Electrical Machinery and Transformer Technology

Richard A. Pearman
Emeritus, Sault College

SAUNDERS COLLEGE PUBLISHING
Harcourt Brace College Publishers
Fort Worth Philadelphia San Diego New York Orlando Austin
San Antonio Toronto Montreal London Sydney Tokyo

Text Typeface: 10/12 Times Roman

Compositor: Weimer Graphics

Production Service: Elm Street Publishing Services, Inc.

Acquisitions Editor: Emily Barrosse

Managing Editor: Carol Field

Manager of Art and Design: Carol Bleistine

Art Director: Robin Milicevic

Cover Designer: Lawrence R. Didona

Director of EDP: Tim Frelick

Production Manager: Joanne Cassetti

Marketing Manager: Monica Wilson

Vice President of Marketing: Marjorie Waldron

Cover Credit: Rick Rusing-Tony Stone Images

Printed in the United States of America

Electrical Machinery and Transformer Technology

ISBN: 0-03-097713-4

Library of Congress Catalog Card Number: 93-40123

4567   042   987654321

This book was printed on paper made from waste paper, containing 10% post-consumer waste and 40% pre-consumer waste, measured as a percentage of total fiber weight content.

To Eileen, our children and grandchildren.

THE SAUNDERS COLLEGE PUBLISHING SERIES IN ELECTRONICS TECHNOLOGY

Bennett ADVANCED CIRCUIT ANALYSIS
ISBN: 0-15-510843-4

Brey THE MOTOROLA MICROPROCESSOR FAMILY: 68000, 68010, 68020, 68030, and 68040; Programming and Interfacing with Applications
ISBN: 0-03-026403-5

Carr ELEMENTS OF MICROWAVE TECHNOLOGY
ISBN: 0-15-522102-7

Carr INTEGRATED ELECTRONICS: OPERATIONAL AMPLIFIERS AND LINEAR IC's WITH APPLICATIONS
ISBN: 0-15-541360-0

Driscoll DATA COMMUNICATIONS
ISBN: 0-03-026637-8

Filer/Leinonen PROGRAMMABLE CONTROLLERS AND DESIGNING SEQUENTIAL LOGIC
ISBN: 0-03-032322-3

Garrod/Borns DIGITAL LOGIC: ANALYSIS, APPLICATION AND DESIGN
ISBN: 0-03-023099-3

Grant UNDERSTANDING LIGHTWAVE TRANSMISSION: APPLICATIONS OF FIBER OPTICS
ISBN: 0-15-592874-0

Greenfield THE 68HC11 MICROCONTROLLER
ISBN 0-03-051588-2

Harrison TRANSFORM METHODS IN CIRCUIT ANALYSIS
ISBN 0-03-020724-X

Hazen EXPLORING ELECTRONIC DEVICES
ISBN 0-03-028533-X

Hazen FUNDAMENTALS OF DC AND AC CIRCUITS
ISBN 0-03-028538-0

Hazen EXPERIENCING ELECTRICITY AND ELECTRONICS, 2/e
Conventional Current Version
ISBN 0-03-076692-3
Electron Flow Version
ISBN 0-03-076691-5

Ismail/Rooney DIGITAL CONCEPTS AND APPLICATIONS, 2/e
ISBN 0-03-097196-9

Laverghetta ANALOG COMMUNICATIONS FOR TECHNOLOGY
ISBN 0-03-029403-7

Leach DISCRETE AND INTEGRATED CIRCUIT ELECTRONICS
ISBN: 0-03-020844-0

Ludeman INTRODUCTION TO ELECTRONIC DEVICES AND CIRCUITS
ISBN 0-03-009538-7

McBride COMPUTER TROUBLESHOOTING AND MAINTENANCE
ISBN: 0-15-512663-6

Oppenheimer SURVEY OF ELECTRONICS
ISBN 0-03-020842-4

Pearman: ELECTRICAL MACHINERY AND TRANSFORMER TECHNOLOGY
ISBN: 0-03-097713-4

Prestopnik DIGITAL ELECTRONICS: CONCEPTS AND APPLICATIONS FOR DIGITAL DESIGN
ISBN 0-03-026757-9

Seeger INTRODUCTION TO MICROPROCESSORS WITH THE INTEL 8085
ISBN: 0-15-543527-2

Spiteri ROBOTICS TECHNOLOGY
ISBN 0-03-020858-0

Wagner DIGITAL ELECTRONICS
ISBN: 0-15-517636-6

Winzer LINEAR INTEGRATED CIRCUITS
ISBN: 0-03-032468-8

Yatsko/Hata CIRCUITS: PRINCIPLES, ANALYSIS, AND SIMULATION
ISBN: 0-03-000933-2

Preface

In this text, every effort has been made to clearly present the fundamental principles behind the operation of electrical machines and transformers, with special emphasis on the basics of power electronics and its application to the control of ac and dc machines. In addition, considerable attention has been devoted to fractional, subfractional, and special-purpose machines, which are being rapidly introduced into automated manufacturing systems.

The author, with many years of engineering experience, has taught electrical machines and control systems for over 20 years at the electrical technology level and has been aware of the lack of an up-to-date text reflecting the latest technologies practiced in the industrial environment. This text is designed to present in a concise format an understanding of the principles and operation of electrical machines and transformers without undue emphasis on detailed mathematical analysis that is standard in many texts. The text is designed to increase the interest and curiosity of the electrical and nonelectrical student of both technology and engineering programs.

The text covers the requirements of introductory-level courses in electrical machines, in both the three-year associate degree and the four-year bachelor of technology level programs. The text is also suitable for survey courses at the undergraduate level for engineering students not in an electrical program.

The SI system of units has been used in the development of all theory because it simplifies the presentation and because it is the worldwide standard. Parallel formulas in the English unit system are also included.

Review of the text will show that Chapter 1 lays the groundwork for the fundamental concepts required in the rest of the text. Chapter 2 provides a survey of solid-state devices and the principles of electronic power conversion techniques used in the control of ac and dc machines. Specific applications are dealt with in subsequent chapters. Chapters 3 through 7 deal with the construction, principles, and characteristics of dc machines. Chapter 8 covers the control of dc motors, both the conventional electromagnetic approach as well as the use of the rapidly developing field of power electronics for dc drive control. Chapter 9 presents an overview of the factors involved in dc machine selection. Chapter 10 lays the groundwork for the study of polyphase ac machines. Chapters 11 and 12 deal with the construction, principles, and characteristics of synchronous generators and motors. Chapter 13 covers transformer theory, testing and application of single- and three-phase transformers, as well as phase-changing techniques. Chapter 14 discusses the principles, characteristics, and control of polyphase induction motors, including automatic starting and speed control by conventional and solid-state methods. Chapter 15 focuses on the theory, characteristics, performance, and control of single-phase induction motors. Chapter 16 is devoted to the principles of operation, characteristics, applications, and control of single-phase ac motors. Chapter 17 concentrates on special-purpose machines such as stepper motors, ac and dc servomotors, switched reluctance motors, speed transducers, and brushless and linear induction motors. Chapter 18 deals with the selection of ac machines.

Acknowledgments

The author would like to acknowledge the assistance and cooperation of the many manufacturers that have enabled this text to be produced. Also appreciated are the efforts of the following reviewers in providing helpful suggestions and comments:

Robert F. Abrams, *Garden State Community College*
S.P. Desai, *University of Southern Indiana*
Warren D. Hill, *University of Southern Colorado*
Richard D. Stoy, *Widener University*
Jim Humphries, *Santa Fe Community College.*

The author also thanks Nancy Shanahan at Elm Street Publishing Services for her hard work and assistance on the book.

Finally, the author wishes to acknowledge his indebtedness to his wife for her patience and support during the many long hours of work involved in preparing this text.

Richard A. Pearman
December 1993

Contents

6 DC MOTORS 193

7 DC MACHINE LOSSES, EFFICIENCY, AND TESTING 230

CHAPTER

1

Fundamental Concepts

1-1 INTRODUCTION

In our day-to-day activities we are measuring most features of the environment in which we live. In nearly all cases the units of measurement are taken for granted, without any appreciation of their origins. Most units of measurement are based on the physical laws of nature and, as a result, do not vary and are reproducible; for example, length is defined in terms of the wavelength of the atomic radiation of krypton 86, time is defined in terms of the atomic radiation of cesium 133. While these basic reproducible reference standards are universally recognized by all countries, the everyday units of measurement are not universal. For example, in the United States length is commonly measured in inches, yards, and miles, while the remaining countries of the world use the millimeter, meter, and kilometer. Also specialized groups, such as astronomers, physicists, and surveyors, use the parsec, angstrom, and foot to measure length. However, all these units of length can be compared accurately because they are based on the wavelength of krypton 86. Internationally accepted standard units of length, mass, and time permit the comparison of the units of measurement of a country or a specific science with those of other countries.

1-2 SYSTEMS OF UNITS

A system of units bears a direct numerical relationship between its members. In the English system the inch, foot, and yard are related by the numbers 12, 3, and 36. In the metric system the millimeter, centimeter, meter, and kilometer are related by the numbers 10, 100, 1,000 and 100,000, that is, the units are related by multiples of 10. An obvious inference is that it is easier to convert meters to centimeters, or kilometers to meters, than it is to convert yards to feet.

The most common systems used in engineering are:

1. The English gravitational system, also sometimes referred to as the U.S. customary system, in North America
2. The international system of units, or SI system

The English System

Approximately 95% of the world has converted to the international system of units (SI system). However, the English system is still in common use in the United States. As a result, engineers and engineering students must be familiar with both systems since most engineering publications throughout the world use SI units. Also, the technical documentation for imported equipment and systems will use the SI system. Therefore this text will stress the use of the SI system, but it will not neglect the English system.

The English and SI systems both define length L, and time T as fundamental dimensions; however, the treatment of force and mass differs. In the English system force is a fundamental dimension and mass is therefore a derived dimension. It should be noted that any system that defines force as a fundamental dimension is known as a gravitational system.

From Newton's second law,

$$F = m\alpha \tag{1-1}$$

where F is the force acting on a rigid body of mass m, which causes it to accelerate with acceleration α. Then, transposing,

$$m = \frac{F}{\alpha} = \frac{F}{L/T^2} = \frac{FT^2}{L} \tag{1-2}$$

The basic dimensions of force, length, and time are pound (lb), foot (ft), and second (s). Therefore it can be seen that the units of mass are

$$m = \frac{FT^2}{L} = \frac{\text{lb} \cdot \text{s}^2}{\text{ft}} \tag{1-3}$$

where lb. s^2/ft is called a *slug*.

The SI system, on the other hand, defines mass as a fundamental unit, and hence, from Newton's second law, the force F is a derived unit,

$$F = m\alpha = \frac{ML}{T^2} = \frac{\text{kg} \cdot \text{m}}{\text{s}^2} \tag{1–4}$$

where kg.m/s^2 is called the newton (N).

The English system leaves us in a quandary since both force and mass are defined as fundamental dimensions. Then a proportionality constant k must be introduced into Eq. (1-1) to maintain a dimensional balance. Thus

$$F = km\alpha$$

Therefore

$$k = \frac{F}{m\alpha} = \frac{F}{ML/T^2} = \frac{FT^2}{ML} \tag{1–5}$$

A further complication occurs because the pound is used for both the unit of force and the unit of mass. Since they are not both the same, it is necessary to differentiate between the two as follows: the pound force (lb_f) and the pound mass (lb_m). By convention if the subscript is omitted, the unit is assumed to be the pound force.

Since in the English system the acceleration due to gravity g is 32.174 ft/s^2, and since

$$F = kmg \tag{1–6}$$

We have

$$1 \text{ lb}_f = k\,(1 \text{ lb}_m)(32.174 \text{ ft/s}^2)$$

$$= k\,(32.174) \text{ lb}_m \cdot \text{ft/s}^2$$

From this it can be seen that a 1-lb force will accelerate a 1-lb mass at a rate of 32.174 ft/s^2. Therefore

$$k = \frac{1}{32.174} \times \frac{\text{lb}_f \cdot \text{s}^2}{\text{lb}_m \cdot \text{ft}}$$

and Eq. (1-6) becomes

$$F = \frac{1}{32.174}\, mg \tag{1–7}$$

Since the acceleration due to gravity does not vary appreciably over the surface of the planet, then

$$F = m \tag{1–8}$$

Therefore, in the English system,

$$1 \text{ lb}_f = (1 \text{ slug})(32.174 \text{ ft/}s^2) \tag{1–9}$$

$$= 32.174 \text{ lb}_m \cdot \text{ft/}s^2$$

and thus, in the English system, a 1-lb force will accelerate a 1-slug mass at 1 ft/s^2, that is,

$$1\ \text{lb}_f = (\text{slug})\ (1\ \text{ft}/s^2) = 1\ \text{slug} \cdot \text{ft}/s^2 \tag{1-10}$$

Equating Eqs. (1-9) and (1-10),

$$32.174\ \text{lb}_m \cdot \text{ft}/s^2 = 1\ \text{slug} \cdot \text{ft}/s^2$$

Therefore

$$1\ \text{slug} = 32.174\ \text{lb}_m$$

The SI System

The international system of units (SI system) is based on the mksa (meter, kilogram, second, and ampere) system. It has been adopted worldwide by standardization bodies, which include the Institute of Electrical and Electronic Engineers (IEEE), the American National Standards Institute (ANSI), and the International Electrotechnical Commission (IEC).

There are seven base units in the SI system: length, mass, time, electric current, thermodynamic temperature, the amount of substance, and luminous intensity. These units and their symbols are listed in Table 1-1.

Furthermore, there are two supplementary SI units which may be considered as either base units or derived units, the plane angle and the solid angle. These units and their symbols are shown in Table 1-2. These seven base units and two supplementary units form the foundation of the SI system, and from them all other SI units are derived.

Definitions of SI Base Units. The following official definitions illustrate the high degree of precision associated with these units:

1. The *meter* (m) is the length equal to 1,650,763.73 wavelengths in vacuum of the radiation corresponding to the transition between the levels $2p_{10}$ and $5d_5$ of the krypton 86 atom.
2. The *kilogram* (kg) is the unit of mass. It is equal to the mass of the international prototype of the kilogram. (*Note*: The prototype is a platinum-iridium cylinder maintained at the International Bureau of Weights and

Table 1-1 SI base units

Quantity	Unit	Symbol
Length	meter	m
Mass	kilogram	kg
Time	second	s
Electric current	ampere	A
Thermodynamic temperature	kelvin	K
Amount of substance	mole	mol
Luminous intensity	candela	cd

Measures in Sèvres, France.) The kilogram is approximately equal to the mass of 1,000 cubic centimeters of water at its temperature of maximum density (3.98°C).

3. The *second* (s) is the duration of 9,192,631,770 periods of the radiation corresponding to the transition between the hyperfine levels of the ground state of the cesium 133 atom.
4. The *ampere* (A) is the constant current that, if maintained in two straight parallel conductors of infinite length and of negligible circular cross section, placed 1 meter apart in vacuum, would produce between these conductors a force equal to 2×10^{-7} newton per meter of length.
5. The Kelvin (K), the unit of thermodynamic temperature, is the fraction 1/273.16 of the thermodynamic temperature of the triple point of water. [Note: The zero of the Celsius scale, 0°C (the freezing point of water), is defined as 0.01 K below the triple point, that is, 273.15 K.]
6. The *mole* (mol) is the amount of substance of a system that contains as many elementary entities as there are atoms in 0.012 kilogram of carbon12.
7. The *candela* (cd) is the unit of luminous intensity, in the perpendicular direction, of a surface of 1/600,000 square meter of a blackbody at the temperature of freezing platinum under a pressure of 101,325 newtons per square meter.

The two supplemental SI units, which may be considered as either base units or derived units, are defined as follows:

8. The *radian* (rad) is the unit of plane angle between two radii of a circle which cuts off an arc on the circumference equal in length to the radius. $1 \text{ rad} = 57.2957° \approx 57.3°$.
9. The *steradian* (sr) is the unit of solid angle which, having its vertex in the center of a sphere, cuts off an area of the surface of the sphere equal to that of a square with sides equal to the radius of the sphere.

Definitions of Derived SI Units. By far the greatest number of SI units used in electrical engineering are derived SI units. Table 1-3 lists the principal electrical quantities in the SI system.

These units are defined as follows:

1. The *siemens* (S) is the conductance of a conductor such that when a constant voltage of 1 volt is applied between its ends, a constant current of 1 ampere will flow through it. It can also be defined as the reciprocal of the ohmic resistance.

Table 1-2 SI supplementary units

Quantity	Unit	Symbol
Plane angle	radian	rad
Solid angle	steradian	sr

Table 1-3 SI derived units used in electrical engineering.

Quantity	Name	Symbol
Capacitance	farad	F
Charge	coulomb	C
Conductance	siemens	S
Potential	volt	V
Resistance	ohm	Ω
Energy	joule	J
Force	newton	N
Frequency	hertz	Hz
Inductance	henry	H
Magnetic flux	tesla	T
Magnetic flux density	weber	Wb
Power	watt	W

2. The volt (V) is the potential difference (PD) between two points of a conducting medium carrying a constant current of 1 ampere when the power dissipated between these points is 1 watt, that is, 1 volt = 1 watt per ampere.
3. The ohm (Ω) is the resistance of a conductor such that a constant current of 1 ampere flowing through it produces a PD of 1 volt between the ends.
4. The *joule* (J) is the work done by a force of 1 newton acting through a distance of 1 meter.
5. The *newton* (N) is the force that imparts an acceleration of 1 meter per second per second to a mass of 1 kilogram.
6. The *hertz* (Hz) is the unit of frequency and is equal to 1 cycle per second.
7. The *henry* (H) is the unit of inductance of a closed circuit in which an electromotive force (emf) of 1 volt is produced when the current changes at the rate of 1 ampere per second.
8. The *tesla* (T) is the unit of magnetic induction and is equal to 1 weber per square meter.
9. The *weber* (Wb) is the magnetic flux that, linking a circuit of one turn, produces in it an electromotive force of 1 volt as it is reduced to zero at a uniform rate in 1 second.
10. The *watt* (W) is the power called for to do work at the rate of 1 joule per second, that is, 1 watt = 1 joule per second.

Other units commonly used in mechanics and thermodynamics are listed in Tables 1-4 and 1-5.

SI Decimal Prefixes. All SI units may be represented by standard prefixes which multiply the appropriate quantity by a power of 10. Table 1-6 lists the standard prefixes, their symbols, and the factor by which the SI unit is multiplied.

Conversion of Units. Because two different unit systems are in common use in the United States, it is necessary to be able to convert from one unit system to the other, that is, from English to SI units or vice versa. The conversion of units is greatly simplified by a conversion table such as the one shown in Table 1-7.

It should be noted that it is always good practice to include the units with each component of an equation when carrying out calculations. Also, when doing conversion calculations, the precision of the input units should be maintained in the answer. Table 1-7 is an abbreviated list of conversion factors. A more comprehensive listing is given in Appendix A. The following examples can be solved by using Table 1-7.

➤ EXAMPLE 1-1

A man is 5 ft 11 in tall and his weight is 195 lb. What are: (**a**) his height in meters; (**b**) his weight in newtons; (**c**) his weight in kilograms?

SOLUTION

(a)

$$5 \text{ ft } 11 \text{ in} = 5 \text{ ft} \times 12 \text{ in/ft} + 11 \text{ in} = 71 \text{ in}$$

Therefore his height H is

$$H = (2.54 \times 10^{-2} \text{ m/in})(71 \text{ in}) = 1.803 \text{ m}$$

(b) His weight W is

$$W = \frac{193 \text{ lb}}{0.2248 \text{ lb/N}} = 867.44 \text{ N}$$

(c) His weight in kilograms is obtained from $W = mg$, where

$$m = \frac{W}{g} = \frac{867.44 \text{ N}}{9.8 \text{ m/s}^2} = 88.51 \text{ kg}$$

➤ EXAMPLE 1-2

A body with a mass of 20 kg is lifted through 10 m. Calculate the work done in joules.

SOLUTION

The force required to lift the body is equal to the weight of the body.

$$\text{Weight of body} = mg = 20 \text{ kg} \times 9.8 \text{ m/s}^2 = 196 \text{ N}$$

$$\text{Work} = \text{force} \times \text{distance moved in direction of force}$$

$$= 196 \text{ N} \times 10 \text{ m} = 1{,}960 \text{ N} \cdot \text{m} = 1{,}960 \text{ J}$$

Table 1-4 SI derived units used in mechanical engineering.

Quantity	Unit	Symbol
Angle	radian	rad
Area	square meter	m^2
Energy (work)	joule	J
Force	newton	N
Length	meter	m
Mass	kilogram	kg
Power	watt	W
Pressure	pascal	Pa
Speed (linear)	meters per second	m/s
Speed (rotational)	radians per second	rad/s
Torque	newton-meter	N·m
Volume	cubic meter	m^3
Volume	liter	L

Table 1-5 SI derived units used in thermodynamics.

Quantity	Unit	Symbol
Heat	joule	J
Thermal power	watt	W
Specific heat	joules per kilogram-kelvin	J/kg · K or J/kg · °C
Temperature	kelvin or °C	K or °C
Temperature difference	kelvin or °C	K or °C
Thermal conductivity	watts per meter-kelvin	W/m · k or W/m · °C

➤ EXAMPLE 1-3

A pump supplies 150 gal of water per minute through a large pipe into a tank 20 ft above its intake. How much work is done in 1 hr? Water weighs 8.34 lb/gal.

SOLUTION

$$\text{Weight of water per hour} = 150 \text{ gal} \times 60 \text{ min} \times 8.34 \text{ lb/gal} = 75{,}060 \text{ lb}$$

$$\text{Work done} = \text{force} \times \text{distance} = 75{,}060 \text{ lb} \times 20 \text{ ft} = 1.5 \times 10^6 \text{ ft} \cdot \text{lb}$$

➤ EXAMPLE 1-4

A 10-kW motor with an efficiency of 91% drives a crane with an efficiency of 45%. With what velocity does the crane lift a 750-lb load if the motor draws 10 kW from the power supply?

SOLUTION

$$\text{Overall efficiency (\%)} = \text{motor efficiency} \times \text{crane efficiency}$$
$$= 91\% \times 45\% = 41\%$$

$$\text{Power available to lift load (kW)} = 0.41 \times 10 \text{ kW} = 4.1 \text{ kW}$$

$$\text{Power output of crane (ft} \cdot \text{lb/s)} = \text{load (lb)} \times \text{velocity (ft/s)}$$

$$4.1 \text{ kW} \times \frac{1.3 \text{ hp}}{1 \text{ kW}} \times \frac{550 \text{ ft} \cdot \text{lb/s}}{1 \text{ hp}} = 750 \text{ lb} \times v\,(\text{ft/s})$$

$$3,021.70 \text{ ft} \cdot \text{lb/s} = 750 \text{ lb} \times v\,(\text{ft/s})$$

Therefore

$$v = \frac{3,021.70 \text{ ft} \cdot \text{lb/s}}{750 \text{ lb}} = 4.03 \text{ ft/s}$$

➤ EXAMPLE 1-5

Find the average power required to lift 10,000 lb through a height of 50 ft in 30 s. Express the answer in horsepower and kilowatts.

SOLUTION

$$\text{Power}\,(\text{ft} \cdot \text{lb/s}) = \frac{\text{work}\,(\text{ft} \cdot \text{lb})}{\text{s}}$$

$$= \frac{10,000 \text{ lb} \times 50 \text{ ft}}{30\text{s}} = 16,666.67 \text{ ft} \cdot \text{lb/s}$$

$$\text{Power}\,(\text{hp}) = 16,666.67 \text{ ft} \cdot \text{lb/s} \times \frac{1 \text{ hp}}{550 \text{ ft/s}} = 30.03 \text{ hp}$$

$$\text{Power}\,(\text{kW}) = 30.03 \text{ hp} \times \frac{746 \times 10^{-3} \text{ kW}}{1 \text{ hp}} = 22.4 \text{ kW}$$

or

$$\text{Power}\,(\text{kW}) = 30.03 \text{ hp} \times 1.341 \text{ hp/kW} = 22.39 \text{ kW}$$

Table 1-6 SI prefixes for decimal factors.

Factor	Prefix	Symbol
10^{18}	exa	E
10^{15}	peta	P
10^{12}	tera	T
10^{9}	giga	G
10^{6}	mega	M
10^{3}	kilo	k
10^{2}	hecto	h
10^{1}	deka	da
10^{-1}	deci	d
10^{-2}	centi	c
10^{-3}	milli	m
10^{-6}	micro	μ
10^{-9}	nano	n
10^{-12}	pico	p
10^{-15}	femto	f
10^{-18}	atto	a

➤ EXAMPLE 1-6

The magnetic core of a reactor is made up of laminations of M-5 grain-oriented steel. The core has an effective volume of 200 in³. The density of the steel is 7.65 g/cm³. Find the weight of the core.

SOLUTION

$$\text{Core weight } W_c = 200 \text{ in}^3 \times \frac{7.65 \text{ g}}{\text{cm}^3} \times \frac{(2.54)^3 \text{cm}^3}{1 \text{ m}^3} = 25.07 \text{ kg}$$

1-3 FUNDAMENTAL POSITIONAL RELATIONSHIPS

Since all electrical machines except transformers and linear motors have rotating shafts, it is necessary to have an understanding of the fundamentals of rotational motion. In this section the concepts of angular position (distance), angular velocity, angular acceleration, torque, Newton's law of rotation, work, power, kinetic energy, and moment of inertia are developed to provide a basic understanding of rotating machine characteristics and performance.

Angular Position

Angular position Θ or angular displacement is the angle turned through by the shaft of a motor. This angle may be expressed in degrees (°), radians (rad), or revolutions (rev). When viewed from the shaft end, the direction of rotation may be classified as clockwise (CW) or counterclockwise (CCW). Counterclockwise rotation is assumed

Table 1-7 Conversion factors.

Quantity	Unit	Factor
Length	1 meter (m)	= 1.093 yd
		= 3.2808 ft
		= 39.37 in
		= 100 cm
Mass	1 kilogram (kg)	= 0.0685 slug
		= 2.2046 lb_m (mass)
		= 1,000 g
Force	1 newton (N)	= 0.2248 lb_f (force)
		= 7.233 pdl (poundals)
		= 0.102 kg (force)
Torque	1 newton-meter (N · m)	= 0.738 lb · ft
		= 141.612 oz · in
Energy	1 joule (J)	= 0.738 ft · lb_f
		= 2.777 kWh
Power	1 watt (W)	= 0.738 ft · lb_f/s
		= 1,341 × 10^{-3}hp
	1 horsepower (hp)	= 746 W
Temperature	°C	= 5 (°F–32)/9
	°F	= [9(°C)/5] + 32
	K	= °C + 273.15

to be positive and clockwise rotation is assumed to be negative. The angular position is analogous to the linear concept of distance in a straight line.

By definition, 1 radian is the angle Θ subtended at the center of a circle by an arc s equal in length to the radius r. Then, as can be seen from Fig. 1-1, Θ is

$$\Theta = \frac{s}{r} \tag{1–11}$$

and

$$s = r\Theta \tag{1–12}$$

Since 1 rev = 360° = 2pπ rad,

$$1 \text{ rad} = \frac{360°}{2\pi} = 57.3°$$

Angular Velocity

The angular velocity ω of a body (shaft) is its time rate of change in the angular position about the axis of rotation. It is usually expressed in radians per second (rad/s),

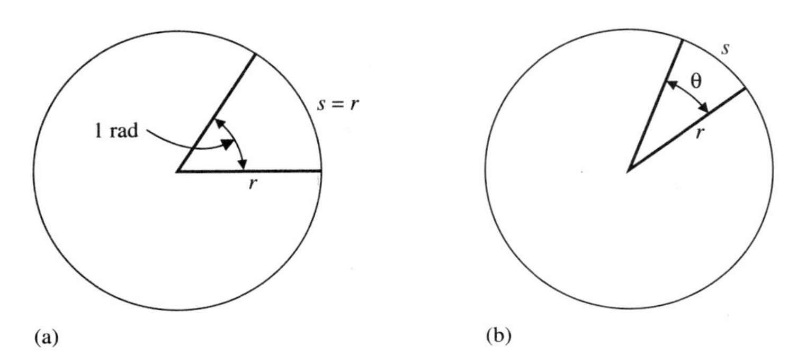

Figure 1-1 Angular position. (a) Definition of radian. (b) Definition of angular displacement.

degrees per second (°/s), or revolutions per minute (r/min).

If a body turns through an angle of Θ rad in t s, its average angular velocity in radians per second is

$$\omega = \frac{\text{angle turned through (rad)}}{\text{time (s) required to turn through this angle}}$$

$$= \frac{\Theta(\text{rad})}{t(\text{s})}\ \text{rad/s} \tag{1–13}$$

Since 1 r/s = 2π rad/s, then

$$\omega\ (\text{rad/s}) = 2\pi \times \text{r/s}$$

Angular Acceleration

The angular acceleration α of a body (shaft) is the time rate of change of its angular velocity. It is usually expressed in radians per second per second (rad/s²).

Assuming the angular velocity of a body changes uniformly from ω_0 to ω_t rad/s in t s, then

$$\alpha = \frac{\text{change in angular velocity (rad/s)}}{\text{time (s) needed for the change}}$$

$$= \frac{\omega_t - \omega_0}{t}\ \text{rad/s}^2 \tag{1–14}$$

➤ EXAMPLE 1-7

Convert: (a) 10 rad to revolutions; (b) 1,200 rev to radians; (c) 1,800 r/min to rad/s.

SOLUTION

(a)

$$10 \text{ rad} = 10 \text{ rad} \times \frac{1 \text{ rev}}{2\pi \text{ rad}}$$

$$= \frac{5}{\pi} = 1.59 \text{ rev}$$

(b)

$$1,200 \text{ rev} = 1,200 \text{ rev} \times \frac{2\pi \text{ rad}}{1 \text{ rev}}$$

$$= 2,400\pi = 7,539.82 \text{ rad}$$

(c)

$$1,800 \text{ r/min} = 1,800 \text{ r/min} \times \frac{1 \text{ min}}{60 \text{ s}} \times \frac{2\pi \text{ rad}}{1 \text{ rev}}$$

$$60\pi = 188.5 \text{ rad/s}$$

➤ EXAMPLE 1-8

A motor accelerates to 900 r/min (30π rad/s) in 10 s from rest. Find the angular acceleration.

SOLUTION

$$\alpha = \frac{\omega_t - \omega_0}{t} = \frac{(30\pi - 0) \text{ rad/s}}{10 \text{ s}}$$

$$= 3\pi \text{ rad/s}^2 = 9.42 \text{ rad/s}^2$$

The following equations for uniformly accelerated angular motion are the analogs of those for linear motion. If v_0 and ω_0 represent the initial linear and angular velocities, respectively, and v_t and ω_t represent the linear and angular velocities, respectively, at time t, then

$$v_t = v_0 + at \qquad s = v_0 t + \tfrac{1}{2} at^2 \qquad v_t^2 = v_0^2 + 2as$$

$$\omega_t = \omega_0 + \alpha t \qquad \theta = \omega_0 t + \tfrac{1}{2} \alpha t^2 \qquad \omega_t^2 = \omega_0^2 + 2\alpha\theta$$

In the case of starting from rest, where $v_0 = 0$ and $\omega_0 = 0$, then

$$v_t = at \qquad s = \tfrac{1}{2} at^2 \qquad v_t^2 = 2as$$

$$\omega_t = \alpha t \qquad \theta = \tfrac{1}{2} \alpha t^2 \qquad \omega_t^2 = 2\alpha\theta$$

Applications are best illustrated by the following examples.

➤ EXAMPLE 1-9

The shaft of a motor is revolving at 1,200 r/min and slows down to 1,000 r/min in 2s. Calculate: (**a**) the angular deceleration of the motor; (**b**) the number of revolutions the shaft makes in this time interval.

SOLUTION

$$\omega_0 = 1{,}200 \text{ r/min} = \frac{1{,}200 \text{ r/min} \times 2\pi \text{ rad}}{60 \text{ s}} = 40\pi \text{ rad/s}$$

$$\omega_t = 1{,}000 \text{ r/min} = \frac{1{,}000 \text{ r/min} \times 2\pi \text{ rad}}{60 \text{ s}} = 33.33\pi \text{ rad/s}$$

(a)

$$\alpha = \frac{\omega_t - \omega_0}{t} = \frac{(33.33\pi - 40\pi)\text{rad/s}}{2 \text{ s}} = -3.33 \text{ rad/s}$$

(b)

$$\begin{aligned}\Theta &= \omega_0 t + \frac{1}{2}\alpha t^2 \\ &= \omega_0 t + \frac{1}{2}\left(\frac{\omega_t - \omega_0}{t}\right) \times t^2 \\ &= \omega_0 t + \frac{1}{2}(\omega_t - \omega_0)t \\ &= \frac{1}{2}(\omega_0 + \omega_t)t \\ &= \frac{1}{2}(40\pi + 33.33\pi) \text{ rad/s} \times 2 \text{ s} = 73.33\pi \text{ rad} = 36.67 \text{ r/min}\end{aligned}$$

➤ EXAMPLE 1-10

A motor whose shaft is turning at 120 r/min has its speed increased to 660 r/min in 9s. Find the angular acceleration in rev/s^2 and rad/s^2.

SOLUTION

$$\omega_0 = 120 \text{ r/min} = \frac{120 \text{ r/min} \times 2\pi \text{ rad}}{60 \text{ s}} = 4\pi \text{ rad/s}$$

$$\omega_t = 660 \text{ r/min} = \frac{660 \text{ r/min} \times 2\pi \text{ rad}}{60 \text{ s}} = 22\pi \text{ rad/s}$$

$$\omega_t = \omega_0 + \alpha t$$

$$\alpha = \frac{\omega_t - \omega_0}{t} = \frac{(22\pi - 4\pi) \text{ rad/s}}{9 \text{ s}} = 2\pi \text{ rad/s}^2 = 1 \text{ r/s}^2$$

Torque

Torque T is the turning force that tends to produce rotation around an axis. It is the product of the applied force F and the perpendicular distance from the axis of rotation to the line of action of the force (Fig. 1-2). In the case of electric motors, torque is the turning force developed by the motor. Countertorque is the opposition, or resistance, offered by the connected motor load to the turning force.

In Fig. 1-2(a) the force F is applied at right angles to the radius r and $\Theta = 90°$. In Fig. 1-2(b) the force is applied at an angle Θ greater than 90° to the radius r. As a result, the perpendicular distance from the axis of rotation to the line of action of the force will be less than r. A general relationship for torque is

$$\begin{aligned} T &= \text{force} \times \text{perpendicular distance from axis of rotation to line of action of force} \\ &= F \times r \sin \Theta \\ &= Fr \sin \Theta \qquad (1\text{–}15) \end{aligned}$$

It should be noted that the torque T is a maximum when $\Theta = 90°$ since $\sin 90° = 1$. Then

$$T = Fr$$

Torque is expressed in lb · ft in the English unit system when the force F is in pounds and the radius r in feet. In the SI system torque T is expressed in N.m when the force F is in newtons and the radius r is in meters. A more meaningful method of expressing torque is as a percentage of rated full-load torque in electric motor applications.

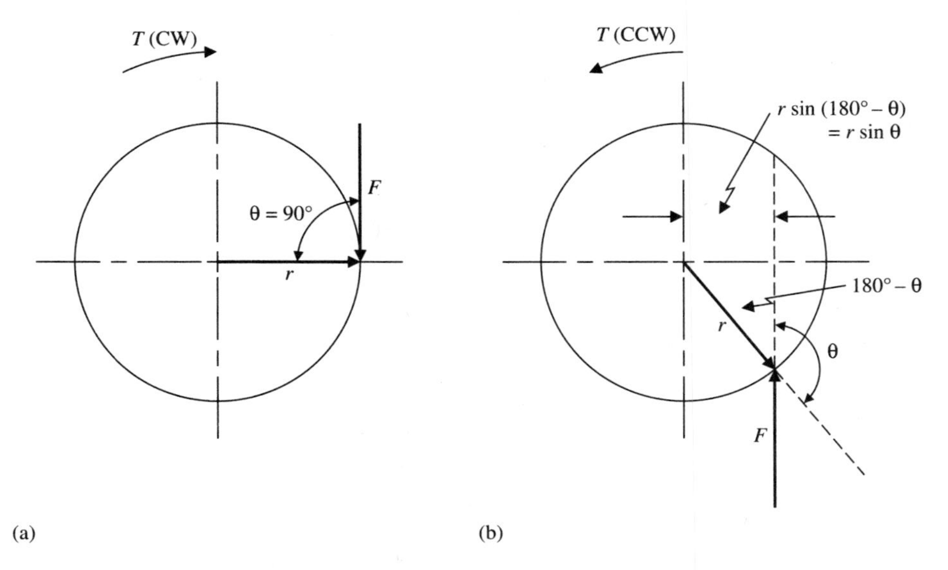

Figure 1-2 Torque produced by a force F. (a) Applied tangentially to a cylinder. (b) Applied at an angle to cylinder surface.

➤ EXAMPLE 1-11

A wheel 3 ft in diameter has a force of 5 lb applied tangentially to its outer surface. Find the torque produced by the force: (a) in lb · ft; (b) in N · m.

SOLUTION

$$\text{Radius } r = \frac{1}{2}(3 \text{ ft}) = 1.5 \text{ ft} = 1.5 \text{ ft} \times 0.3048 \text{ m/ft} = 0.46 \text{ m}$$

$$\text{Force } F = 5 \text{ lb} = 4.448 \text{ N / lb} \times 5 \text{ lb} = 22.24 \text{ N}$$

(a) $$T = Fr = 5 \text{ lb} \times 1.5 \text{ ft} = 7.50 \text{ lb} \cdot \text{ft}$$

(b) $$T = Fr = 22.24 \text{ N} \times 0.46 \text{ m} = 10.23 \text{ N} \cdot \text{m}$$

Work

The work done on a rotating body by a constant torque is equal to the product of the torque T and the angular displacement Θ,

$$W = T\Theta \tag{1–16}$$

and

$$W\,(\text{ft} \cdot \text{lb}) = T\,(\text{lb} \cdot \text{ft}) \times \Theta\,(\text{rad}) \tag{1–16)(E}$$

or

$$W\,(J) = T\,(\text{N} \cdot \text{m}) \times \Theta\,(\text{rad}) \tag{1–16)(SI}$$

Power

Power P is the rate of doing work, or the increase in work per unit time. Power is expressed as

$$P = \frac{W}{t} \tag{1–17}$$

and

$$P\left(\text{ft} \cdot \text{lb/s}\right) = \frac{W(\text{ft} \cdot \text{lb})}{t(\text{s})} \tag{1–17)(E}$$

or

$$P\,(\text{W}) = \frac{W(J)}{t\,(\text{s})} \tag{1–17)(SI}$$

In the English system we refer to work in units of horsepower (hp), where

$$1 \text{ hp} = \frac{33,000 \text{ ft} \cdot \text{lb}}{\text{min}} = \frac{550 \text{ ft} \cdot \text{lb}}{\text{s}} = 746 \text{ W}$$

The SI unit of power is the *watt* (W) = N · m/s = J/s. Usually the power input to or the power output from a motor is expressed in kilowatts (kW), which is equal to 1,000 W.

The power in rotational motion is equal to the product of the torque T and the angular velocity ω with which the torque T is applied,

$$P = T\omega \tag{1–18}$$

and

$$P(\text{ft} \cdot \text{lb/s}) = \text{T}(\text{lb} \cdot \text{ft}) \times \omega\,(\text{rad/s}) \tag{1–18)(E}$$

or

$$P(\text{W}) = \text{T}(\text{N} \cdot \text{m}) \times \omega\,(\text{rad/s}) \tag{1–18)(SI}$$

Equation (1-18) is the valid relationship between power, torque, and speed if power is measured in watts, torque in newton-meters, and speed in radians per second. In engineering applications in the United States torque is usually measured in pound-feet, speed in revolutions per minute, and power in horsepower. Hence it is necessary to be able to convert from one unit system to the other. Equation (1-18) in terms of the English unit system is then expressed as

$$P(\text{W}) = \frac{T(\text{lb} \cdot \text{ft}) \times S\ (\text{r/min})}{7.04} \tag{1–19}$$

and

$$P(\text{hp}) = \frac{T(\text{lb} \cdot \text{ft}) \times S\ (\text{r/min})}{5252} \tag{1–20}$$

➤ EXAMPLE 1-12

An electric motor raises an elevator of mass 1,000 kg through a height of 25 m in 12 s. Find the output power developed by the motor in kilowatts and in horsepower.

SOLUTION

The cable tension F is

$$F = mg = 1{,}000\ \text{kg} \times 9.8\ \text{m/s}^2 = 9{,}800\ \text{N}$$

The work done is

$$W = Fd = 9{,}800\ \text{N} \times 25\ \text{m} = 245{,}000\ \text{J}$$

Then the output power in kilowatts is

$$P = \frac{W}{t} = \frac{245{,}000\ \text{J}}{12\ \text{s}} = 20{,}416.67\ \text{W} = 20.42\ \text{kW}$$

The output power in horsepower is

$$P = \frac{20{,}416.67\ \text{W}}{746\ \text{W/hp}} = 27.37\ \text{hp}$$

Torque and Angular Acceleration

Newton's second law describes in the case of linear motion the relationship between the force applied to an object and the resulting acceleration.

There is a similar relationship describing the effect of a torque applied to an object and the resulting acceleration. This relationship, which is called Newton's law of rotation, is

$$T = J\alpha \tag{1-21}$$

and

$$T\,(\text{lb} \cdot \text{ft}) = J\,(\text{slug} \cdot \text{ft}^2) \times \alpha\,(\text{rad/s}^2) \tag{1-21)(E}$$

or

$$T\,(\text{N} \cdot \text{m}) = J\,(\text{kg} \cdot \text{m}^2) \times \alpha\,(\text{rad/s}^2) \tag{1-21)(SI}$$

The J term, called the *moment of inertia,* is the measure of the resistance offered by a rotating body to any change in angular velocity. The moment of inertia J depends on the mass and shape of the body and, as will be seen, plays a very important part in the performance of rotating machines.

Kinetic Energy

The kinetic energy of a body is its ability to do work because of its motion. We must consider two forms of kinetic energy—that possessed by a body moving in a straight line and that possessed by a rotating body.

In the case of linear motion the kinetic energy is

$$E = \frac{1}{2}mv^2 \tag{1-22}$$

and

$$E(\text{ft} \cdot \text{lb}) = \frac{1}{2}m\,(\text{slug}) \times v^2(\text{ft/s}^2) \tag{1-22)(E}$$

or

$$E(J) = \frac{1}{2}m\,(\text{kg}) \times v^2(\text{m/s}^2) \tag{1-22)(SI}$$

In the case of rotary motion the kinetic energy is

$$E = \frac{1}{2}J\omega^2 \tag{1-23}$$

and

$$E(\text{ft} \cdot \text{lb}) = \frac{1}{2}J(\text{slug} \cdot \text{ft}^2) \times \omega^2(\text{rad/s})^2 \tag{1-23)(E}$$

or

$$E(J) = \frac{1}{2} J(\text{kg} \cdot \text{m}^2) \times \omega^2(\text{rad/s})^2 \qquad \textbf{(1–23)(SI)}$$

Equation (1-23)(SI) can be simplified as follows:

$$E = \frac{1}{2} J\left(\text{kg} \cdot \text{m}^2\right) \times \omega^2(\text{rad/s})^2$$

$$= \frac{1}{2} J\left(\text{kg} \cdot \text{m}^2\right) \times \left[\frac{2\pi\, S(\text{r/min})}{60}\right]^2$$

$$= \frac{1}{2} J\left(\text{kg} \cdot \text{m}^2\right) \times \left[\frac{4\pi^2\, S^2(\text{r/min})}{3{,}600}\right]$$

$$= J\left(\text{kg} \cdot \text{m}^2\right) \times \left[\frac{\pi^2\, S^2(\text{r/min})}{1{,}800}\right]$$

$$= 5.48 \times 10^{-3} \times J\left(\text{kg} \cdot \text{m}^2\right) \times S^2(\text{r/min}) \qquad \textbf{(1–24)}$$

From Eq. (1-23) it can be seen that the kinetic energy of a rotating body is proportional to the moment of inertia and the square of the angular velocity. The moment of inertia J is determined by the shape and mass of the rotating body. Figure 1-3 illustrates the moments of inertia of a few but important body shapes. It should be noted that it is not intended to develop moments of inertia for various body shapes. However, it is felt absolutely necessary that the significance of J in the control and operation of rotating machines be appreciated.

➤ EXAMPLE 1-13

A solid cylindrical rotor has a mass of 100 kg and a radius of 12 cm. Calculate: **(a)** the moment of inertia; **(b)** the kinetic energy when rotating at 1,200 r/min.

SOLUTION

From Fig. 1-3(b)

$$J = \frac{1}{2} mr^2$$

$$= \frac{1}{2} \times 100 \text{ kg} \times \left(\frac{12 \text{ cm}}{100 \text{ cm/m}}\right)^2$$

$$= 0.72 \text{ kg} \cdot \text{m}^2$$

From Eq. (1-24), the kinetic energy of rotation is

$$E = 5.48 \times 10^{-3} \times J\left(\text{kg} \cdot \text{m}^2\right) \times S^2(\text{r/ min})$$
$$= 5.48 \times 10^{-3} \times 0.72 \text{ kg} \cdot \text{m}^2 \times 1,200^2$$
$$= 5,681.66 \text{ J}$$

➤ EXAMPLE 1-14

Repeat Example 1-13 with a radius of 6 cm.

SOLUTION

$$J = \frac{1}{2}mr^2$$
$$= \frac{1}{2} \times 100 \text{ kg} \times \left(\frac{6 \text{ cm}}{100 \text{ cm/m}}\right)^2$$
$$= 0.18 \text{ kg} \cdot \text{m}^2$$
$$E = 5.48 \times 10^{-3} \times J\left(\text{kg} \cdot \text{m}^2\right) \times S^2(\text{r/min})$$
$$= 5.48 \times 10^{-3} \times 0.18 \text{ kg} \cdot \text{m}^2 \times 1,200^2$$
$$= 1,420.42 \; J$$

As can be seen from these two examples, if the diameter of the rotating cylinder is halved, then with the mass and angular velocity remaining constant, the moment of inertia and the kinetic energy are each one-quarter of their original values. This means that if it is necessary to accelerate or decelerate a rotating cylindrical body rapidly, then the moment of inertia must be minimized. There are many industrial and military applications where motors must be capable of rapid speed changes or reversals; these motors are characterized by having long small-diameter rotors, and sometimes the iron in the rotor magnetic circuit has been eliminated to reduce the moment of inertia, for example, moving-coil servomotors.

In applications where a sudden load demand causes large speed fluctuations, such as in a punch press, or where it is necessary to minimize speed fluctuations, as in a reciprocating engine-driven alternator, these effects may be minimized by using a flywheel, which has a large moment of inertia.

The moment of inertia is usually given in the manufacturer's data in the form of

$$J = mk^2 \qquad \textbf{(1–25)(SI)}$$

or

$$J = Wk^2 \qquad \textbf{(1–25)(E)}$$

where J is expressed in slug · ft^2 in the English unit system, or in kg · m^2 in the SI system,

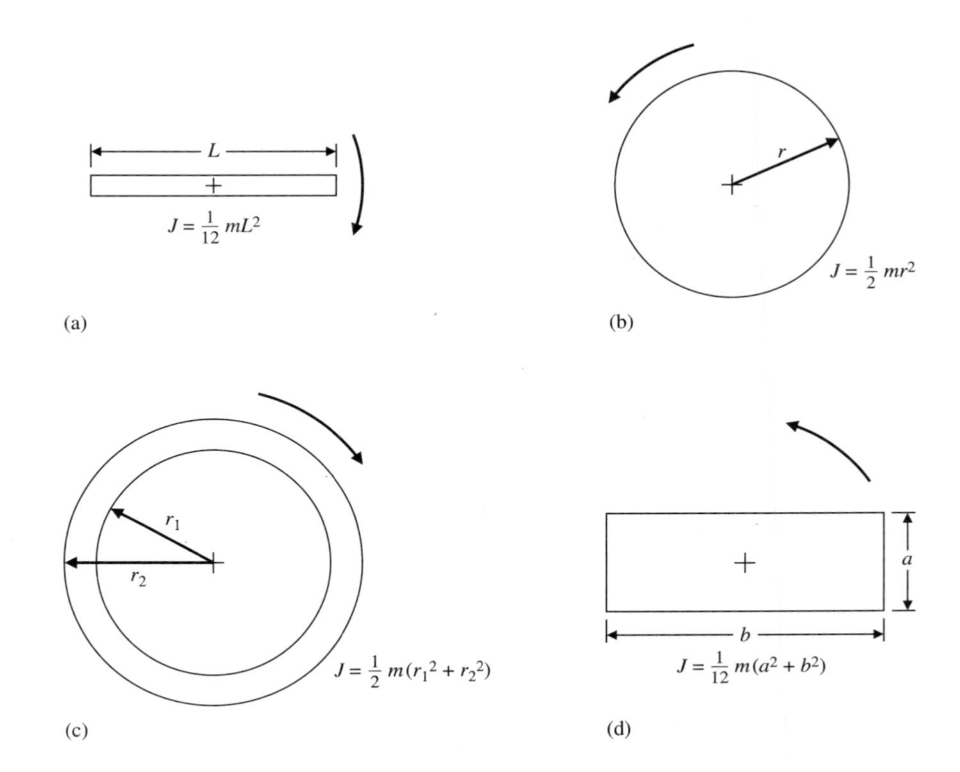

Figure 1-3 Moments of inertia of various shaped bodies. (a) Slender rod of mass m rotating about its center. (b) Solid cylinder of mass m rotating about its center. (c) Hollow cylinder of mass m rotating about its center. (d) Rectangular plate of mass m rotating about its center.

and k is known as the *radius of gyration* and is the distance from the axis of rotation to the point at which the total mass may be concentrated without changing the moment of inertia.

1-4 MAGNETISM

Magnetic fields are an essential element in the conversion of mechanical energy to electrical energy or vice versa, that is, all generators and motors depend on a magnetic field for their operation.

Magnetic Fields

Magnetic fields can occur naturally, for example, the lodestone which was used by the Chinese 2,000 years ago as a navigational aid, or they may be produced by electric current flowing in a coil. They all have certain common characteristics.

1. Like magnetic poles repel each other; unlike magnetic poles attract each other.

2. Magnetic lines of force form closed loops.
3. Magnetic lines of force possess direction.
4. Magnetic lines of force are in a state of tension, which causes them to be as short as possible.
5. Magnetic lines of force repel each other.
6. Magnetic lines of force never intersect.
7. Magnetic lines of force always arrange themselves so that the maximum number of lines of force are established.

These properties are shown in Fig. 1-4.

Electromagnetism

While the magnetic field characteristics described in the foregoing apply to all magnetic fields, they are characteristic of fields produced by permanent magnets, that is, they are constant. In rotating-machine applications, solenoid-operated devices, electromagnets, and so on, it is necessary to be able to control the magnitude and the presence of a magnetic field. This is accomplished by using magnetic fields created by an electric current flowing through an electrical conductor, which, of course, is controllable.

A brief review of the production of magnetic fields by current-carrying conductors is in order. But first we shall use the conventional current flow notation. Recall that this means that in the external circuit current flow is from the positive terminal to the negative terminal through the external load.

Magnetic Fields Surrounding a Current-Carrying Conductor

Figure 1-5 shows concentric magnetic fields that surround current-carrying conductors. The direction of the magnetic field can be determined by using the right-hand grasp rule, which states that if the conductor is grasped in the right hand with the thumb pointing in the direction of the current flow, the fingers will be pointing in the direction of the magnetic field.

If this concept is extended to a multiturn coil wound over a nonmagnetic former, then, as can be seen in Fig. 1-6, a magnetic field is produced which is identical to that of the bar magnet shown in Fig. 1-4(c).

Figure 1-6(a) shows a section of a coil and the magnetic fields produced by the current flow in the coil turns. It can be seen that the fields cancel in the space between adjacent turns, but reinforce each other in the center and the outside of the coil. It should also be noted that there is an attractive force between adjacent turns because the current flows are in the same direction. Figure 1-6(b) shows the production of the magnetic field by a multiturn coil.

A simple rule that can be used to determine the magnetic polarity of a direct-current (dc) energized coil is easily developed from Fig. 1-6(b), namely, grasp the coil in the right hand so that the fingers point in the direction of the current flow, then the thumb will point to the N pole.

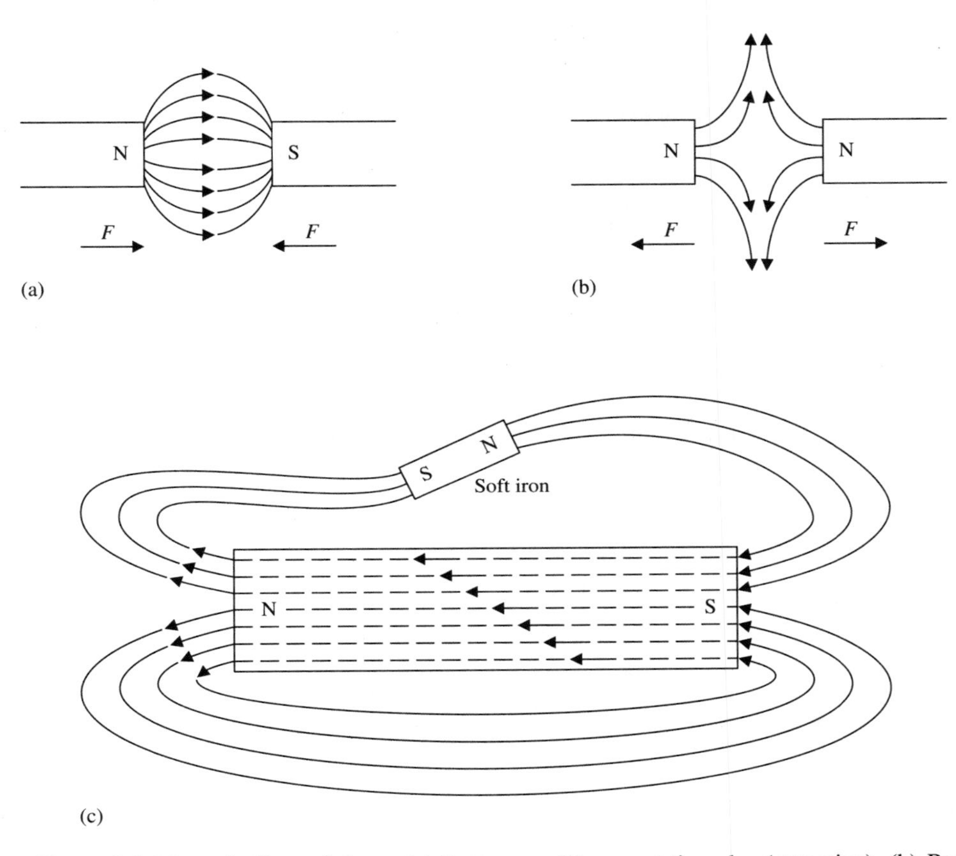

Figure 1-4 Magnetic lines of force. (a) Between unlike magnetic poles (attraction). (b) Between like poles (repulsion). (c) Around a bar magnet (the presence of a magnetic material concentrates the magnetic lines of force).

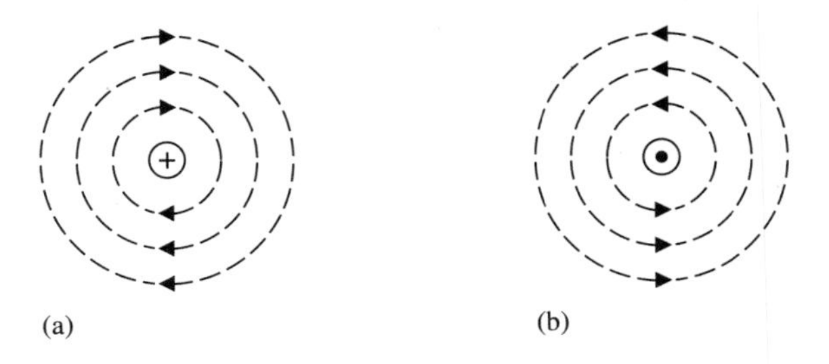

Figure 1-5 Magnetic fields surrounding a current-carrying conductor. (a) Current flowing into paper. (b) Current flowing out of paper.

Magnetic Flux and Magnetic Flux Density

The total number of magnetic lines of force present in a magnetic field is termed the *magnetic flux* Φ. The *weber* (Wb), the SI unit of magnetic flux, is that magnetic flux which, linking a single-turn coil, induces an electromotive force of 1 V in it when the flux is reduced to zero at a uniform rate in 1 s.

Magnetic flux density B is a measure of the concentration of the magnetic flux Φ per unit area. The *tesla* (T) the SI unit of magnetic flux density, is the flux in a magnetic field when 1 Wb of flux occurs in a plane of 1 m^2, that is, 1 T=1 Wb/m^2.

Flux and flux density are related by

$$B = \frac{\Phi}{A} \qquad (1\text{–}26)$$

where B is the flux density in tesla, Φ the total flux in webers, and A the cross-sectional area in meters squared.

It should be noted that the flux density is not necessarily uniform in all cross-sections of a magnetic field. From Figs. (1-4) and (1-6) it can be seen that the magnetic lines of force are concentrated at the pole faces and are less dense at all other positions. This means that the flux density is greatest at the pole faces and internally in the magnets.

➤ EXAMPLE 1-15

The total magnetic flux at the pole face of a bar magnet is 3 × 10–4 Wb. The bar magnet is rectangular and has a cross-sectional area of 2 cm^2. What is the flux density within the magnet?

SOLUTION

$$B = \frac{\Phi}{A} = \frac{3 \times 10^{-4}\,\text{Wb}}{2 \times 10^{-4}\,\text{m}^2} = 1.5\ \text{T}$$

Magnetomotive Force

In the electric circuit current flows when an electromotive force is applied to a complete circuit. The magnitude of the current is determined by Ohm's law. Similarly, in a magnetic circuit magnetic flux is created when a magnetomotive force (mmf) F_m acts on the circuit. Since flux is produced only when a current I flows in the associated electric circuit, it is possible to create a unit of magnetomotive force related to the current flow through a single-turn coil of wire. This relationship is

$$F_m = NI\ A \qquad (1\text{–}27)$$

where F_m is the magnetomotive force in amperes, N the number of coil turns, and I the current flowing through the coil in amperes.

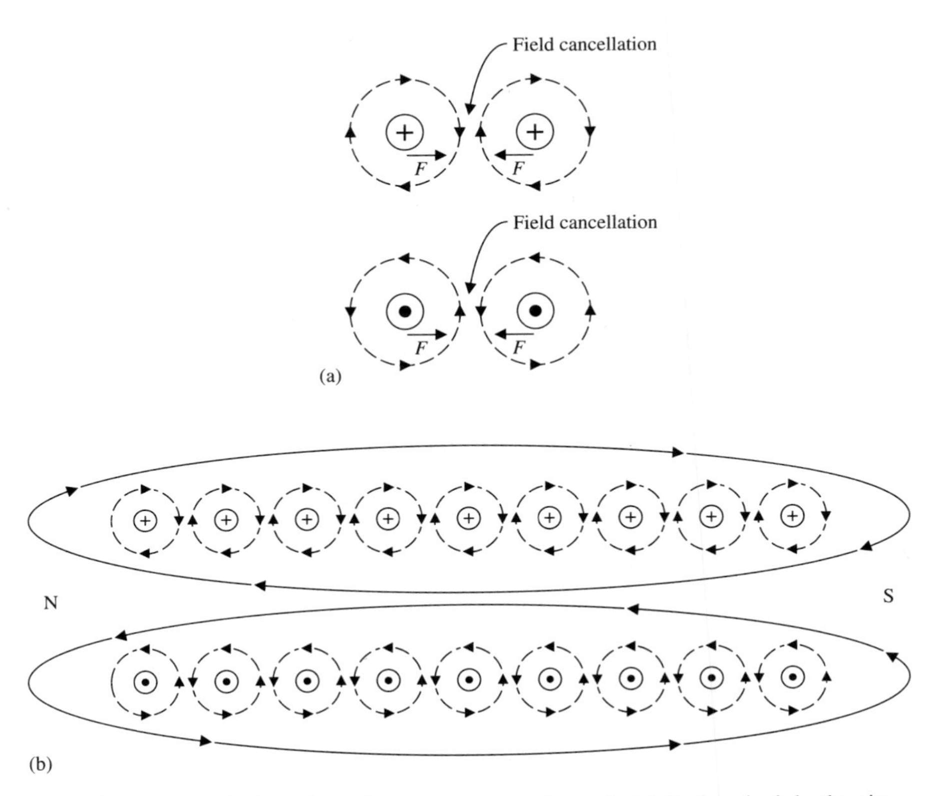

Figure 1-6 Magnetic field produced by a current-carrying coil. (a) Basic principle showing field cancellation and reinforcement. (b) Magnetic field produced by multiturn coil.

Reluctance

In the electric circuit the resistance R of the circuit is the opposition offered by the circuit to the passage of current. In the magnetic circuit there is an analogous unit called *reluctance* R_m, which is the opposition offered by the magnetic circuit to the establishment of the magnetic flux F by the magnetomotive force F_m. These are related as

$$R_m = \frac{F_m \text{ A}}{\Phi \text{ Wb}} \qquad (1\text{–}28)$$

where R_m is the reluctance in A/Wb.

Permeance

Just as in the analysis of parallel electric circuits it is more convenient to think in terms of conductance, similarly, in parallel magnetic circuits it is more convenient to carry out calculations in terms of permeance Pm, where permeance is

$$P_m = \frac{1}{R_m} \tag{1–29}$$

It is expressed as webers per ampere (Wb/A). The SI unit of permeance is the *henry* (H).

Permeance may be defined as the measure of the ability of a magnetic circuit to permit the establishment of a magnetic field.

Permeability

In the electric circuit we compare conductor materials by means of resistivity r, or specific resistance, which is the resistance of a unit length and unit cross section of a conductor material. Then

$$\rho = \frac{\text{volt}}{\text{meter}} \times \frac{\text{meter}^2}{\text{ampere}} = \frac{\text{volt}}{\text{ampere} \times \text{meter}}$$

$$= \text{ohm-meter} = \Omega \cdot \text{m} \tag{1–30}$$

In turn the resistance of the conductor in terms of its physical characteristics is

$$R = \frac{\rho l}{A}\ \Omega \tag{1–31}$$

where R is the conductor resistance, l the conductor length in meters, A the conductor cross-sectional area in meters squared, and ρ the resistivity of the conductor material in ohm-meters at 20° C.

Similarly, in the magnetic circuit the magnetic properties of materials are compared by means of the permeance per unit length and cross-sectional area. The permeance per unit length and cross-sectional area of a magnetic field is called *permeability*.

The permeability of a magnetic material is the measure of the ease with which a magnetic field may be established. Permeability μ is expressed as

$$\mu = P_m \frac{l}{A} = \frac{l}{R_m A} \tag{1–32}$$

where μ is the permeability of the magnetic material, P_m the permeance in Wb/A, R_m the reluctance in A/Wb, l the length of the magnetic circuit in meters, and A the cross-sectional area of the magnetic circuit in meters squared. In the SI system permeability is expressed in henrys per meter (H/m).

Often the material surrounded by a conductor or coil is nonmagnetic, such as air, plastic, or even a vacuum. The permeability in these cases is constant and is termed the *permeability of free space* μ_0,

$$\mu_0 = 4\pi \times 10^{-7}\ \text{Wb/A} \cdot \text{m} \tag{1–33}$$

The permeability of nonmagnetic materials, such as air, glass, wood, and brass, is approximately equal to that for a vacuum. However, magnetic materials such as iron, steel, cobalt, and specialty alloys have permeabilities very much greater than that of free space. The improvement in the ease of establishing magnetic fields in these materials, called *ferromagnetic materials,* is quantified by assigning a relative permeability μ_r to each material. Then

$$\mu = \mu_0 \mu_r \qquad \text{(1–34)}$$

where $\mu_r = 1$ for air and nonmagnetic materials and can range from 500 to 2,500 for irons and steels used with rotating machines and transformers. It should be noted that μ_r is dimensionless. Also μ is not constant for most ferromagnetic materials and therefore must be obtained from experimental data.

Magnetic Field Intensity

Magnetic fields exist around current-carrying conductors and permanent magnets, but so far we do not have a relationship between the strength of the magnetic field and the coil current producing it. The magnetic field depends on the current and the material in which the magnetic field exists. In addition, the relationship is nonlinear for most materials, air being the exception.

Since permeability is based on a unit volume of the magnetic material, it is of little importance to know the magnetomotive force for the entire magnetic circuit. What is of greater importance, is to know the magnetomotive force to create a given flux density in a unit length of the magnetic circuit. The magnetomotive force per unit length is known as the *magnetic field intensity H* and is defined as

$$H = \frac{F_m}{l} \qquad \text{(1–35)}$$

where H is the magnetic field intensity in A/m, F_m the magnetomotive force in amperes, and l the length of the magnetic circuit in meters.

Earlier we defined

$$R_m = \frac{F_m}{\Phi} \text{ and } P_m = \frac{\Phi}{F_m}$$

substituting in Eq. (1-32),

$$\mu = \frac{\Phi}{F_m} \times \frac{l}{A} = \frac{\Phi}{A} \times \frac{l}{F_m}$$

But from Eqs. (1-26) and (1-35)

$$B = \frac{\Phi}{A} \text{ and } H = \frac{F_m}{l},$$

we have

$$\mu = \mu_0 \mu_r = \frac{B}{H} \qquad \text{(1–36)}$$

➤ EXAMPLE 1-16

What are the magnetic field intensity and the magnetomotive force needed to produce a flux of 2×10^{-4} Wb in a steel ring whose mean circumferential length is 100 cm and which has a cross-sectional area of 5 cm²? Assume $\mu_r = 500$.

SOLUTION

$$\Phi = BA = \mu_0 \mu_r HA = \mu_0 \mu_r \frac{NI}{l} A$$

$$F_m = NI = \frac{\Phi l}{\mu_0 \mu_r A} = \frac{2 \times 10^{-4}\,\text{Wb} \times 100 \times 10^{-2}\,\text{m}}{4\pi \times 10^{-7} \times 500 \times 5\,\text{cm}^2 \times 10^{-9}}$$

$$= 636.62\ \text{A} \approx 637\ \text{A}$$

$$\text{H} = \frac{F_m}{l} = \frac{637\ \text{A}}{100 \times 10^{-2}\,\text{m}} = 637\ \text{A/m}$$

Magnetic Moments

From earlier studies of the Bohr atom we know that electrons orbit the nucleus of an atom. The atom is neutrally charged, and when a charge moves in an orbit around a space in a plane, a magnetic moment is generated. The magnetic moment is a magnetic flux which will try to align itself with any external magnetic field that is present. Since all matter has orbiting electrons, obviously all materials have the potential to exhibit magnetic properties. However, as we know, not all materials are magnetic. The answer lies in the way that the magnetic moments of the atoms, molecules, and crystals align themselves. For example, if two equally strong magnetic moments directly oppose each other, they will cancel each other. If partial cancellation happens, there will be a small resultant moment; if they are additive there will be a strong magnetic moment.

Magnetic Domains

All solid materials consist of atoms and molecules formed into regular geometric shapes called crystals. These crystalline structures have in them areas in which the magnetic moments are additive. These areas are called *domains*. The domain has a net magnetic field, which when subjected to an external magnetic field will orient itself with that field.

Consider a bar of ferromagnetic material, that is, a material that can be magnetized, such as steel. In the demagnetized state the magnetic domains are oriented randomly, and the bar will not have an external magnetic field present. This situation is represented in Fig. 1-7(a). It also will be the point 0 of the B–H curve, in Fig. 1-7(d), which describes the state of the magnetic field as H is increased from 0. As the magnetizing force H increases, the flux density also increases, [line $0a$ in Fig. 1-7(d) and Fig. 1-7(b)]. During this stage a number of the domains have aligned themselves with the external magnetic field and the flux density B is increasing lin-

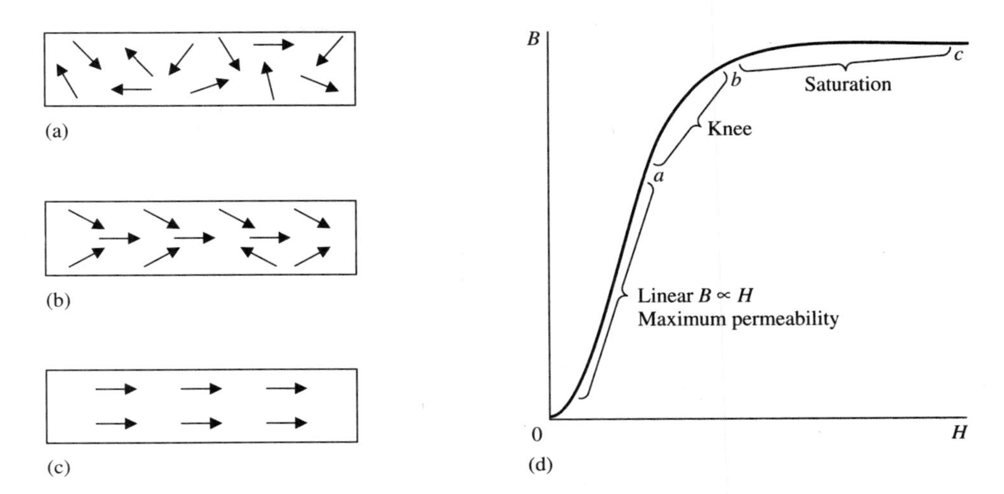

Figure 1-7 Magnetic field domains. (a) Random. (b) Partially oriented. (c) Fully oriented. (d) *B-H* curve.

early as H is increased. Simultaneously the permeability is at a maximum. As H is increased further, a point is reached where the easily aligned domains have already been lined up with the external magnetic field. At this stage, called the *knee* of the *B–H* curve, [curve *ab* in Fig. 1-7(d)], the rate of increase of the flux density B is no longer proportional to the magnetizing force H. This is caused by a small number of domains calling for a greater force to move them into alignment with the external magnetic field. When nearly all the domains have been aligned, further increases in H produce a negligible increase in the flux density B. This region [*bc* in Fig. 1-7(d)] is called *saturation*.

Magnetization Curves

Magnetic flux Φ is usually produced by passing an electric current through a multi-turn coil which creates a magnetomotive force F_m. Of particular importance is the relationship between the flux and magnetomotive force. To have effective comparisons between materials, we compare them in terms of unit quantities, that is, flux density B (Wb/m^2 or T) and magnetizing force H (A/m). The plots of experimental results are called *magnetization*, or *B–H curves*. There is a unique *B–H* curve for each material. Representative curves are shown in Fig. 1-8.

The straight-line graph shows B proportional to H, and the slope is $B/H = \mu = 4\pi \times 10^{-7}$, or the permeability of free space. This curve clearly shows that the magnetic field produced by a coil with an air core or with materials such as wood, glass, fabrics, or plastics is small when compared against the *B–H* curves for ferromagnetic materials, such as cast iron, cast steel, and sheet steel.

However, since the slope of the *B–H* curve is equal to the permeability, it can be

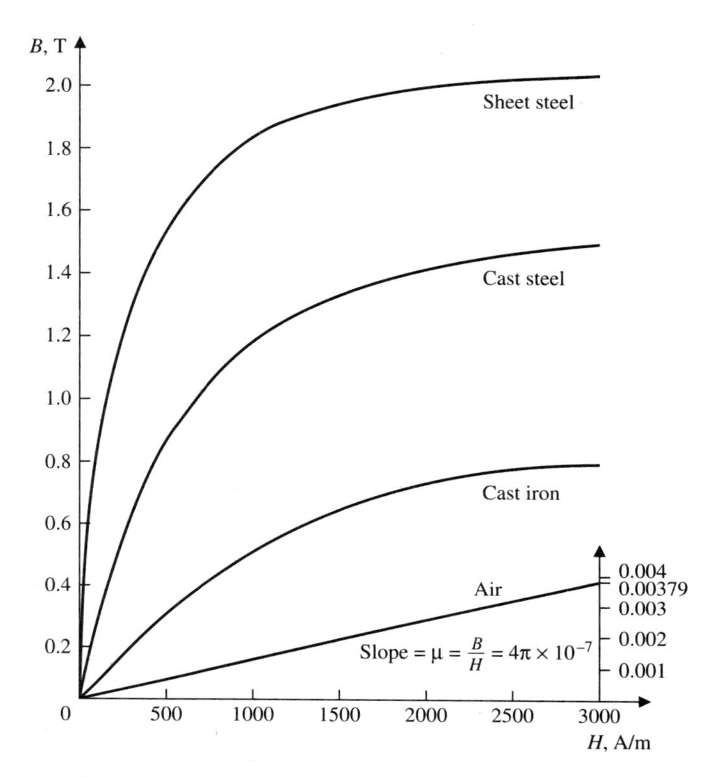

Figure 1-8 Magnetization or *B-H* curves.

seen that the ferromagnetic materials have varying permeabilities which depend on the magnetizing force *H*.

Consider the hypothetical *B–H* curve shown in Fig. 1-9. It can be divided into four regions. As *H* is increased slowly from 0, the *lower knee* 0*a* indicates the alignment of the magnetic domains whose axes are nearly parallel to the applied magnetic field. The steepest region *ab* shows the movement of most domains into alignment with the applied magnetic field. In this region the increase in flux density *B* is proportional to *H*. Since it is also linear, the permeability is constant and at its maximum. The region *bc* is called the *upper knee* and indicates the increasing magnetizing force that must be applied to force the alignment of these domains with the applied magnetic field instead of with their crystal axes. In this region the flux density *B* is no longer proportional to *H*, and the permeability is decreasing. The region *cd* is known as *saturation* and represents the region where most of the domains have been aligned with the external magnetic field. This region represents the practical limit to any further attempts to increase the flux density. In this region the permeability of the material is approaching 0. The variations of permeability μ versus the flux density *B* are shown in Fig. 1-10. It should also be noted that the permeability of ferromagnets is affected by temperature increases. At a temperature called the *Curie temperature* the material will act as a paramagnetic material. This temperature varies for different ferromagnetic materials and is thought to be due to thermal agitation, which causes the magnetic domains to cease to exist.

Hysteresis

In Fig. 1-9 a previously unmagnetized ferromagnetic sample was subjected to the influence of an increasing positive magnetizing force. Figure 1-11 shows the variations of flux density as the magnetizing force is varied from 0 to a positive maximum, to 0, to a negative maximum, through to a positive maximum.

As H is increased from 0 to its positive maximum, the corresponding changes of B are shown by the curve $0a$. As H is decreased to 0 from its positive maximum, B changes as illustrated by the segment ab. This shows that B decreases, but not as rapidly as the decrease in H. This is the result of some domains remaining aligned with the axis of the magnetic field. When H is 0, there is a significant retained flux, which is represented by $0b$. This property of retaining some magnetic field is known as *retentivity* or *residual magnetism*.

If H is increased from 0 to a negative maximum, the flux density decreases to 0 and then increases negatively to its negative maximum along the path bd. Where the flux density curve crosses the 0 axis point c, the sample does not display any external magnetic field. The distance $0c$ represents the opposing magnetizing force that is needed to reduce the retained or residual magnetic field $0b$ to 0. This is known as the *coercive force*.

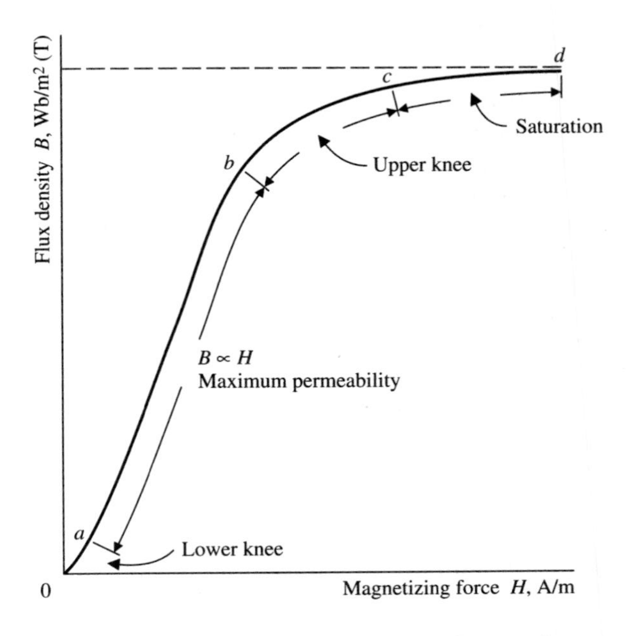

Figure 1-9 B-H curve for ferromagnetic material showing significant regions.

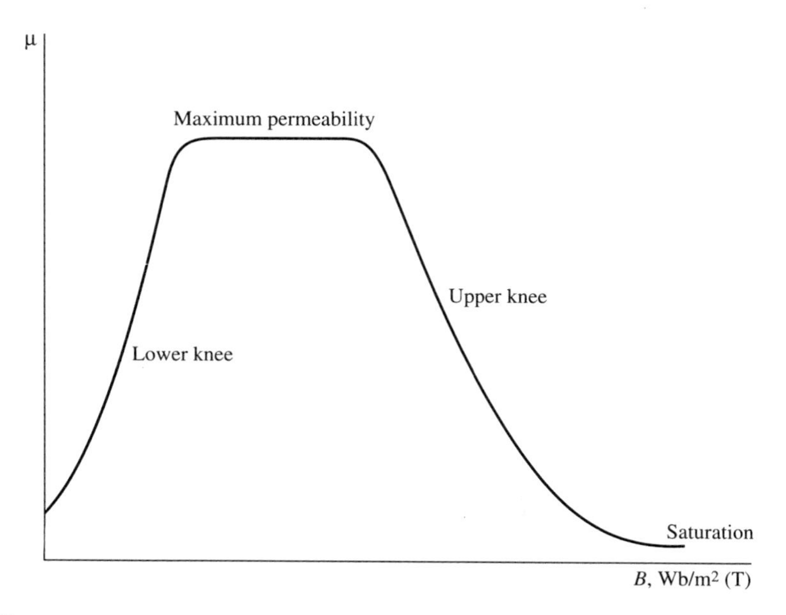

Figure 1-10 Plot showing variation of permeability with increasing flux density for ferromagnetic material.

As H is decreased from its negative maximum to 0 and then to its positive maximum, the flux density B decreases to 0 and then increases to its positive maximum along the curve *defa*. When H is 0, the retained magnetic field is 0*e*, which is approximately equal to 0*b* but of reversed polarity. Similarly, as the flux density crosses the 0 axis at point *f*, the reversed coercive force 0*f* is approximately equal to 0*c*. As H is increased to its positive maximum, the flux density increases to its positive maximum at point *a*.

If H is varied cyclically between $+H$ and $-H$, the flux density B will vary cyclically along the path *abcdefa*. This path is called the *hysteresis loop*.

Each type of ferromagnetic material has its own distinct hysteresis loop. The property by which B lags behind H is called *hysteresis* and is the property that permits us to make permanent magnets. However, as can be seen, energy must be expended to reverse the magnetic field as a material is subjected to cyclic variations of H, which is the case in alternating-current (ac) applications such as single- and polyphase motors and transformers. The area enclosed by the hysteresis loop represents the power loss involved in overcoming the retained magnetic field. The hysteresis loss is

$$P_h = K_h f B_m^n \text{ W} \tag{1–37}$$

where K_h is a constant determined by the ferromagnetic material and its dimensions, f the frequency in hertz, B_m the maximum flux density, and $1.5 \leq n \leq 2.5$, n being an empirically determined constant.

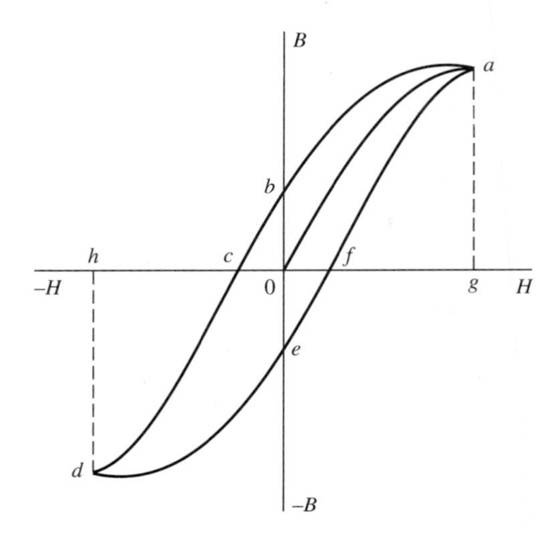

Figure 1-11 Typical hysteresis loop.

Hysteresis losses also occur in rotating machines. This is caused by the reversal of the magnetic domains as a rotating ferromagnetic rotor rotates past a fixed magnetic field. In transformer and rotating-machine applications it is essential that steels having a hysteresis loop with a minimum enclosed area be chosen. The hysteresis loss is classified as an *iron loss* and appears as heat.

Typical hysteresis loops for cast iron, sheet steel, and ferrites are shown in Fig.1-12. The ferrite material is used for magnetic memories, and, as can be seen, B remains approximately constant until H becomes equal to the coercive force in the reverse direction at which point B reverses.

Eddy Currents

Voltages are induced in a closed conductor system when there is relative motion between a magnetic field and a conductor. The relative motion can be caused by a closed conductor moving past a stationary magnetic field, or vice versa. An alternating magnetic field linking a closed conductor system will also induce a voltage in the conductor system. The latter is the main cause of eddy currents in the magnetic circuits of transformers and rotating machines.

Figure 1-13(a) shows a solid ferromagnetic core surrounded by a coil through which an alternating current is flowing. In turn, this coil produces an alternating flux along the coil axis. This alternating flux induces an alternating voltage in the core, which gives rise to circulating currents called *eddy* or *Foucault currents*. Since the core material has a low ohmic resistance, these currents can be quite significant and as a result cause the core material to heat because of the I^2R losses.

Eddy currents can be reduced by increasing the path resistance. This is achieved by splitting the core into thin sheets parallel to the core flux. These sheets are called

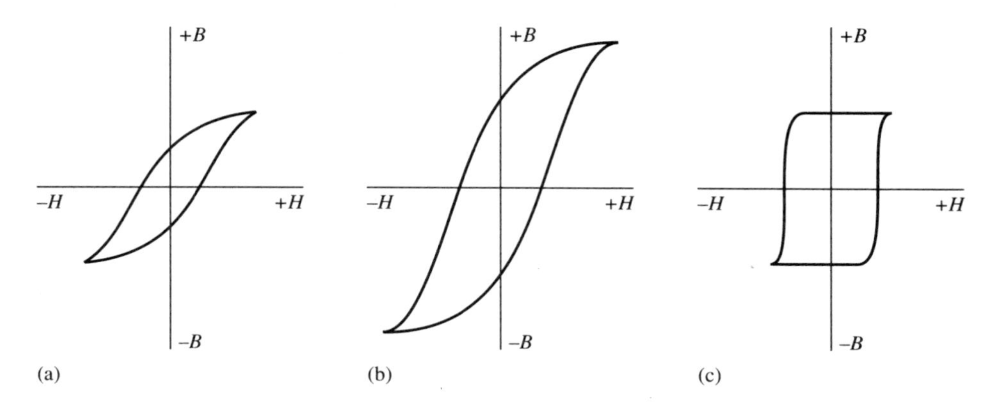

Figure 1-12 Typical hysteresis loops. (a) Cast iron. (b) Sheet silicon steel. (c) Ferrite.

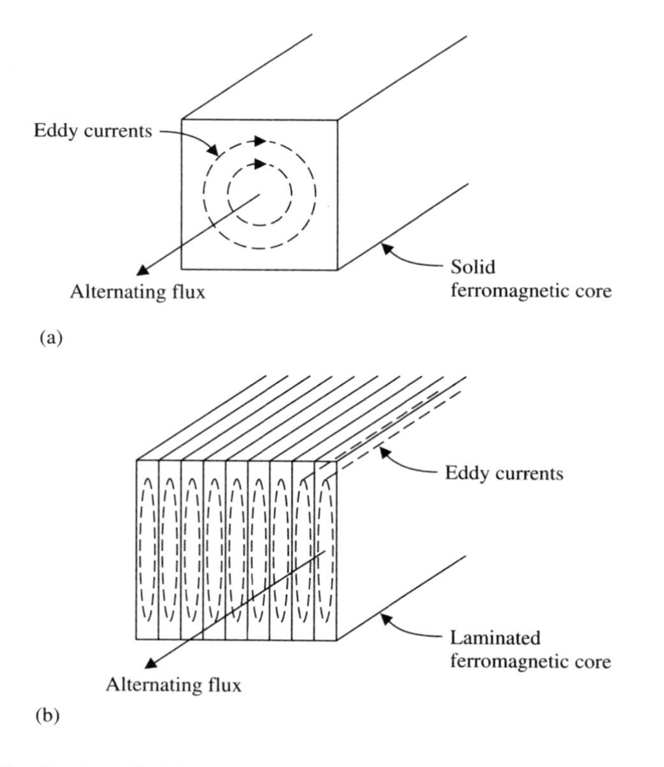

Figure 1-13 Production of eddy current. (a) In a solid ferromagnetic core. (b) In a laminated ferromagnetic core.

laminations [Fig. 1-13(b)]. To prevent an electrical path between the laminations, they are coated with a thin surface skin of oxide supplemented by an insulating varnish. Typical lamination thicknesses are 0.014 in (0.36 mm), 0.0185 in (0.47 mm), and 0.025 in (0.64 mm). Laminations are cut or punched to the desired shape and stacked on top of each other until the desired depth is obtained. Then they are tightly clamped to prevent movement.

The eddy current loss is

$$P_e = K_e B_{max}^2 f^2 t^2 \text{ W} \tag{1–38}$$

where K_e is a constant determined by the ferromagnetic material and the core dimensions, B_{max} is the peak flux density, f the frequency, and t the lamination thickness.

Laminated cores are effective in minimizing eddy currents to 50 kHz, although handling laminations as thin as 25 μ in can be a problem. At higher frequencies the core usually consists of powdered ferromagnetic material bonded together by a nonconducting material.

Eddy currents can also be produced if there is a variation of flux density. They also have the effect of reducing the rate of change of flux density. In dc machines the armature, which is an integral part of the magnetic circuit, rotates past the stationary-field pole system. If the armature were made of solid steel, an induced voltage would be produced which would be acting parallel to the axis of rotation. In turn this voltage would cause eddy currents to flow parallel to the shaft. These eddy currents, because of the low path resistance, would create a large I^2R loss which appears as a heat loss. This problem is solved by fabricating the armature from insulated punched laminations keyed to the shaft to prevent rotation.

Classification of Magnetic Materials

Any substance or material can be classified into one of the five following categories, depending on the type of magnetic behavior it shows.

Nonmagnetic Materials. *Nonmagnetic* materials have the same permeability as free space and hence have no effect on magnetic fields. Typical nonmagnetic materials are wood, glass, and rubber.

Diamagnetic Materials. *Diamagnetic* materials have a permeability slightly less than that of free space and exhibit a slight repulsion to magnetic lines of force. The diamagnetic effect is very small and can only be detected by very sensitive instruments. Examples of diamagnetic materials are copper, mercury, boron, silicon, phosphorus, inert gases, such as argon and helium, and most organic compounds.

Paramagnetic Materials. *Paramagnetic* materials have a permeability that is slightly greater than that of free space, that is, the materials are very slightly magnetic. Examples of paramagnetic materials are manganese, chromium, cobalt, nickel, platinum, and silver. Ferromagnetic materials can be classified as paramagnetic, but they are normally treated as a separate group because of their strong magnetic properties.

Ferromagnetic Materials. *Ferromagnetic* materials have a permeability many times greater than that of free space. They are strongly magnetic and are of special interest to the electrical industry. Some pure elements such as iron, nickel, cobalt, and rare-earth materials are strongly ferromagnetic, although most of the ferromagnetic materials used in the electrical industry contain varying percentages of other elements.

Ferromagnetic materials used in the electrical industry can be divided into two main groups, magnetically *soft* and magnetically *hard* materials. Soft magnetic materials, or electrical steels, in either sheet or strip form are used extensively as magnetic core materials. They are easily magnetized or demagnetized and have high permeabilities, low coercive forces, and low core losses, that is, the combined hysteresis and eddy current losses. These characteristics, combined with their low cost, have led to their use for stacked or wound magnetic cores for transformers, motors, and reactors operating at power-line frequencies. Hard magnetic materials possess a high energy content, that is, they are not easily magnetized or demagnetized, have a high retentivity, and are principally used as permanent magnets. Typical *B-H* curves of commercial magnetic materials are shown in Fig. 1-14.

Ideally iron would be the perfect magnetic material if it were not for eddy currents. Silicon is added to increase its volume resistivity, which reduces the magnitude of the eddy currents and therefore the eddy current loss, but it also affects the grain

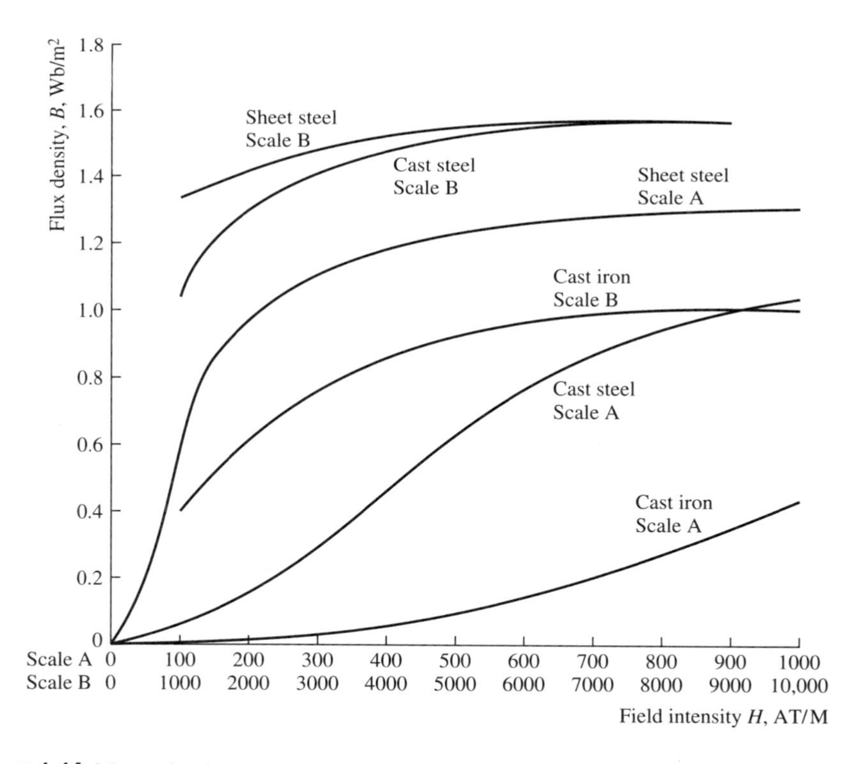

Figure 1-14 Magnetization curves.

structure, which results in a reduction of the hysteresis loss. It also reduces the permeability. This means that the silicon content must be chosen to produce an acceptable compromise between the reduction of the core losses and the reduction of the permeability. Other elements, such as manganese and aluminum, are added to improve the metallurgical properties and are usually restricted to amounts of between 0.1 to 0.5%. Carbon is always present in steels, but it is usually kept to less than 0.01%, because it increases the coercive force and decreases the permeability. The carbon content is reduced by special heat treatment techniques at the steel mill for fully processed steels, and in semi-processed steels by annealing by the electrical manufacturer.

Electrical steels are generally broken down into five classifications:

1. *Nonoriented steels,* in which the magnetic properties are about the same in all directions.
2. *Grain-oriented steels,* in which the grains have been aligned by the rolling process to give the highest permeability in the direction of rolling.
3. *Thin electrical steel,* typically 0.001–0.007 in (0.03–0.18 mm), which are supplied in either the nonoriented or the grain-oriented form and are mostly suitable for the 400–3,000-Hz frequency range.
4. *Fully processed steels,* in which the full magnetic characteristics have been developed by the steel mill. They are usually heat treated by the electrical manufacturer to relieve manufacturing stresses.
5. *Semi-processed steels,* which are shipped from the steel mill rolled to their final thickness. The final annealing to fully develop their full magnetic properties is done by the electrical manufacturer.

The choice of whether to use nonoriented or grain-oriented steels is dependent on their application. In nonoriented steels there is little difference between the magnetic properties in the direction of rolling. This property is preferred for induction motor stator and rotor laminations because there are a number of flux paths. Grain-oriented steels have a marked difference in the permeabilities and core losses with and across the grain. The difference between exciting currents with and across the grain can be as much as 20 times greater than that for nonoriented steels. Therefore grain-oriented steels are used primarily where the flux flows in well-established paths, such as in large power transformers. The use of grain-oriented steels in transformer applications results in smaller cores and lower costs and increases the power ratings that can be manufactured and shipped over highways or railroads. Another application of grain-oriented steels is in the rotors of large two-pole turboalternators.

Since the core loss is a major factor in the selection of magnetic core materials, electrical steels are graded in terms of their core losses. The system in most common use has been established by the American Iron and Steel Institute (AISI). Originally the AISI grading system used the letter M (magnetic material) followed by a number, which represented the core loss of that grade. However, because of the improvement in core losses resulting from technological changes, the number no longer is truly representative of the actual core losses. Nevertheless, the numbers still represent the relative core losses between different grades of steel. Table 1-8 shows the general characteristics and applications of nonoriented and grain-oriented steels.

From Eq. (1-38) the eddy current loss is proportional to the square of the lamination thickness. This therefore limits the practical thickness of laminations at 60 Hz to approximately 0.025 in (0.64 mm). This gauge of laminations is used primarily for small motors and intermittently operated devices such as relays, where a high core loss is acceptable. To ensure a more moderate core loss at 60 Hz, 0.0185 in (0.47-mm) thick laminations are used. To reduce core losses even further, in the 60–200-Hz range 0.012–0.14 in (0.30–0.36 mm) thick laminations are used. However, the cost of manufacturing laminations thinner than this increases rapidly, as does the cost of punching the laminations. For frequencies of between 400 and 3,000 Hz laminations between 0.001 and 0.007 in (0.025 and 0.178 mm) are used sometimes, but only when it is not practical to increase the volume resistivity of the material by increasing the alloy content.

Because of the *skin effect* caused by the eddy currents, crowding the magnetic flux from the center of the laminations under ac operating conditions, the effective thickness of the laminations is less than under dc operating conditions. Hence for ac operation the permeability is also less than under dc operating conditions.

Since laminations are punched to form quite intricate shapes and because there can be interlaminar losses, laminations are usually coated with organic or inorganic insulating coatings. These coatings have been assigned AISI designations as follows:

1. *C-0*: Natural oxide coating formed during heat treatment. It is suitable for fractional-horsepower motors, relay armatures, pole pieces, and small power transformers.
2. *C-1*: An organic varnish coating used in non-oil-immersed applications.
3. *C-2*: An inorganic ceramic glasslike film suitable for cores operating in an air or oil-immersed environment. Typically this is used for distribution transformers and saturable reactors.
4. *C-3*: An enamel coating suitable for both air-cooled and oil-immersed cores. Typical applications are medium-size power and distribution transformers and continuous-duty high-efficiency rotating machines.
5. *C-4*: A chemically treated surface coating suitable for both air-cooled and oil-immersed cores. Typical applications are medium-size power and distribution transformers and continuous-duty high-efficiency rotating machines.
6. *C-5*: An inorganic high-resistance coating similar to the C-4 coating, but with more fillers, which give it a very much higher insulation resistance. Typical applications are large power and distribution transformers and large rotating machines.

Specialty Alloys. Besides the electrical steels many specialty magnetically soft materials have been developed to satisfy specialized niches in the magnetic spectrum. One group is the nickel-iron alloys, where as much as 78% nickel combined with specialized heat treatment produces materials with permeabilities as high as

Table 1-8 Characteristics and applications of electrical steel laminations.

Type	Characteristics	Applications
M4, M5, M6	Grain-oriented, very low core loss and high permeability in direction of rolling	High-performance power and distribution transformers with minimum weight/kVA
M7, M8	As above	Large generators and power transformers
M15	Lowest core loss, nonoriented steel, high permeability at low flux densities	Large power transformers and high-efficiency rotating machines
M19, M22, M27	Low core loss and good permability at low and medium flux densities	Reactor cores, communication transformers and reactors, and high-efficiency rotating machines
M36, M43	Very suitable for punched laminations, moderate core loss and good permeability at medium to high flux densities	High-efficiency continuous-duty rotating machines, voltage regulators, chokes, and appliance motors
M45, M47	Especially suitable for punched laminations, ductile, high core loss and good permeability at high flux densities	Fractional-hp motors, appliance motors, relays, and ballasts

900,000, very low coercive forces, resistivities of 86 μΩ/cm, and high saturation flux densities. A typical example of these alloys is Deltamax, which is a 50% nickel alloy with a saturation flux density of 1.5 T and a permeability of 85,000. It is used for the cores of pulse transformers and magnetic amplifiers.

Another family of alloys are the iron-nickel-copper-chromium alloys, in which the addition of copper and chromium to high-percentage nickel-iron alloys has been found to produce high permeabilities at low flux densities. Typical members of this grouping are marketed under the names of Mumetal, 1040 alloy, and Hymu 80. These materials are principally used in the cores of magnetic amplifiers and for magnetic shielding.

To minimize waveform distortion a number of constant-permeability alloys have been developed. These alloys, such as Conpernik and Isoperm, usually have a nickel content of between 40 and 55% and are cold worked.

From Eqs. (1-37) and (1-38) the hysteresis loss is proportional to B^n_{max}, where n is usually about 1.5, while the eddy current losses are proportional to B^2_{max}, f^2, and t^2. It is therefore obvious that in high-frequency applications great care is needed in the selection of ferromagnetic materials to minimize iron losses, especially the eddy current loss. In addition, high resistivities will also reduce the magnitude of the eddy currents. These requirements are met by using high-permeability alloys, either as very thin tape, as compressed powdered iron alloys, or by sintered ferrites.

Tape applications usually are supplied by high-permeability nickel-iron alloys with thicknesses of between 0.001 and 0.010 in (0.0254 and 0.254 mm). They are most suitable for frequencies between 100 Hz to 100 kHz.

Compressed powdered iron alloy cores consist of powdered iron grains about 10 μm (10^{-3} cm) in diameter, coated with about a 1 μm film of insulating varnish. These small beads are mixed with a phenol resin binder, and the combination is then subjected to high pressure in a mold and baked. This treatment produces a magnetically stable ferromagnetic core, which is approximately 90 wt% pure iron. It has also the additional advantage that it can be machined. These cores are used in high-volume applications where cost is important. Typical applications include dc filter inductors and electromagnetic interference inductors (EMI) and are characterized by low core losses.

Ferrites, although similar to ferromagnetic materials, have a different magnetic mechanism and are classified as *ferrimagnetic*. Ferromagnetic materials were considered in terms of the atom, that is, pure materials. In ferrimagnetic materials the molecule possesses a magnetic moment. Also, ferrimagnetic materials are mixtures of two different compounds, each with its own different magnetic moments. As a result, coupling forces exist between the molecules of the compounds that hold the magnetic moment of like molecules in parallel and those of dissimilar molecules in a nonparallel arrangement. Therefore there is a net magnetic moment caused by the unbalance of the magnetic moments.

Ferrite cores consist of a mixture of metallic oxide powders where two oxides, such as Fe_2O_3 and NiO, are mixed in the correct proportions, then ground together, molded, and sintered at about 1,300° C for several hours. The end product is similar to ceramic materials, that is, brittle and hard. The oxygen ions in the ferrites insulate the metallic ions, and hence ferrites have resistances in the range of those of semiconductor materials. Typical resistivities are 1×10^3 to 1×10^8 μΩ . cm. As a result of the nonparallel alignment of the magnetic moments, saturation densities are usually less than 0.5 T, with permeabilities as high as 5,000.

Magnetic Circuits

Magnetic circuits are analogous to electric circuits, and similar techniques can be used in their solution. Magnetomotive force F_m, flux Φ, and reluctance R_m are the magnetic circuit equivalents of electromotive force V, current I, and resistance R.

Series Magnetic Circuits. The series magnetic circuit is similar to the series electrical circuit. Just as in the electric circuit, where the current I is common to all electric elements in series, the flux Φ is common to all magnetic elements in series. The law for magnetic circuits is similar to Kirchhoff's voltage law and is called *Ampere's circuital law*, which states: "The algebraic sum of the magnetomotive force drops around the magnetic circuit is equal to the applied magnetomotive force."
This may be stated for the general case as follows:

$$F_m = NI = H_1 l_1 + H_2 l_2 + \cdots + H_n l_n \tag{1–39}$$

$$= F_{m1} + F_{m2} + \cdots + F_{mn} \tag{1–40}$$

➤ EXAMPLE 1-17

A flux of 0.2 Wb/m^2 is to be produced in a toroid of cast steel (Fig. 1-15). The mean diameter of the toroid is 10 cm. Calculate the required current and the relative permeability of the cast steel if a coil of 1,000 turns is wound around the toroid.

SOLUTION

The length of the mean magnetic path l is

$$l = \pi \times 0.1 \text{ m} = 0.314 \text{ m}$$

From Fig. 1-14 for $B = 0.2$ Wb/m^2, H for cast steel (scale A) = 250 A/m.
Then from Eq. (1-39),

$$F_m = NI = Hl$$

Therefore

$$I = \frac{Hl}{N} = \frac{250 \text{ A/m} \times 0.314 \text{ m}}{1{,}000 \text{ turns}} = 0.0785 \text{ A} = 78.5 \text{ mA}$$

From Eq. (1-36),

$$\mu = \mu_0 \mu_r = \frac{B}{H}$$

Then

$$\mu_r = \frac{B}{\mu_0 H} = \frac{0.2 \text{ Wb/m}^2}{4\pi \times 10^{-7} \times 250 \text{ A/m}} = 637$$

➤ EXAMPLE 1-18

Repeat Example 1-17 with $B = 1.5$ Wb/m^2

SOLUTION

From Fig. 1-14 with B = 1.5 Wb/m2, H for cast steel (scale B) = 4,000 A/m. Then

$$I = \frac{Hl}{N} = \frac{4{,}000 \text{ A/m} \times 0.314 \text{ m}}{1{,}000 \text{ turns}} = 1.26 \text{ A}$$

$$\mu_r = \frac{B}{\mu_0 H} = \frac{1.5 \text{ Wb/m}^2}{4\pi \times 10^{-7} \times 4{,}000 \text{ A/m}} = 298$$

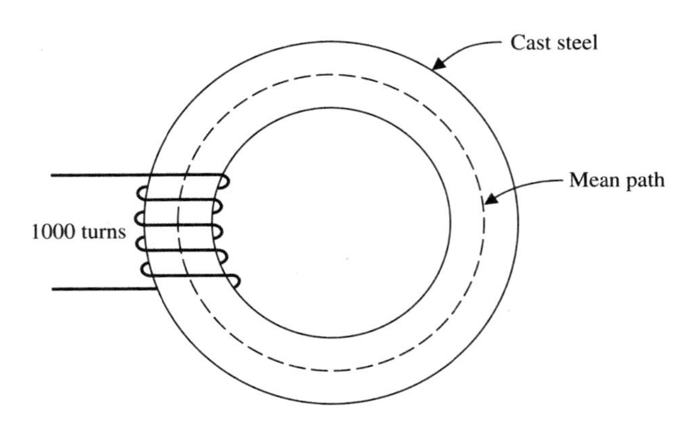

Figure 1-15 Toroid for Examples 1-17 and 1-18.

From Examples 1-17 and 1-18 it can be seen that the relative permeability is not constant.

The previous examples dealt with a magnetic circuit with one type of material. Many magnetic circuits have two or more ferromagnetic materials including air gaps.

➤ EXAMPLE 1-19

Find the current needed to create a flux of 1.5×10^{-4} Wb in the magnetic circuit shown in Fig. 1-16.

SOLUTION

$$B = \frac{\Phi}{A} = \frac{1.5 \times 10^{-4}\ \text{Wb}}{1.25 \times 10^{-4}\ \text{m}^2} = 1.20\ \text{Wb/m}^2$$

From Fig. 1-14, for $B = 1.20\ \text{Wb/m}^2$

Sheet steel, $H_{ab} = 400$ A/m (scale A)
Cast steel, $H_{bcda} = 1{,}500$ A/m (scale B)
Mean path length of sheet steel, $l_{ab} = 0.15$ m
Mean path length of cast steel, $l_{bcda} = 0.35$ m

From Eq. (1-39),

$$\begin{aligned} NI &= H_{ab}l_{ab} + H_{bcda}l_{bcda} \\ &= 400\ \text{A/m} \times 0.15\ \text{m} + 1,500\ \text{A/m} \times l_{bcda} \\ &= 60\ \text{A} + 525\ \text{A} = 585\ \text{A} \end{aligned}$$

$$I = \frac{NI}{N} = \frac{585\ \text{A}}{100\ \text{turns}} = 5.85\ \text{A}$$

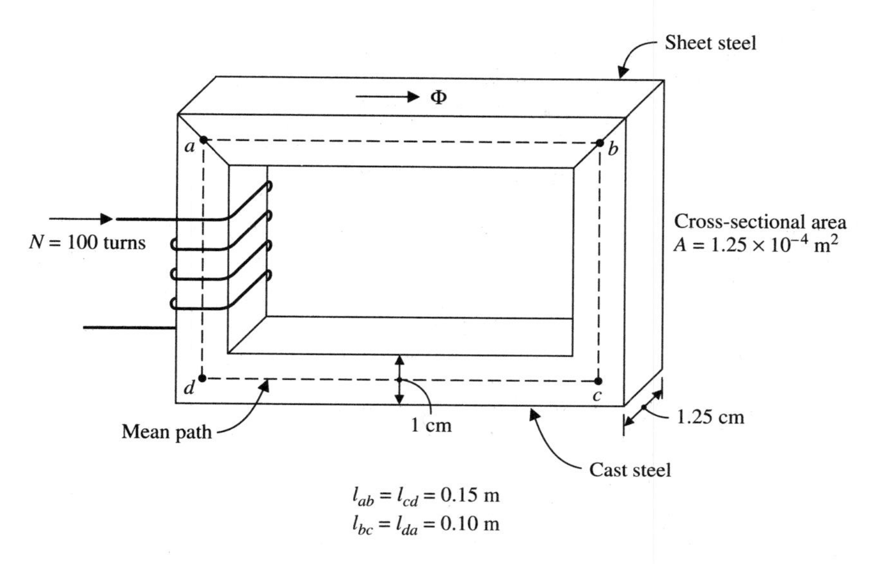

Figure 1-16 Magnetic circuit for Example 1-19.

Air Gaps and Their Effects in the Magnetic Circuit. In many applications the flux path is not solely confined to the high-permeability path of the ferromagnetic materials. In these cases the magnetic flux must cross one or more air gaps in the magnetic circuit, for example, the air gaps between stationary (stator) and rotating (rotor) parts of an ac or dc motor or generator (Fig. 1-17). Since the relative permeability of air is $\mu_r = 1$, the effect of an air gap is to increase the total reluctance of the magnetic circuit as all the flux must pass through the air gap.

As the magnetic lines of force cross the air gap, they spread out because the individual lines repel each other. This spreading-out effect, known as *fringing,* also results in the air-gap flux density being less than that in the immediately adjacent ferromagnetic material. Figure 1-18 shows the fringing effect and the method of calculating the effective cross-sectional area of the air gap. The effect of the air gap in a magnetic circuit is illustrated in the following example.

➤ EXAMPLE 1-20

Repeat Example 1-19, assuming that there is a 2 mm air gap between *a* and *b*.

SOLUTION

The effective cross-sectional area of the air gap is

$$A_{ag} = (1 + 0.2)(1.25 + 0.2) = 1.74 \text{ cm}^2 = 1.74 \times 10^{-4} \text{ m}^2$$

Therefore the air-gap flux density is

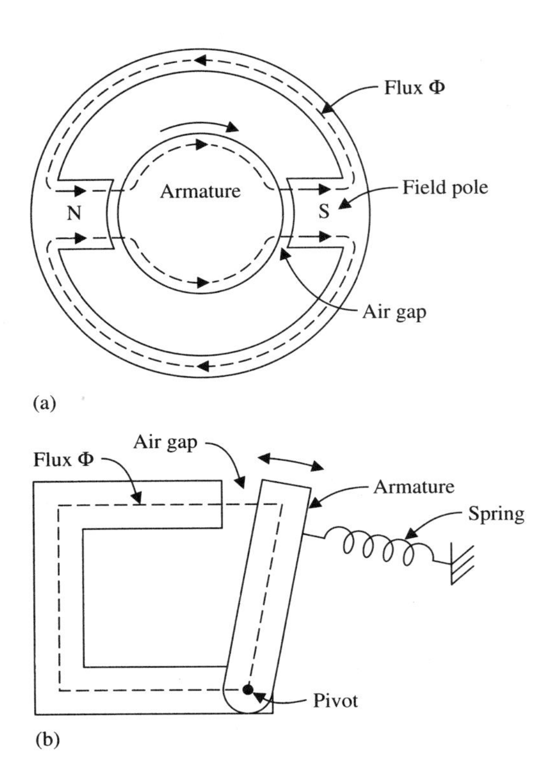

Figure 1-17 Air gaps. (a) Dc machine. (b) Electromagnetic relay.

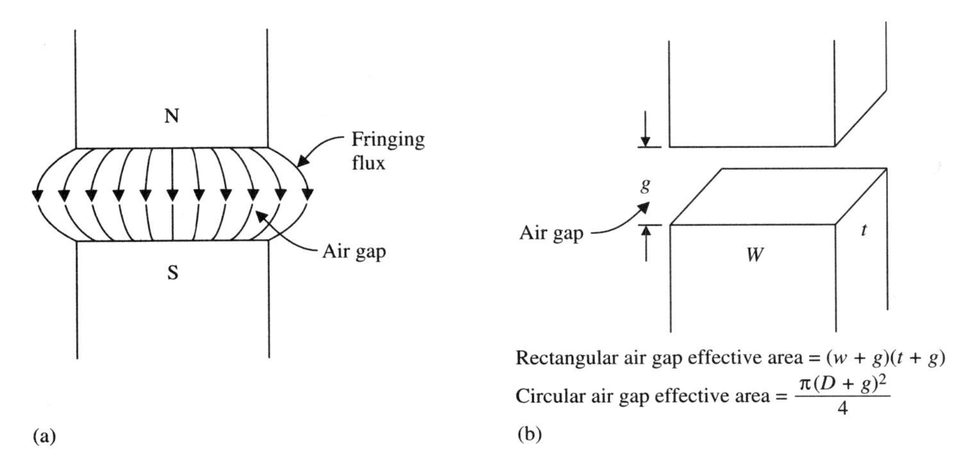

Figure 1-18 Fringing. (a) Effect of air gap. (b) Effective cross-sectional areas.

$$B_{ag} = \frac{\Phi}{A_{ag}} = \frac{1.5 \times 10^{-4}\ \text{Wb}}{1.74 \times 10^{-4}\ \text{m}^2} = 0.86\ \text{Wb/m}^2$$

The flux density in the ferromagnetic material is

$$B = \frac{\Phi}{A} = \frac{1.5 \times 10^{-4}\ \text{Wb}}{1.25 \times 10^{-4}\ \text{m}^2} = 1.20\ \text{Wb/m}^2$$

From Fig. 1-14 for $B = 1.2\ \text{Wb/m}^2$:

Sheet steel, $H_{ab} = 400$ A/m
Cast steel, $H_{bcda} = 1{,}500$ A/m
Magnetic intensity H_{ag} for air gap,

$$H_{ag} = \frac{B_{ag}}{\mu_0} = \frac{0.86\ \text{Wb/m}^2}{4\pi \times 10^{-7}} = 6.84 \times 10^5\ \text{A/m}$$

Mean path length of sheet steel, $l_{ab} = 0.15 - 0.002 = 0.148$ m
Mean path length of cast steel, $l_{bcda} = 0.35$ m
Mean path length of air gap, $l_{ag} = 0.002$ m

From Eq. (1-39)

$$NI = H_{ab}l_{ab} + H_{bcda}l_{bcda} + H_{ag}l_{ag}$$

$$= 400\ \text{A/m} \times 0.148\ \text{m} + 1{,}500\ \text{A/m} \times 0.35\ \text{m} + 6.84 \times 10^5\ \text{A/m} \times 0.002\ \text{m}$$

$$= 59.20\ \text{A} + 525\ \text{A} + 1{,}368.73\ \text{A} = 1{,}952.93\ \text{A}$$

Therefore

$$I = \frac{NI}{N} = \frac{1{,}952.93\ \text{A}}{100\ \text{turns}} = 19.53\ \text{A}$$

As can be seen, the effect of even a 2 mm air gap is quite dramatic. The major part of the magnetomotive force is needed to establish the desired flux in the air gap. Conversely, if a constant magnetomotive force is maintained, then the effect of an air gap is to reduce the overall flux density. This technique will prevent saturation in a magnetic circuit.

Parallel Magnetic Circuits. In the magnetic circuits considered so far, the magnetic flux has been common to all parts of the circuit. As can be seen in Fig. 1-17(a), the magnetic flux produced by the field-pole magnetomotive forces divides equally between the two paths. This is similar to the parallel electric circuit, where the current divides in direct proportion to the conductances of the parallel paths. In the parallel magnetic circuit the flux divides in proportion to the permeance of the parallel paths. The magnetic equivalent of Kirchhoff's current law is

$$\Phi_T = \Phi_1 + \Phi_2 + \Phi_3 + \ldots + \Phi_n \qquad (1\text{–}41)$$

➤ EXAMPLE 1-21

In the parallel magnetic circuit of Fig. 1-19(a) calculate the coil current if the coil has 50 turns to obtain a flux of 1.0×10^{-4} Wb in the path *bcde*.

SOLUTION

The cross-sectional area A is

$$A = (2 \text{ cm} \times 2 \text{ cm}) \times 10^{-4} \text{ m}^2$$

Then the flux density in the Φ_1 path *bcde* is

$$B_{bcde} = \frac{\Phi_1}{A} = \frac{1.0 \times 10^{-4} \text{ Wb}}{4 \times 10^{-4} \text{ m}^2} = 0.25 \text{ Wb/m}^2 \text{ (T)}$$

From Fig. 1-14 for cast steel (scale A), $H_{bcde} = 280$ A/m. The length of the Φ_1 path is

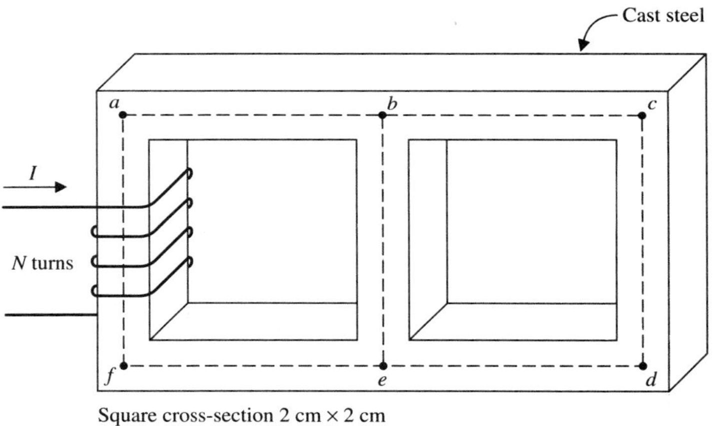

(a)

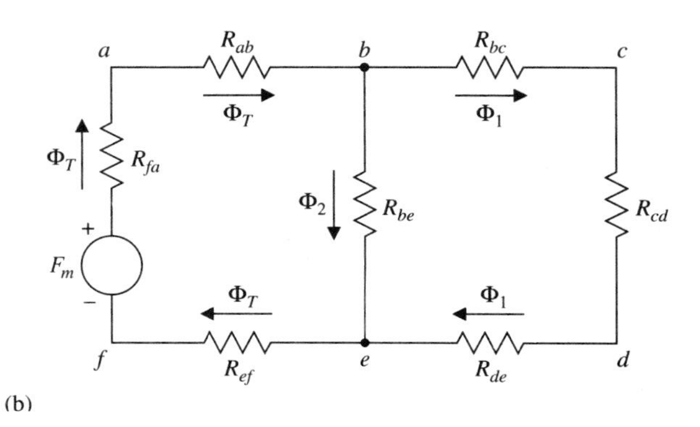

(b)

Figure 1-19 Parallel magnetic circuit. (a) Isometric view. (b) Equivalent electrical analog.

$$l_{bc} + l_{cd} + l_{de} = 10 \text{ cm} + 10 \text{ cm} + 10 \text{ cm} = 30 \text{ cm} = 0.30 \text{ m}$$

Therefore the required magnetomotive force is

$$H_{bcde}\, l_{bcde} = 280 \text{ A/m} \times 30 \times 10^{-2} \text{ m} = 84 \text{ A}$$

Using Ampere's circuital law around the loop *bcdea*, we find $H_{be} l_{be}$. Then since

$$\sum_{bcdea} F_m = 0$$

we have for the assumed flux directions,

$$H_{bcde} l_{bcde} - H_{be} l_{be} = 0$$

$$84 \text{ A} - H_{be} l_{be} = 0$$

$$H_{be} l_{be} = 84 \text{ A}$$

The magnetizing force for the path *be* is

$$H_{be} = \frac{84 \text{ A}}{0.10 \text{ m}} = 840 \text{ A/m}$$

From Fig. 1-14 for H_{be} = 840 A/m (scale *A*), B_{be} = 0.97 T. Therefore Φ_2 is

$$\Phi_2 = B_{be} A = 0.97 \times 4 \times 10^{-4} = 3.88 \times 10^{-4} \text{ Wb}$$

From Eq. (1-41),

$$\Phi_T = \Phi_1 + \Phi_2$$

$$= 1.0 \times 10^{-4} + 3.88 \times 10^{-4} = 4.88 \times 10^{-4} \text{ Wb}$$

The flux density in the path *efab* is

$$B_{efab} = \frac{\Phi_T}{A} = \frac{4.88 \times 10^{-4} \text{ Wb}}{4 \times 10^{-4} \text{ m}^2} = 1.22 \text{ T}$$

From Fig. 1-14 for cast steel (scale *B*), when B_{efab} = 1.22 T, H_{efab} = 1,550 A/m. Therefore the magnetomotive force for the path *efab* is

$$H_{efab} l_{efab} = 1{,}550 \text{ A/m} \times 30 \times 10^{-2} \text{ m} = 465 \text{ A}$$

Using Ampere's circuital law around the path *abefa*, the total magnetomotive force needed to produce the desired flux is

$$NI = H_{efab} l_{efab}$$

$$= 465 \text{ A}$$

Hence the desired current is

$$I = \frac{NI}{N} = \frac{465 \text{ A}}{50 \text{ turns}} = 9.3 \text{ A}$$

It can be seen that because of the repetitive nature of this type of calculation, the use of a programmable calculator or a personal computer will relieve the tedium of repeated calculations. In actual practice, all rotating machines and transformers are designed almost exclusively using computer-aided design (CAD).

Magnetic Unit Systems. So far in this section we have used the SI system of magnetic units. The English system of magnetic units is still used extensively in the United States. In the English system of magnetic units, flux is in lines, flux density is in lines per square inch, magnetomotive force is in ampere-turns, and field intensity is in ampere-turns per inch. Irrespective of which system is used, it is essential that all units in a specific problem be expressed in the same system of units (see Table 1-9). Tables 1-10 and 1-11 show the conversion factors between SI and English magnetic units.

As should be suspected, the permeabilities of free space differ for the two systems. Recall

$$H = \frac{B}{\mu}$$

Therefore

$$\mu = \frac{B\left(\text{Wb/m}^2\right)}{H\ (\text{A}\cdot\text{m})}$$

Assuming $\mu_r = 1$, then

$$\mu_0 = 4\pi \times 10^{-7} \times \frac{\text{Wb}}{\text{A}\cdot\text{m}} \times \frac{10^8\ \text{lines}}{\text{Wb}} \times \frac{1\ \text{At}}{1\ \text{A}} \times \frac{1\ \text{m}}{39.35\ \text{in}}$$

$$= 4\pi \times 10^{-7} \times \frac{10^8\ \text{lines}}{39.35\ \text{At}\cdot\text{in}} = 3.19\ \text{lines/At}\cdot\text{in}$$

for the English system of magnetic units.

Permanent Magnets

Permanent magnets are *magnetically hard*, that is, they are difficult to magnetize but have a high residual flux density B_r and a high coercive force H_c as compared to the magnetically soft materials discussed previously. In order that the effectiveness of permanent magnets may be assessed, it is necessary to understand their operating characteristics.

Permanent magnets operate on the demagnetization part of the hysteresis curve (Fig. 1-20). The curve B_r–H_c shows the flux density that exists for any given value of external demagnetizing force less than H_c. The value of B_r shown exists only if the permanent magnet forms part of a closed magnetic circuit. However, in all practical applications one or more air gaps are present. These air gaps act on the permanent magnet as demagnetizing forces. As a result the actual flux density will be less than B_r.

Table 1-9 Comparison of SI and English magnetic units.

Magnetic Term	SI	English
Flux Φ	Wb	lines
Flux density B	Wb/m² (T)	lines/in²
Magnetomotive force F_m	A	At
Field intensity H	A/m	At/in

Table 1-10 Conversion of SI to English magnetic units.

SI Unit	English Unit
1 Wb (Φ)	10^8 lines
1 Wb/m² (B)	64.52×10^3 lines/in²
1 A (F_m)	1 At
1 A/m (H)	2.54×10^{-2} At/in

Table 1-11 Conversion of English to SI units.

English Unit	SI Unit
1 lines (Φ)	10^{-8} Wb
1 lines/in² (B)	1.552×10^{-5} Wb/m²
1 At (F_m)	1 A
1 At/in (H)	39.4 A/m

The demagnetization curve B_r–H_c can be thought of as having two components, the first strictly due to the magnetic material, called the *intrinsic induction*, and the second resulting from the flux density of free space with the magnetic material removed, called the *normal curve*. From Fig. 1-20 it can be seen that the intrinsic and normal curves have the same flux density at B_r, but at H_c, the *intrinsic coercivity* is much greater than the normal coercivity H_c. This is very significant since it is a measure of the ability of a permanent-magnet material to oppose any demagnetizing forces that may be present, and it is an important parameter in assessing the performance of a permanent magnet.

Another method of assessing permanent-magnet materials is in terms of the *energy product,* which is the product BH, whose units are kJ/m³. The curve of the energy product is also shown in Fig. 1-20.

In permanent-magnet applications the usual problem is to establish a given flux Φ in an air gap of a specific length. Then

$$V = \frac{2W}{BH} \tag{1–42}$$

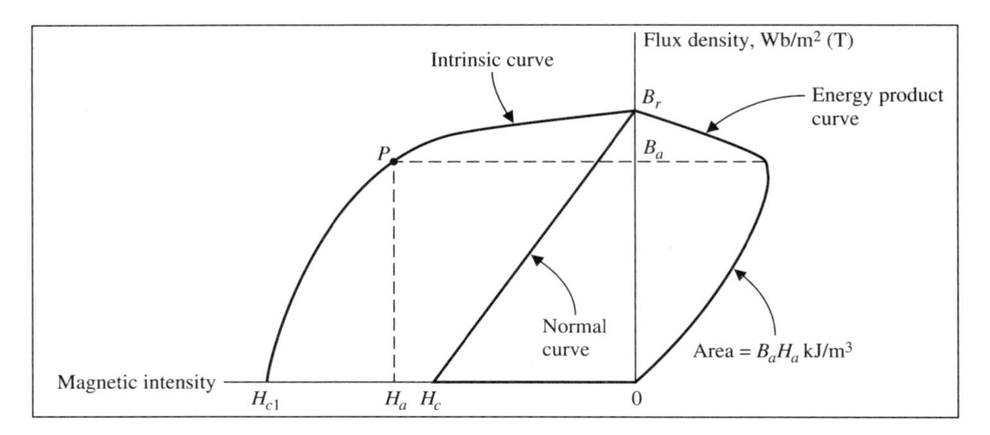

Figure 1-20 Characteristics of permanent-magnet material.

where V is the volume of the permanent-magnet material in cubic meters, W the magnetic energy in the air gap in joules, B the flux density in Wb/m^2, and H the magnetic field intensity produced by the magnet in A/m.

It can be seen from Eq. (1-42) that for a given magnetic energy in the air gap, the product BH must be as large as possible to minimize the volume of the magnet.

It is possible to manufacture permanent magnets from almost any steel that can be hardened by suitable heat treatment. However, the best results are obtained if they are manufactured from specially selected materials. These materials can be grouped into four classes:

1. Precipitation-hardened alloys
2. Quench-hardened alloys
3. Ceramic magnet materials
4. Rare-earth materials

Precipitation-Hardened Alloys. The most prominent members of this group are the cast alloys or *Alnico* (aluminum-nickel-cobalt), which were discovered by Mishma in 1932. The characteristics of the Alnico family are that they have the highest energy product of any permanent-magnet material. For example, Alnico V-7 has an energy product of 56–64 kJ/m^3, as compared to 0.005 kJ/m^3 for silicon steel, a higher coercivity (110–160 kA/m for Alnico V-7), lower retentivity than electrical steels, and it is only slightly affected by temperature. This latter feature, where B decreases 0.01–0.02% per °C increase in temperature, makes these alloys especially suitable for use in rotating machines.

These magnets are produced by casting or sintering. However, they are mechanically weak, hard, and brittle, and are usually machined by grinding. Also the castings shrink significantly when being cooled. They must be subjected to a strong magnetic field intensity H of about 159–238 kA/m to achieve the required magnetization.

A family of Alnico alloys consisting of varying percentages of aluminum, nickel, and cobalt, and small percentages of elements such as titanium, with the balance being iron, are used to obtain permanent magnets with different retentivities, coercivities, and energy products to meet a wide range of applications. Typical demagnetization curves are shown in Fig. 1-21.

The curves for Alnico V and V-7 with residual flux densities of 1.2 and 1.3 T, respectively, make these magnets particularly suitable for small high-torque permanent-magnet (PM) motors. There are several disadvantages to their use in motor applications, such as cost and their moderate coercivity, which can lead to demagnetization of the magnetic field system under locked-rotor conditions. This can be minimized, however, by using current-limit protection.

Other members of the precipitation-hardened family include Remalloy, a cobalt (12%) and molybdenum (17%) alloy that is cast and hot-rolled at 1,200–1,300°C (2,192–2,372°F) before machining and then aged for 1 h at 650–700°C (1,202–

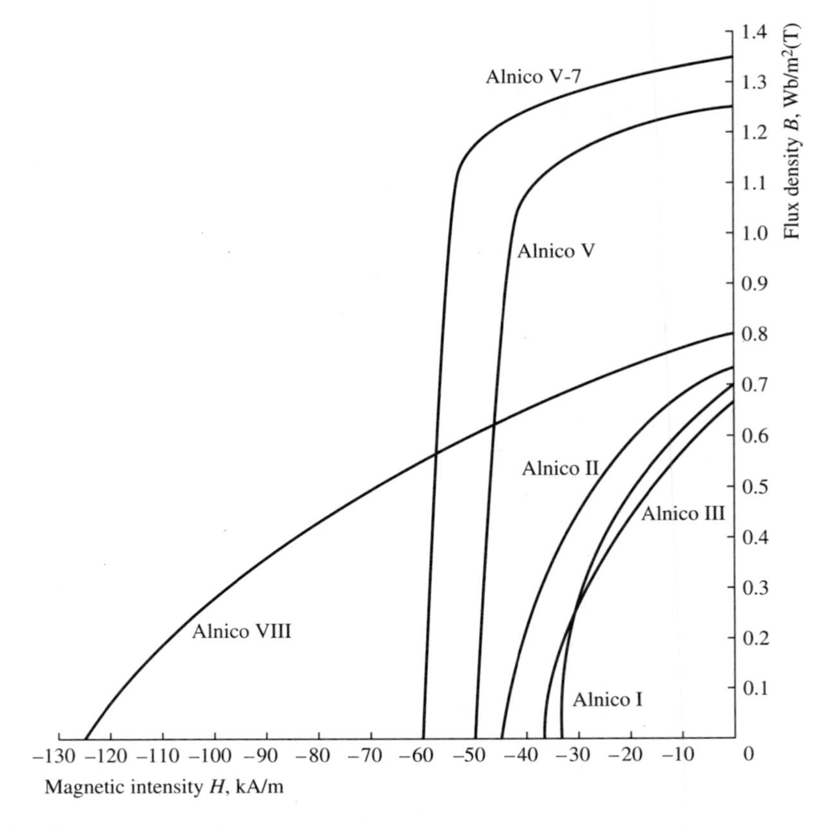

Figure 1-21 Demagnetization curves for Alnico permanent magnets.

1,292°F); and Cunife, a copper-nickel-iron alloy that has the advantage of being malleable and ductile as well as possessing good machining properties. Its retentivity, however, drops significantly with temperature increases and it becomes nonmagnetic at about 400°C.

Quench-Hardened Alloys. Quenched carbon steel has the peculiar property that both the coercivity and the retentivity increase as the carbon content increases to 0.85%. If the carbon content is increased further, the retentivity decreases.

Slight increases in both coercivity and retentivity can be obtained with tungsten steel containing about 5% tungsten, 0.5% manganese, and 0.6–0.8% carbon. A less expensive substitute for tungsten steel is chrome magnetic steel, which has the same ingredients as tungsten steel plus 2–6% chromium.

Cobalt-chrome magnet steel (11% cobalt and 9% chromium) and cobalt magnet steel (36% cobalt, 4% tungsten, and 6% chromium), while possessing good working and machining properties, are expensive because of the cobalt content.

Platinum-iron alloys (60–90% platinum) possess very high coercivities as a result of specialized heat treatment. Platinum-cobalt alloys achieve energy products comparable to Alnico V. However, due to the high cost of platinum, their use is restricted to high-temperature applications because of their high Curie temperature of about 1,100°C (2,012°F).

Ceramic Magnet Materials. Ceramic magnet materials have resistivities approaching those of insulators. They usually consist of metals such as manganese (Mn), nickel (Ni), zinc (Zn), barium (Ba), magnesium (Mg), copper (Cu), and iron (Fe) alloyed with ferric oxide (Fe_2O_3), the exact combinations and proportions usually being proprietary information. The ingredients are ground to a fine powder, mixed, formed in dies under heavy pressure, into the desired shapes, in the presence of a strong magnetic field, and sintered. Usually the parts are made as much as 40–60% oversize to allow for volume shrinkage. Machining is usually carried out by grinding because of the very abrasive nature of the material.

Ceramic magnets have lower retentivities than Alnico magnets, but their coercivities are much greater, and as a result the energy product is about the same as that for the Alnico family.

Ceramic magnets have a nearly linear demagnetization curve and a temperature sensitivity of 0.2%/°C. The magnets are usually magnetized in their final location with magnetizing forces as great as 788 kA/m, the magnetizing current being supplied from a capacitor-discharge or dc electromagnet system.

Ceramic magnets consisting of oriented barium ferrites in a rubber binder are in common use as door gaskets. Recently, flexible magnets with the fields oriented perpendicular to the top and bottom surfaces of the strip have been used in permanent-magnet motors.

Rare-Earth Metals. A number of rare earths, such as samarium, lanthanum, cerium, praseodymium, and misch metal, which occurs naturally when mixed with cobalt, exhibit good retentivity $B_r = 0.9$ T, very high coercivities of 675 kA/m, and very high energy products. To give a relative comparison between energy products, the Alnicos typically range from 11 to 56 kJ/m^3, with the majority between 36 and 40 kJ/m^3, ferrites usually have energy products of between 11 and 36 kJ/m^3, while samarium-cobalt permanent magnets have energy products of between 120 and 175 kJ/m^3.

As can be seen, the rare-earth–cobalt combinations raise the possibility of very significant reductions in weight and volume of permanent magnets. The major drawback is cost, first because cobalt is not in abundant supply, and second because the amount and cost of processing the materials to achieve the desired magnetic properties are high.

General Motors has substituted iron for cobalt and produced a ferrite–rare-earth magnet composed of neodymium-iron-boron ($Nd_2Fe_{14}B$) with a high magnetocrystalline anistropy, that is, the ability to resist rotation of the magnetic moment. This material is produced by heating the iron, boron, and neodymium in an argon atmosphere at 1,000°F (537.8°C) in a quartz crucible. The molten alloy is then ejected against a spinning copper disk which rotates at approximately 19 m/s. The alloy flies off the disk and forms particles which cool at about $1 \times 10^{6\circ}$C/s. Cooling at this rate forms grains of about 50 nm (5×10^{-10} m). It has been found that keeping the grain size to 50 nm or less permits the formation of single magnetic domains, which will reverse when subjected to a reversing magnetic field. This material has been called Magnequench by General Motors and has an energy product of 239 kJ/m^3, which matches that of the most powerful permanent magnets, but at reduced cost. This permanent magnet has been developed to produce more efficient, lighter, and less expensive motors in the low and subfractional horsepower (kilowatt) range.

1-5 FARADAY'S LAW OF ELECTROMAGNETIC INDUCTION

In 1831 Faraday discovered what was to become one of the most important advances in electromagnetism and is known as *Faraday's law of electromagnetic induction.* This law establishes the relationship between the magnetic flux and the voltage in a circuit. It states that the amplitude of the induced voltage in a coil is proportional to the time rate of change of the magnetic flux ($\Delta\Phi/\Delta t$) linking the coil, and to the number of coil turns N. Recalling the SI definition of the weber (Wb), which states that a voltage of 1 V will be produced when the magnetic flux linking a 1-turn coil varies at the rate of 1 Wb/s, the voltage induced in a 1-turn coil is

$$e = -\frac{\Delta\Phi}{\Delta t} \tag{1–43}$$

Since in an N-turn coil the same flux links with all the turns, the voltage induced in the whole coil is

$$e = -N\frac{\Delta\Phi}{\Delta t} \tag{1–44}$$

where e is the voltage induced in an N-turn coil, $\Delta\Phi$ the change of flux linking the coil in webers, and Δt the time interval in seconds during which the flux changes.

Faraday's law relates the amplitude of the induced voltage to the rate of change of flux, and is fundamental to the production of an electromotive force in all rotating machines and transformers. The induced electromotive force is produced by one of the following means:

1. The movement of a closed conductor system with respect to a stationary magnetic field. This is the principle of operation of all dc machines.
2. The movement of a magnetic field with respect to a stationary conductor system. This technique is used in all hydroelectric and steam turbine driven generators (alternators).
3. The linking of a time-varying magnetic field with a stationary conductor system, the basis of transformer action.

From Eq. (1-44) there are a number of factors that affect the magnitude of the induced voltage. They are as follows:

1. The induced voltage e is directly proportional to the rate of change of the magnetic flux $\Delta\Phi/\Delta t$, that is, the greater or lesser the rate of change, the greater or lesser is the induced voltage.
2. The induced voltage e is directly proportional to the number of coil turns N linked by the changing flux.

➤ EXAMPLE 1-22

The magnetic field linking a 100-turn coil changes from 5×10^{-3} to 3×10^{-3} Wb in 0.02 s. What is the amplitude of the induced voltage?

SOLUTION

$$\Delta\Phi = (5 \times 10^{-3} - 3 \times 10^{-3})\ \text{Wb}$$

$$= 2 \times 10^{-3}\ \text{Wb}$$

$$\Delta t = 0.02\ \text{s}$$

Therefore

$$e = -N\frac{\Delta\Phi}{\Delta t} = -100 \times \frac{2 \times 10^{-3}\ \text{Wb}}{0.02\ \text{s}} = 10\ \text{V}$$

➤ EXAMPLE 1-23

Repeat Example 1-22: (**a**) with a 1,000-turn coil; (**b**) with $\Delta t = 0.01$ s.

SOLUTION

(a)
$$e = -1{,}000 \times \frac{2 \times 10^{-3}\ \text{Wb}}{0.02\ \text{s}} = 100\ \text{V}$$

(b)
$$e = -100 \times \frac{2 \times 10^{-3}\ \text{Wb}}{0.01\ \text{s}} = 20\ \text{V}$$

1-6 LENZ'S LAW

The minus sign in Eq. (1-44) shows that the direction of the induced electromotive force is such as to create, in a closed conductor system, a current that produces a flux opposing the flux change ΔF. This is known as Lenz's law. The induced current flows in such a direction as to oppose, by electromagnetic action, the change causing it. For example, if the induced current is caused by a decrease in the magnetic flux linking a coil, then the induced current is in such a direction as to create a magnetic field opposing the decreasing flux. In another example, which occurs in dc machines when there is an increase in load, the load current creates a magnetic field that produces a countertorque which reduces the angular velocity (r/min) of the machine.

1-7 VOLTAGE INDUCED IN A CONDUCTOR

Figure 1-22 shows the movement of a single conductor in three possible directions with respect to a stationary magnetic field. In Fig. 1-22(a) the conductor is moving with velocity v parallel to the field. Since it does not link the magnetic field, that is, the conductor does not experience a change of flux, no electromotive force is induced in the conductor.

Figure 1-22(b) shows the conductor moving with velocity v at right angles to the magnetic field. As a result the magnetic field is cut at the maximum rate by the conductor. The induced electromotive force will be a maximum. Figure 1-22(c) shows the conductor moving with velocity v at an angle Θ with respect to the magnetic field. The magnitude of the induced electromotive force is proportional to the velocity component perpendicular to the magnetic field.

The magnitude of the induced electromotive force in all these cases is

$$e = Blv \sin Q \qquad \textbf{(1-45)(SI)}$$

or

$$e = Blv \sin \Theta \times 10^{-8} \qquad \textbf{(1-45)(E)}$$

where e is the induced voltage in volts, B the flux density in Wb/m2 in SI units or lines/in2 in English units, l the conductor length in the magnetic field in meters (SI) or inches (English units), v the conductor velocity in m/s (SI) or in/s (English units), and Q is in degrees.

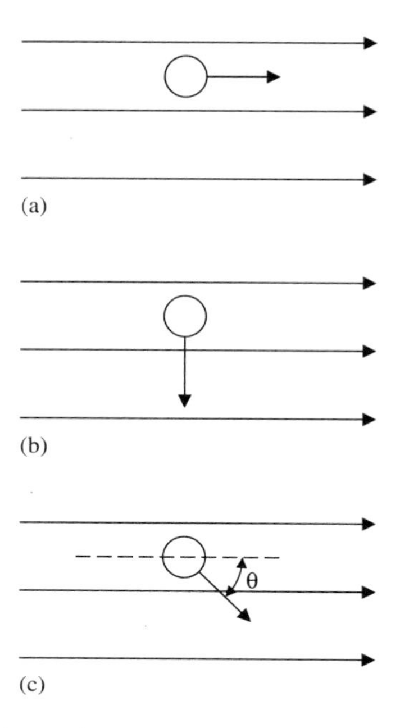

Figure 1-22 Voltage induced in conductor. (a) Moving parallel to magnetic field. (b) Moving at right angles to magnetic field. (c) Moving at angle Θ to magnetic field.

➤ EXAMPLE 1-24

A conductor 12 in (0.305 m) long is located on the periphery of the armature of a dc generator 18 in (0.457 m) in diameter, which is revolving at 1,200 r/min (125.66 rad/s). If the flux density under the field poles is 7,500 lines/in^2 (0.116 Wb/m^2), calculate the induced electromotive force in the conductor.

SOLUTION

(a) In SI units,

$$v = \pi d \times \text{r/s} = \pi \times 0.457 \text{ m} \times \frac{1{,}200 \text{ r/min}}{60 \text{ s}} = 28.71 \text{ m/s}$$

$$e = Blv \sin \Theta$$

$$= 0.116 \text{ Wb/m}^2 \times 0.305 \text{ m} \times 28.71 \text{ m/s} \times \sin 90°$$

$$= 1.018 \text{ V}$$

(b) In English units,

$$v = \pi d \times \text{r/s} = \pi \times 18 \text{ in} \times \frac{1{,}200 \text{ r/min}}{60 \text{ s}} = 1{,}130.97 \text{ in/s}$$

$$e = Blv \sin \Theta \times 10^{-8}$$

$$= 7{,}500 \text{ lines/in}^2 \times 12 \text{ in} \times 1{,}130.97 \text{ in/s} \times \sin 90° \times 10^{-8}$$

$$= 1.018 \text{ V}$$

The direction of the induced electromotive force in the conductor is given by Fleming's right-hand rule, (Fig. 1-23), which states that if the thumb and the index and middle fingers are mutually extended at right angles to each other, then if the index finger is pointed in the direction of the magnetic field and the thumb in the direction of motion of the conductor, the middle finger will point in the direction of the induced voltage.

1-8 VOLTAGE INDUCED IN A COIL

Equations (1-45)(SI) and (E) determine the instantaneous value of the induced voltage in a conductor. However, we are more interested in the average value of the induced electromotive force. From Fig. 1-24(a) it can be seen, assuming that the conductor is rotating clockwise at a constant angular velocity, that at point *a* (0°) the conductor is traveling parallel to the magnetic field. Since it is not being linked by the field, it will not have an electromotive force induced in it. From point *a* to point *b* (0° to 90°) the conductor is cutting the magnetic lines of force at an increasing rate until, at point *b*, it is traveling at right angles to the magnetic field. Therefore, as the conductor has traveled from point *a* to point *b*, the induced voltage has increased from 0 to Vmax. This increase is plotted in Fig. 1-24(b). From point *b* to point *c* (90° to 180°) the voltage has decreased to 0. As the conductor travels from point *c* to point *d* (180° to 270°), through point *d*, and back to point a (270° to 360° or 0°), the induced electromotive force has increased from 0 to the negative maximum $-V_{max}$, and then decreased to 0 at point *a*, at which point the cycle repeats the process.

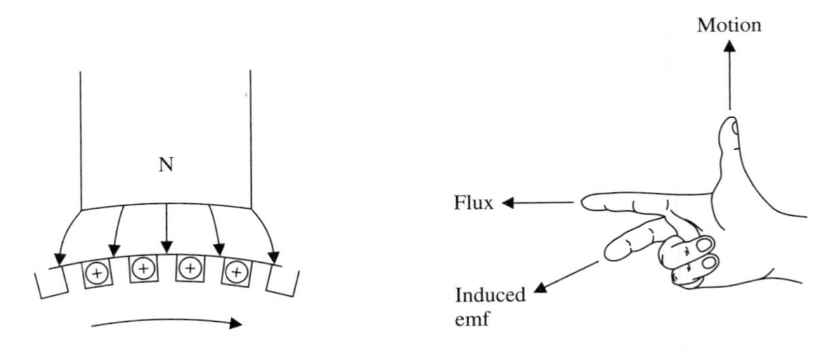

Figure 1-23 Fleming's right-hand rule.

The graph of the instantaneous induced electromotive force for 1 revolution is shown in Fig. 1-24(b). This graph is a sine curve whose amplitude at any instant t is

$$v = V_{max} \sin \omega t \tag{1–46}$$

$$= V_{max} \sin 2\pi ft \tag{1–47}$$

$$= V_{max} \sin \Theta° \tag{1–48}$$

where v is the instantaneous value of the induced voltage at time t s, f is the frequency in Hz, $\Theta = 2\pi ft = \omega t$, or the angle turned through in t s, and V_{max} is the maximum value of the instantaneous voltage.

From Fig. 1-24 it can be seen that the conductor moves from position a to position b in one quarter of a revolution, that is, from a position of 0 flux linking the conductor to the position of maximum flux linking the conductor. The average induced electromotive force E_{AV} in a single conductor is

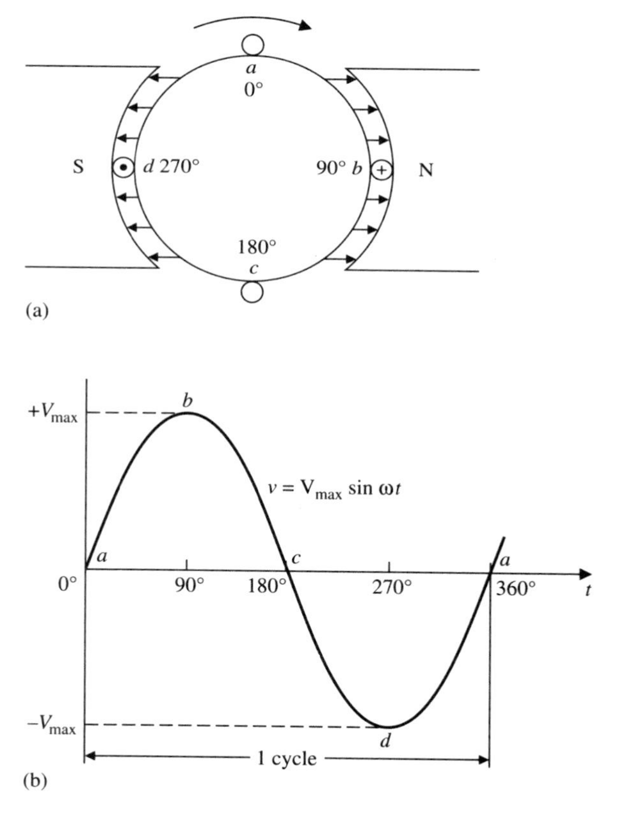

Figure 1-24 Voltage production by rotating coil. (a) Coil rotating in magnetic field. (b) Sine curve of instantaneous voltages.

$$E_{AV} = \frac{1}{2} \times \frac{\Phi}{t} \text{ V} \qquad \textbf{(1–49)(SI)}$$

or

$$E_{AV} = \frac{1}{2} \times \frac{\Phi}{t} \times 10^{-8} \text{ V} \qquad \textbf{(1–49)(E)}$$

For a single-turn coil consisting of two sides,

$$E_{AV} = \frac{\Phi}{t} \text{ V/turn} \qquad \textbf{(1–50)(SI)}$$

or

$$E_{AV} = \frac{\Phi}{t} \times 10^{-8} \text{ V/turn} \qquad \textbf{(1–50)(E)}$$

Since the time t for the coil side to travel one quarter of a revolution is $1/4s$, where s is the number of revolutions per second, then substituting for t,

$$E_{AV} = 4s\Phi \times 10^{-8} \text{ V/turn} \qquad \textbf{(1–51)(E)}$$

or

$$E_{AV} = \frac{4\omega\Phi}{2\pi} \text{ V/turn} \qquad \textbf{(1–51)(SI)}$$

$$= 0.637\omega\Phi \text{ V/turn} \qquad \textbf{(1–52)(SI)}$$

Therefore for a coil consisting of N turns, Eqs. (1-51)(E) and (1-51)(SI) become

$$E_{AV} = 4sN\Phi \times 10^{-8} \text{ V/coil} \qquad \textbf{(1–53)(SI)}$$

and

$$E_{AV} = \frac{4\omega\Phi N}{2\pi}$$

$$= 0.637\ \omega\Phi N \text{ V/coil} \qquad \textbf{(1–53)(SI)}$$

These equations represent the electromotive force induced in a coil under a pair of poles. Since a one-turn coil has two sides or conductors Z, an N turn coil has

$$Z = 2N \text{ conductors} \qquad \textbf{(1–54)}$$

In addition, we have only considered the voltage produced under one pole. Then for a P-pole machine we must consider the induced electromotive force in all the conductors as if they were in series under the P poles. Then Eqs. (1-53) become

$$E_{AV} = \frac{\Phi ZSP}{60} \times 10^{-8} \text{ V/coil} \qquad \textbf{(1–55)(E)}$$

or

$$E_{AV} = \frac{\Phi Z\omega P}{2\pi} \text{ V/coil} \qquad \textbf{(1–55)(SI)}$$

where Z is the number of conductors in the coil and P is the number of magnetic field poles.

➤ EXAMPLE 1-25

In a 6-pole dc machine a coil containing 100 turns revolves at 1,000 r/min (104.72 rad/s). The flux per pole is 250×10^3 lines (2.5×10^{-3} Wb). Calculate the average voltage per coil.

SOLUTION

In English units,

$$Z = 2N = 2 \times 100 \times = 200 \text{ conductors}$$

Therefore

$$\begin{aligned} E_{AV} &= \frac{\Phi ZSP}{60} \times 10^{-8} \\ &= \frac{250 \times 10^3 \text{ lines} \times 200 \text{ conductors} \times 1,000 \text{ r/min} \times 6 \text{ poles}}{60} \times 10^{-8} \\ &= 50.0 \text{ V/coil} \end{aligned}$$

In SI units,

$$\begin{aligned} E_{AV} &= \frac{\Phi Z\omega P}{2\pi} \\ &= \frac{2.5 \times 10^3 \times 200 \text{ conductors} \times 104.72 \text{ rad/s} \times 6 \text{ poles}}{2\pi} \\ &= 50.0 \text{ V/coil} \end{aligned}$$

Equations (1-55) can be modified as follows:

$$E_{AV} = KS \text{ V/coil} \qquad \textbf{(1–56)(E)}$$

where

$$K = \frac{ZP}{60} \times 10^{-8}$$

or

$$E_{AV} = k\,\omega\,\Phi \text{ V/coil} \qquad \textbf{(1–56)(SI)}$$

where

$$k = \frac{ZP}{2\pi}$$

These forms of the equations for average voltage per coil are very useful, because after the fixed terms have been replaced with a constant (K or k), they clearly show that the average voltage is proportional to the speed (angular velocity) and flux.

1-9 ELECTROMAGNETIC FORCE AND TORQUE

An extremely important effect, which is the basis of all motor action, is that caused by the interaction of the magnetic field produced by a current-carrying conductor within a magnetic field.

Biot-Savart's Law

Biot-Savart's law relates the force F that is produced on a conductor of length l carrying a current I in a magnetic field with a flux density B (Fig. 1-25). Expressed in English units, Biot-Savart's law is

$$F = \frac{BIl}{1.13} \times 10^{-7} \text{ lb} \qquad \textbf{(1–57)(E)}$$

where F is the force acting on the conductor in pounds, B the magnetic field flux density in lines/in^2, I the current in amperes, and l the conductor length in the magnetic field in inches. The constant $10^{-7}/1.13$ is the conversion factor from the cgs unit system to the English system.

In the SI system Eq.(1-57)(E) becomes

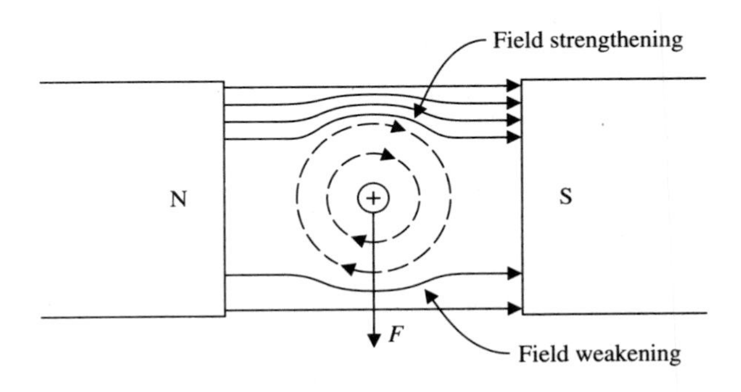

Figure 1-25 Biot-Savart's law

$$F = BIl\text{N} \qquad \textbf{(1–57)(SI)}$$

where F is the force acting on the conductor in newtons, B the magnetic field flux density in Wb/m^2, I the current in amperes, and l the conductor length in the magnetic field in meters.

➤ EXAMPLE 1-26

Each conductor on the surface of a dc generator armature is under the influence of the field pole magnetic field for 12 in (0.305 m) of its axial length. Under load conditions a current of 55 A flows in the conductor. What force in English and SI units is exerted on the conductor when the field pole flux density is 85,000 lines/in^2 (1.32 T)?

SOLUTION

In English units, from Eq. (1-57)(E),

$$F = \frac{BIl}{1.13} \times 10^{-7} \text{ lb}$$

then

$$F = \frac{85,000 \text{ lines/in}^2 \times 55 \text{ A} \times 12 \text{ in} \times 10^{-7}}{1.13}$$

$$= 4.96 \text{ lb}$$

In SI units from Eq. (1-57)(SI)

$$F = BIlN$$

$$= 1.32 \text{ T} \times 55 \text{ A} \times 0.305 \text{ m} = 22.14 \text{ N}$$

Check: There are 4.4476 N/lb; then

$$4.96 \text{ lb} \times 4.4476 \text{ N/lb} = 22.06 \text{ N}$$

It should be noted that the Biot-Savart law is not solely confined to motors or deflection-type measuring instruments, but is equally applicable to generators. The effect of current being carried by the armature coils is to produce a force that is less than but directly opposes the torque applied to the generator by the prime mover. This is a physical expression of Lenz's law.

Force Produced by a Conductor

In Fig. 1-25 a current-carrying conductor is shown in a magnetic field. When the current in the conductor flows into the paper, concentric magnetic lines of force are established about the conductor; these lines of force have a clockwise orientation. The lines of force at the top of the conductor add to those produced by the magnetic field poles, while at the bottom of the conductor they are acting in opposition to the main field. As a result, in this region the field is reduced. The unbalance in the magnetic fields is directly proportional to the conductor current; this creates a force on the conductor which, in this case, moves the conductor down.

The direction of motion of the conductor can be determined by Fleming's left-hand rule, which states that if the thumb and the index and middle fingers of the left hand are mutually extended at right angles to each other, and the middle finger is pointed in the direction of the current flow in the conductor, then the thumb will point in the direction of motion of the conductor (Fig. 1-26).

Torque Production

Figure 1-27(a) shows a cylinder with a single current-carrying conductor mounted at a radius r. The cylinder is free to rotate about an axis at its center, between the magnetic poles. When current flows through the conductor in the direction shown, a force F is created that will cause the cylinder to turn in a clockwise direction. When a force is applied to a body that is free to rotate about an axis, a torque is developed. The torque is the product of the force and the perpendicular distance from the axis of rotation to the line of action of the force (see Section 1-3 and Fig. 1-2). The general relationship for torque is

$$T = Fr \sin \Theta \tag{1-58}$$

where the torque T is a maximum when $Q = 90°$.

The torque T is in lb · ft when the force F is in pounds, the radius r in feet for the English system. In the SI system T is in newton-meters (N · m) when the force F is in newtons and the radius r in meters.

If Eqs. (1-57) are substituted in Eq. (1-58), then the torque equation becomes

$$T = \frac{BIlr}{1.13} \times 10^{-7} \text{lb} \cdot \text{ft}$$

$$= 0.885BIlr \times 10^{-7} \text{lb} \cdot \text{ft} \tag{1-59)(E}$$

or

$$T = BIlr \text{ N} \cdot \text{m} \tag{1-59)(SI}$$

where T is in lb · ft, B in lines/in^2, I in amperes, l in inches, and r in feet in the

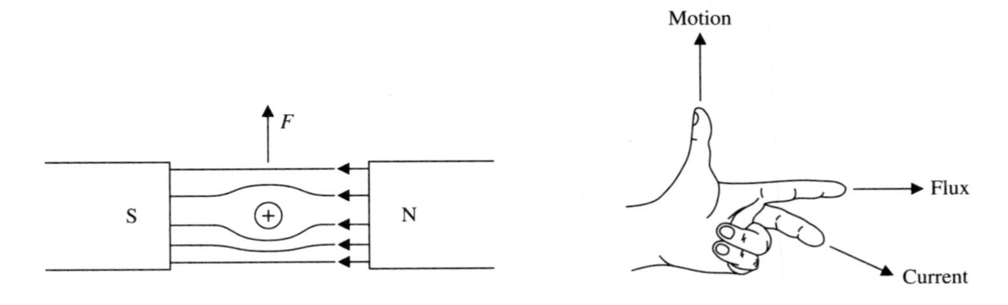

Figure 1-26 Fleming's left-hand rule.

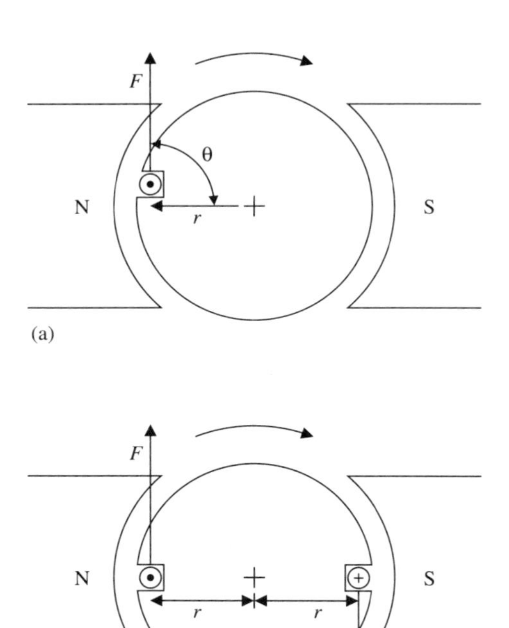

Figure 1-27 Production of torque. (a) By conductor. (b) By coil.

English system. In the SI system T is in N · m, B in tesla, I in amperes, l in meters, and r in meters.

The torque produced by a coil as shown in Fig. 1-27(b) is the sum of the individual torques produced by the two coil sides. Therefore the torque produced by a single-turn coil carrying a current I is

$$T = 2 \times 0.885\, BIlr \times 10^{-7}\ \text{lb} \cdot \text{ft}$$
$$= 1.77\, BIlr \times 10^{-7}\ \text{lb} \cdot \text{ft} \qquad \textbf{(1–60)(E)}$$

or

$$T = 2BIlr\ \text{N} \cdot \text{m} \qquad \textbf{(1–60)(SI)}$$

where T is in lb · ft, B in lines/in^2, I in amperes, l in inches, and r in feet in the English system. In the SI system T is in N · m, B in Wb/m^2, I in amperes, l in meters, and r in meters.

➤ EXAMPLE 1-27

A single-turn armature coil wound on a steel-cored armature 9 in (0.229 m) in radius and 12 in (0.305 m) long is subjected to a field pole flux whose density is 1.5 T

(96,780 lines/in^2). If the coil carries a current of 45 A, calculate the torque produced in lb · ft and N · m.

SOLUTION

From Eq. (1-60)(E),

$$T = 1.77\, BIlr \times 10^{-7}\ \text{lb} \cdot \text{ft}$$

$$= 1.77 \times 96{,}780\ \text{lines/in}^2 \times 45\text{A} \times 12\ \text{in} \times 0.75\ \text{ft} \times 10^{-7}\ \text{lb} \cdot \text{ft} = 6.94\ \text{lb}$$

or in SI units, from Eq. (1-60)(SI),

$$T = 2\, BIlr\ \text{N} \cdot \text{m}$$

$$= 2 \times 1.5\ \text{T} \times 45\text{A} \times 0.305\ \text{m} \times 0.229\ \text{m} = 9.413\ \text{N}$$

QUESTIONS

1-1 What are the seven base units in the SI system? Specify each unit and unit symbol.

1-2 What are the supplemental SI units?

1-3 Why is it advisable to be familiar with both the SI and the English unit systems?

1-4 What is meant by the terms *derived* and *base* units?

1-5 List by name and symbol the SI-derived units most commonly used in electrical engineering.

1-6 List by name and symbol the SI-derived units most commonly used in mechanical engineering.

1-7 SI units are represented by standard prefixes. Specify the prefix, the symbol, and the factor for the most commonly used prefixes.

1-8 Explain what is meant by mass and weight.

1-9 Explain what is understood by the following terms: **(a)** angular position; **(b)** angular velocity; **(c)** angular acceleration; **(d)** kinetic energy; **(e)** moment of inertia.

1-10 Explain what is meant by torque.

1-11 What is the work done on a rotating body?

1-12 What is meant by power?

1-13 Explain Newton's law of rotation.

1-14 What is meant by the term *moment of inertia*? What determines its magnitude, and what is its effect on a rotating body?

1-15 What is meant by the kinetic energy of a rotating body?

1-16 What is meant by the term *radius of gyration*?

1-17 What are the characteristics of a magnetic field?

1-18 How can the magnetic polarity of a solenoid be determined?

1-19 Define the following terms: **(a)** weber; **(b)** tesla; **(c)** flux density; **(d)** magnetomotive force; **(e)** reluctance; **(f)** permeability; **(g)** magnetic field intensity.

1-20 What is understood by the term *magnetic moment*?

1-21 What is a magnetic domain?

1-22 Discuss with the aid of a sketch the *B–H* curve of a ferromagnetic material.

1-23 What is the significance of hysteresis in a magnetic material?

1-24 Explain the terms *retentivity* and *coercive force*.

1-25 Explain the significance of a hysteresis loop. What conclusions can be drawn from its shape and the area enclosed by the loop?

1-26 What factors determine the magnitude of the hysteresis loss?

1-27 How can hysteresis loss be minimized?

1-28 What are the causes of hysteresis losses in static and rotating machines?

1-29 In what form does the hysteresis loss appear?

1-30 What are eddy currents? How are they produced in static and rotating machines?

1-31 What factors determine the magnitude of the eddy current loss?

1-32 How can the effects of eddy currents be reduced?

1-33 Discuss what is meant by the following: **(a)** diamagnetic; **(b)** paramagnetic; **(c)** ferromagnetic. Give examples of typical materials with these properties.

1-34 What are soft magnetic materials? What are their characteristics? Where are they used in electrical equipment?

1-35 What are laminated sheet steels? What are their characteristics? Where would they be used in electrical equipment?

1-36 What is the advantage of using grain-oriented steels in electrical equipment?

1-37 Where would nonoriented steels be used?

1-38 What is meant by specialty alloys? What are typical characteristics and applications?

1-39 What are ferrites? Where would they be used?

1-40 What is meant by ferrimagnetic?

1-41 What effect does an air gap have in a magnetic circuit?

1-42 What is meant by the term *fringing*?

1-43 What are the characteristics of permanent magnets?

1-44 What is meant by the energy product of a permanent magnet?

1-45 Discuss the four classes of permanent-magnet materials.

1-46 What are the characteristics of Alnico permanent magnets?

1-47 What are the characteristics of ceramic permanent magnets?

1-48 What are the characteristics of rare-earth permanent magnets?

1-49 Discuss Faraday's law of electromagnetic induction.

1-50 What are the three methods of producing an electromotive force in electrical equipment?

1-51 What factors affect the magnitude of the induced voltage?

1-52 Explain Lenz's law and discuss its effects in electrical equipment.

1-53 State Fleming's right-hand rule and where would it be used?

1-54 What is the significance of Biot-Savart's law?

1-55 State Fleming's left-hand rule and where would it be used?

PROBLEMS

1-1 Express the following as numbers between 1 and 10 with cnrrect SI prefix: **(a)** $2 \times 10^6 \Omega$; **(b)** 1×10^{-6}F; **(c)** 2×10^{-3}H; **(d)** 2,500 g; **(e)** 65,000 Pa; **(f)** 5 × 106 W; **(g)** 500,000 V; **(h)** 0.000,002 A; **(i)** 2,500 J; **(j)** 0.000,000,001 F.

1-2 In the following list of SI units, identify those that are base or fundamental units and those that are derived units: **(a)** time; **(b)** angular velocity; **(c)** temperature; **(d)** mass; **(e)** pressure; **(f)** acceleration; **(g)** force; **(h)** volume; **(i)** luminous intensity.

1-3 Correct the following in accordance with the SI rules:**(a)** 52.3 N meters; **(b)** 25.4 millimeters; **(c)** 14.2 amps; **(d)** 16 sec; **(e)** 23 meters; **(f)** 9.81 m/s/sec.

1-4 Determine the weight of a body whose weight is **(a)** 1 kg; **(b)** 15 g **(c)** 1 slug.

1-5 What is the mass of a body whose weight is **(a)** 22.4 N; **(b)** 35 lb.

1-6 Convert the following quantities as specified: **(a)** 6 in to mm; **(b)** 74.2 N to lb; **(c)** 50 lb · ft to N · m; **(d)** 85.6 N · m to lb · ft; **(e)** 10.56 kPa to lb/in^2; **(f)** 525°C to °F; **(g)** - 200°F to K; **(h)** 90 km/h to m/s; **(i)** 24.2 in^2 to m^2; **(j)** 10.25 ft^3 to m^3; **(k)** 6.25 gal (U.S.) to L.

1-7 Calculate the work done in pumping 1,000 imperial gallons of water into a tamk 100 ft above the intake. Water weighs 62.3 lb/ft^3, 1 imperial gallons = 277.3 in^3.

1-8 Convert the following quantities as specified: **(a)** 250° to rad; **(b)** 2.5 rad to degrees; **(c)** r/s to rad/s; **(d)** 1,800 r/m to rad/s; **(e)** 100 rad/s to r/min.

1-9 A motor accelerates from rest to 1,760 r/min in 5 s. Calculate the angular acceleration in rad/s^2

1-10 The shaft of a motor decelerates from 1,750 to 1,200 r/min in 5 s. Calculate the angular deceleration in rad/s^2 and r/s^2.

1-11 A force of 10 lb acts tangentially at the rim of a flywheel whose diameter is 14 in. Calculate the torque in lb · ft and N · m.

1-12 A flywheel 2 ft in diameter has a force of 10 lb applied at an angle of 60° to the radius. Calculate the torque in lb · ft and N · m.

1-13 An electric motor raises a cage of mass 500 kg through a height of 50 m in 20 s. If the motor is 80% efficient, what is the input power required by the motor?

1-14 A flywheel has a mass of 10 kg and a radius of gyration of 25 cm and rotates at 500 r/min. Calculate the moment of inertia and the kinetic energy in SI and English units.

1-15 Calculate the moment of inertial of a flywheel whose angular avelocity increases from 100 to 300 r/min when 250 J of work is done.

1-16 An electric motor develops a torque of 250 lb · ft at 1,750 r/min. **(a)** Calculate the output power in hp and kW. **(b)** If the motor develops 150 hp at 3,450 r/min, what is the torque of lb · ft and N · m.

1-17 A solenoid 20 cm long has a cross-sectional area of 10 cm^2 and is wound with 250 turns of wire carrying a current of 2.5 A. Given that the permeability of the core is 950, calculate for the core material: **(a)** H; **(b)** B **(c)** Φ

1-18 A magnetic intensity $H = 375$ A/m produces a flux density B = 0.375 Wb/m2. Calculate for the core material: **(a)** μ; **(b)** μ_r.

1-19 A circular ring of cast steel has a cross-sectional area of 3 cm^2 and a mean diameter of 12 cm, and is wound with 500 turns. The core permeavility is 2,000. Calculate: **(a)** the core reluctance: **(b)** the magnetomotiave force when the coil current is 1 A; **(c)** B; **(d)** Φ

1-20 A soft iron ring has a cross-sectional area of 10 cm^2, and a mean diemeter of 12.5 cm is wound with 500 turns. An air gap 2 mm long is cut in the ring, and the relative permeability of the core is 650. Calculate the current required to produce a flux of 0.85×10^{-4}Wb.

In Problems 1-21 to 1-26, use the magnetization curves given in Fig. 1-14.

1-21 A sheet steel toroid of circular cross section has an inside diameter of 4 in and an outside diameter of 5 in and is wound with 500 turns. Calculate the required current and the relative permeability if a flux density of 0.3 Wb/m^2 is to be produced.

1-22 Determine the current required to create a flux of 1.2×10^{-4} Wb in the magnetic circuit shown in Fig. 1-16.

1-23 Determine the current required to create a flux of 1.2×10^{-4} Wb in the magnetic circuit shown in Fig. 1-16 if the relative positions of the cast and sheet steel are reversed.

1-24 Repeat Problem 1-23 assuming that there is a 1.5 mm air gap between b and c. Neglect fringing.

1-25 In the magnetic circuit of Fig. 1-19 **(a)** calculate the coil current required to obtain a flux of 1.2×10^{-4} Wb in path be if $N = 100$ turns. Assume the material is sheet steel.

1-26 Repeat Problem 1-25, except that there is now a 3 mm air gap between c and d. Neglect fringing.

1-27 Convert the following quantities as specified: **(a)** 1.75 Wb/m^2 to $lines/in^2$; **(b)** 5.25×10^{-2} At/in to A/m; **(c)** 75,000 lines to Wb; **(d)** 55,000 $lines/in^2$ to T.

1-28 The flux linking a field pole winding with 1,000 turns collapses from 5×10^{-3} Wb to 0 in 0.002 s. Calculate the induced voltage.

1-29 In a four-pole dc machine a 20-turn coil rotates at 1,750 r/min (183.26 rad/s) between two magnetic poles. The field pole flux per pole is 300×10^{3} lines (3.0×10^{-3} Wb). Calculate the average induced voltage.

1-30 A single-turn coil wound on a steel core 10 in in diameter and 8 in long is subjected to a magnetic field whose flux density is 1.2 T. If the coil carries a current of 25 A, calculate the torque produced in lb · ft and N · m.

CHAPTER

2

Introduction to Power Electronics

2–1 INTRODUCTION

Since 1957, with the introduction of the silicon-controlled rectifier (SCR) combined with the rapid development of integrated circuits (ICs), digital logic control techniques, microprocessors, and programmable logic controllers (PLCs), the application of high- and low-power solid-state devices has made considerable inroads into the methods of controlling ac and dc rotating machines and industrial processes. The result of these and ongoing developments has been to create a new field of activity in the electrical industry—power electronics. Power electronics is defined as the application of high-power solid-state devices and associated control devices to the conversion, control, and conditioning of electric power.

The need for reliable, versatile, and accurate control of dc and ac motor drive systems has always been with us. The first method developed in the late nineteenth century was the Ward-Leonard system of controlling dc drives. The Ward-Leonard system has been modified and improved continuously, using gaseous-discharge devices such as the thyratron, the mercury pool rectifier, the ignitron, and lastly the excitron during the 1940s. During the 1950s, with the introduction of the transistor and the magnetic amplifier, improved closed-loop control methods were applied to the control of dc drives. In the 1960s the gaseous-discharge devices were beginning to be replaced in new systems by the SCR as the source of variable dc voltage for armature control of dc motors. Concurrently, and into the 1970s, the operational amplifier, digital ICs, and the microprocessor led to equally dramatic improvements in the low-power control of thyristors.

Modern phase-controlled converters using thyristors, digital ICs, and microprocessors are being used for speed and position control of adjustable-speed dc drives ranging from subfractional horsepower (watt) ratings to drives of more than

20,000 hp (14,920 kW) in such applications as machine tools, rolling mills in the metalworking and papermaking industries, mine hoists, electric trains and streetcars, cranes, elevators, and ship-propulsion systems.

Modern variable-frequency ac drives using thyristors, gate-turn-off thyristors (GTOs), or power transistors are rapidly supplanting traditional starters and speed-control techniques such as pole changing and pulse-amplitude modulation in power outputs ranging from subfractional to 20,000 hp (14,920 kW).

The major reasons for these important developments have been the need to reduce energy and maintenance costs as well as the continuing improvements being made to solid-state devices. It is not the aim of this text to provide a detailed study of power electronics and associated devices, which is best left to texts solely devoted to the subject. However, there are a number of high-power solid-state devices together with their application to dc and ac motor drive systems which must be included for a thorough review of rotating-machine control methods.

2–2 POWER ELECTRONIC DEVICES

Power semiconductors used in rectifiers (ac-to-dc power converters), inverters (ac-to-dc to variable-frequency ac converters), and choppers (dc-to-dc converters) operate in the switching mode, which reduces device losses and leads to higher conversion efficiencies. The penalties incurred include the production of harmonics, increased complexity, and, in thyristor phase-controlled converters, low power factors on the ac side when operated at low output voltages.

The most important power semiconductor devices used in dc and ac drive systems are diodes, thyristors, power transistors, and power MOSFETs.

Diodes

The semiconductor diode is a two-terminal *p-n* junction device which acts as a high-speed electronic switch, with a low impedance when forward biased and a high impedance when reverse biased. Most diodes are diffused-junction silicon devices. Silicon is preferred to germanium because of the higher temperature and its current-handling capability.

Silicon diodes are available in two types:

1. The *general-purpose diode*, with current ratings from 1 to 2,200 A and voltage ratings from 600 to 4,000 V, although not necessarily in the same device. Typical applications include motor controls, cranes, hoists, electric-discharge machining, arc furnaces, forklift trucks, battery chargers, and electroplating.
2. The *fast-recovery diode*, which is available in current ratings from 6 to 1,400 A and in voltage ratings from 600 to 3,200 V, with reverse recovery times from 200 ns to 5 μs. Typical applications are uninterruptible power supplies (UPSs), variable-frequency inverters, choppers, and as bypass diodes.

The general-purpose diode is supplied in axial lead mounting packages, and the general-purpose and fast-recovery diodes are available in stud mount (both standard and reverse polarity) and in disk or hockey-puck packages. The device symbol and the steady-state voltage-current (*V–I*) characteristic of a diode are shown in Fig. 2-1.

When forward biased, that is, the anode potential is positive with respect to the cathode, forward current I_D flows and there is a small forward voltage drop. The magnitude of the forward voltage drop is usually 0.7 V or less and is determined by the amount of doping and the operating temperature of the device. It should be noted that the maximum acceptable junction temperatures are usually from 170 to 200°C. When reverse biased, the reverse current is usually in the micro- or milliampere range and increases slightly as the reverse-bias voltage increases, until it goes into avalanche conduction at the zener voltage.

The essential diode ratings are voltage, current, and junction temperature. The voltage rating is determined by the thickness of the silicon die, while the current rating is expressed in terms of the maximum permissible forward current and is determined by the geometry of the silicon die and the method of dissipating excess junction heat. The maximum permissible junction temperature usually is not more than 200°C. The major causes of junction heating are:

1. The forward power loss, which is the product of the forward voltage drop (about 0.7 V for a silicon diode) and the forward current. Since the forward voltage drop is essentially constant, it is therefore proportional to the forward current.
2. The reverse power loss, which is the product of the reverse voltage and the reverse leakage current. The reverse leakage current increases rapidly as the junction temperature increases.

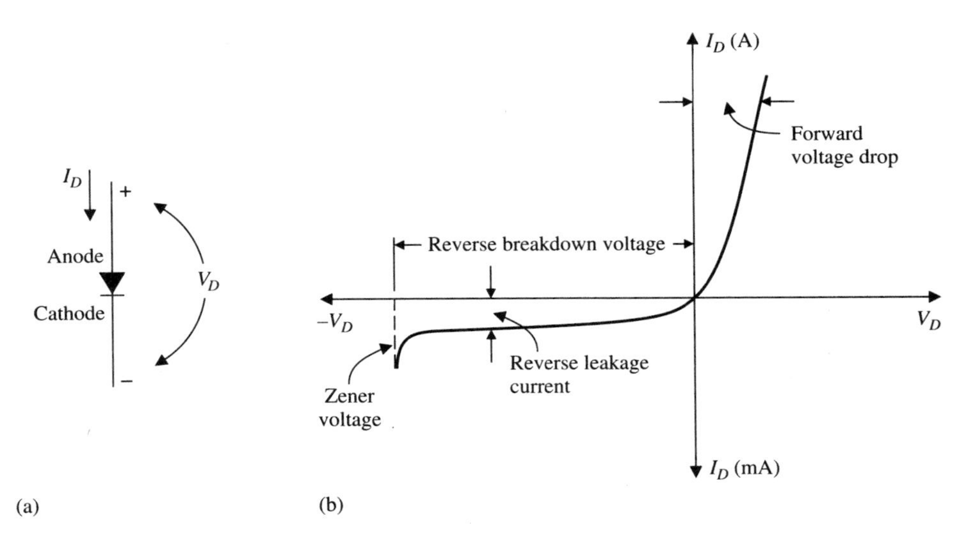

Figure 2-1 The diode. (a) Device symbol. (b) *V–I* characteristics.

3. The high-frequency switching loss, which is negligible at power-line frequencies, but increases rapidly as the switching frequency increases.

Normally the turn-on time of a diode is small, usually in the nanosecond range. However, during turn-off an appreciable time interval is needed to remove the charge carriers from the junction region before the reverse blocking capability is completed. The amount of stored energy is determined by the diode design and the magnitude of the forward current. This time interval is known as the *reverse recovery time* t_{rr} and is usually between 200 ns and 5 μs for fast-recovery diodes, between 2 and 20 μs for general-purpose diodes.

Junction temperatures are usually controlled by heat sinks, which can be natural, forced-air cooled, or water cooled, depending on the amount of heat to be removed.

Thyristors

Thyristors include a family of about 25 semiconductor devices. The thyristor is a four-layer *p-n-p-n* device. The most prominent member, because of its power-handling capability, is the *silicon-controlled rectifier* (SCR). Another member of the family which is used extensively in lower power-handling applications, is the bidirectional thyristor (TRIAC). It is used in full-wave ac heating and motor controls.

The SCR is a three-terminal device, sometimes called a reverse blocking triode thyristor. It is a fast switching unidirectional power switch which has a high impedance in the off-state and a low impedance in the on-state. The SCR is switched on by applying a low-level positive signal to the gate terminal when the device is forward biased. Once turned on, the SCR will remain in conduction until the anode current falls to a low value, called the *holding current* I_H. The device symbol and the steady-state V–I characteristics are shown in Fig. 2-2.

From Fig. 2-2(b) it can be seen that the SCR will not turn on if the gate current $I_G = 0$. However, if the anode-cathode voltage V_{AK} is increased sufficiently, the SCR will turn on. This voltage level is known as the *forward break-over voltage* V_{FBO}. It should be noted that this is not the recommended way of achieving turn-on. Once turn-on has been completed, the anode current I_A will increase to a value determined by the load impedance.

The mere act of applying a positive gate current I_G will not complete a turn-on unless V_{AK} has reached a large enough amplitude. As I_G is increased in amplitude, the SCR can achieve turn-on with smaller values of V_{AK}.

When the SCR is reverse biased, a small leakage current will flow for values of reverse voltage less than V_{RB}, the repetitive peak reverse voltage. If the reverse voltage increases to V_{RBT}, the nonrepetitive peak reverse voltage, there will be a significant increase in the reverse leakage current. The anode current will stop if the current drops to or below the holding current I_H.

The SCR can be turned on in several ways:

1. *Gate current* I_G: The application of a positive pulse of sufficient amplitude and duration. Care must be taken not to exceed the safe power level, or overheating will occur.

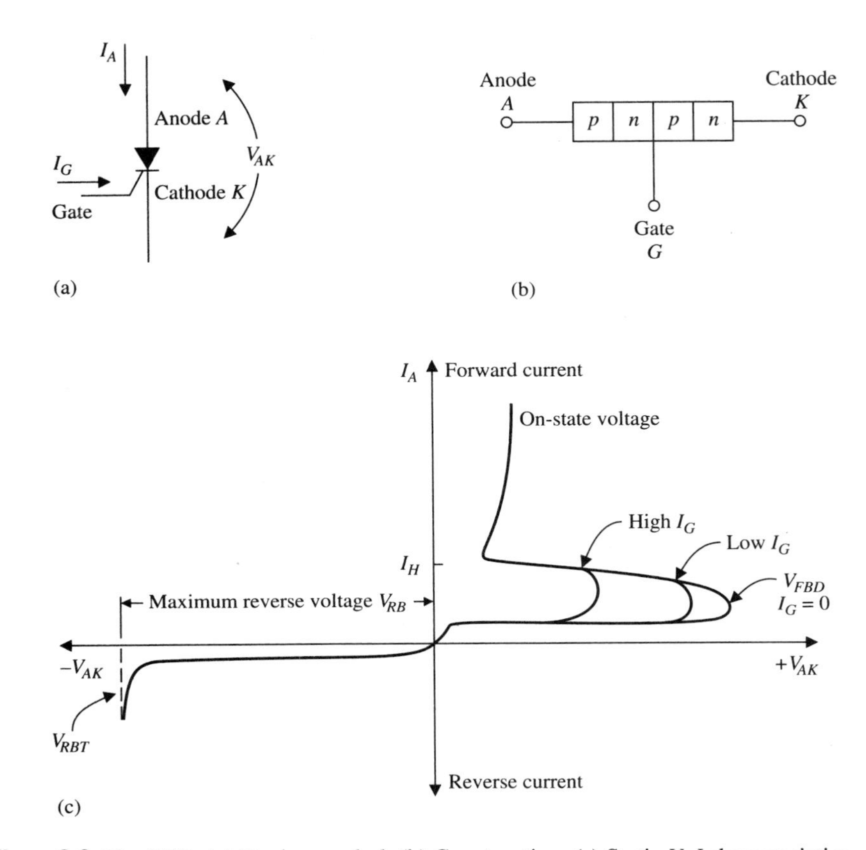

Figure 2-2 The SCR. (a) Device symbol. (b) Construction. (c) Static *V–I* characteristics.

2. *Overvoltage*: A forward V_{AK} greater than V_{FBO} will start turn-on. This method is not recommended.
3. *dv/dt*: Since the *p-n* junctions are capacitive during blocking because of the depletion layer, the application of a fast-rise-time anode-cathode voltage *dv/dt* will produce a capacitive charging current sufficient to initiate turn-on. This effect may be reduced by installing a *snubber*, that is, a series *RC* circuit in parallel across the device.
4. *Thermal*: An increase in the device junction temperature T_J will cause turn-on at lower forward anode-cathode potentials. Since this is an uncontrolled turn-on, the device must be kept within acceptable temperature limits by heat sinking, that is, by removing the excess heat.
5. *Light or radiation*: Photons, gamma rays, neutrons, protons, electrons, and hard and soft X rays, if permitted to hit an unshielded thyristor, will initiate turn-on. It should be noted that some specialized SCRs are turned on by light energy signals applied to the silicon wafer via fiber optics.

The SCR is available in several different configurations, such as stud-mounted, disk or hockey-puck, and molded power modules. There are two classes of SCRs, the phase-controlled SCR, which is used in thyristor phase-controlled converters where the applied anode-cathode *dv/dt* is usually less than 200 V/μs, and the inverter-grade SCRs, where the permissible *dv/dt* may be as great as 1,000 V/μs.

The Thyristor Family

The SCR is still the most commonly used member of the thyristor family, but its basic structure has been modified to reduce limits in its ratings and to improve the overall performance. Some of the more prominent members of the thyristor family are briefly described here. However, with the exception of the TRIAC and the GTO, none of these devices has the power-handling capability of the SCR.

Amplifying Gate SCR. Many gate structures have been developed to be able to use high-current fast-rise-time gate pulses to minimize the turn-on loss for fast-turn-on thyristors in high *di/dt* applications such as inverters and choppers. To minimize the input gate power requirements, an auxiliary SCR is built into the thyristor structure. The auxiliary SCR acts as a triggering device to control the gate current to the main SCR. The main SCR gate current is obtained from the anode circuit. The concept of the amplifying thyristor is shown in Fig. 2-3.

From Fig. 2-3, when a low-level positive gate pulse is applied to the gate of the pilot SCR, it turns on, and current is drawn from the anode of the main SCR, which is assumed to be forward biased, and applied to the gate of the main SCR. As the gate current flows through to the cathode of the main SCR, it rapidly promotes the buildup of the conduction area for the main current flow. The major advantage of this type of thyristor is that the gate signal to the pilot SCR remains unchanged irrespective of the rate of change of the load current *di/dt* in the main SCR and, hence, reduces the need for the complicated gate geometries called for in conventional high *di/dt* thyristors.

Asymmetrical Thyristor. The asymmetrical silicon-controlled rectifier (ASCR) is used in inverter and induction heating applications. Here the thyristor is not required to block reverse voltage since this is done by feedback diodes in antiparallel with the device. The ASCR is constructed so that it has a limited reverse voltage blocking

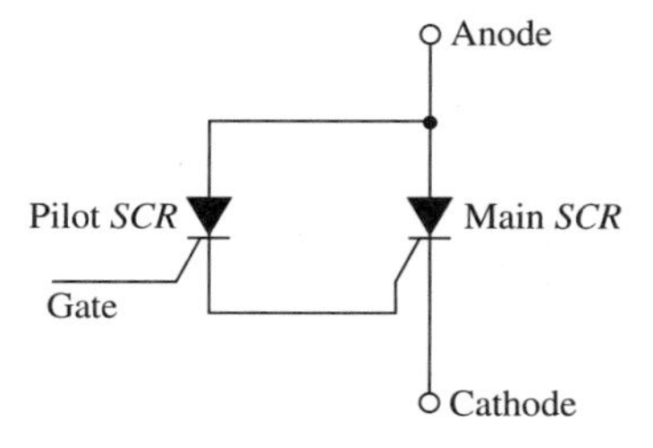

Figure 2-3 Amplifying gate SCR.

capability, typically about 30 V. This in turn produces a reduction in turn-on and turn-off times, including reducing the on-state voltage.

Reverse Conducting Thyristor. The reverse conducting thyristor (RCT) is an ASCR without any reverse blocking capability. The RCT is an integrated device having a diode connected in antiparallel with the thyristor. Just as with the ASCR, higher voltage faster switching devices with low forward-voltage drops can be manufactured. This type of device is being used in electric traction applications.

Gate Turn-Off Thyristor. The GTO is a thyristor which is turned on by a low-level positive signal, just as with the SCR. However, it can be turned off by a high-level negative signal, typically 20–30% of the load current.

The GTO has a high-voltage blocking capability, high surge current capability, and a high gain, with the added advantages over the conventional SCR of improved efficiency and reduced turn-off times, and, when used in inverters and choppers, it ends the need for complicated forced-commutation circuits and components.

The TRIAC and DIAC. Unlike the SCR, which conducts only in one direction, the TRIAC conducts in both directions. It functionally consists of two phase-controlled SCRs with a common gate connected in inverse parallel on the same silicon wafer [Fig. 2-4(a)].

The TRIAC is normally turned on by applying either a low-power positive gate pulse when MT2 is positive with respect to MT1 in quadrant I; or a low-power negative gate signal when MT2 is negative with respect to MT1 in quadrant III [Fig. 2-4(c)].

Since the TRIAC can conduct on both halves of the ac cycle, it is especially useful for ac phase-control applications such as heating control and variable-speed hand-held power tools. It is also used extensively in solid-state relays (SSRs).

The TRIAC is usually limited to applications where the switching frequency is less than 400 Hz. As compared to the SCR, the TRIAC has a longer turn-off time, the gate is less sensitive, and the device has a low reapplied dv/dt capability. It should be noted that if a TRIAC is switched on in the middle of an ac half-cycle with an RL load, it can cause sudden current changes, which are the source of *electromagnetic interference* (EMI), which in turn can cause interference with other sensitive electronic equipment.

The DIAC, which is essentially a TRIAC without a gate connection, is used as a trigger device for TRIACs and SCRs. It is available in trigger voltages from ± 15 to ± 50 V.

Power Transistors

The major strengths of the SCR are its ability to withstand reverse voltage and its high-power low-frequency (less than 50-kHz) switching applications. The major disadvantage of the SCR is its inability to be turned off by other than forced commutation methods.

The large-signal or *power transistor* used in the switching mode is now available in voltage ratings of $V_{CEO} = 250$–800 V and in current ratings of $I_C = 20$–200 A with dc current gains of 5–10. These ratings, combined with low forward voltage drops of

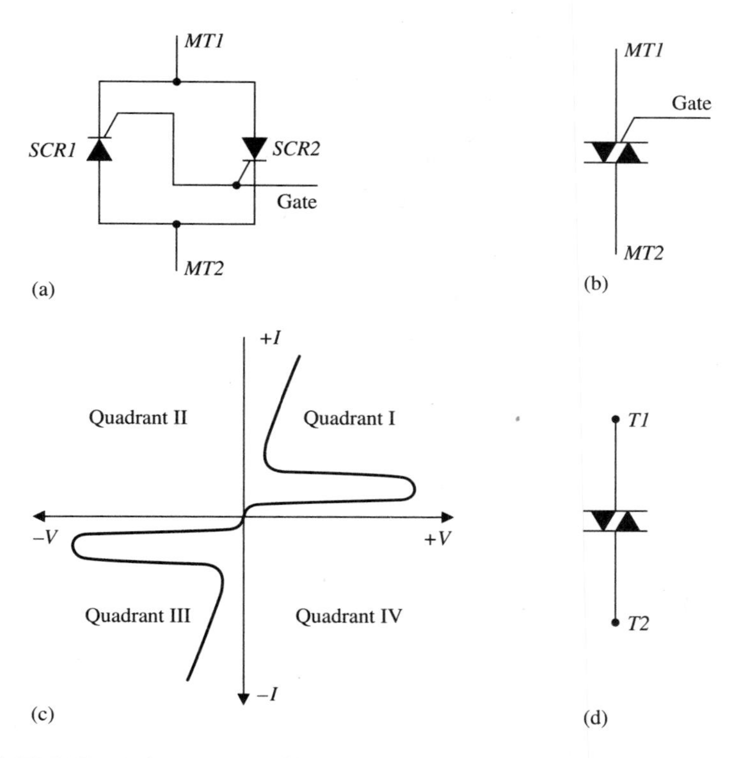

Figure 2-4 TRIAC. (a) Concept. (b) Circuit symbol. (c) Static *V–I* characteristic. (d) DIAC circuit symbol.

0.3–0.8 V as compared to the SCR voltage drops of 1.2–2.0 V, reduced turn-on time, and almost nonexistent turn-off problems have made the power transistor extremely competitive in low- and medium-power dc voltage-fed applications such as variable-frequency inverters and choppers.

The power transistor is a current-controlled device where the base current I_B controls the output or collector current I_C. If the base current $I_B = 0$, the transistor acts as an open circuit; if the base current is large enough, the transistor is driven into saturation and acts as a short circuit [Fig. 2-5(c)].

It is necessary to apply a continuous base current for the duration of the conduction period. Since the dc gain $h_{FE} = I_C/I_B$ is usually no greater than 10, it can be seen that if, for example, $I_C = 100$ A, a base current $I_B = 10$ A is called for to maintain the transistor in conduction. Hence the power loss in the base circuit can be quite significant.

The removal of the base signal I_B will begin turn-off. The turn-off time is appreciable, but it may be reduced by reverse biasing the base during turn-off. This has the effect of speeding up the removal of the extra charge carriers injected by the base current.

As the demand for devices having increased power-handling capability has risen, it also became more difficult to produce the higher base currents. This problem has been overcome with the introduction of the *n-p-n Darlington* power transistors. The Darlington power transistor has current gains of between 50 and 100, or even higher in high-current Darlingtons, with current ratings of 300–800 A and V_{CEO} ratings of 400–500 V being currently available (Fig. 2-6).

The current rating of the Darlington can be increased with two paralleled matched transistors Q2 in the output stage [Fig. 2-6(b)].

The most important transistor parameters are the dc gain h_{FE}, the collector-emitter breakdown voltage with the base circuit open V_{CEO}, the switching time, and the designated safe operating area (SOA).

The power-handling capability is determined by the junction temperature, which in turn is controlled by adequate heat sinking, and by secondary or *second breakdown*, which can occur at any temperature, is caused by localized heating at the emitter-base junction, and is especially noticeable when switching inductive loads. Typical SOA charts contain three regions: (1) bonding wire limit, (2) thermal limit, and (3) second breakdown, as illustrated in Fig. 2-7. The effect of second break-

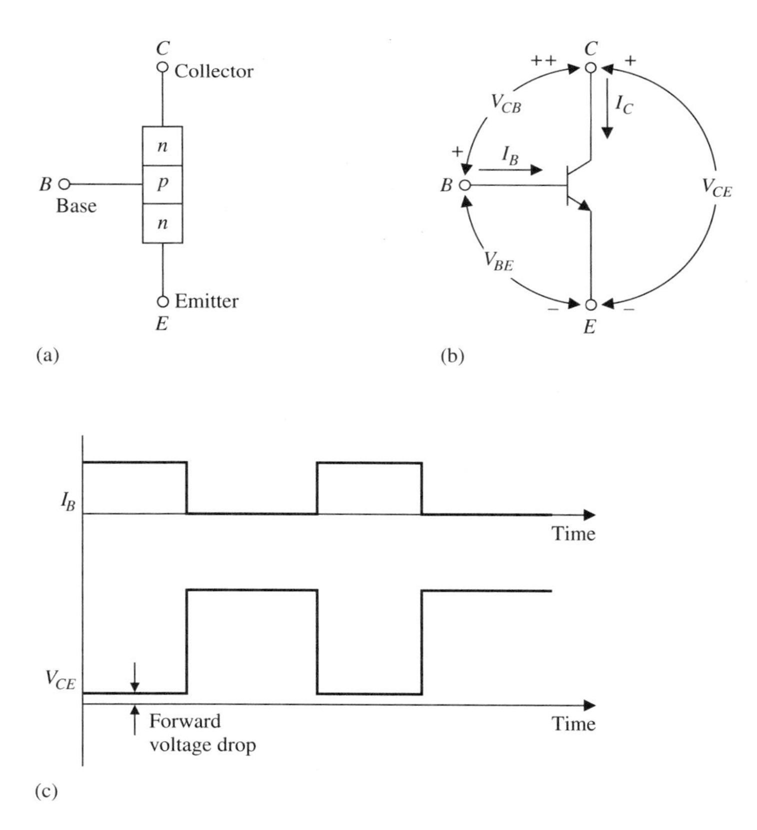

Figure 2-5 *n-p-n* power transistor. (a) Basic structure. (b) Circuit symbol. (c) Switching waveforms.

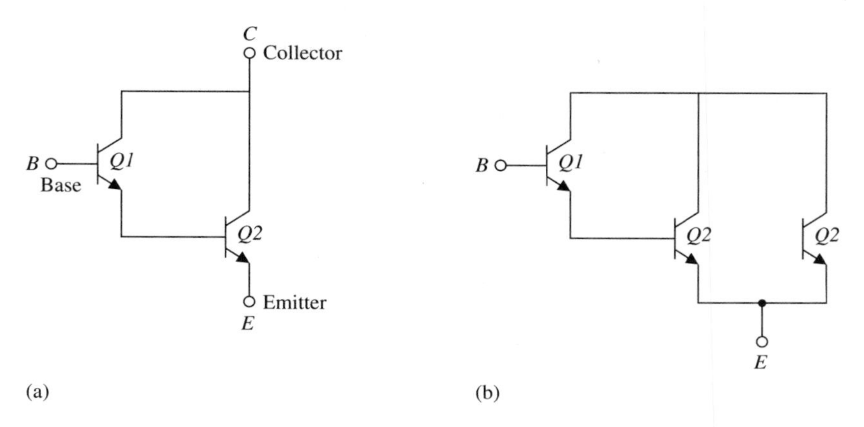

Figure 2-6 Power Darlington. (a) Low-current. (b) High-current.

down, which is a thermal runaway condition, can be eliminated by observing the parameters established by the manufacturer's designated SOA graphs for forward- and reverse-bias operating conditions.

As can be seen from Fig. 2-7 for a 140-V 30-A power transistor, it cannot control 30 A at 140 V, since the power dissipation limits the current at 140 V to about 0.6 A for a continuous base drive. As the pulse duration or the duty cycle of the base drive signal decreases, the magnitude of the collector current that can be carried for the same power dissipation increases.

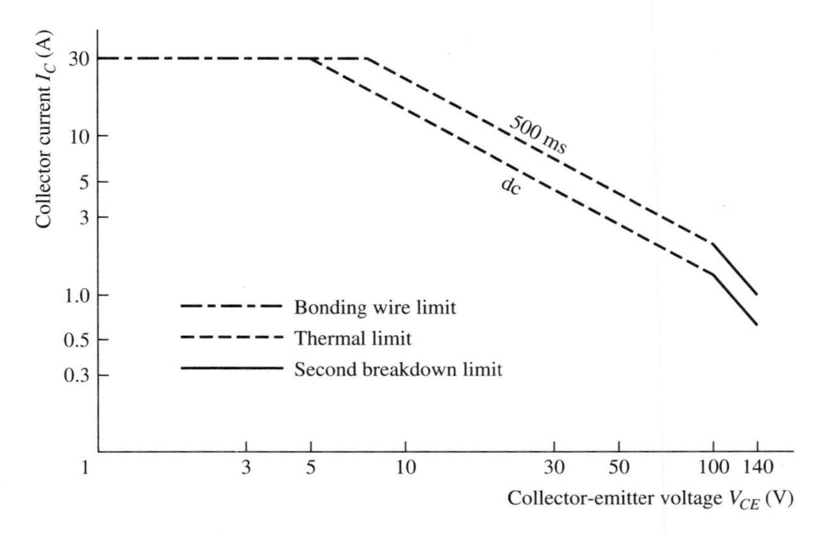

Figure 2-7 Typical maximum forward-bias safe operating curves for power transistor.

Power MOSFETs

Metal-oxide-semiconductor field-effect transistors (MOSFETs) were originally developed for use in large-scale-integration (LSI) and very-large-scale-integration (VLSI) applications. The first MOSFETs were limited in their applications by having a low current-handling capability and a drain-source breakdown voltage of 10–15 V. The power MOSFET was originally developed to meet the need for a power transistor that could be controlled by lower control signal power levels than were needed by bipolar transistors.

The power transistor and power MOSFET use two different technologies, since the *n-p-n* bipolar transistor is a minority carrier (electron) device and the power MOSFET is a majority carrier (electrons for an *n*-channel FET) device. The amount of charge needed to achieve turn-on is determined by the carrier type, which in turn determines the device characteristics. The power MOSFET is a voltage-controlled majority carrier device, and the power transistor is a current-controlled minority carrier device. The differences in control methods are illustrated in Fig. 2-8.

The basic structure of a HEXFET, which is a power MOSFET manufactured by International Rectifier Corporation, is shown in Fig. 2-9(a). In Fig. 2-9(b), when a positive voltage with respect to the source is applied to the gate, an *n* channel is induced in the device, which permits current to flow through the device. Since an extremely thin layer of silicon dioxide (SiO_2), which acts as an insulator, is between the gate electrode and the channel of the MOSFET (the device has a very high input impedance, approximately 10^9 Ω), it is possible to forward bias the gate without initiating a current flow. This characteristic permits a MOSFET to be driven directly from TTL and CMOS logic. It should also be noted that there is an integrated feedback diode built into the device structure, which permits it to be used in inductive circuit applications.

The power MOSFET has several advantages compared to the bipolar transistor. They are summarized as follows:

1. Because of the high input resistance combined with being voltage controlled, the power levels of the gate and control circuits are much less than for a bipolar transistor.

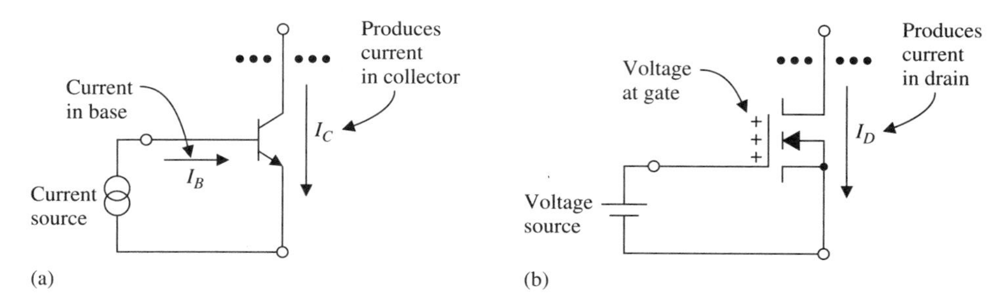

Figure 2-8 Control methods. (a) Base current drive for a bipolar transistor. (b) Gate voltage drive for a power MOSFET. (Courtesy of International Rectifier Corporation)

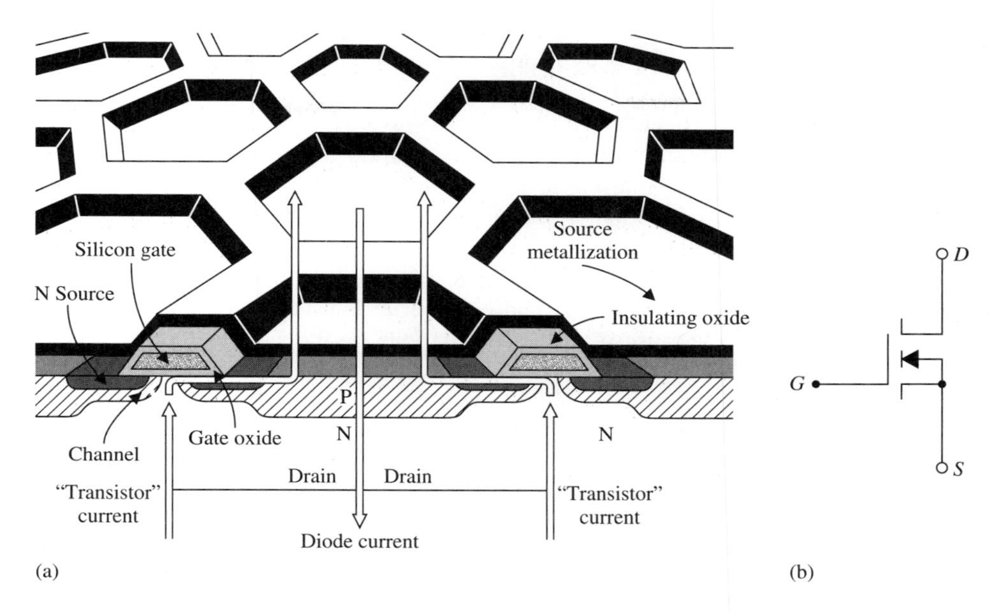

Figure 2-9 HEXFET. (a) Structure. (b) Circuit symbol. (Courtesy of International Rectifier Corporation)

2. Because there are no stored charges in the MOSFET, its switching times are significantly less than for the bipolar transistor.
3. Since the MOSFET has a positive resistance temperature coefficient, the possibility of local hot spots in the device has been almost completely eliminated, that is, unlike with the bipolar transistor, there is a very small possibility of thermal breakdown.
4. The power MOSFET does not suffer from the second breakdown effect of the power transistor, since the conduction area is evenly distributed over the whole silicon die. As a result, the only requirement for safe operation is to ensure that MOSFETs are adequately heat sinked.
5. Because the positive resistance temperature coefficient promotes current sharing, power MOSFETs can be used in parallel to increase their current-handling capability.

2–3 THYRISTOR PHASE-CONTROLLED CONVERTERS

Thyristor phase-controlled converters are naturally or line commutated ac-to-dc power converters that produce a variable dc output voltage whose amplitude is varied by phase-control techniques, that is, the duration of the conduction period is controlled by varying the point at which a positive gate signal pulse is applied to a forward-biased thyristor.

In most converters the power flow is from the ac source to the dc load, that is, the rectification mode. When an active load or source of negative voltage such as a dc motor or a battery bank is present, then if the load voltage is greater than the converter output voltage, it is possible for power flow to take place from the dc load to the ac source. This is known as *synchronous inversion*.

Since phase-controlled converters contain no moving parts, except sometimes cooling fans, which are used to remove heat from the heat sinks, they have a high efficiency (usually 95% or better) because of the low power losses in the thyristors, gate, and control circuits.

Phase-controlled converters may be supplied from single- or three-phase ac sources, although the single-phase converter is usually limited to a maximum of 7.5 hp (5.6 kW) because of current limitations on the ac side. Three-phase converters are available in output ranges of up to 20,000 hp (14,920 kW).

Thyristor phase-controlled converters are divided into one-, two-, and four-quadrant or dual converters (Fig. 2-10). The one-quadrant, half-controlled, or semiconverter only operates as a controlled rectifier, that is, the polarity of the output

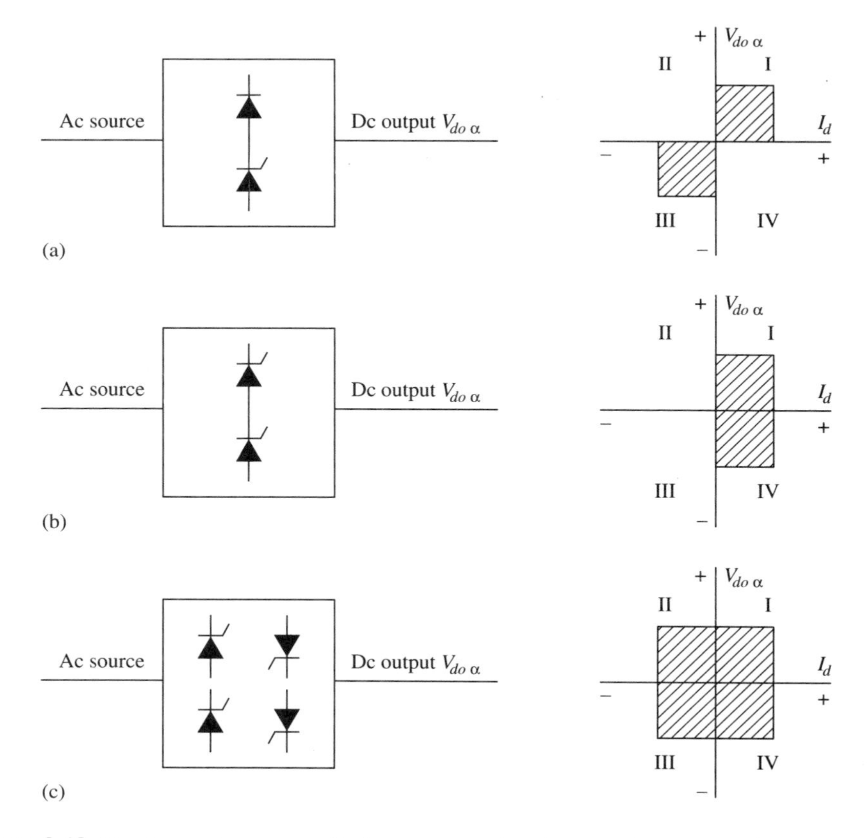

Figure 2-10 Thyristor phase-controlled converters. (a) One-quadrant or semiconverter. (b) two-quadrant or full converter. (c) Four-quadrant or dual converter.

voltage and the direction of the output current are constant. These converters use SCRs in half the positions and rectifier diodes in the remaining positions, and they cannot operate in the synchronous inversion mode [Fig. 2-10(a)].

Two-quadrant or full converters can operate with voltage polarity reversal if supplied from an active load. However, the current flow remains unidirectional. They operate in quadrant I as rectifiers and in quadrant IV as synchronous inverters. These converters use SCRs in all positions [Fig. 2-10(b)].

Four-quadrant or dual converters are essentially two two-quadrant converters connected back to back and can operate as controlled rectifiers in quadrants I and III, and in the synchronous inversion mode in quadrants II and IV.

Phase-Control Principles

The concept of phase control is best illustrated by considering a single-phase half-wave controlled thyristor converter with resistive and resistive-inductive loading, as shown in Fig. 2-11.

During the negative half-cycle of the ac source voltage the SCR is reverse biased and no voltage or current will be supplied to the load. During the positive half-cycle of the ac source voltage the SCR is forward biased, but will not conduct until the SCR is gated on.

If a gate pulse is applied to the SCR at time t_1, load current will flow and the ac supply voltage minus the device on-state voltage drop will be applied to the load [Fig. 2-11(b)]. If the load is purely resistive, the SCR becomes reverse biased at time t_2, and neither voltage nor current will be applied to the load until the SCR is again forward biased and turned on at time t_4. The amplitude of the mean dc voltage $V_{do\,\alpha}$ applied to the load is controlled by varying the firing delay angle α. Decreasing α increases the output voltage; increasing α decreases the output voltage. It should be noted that with a resistive load the output current stops at the point where the SCR becomes reverse biased.

In the case of a resistive-inductive (*RL*) load [Fig. 2-11(c)], although current commences at time t_1, at time t_2, when the SCR is reverse biased, load current continues to flow because the energy stored in the magnetic fields of the load is being returned to the system, and it will continue to flow until time t_3. At time t_3 the current has decayed to less than the holding current and the SCR turns off. Between times t_3 and t_5 neither voltage nor current is applied to the load.

This circuit provides a good insight into the principles of phase control. However, it can be seen that the output voltage has a very high ripple content, and a series of current pulses are supplied to the load. Hence very little use is made of this circuit.

Two-Quadrant Converters

To get two-quadrant operation, that is, unidirectional current flow and both polarities of dc voltage, the prime requirement is to have SCRs in all positions in the power circuit of the converter.

To simplify our discussions on phase-controlled converters, the following assumptions are made:

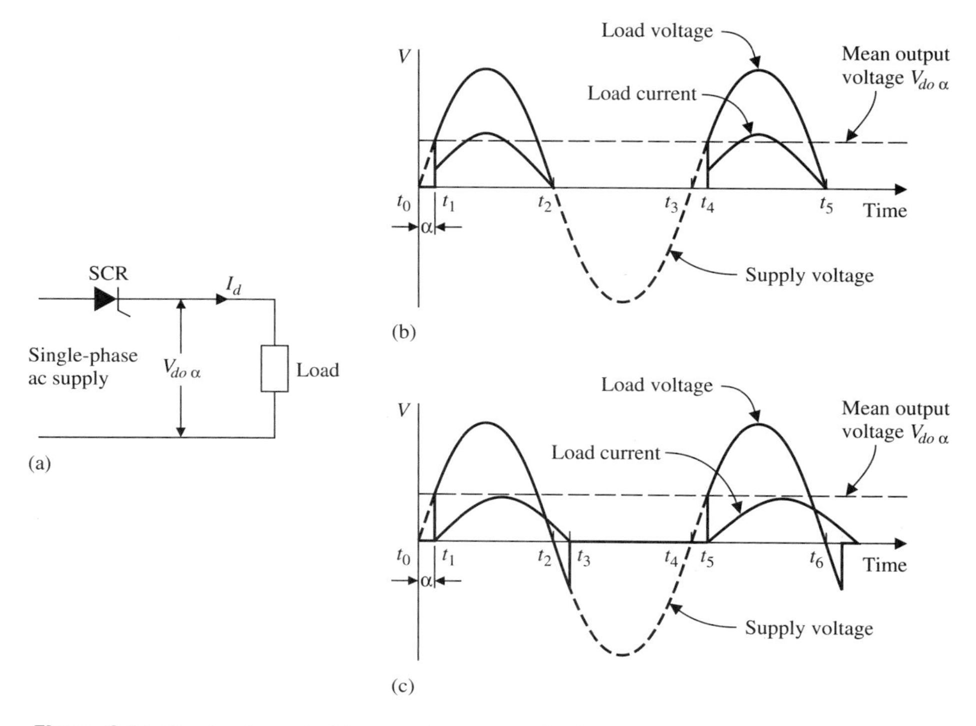

Figure 2-11 Single-phase, half-wave phase-controlled converter. (a) Schematic. (b) Waveforms for a resistive load. (c) Waveforms for a resistive-inductive load.

1. The voltage drop across a conducting SCR is negligible.
2. There is no leakage current through the device when it is not conducting.
3. The SCR turns on and off instantly.
4. The load contains an infinite inductance, that is, an ideal filter, and hence the output current I_d is constant and ripple free.

There are two possible types of two-quadrant converters:

1. Midpoint or half-wave converters, which are supplied from a transformer.
2. Bridge or full-wave converters, which may be line fed or transformer fed.

Two-Pulse Midpoint Converter. Figure 2-12(a) shows the basic arrangement of a single-phase two-pulse midpoint converter. The center-tapped transformer secondary produces two equal voltages v_1 and v_2 180° out of phase with each other relative to the center-tapped neutral point N.

During the positive half-cycle of the ac supply voltage, SCR1 is forward biased and when gated on, current flows from S1 through SCR1 and the load back to the neutral point N of the transformer secondary. At the same time SCR2 is reverse biased. During the negative half-cycle of the ac supply voltage, SCR2 is forward

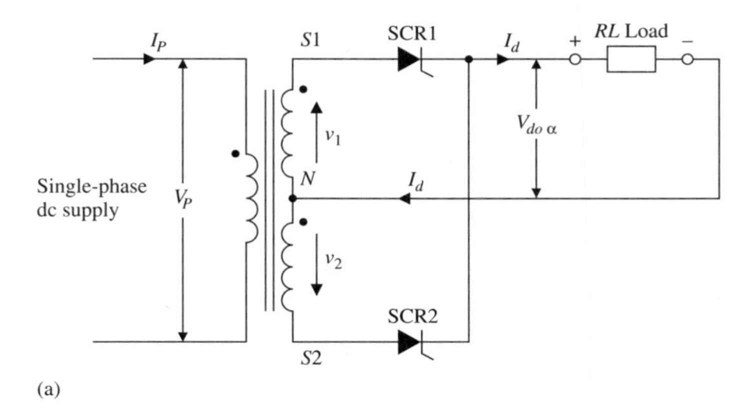

Figure 2-12a Single-phase two-pulse midpoint converter. (a) Schematic.

biased and when gated on, current flows via S2, SCR2, and the load back to the neutral point *N*. It should be noted that the current flow through the load is unidirectional.

From Fig. 2-12(b) SCRs 1 and 2 are gated on by the gate firing pulses i_{g1} and i_{g2}. Their position relative to the point where the anode potentials of the SCRs are positive with respect to their cathodes is determined by the gate firing logic circuits of the converter.

When the firing delay angle $\alpha = 0°$, the converter acts as an uncontrolled diode rectifier and the output voltage of the converter is at its positive maximum. It consists of a dc component V_{do} and a superimposed ac ripple voltage whose frequency is twice that of the source frequency, which accounts for the name of two-pulse converter.

The load current I_d is supplied alternately by SCR1 and SCR2 and is assumed to be square (assumption 4). When $\alpha = 0°$, the load current flowing in the secondary is in phase with the secondary voltage, that is, the converter is drawing power from the ac source at unity power factor.

As the firing delay angle is increased, there is a decrease in the average dc voltage $V_{do\,\alpha}$. Since it is assumed that the load is inductive, the previously conducting SCR will remain in conduction, even though reverse biased until the incoming SCR is gated on, that is, each SCR conducts for 180° and supplies load current for 180°. Therefore as the firing delay angle is increased, the load current is phase shifted by approximately the same angle as the firing delay angle, that is, the input power factor is worsening.

When $\alpha = 90°$, the average dc voltage $V_{do\,\alpha}$ is zero, and the load voltage contains only the double frequency ac ripple component. The load currents in the SCRs are now lagging their anode-cathode voltages by 90° or, in other words, the primary supply current now lags the supply voltage by 90° and no power is transferred from the ac source to the dc load.

To summarize, in dc motor speed control applications, where the speed is proportional to the voltage supplied to the armature circuit, the speed is a maximum when $\alpha = 0°$, and zero when $\alpha = 90°$, but the power factor of the converter as seen from the ac source worsens as the firing delay angle increases.

In certain conditions, for example, when a dc motor is operating under overhauling load conditions, the load is accelerating the armature and presents a negative voltage to the converter output terminals. Under these conditions the firing delay angle can be increased beyond 90°, for example. If $\alpha = 135°$, the average dc voltage $V_{do\alpha}$ is negative. The load current will still flow through each SCR for 180° in its original direction. However, since the voltage has reversed polarity, the power flow is now from the dc load to the ac source and the converter is operating in the synchronous inversion mode.

As the firing delay angle is increased, the amplitude of the negative dc voltage $V_{do\alpha}$ increases, reaching theoretically a maximum at $\alpha = 180°$. Since the SCRs remain in conduction for 180°, the supply current lags the supply voltage by 180°, and the power transferred to the ac source is a maximum.

In actual practice this condition cannot be met, since the period of reverse bias of

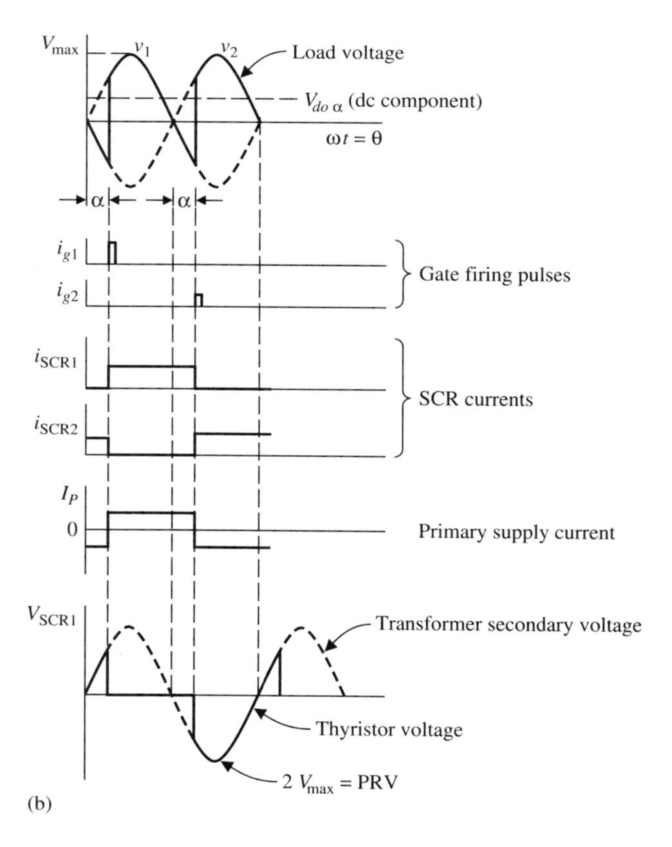

Figure 2-12b Single-phase two-pulse midpoint converter. (b) Waveforms for $\alpha = 45°$.

the thyristors must be great enough to allow for the thyristors to turn off and regain their forward blocking capability before forward voltage is reapplied. For 60-Hz systems the practical limit for the maximum firing delay angle is approximately 160 to 170°.

When operating in the synchronous inversion mode, the inversion region is specified in terms of the inverter advance angle β, where $\beta = 180° - \alpha$.

The value of the mean dc voltage is given by

$$V_{do\,\alpha} = \frac{2V_{max}}{\pi}\cos\alpha \tag{2–1}$$

where V_{max} is the maximum value of the transformer secondary voltages v_1 and v_2, and α is the firing delay angle.

Each SCR is subjected to a peak reverse voltage (PRV) of $2V_{max}$, or $2.8V$, where V is the root mean square (rms) value of V_{max}. The maximum average current carried by each SCR is $V_{max}/\pi R$, where R is the resistance of the load.

Summarizing the characteristics and performance of the two-pulse midpoint converter:

1. The converter operates as a rectifier for firing delay angles less than 90° and as an inverter for firing delay angles greater than 90° and less than 180°.
2. In the rectifying mode power is transferred from the ac source to the dc load. In the synchronous inversion mode power is transferred from the dc load to the ac source if there is a source of negative voltage at the load.
3. As the firing delay angle increases, so does the phase angle between supply voltage and supply current, that is, the power factor worsens.
4. The mean dc voltage $V_{do\,\alpha}$ decreases from a positive maximum at $\alpha = 0°$ to zero at $\alpha = 90°$ to a negative maximum at $\alpha = 180°$. At the same time the applied ac ripple voltage increases from a minimum at $\alpha = 0°$ to a maximum at $\alpha = 90°$, and then decreases to a minimum at $\alpha = 180°$.
5. Depending on the figure of merit X_L/R, the thyristors will remain in conduction for 180° if the load is sufficiently inductive, that is, I_d is continuous. For a purely resistive load, since conduction will stop as soon as the thyristor becomes reverse biased, the load current will become discontinuous.
6. When choosing thyristors, their voltage rating must be at least equal to twice the maximum applied ac voltage.
7. Midpoint converters only need half the thyristors required by a bridge converter, with a corresponding savings in devices and control circuit components and circuitry, which is offset to some extent by the increased cost of the higher voltage rated thyristors.
8. Midpoint converters are transformer fed and are mainly used where electrical isolation is required.
9. Single-phase midpoint converters are not in common use, being principally restricted by the capacity of the single-phase source. Their major advantage

is that by increasing the ripple frequency, the amplitude of the ripple component is reduced.

Two-Pulse Bridge Converter. The two-pulse bridge converter shown in Fig. 2-13 produces exactly the same output waveforms as the two-pulse midpoint converter, but has the advantage that it is not transformer fed.

Diagonally opposite pairs of SCRs conduct and commutate together. This is accomplished by simultaneously applying gate pulse firing signals to SCRs 1 and 4 or SCRs 2 and 3. Again, the mean dc voltage $V_{do\,\alpha}$ can be varied from maximum positive to zero, the rectifying mode, and from zero to maximum negative (if there is a source of negative voltage present at the load) by varying the firing delay angle from 0 to 180°.

The mean dc voltage $V_{do\,\alpha}$ is given by

$$V_{do\,\alpha} = \frac{2V_{\max}}{\pi}\cos\alpha \tag{2–2}$$

while the average SCR current is

$$I_d = \frac{V_{\max}}{\pi R} \tag{2–3}$$

The maximum PRV that is applied to an SCR is $V_{\max} = \sqrt{2}V$ where V is the rms voltage of the source.

The major differences between the two-pulse bridge converter and the two-pulse midpoint converter are:

1. There are always two thyristors in conduction simultaneously in the bridge converter, which means that the device voltage drop is twice that of the midpoint converter.
2. The maximum voltage from the source is the maximum peak voltage applied to each thyristor.
3. The control circuitry is slightly more complex, with a resulting increase in cost. However, this is more than offset by not having to use a transformer.

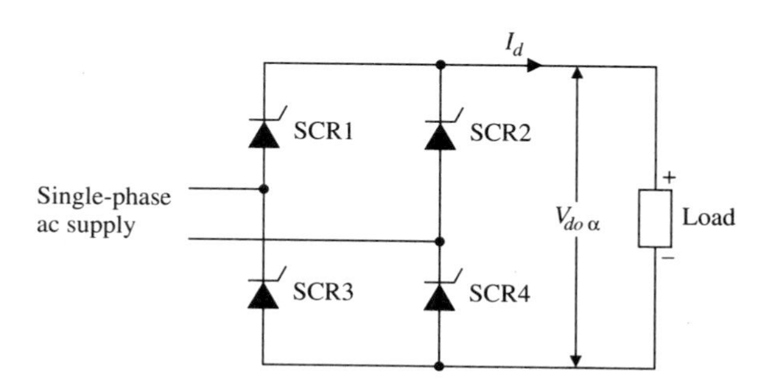

Figure 2-13 Two-pulse bridge converter.

Three-Pulse Midpoint Converter. To reduce the amplitude of the ac ripple component in the output dc voltage, and to simultaneously remove the power limitations imposed by single-phase sources, the three-phase three-pulse midpoint converter was developed.

The simplest practical version of the three-pulse midpoint converter is shown in Fig. 2-14(a). The input transformer is star-zigzag connected. The zigzag or interconnected star secondary consists of two separate identical secondary windings per phase. This arrangement will prevent dc magnetization of the transformer core by the dc component in the ac supply, which occurs if a star-connected secondary winding is used. This is achieved by permitting equal and opposite currents to flow in each half of the secondary phase windings.

From Fig. 2-14(b) it can be seen that each thyristor may be turned on by a gate pulse after the thyristor becomes forward biased 30° after the phase voltage crosses the zero axis; at this point $\alpha = 0°$. Only the thyristor with the highest instantaneous voltage applied to it can conduct when a gate pulse is applied. This produces the load voltage waveforms shown. Each thyristor conducts for 120° and blocks reverse voltage for 240°, the maximum blocking voltage being equal to the secondary line-to-line voltage. The frequency of the ac ripple component is $3f$, where f is the ac source frequency. Similarly the amplitude of the ripple component is 18.3% as compared to 48% for the two-pulse midpoint and two-pulse bridge converters. Since the thyristor conducts for 120°, the duration of the thyristor current flow is 120° and its magnitude is $I_d/3$.

Just as with the two-pulse midpoint and bridge converters for firing delay angles $\geq 90°$, it is necessary to have a negative dc voltage source present at the load terminals, so that operation in the synchronous inversion mode can be achieved.

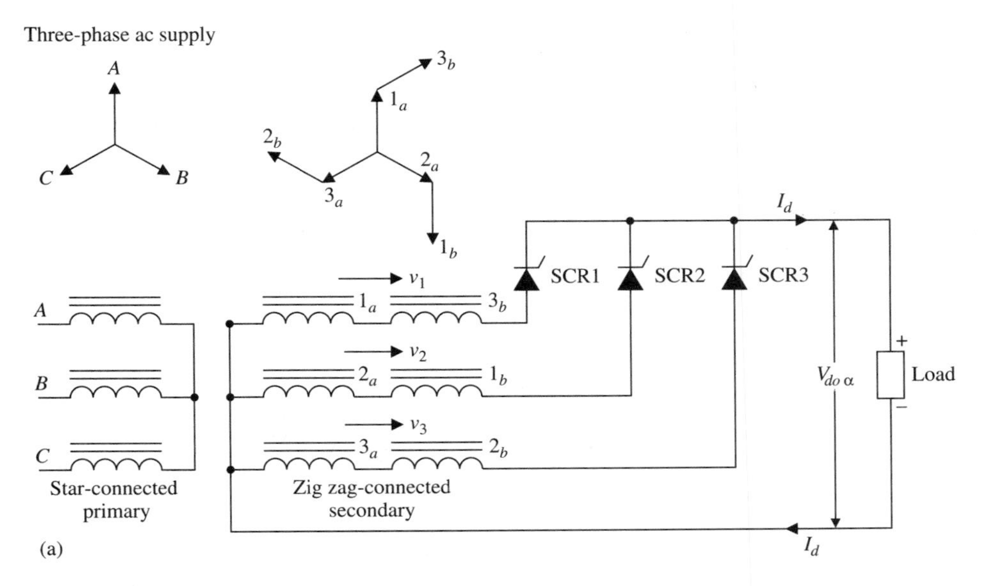

Figure 2-14a Three-phase three-pulse midpoint converter. (a) Schematic.

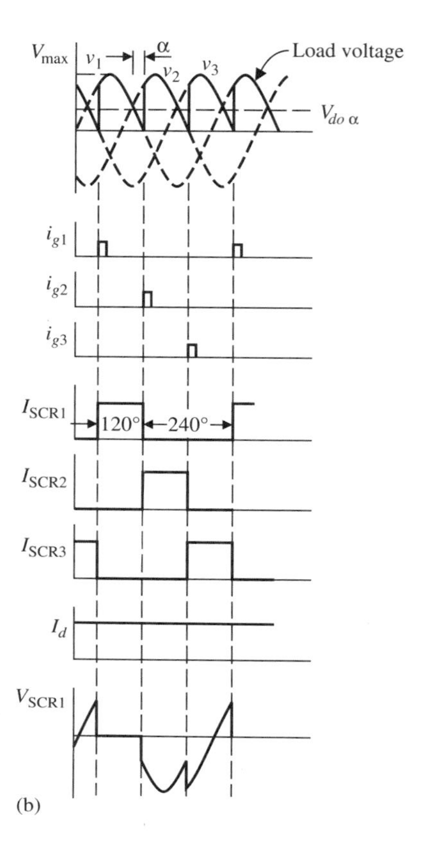

Figure 2-14b Three-phase three-pulse midpoint converter. (b) Waveforms for α = 45°.

The value of the mean dc voltage is

$$V_{do\alpha} = \frac{3\sqrt{3}V_{\max}}{2\pi} \cos\alpha \qquad \textbf{(2–4)}$$

where $V_{\max}$ is the maximum value of the instantaneous secondary phase voltage and α is the firing delay angle.

Each thyristor is subjected to a peak reverse voltage of $\sqrt{3}V_{\max}$, or $2.45V$, where V is the rms value of $V_{\max}$ and the average thyristor current is $0.33I_d$.

The three-pulse midpoint converter is supplied from a three-phase ac source. The amplitude of the ac ripple component can be further reduced by increasing the pulse number. For example, a six-phase supply can be obtained by using a delta–double-star-connected three-phase transformer connected to a six-pulse midpoint converter with an interphase reactor, as shown in Fig. 2-15.

This converter arrangement consists of two three-pulse midpoint converters operating in parallel through the interphase reactor. Each of the converters operates independently of the other, and each converter supplies half the load current, with

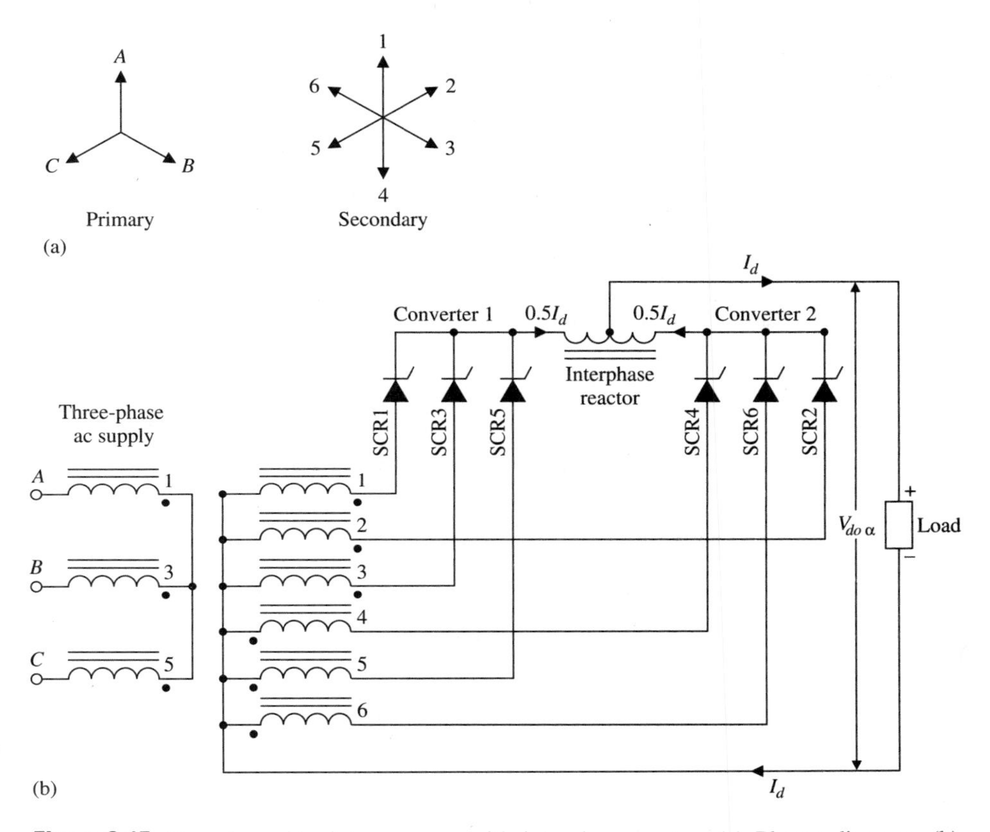

Figure 2-15 Six-pulse midpoint converter with interphase reactor. (a) Phasor diagram. (b) Schematic.

the interphase reactor absorbing the differences between the instantaneous voltages. The ac ripple component of each converter has a frequency of $3f$, but because there is a phase difference between the ac ripple components of each converter, the ripple frequency present in the dc output at the center tap of the interphase reactor is $6f$, and the amplitude of the ac ripple component is now 4.2%. The load is supplied with six segments of dc voltage, and each thyristor conducts for 120°.

It is essential that the interphase reactor carry enough dc current to maintain an adequate core flux level, or under light load conditions the converter will operate as a six-pulse midpoint converter without an interphase reactor. As a result the thyristors will only conduct for 60° and the output current will be discontinuous with a significant drop in the amplitude of the mean output dc voltage. This problem is cured by connecting a bleeder resistance across the output terminals to maintain a minimum dc current through the interphase reactor.

A twelve-pulse midpoint converter can be obtained by connecting two six-pulse midpoint converters with interphase reactors in parallel with each other through a third interphase reactor. The result is a smoother output dc voltage with a $12f$ ac ripple frequency. In practice it is unusual to extend the concept beyond 12-pulse converters because of the complexity of the transformer connections and the electronic firing circuitry.

Six-Pulse Bridge Converter. The six-pulse bridge converter consists essentially of two three-pulse midpoint converters connected in series [Fig. 2-16(a)]. It can be seen that the individual output currents of the two converters are acting in opposition in the neutral line, and as a result the neutral line is not required. Figure 2-16(b) shows the result of eliminating the neutral line.

The advantages of using the six-pulse bridge converter as compared to the three-pulse converter are:

1. The mean dc voltage has doubled using the same ac supply voltage.
2. The amplitude of the ac ripple component has decreased from 18.3 to 4.2 %, and the ripple frequency has doubled to 6*f*.

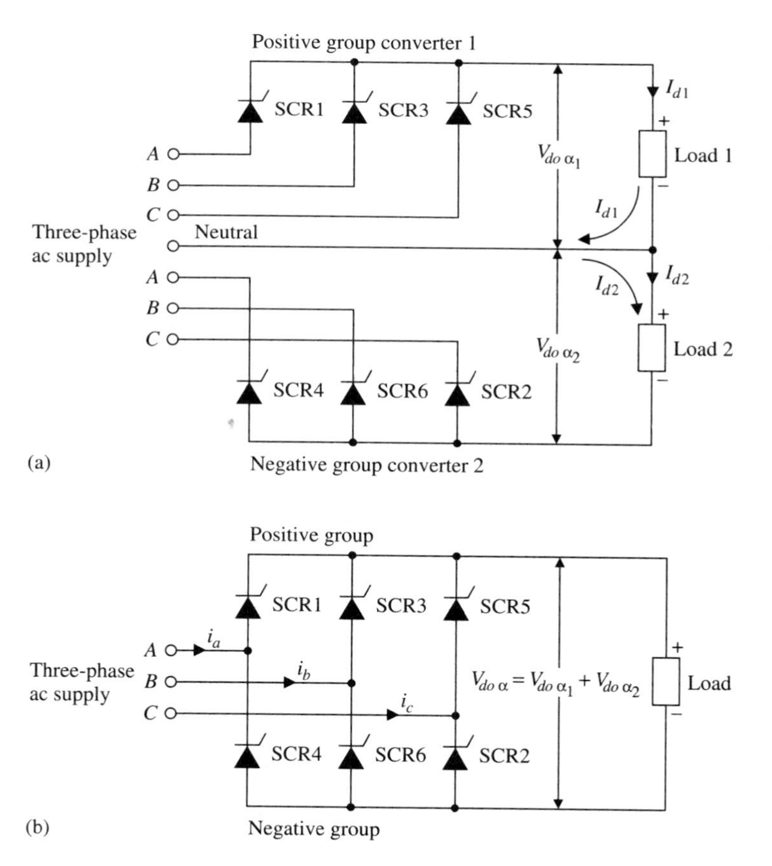

Figure 2-16a and b Six-pulse bridge converter. (a) Two three-pulse midpoint converters in series. (b) Basic circuit.

3. The converter can be supplied directly from the three-phase ac source, eliminating the need for an input transformer, although a transformer may be needed to match the source voltage to the load voltage.

From Fig. 2-16(c), when $\alpha = 45°$, the output dc voltage $V_{\text{do}\,\alpha}$ is the voltage sum of the two series-connected three-pulse midpoint converters. The positive group with the common cathode connection will have the SCR with the most positive anode conducting, namely, SCR1. Similarly, the SCR with the most negative cathode, SCR2, will be conducting, making the common anode connection negative. The mean output dc voltage $V_{\text{do}\,\alpha}$ is the algebraic sum of the instantaneous voltages of the positive and negative converter groups, and consists of segments of the three-phase line-to-line voltages. The amplitude of $V_{\text{do}\,\alpha}$ is double that of the individual three-pulse midpoint converters. The frequency of the ac ripple component is $6f$. Each SCR conducts and carries the load current I_d for 120° and blocks for 240°. From the last waveform showing the supply voltage and currents flowing through SCRs 1 and 4, it can be seen that the ac line current is symmetrical, that is, there is no dc compo-

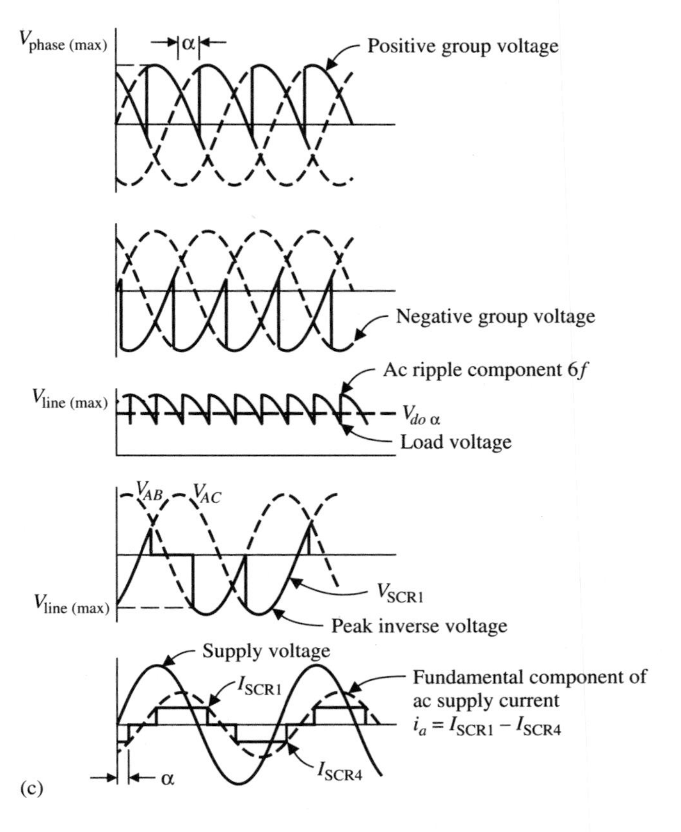

Figure 2-16c Six-pulse bridge converter. (c) Waveforms for $\alpha = 45°$.

nent, and the fundamental component of the ac line current lags the supply voltage by about the same amount as the firing delay angle α.

As α increases, the mean output dc voltage $V_{\mathrm{do}\,\alpha}$ decreases until, when $\alpha = 90°$, $V_{\mathrm{do}\,\alpha} = 0$ V. Assuming the load is a dc motor under overhauling load conditions, then as α is increased beyond 90°, the mean output dc voltage $V_{\mathrm{do}\,\alpha}$ becomes increasingly more negative, and the converter is operating in the synchronous inversion mode, returning power back to the ac source.

Assuming the ac supply voltage phase sequence is *ABC*, the firing order of the thyristors is SCRs 1 and 2, SCRs 2 and 3, SCRs 3 and 4, SCRs 4 and 5, SCRs 5 and 6, with commutations occurring alternately in the positive and negative groups every 60°. This is necessary so that a complete path is available for current flow, that is, the SCRs must be gated on by supplying sequential gate pulses spaced 60° apart. This is adequate provided there is a continuous current flow. In the case of discontinuous current flow it is possible for SCR2 to have stopped conducting before SCR3 is turned on. Then it is necessary to apply a gate pulse simultaneously to SCRs 2 and 3. This problem is overcome by *double pulsing*, that is, supplying two pulses per cycle, spaced 60° apart, to each SCR. Alternatively a long pulse, greater than 60°, will overcome the problem, especially when supplying an inductive load. It is difficult to produce long pulses. So usually a pulse train with a frequency of about 10 kHz is used to simulate a long pulse. This has the added benefit that it helps to reduce heating at the gate-cathode junction.

The performance and characteristics of the six-pulse bridge converter may be summarized as follows:

1. For firing delay angles ≤90° the converter is operating in the rectifying mode. If a source of negative dc voltage is present in the load, the converter operates in the synchronous inversion mode for firing delay angles between 90° and 180°.
2. The peak inverse voltages applied to the thyristors are half those in the six-pulse midpoint converter, and the load current carried by the individual thyristors is double that of the six-pulse midpoint converter.
3. The ripple frequency is $6f$, where f is the ac supply frequency.
4. There is no dc component present in the ac line current, and the ac line current lags the ac supply voltage by approximately the same angle as the firing delay angle.
5. The peak forward and reverse voltages applied to the SCRs are $\sqrt{2}V$, where V is the rms line-to-line voltage. The mean output dc voltage $V_{\mathrm{do}\,\alpha}$ is

$$\begin{aligned} V_{do\,\alpha} &= \frac{3\sqrt{2}V}{\pi}\cos\alpha \\ &= 1.35V\cos\alpha \end{aligned} \tag{2–5}$$

the average thyristor current is

$$I_{T(AV)} = I_d$$

and the rms current carried by each thyristor is

$$I_{T(RMS)} = \frac{I_d}{\sqrt{3}} \quad (2\text{–}6)$$

One-Quadrant Converters

Two-quadrant converters operate with positive and negative load voltages and, when rectifying, convert ac power to dc power. In the synchronous inversion mode they remove dc power from an active load and return it to the ac source as ac power.

There are many applications where it is only necessary to supply variable dc power to the load, that is, operation in the rectifying mode only. Thyristor phase-controlled converters that meet this requirement are called one-quadrant converters, or half-controlled or semiconverters. These converters only need thyristors in half the active positions. The remaining positions are filled with rectifiying diodes. The result is a cheaper converter since diodes are not as expensive as thyristors, and only half as many gate firing control circuits are needed.

Two-Pulse Half-Controlled Bridge Rectifiers. The principle of one-quadrant converters is most easily shown by considering the two-pulse half-controlled converter illustrated in Fig. 2-17.

It can be seen that the half-controlled bridge converter is made up of two two-pulse midpoint converters connected in series. The positive group has its output voltage controlled by phase shift control of SCRs 1 and 2. The negative group consists of the uncontrolled diodes D1 and D2. As the firing delay angle α is increased from 0 to 180°, the mean output dc voltage $V_{\text{do}\,\alpha}$ varies from a positive maximum to nearly 0 V.

When $\alpha = 0°$, the anode of SCR1 is positive with respect to N, and the cathode of D2 is negative with respect to N, current will flow through SCR1 and the load, and will return to the ac line via D2. In the next half-cycle, SCR2 and D1 are the conducting devices. As the firing delay angle increases, the outgoing SCR commutates

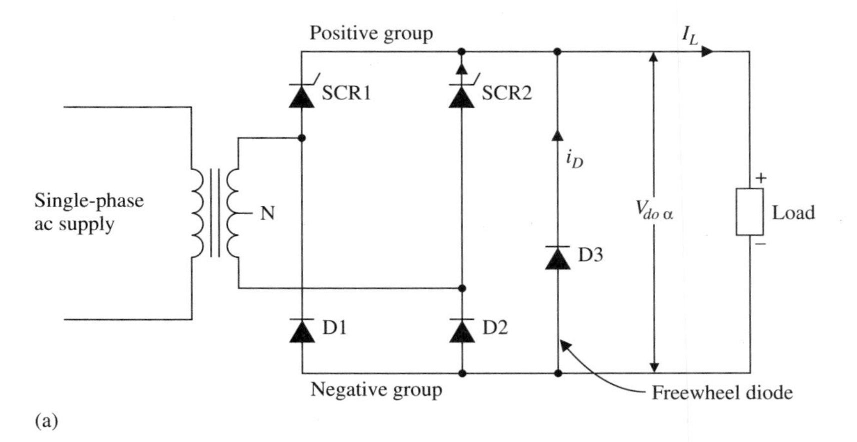

Figure 2-17a Two-pulse half-controlled converter. (a) Basic circuit.

off as the incoming SCR is gated on. The diodes turn on and commutate off at the zero crossing points of the voltage waveforms. As can be seen, the peak amplitude of the load voltage waveform is twice that of either converter group. As the firing delay angle is increased, the duration and amplitude of the voltage pulses forming the load voltage decrease but never become negative. However, the output voltage will never be completely zero because of the contribution from the diode converter.

The freewheel diode D3 will carry current i_D during the interval between the point that the decreasing supply voltage crosses the zero axis and the point where the next SCR is gated on. The current i_D is produced by the stored magnetic energy in the inductive components of the load being returned through D3, instead of through either SCR1 and D1 or SCR2 and D2, and ensures that the outgoing SCR has turned off and regained its forward blocking state.

The mean output dc voltage is

$$V_{do\,\alpha} = \frac{V_{\max}}{\pi}(1 + \cos\alpha) \tag{2–7}$$

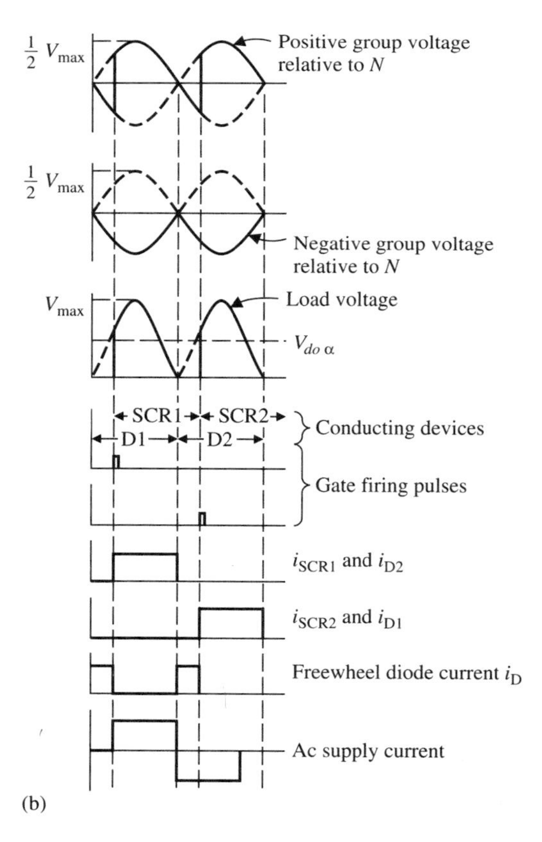

Figure 2-17b Two-pulse half-controlled converter. (b) Waveforms at $\alpha = 45°$.

The peak forward and reverse voltage applied to the SCRs and diodes is V_{max}, and the average current in the SCRs and diodes is

$$I_{T(AV)} = \frac{V_{max}}{\pi R} \tag{2–8}$$

where R is the load resistance.

The current carried by the freewheel diode D3 is

$$i_D = \frac{V_{max}}{\pi R} \tag{2–9}$$

for resistive loads and

$$i_D = \frac{0.42 V_{max}}{\pi R} \tag{2–10}$$

for inductive loads.

Three-Pulse Half-Controlled Bridge Converters. The basic circuit of the three-pulse half-controlled bridge converter is shown in Fig. 2-18(a). It consists of a positive controlled three-pulse midpoint converter and an uncontrolled negative three-pulse midpoint converter. It should be noted that the relative positions of the two midpoint converters is unimportant, and their positions can be reversed without affecting the performance of the bridge converter.

The operation of the converter can be seen clearly from the three three-pulse waveforms shown in Fig. 2-18(b), (c), and (d). In Fig. 2-18(b), with $\alpha = 0°$, the upper waveform shows the output of the controlled converter and the lower the uncontrolled diode converter output. The sum of the instantaneous values of the two waveforms is the load voltage waveform. The mean output dc voltage $V_{do\alpha}$ is at its maximum and the ac ripple component has a frequency of $6f$. Each thyristor and diode conducts for 120° and blocks for 240°, and the ac supply current has alternat-

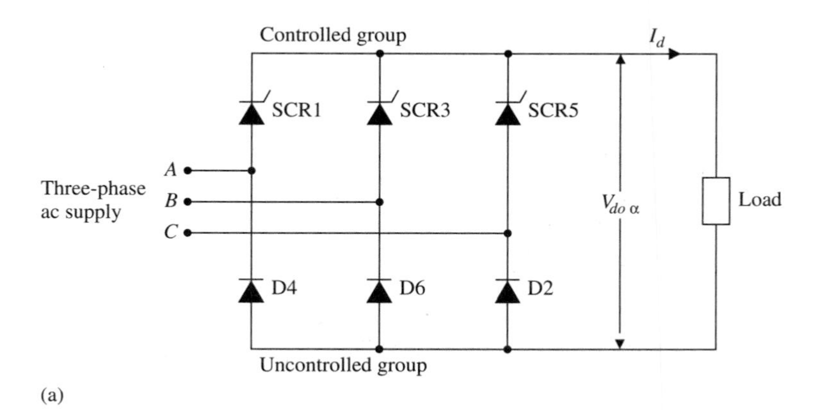

Figure 2-18a Three-pulse half-controlled bridge converter. (a) Schematic.

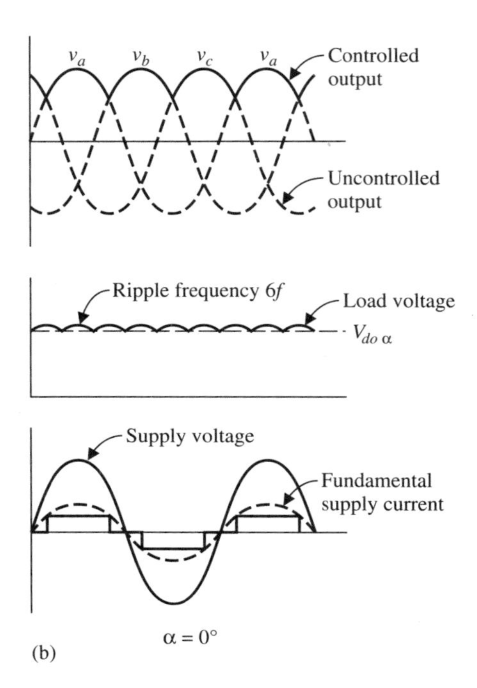

Figure 2-18b Three-pulse half-controlled bridge converter. (b) Waveforms for $\alpha = 0°$.

ing components 120° in duration symmetrically displaced from each other and in phase with the ac voltage.

As the firing delay angle is increased to $\alpha = 60°$ [Fig. 2-18(c)], the mean output dc voltage decreases, and the ripple frequency of the ac component has decreased to $3f$ as soon as α becomes greater than 0°. The ac current supplied to the controlled converter has been delayed by 60° and then it flows for 120°. However, the ac current drawn by the uncontrolled converter has remained in phase with the ac source voltage and flows for 120°. The net current drawn from the ac source is alternating, but the phase shift is less than that associated with a three-pulse fully controlled converter.

Increasing the firing delay angle past 60° will cause the thyristor and the diode in the same leg to be conducting simultaneously, that is, the load current is freewheeling through a diode and a thyristor, and during these intervals the load voltage will be zero [Fig. 2-18(d)]. Simultaneously the amplitude of the ac supply current is also decreasing. The penalty imposed by the decrease in frequency of the ac ripple component is that there is a greater need to filter the output voltage.

Provided that the load is highly inductive, a possibility exists of the converter *half-waving* at small firing delay angles. This is overcome by placing a freewheel diode across the converter output terminals, which also aids the thyristor in regaining its forward blocking capability.

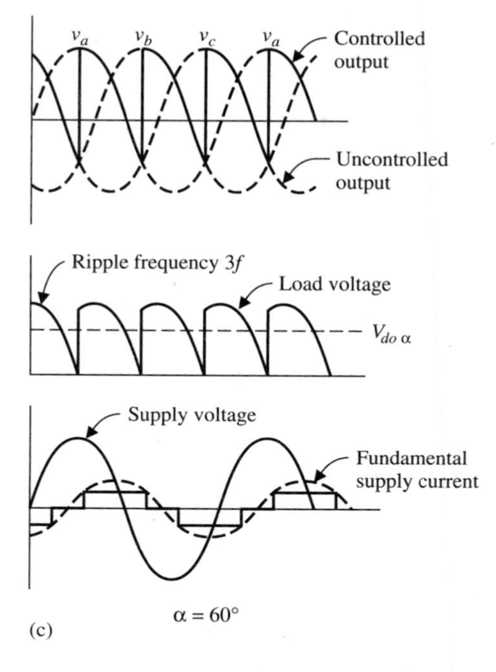

Figure 2-18c Three-pulse half-controlled bridge converter. (c) Waveforms for $\alpha = 60°$.

The peak forward and reverse voltage applied to the diodes and thyristors is $\sqrt{2}V$, where V is the rms line-to-line voltage.

The mean output dc voltage is

$$\begin{aligned} V_{do\,\alpha} &= \frac{3\sqrt{2}V}{\pi}(1+\cos\alpha) \\ &= 1.35V(1+\cos\alpha) \end{aligned} \tag{2–11}$$

and the thyristor current is $I_d/3$. The rms current carried by the devices is

$$I_{T(rms)} = \frac{I_d}{\sqrt{3}} \tag{2–12}$$

and the current carried by the freewheel diode is

$$I_{FWD} = 0.33I_d \tag{2–13}$$

To summarize the advantages of using half-controlled converters as compared to full controlled converters for one-quadrant applications:

1. They are less costly since diodes are cheaper than thyristors.
2. They are less complex since only half the devices require firing circuits.
3. The amplitude of the ac ripple component is less since there are no negative components in the output voltage waveform.

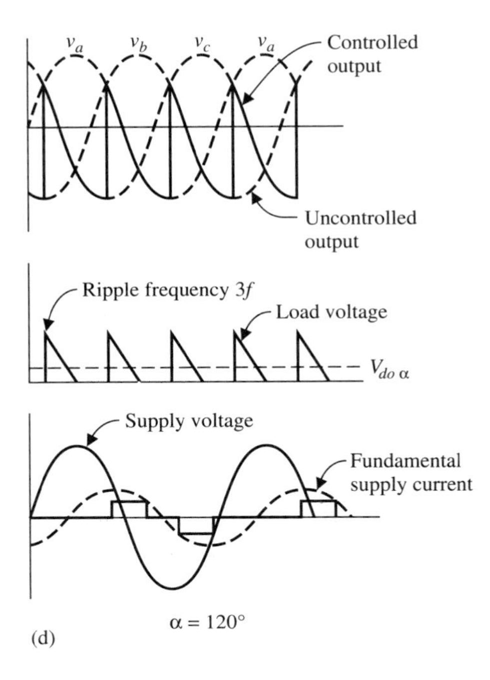

Figure 2-18d Three-pulse half-controlled bridge converter. (d) Waveforms for $\alpha = 120°$.

4. The ripple frequency is half that of the fully controlled converter. However, this may mean that additional filtering is required.
5. The output dc voltage $V_{do\alpha}$ varies from a positive maximum to nearly zero as the firing delay angle varies from 0 to 180°.
6. Using a freewheel diode across the converter output terminals of a two-quadrant converter will cause it to operate as a one-quadrant converter.
7. The input power factor is improved.

2–4 FIRING CIRCUITS

So far no mention has been made of how gate firing pulses are produced, or how they are applied to the proper device at exactly the right time. The firing circuit is the most important section of any converter, whether it is used for ac to dc, dc to dc, or variable-frequency control.

There are many methods of producing gate firing pulses, either using discrete components for analog circuits, or digital circuits using TTL or CMOS ICs. It is our aim to introduce the reader to the basic concepts, so that the principles of firing control are understood.

To obtain the desired operation of any thyristor circuit it is necessary to ensure that unwanted or spurious gate signals are not supplied to the gate circuit. Since a thyristor responds to gate signals as low as 0.25 V with a pulse duration of less than 50 μs, it is essential that all unwanted signals be screened out of the control circuits.

Electrical Noise

Electrical noise is due to fast-rise-time electrical transients present in power and control wiring. The industrial environment has many sources of electrical noise caused by high rates of change of current or voltage. Some major noise sources are the operation of load-carrying switches and relays, brush sparking in dc machines, inductive devices such as relay coils and solenoids, induction heating systems, and thyristor switching. Noise can be transmitted by capacitive or inductive coupling, conduction, radiation, or common line injection. It is extremely difficult to eliminate noise, so most of our efforts are concentrated on reducing the effects of electrical noise by shielding, screening, isolation (physically and electrically), and filtering.

The most productive methods of suppressing noise are:

1. Suppression at the source, such as replacing electromagnetic relays with solid-state relays (SSRs) that incorporate zero voltage switching (ZVS), and connecting capacitors across the relay contacts to reduce arcing. Diodes, varactors, zeners, and *RC* snubbers connected in parallel with inductive elements such as coils reduce transients.
2. Desensitizing the control circuitry, for example, in logic-controlled systems, by using high threshold logic (HTL) so that the voltage level of the enabling signals exceeds the voltage level of the noise signals.
3. Reducing the possibility of coupling between the control circuits and the noise sources. Some common methods incorporated during manufacture or installation are:
   a. Reducing inductive and capacitive coupling by maintaining the greatest physical distance between power and control circuits
   b. Keeping wiring runs as short as possible and avoiding grouping wiring runs together
   c. Reducing capacitive coupling by using shielded cable, with the shielding grounded at only one end
   d. Reducing magnetic coupling by twisting conductors together
   e. Always grounding the control system to the board common connection, never to the electrical common
   f. Providing a grounded screen between the windings of all control system transformers to reduce the effects of interwinding capacitance

Requirements of a Gate Firing Circuit

All firing circuits should meet the following requirements:

1. The amplitude and duration of the gate firing pulse must be great enough to ensure that the thyristor is turned on.
2. There is adequate control of the firing point over the desired output range of the converter.
3. In a closed-loop control system the response must be fast enough to correct a detected error in the controlled variable (usually speed) without the system becoming unstable.

4. There is a linear relationship between the firing delay angle and the converter output.
5. In polyphase systems the phase relationship between the firing pulses must be maintained accurately and synchronized to the ac supply.

Firing Pulses

The shape, duration, and amplitude of the firing pulse is determined by the gating requirements of the thyristor and the connected load. In high *di/dt* applications the gate voltage can vary from 3 to 20 V and the gate current from 0.25 to 4 A, with a rise time of less than 1 μs to ensure rapid spreading of the conduction area in the thyristor. It must be long enough for the thyristor to latch.

A large fast-rise-time gate firing pulse, if applied for too long a time, may cause overheating of the gate-cathode junction and cause the destruction of the thyristor. This can be minimized by using a two-level pulse, as shown in Fig. 2-19.

Double Pulsing and Long Pulsing

As was mentioned when we considered the six-pulse bridge converter, it is necessary for two or more thyristors to be conducting simultaneously to provide a complete current path. If the load is inductive or the load current discontinuous, one short-duration gate firing pulse will not guarantee that all thyristors will be turned on. One commonly used solution is to apply two gate pulses per cycle to each thyristor. These pulses are separated by an angle equal to 2π/pulse number. For example, for a six-pulse bridge converter the spacing is $2\pi/6 = 60°$ [Fig. 2-20(a)]. This method is known as *double pulsing*. An alternative method is to apply *long pulses*, which are usually greater than 60° [Fig. 2-20(b)]. The major problem with long pulses is that provided the load is highly inductive, there will be a delay in current buildup in the thyristors, and by the end of the gate pulse the thyristor may not be latched. A solution is to make the gate pulse 120°, or equal to the conduction period. However,

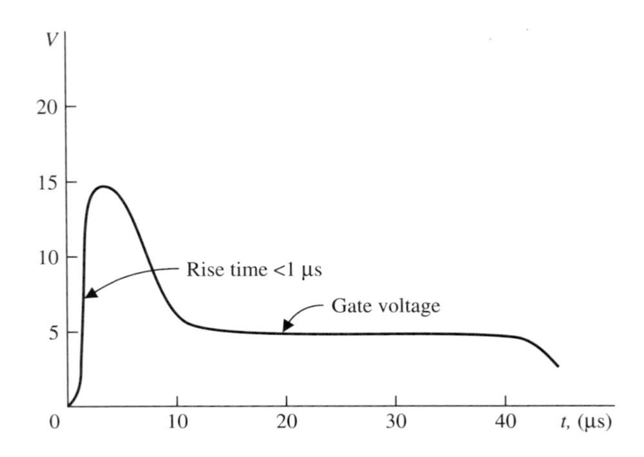

Figure 2-19 Two-level gate pulse for high *di/dt* applications.

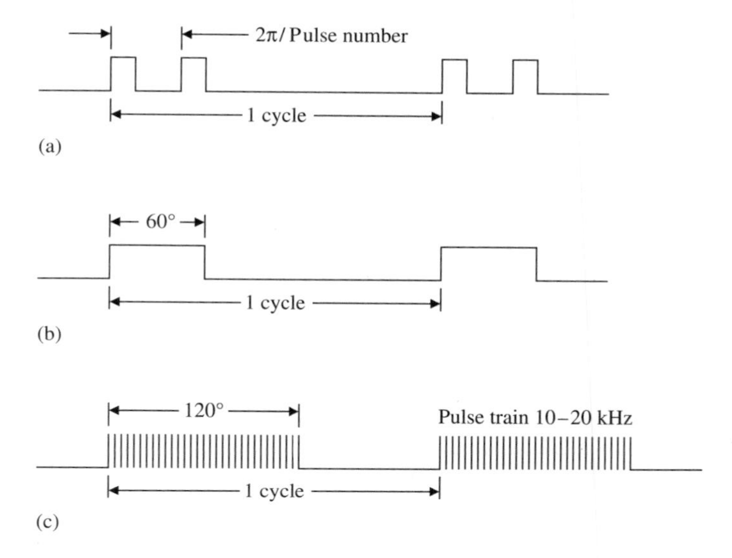

Figure 2-20 Gate firing pulses. (a) Double pulses. (b) Long pulses. (c) Pulse train.

there are two problems. First, it is difficult to produce a pulse of 120° duration, and second, the gate-cathode junction will most probably overheat. This problem can be overcome by using a long firing pulse modulated at a frequency of 10–20 kHz, as shown in Fig. 2-20(c). The pulse train should have a duty cycle of approximately 70–75%.

Pulse Isolation

In all power electronic converters there are potential differences between the gates, and between the individual thyristors and the *gate pulse generators* (GPGs). As a result it is necessary to provide electrical isolation between the gate firing circuits and the thyristors.

Pulse Transformers. The most common method of providing electrical isolation is by the use of a pulse transformer. The pulse transformer normally has a primary winding and one or more secondary windings, which permits simultaneous gate firing pulses to be applied to series- and parallel-connected thyristors. The pulse transformer windings are tightly coupled to reduce leakage inductance, which ensures that the output pulse will have a fast rise time. They are insulated from each other and tested to withstand 2,500 V between windings.

Optocouplers. *Optocouplers* consist of two parts, an infrared light source supplied by a light-emitting diode (LED) and a junction-type photoconductor such as a phototransistor, a photo-Darlington, a photo-SCR, or a photo-FET (Fig. 2-21). These components are assembled into an IC package and arranged so that the light output

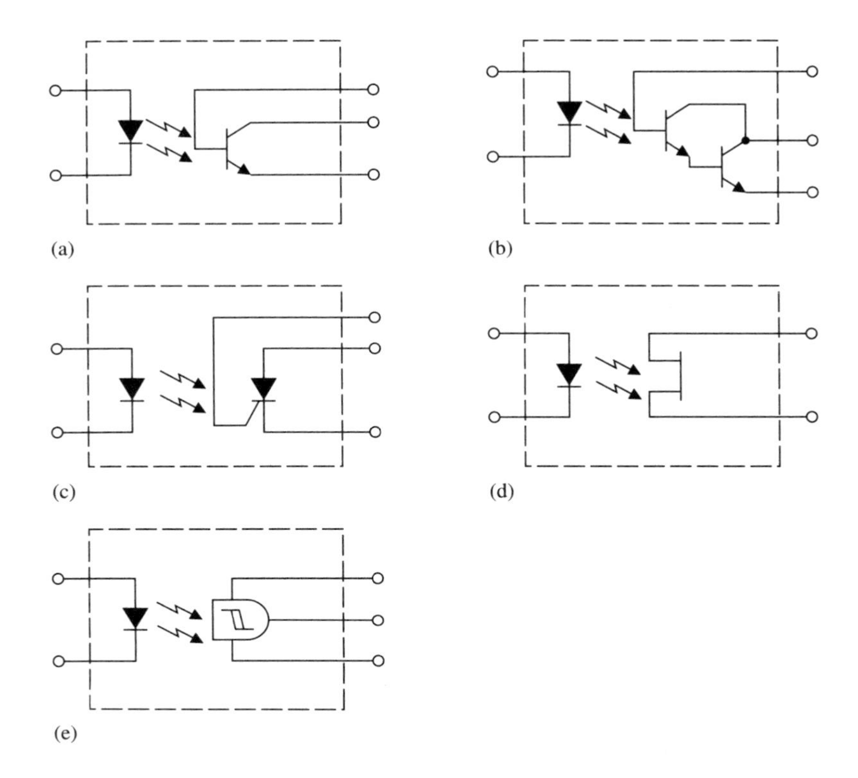

Figure 2-21 Typical optocouplers. (a) Phototransistor. (b) Photo-Darlington. (c) Photo-SCR. (d) Photo-FET. (e) Photo-Schmitt.

from the LED strikes the light-active region of the junction-type photoconductor, which is then turned on. The optocoupler needs a power source on the load side, which may be obtained from either a separate power supply or a voltage divider connected across the thyristor input. The latter method has the disadvantage of providing a leakage path around the thyristor when it is not conducting.

The type of optocoupler that is chosen is mainly determined by the power requirements of the gate circuit. Typical operating speeds of these devices are 100–500 kHz for phototransistors, 2.5–10 kHz for photo-Darlingtons, and 2–20-μs turn-on time for photo-SCRs.

Firing Delay Angle Control. The output of phase-controlled thyristor converters is controlled by varying the firing delay angle α with respect to the point where the incoming thyristor is becoming forward biased. There are two basic methods, linear and cosine angle control of the firing delay angle.

Linear angle control. This method is illustrated in Fig. 2-22 for a single-phase full converter. Figure 2-22(b) shows the block diagram for one-half of the linear angle control scheme. The synchronizing signals v_1 and v_2 are obtained from a center-tapped step-down transformer connected across the ac input lines. Considering the production of the firing delay angle control for SCRs 1 and 4, the synchronizing signal v_1 is squared by the square-wave generator whose output voltage v_S is in turn supplied to the ramp generator, which produces the positive-going ramp voltage v_R. The next step is to compare the ramp voltage v_R against the variable dc control voltage V_C by means of the comparator. The comparator produces a square-wave output voltage v_c whenever $V_C \geq v_R$. The positive-going edge of the comparator voltage v_c marks the end of the firing delay angle α. The firing delay angle is increased by increasing the dc control voltage V_C, or vice versa, which in turn advances or retards the intersection point of the ramp and control voltages. The comparator output voltage v_c is then applied to the gate pulse generator, which in turn applies a pulse to the pulse transformer primary, which amplifies the GPG signal and produces the thyristor gate signal at the secondary terminals for SCRs 1 and 4. A similar arrangement using v_2, which is 180° out of phase with v_1, produces the gating signals for SCRs 2 and 3. The firing delay angle is given by

$$\alpha = kV_C \tag{2–14}$$

From Eq. (2-2) the mean output dc voltage is

$$\begin{aligned} V_{do\,\alpha} &= \frac{2V_{max}}{\pi}\cos\alpha \\ &= \frac{2V_{max}}{\pi}\cos(kV_C) \end{aligned} \tag{2–15}$$

As can be seen, $V_{do\,\alpha}/V_C$ is not a linear relationship, that is, for equal changes of V_C, $V_{do\,\alpha}$ does not change equally. However, in many applications the lack of linearity is acceptable.

Cosine angle control. Unlike linear angle control, the cosine angle control scheme produces a linear relationship between the variable dc reference voltage V_R and the mean dc output voltage $V_{do\,\alpha}$. The basic concept of cosine angle control is shown in Fig. 2-23.

The cosine timing signal is obtained from a step-down transformer connected across the ac input, and the output voltage is phase shifted 90° by an *RC* network to produce the cosine timing signal with a peak amplitude of usually ± 10 V. The variable dc reference signal V_R is the error signal generated by comparing a variable dc signal (usually 0–10 V), representing the desired output speed of the dc motor, to the negative feedback signal, representing the actual motor speed. This signal may be obtained either from a voltage divider connected across the motor terminals or from a tachogenerator coupled to the motor shaft. The firing delay signal is obtained by comparing the positive or negative dc error signal against the cosine timing wave. As V_R becomes more positive, indicating low speed, the firing delay angle is reduced and the converter output voltage increased. Similarly, if V_R increases negatively, indicating that the motor is overspeeding, the firing delay angle is increased, thus reduc-

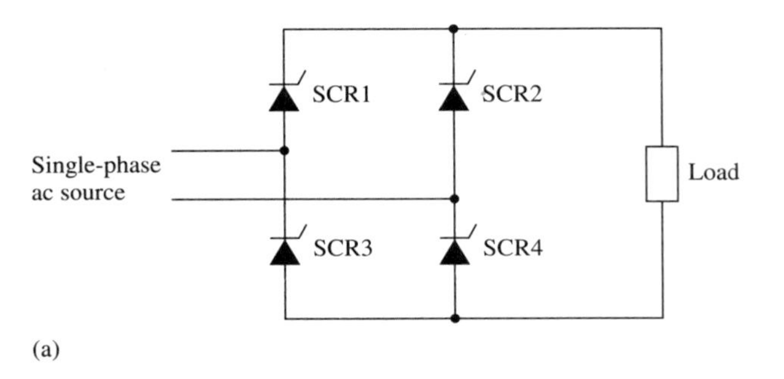

Figure 2-22a Linear angle control. (a) Schematic.

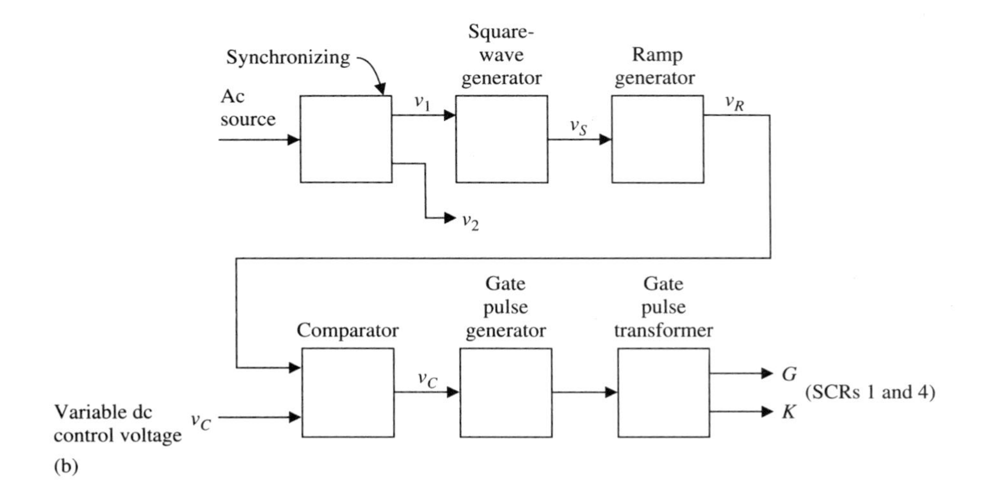

Figure 2-22b Linear angle control. (b) Block diagram.

ing the converter output.

The cosine angle or cosine-crossing method of controlling the firing point of the thyristors is in common use. The system is sensitive to changes in the ac source voltage, since a decrease or increase of the source voltage will produce a corresponding change in the amplitude of the cosine timing wave, and since the dc error signal V_R is independent of the ac source voltage, the firing delay angle will be increased or decreased, and therefore the output voltage will remain constant. The scheme is also sensitive to variations in the source frequency because the *RC* phase-shifting network is frequency sensitive. The firing delay angle α is

$$\alpha = \text{arc cos}\left[\frac{V_R}{v_{\max}}\right] \tag{2–16}$$

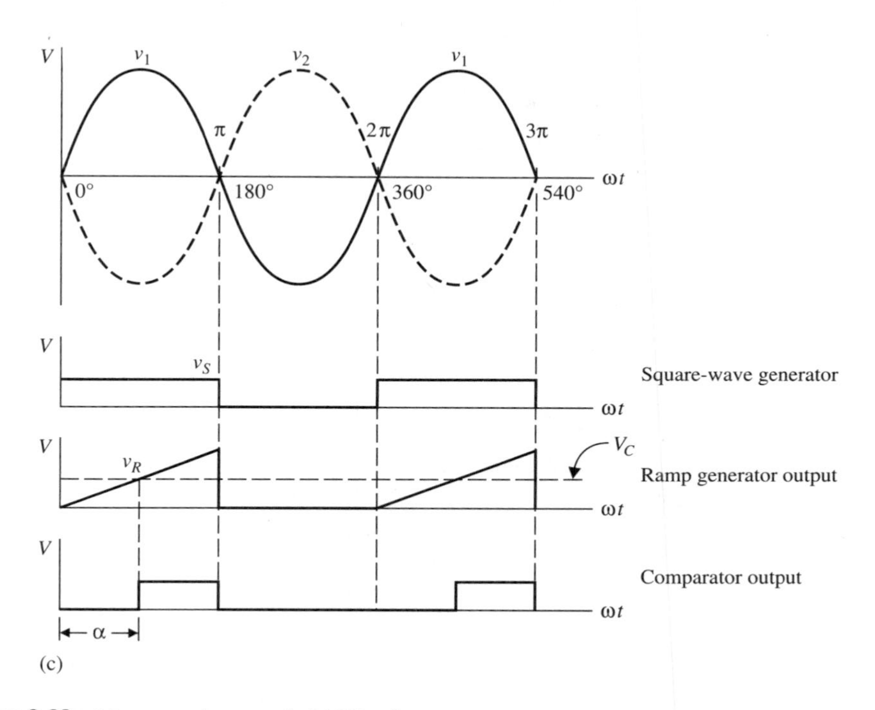

Figure 2-22c Linear angle control. (c) Waveforms.

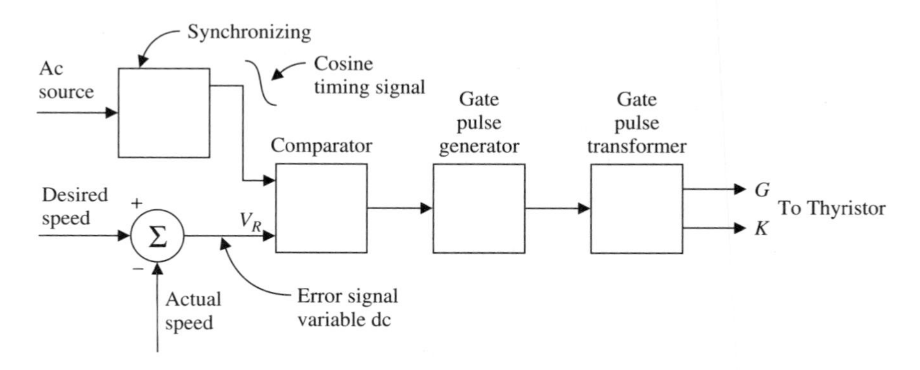

Figure 2-23 Block diagram of cosine angle control applied to dc motor control.

where v_{max} is the maximum amplitude of the cosine timing wave. Then for a two-pulse bridge converter,

$$\begin{aligned} V_{do\,\alpha} &= \frac{2V_{max}}{\pi}\cos\alpha \\ &= \frac{2V_{max}}{\pi}\cos\left(\text{arc}\cos\frac{V_R}{v_{max}}\right) \\ &= \frac{2V_{max}}{\pi v_{max}}V_R \\ &= kV_R \end{aligned} \tag{2–17}$$

Equation 2-17 shows that the cosine control provides a linear relationship between the output voltage $V_{do\,\alpha}$ and the error signal voltage V_R.

2–5 AC THYRISTOR CONTROLLERS

The thyristor can be used with equal ease at power-line frequencies to control the AC power supplied to a load. It is a unidirectional current-carrying device. However, provided thyristors are used in pairs connected in inverse parallel, they can be used in ac applications, although the currents will be alternating and reversing every cycle, that is, the thyristors will be turning off as the current decreases and reverses direction.

The major applications of ac thyristor controllers are industrial heating, single- and polyphase motor speed control, on-load transformer tap changing, and spot welding control.

Single-Phase AC Controllers

The concept of ac phase control is most readily understood by considering the simple single-phase full-wave ac controller shown in Fig. 2-24. The voltage applied to the load is controlled by varying the firing delay angle α, the gate pulses being applied alternately to the SCRs. Theoretically a TRIAC may be used instead of the inverse-parallel SCR arrangement. However, because of the low *dv/dt* capability and current restrictions, normally inverse-parallel connected SCRs are used.

With a resistive load [Fig. 2-24(b)], applying a gate pulse at α will cause the forward-biased SCR to turn on, and the remaining portion of the ac half-cycle of voltage is transferred to the load. The load current waveform will have exactly the same shape as the voltage waveform, its amplitude being determined by the load resistance. Conduction will cease in the conducting SCR as it becomes reverse biased.

With an *RL* load, as the firing delay angle is increased, the flow of load current will be continuous, but displaced with respect to the load voltage until the firing delay angle equals the phase angle ϕ of the load, where ϕ is the angle by which the load current lags the load voltage,

$$\phi = \arctan \frac{\omega L}{R} \tag{2–18}$$

where ωL is the inductive reactance and R the resistance of the load.

When α is less than ϕ, each SCR will conduct for 180°. The current will not flow at the point where the SCR is gated on, however, but starts flowing at the point where the load current waveform crosses the zero axis going positive [Fig. 2-24(c)]. This means that for $\alpha < \phi$, any increase in α will not cause a reduction in the load current. For $\alpha > \phi$ [Fig. 2-24(d)], load current flow is discontinuous, and the load current is reduced.

Under resistive loading conditions, and for $\alpha > \phi$, for an *RL* load there is a rapid rise in the voltage across the SCRs at the end of each current pulse. The rate of voltage change *dv*/*dt* across the SCRs may exceed the capability of the SCR. This prob-

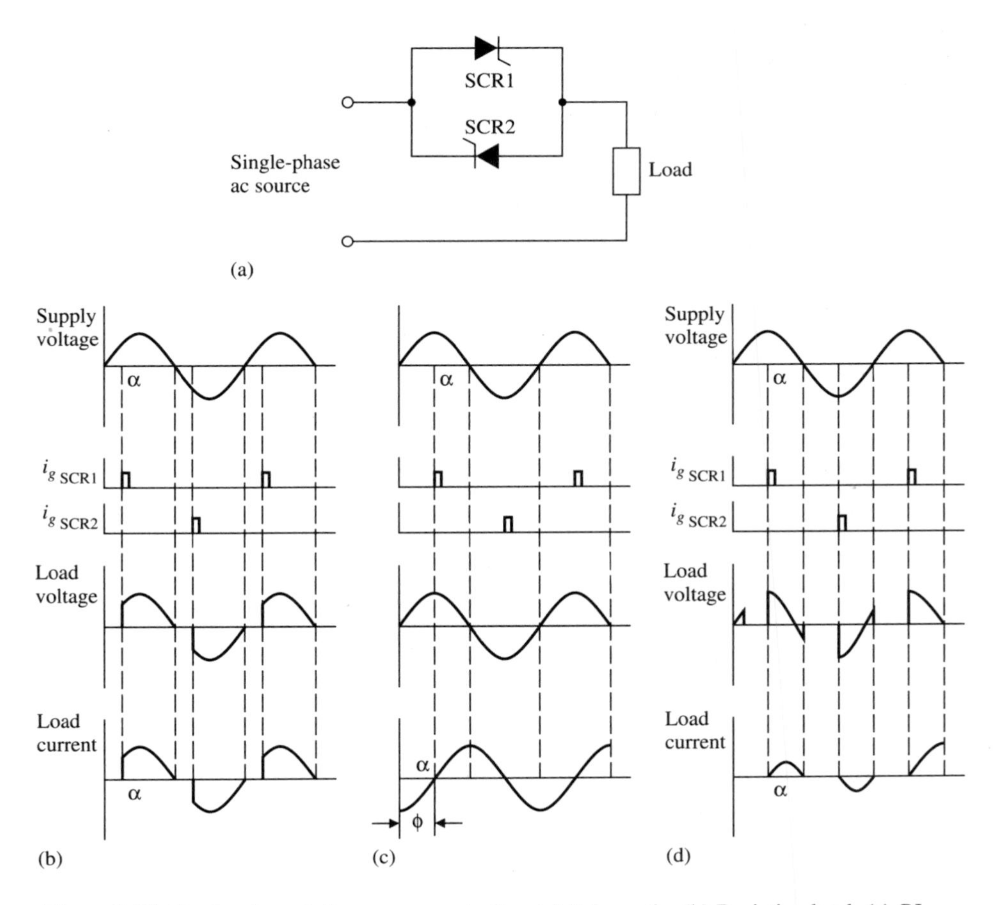

Figure 2-24 Single-phase full-wave ac controller. (a) Schematic. (b) Resistive load. (c) *RL* load $\alpha \leq \phi$. (d) *RL* load $\alpha \geq \phi$.

lem may be reduced by connecting an *RC* snubber across the SCRs.

It should be noted that it is advisable to use long pulse train gating signals to ensure that the incoming SCR is gated on before the current reverses, or the previously conducting SCR will remain in conduction until the current reverses, that is, the incoming SCR will not be turned on and load current will not flow until the next SCR is fired.

Other combinations of single-phase ac controllers are shown in Fig. 2-25. The TRIAC circuit in Fig. 2-25(a) is equivalent to the inverse-parallel SCR circuit just considered. However, its use is limited by the current-carrying capability of the TRIAC. In Fig. 2-25(b) SCR1 and D2 conduct for one half-cycle, and SCR2 and D1 conduct for the other half-cycle. It also has a simplified firing circuit since the SCRs have a common cathode. This circuit is suitable for resistive and inductive loads, and has a control range from 0 to 100%. Figure 2-25(c) consists of an SCR and a diode in inverse parallel, and is only suitable for light resistive loads such as heating and light level control of incandescent lamps. The control range is from 50 to 100% since the diode conducts for a full half-cycle. In Fig. 2-25(d) both halves of the ac cycle are controlled by connecting the SCR across the diode bridge. The load current is alternating, but the SCR current is direct current. This circuit is suitable for resistive or low resistive-inductive loads such as heating, incandescent lamp control and universal motor speed control, and has a control range from 0 to 100%. However, the relative power output is 70% of the input.

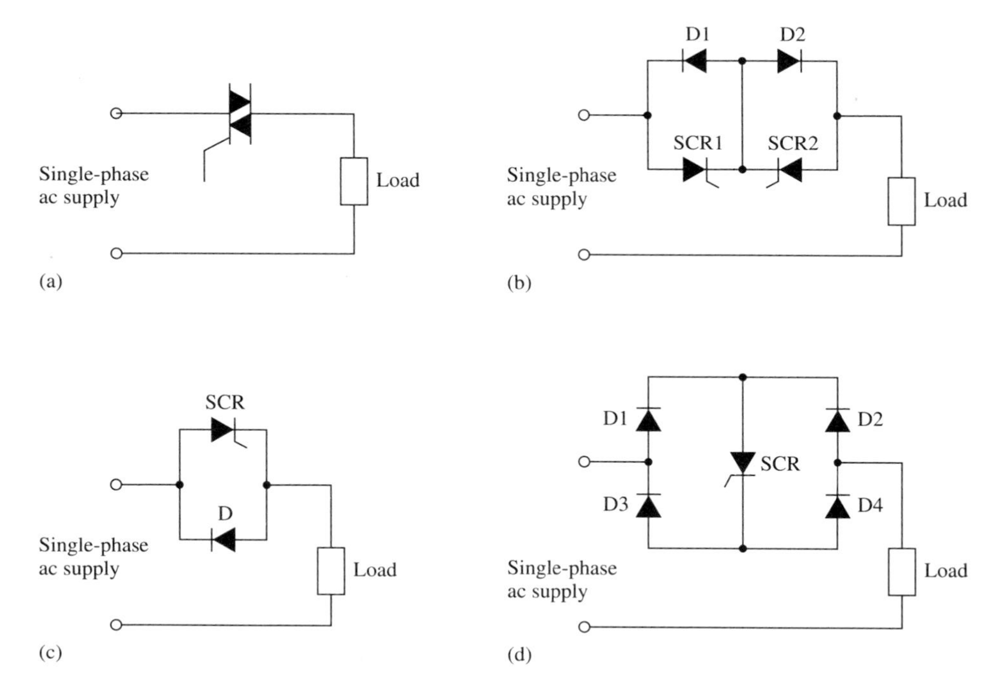

Figure 2-25 Single-phase ac power controllers. (a) TRIAC. (b) With common cathode. (c) Inverse-parallel SCR-diode. (d) SCR-diode bridge.

Control Methods

The ac controller can be controlled by phase-control methods, as discussed earlier, and it can also be used as a switch, where the SCRs are triggered at the voltage crossing points to minimize radio frequency interference (RFI).

The second method of control is known as *integrated cycle* or *pulse burst modulation*. It consists of a number of cycles ON followed by a number of cycles OFF. The power transferred to the load is

$$P_L = \frac{V_{\text{max}}^2}{2R} \frac{\text{no. of cycles ON}}{\text{no. of cycles ON} + \text{no. of cycles OFF}} \tag{2–19}$$

The principle of integral cycle control is shown in Fig. 2-26. It can also be combined with phase control.

Integral cycle control has the advantage of eliminating harmonic currents that occur with phase control. It is most suitable for heating control and control of the current to spot welder electrodes during the heat cycle. However, because of the relatively low frequency of the integral cycles, integral cycle control is not suitable for motor control, and if used for lighting control, it has an annoying flicker.

Three-Phase AC Controllers

The concept of the single-phase ac controller is easily extended to the control of three-phase loads, as shown in Fig. 2-27. Usually inverse-parallel-connected SCRs are placed in series with the ac supply lines supplying star- or delta-connected loads.

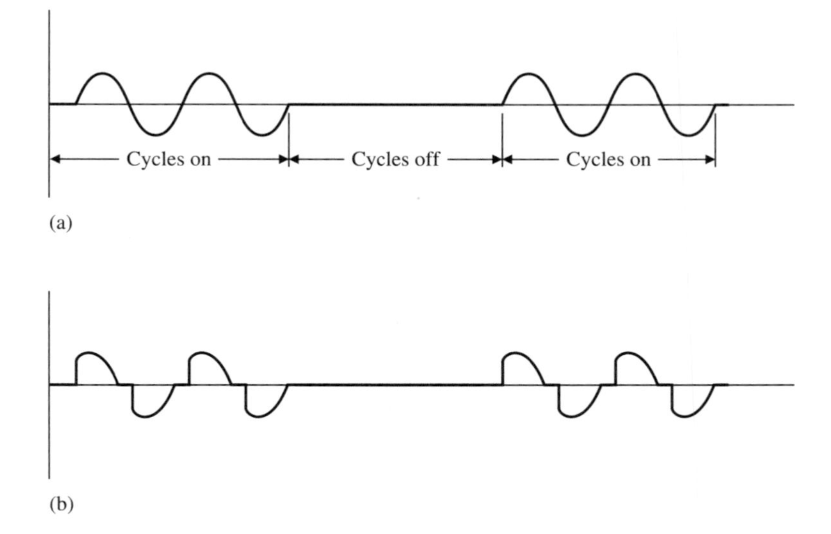

Figure 2-26 Integral or pulse modulation control. (a) Integral. (b) Integral combined with phase control.

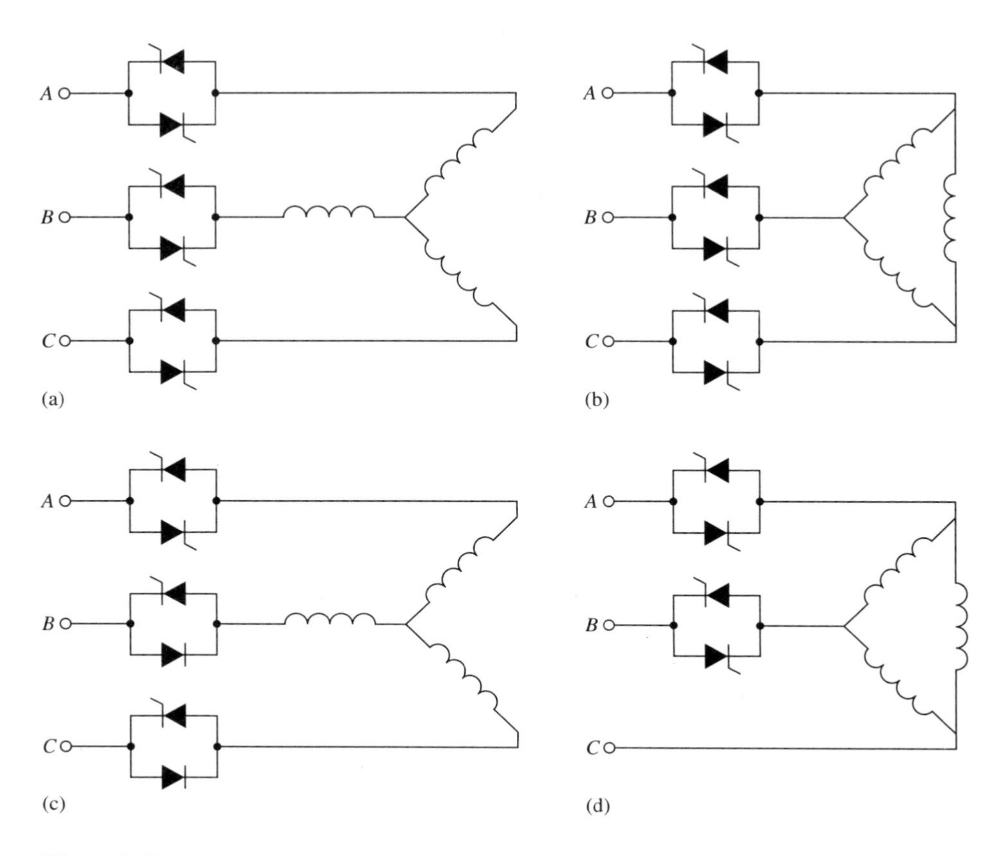

Figure 2-27 Three-phase ac controllers. (a) Fully-controlled star connected load. (b) Fully-controlled delta-connected load. (c) Half-controlled star-connected load. (d) Delta-connected load two-phase control.

In low-current applications TRIACs may be substituted for the inverse-parallel-connected SCRs.

Figure 2-27(a) and (b) shows a fully controlled ac controller connected to star- and delta-connected loads. The controller may be operated as an ac controller by applying a continuous series of gate pulses at $\alpha = 0°$ for the required operational time. Provided the controllers are being phase controlled and the load is partly inductive, as would be the case with the stator of a polyphase induction motor, it is necessary to double-pulse the SCRs, especially for firing delay angles in excess of 90°.

Figure 2-27(c) shows an SCR-diode combination applied to the control of a star-connected load, although it could be a delta-connected load as well. This method is best suited to integral control, as the SCRs and the diodes will be fully on or fully off. If phase control is used, harmonics will be present in the load current.

If unbalanced voltage operation is acceptable, the total cost can be reduced by using the configuration shown in Fig. 2-27(d).

One of the major problems of phase-controlled ac controllers is the production of harmonics, which can be reduced by using integral cycle control, but at the cost of low-frequency voltage fluctuations. Changes of voltage level can be made by using a thyristor-controlled tap-changing transformer or autotransformer. A two-voltage level tap-changing transformer arrangement is shown in Fig. 2-28(a). When SCRs 1 and 2 are gated on, the load is connected to the full secondary voltage; when SCRs 3 and 4 are fired and SCRs 1 and 2 are off, the load is connected to the reduced voltage tap T2. As long as a pair of SCRs is gated on at $\alpha = 0°$, the appropriate voltage is connected to the load, with the added benefits that no harmonics, radio interference, or reduction in the input power factor are introduced, as would be the case with phase control. The concept can be extended to more than two voltage levels, but the control complexity also increases.

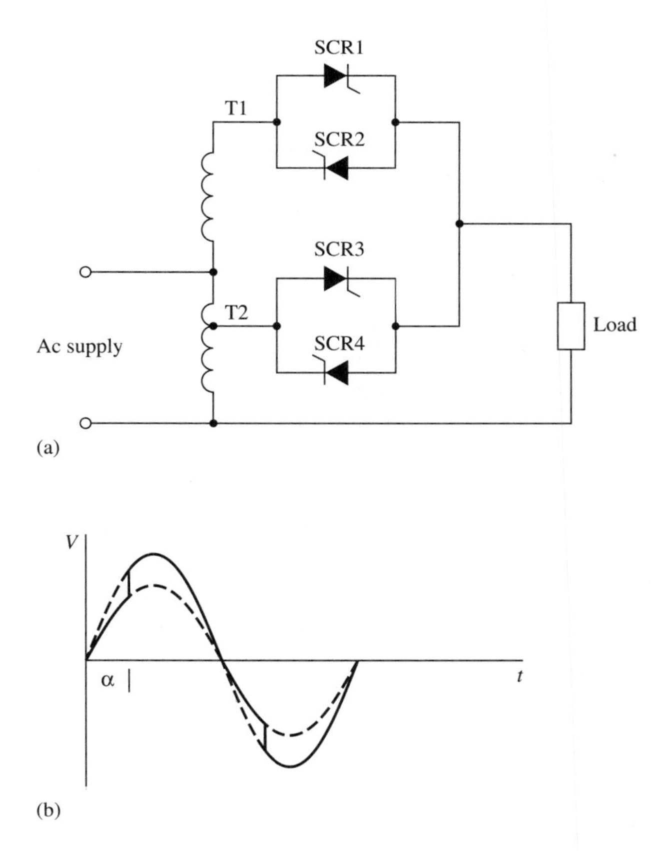

Figure 2-28 Thyristor-controlled transformer tap changer. (a) Circuit. (b) Output waveform of a synchronous tap changer.

The same circuit can be used as a synchronous tap changer, that is, a combination of tap changing by switching from the lower tap to the higher tap at any point during a half-cycle. The resulting waveform is stepped with two voltage levels, as shown in Fig. 2-28(b).

2-6 DC-TO-DC OR CHOPPER CONTROL

The chopper or dc-to-dc converter varies the average output dc voltage at the converter terminals by rapidly switching a constant input dc voltage on and off. The mean load voltage V_{do} can be varied by using any of the following operating modes (Fig. 2-29):

1. *Pulse frequency modulation*: t_{on} constant and t_{off} variable, that is, the periodic time T is variable.
2. *Pulse-width modulation*: t_{on} variable, t_{off} variable, and the periodic time T constant.
3. Some combination of modes 1 and 2.

The implementation of these operating schemes has presented some difficulties in the past. However, with the increasing use of IC digital logic it is now easily achieved. Pulse-width modulation (PWM) combined with devices such as the power transistor and the GTO has eliminated the commutation problems associated with the SCR, and is in common use in choppers and variable-frequency inverters.

The output voltage V_{do} obtained by using PWM or pulse frequency control is

$$V_{do} = \frac{t_{\text{on}}}{t_{\text{on}} + t_{\text{off}}} V_d \tag{2–20}$$

Choppers fed from a constant dc voltage source V_{d} are used to control dc traction motors in streetcars, trolley buses, and subway trains; switching power supplies; and battery-operated equipment such as forklift trucks and delivery vehicles. In these applications the major advantages are:

1. Rheostatic controls and the consequent power loss are eliminated in dc motor control applications.
2. They permit smooth lossless acceleration, thus conserving battery power in electric vehicle applications.
3. Power can be returned back to the source by using regeneration.
4. Rectifier supplied choppers can control the output voltage without the power factor variations that occur with thyristor phase-controlled converters.
5. They are used for one-, two-, and four-quadrant control of dc motors up to 3,000 hp (2,249 kW).

Using thyristors as the switching devices presents a problem, namely, a forward-biased thyristor once turned on and supplying more load current than the holding current will remain in conduction. It is necessary, after the conduction period, to apply to the SCR a reverse bias of sufficient magnitude and duration to ensure that the device has regained its forward blocking capability before the next gate pulse is applied. This technique is called *forced commutation*. Forced-commutation circuits usually get the reverse-bias voltage from a charged capacitor. The advantage of using the power transistor is that it turns off when the base drive signal is removed. Similarly, the GTO turns off when a strong negative pulse is applied to its gate. Therefore neither of these devices requires forced commutation.

Basic Step-Down Chopper

The basic step-down one-quadrant chopper is shown in Fig. 2-30(a). It is called step-down because the average output voltage is less than the source voltage.

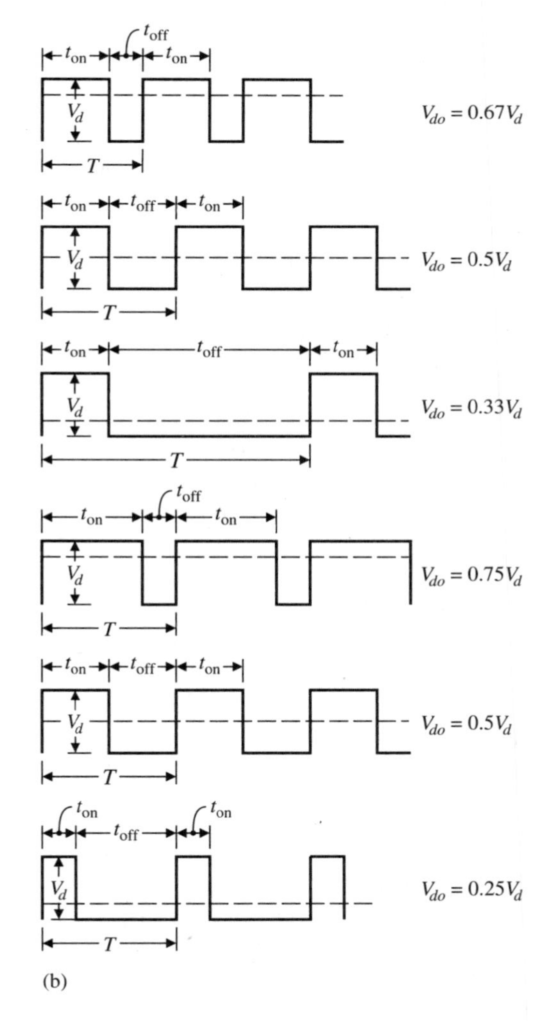

Figure 2-29 Chopper or dc-to-dc control. (a) Pulse-frequency modulation. (b) Pulse-width modulation.

The term one-quadrant shows that it only operates with positive voltage and current, that is, it cannot return power to the source.

When SCR1 is gated on, the constant source voltage V_d minus the device voltage drop is applied to the motor armature, and the freewheel diode D1 is reverse biased. Provided SCR1 remains in conduction, the armature current I_a will increase exponentially because of the filter and armature inductance,

$$I_a = \frac{V_{do} - E_c}{R_a} \tag{2–21}$$

where E_c is the motor counterelectromotive force and R_a is the armature resistance. However, under starting conditions $E_c = 0$, and

$$I_a = \frac{V_{do}}{R_a} \tag{2–22}$$

Since the armature resistance of integral horsepower (kW) motors is usually less than 1 Ω, I_a will become very large.

The armature current during starting is controlled by commutating SCR1 off before the armature current has reached an excessive value. After SCR1 is turned off, the decreasing current will flow clockwise through the motor armature, the freewheel diode D1, and the filter inductance, and will decay to either zero or some value less than that achieved when SCR1 was conducting. By turning SCR1 on and off rapidly, the average value of I_a can be controlled easily [Fig. 2-30(b)].

The effect of the filter inductance is determined by the switching frequency of SCR1. High switching frequencies, usually from 100 to 1 kHz, are normal, permitting relatively small filter inductances to be used to smooth the output voltage to acceptable levels. At low frequencies, that is, less than 400 Hz, it may be necessary to use an *LC* filter to get the desired reduction in the ripple component.

Regenerative Chopper

The regenerative or bidirectional chopper shown in Fig. 2-31 is capable of returning power to the source, that is, *regeneration* or *regenerative braking* can be achieved provided there is a source of electromotive force in the load. This is a bidirectional or two-quadrant chopper, in which the load current may flow in either direction, but the output voltage is always positive.

Obviously SCRs 1 and 2 must never be turned on together, or a short circuit will be placed across the source terminals. Also there should be an interval of at least 100 μs between commutating off the outgoing SCR and applying a gate signal to the incoming SCR. This interval is necessary to ensure that the outgoing SCR has regained its forward blocking capability.

For first-quadrant operation, SCR1 is turned on, which causes the armature current I_a to flow through SCR1, the filter inductor, and the motor armature to the negative terminal of the source. When SCR1 is commutated off, the armature current freewheels through D1, the filter inductor, and the motor armature. When the motor counterelectromotive force E_c is greater than zero and SCR2 is fired, then the arma-

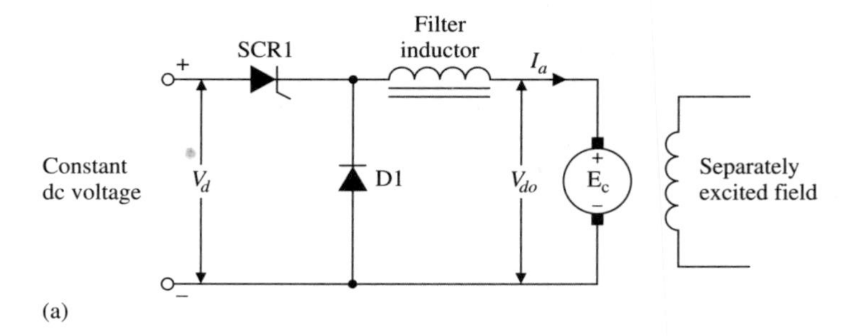

Figure 2-30a Basic step-down chopper. (a) Circuit without forced-commutation circuits.

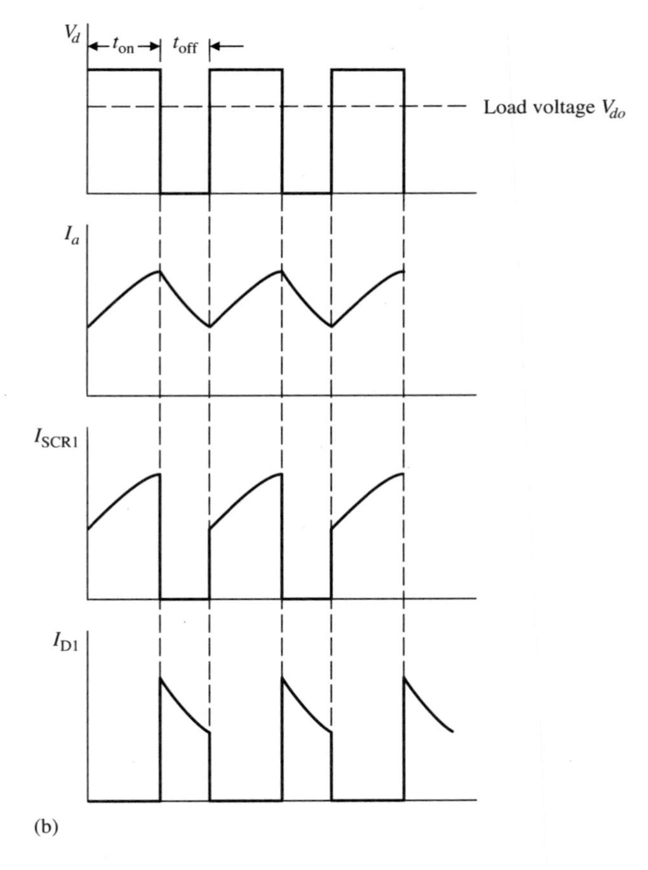

Figure 2-30b Basic step-down chopper. (b) Waveforms.

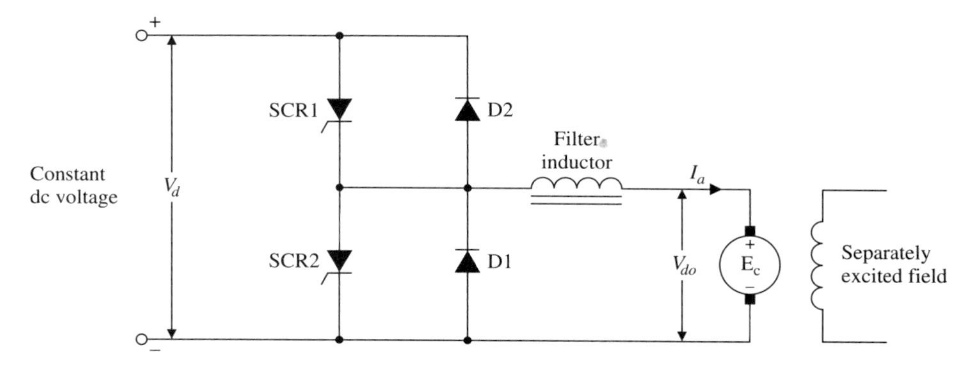

Figure 2-31 Regenerative or bidirectional chopper.

ture current is reversed. Provided SCR2 is commutated off, the armature current flows from the motor armature through the filter inductor and D2 to the source, thus returning power from the motor and braking the motor.

Four-Quadrant Chopper

The four-quadrant chopper is an extension of the two-quadrant chopper shown in Fig. 2-31. This extension permits the chopper to operate with both directions of current and both polarities of voltage.

As can be seen from Fig. 2-32(a), if either SCRs 1 and 2 or SCRs 3 and 4 are turned on together, they will place a short circuit across the source terminals. Turning on SCRs 1 and 4 will cause armature current to be supplied via SCR1, the motor armature, the filter inductor, and SCR4. If SCR1 is turned off and SCR4 remains in conduction, the current will decay exponentially, either to zero, provided SCR1 is not turned on again, or to some lower value if SCR1 is turned on again before the current has reached zero. The motor current freewheels through the

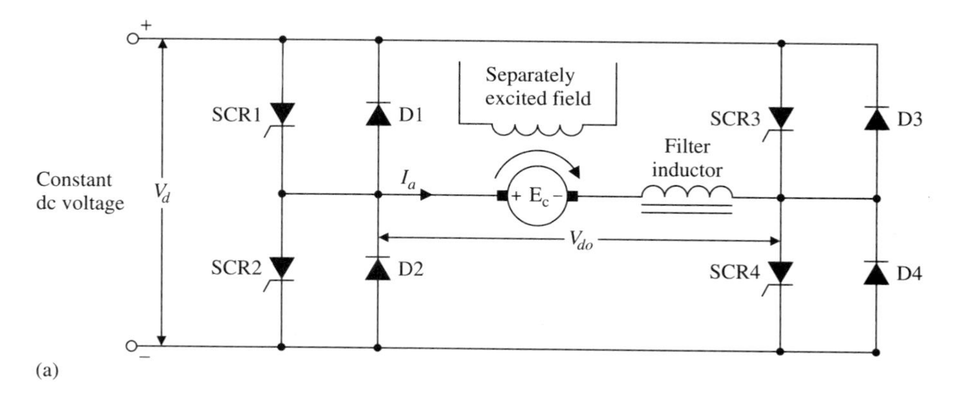

Figure 2-32a Four-quadrant chopper. (a) Circuit.

path consisting of D2, the motor armature, the filter inductor, and SCR4. This first-quadrant operation is illustrated in Fig. 2-32(b).

When SCR2 is turned on with SCR4 remaining on at all times, energy is stored in the armature and filter inductances. When SCR2 is turned off, the stored energy in the inductances is returned to the source via diode D1. This is a second-quadrant operation.

Referring to Fig. 2-32(c), where SCR1 is off and SCR2 is on continuously, turning on SCR3 will cause current to flow in the opposite direction through the filter inductor, the armature, and SCR2. Turning off SCR3 causes the decaying current to freewheel through D3. Turning SCR3 on and off cyclically will cause the stored energy in the inductances to be returned to the source via D4 as SCR3 turns off. The chopper is operating in the third and fourth quadrants.

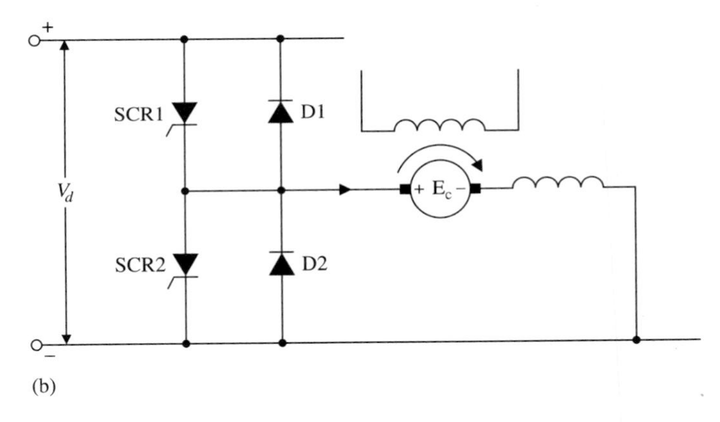

Figure 2-32b Four-quadrant chopper. (b) Operation in quadrants 1 and 2.

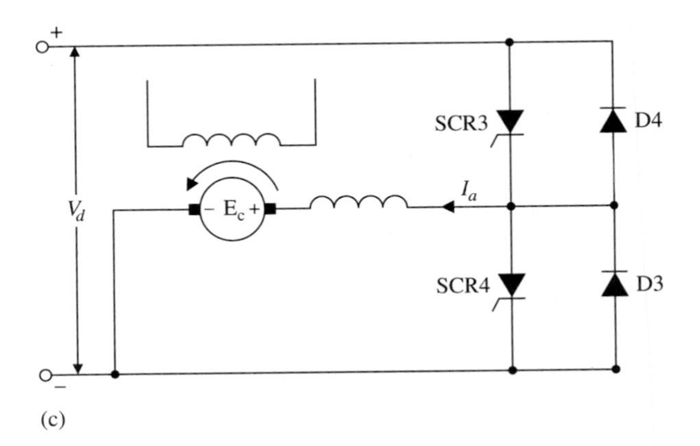

Figure 2-32c Four-quadrant chopper. (c) Operation in quadrants 3 and 4.

2–7 STATIC FREQUENCY CONVERSION

The key to the speed control of polyphase induction and synchronous motors is that the speed for a given motor is proportional to frequency. The major means of producing variable frequency is by the *force-commutated thyristor inverter*, which is basically a square-wave generator whose output is obtained from alternately switching a dc voltage source across the ac load. Modern solid-state power devices, such as the power transistor and the GTO, enable inverters to be simplified by eliminating the requirement for forced commutation.

Single-Phase Bridge Inverter

The basic concept of inverter operation is most easily understood by considering the single-phase bridge inverter shown in Fig. 2-33(a). As can be seen, this circuit is basically a four-quadrant chopper. As before, only one thyristor at a time can be in conduction in a leg of the inverter, that is, SCR1 or SCR2 and SCR3 or SCR4, or a short circuit will be placed across the dc source. Diodes D1 through D4 provide a feedback path for the return of the energy stored in the magnetic field of the load. For simplicity, the forced commutation circuits are not shown.

The inverter operation is as follows. The SCRs are turned on and off for a time corresponding to 180° of the desired output ac cycle. Gating SCR1 on makes point A of the load positive with respect to the negative bus. Turning SCR1 off and turning SCR2 on makes point A negative. Similarly, turning SCR3 on makes point B positive with respect to the negative bus; turning SCR3 off and SCR4 on makes point B negative. Alternately turning SCRs 1 and 2 on and off produces a series of positive square voltage pulses V_{AN}. Similarly, cycling SCRs 3 and 4 on and off produces a series of positive square pulses V_{BN} displaced 180° from V_{AN}.

The alternating voltage across the load V_{AB} is produced by turning on SCRs 1 and 4 simultaneously for 180° and then turning them off, then repeating the process with SCRs 2 and 3, and then repeating the cycle. The frequency of V_{AB} is solely determined by the switching rates of the SCRs.

In addition to the ability to control the output frequency of a variable-frequency

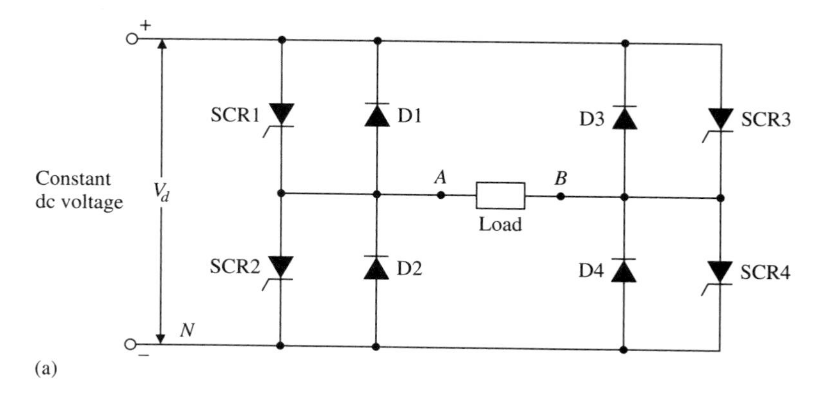

Figure 2-33a Single-phase bridge inverter. (a) Circuit.

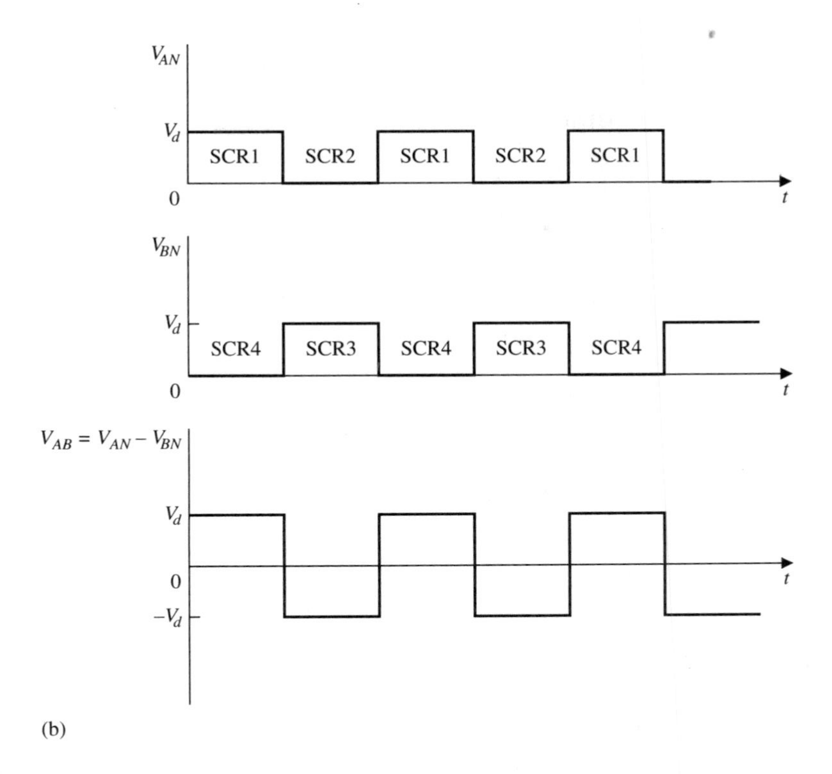

Figure 2-33b Single-phase bridge inverter. (b) Voltage waveforms.

ac drive, it is also necessary to be able to control the output torque. To produce a constant torque output at reduced frequencies it is necessary to maintain a constant voltage-to-frequency (volts/Hz) ratio so that the air-gap flux remains constant. This is done by controlling the inverter voltage. There are several ways of doing this. Provided the source voltage is ac, the dc input voltage to the inverter can be controlled by using a thyristor phase-controlled converter. If the source voltage is dc, then the inverter input voltage may be controlled by using a chopper. These methods are called *variable-input voltage control.* There are several drawbacks to their use:

1. Controlling the dc input voltage to the inverter increases the control complexity.
2. The input dc voltage must be smooth, which means that it must be filtered.
3. The current commutating capability of the inverter varies with the dc voltage level. As a result, at low frequencies, which also mean slow dc voltage levels, the inverter may fail to commutate.

Modern practice is to use voltage control within the inverter, that is, controlling the ratio between the input dc voltage and the output ac voltage. The most common method of voltage control within the inverter is by *pulse-width modulation* (PWM) (Fig. 2-34).

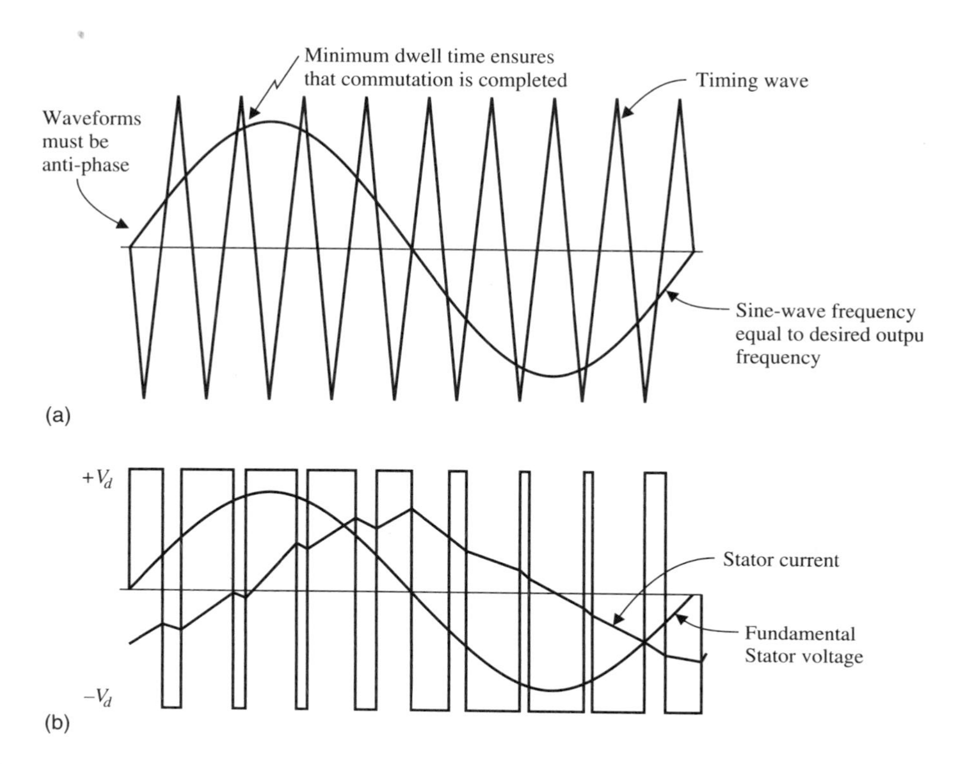

Figure 2-34 Pulse-width modulation. (a) Generation of switching points. (b) Resulting output waveforms.

Comparison of the triangular timing wave against a variable-amplitude variable-frequency sine wave determines the turn-on and turn-off points of the switching thyristors [Fig. 2-34(a)]. The resulting waveform is shown in Fig. 2-34(b), which is a series of constant-amplitude rectangular pulses of varying widths, symmetrical about the center point of the output half-cycles. This waveform effectively varies the instantaneous voltage values by control of the pulse width. The result is a fundamental sinusoidal voltage with a high harmonic content and a nearly sinusoidal current waveform. The harmonic content can be reduced by increasing the switching rate of the devices, which in turn leads to high switching losses if SCRs are used. However, using power transistors in low- to medium-current applications and GTOs in medium- to high-current applications will significantly reduce the switching losses.

Three-phase inverters are obviously the most commonly used type of inverter, and Fig. 2-35 illustrates a typical arrangement. This arrangement is known as a *dc link converter*. The firing and commutation circuits have been omitted to simplify the explanation. The input three-phase power is rectified by the diode bridge. A diode bridge is used to present a high input power factor to the ac source, as well as simplifying the control of the system. The reactor L and the capacitor C smooth the fixed dc output voltage of the diode bridge. The thyristors in the inverter section are

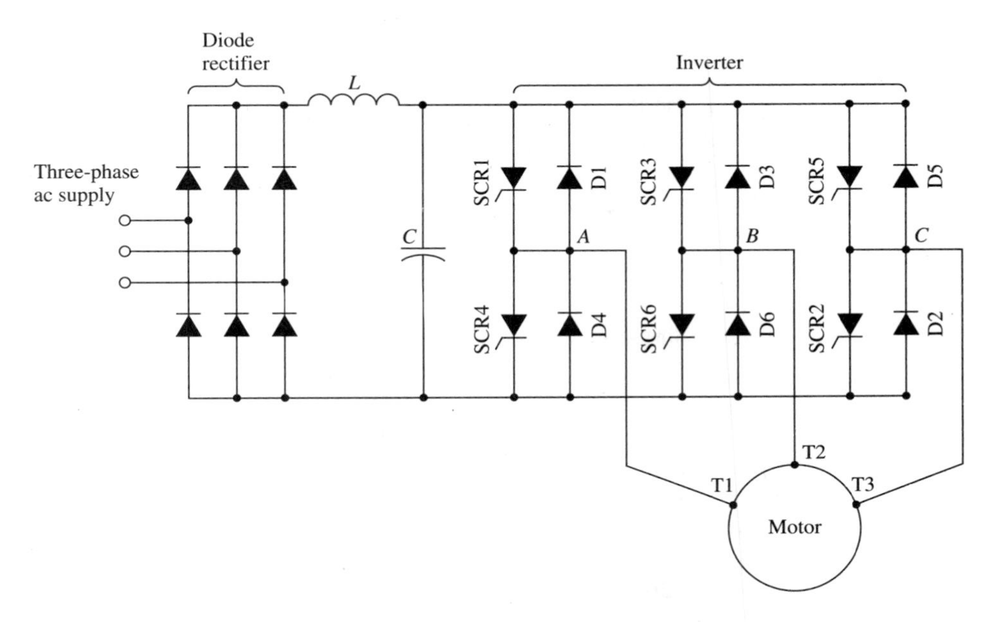

Figure 2-35 Voltage source dc link converter.

numbered in the firing order required to produce positive phase sequence voltages V_{AB}, V_{BC}, and V_{CA} at output terminals A, B, and C of the inverter. Since the inverter is supplied from a constant dc voltage source, the inverter output voltage must be varied to maintain a constant air-gap flux by maintaining a constant voltage-to-frequency (volts/Hz) ratio at reduced frequencies by PWM.

Three-phase inverters are in common use in commercial and industrial applications, where they are displacing dc drives in most cases. There are a number of advantages to their use. For example, they eliminate traditional reduced-voltage starting methods, since the motor can be started at low frequencies and accelerated to the desired speed electronically. Also motor reversal only requires a change in the firing sequence of the switching devices.

2-8 CYCLOCONVERTERS

The cycloconverter is a frequency converter that converts ac power at one frequency to ac power at another frequency in one step without the need for an intermediate power conversion.

Usually cycloconverters step down the frequency. For example, a three-phase 60-Hz input would be varied by the cycloconverter to produce a three-phase output with a frequency that can be varied between 0 and 20 Hz. Cycloconverters are normally operated such that the output frequency is no more than one-third of the input frequency, so that the harmonics in the output voltage are within acceptable limits.

Cycloconverters are used with polyphase induction and synchronous motors for low-speed high-power output applications, such as cement kilns and mine hoists. The advantages of using cycloconverters are:

1. They eliminate the requirement for an intermediate power conversion by a dc link converter, so the overall efficiency is higher.
2. Voltage and frequency control are done entirely by solid-state electronics.
3. The thyristors are line commutated, so there is no requirement for forced commutation and its associated controls.

Single-Phase-to-Single-Phase Cycloconverter

The concept and operation of a cycloconverter are most easily understood by considering the single-phase-to-single-phase cycloconverter shown in Fig. 2-36.

The cycloconverter in Fig. 2-36(a) consists of two two-pulse midpoint converters connected back to back to form a positive group and a negative group. The two groups operate as a dual converter with four-quadrant capability. The firing delay angles of each group, α_N and α_P, are modulated sinusoidally at the desired output frequency so that segments of the secondary voltages v_1 and v_2 form the output voltage waveform [Fig. 2-36(b)]. The amplitude of the sinusoidal modulated control voltage and its frequency are readily adjusted to produce the required variable-frequency variable-voltage output. If the output frequency is increased above one-third of the input frequency, the harmonic content becomes unacceptable.

From Fig. 2-36(b) the fundamental load current lags the fundamental load voltage by the phase angle ϕ. Referring to Fig. 2-36(c), when both waveforms are positive, positive load current is being supplied by the positive converter operating as a rectifier. Similarly, when both waveforms are negative, negative load current is supplied by the negative converter acting as a rectifier. Also the positive group with negative voltage and positive current, and the negative group with positive voltage and negative current, will invert. This means that power can flow into the load, that is, rectification; or power can flow from the load to the source, that is, inversion. It should be noted that despite the direction of power flow, the positive group will only carry positive current, and the negative group will only carry negative current. A cycloconverter can handle all loads varying from zero power factor lagging to zero power factor leading.

From Fig. 2-36(b) it was observed that the output voltage waveform had a high-amplitude ripple content. As was the case for phase-controlled converters, the amplitude of the ripple component can be reduced by increasing the pulse number.

Three-Phase-to-Three-Phase Cycloconverter

Normally cycloconverters are used to provide speed control of high-output polyphase motors. Therefore extending the single-phase concept to three-phase applications will produce an immediate increase in the pulse number. Figure 2-37 shows a four-quadrant six-pulse bridge cycloconverter with a minimum of 36 thyristors. This number can increase very rapidly when series and parallel arrangements of SCRs are needed for high-voltage and high-current applications, with a subsequent increase in control complexity.

Envelope Cycloconverters

When it is only necessary to operate at one or more reduced frequencies, that is, ratios of output frequency to input frequency, such as 2:1, 3:1, or 4:1, it is relatively

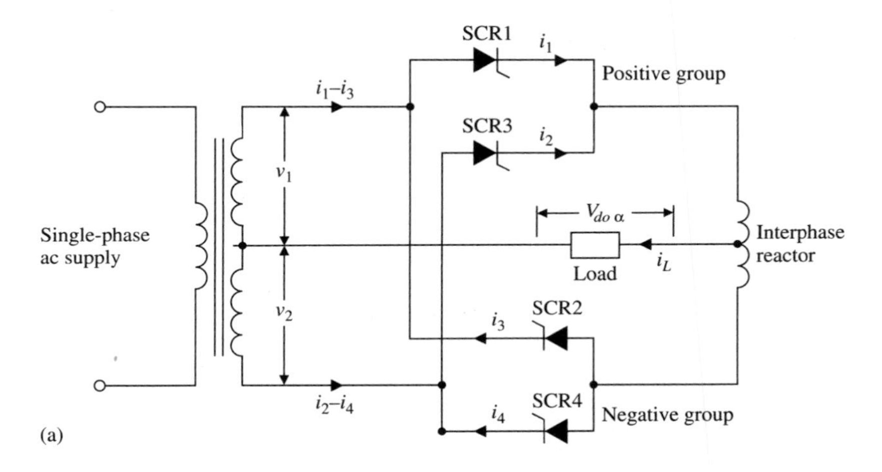

Figure 2-36a Single-phase-to-single-phase cycloconverter. (a) Circuit.

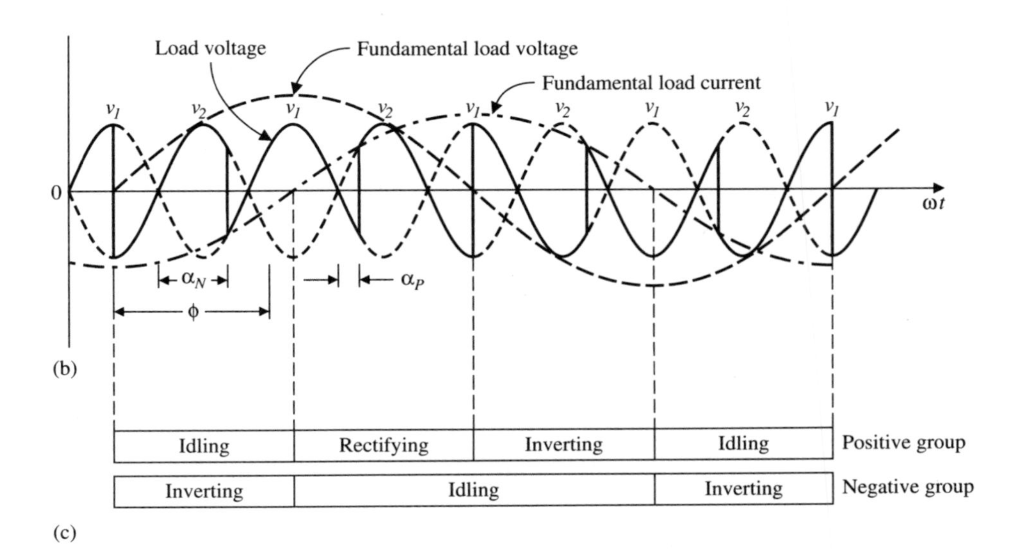

Figure 2-36b Single-phase-to-single-phase cycloconverter. (b) Load voltage waveform. (c) Positive and negative group operating cycles.

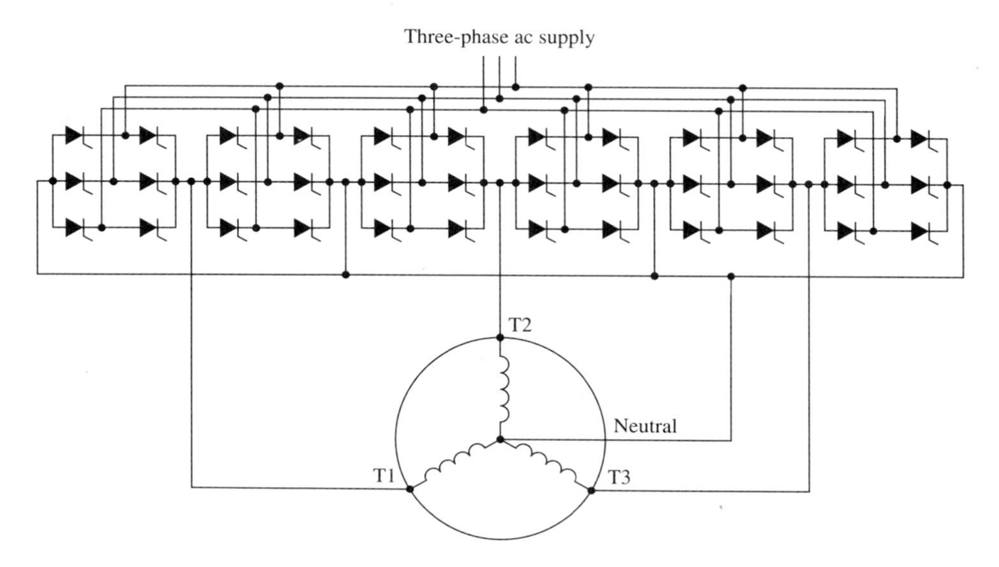

Figure 2-37 Three-phase-to-three-phase six-pulse bridge cycloconverter.

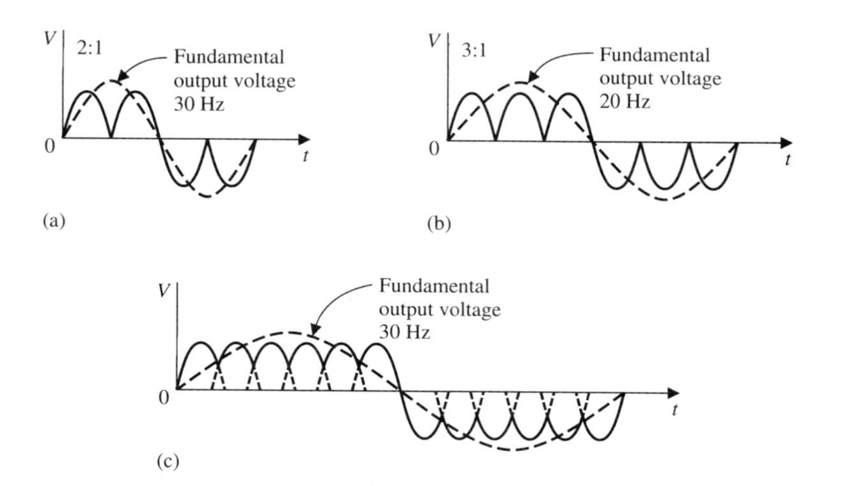

Figure 2-38 Envelope cycloconverter. (a) Single-phase outputs. (b) Three-phase outputs.

easy, using digital logic, to produce outputs as shown in Fig. 2-38. These waveforms are achieved by integral cycle control, that is, firing the positive group continuously for the proper duration of the positive half-cycle of the desired output frequency, with the negative group completely off, followed by similar action for the negative group.

The control circuits for this type of cycloconverter are much simpler than for the phase-controlled cycloconverter. However, the output voltage envelope tends to be rectangular, that is, it has a high harmonic content. Also, since the two converter groups operate only as rectifiers, that is, they cannot invert, operation is limited to unity or near-unity power factor loads.

The objection to a rectangular output waveform can be minimized by using a three-phase star–double-star transformer with four secondary voltages. With a six-pulse diode-thyristor converter, the resulting waveform is an acceptable approximation to a sine wave (Fig. 2-39).

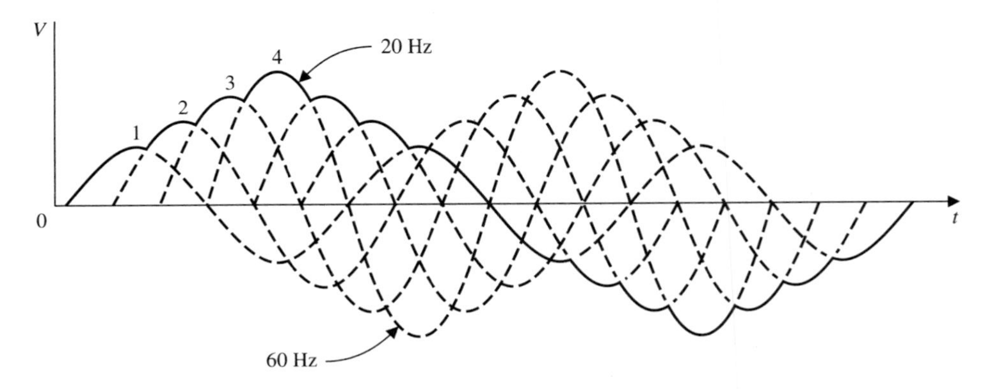

Figure 2-39 Improved envelope cycloconverter with output waveform synthesized from a six-pulse converter having a star–double-star transformer with four different output voltages.

QUESTIONS

2-1 Explain what you understand by the term *power electronics.*

2-2 Briefly describe the function and operation of a diode. Draw the device symbol and the static V–I characteristic.

2-3 What is a general-purpose diode? What is a fast recovery diode? Where would each be used?

2-4 What are the essential diode ratings? What factors affect each of them?

2-5 What is a thyristor?

2-6 What is an SCR? Draw the device symbol and the static V–I characteristic.

2-7 Discuss the ways by which an SCR may be turned on. Which is the preferred method?

2-8 What is the function of a snubber?

2-9 Discuss the differences between a phase-controlled SCR and an inverter grade SCR.

2-10 What is an amplifying gate SCR? Draw the device symbol.

2-11 Discuss the advantages of using an amplifying gate SCR.

2-12 What is an asymmetrical thyristor? What are its advantages?

2-13 What is a reverse conducting thyristor and where would it be used?

2-14 What is a gate turn-off thyristor and where would it be used?

2-15 What is a TRIAC? Draw the device symbol and the static V–I characteristic.

2-16 Where would a TRIAC be used?

2-17 What is a DIAC and where would it be used?

2-18 Why would a power transistor be used in preference to an SCR?

2-19 How is an *n-p-n* power transistor controlled?

2-20 Draw the device symbol of an *n-p-n* power transistor.

2-21 What is an *n-p-n* Darlington power transistor? When would it be used and why? Draw the device symbol.

2-22 What determines the power-handling capability of power transistors?

2-23 What is meant by an SOA graph? What are the three regions of an SOA graph?

2-24 What is a power MOSFET? Draw the device symbol.

2-25 What are the advantages and disadvantages of using power MOSFETs?

2-26 What is a thyristor phase-controlled converter?

2-27 What is synchronous inversion, and under what conditions can it occur?

2-28 Explain what you understand by the terms *one-*, *two-*, and *four-quadrant converters.*

2-29 Explain the phase-control principle as applied to a thyristor phase-controlled converter.

2-30 Explain with the aid of a sketch the operation of a two-pulse midpoint converter.

2-31 What are the characteristics of a two-pulse midpoint converter?

2-32 With the aid of a sketch explain the operation of a two-pulse bridge converter.

2-33 Discuss the major differences of a two-pulse bridge converter and a two-pulse midpoint converter.

2-34 Why would you choose to use a three-pulse midpoint converter?

2-35 Explain with the aid of a sketch the operation of a three-pulse midpoint converter.

2-36 What is the purpose of the interphase reactor in a six-pulse midpoint converter?

2-37 Sketch a six-pulse bridge converter, and explain its waveforms for a 45° firing delay angle.

2-38 What are the advantages of using six-pulse bridge converters?

2-39 What are the characteristics of a six-pulse bridge converter?

2-40 When would a one-quadrant converter be used and why?

2-41 Explain with the aid of a sketch and waveforms the operation of a two-pulse half-controlled bridge converter.

2-42 Explain with the aid of a schematic and waveforms the operation of a three-pulse half-controlled bridge converter.

2-43 What are the advantages and disadvantages of using three-pulse half-controlled converters?

2-44 What is electrical noise? What are the causes of electrical noise? What effects can it have on power electronic equipment?

2-45 How can electrical noise be suppressed?

2-46 What are the requirements that must be met by a gate firing circuit?

2-47 What is meant by double pulsing? Why is it necessary?

2-48 Why is pulse isolation necessary?

2-49 What is the function of a pulse transformer?

2-50 What is the function of an optocoupler?

2-51 Discuss linear angle control of a thyristor phase-controlled converter.

2-52 Discuss cosine angle control of a thyristor phase-controlled converter.

2-53 Explain with the aid of a schematic and waveforms the operation of a single-phase ac controller with resistive and resistive-inductive loading.

2-54 With the aid of schematics briefly discuss the performance of the following types of single-phase ac controllers: **(a)** TRIAC; **(b)** SCR-diode combination with a common cathode; **(c)** inverse-parallel SCR-diode; **(d)** SCR-diode bridge.

2-55 Discuss the relative merits of phase control and integral cycle control of ac controllers.

2-56 What is meant by chopper control?

2-57 Explain what is meant by: **(a)** pulse-frequency-modulation control of chopper; **(b)** pulse-width-modulation control of a chopper.

2-58 What benefits are obtained by using chopper control?

2-59 Why is forced commutation necessary when using SCRs supplied from a dc source?

2-60 Explain with the aid of a schematic the operation of a step-down chopper.

2-61 Explain with the aid of a schematic the operation of a regenerative chopper.

2-62 Explain with the aid of a schematic the operation of a four-quadrant chopper.

2-63 Explain with the aid of a schematic and waveforms the production of an ac voltage by a single-phase bridge converter.

2-64 Discuss the methods by which the output voltage of an inverter may be varied.

2-65 Discuss pulse-width-modulation control of an inverter.

2-66 With the aid of a block diagram explain the operation of a voltage source dc link converter.

2-67 What are the advantages of using three-phase pulse-width-modulated inverters, and where would they be used?

2-68 What is a cycloconverter and where would it be used?

2-69 What are the advantages and disadvantages of using three-phase cycloconverters?

2-70 With the aid of a schematic and waveforms explain the operation of a single-phase cycloconverter.

2-71 What is an envelope cycloconverter? Where would it be used?

CHAPTER

3

DC Machine Construction

3–1 INTRODUCTION

Even though the polyphase ac induction motor, especially in sizes greater than 5 hp (3.73 kW), accounts for approximately 93% of the connected horsepower in North America, dc machines, and in particular the dc motor, play a vital role in modern industry and transportation. The ability of the dc motor to operate at its designed operating range and to maintain any desired speed within this range, still makes it the preferred machine in rolling operations in the papermaking and steel industries, electric traction, and hoisting applications.

The dc machine consists of the following parts:

1. The magnetic circuit, which provides a minimum-reluctance path for the magnetic flux. The magnetic circuit is made up of the following ferromagnetic parts: the main frame or yoke, the main and commutating poles, and the armature core.
2. The field pole windings, which produce the magnetic fields and include the main, commutating, and compensating windings.
3. The armature assembly, whose function it is to support the armature windings so that they move relative to the stationary magnetic fields produced by the field pole windings. Also mounted on the armature shaft is the commutator, whose function it is to rectify the ac voltages produced in the armature windings. Another function of the armature is to provide a low-reluctance path for the magnetic flux.
4. The brushes and brushgear, which provide the means of removing or applying electric energy from or to the machine.
5. The mechanical structure, which holds all the parts together as well as supporting the armature bearings and brush rigging. The mechanical structure must also be designed to meet various mounting requirements such as hori-

zontal and vertical mounting, mechanical stresses, and cooling requirements, as well as protecting the motor against various environmental conditions and, last but not least, protecting personnel against electrical and mechanical hazards.

3–2 MAGNETIC CIRCUIT

Figure 3–1 shows the magnetic flux paths of a four-pole machine. The flux is produced by prewound field coils placed on the field poles. The armature, which is centered between the field poles, performs two tasks: it supports the armature coils as they pass through the magnetic fields, and it provides a low-reluctance path between the field poles to complete the magnetic circuit. The air gap is kept to the minimum consistent with the mechanical limitations of the machine, that is, there must not be any possibility of the armature touching the field poles under all operating conditions. Air gaps normally vary from 0.15 in (3.81 mm) for small precision dc machines to 0.25 in (6.35 mm) for the largest machines.

3–3 FIELD POLES AND WINDINGS

The main poles and the *interpole* or *commutating* poles are rectangular in cross section and are formed from laminations that are riveted together. The laminations are punched out from high-permeability electrical sheet steel. These laminations are normally not as thin as those used in the armature core. Usually they are between 0.062 and 0.125 in (1.57 and 3.18 mm) thick since they are subjected to a constant or slowly varying flux change. The laminated structure is used to reduce to a minimum the eddy currents produced in the *pole shoes* by the high-frequency flux variations that they are experiencing as a result of the movement of the toothed periphery of the armature past the pole shoes. The ends of the poles are shaped to the same radius as the inside of the main frame to ensure a minimum-reluctance joint. The poles are bolted to the main frame. Typical field pole and interpole laminations are shown in Fig. 3 –2(a) and (b).

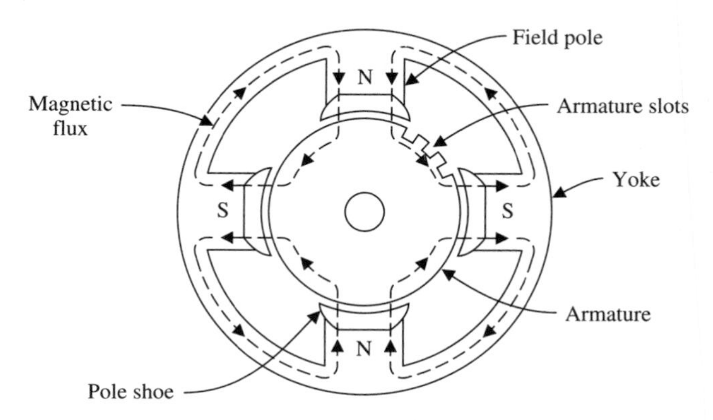

Figure 3-1 Magnetic circuit of a 4-pole dc machine.

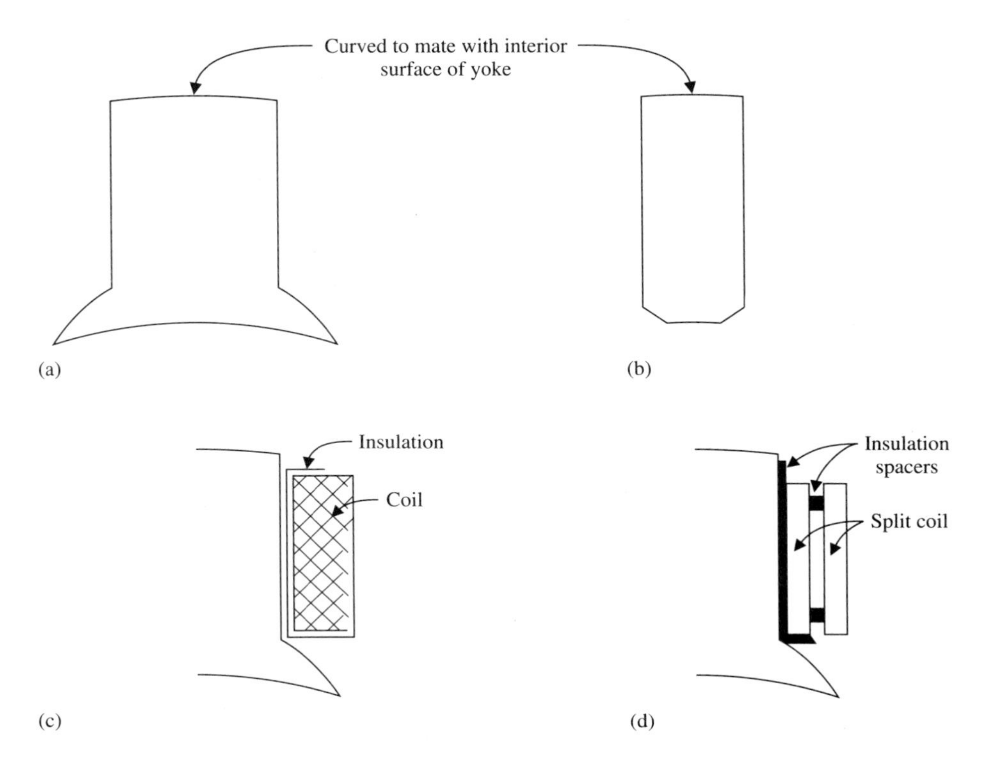

Figure 3-2 Field and interpole assemblies. (a) Field pole lamination. (b) Interpole lamination. (c) Insulated main field coil mounting. (d) Insulated main field coil split to improve cooling.

There are three types of main field coils. For small machines, the coils are wound either on an insulated spool or on an insulated steel spool to improve heat transfer to the field pole. For medium-size to large machines the field coil is wound in two sections. The inner section is wound tightly over the insulated field pole to assist in heat conduction; the outer section is wound over spacers to provide air passages for the cooling air [Fig. 3–2(c) and (d)]. It should be noted that shunt field coils contain a large number of turns of small-gauge enameled copper wire, and that series or interpole windings contain a few turns of large-gauge or rectangular enameled copper wire, the latter usually being edge wound.

The insulating material between the coils and the field pole must be able to withstand high *hot-spot temperatures* caused by the I^2R losses of the coil and the heat radiated from the armature windings and the commutator. The maximum temperatures that can be accepted are determined by the temperature limits imposed by the class of insulation used. There are four classes of insulation that are commonly used in insulating electric machines:

1. *Class A* consists of organic materials such as cotton, paper, cellulose-acetate films, enamel-coated wire impregnated with varnishes, and syn-

thetic resins as binders. This class of insulation materials, sometimes called class 105, indicating designed hot-spot temperatures of 105°C, is rarely used today.

2. *Class B* includes inorganic materials such as mica, glass fiber, asbestos, and synthetic films with suitable binding materials possessing the desired thermal stability. Class B, also sometimes called class 130, indicating a designed hot-spot temperature of 130°C, is used very frequently.
3. *Class F* includes all the materials used in class B, but with binder materials selected to be able to withstand hot-spot temperatures of 155°C (class 155). Class F materials are being used more frequently.
4. *Class H*, a class 180 insulating system, includes mica, glass fibers, asbestos, silicone elastomers, and high-temperature binding materials. Class H insulation systems are principally used in applications where a high ambient temperature exists, or where a small motor is required to supply a high output. Class H materials, because they use silicone varnishes, must be used with discretion in any application where there is a stationary-to-rotary interface involving carbon brushes. The silicone vapors poison the interface between the carbon brushes and the commutator or slip rings and cause sparking and rapid wear of the contacting surfaces.

Modern practice is to use an epoxy-capsulated field bonded to the field pole: this arrangement permits heat transfer directly to the pole and then to the yoke. This technique also reduces maintenance costs by eliminating coil movement and the resulting chafing of the coil against the metal laminations. The interpole or commutating poles are placed midway between the main field poles. These poles are also laminated but have smaller cross sections than the main poles. They also do not have any pole shoes. Small integral-horsepower motors up to 20 hp (14.93 kW) may have only half as many interpoles as main poles as a cost cutting measure.

Some motors up to 200 hp (149.20 kW) are now using permanent-magnet main field poles. This arrangement has the advantage of reducing the field copper losses and the cooling requirements.

3–4 ARMATURE ASSEMBLY

The armature rotates between the field poles and has a laminated core to minimize hysteresis and eddy current losses, with slots for the armature windings around the periphery. The laminations are stamped out from high-permeability electrical sheet steel. They are usually between 0.017 and 0.025 in (0.43 and 0.64 mm) thick and are coated with an insulating film. Laminations for small machines are stamped out in one piece and are mounted on and keyed to the shaft. Laminations for armatures in excess of 45 in (1.14 m) are stamped out in segments and mounted on iron or steel spiders, the final assembly looking somewhat like a wagon wheel.

The armature laminations have slots punched out around the periphery, in which the armature windings are placed. The armature windings are designed to have as many poles as there are main poles. In small-size machines up to 10 hp (7.46 kW) partially enclosed slots are used, and the coils are wound directly into the slots, either manually or by machine. Larger size machines use open slots and form-wound preinsulated armature coils, which are inserted into the slots. In addition, coils in open slots have a lower reactance.

In the case of armatures with axial lengths in excess of 8 in (20.32 cm), radial ventilating ducts about 0.25–0.375 in (6.35 –9.53 mm) wide are provided about every 2.5–3.0 in (6.35 –7.62 cm). Axial ventilating ducts are also provided in armatures made up from single stampings.

The armature laminations are usually clamped between two steel plates. These end plates also provide support for the overhanging coil ends. The armature coils are tightly wedged into the slots. Glass fiber bands under tension are wound over the overhanging coil ends, and in many cases also in recesses on the surface of the armature core, to prevent coil movement caused by centrifugal forces. Dc motors operating under low-speed conditions quite often have the coil slots skewed, usually by one slot pitch, to prevent torque pulsations. This produces an armature with an almost constant reluctance and uniform torque when operating with a constant armature current. Another benefit is the reduction of noise. It should also be recalled from the section on kinetic energy in Chapter 1 that in applications requiring rapid speed changes the moment of inertia is reduced by decreasing the diameter and increasing the length of the armature.

The rotating parts of a dc machine, namely, the armature assembly including the commutator, are also commonly called the *rotor.* Similarly, the stationary parts of the machine are called the *stator.*

Armature Windings

There are two basic types of armature windings used in dc machines: the *lap winding* and the *wave winding.* Sometimes another armature winding known as *frog-leg winding* is used, but it is a combination of lap and wave windings. These windings, regardless of whether they are lap or wave, are prewound and insulated [Fig. 3–3(a)] and then inserted into the slots around the armature core. The coils consist of a number of turns of enamel-insulated copper wire, the whole assembly being taped

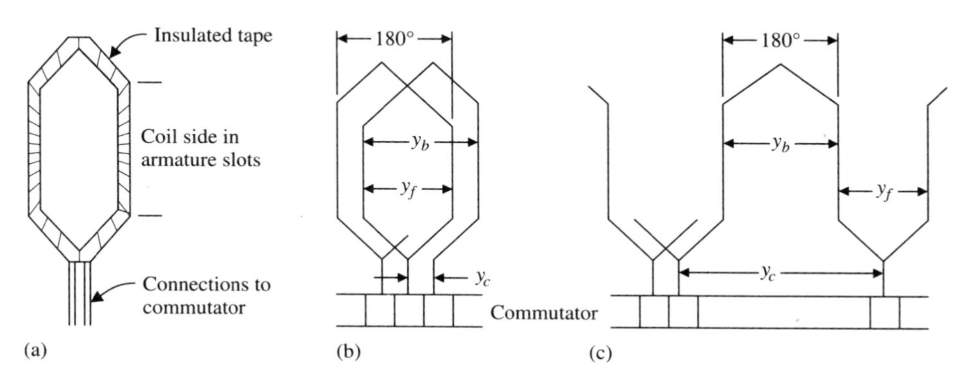

Figure 3-3 Armature coils. (a) Preformed coil. (b) Lap winding. (c) Wave winding.

with insulating material. The difference between the two types of windings shown in Fig. 3–3(b) and (c) consists mainly in the way the coil ends are brought out to the commutator.

The *coil span,* that is, the distance between the two coil sides of a preformed coil, is usually 180° electrical, or equal to the *pole pitch,* that is, the angle from the center of one main pole to the center of the next main pole of opposite polarity. Coils whose span is equal to the pole pitch are termed *full-pitch coils,* and those that have a span less than a pole pitch are called *fractional-pitch coils.* Fractional-pitch coils use slightly less copper than full-pitch coils, and they also improve commutation.

Referring to Fig. 3–3(b) and (c), the coil connections to the commutator and to each other can be defined in terms of the back pitch y_b, the front pitch y_f of the coil, and the commutator pitch y_c . The back pitch of any dc armature winding will be approximately equal to the pole pitch, that is, the back pitch for a P-pole machine with S slots will be an integer number equal to or less than P/S. The commutator pitch will be the algebraic sum of the front and back pitches,

$$y_c = y_b - y_f \quad \text{(lap winding)} \tag{3–1}$$

$$y_c = y_b + y_f \quad \text{(wave winding)} \tag{3–2}$$

The commutator pitch of a simplex lap winding will always be one commutator segment, as shown in Fig. 3–3(b). However, in the case of multiplex lap windings of multiplicity m the commutator pitch will be m commutator segments.

➤ EXAMPLE 3-1

A four-pole dc machine has 24 slots and is to be lap wound. What are the front and back pitches?

SOLUTION

Since there are 24 slots, the number of slots per pole is $24/4 = 6 =$ pole pitch. The front and back pitches must be approximately the same as the pole pitch, and they must be odd numbers. Therefore assuming $y_b = 7$ and $y_f = 5$ meets this requirement.

Using these values, the completed armature winding can be developed as shown in Fig. 3–4. If the directions of the induced electromotive forces in the coils are inserted, then it can be seen that the positive brushes should be making contact at commutator segments 2 and 8, and the negative brushes at commutator segments 5 and 11. If the developed winding is studied carefully, it will be seen that the armature windings form four paths in parallel. There must be as many brushes as there are paths in parallel. Brushes of the same polarity are joined together. This is an important characteristic of lap windings. There are as many paths in parallel in the armature circuit as there are field poles, that is, the total current in the external circuit is equal to the sum of the currents in the parallel paths. This leads to the conclusion that a lap-wound machine is best suited for heavy-current applications.

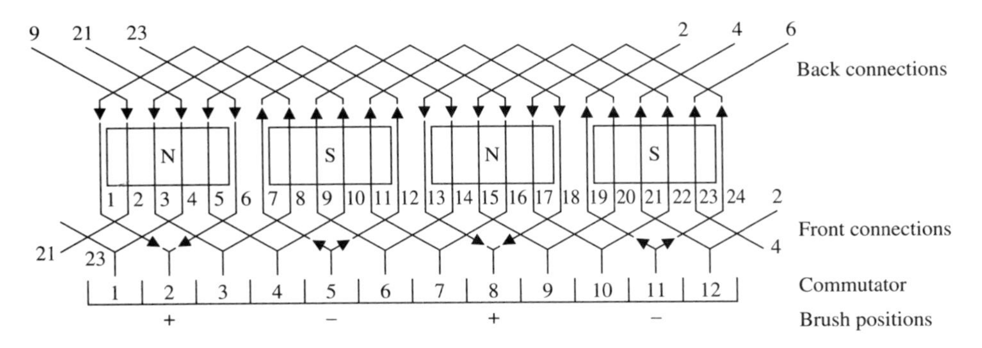

Figure 3-4 Developed simplex lap winding for a 4-pole dc machine with 24 slots.

The wave, or, as it is sometimes called, the series winding differs from the lap winding in that each coil, instead of being connected to the next coil lying under the same pair of poles, is connected to a coil lying in approximately the same position under the next pair of poles. The commutator pitch for a simplex wave winding is

$$y_c = \frac{N \pm 1}{\text{no. of pole pairs}} \tag{3–3}$$

where N, which must be odd, is the number of commutator segments, and y_c must be a whole number.

The coil ends are connected to commutator segments approximately two pole pitches apart, that is, each time a coil is passed through, an advance of two pole pitches around the commutator has been made. For example, in the case of a six-pole machine there are three pairs of poles. So after passing through the three coils, the finishing point will be made to commutator segment 2 before or after the one at which the winding started. Repeating this process for all coils, the winding will eventually close on itself.

➤ EXAMPLE 3-2

A four-pole dc machine with 19 commutator slots is to be wound with a simplex wave winding. What are the commutator, forward, and back pitches?

SOLUTION

From Eq. (3-3), y_c is $(19 \pm 1)/2 = 9$ or 10. Assuming 9, then the back pitch is $19/4 = 4.75$, or 5 slots. The front pitch $y_f = y_c - y_b = 9 - 5 = 4$ slots.

Figure 3–5(a) shows the developed winding of Example 3–2. The brush positions can be found by drawing the equivalent ring diagram [Fig. 3–5(b)]. Point A is the point where the electromotive forces separate, and since it is at the back of the winding, the negative brush connection is made to commutator segment 10. Similarly, point *B* is the point where the two electromotive forces join together. Once again it is at the back of the winding. Therefore the positive brush connection is made to commutator segment 5. Since the wave winding consists of two parallel paths irrespective of the number of poles, only two brushes are required. The electromotive force produced by the wave winding is equal to that produced by half the armature conductors connected in series, and the total current in the external circuit is twice that in either armature path. The wave connection is used in dc machines that operate at high voltages and low currents. In general, the lap-wound machine is usually found in the larger ratings, and the wave-wound machine in small to medium-size machines.

Commutator

The commutator is an integral part of the armature assembly and is unique to dc machines. It is made up of a number of tapered segments of hard-drawn copper, sometimes containing a small percentage of silver to reduce the specific resistance. The copper segments are insulated from each other by mica segments, usually 0.02 – 0.05 in (0.51 – 1.27 mm) thick. The commutator and mica segments are clamped between two mica-insulated metal V rings [Fig. 3–6(a)]. A more modern technique molds the commutator segments together with epoxy into a solid mass, with glass fiber bands being applied under tension in grooves around the commutator ends and the whole assembly held rigidly against an insulated steel cylinder [Fig. 3–6(b)].

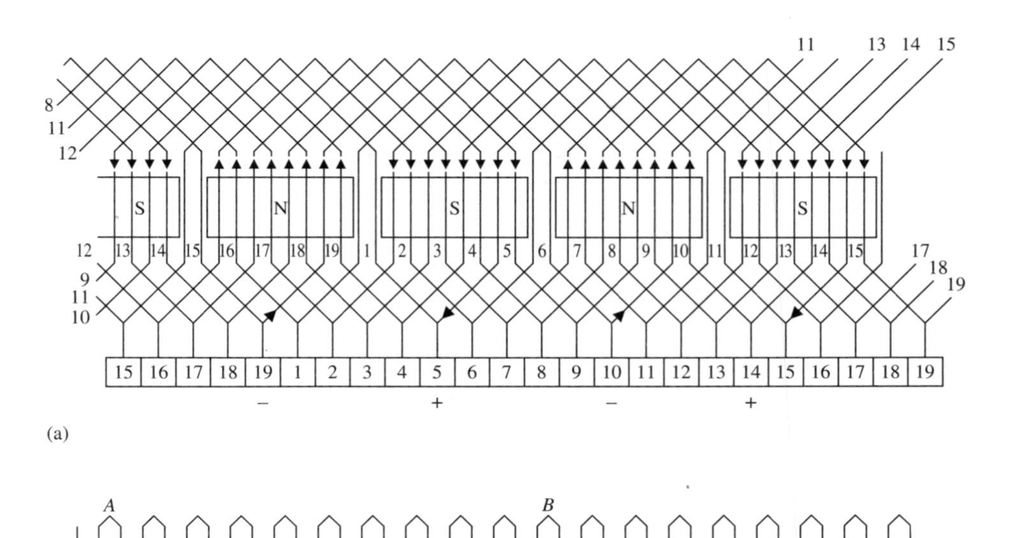

Figure 3-5 Developed simplex wave winding. (a) For a 4-pole dc machine with 19 slots. (b) Equivalent ring diagram.

This type of construction is lighter and more rigid; however, it cannot be repaired. The armature coil ends are connected to slots in the commutator segment risers and either soldered or tungsten inert gas (TIG) welded. The TIG process produces a copper-to-copper connection, which eliminates the problems associated with low-melting-point solders under overload conditions.

Cooling Fan

Another important item that forms part of the armature assembly is the cooling fan. The fan is keyed to the shaft and provides the airflow to cool the field coils, armature core, and commutator.

3–5 BRUSHES AND BRUSH RIGGING

Carbon and graphite brushes are used to remove or apply electric energy in dc machines. Originally natural graphite was chosen because of its natural lubricating properties. However, it also has a number of properties that make it the ideal material for sliding electric contact.

1. It is resistant to changes caused by local high temperatures as occur with arcing brushes in dc machines.

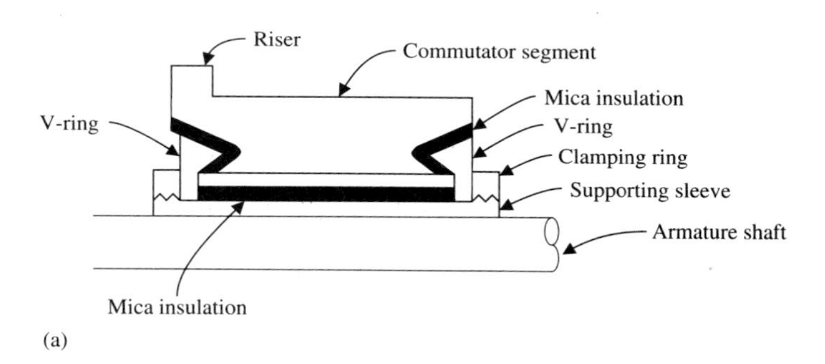

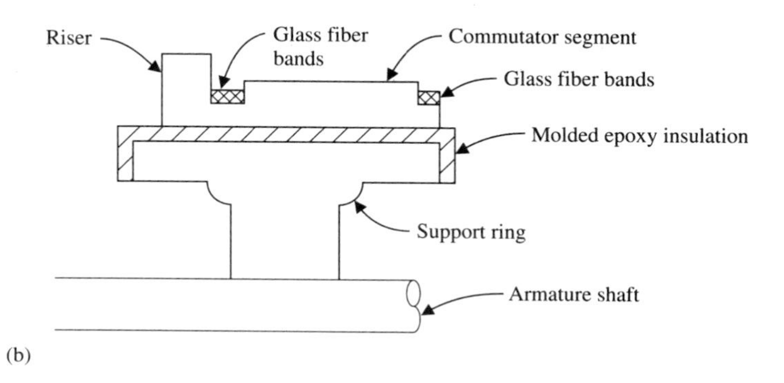

Figure 3-6 Commutator construction. (a) Clamped type. (b) Molded epoxy type.

2. Because of its low density, it has a low inertia and will therefore follow any surface irregularities such as an eccentric commutator.
3. Carbon will not weld to metals during arcing.
4. Carbon has a low coefficient of friction and therefore will minimize friction between the brush and the commutator as well as between the brush and the brush holder.

Most carbon used in electrical applications consists of powdered carbon or graphite or a mixture of both (lampblack or petroleum coke), mixed into a homogeneous mass with binders and baked at approximately 900°C (1652°F) in the absence of air. The volatile binder components are driven off, and the remainder is carbonized. This product can be converted into electrographite by baking at temperatures in excess of 2200°C (3992°F) in the absence of oxygen.

Carbon and graphite brushes used in dc machines must perform two distinct tasks. First, they must provide the current path into or out of the commutator. Second, during the commutation process they must carry all or part of the current as it reverses during the time that the armature coils are short-circuited by the brush. The latter requires that there be an appreciable resistance in the brush-commutator interface.

There are four classes of brushes in common use.

1. *Natural graphite brushes* have a high self-lubricating property, good thermal conductivity, and a high contact resistance. They are used with high-commutator-speed high-voltage machines.
2. *Hard carbon brushes* are mechanically strong and have low electrical and thermal conductivity. They are used mainly in low-surface-speed and low-current-density applications.
3. *Electrographitic brushes* do not contain any natural graphite and are made from various types of carbon. They are converted into artificial graphite at temperatures in excess of 2,500°C (4,532°F). This produces a material that has the properties of both carbon and graphite, that is, it has good thermal and electrical conductivity, and a high contact drop with good lubricating properties. At the same time it has maintained its mechanical properties, but is softer than before graphitization. This type of brush is especially suitable for applications where it is subjected to mechanical shock and sparking such as in traction motors.
4. *Metal graphite brushes* contain a very fine mixture of copper or silver combined with electrographitic material. They have a very low contact drop, excellent thermal conductivity, and a high current-carrying capability. These brushes are used in very-low-voltage high-current applications, such as welding machines.

The correct type of brush, the permissible current densities, the brush angle to the commutator surface, and the brush pressure are usually specified by the machine manufacturer.

The brushes are housed in brush holders. The most common type of brush holder is the slide type, where the brush fits inside a box-shaped guide made of brass, steel, bronze, or a copper-based alloy. Pressure is applied to the top of the brush by a spring, the actual pressure being controlled by the spring pressure. The electrical connection between the brush and the holder is usually made by a plaited or twisted copper flexible connector or *pig tail.* The pig tail is inserted into the brush material during the fabrication process.

The brush holders are mounted on brush yokes or rockers, which in turn are attached to the end bell. The brush rocker has a limited arc of travel so that the brushes can be adjusted to the brush neutral position.

3–6 MECHANICAL STRUCTURE

The *main frame* or *yoke* provides the return path for the magnetic flux and is usually fabricated from rolled mild steel plate. In applications where there are rapidly changing load conditions, the yoke may be laminated. The cross-sectional area of the yoke must also be large enough to ensure that the magnetic circuit is not saturated.

After rolling, the two ends of the yoke are butt-welded. The inner surface and the ends are machined to ensure that the inner diameter is cylindrical and that the end faces are perpendicular to the horizontal centerline of the yoke. The inner surface is machined to ensure that when the main field poles and the interpoles are bolted to it, there will be a minimum-reluctance joint between them and the yoke. The end faces are machined, drilled, and tapped to ensure that the end bells, which house the bearings and support the brush rigging, can be accurately placed. Mounting feet are welded or bolted to the outside of the yoke. On large machines the bearings are usually oil-ring lubricated sleeve-type babbitt bearings located in separate pedestal bearing housings external to the machine. This type of bearing, especially in low-speed applications, may be supplied from a forced-oil lubrication system. Antifriction ball or roller bearings are commonly used in small and medium-size machines. Machines that are operated in vertical or inclined positions use thrust-type antifriction bearings. It should be noted that antifriction bearings are noisier than sleeve-type bearings.

Some large dc machines and mill motors have the yoke split and sometimes are hinged along the horizontal centerline of the motor. This greatly improves the installation, inspection, maintenance, and removal of the armature. A typical mill motor is shown in Fig. 3–7.

In most industrial installations dc motors will be equipped with one or more of the following accessories:

1. Overspeed trip systems are used as safety devices to prevent overspeeding, usually in excess of 115% of maximum rated speed, caused by the loss of the shunt field or under overhauling load conditions.
2. Tachogenerators provide a speed indication, or are used in closed-loop control systems to achieve accurate speed control.
3. Imbedded temperature detectors are installed in main or interpole windings of large machines to warn of overheating. They are connected to indicating meters, alarms, or recorders.

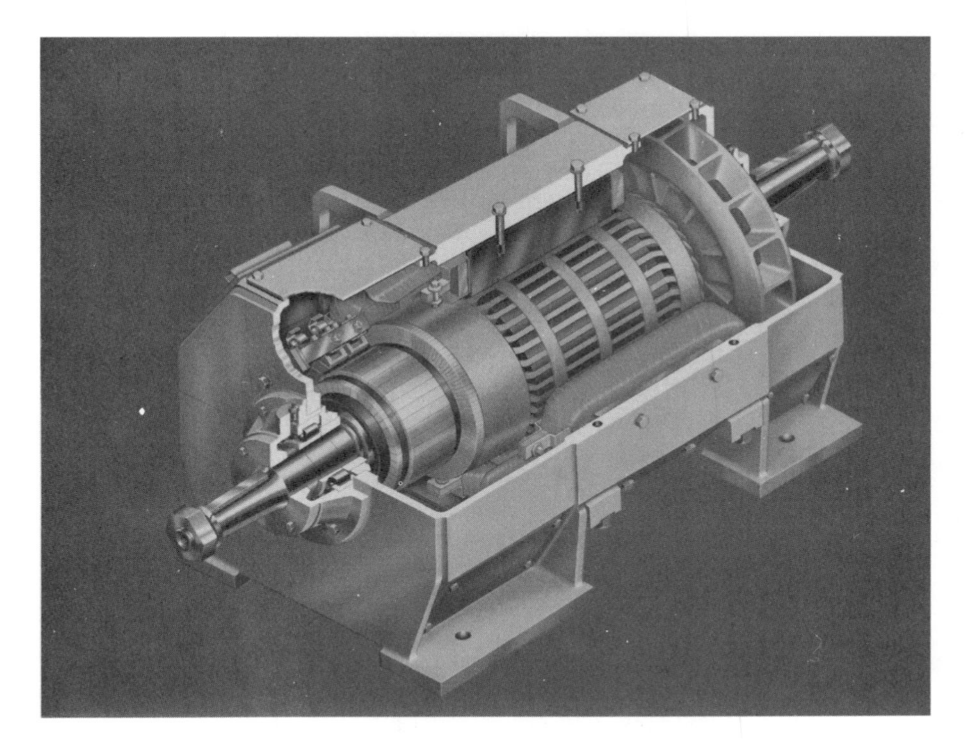

Figure 3-7 DC mill motor. Source: Courtesy of General Electric.

4. Bearing temperature detectors are very often used in babbitted sleeve bearings. They are connected to alarms to give visual or audible warning of an overheated bearing.
5. Magnetically operated disk or shoe brakes are used to hold the motor shaft stationary in hoist applications. These units are fitted either integral to the machine or on a shaft extension.
6. Space heaters are very often installed in machines where high humidity or condensation occurs. They serve to prevent deterioration of the insulation integrity of the machine.

3–7 ENCLOSURES

Machines are enclosed to protect the vital parts such as windings, commutator, and bearings from contamination by airborne contaminants. Another function is to protect personnel from operating hazards. Enclosures basically take one of two forms, ventilated or totally enclosed. The enclosure type varies to suit specific environmental conditions. Briefly, enclosures are defined as follows:

1. A drip-proof enclosure is the most common enclosure for small and medium-size machines. It is constructed so that any particles or liquids that fall from an angle within 15° of the vertical cannot enter the machine.

2. A splash proof enclosure is an open machine in which the ventilating openings prevent the entry of liquids or solids at any angle between the vertical and 10° below the horizontal.
3. Semiguarded machines are open machines in which the upper ventilating openings are fitted with screens to prevent objects greater than 0.5 in (1.27 cm) in diameter from entering the machine.
4. Guarded machines are open machines where the access to live or rotating parts is restricted by screens so that a cylindrical rod 0.5 in (1.27 cm) in diameter cannot reach any parts that are within 4 in (10.16 cm), or a cylindrical rod 0.75 in (1.91 cm) in diameter cannot reach parts that are more than 4 in (10.16 cm) from the guarded opening.
5. An open pipe-ventilated machine is an open machine with ventilating openings designed to permit the direct connection of inlet ducts or pipes. The ventilating air is circulated either by an integral fan or by external means.
6. An open externally ventilated machine has the ventilation air driven through the machine by a blower mounted externally on the machine enclosure.
7. There are two types of weather-protected machines:
   Type I is an open machine with ventilating passages designed to minimize the entrance of snow, rain, and airborne particles to the electrical parts. The ventilating openings must not permit the passage of a cylindrical object greater than 0.75 in (1.91 cm) in diameter.
   Type II enclosure has the inlet and discharge ventilating passages arranged so that high-velocity air and airborne particles entering the machine can be discharged without entering the internal ventilating passages leading directly to the electrical parts of the machine. The incoming ventilating air passes through baffles or separate housings with at least three abrupt changes of direction, none of which is less than 90°. Also a zone of low-velocity airflow not exceeding 600 ft/min (182.88 m/min), must be provided in the intake path to reduce the possibility of airborne moisture or dirt being carried over to the electrical parts.
8. Totally enclosed nonventilating machines are not equipped for cooling by means external to the enclosing parts.
9. Totally enclosed fan-cooled machines are totally enclosed machines equipped for exterior cooling by means of a fan or fans integral with the machine, but external to the enclosing parts.
10. Explosion-proof machines are totally enclosed machines whose enclosure is designed and constructed to withstand an explosion of a specified gas or vapor that may occur within it, and to prevent ignition of the specified gas or vapor surrounding the machine by sparks, flashes, or explosions of the specified gas or vapor that may occur within the machine casing.
11. Dust-ignition-proof machines are totally enclosed machines whose enclosure is designed and constructed to exclude ignitable amounts of dust and to prevent arcs, sparks, or motor heat from causing the ignition or explosion of an ambient atmosphere of the specific dust or from causing the ignition of dust accumulated on or around the machine.

12. Waterproof machines are totally enclosed machines that will not permit the entry of water when applied as a stream from a hose, except at the shaft, provided that the water is prevented from entering the coil reservoir and that provision is made for automatically draining the machine.
13. Totally enclosed pipe-ventilated (TEPV) machines are similar to the open pipe-ventilated machine except that inlet and outlet ducts or pipes may be connected to them for the entry and discharge of the ventilating air. Usually fans integral with the machine draw ventilating air from the ducts; if the ducting is over 20 ft (6.09 m), external blowers may be added. The machine is then known as a separately ventilated or forced-ventilated machine.
14. A totally enclosed water-cooled (TEWC) machine is a totally enclosed machine that is cooled by circulating water, the water or water piping coming in direct contact with the actual machine parts.
15. A totally enclosed water–air-cooled machine is a totally enclosed machine in which the internal circulating air is cooled by circulating water in a heat exchanger.
16. A totally enclosed air-to-air-cooled machine is a totally enclosed machine in which the internal circulating air is cooled by circulating it through an air-cooled exchanger.
17. A totally enclosed fan-cooled guarded machine is basically a totally enclosed fan-cooled machine with the exterior of the frame ribbed to assist heat transfer. All access to the fan is protected by grills that will not permit entry by an object greater than 0.75 in (1.9 cm).

QUESTIONS

3–1 Why is the dc motor the preferred motor in a number of industrial applications?

3–2 With the aid of a sketch discuss the magnetic circuit of a six-pole dc machine.

3–3 What is the function of the armature assembly?

3–4 Discuss the construction of an armature.

3–5 Why are the slots of some armatures skewed?

3–6 Why are the main field poles laminated?

3–7 Discuss three types of main field pole windings.

3–8 Discuss the types of insulating materials that are used for dc machine winding insulation.

3–9 Why is it not good practice to use class H insulation in a dc machine?

3–10 What is meant by hot-spot temperature?

3–11 Why are epoxy-encapsulated field poles used?

3–12 Why is the armature laminated?

3–13 How is a long armature ventilated?

3–14 With the aid of sketches describe the two types of armature windings that are used in dc machines.

3–15 Why are there as many brushes as poles in a lap winding?

3–16 What is meant by fractional-pitch and full-pitch coils? What are their advantages?

3–17 Why are wave-wound machines classified as high-voltage low-current machines?

3–18 What is meant by commutator, forward, and back pitches?

3–19 Why are only two brush sets required for a wave-wound machine?

3–20 What is the function of the commutator?

3–21 With the aid of sketches describe two types of commutator construction.

3–22 What is the function of carbon brushes in a dc machine?

3–23 Why is carbon chosen for brushes?

3–24 What is the function of a brush holder, and how is it constructed?

3–25 What is the main frame or yoke, and how is it constructed?

3–26 Why are the inside and the ends of the yoke machined?

3–27 When would oil-ring sleeve-type bearings be used?

3–28 What is the function of the end bells?

3–29 What is meant by a mill motor? What are the advantages of using a mill motor?

3–30 Discuss the accessories that are commonly used with dc motors.

3–31 Why are enclosures required for dc machines?

CHAPTER

4

DC Machine Principles

4–1 INTRODUCTION

The modern dc machine hinges on the discovery by H. C. Oersted in 1819 of the production of magnetic fields by current-carrying conductors. This led Michael Faraday in 1821 to demonstrate that rotation could be produced by electromagnetic means, that is, the first electric motor. In 1831 Faraday showed that electric energy could be produced by moving a conductor through a magnetic field. Both these rudimentary machines are classified as *homopolar* machines, which are low-voltage high-current machines.

The next problem to be overcome before the dc machine as we know it was produced, was *commutation,* which was solved in 1832 by H. Pixii in Paris and W. Ritchie in London. Even though the first dc motors were built in 1837, it was necessary at this point to overcome the economic advantages of steam power. It was not until 1873, when Z. T. Gramme showed that dc machines could operate as either motors or generators, that the dc motor was used commercially. Even though the induction motor was developed by Nikola Tesla in 1888, the dc motor has still retained its edge because of the ease with which it can be controlled, as well as for its ability to operate from storage batteries.

4–2 GENERATOR PRINCIPLES

Recalling from Section 1–5, Faraday's law of electromagnetic induction stated that

$$e = -\frac{\Delta\Phi}{\Delta t} \tag{1–43}$$

that is, the instantaneous voltage of a single-turn coil is equal to the rate of change of the magnetic flux linking the coil. The voltage is increased if the number of turns in the coil is increased. Therefore

$$e = -N\frac{\Delta\Phi}{\Delta t} \quad \textbf{(1–44)}$$

In terms of a conductor rotating with respect to a stationary magnetic field, the instantaneous voltage is

$$e = Bl\upsilon \sin\theta \quad \textbf{(1–45)(SI)}$$

or

$$e = Bl\upsilon \sin\theta \times 10^{-8} \quad \textbf{(1–45)(E)}$$

with the direction of the electromotive force in the conductor being given by Fleming's right-hand rule.

However, of far greater importance is the voltage developed by a coil. In Section 1–8 it was shown that the voltage produced by a single-turn coil is

$$\upsilon = V_{\max}\sin\omega t \quad \textbf{(1–46)}$$

$$= V_{\max}\sin 2\pi f t \quad \textbf{(1–47)}$$

$$= V_{\max}\sin\theta \quad \textbf{(1–48)}$$

It should be noted at this point that the voltage produced in the coil is an ac voltage.

Finally it was shown that the average voltage produced by an N-turn coil rotating in a P-pole machine at S r/min or ω rad/s is

$$E_{AV} = \frac{\Phi ZSP}{60} \times 10^{-8}\ \text{V/coil} \quad \textbf{(1–55)(E)}$$

or

$$E_{AV} = \frac{\phi Z\omega P}{2\pi}\ \text{V/coil} \quad \textbf{(1–55)(SI)}$$

Equations (1–55) only permit the calculation of the average voltage generated in an N-turn coil. Recall that $Z = 2N$ when it is rotating at S r/min or ω rad/s in a P-pole machine with a flux per pole of Φ lines or webers, as appropriate. From Chapter 3 we know that the armature winding consists of a number of such coils connected in series. The number of coils in series depends on whether the winding is lap or wave, and also on the multiplicity m of the winding.

Equations (1–55) can be modified to represent the voltage produced by the complete armature winding, that is,

$$E = \frac{\Phi ZSP}{60a} \times 10^{-8}\ \text{V} \quad \textbf{(4–1)(E)}$$

or

$$E = \frac{\phi Z \omega P}{2\pi a} \text{ V} \tag{4–1)(SI}$$

where a is the number of current paths in the armature circuit. There are simple rules for determining the number of current paths a.

1. Simplex lap windings have as many paths in parallel as there are main field poles. Therefore

$$a = P \tag{4–2}$$

2. Simplex wave windings only have two parallel paths, irrespective of the number of main field poles. Therefore

$$a = 2 \tag{4–3}$$

As was mentioned briefly in Chapter 3, lap and wave windings can be modified to obtain higher voltages or currents. One method is the use of *multiplex* windings, which consist of sets of independent and completely closed windings. If there is a set of coils forming only one closed winding, it is called a *simplex* winding; if there are two sets of coils forming two closed windings, it is called a *duplex* winding. Similarly for a *triplex* winding, and so on. The multiplicity m increases the number of possible armature paths, with the obvious result that the output voltage rating is decreased and at the same time the current rating is increased.

The number of parallel paths in a multiplex winding of multiplicity m is

3.

$$a = mP \text{ (lap winding)} \tag{4–4}$$

4.

$$a = 2m \text{ (lap winding)} \tag{4–5}$$

➤ EXAMPLE 4-1

A six-pole dc machine is wound with a duplex winding. How many parallel paths exist?

SOLUTION

$$\begin{aligned} a &= mP \\ &= 2 \times 6 = 12 \text{ paths} \end{aligned}$$

➤ EXAMPLE 4-2

A six-pole dc machine is wound with a duplex wave winding. How many parallel paths exist?

SOLUTION

$$a = 2m$$
$$= 2 \times 2 = 4 \text{ paths}$$

➤ EXAMPLE 4-3

Calculate the voltage generated by each of the above machines if the flux per pole is 5.0×10^6 lines (0.05 Wb) and the generator speed is 120 r/min (12.57 rad/s). There are 200 armature coils and each coil has 15 turns.

SOLUTION

In English units,

$$Z = 200 \text{ coils} \times 2 \text{ conductors/ turn} \times 15 \text{ turns/coil}$$
$$= 6{,}000 \text{ conductors}$$

From Eq. (4–1)(E),

$$E = \frac{\Phi ZSP}{60a} \times 10^{-8}\,\text{V}$$

Therefore for the six-pole duplex lap winding,

$$E = \frac{5 \times 10^{-6} \times 6{,}000 \text{ conductors} \times 120 \text{ r/min} \times 6 \text{ poles}}{60 \times 12 \text{ paths}} \times 10^{-8}\,\text{V}$$
$$= 300\,\text{V}$$

For the six-pole duplex wave-wound machine,

$$E = \frac{5 \times 10^{6} \times 6{,}000 \text{ conductors} \times 120 \text{ rpm} \times 6 \text{ poles}}{60 \times 4 \text{ paths}} \times 10^{-8}\,\text{V}$$
$$= 900\,\text{V}$$

In SI units, from Eq. (4–1)(SI),

$$E = \frac{\phi Z \omega P}{2\pi a}\,\text{V}$$

For the six-pole duplex lap winding,

$$E = \frac{0.05 \text{ Wb} \times 6{,}000 \text{ conductors} \times 12.57 \text{ rad/s} \times 6 \text{ poles}}{2\pi \times 12 \text{ paths}}\,\text{V}$$
$$= 300\,\text{V}$$

For the six-pole wave-wound machine,

$$E = \frac{0.05 \text{ Wb} \times 6{,}000 \text{ conductors} \times 12.57 \text{ rad/s} \times 6 \text{ poles}}{2\pi \times 4 \text{ paths}}\,\text{V}$$
$$= 900\,\text{V}$$

4–3 COMMUTATOR ACTION

A sinusoidal alternating voltage is produced when a closed conductor system revolves in a magnetic field [Fig. 4–1(a)]. However, the requirement is to produce a constant-polarity or dc voltage. This is achieved by using a commutator.

Figure 4–1(b) shows a simple loop coil revolving counterclockwise in a magnetic field. Previously the coil ends were connected to two slip rings, and the ac output was taken from the slip rings by two carbon brushes and supplied to the load. In Fig. 4–1(b) one slip ring has been removed. The other slip ring has been split into two semicircular segments 1 and 2 which are electrically insulated from the shaft and each other. The two ends of the coil have each been connected to a segment. This is the simplest form of a commutator.

Refer to Fig. 4–1(c) with the brushes positioned as shown and the current directions as marked on coil sides *A* and *B*. In position 1 the current flow is out of coil side *A* and into commutator segment 1, that is, the brush making contact with commutator segment 1 is positive, and similarly, the brush making contact with commutator segment 2 is negative. When the armature coil has rotated 90° counterclockwise to position 2, coil sides *A* and *B* are moving parallel to the magnetic field and no

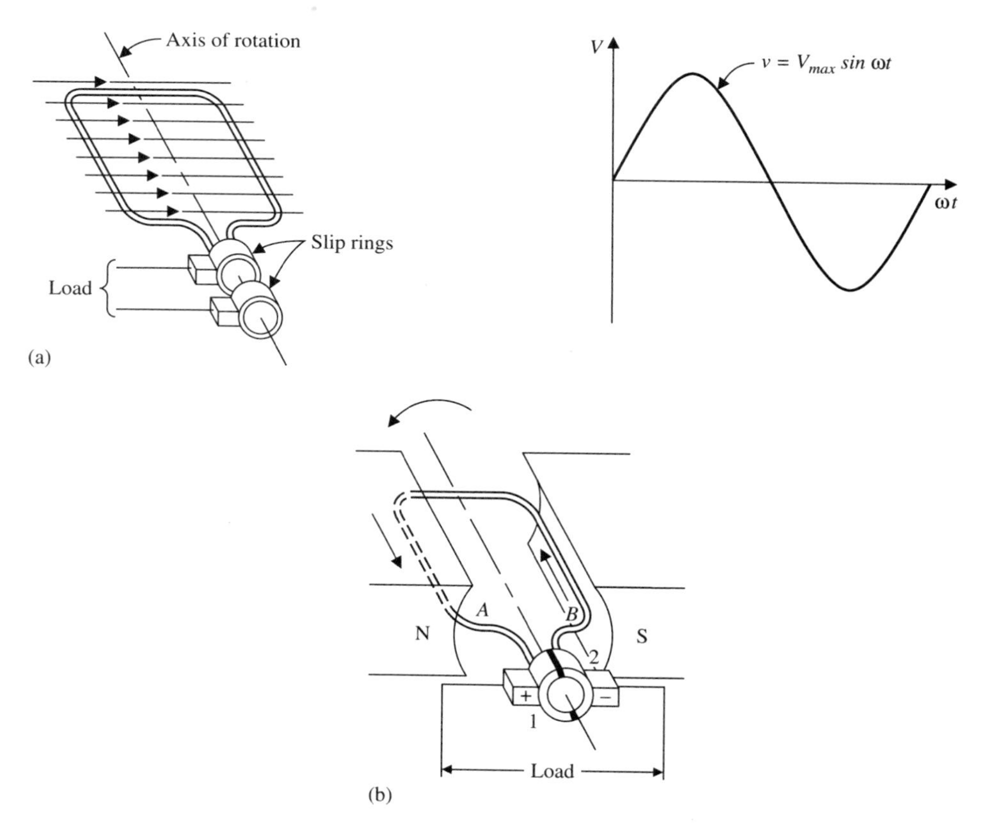

Figure 4-1 a and b Generation of an emf. (a) Single loop alternator. (b) 2 segment commutator.

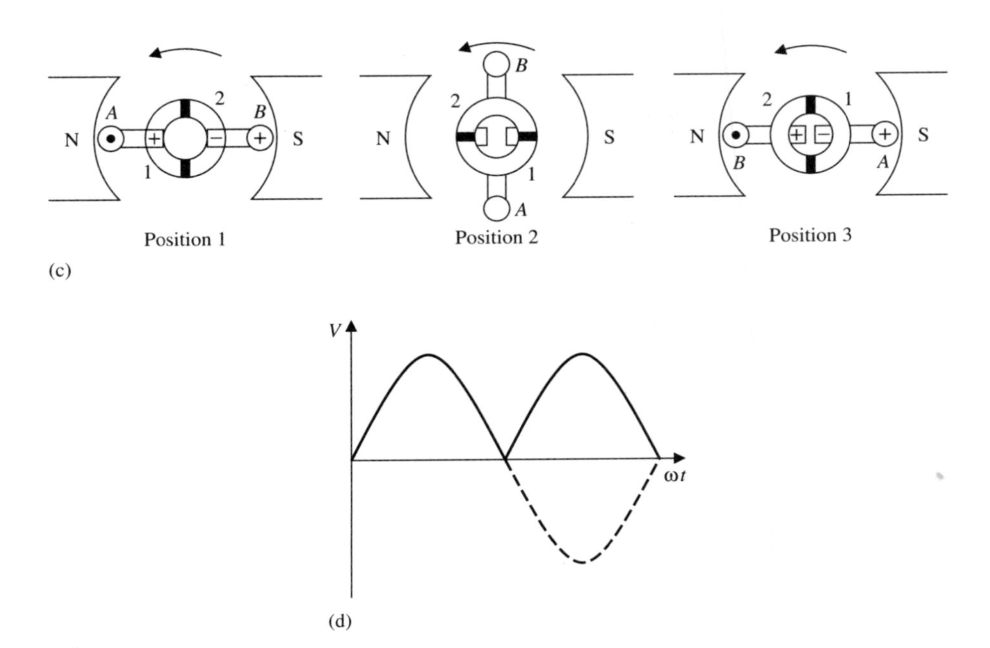

Figure 4-1 c and d Generation of an emf. (c) Commutator action. (d) Rectified output.

voltage is induced in the coil. At this point the coil undergoing commutation is short-circuited by the brushes. The brushes are lying on what is termed the *geometric neutral axis* (GNA). After turning another 90° counterclockwise to position 3, the two coil sides *A* and *B* and the commutator segments have reversed their positions. However, the left-hand brush now making contact with commutator segment 2 is still positive, and the brush making contact with commutator segment 1 is still negative. The waveform in Fig. 4–1(d) shows the output voltage across the brushes. As can be seen, the effect of the commutator is that it has reversed the negative half-cycle voltage produced by the armature conductors, that is, the output voltage has been rectified by the commutator.

The output voltage meets the requirements of a dc voltage, but it has a large ripple content, and the magnitude of the output voltage is relatively small. The amplitude of the output voltage is increased by increasing the number of turns per coil. The ripple component amplitude is reduced by spacing multiturn coils evenly around the periphery of the armature core, as was discussed in Chapter 3 for lap and wave windings. The combined effect of these two measures is to produce a greater amplitude dc voltage with a high-frequency low-amplitude ac ripple content, as illustrated in Fig. 4–2.

4–4 ARMATURE REACTION

Armature reaction is the effect of the magnetic field produced by current-carrying armature conductors on the main field pole magnetic flux. Referring to Fig. 4–3(a), a

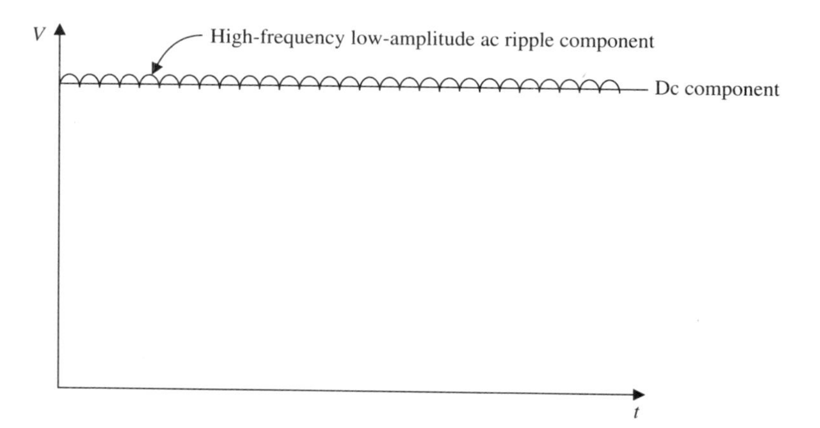

Figure 4-2 Output voltage waveform of a multiturn multicoil dc armature winding.

two-pole dc generator is considered for simplicity. The brushes are assumed to be making contact with the conductors as they pass through the GNA. All the conductors to the left of the GNA carry currents whose direction is inward, while the conductors to the right of the GNA carry currents directed outward. With these current directions the armature can be considered as magnetically equivalent to a solenoid. From the right-hand grasp rule, the armature conductors are producing a cross-magnetic field ϕ_{ar}, which is directed vertically down along the GNA. The magnitude of this field is directly proportional to the magnitude of the armature currents. Figure 4–3(b) shows the distribution of the direct-axis flux ϕ_{da} produced by the main field poles with no current flow in the armature conductors. As can be seen, the magnetic flux is distributed uniformly in the air gaps and through the armature core.

Under load conditions both the armature reaction flux ϕ_{ar} and the main field pole flux ϕ_{da} are present. The resulting flux distribution is a combination of both these fluxes. From Fig. 4–3(a) it can be seen that the armature flux crosses the air gap at the top of the N pole and the air gap at the bottom of the S pole in the same direction as the main field flux, that is, it is strengthening the air-gap flux at these points. Similarly, the air-gap fluxes are weakened at the bottom of the N pole and at the top of the S pole by the armature flux. If the armature teeth are not saturated, then the strengthening of the flux under the trailing pole tips is equal to the weakening of the flux under the leading pole tips, that is, the total pole flux remains unchanged. However, if the armature teeth are saturated, the flux density under the trailing pole tips will remain unchanged, but the flux density under the leading pole tips will have been reduced, that is, the armature reaction has distorted and reduced the total pole flux. In general it can be stated that the effect of armature reaction is to reduce the flux density under the leading pole tips, and to strengthen the flux density under the trailing pole tips of a dc generator. The result of armature reaction is to distort the flux distribution under the main field poles, and it effectively moves the axis of the resulting magnetic field forward in the direction of rotation [Fig. 4–3(c)]. This effect can be better appreciated by considering the vector diagram in Fig. 4–3(d). The vec-

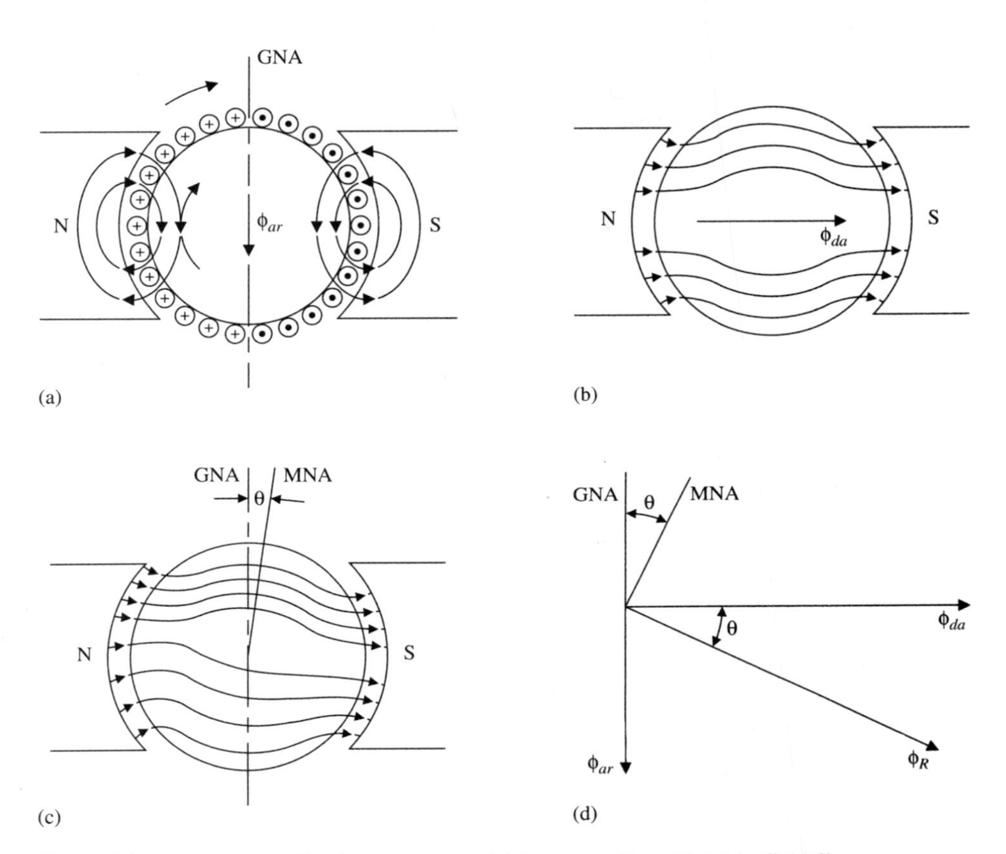

Figure 4-3 Armature reaction in a generator. (a) Armature flux. (b) Main field flux. (c) Resultant flux. (d) Vector diagram showing shift of MNA.

tor ϕ_{da} represents the main pole flux, and in most applications will remain relatively constant. The vector ϕ_{ar} represents the armature flux, and is directly proportional to the magnitude of the armature current I_a and, as a result, is continually changing as the load changes. The resultant flux ϕ_R has advanced in the direction of rotation by the angle θ. The *magnetic neutral axis* (MNA), which is perpendicular to the resulting field, has been moved forward in the direction of rotation. The amount of shift experienced by the MNA is dependent on the magnitude of ϕ_{ar}.

The brushes should be at the point where the electromotive force in the coil undergoing commutation is reversing, namely, the MNA. As can be seen, the effect of armature reaction in a generator is to advance the MNA in the direction of rotation. Therefore for correct commutation the brushes should also be advanced by the same amount. This process, while theoretically correct, is not very practical since it would require that the brushes be repositioned after every load change.

To be complete, we must also discuss the effect of armature reaction in a dc motor. From Fig. 4–4(a) and by the use of Fleming's left-hand rule it can be seen that with the same main pole polarity and direction of rotation, the direction of the cur-

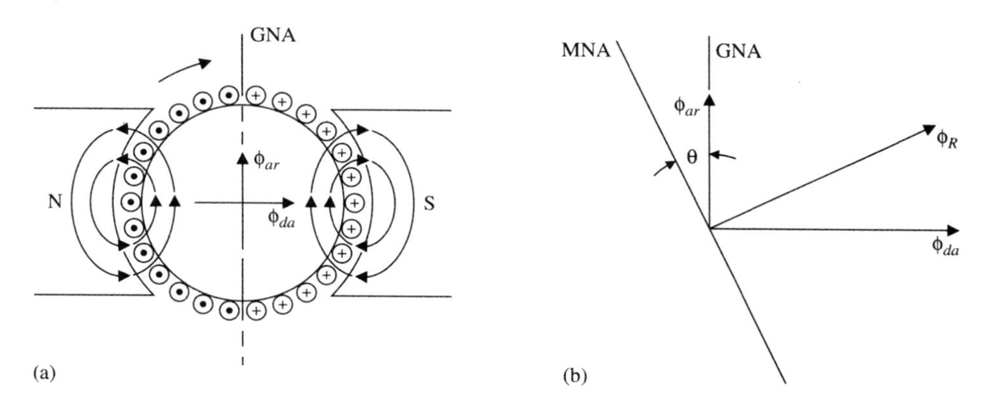

Figure 4-4 Armature reaction in a dc motor. (a) Armature cross magnetization flux. (b) Vector diagram of fluxes.

rent flow in the armature conductors is opposite to that of a dc generator. The armature cross-magnetizing field ϕ_{ar} acts vertically up along the GNA. Also the main pole flux is strengthened under the leading pole tips and weakened under the trailing pole tips. From the vector diagram [Fig. 4–4(b)] it can be seen that the MNA has moved back against the direction of rotation. This suggests that for good commutation the brushes should also be moved back by the same amount.

Since dc machines can be operated as either a motor or a generator, it is obviously not very practical to rely on moving the brushes as a method of reducing the effects of armature reaction. Also the amount of brush movement would have to be proportional to the magnitude of the armature current I_a.

4–5 COMMUTATION

At this point it is necessary to study the commutation process, that is, the reversal of both the voltage and the current in the short-circuited coil, in more detail. Immediately prior to the instant of short circuit, the armature current is flowing in the coil in one direction. Immediately after the short circuit, ideally the armature current has reached its full value in the reverse direction. The difference will appear as an arc between the commutator and the brush. This process is illustrated in Fig. 4–5.

Figure 4–5(a) shows the initial condition, with commutator segment *b* being supplied with current from coils *B* and *C*. After the commutator has moved half a segment to the right, coil *B* is short-circuited by the brush bridging commutator segments *a* and *b* [Fig. 4–5(b)]. During this interval the current in coil *B* should decay to zero, reverse, and build up to its normal value in the opposite direction. As shown in Fig. 4–5(c), the commutator has moved a further half-segment to the right, and the brush is making contact with commutator segment *a* . Under ideal conditions the current flow in coil *B* has been reversed, and the current in coil *A* is just about to be reversed. In less than ideal conditions, the current in coil *B* has not completely reversed and the difference in currents in coils *B* and *C* appears as an arc jumping

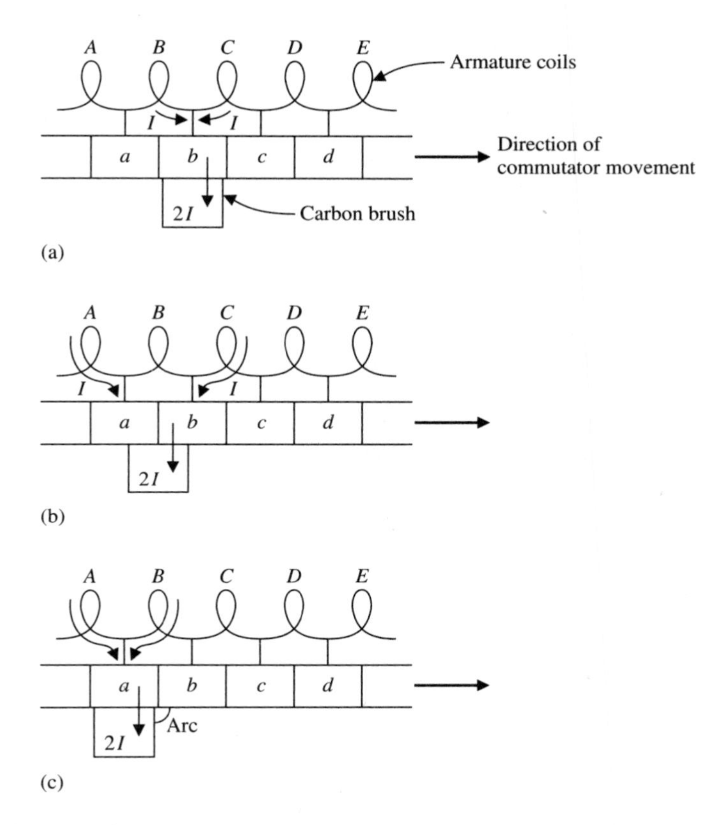

Figure 4-5 Commutation process.

from segment b to the brush making contact with segment a.

The failure of the current in coil B to completely reverse during the short-circuit interval is the direct result of a *reactance voltage* being developed, which opposes the current reversal. The reactance voltage is a counterelectromotive force which is caused by the rapid rate of change of the coil current during reversal, namely, from $+I$ to $-I$ or $2I$ during the interval of short circuit of t s. It is

$$e = \frac{2IL}{t} \text{ V} \tag{4-6}$$

where L is a combination of the coil's self-inductance and its mutual inductance with adjacent coils. The inductance is caused by the fact that the coil sides, which form a significant proportion of the coil, are lying in narrow slots in the steel of the armature core, and consequently possess a significant inductance. The self-inductance of the coils can be minimized by keeping the number of turns per coil to a minimum. In fact in large-capacity machines single-turn armature coils are often used.

The arcing between the commutator and the brushes damages both the brush and the commutator surface. If not corrected it will lead to expensive repairs to the commutator. Arcing can be minimized (1) by using high contact resistance brushes or (2) by counteracting the reactance voltage in the short-circuited coil.

High Contact Resistance Brushes

By increasing the resistance in the path of the short-circuit current it is possible to assist current reversal in the short-circuited coil. Fig. 4–6 shows the principle.

For the moment let us neglect the coil self-inductance and assume that the contact resistance of the brush is much greater than the coil resistance. Then the current from coil C, when it reaches commutator segment b, has two available paths: (1) through the segment to the brush, and (2) through the short-circuited coil B, that is, in the opposite direction to the original current direction in coil B, and then to the portion of the brush in contact with segment a. As can be seen, the greater the brush contact resistance, the greater is the proportion of current diverted through the coil undergoing commutation. The actual contact resistance with segment b varies, being a minimum when the brush is only contacting segment b and increasing progressively as the brush area in contact with segment b decreases.

This method is only partially successful because the reactance voltage resulting from coil self-inductance is always present.

Commutating Poles or Interpoles

Interpoles are narrow field poles set midway between the main field poles, and their windings are in series with the armature circuit, that is, the interpole flux is directly proportional to the load current. They have two functions: (1) to produce a flux which directly counteracts the armature reaction or cross-magnetization flux ϕ_{ar}, and (2) to produce an induced voltage in the coil undergoing commutation, which neutralizes the reactance voltage.

The interpoles are placed on the MNA and the brushes must also be set on the *neutral position*, that is, on the MNA. It is possible to overexcite the interpoles, and as a result the reversed current in the short-circuited coil will be greater than it should be. Once again arcing will occur, but in the reverse direction. This is known as *overcommutation.*

From Fig. 4–3(a) for a generator it can be seen that it is necessary to produce a cross-magnetic field directly opposing ϕ_{ar}. This is accomplished by arranging the interpole polarity to be as shown in Fig. 4–7(a), that is, the polarity of the interpoles

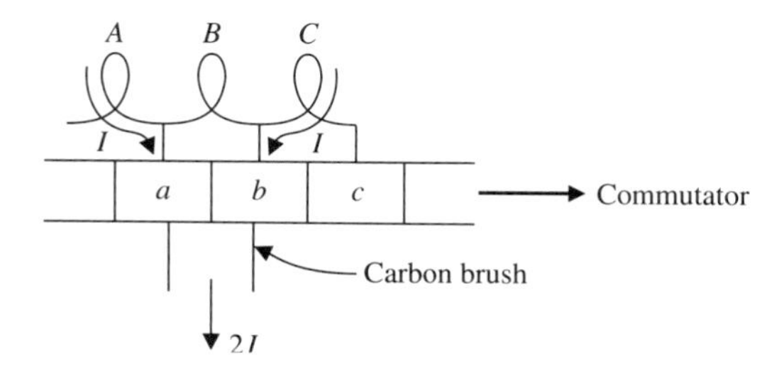

Figure 4-6 The effect of brush contact resistance on commutation.

for a generator is the same as that of the next main field pole ahead in the direction of rotation.

In the case of a dc motor, it can be seen from Fig. 4–7(b) that with the same direction of rotation and main pole polarities, the direction of current flow in the armature is reversed. Also the direction of the armature cross-magnetization field is reversed. Since the interpoles are connected in series with the armature, then as the armature currents reverse when the machine is operated as a motor, the interpole polarity is also reversed [Fig. 4–7(b)], thus maintaining the armature cross-magnetizing field. In the case of a motor the interpole polarity then is the same as the polarity of the preceding main pole in the direction of rotation.

If a machine is designed to operate as either a motor or a generator, for example, under overhauling load conditions, then the interpole polarity will be correct for either mode of operation, and satisfactory commutation will be obtained provided the brushes are initially set on the MNA. As can be seen from the associated vector diagrams, the magnitude of the interpole flux vector ϕ_{ip} is greater than ϕ_{ar}. This difference is required to counteract the effect of the reactance voltage. In actual practice the interpole magnetomotive force is usually about 30% greater than the armature cross-magnetization

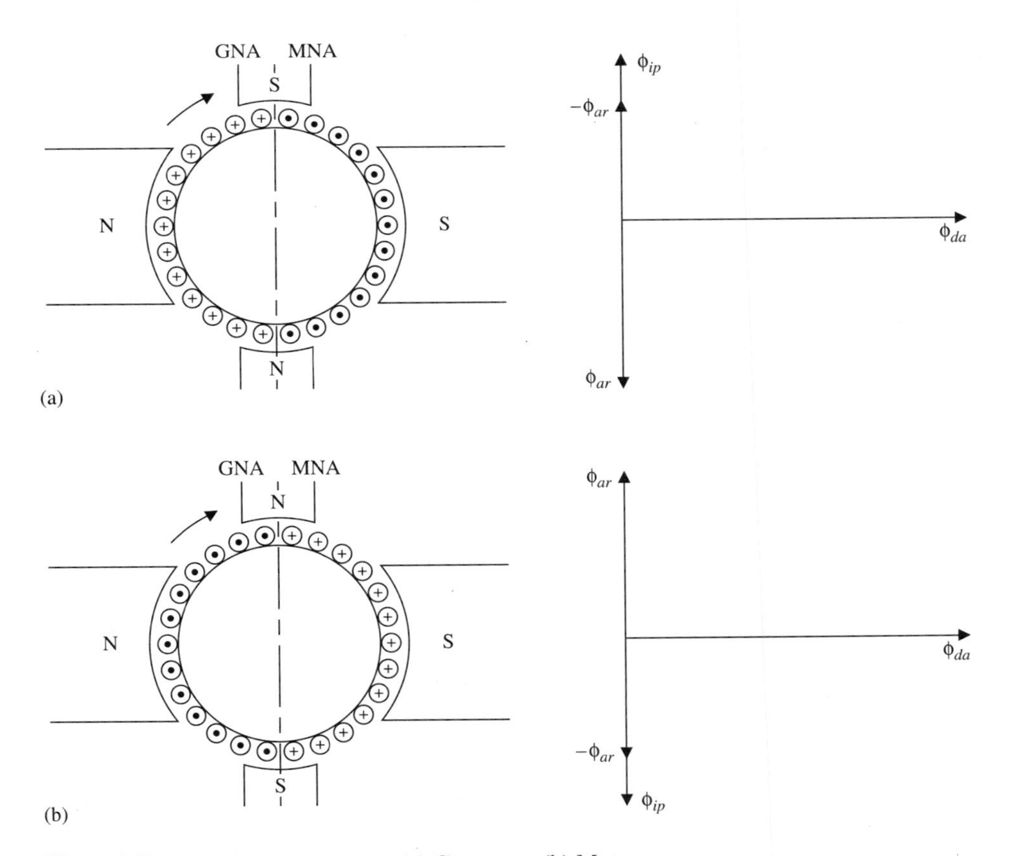

Figure 4-7 Interpole arrangements. (a) Generator. (b) Motor.

magnetomotive force, with the interpole magnetomotive force being adjusted by a diverter resistance to give satisfactory commutation at full load. This results in the interpole being slightly saturated under light loading and slightly weak under overload conditions. This situation may be corrected by the brush contact resistance.

Many dc motors are supplied with rectified ac power from thyristor phase-controlled converters. This rectified power input has a high input ripple content, which in turn creates eddy currents in the main frame of the machine and reduces the effectiveness of the interpoles by producing a phase shift or delay between the interpole flux and the armature cross-magnetization flux. This problem is reduced by laminating the yoke of the machine. As a final point, even though interpoles are effective in eliminating the shift of the MNA, there is still distortion of the magnetic field under the main poles. However, in machines that operate under stable load conditions this is acceptable.

Compensating Windings

In certain applications, such as reversing drives in rolling operations in steel mills or in mine hoist motors, the motor is subjected to rapid reversals and sudden load changes. In these cases the armature currents create a significant distortion of the air-gap flux, which is neutralized by using a *compensating winding*. As the machine is subjected to sudden load changes, there is a corresponding change in the flux linking the armature conductors, and this in turn produces a statically induced electromotive force in the armature conductors. The magnitude of the induced electromotive force is dependent on the rate of change of the load current. If great enough, the induced electromotive force between adjacent commutator segments will cause the commutator to flash over, effectively short-circuiting the entire armature.

This problem can be reduced by inserting a compensating winding in series with the armature into slots in the main pole faces. The compensating winding is designed so that its magnetomotive force is equal and opposite to that produced by the armature conductors under the main pole faces (Fig. 4–8).

Since the compensating winding is in series with the interpoles and the armature, correction will be made automatically for both directions of rotation. Compensating windings also assist in counteracting the demagnetization effects of the armature reaction or cross-magnetization flux by concentrating on the area under the main pole faces. Compensating windings will only be found in large machines because of the high costs involved.

4–6 MOTOR ACTION

Motor action depends on the production of a force by current-carrying conductors interacting with a magnetic field. In Section 1–9 and Fig. 1–25 it was shown that the force F produced on a conductor of length l carrying a current I in a magnetic field with a flux density B is

$$F = \frac{BIl}{1.13} \times 10^{-7} \text{ lb} \qquad \textbf{(1–57)(E)}$$

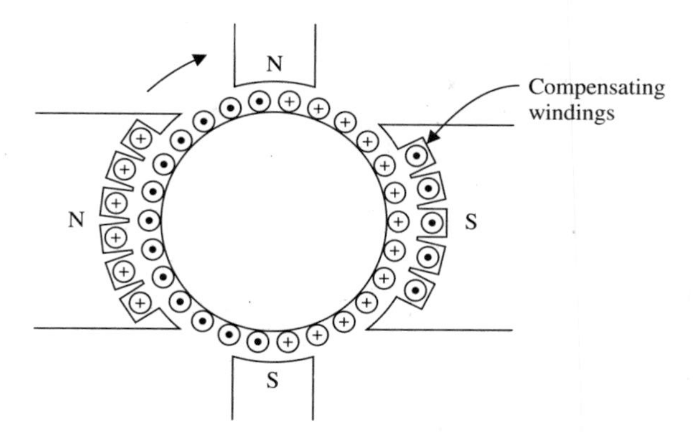

Figure 4-8 Compensating winding for a dc motor.

or

$$F = BIl\,\text{N} \qquad \textbf{(1–57)(SI)}$$

Figure 4–9(a) shows a single current-carrying conductor in a slot on the periphery of an armature, with the armature free to rotate about a shaft at its center. When current flows through the conductor in the direction shown, a force F is produced which causes the armature to move in a clockwise direction, that is, a torque-T is developed which is the product of the force F and the perpendicular distance r from the axis of rotation to the line of action of the force. The torque is

$$T = Fr \sin \theta \qquad \textbf{(1–58)}$$

and T is a maximum when $\theta = 90°$. The torque T is in lb · ft when the force F is in pounds and the radius r in feet, for the English system. In the SI system T is in newton-meters (N · m) when the force is in newtons and the radius r in meters.

Substituting Eq. (1–57) in Eq. (1–58) yields

$$T = \frac{BIlr}{1.13} \times 10^{-7}\ \text{lb} \cdot \text{ft} \qquad \textbf{(1–59)(E)}$$

or

$$T = BIlr\,\text{N} \cdot \text{m} \qquad \textbf{(1–59)(SI)}$$

The torque produced by a single-turn coil [Fig. 4–9(b)] is the sum of the individual torques produced by both coil sides. Therefore the torque produced by a single-turn coil carrying a current I is

$$\begin{aligned} T &= 2 \times 0.885 BIlr \times 10^{-7}\ \text{lb} \cdot \text{ft} \\ &= 1.77 BIlr \times 10^{-7}\ \text{lb} \cdot \text{ft} \end{aligned} \qquad \textbf{(1–60)(E)}$$

or

$$T = 2BIlr\,\text{N} \cdot \text{m} \qquad \textbf{(1–60)(SI)}$$

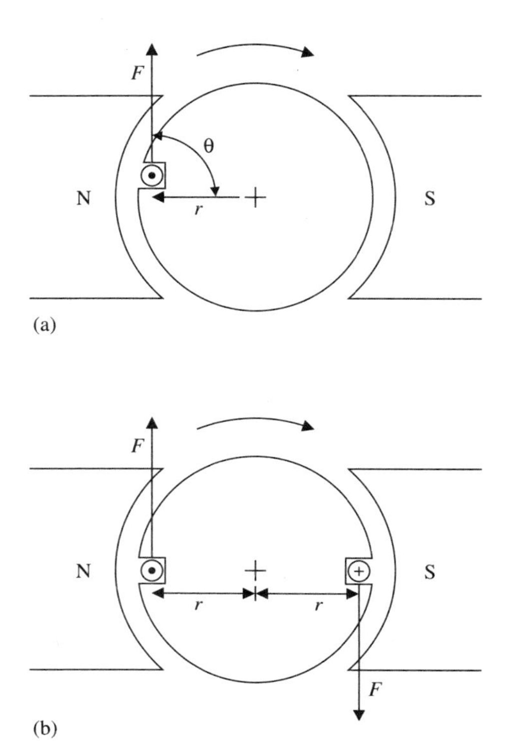

Figure 4-9 Torque production: (a) by a conductor; (b) by a coil.

In a dc motor electric energy is supplied from an external source and torque is produced as a result of the magnetic field produced by current flow in the armature conductors interacting with the field pole flux. As a result the armature rotates and produces a mechanical power output. The commutator performs exactly the same function as it does in a generator, that is, it transfers each coil to the next armature circuit as the coil sides pass between the main poles. This action causes current reversal in each conductor as it comes within the influence of the next pole. The result is that the force on the conductor will always produce a torque in the same direction, and hence the armature will rotate.

4–7 COUNTER- OR BACK ELECTROMOTIVE FORCE

When power is applied to a dc motor, the armature rotates. As a result the armature conductors cut the magnetic flux under the main field poles, and an electromotive force is induced in the armature conductors. The direction of the induced electromotive forces can be determined by Fleming's right-hand rule, which shows that they oppose the current flowing in the armature conductors. This induced electromotive force is known as the *counter electromotive force* E_c or *back electromotive force.* The applied terminal voltage V has to do work against the counter electromotive force.

Obviously in normal operation the applied terminal voltage V is greater than the counter electromotive force E_c. This leads us to conclude that if the generated or induced electromotive force is opposite in direction to the armature current, then the machine is motoring. If the generated electromotive force is in the same direction as the armature current, the machine is generating. The counter electromotive force is

$$E_c = \frac{\Phi ZSP}{60a} \times 10^{-8}\,\text{V} \qquad \textbf{(4–7)(E)}$$

or

$$E_c = \frac{\phi Z\omega P}{2\pi a}\,\text{V} \qquad \textbf{(4–7)(SI)}$$

This is a *speed voltage* and is exactly the same as Eqs. (4–1).

Applying Ohm's law it can be seen that the armature current I_a in a dc motor is

$$I_a = \frac{V - E_c}{R_a} \qquad \textbf{(4–8)}$$

where R_a is the armature resistance.

From Eq. (4–8) it can be concluded that E_c limits the armature current I_a. As a result, when the machine is motoring, the counter electromotive force E_c will always be less than the applied voltage V.

Rearranging Eq. (4–8) we obtain

$$V - E_c = I_a R_a \qquad \textbf{(4–9)}$$

Then multiplying by I_a produces

$$VI_a - E_c I_a = I_a^2 R_a \qquad \textbf{(4–10)}$$

which in turn, after further rearrangement, yields

$$VI_a - I_a^2 R_a = E_c I_a \qquad \textbf{(4–11)}$$

Where VI_a is the electric power supplied to the armature, $I_a^2 R_a$ the power dissipated as heat in the copper of the armature, and $E_c I_a$ the electric power converted into mechanical power.

It can be shown that the maximum mechanical power output is developed when

$$\frac{V}{2} = I_a R_a \quad \text{or} \quad E_c = \frac{V}{2} \qquad \textbf{(4–12)}$$

or, expressed another way, the maximum mechanical power output is obtained when

$$I_a = \frac{V - E_c}{R_a} = \frac{V - V/2}{R_a} = \frac{V}{2R_a} \qquad \textbf{(4–13)}$$

This condition is not normally achievable in practice since it would require higher armature currents than are generally considered desirable. In addition, the armature copper losses would be excessive, leading to an overall efficiency of less than 50%. In actual practice the counter electromotive force is usually on the order of 80–95% of the applied terminal voltage.

4–8 TORQUE EQUATION

Equation (4–11) showed that the total power converted into mechanical power output was $E_c I_a$, that is, the product of the counter electromotive force and the armature current.

The gross mechanical power developed, namely, the shaft power plus the windage and friction losses, is

$$P_g = E_c I_a = T\omega \text{ W}$$

$$T = \frac{P_g}{\omega} = \frac{E_c I_a}{\omega} \tag{4–14}$$

Substituting for E_c,

$$T = \frac{\phi Z P I_a}{2\pi a}$$

$$= \frac{1}{2\pi}(\phi Z I_a) \times \frac{P}{a} \text{ N} \cdot \text{m} \tag{4–15)(SI}$$

$$= k_m \phi I_a \text{ N} \cdot \text{m} \tag{4–16)(SI}$$

where $k_m = ZP/2\pi a$ and is called the *motor constant.*

Since 1 lb · ft = 1.356 N · m, then in English units

$$T = \frac{1}{2\pi \times 1.356} \times \Phi Z I_a \times \frac{P}{a}$$

$$= 0.118(\Phi Z I_a) \times \frac{P}{a} \text{ lb} \cdot \text{ft} \tag{4–16)(E}$$

$$= K_m \Phi I_a \text{ lb} \cdot \text{ft}$$

where $K_m = ZP/2\pi \times 1.356 \times a = ZP/8.52a$.

The motor constants k_m and K_m show that the torque developed by the armature is proportional to the product of the flux per pole and the armature current, that is,

$$T \propto \Phi I_a \tag{4–17}$$

➤ EXAMPLE 4-4

An eight-pole lap-wound armature has 500 conductors. The flux per pole is 0.05 Wb, the armature current is 400 A, and the armature rotates at 600 r/min. Calculate the gross torque and the gross mechanical power output.

SOLUTION

From Eq. (4–15)(SI),

$$T = \frac{1}{2}\pi(\Phi Z I_a) \times \frac{P}{a}\ \text{N}\cdot\text{m}$$

$$= \frac{1}{2}\pi(0.05\ \text{Wb} \times 500\ \text{conductors} \times 400\ \text{A}) \times \frac{8\ \text{poles}}{8\ \text{paths}}$$

$$= 1{,}591.55\ \text{N}\cdot\text{m}$$

$$= \frac{1{,}591.55\ \text{N}\cdot\text{m}}{1.356} = 1{,}173.71\ \text{lb}\cdot\text{ft}$$

$$P_g = T\omega$$

$$= 1{,}591.55\ \text{N}\cdot\text{m} \times \frac{2\pi \times 600\ \text{r/min}}{60}$$

$$= 100\ \text{kW}$$

$$= \frac{100\text{kW}}{0.746\ \text{kW/hp}} = 134.05\ \text{hp}$$

QUESTIONS

4–1 What is understood by the term *multiplicity*?

4–2 What is commutation?

4–3 With the aid of a sketch explain commutator action.

4–4 What is the geometric neutral axis?

4–5 How is the ripple content of the output voltage of a dc generator reduced?

4–6 What is armature reaction?

4–7 What are the effects of armature reaction?

4–8 With the aid of sketches explain the process of commutation.

4–9 What is meant by reactance voltage?

4–10 What is the effect of reactance voltage on commutation?

4–11 How can brush arcing be reduced?

4–12 How does the use of high contact resistance brushes assist commutation?

4–13 What are the functions of interpoles?

4–14 What are the rules defining interpole polarities in a dc generator and a dc motor?

4–15 How does the use of rectified power sources affect interpole performance in a dc motor, and how can the problem be minimized?

4–16 What are compensating windings? Why are they necessary?

4–17 Explain what is meant by motor action.

4–18 What is meant by the term *counter electromotive force?*

4–19 What is meant by the term *gross power?*

4–20 Why is the gross power output greater than the shaft output?

PROBLEMS

4–1 A dc motor has a rated armature current of 500 A, and has eight poles. Calculate the current per path if the armature is: (**a**) simplex lap wound; (**b**) simplex wave wound.

4–2 A four-pole 500-V dc generator has a simplex lap-wound armature with 1,000 conductors. Find the required flux per pole when it is driven at 1,500 r/min.

4–3 An eight-pole dc generator has a wave-wound armature with 96 slots and 24 conductors per slot. If the flux per pole is 0.2 Wb, calculate the generated electromotive force when the machine is running at 2,500 r/min.

4–4 A 50-hp 600-V four-pole 600-r/min dc motor has a flux per pole of 0.02 Wb. The armature has 784 conductors and is simplex wave wound, and the full-load armature current is 95 A. Calculate the gross torque in N · m and lb · ft.

CHAPTER

5

DC Generators

5–1 INTRODUCTION

The role of the dc generator as a major contender in the power generation field has declined rapidly over the years. The rate of decline accelerated with the introduction of high-power solid-state rectifiers and thyristor phase-controlled converters, which now form the prime sources of fixed and variable dc voltage. However, the principles involved in the dc generator are still important and should be clearly understood.

5–2 EQUIVALENT GENERATOR CIRCUIT

We established the generator voltage equations in Chapter 4, which are repeated here:

$$E_g = \frac{\Phi ZSP}{60a} \times 10^{-8}\ \text{V} \qquad \textbf{(4–1)(E)}$$

or

$$E_g = \frac{\phi Z\omega P}{2\pi a}\ \text{V} \qquad \textbf{(4–1)(SI)}$$

If we examine these equations in terms of an existing dc generator, it is clear that there are only two variables, namely, the air-gap flux per pole Φ and the speed of rotation S or the angular velocity ω. Let

$$K_g = \frac{ZP}{60a} \times 10^{-8} \qquad \textbf{(5–1)(E)}$$

or

$$k_g = \frac{ZP}{2\pi a} \qquad \textbf{(5–1)(SI)}$$

where K_g and k_g are the generator constants they are dependent on the individual machine and can be readily determined experimentally. Then the generated voltage equations [Eqs. (4–1)] may be expressed as

$$E_g = K_g \Phi S \text{ V} \tag{5–2)(E}$$

or

$$E_g = k_g \phi \omega \text{ V} \tag{5–2)(SI}$$

The field pole flux Φ varies with the field current I_f and is proportional to I_f as long as the machine is operating on the straight-line portion of the magnetization curve or no-load saturation curve. It is possible to analyze the performance of a dc machine even if the magnetic circuit is saturated by using straight-line approximations of the magnetization curve (Fig. 5 –1). Therefore Eqs. (5–2) can be rewritten as

$$E_g = K_g I_f S \text{ V} \tag{5–3)(E}$$

or

$$E_g = k_g I_f \omega \text{ V} \tag{5–3)(SI}$$

A dc generator can be modeled by two simple circuits as illustrated in Fig. 5–2. (1) The armature circuit consists of an ideal voltage source E_g in series, with R_a

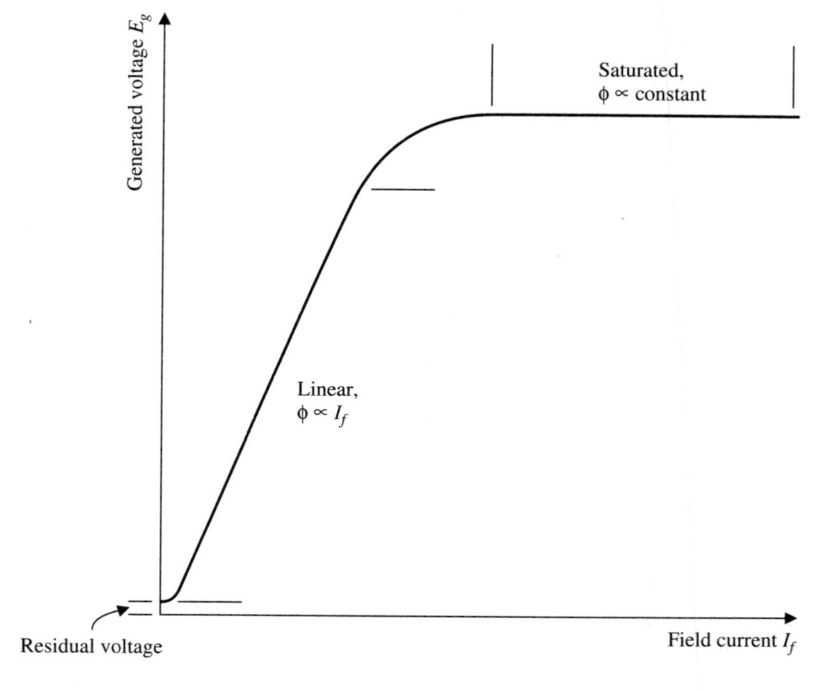

Figure 5-1 Magnetization curve of dc generator.

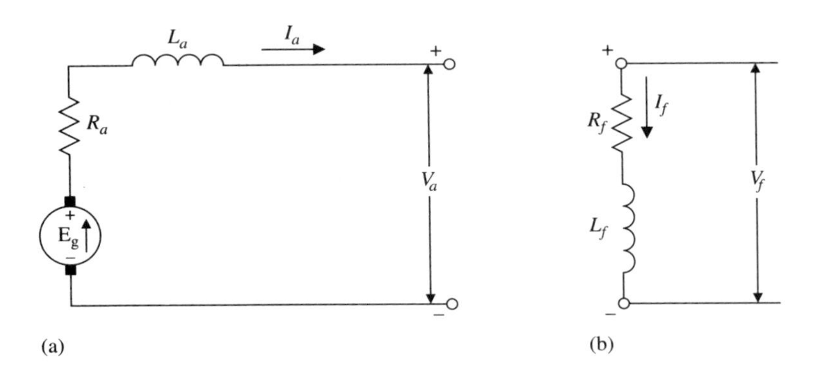

Figure 5-2 Generator circuit models. (a) Armature circuit. (b) Field circuit.

representing the resistance of the armature windings. The armature inductance L_a is included, but is not involved in steady-state analysis. To be complete, the brush resistance, which is nonlinear and varies with humidity, pressure, and current density, is normally treated as a voltage drop V_B and must be included. Typical values are between 1.5 and 5.0 V and remain essentially constant under all load conditions [Fig. 5–2(a)]. (2) The field circuit consists of a voltage source V_f, the field resistance R_f, and the field inductance L_f [Fig. 5–2(b)].

From Fig. 5–2(a), neglecting L_a which only affects the circuit under rapidly changing load conditions, we have

$$V_a = E_g - I_a R_a \tag{5–4}$$

Then multiplying both sides by I_a yields

$$V_a I_a = E_g I_a - I_a^2 R_a \tag{5–5}$$

which in turn can be expressed as

$$P_d = P_{\text{in}} - I_a^2 R_a \tag{5–6}$$

where P_d is the gross power output of the armature, P_{in} the mechanical power input expressed in watts, and $I_a^2 R_a$ the armature copper loss, which is a heat loss.

From Fig. 5–2(b), neglecting L_f, we obtain

$$V_f = I_f R_f \tag{5–7}$$

$$I_f = \frac{V_f}{R_f} \tag{5–8}$$

$$P_f = I_f^2 R_f = \frac{V_f^2}{R_f} \tag{5–9}$$

where P_f is the field copper loss, V_f the applied field voltage, R_f the field resistance, and I_f the field current.

5–3 DC GENERATOR CONNECTIONS

Dc machines are classified by the method of exciting their main field windings. There are two basic methods: separately excited and self-excited. Separately excited machines rely on a dc power source separate from the machine to provide the field current. Typical sources are a battery bank, a solid-state power supply, or another generator. Self-excited machines have one or more of their field windings connected to the armature circuit and depend on the residual magnetic field in the field poles to initiate a voltage buildup.

Separately Excited DC Generator

Fig. 5–3 shows the schematic of a separately excited dc generator. The terminal voltage V_t is equal to the armature voltage V_a. Hence

$$V_t = E_g - I_a R_a \tag{5–10}$$

where

$$E_g = K_g I_f S$$

or

$$E_g = k_g I_f \omega$$

From these relationships it is clear that the terminal voltage V_t is controlled by varying either I_f or $S(\omega)$ or both, but since the prime mover is usually operated at constant speed, normally I_f is used to control the output voltage.

The power relationships are also shown in Fig. 5–3 and include the power supplied to the field P_f, the mechanical power input P_{in}, and the power output P_{out}. We must also consider the power losses, which include the armature copper loss P_a, the field copper loss P_f, and the friction (bearings and brushes), windage, and core (hysteresis and eddy current) losses P_{fwc}. These losses are related by the law of conservation of energy by

$$\begin{aligned} P_f + P_{\text{in}} &= P_{\text{out}} + \sum \text{losses} \\ &= P_{\text{out}} + P_f + P_a + P_{fwc} \end{aligned} \tag{5–11}$$

Canceling P_f from both sides yields

$$\begin{aligned} P_{\text{in}} &= P_{\text{out}} + P_a + P_{fwc} \\ &= V_t I_L + I_a^2 R_a + P_{fwc} \end{aligned} \tag{5–12}$$

Since $I_L = I_a$,

$$P_{\text{in}} = (V_t + I_a R_a) I_a + P_{fwc} \tag{5–13}$$

$$= E_g I_a + P_{fwc} \tag{5–14}$$

$$= P_d + P_{fwc} \tag{5–15}$$

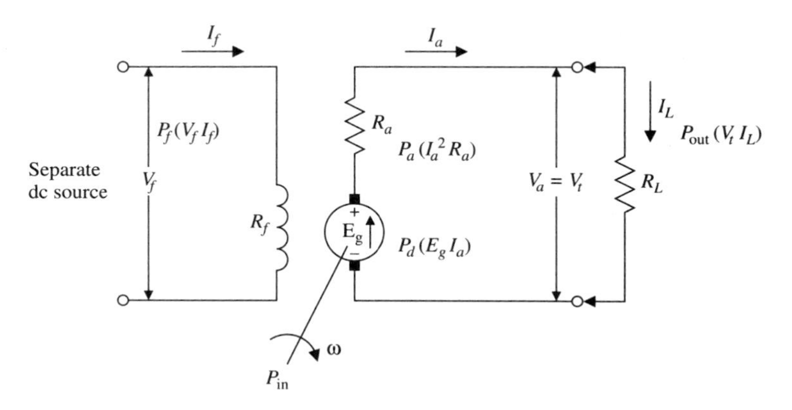

Figure 5-3 Separately excited dc generator.

The cost of operation is always of great importance in assessing machine performance. We use efficiency as the yardstick to access a machine. Efficiency η is the ratio of power output to power input and is usually expressed as a percentage. In general,

$$\eta = \frac{P_{out}}{P_d + P_f + P_{fwc}} \times 100\% \tag{5-16}$$

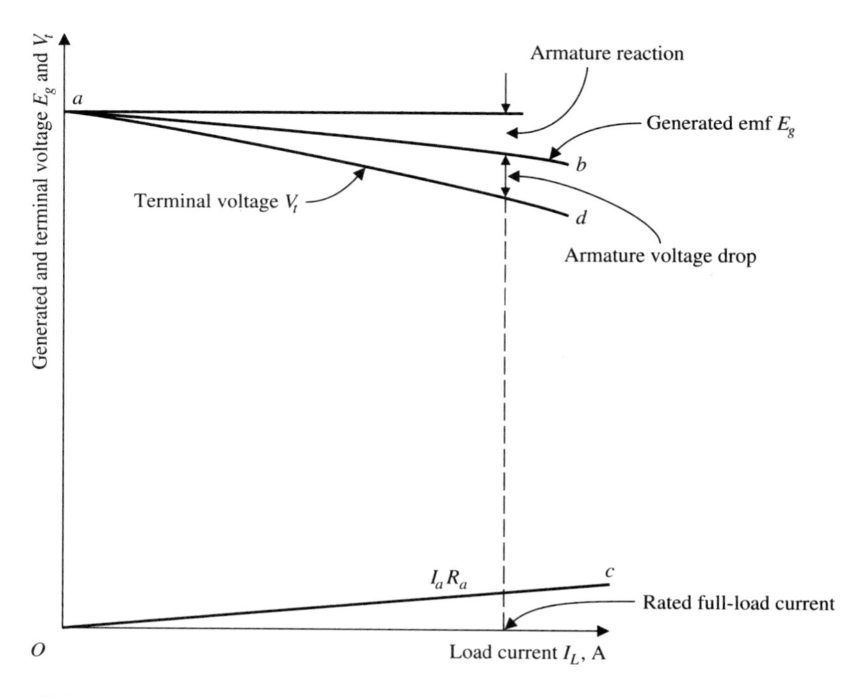

Figure 5-4 Load characteristic of separately excited dc generator.

The output voltage V_t of a separately excited generator drops as the load current increases, even though the field current I_f is maintained constant. This drop is caused by two factors: (1) a slight decrease in the generated voltage E_g because of armature reaction, which is shown by the slightly drooping curve *ab* in Fig. 5–4, and (2) the armature voltage drop I_aR_a (curve *c*), which is proportional to the load current $I_L(I_a = I_L)$. Subtracting the armature voltage drop curve from the generated electromotive force curve yields the load characteristic curve (curve *ad*), or V_t versus I_L curve.

Voltage Regulation. As can be seen from Fig. 5–4, with a constant field current the terminal voltage decreases as the connected load increases. The term *voltage regulation* is used to assess the performance of a generator under load conditions. If there is little change in the terminal voltage between no load and full load, the generator is said to have a good voltage regulation. Conversely, if there is a large change of terminal voltage between no load and full load, the generator is said to have a poor regulation. The American National Standards Institute (ANSI) defines voltage regulation as

$$\%\text{ voltage regulation} = \frac{E_g - V_t}{V_t} \times 100\% \qquad (5\text{–}17)$$

where E_g is the generated voltage and V_t the terminal voltage.

➤ EXAMPLE 5-1

A separately excited generator has a generated electromotive force of 200 V at a speed of 1,200 r/min (125.56 rad/s) with a field current of 2.5 A. Calculate: (**a**) the generated voltage at 1,600 r/min (167.55 rad/s) with a field current of 2.5 A; (**b**) the generated voltage at 1,000 r/min (104.72 rad/s) with a field current of 3.0 A.

SOLUTION

(**a**) In English units, $E_{g1} = 200$ V, $I_{f1} = 2.5$ A, $S = 1{,}200$ r/min, and

$$E_{g2} = E_{g1} \times \frac{S_2}{S_1} \times \frac{I_{f2}}{I_{f2}}$$
$$= 200\text{ V} \times \frac{1{,}600\text{ r/min}}{1{,}200\text{ r/min}} \times \frac{2.5\text{ A}}{2.5\text{ A}}$$
$$= 266.67\text{ V}$$

In SI units, $E_{g1} = 200$ V, $I_{f1} = 2.5$ A, $\omega_1 = 125.56$ rad/s, and

$$E_{g2} = E_{g1} \times \frac{\omega_2}{\omega_1} \times \frac{I_{f2}}{I_{f2}}$$
$$= 200\text{ V} \times \frac{167.55\text{ rad/s}}{125.56\text{ rad/s}} \times \frac{2.5\text{ A}}{2.5\text{ A}}$$
$$= 266.68\text{ V}$$

(**b**) In English units, $E_{g1} = 200$ V, $I_{f1} = 2.5$ A, $S_1 = 1{,}200$ r/min, and

$$E_{g2} = E_{g1} \times \frac{S_2}{S_1} \times \frac{I_{f2}}{I_{f1}}$$

$$= 200\ \text{V} \times \frac{1{,}000\ \text{r/min}}{1{,}200\ \text{r/min}} \times \frac{3.0\ \text{A}}{2.5\ \text{A}}$$

$$= 200\ \text{V}$$

In SI units, $E_{g1} = 200$ V, $I_{f1} = 2.5$ A, $\omega_1 = 125.56$ rad/s, and

$$E_{g2} = E_{g1} \times \frac{\omega_2}{\omega_1} \times \frac{I_{f2}}{I_{f1}}$$

$$= 200\ \text{V} \times \frac{104.72\ \text{rad/s}}{125.56\ \text{rad/s}} \times \frac{3.0\ \text{A}}{2.5\ \text{A}}$$

$$= 200\ \text{V}$$

➤ EXAMPLE 5-2

A separately excited dc generator has a field resistance of 50 Ω, an armature resistance of 0.125 Ω, and a brush drop of 2 V. At no load the generated voltage is 275 V and the full-load current is 95 A. The field excitation voltage is 120 V, and the friction, windage, and core losses are 1,500 W. Calculate: (**a**) the rated terminal voltage and power output; (**b**) the efficiency at full load.

SOLUTION

(**a**)

$$V_t = E_g - V_B - I_a R_a$$

$$= 275 - 2 - 95 \times 0.125$$

$$= 261.13\ \text{V}$$

$$P_{\text{out}} = V_t I_L$$

$$= 261.13 \times 95 = 24.81\ \text{kW}$$

(**b**)

$$\eta = \frac{P_{\text{out}}}{P_{\text{in}}} \times 100\%$$

$$= \frac{P_{\text{out}}}{P_d + P_f + P_{fwc}}$$

$$= \frac{24.81\ \text{kW}}{275 \times 95 + 125^2 / 50 + 1{,}200} \times 100\%$$

$$= \frac{24.81\ \text{kW}}{26.13\ \text{kW} + 0.313\ \text{kW} + 1.2\ \text{kW}} \times 100\%$$

$$= \frac{24.81\ \text{kW}}{27.64\ \text{kW}} \times 100\% = 89.75\%$$

➤ EXAMPLE 5-3

A 100-kW 250-V separately excited generator has an armature resistance of 0.125 Ω. If the generator is operating at rated voltage and power output, calculate the armature current and generated voltage.

SOLUTION

$$I_a = I_L = \frac{100 \text{ kW}}{250 \text{ V}} = 400 \text{ A}$$

$$E_g = V_t + I_a R_a$$

$$= 250 + 400 \times 0.125 = 300 \text{ V}$$

Self-Excited DC Generators

Self-excited dc generators do not require an independent field excitation, but depend instead on the presence of a small residual magnetic flux, which is always present in a machine that is used frequently. Although it may be necessary, especially for the first time, the machine is run to *flash* the field from a separate dc source. As the armature rotates, the residual flux generates a small electromotive force, which produces a small field current. This in turn causes an increase in the field pole flux, provided that this flux adds to the residual flux. Each increase in generated electromotive force increases the field current, which in turn increases the field pole flux, the process repeating itself until the generated voltage has fully built up to its set value. This process is illustrated in Fig. 5–5.

If the magnetization or open-circuit characteristic were a straight line as represented by *OA*, the generated voltage E_g could assume any value along that curve, and the generator operation would be unstable. The slope of the line is $E_g/I_f = R_c$, where R_c is the *critical resistance*. If the resistance of the field circuit, that is, the field resistance and any adjusting resistances, is equal to the critical resistance, then the output voltage will be unstable. If the resistance of the field circuit is greater than the critical resistance, then the generator voltage will not build up. This condition is represented by line *OC*. If the total resistance in the field circuit is less than the critical resistance with the generator running at rated speed, it will self-excite to the voltage value represented by the intersection of the resistance line *OB* and the magnetization curve. If the rotational speed of the generator is increased from ω_1 to ω_2, the open-circuit curve will increase as shown by the dashed curve, and the new generated voltage will be at the intersection of the open-circuit curve at velocity ω_2 and the resistance line *OB* extended to *D*. Conversely, if the speed decreases, the open-circuit voltage will decrease and the machine will produce a new lower generated voltage.

Conditions for Voltage Buildup. There are a number of conditions that must be met before a self-excited dc generator will build up voltage.

1. There must be residual field present. If the residual magnetic field is insufficient or not present, the machine will fail to build up voltage. This condi-

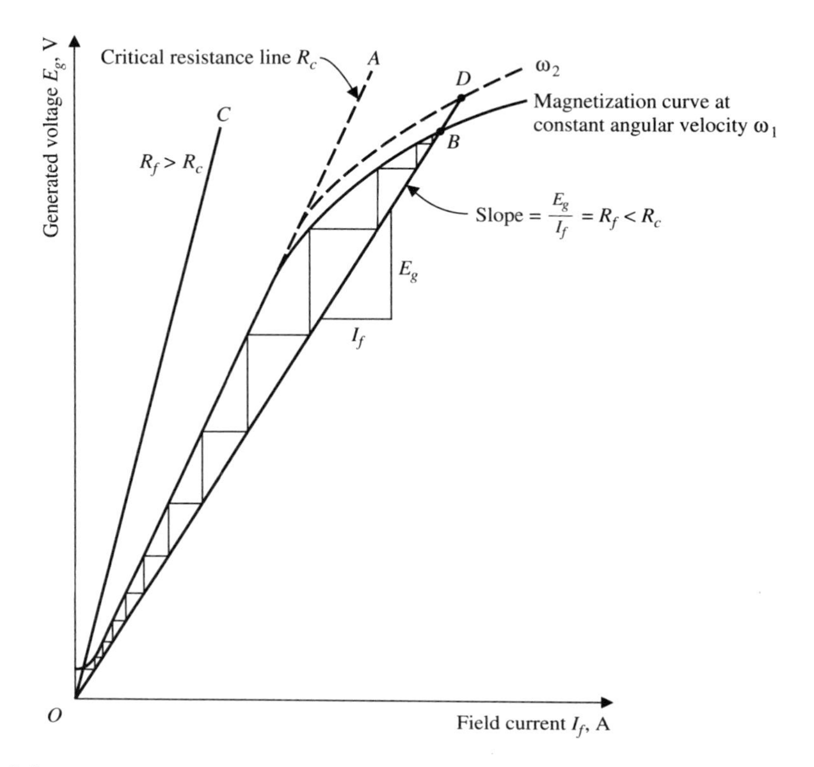

Figure 5-5 Voltage buildup in self-excited generator.

tion can be caused by excessive heat, by vibration, or because the machine has not been used for a considerable period of time. The residual magnetic field can be restored by *flashing*, that is, connecting the field momentarily to a dc source.

2. The field connections must be correct. If the connections are reversed, the field current will create a magnetomotive force which opposes the residual field magnetomotive force, and as result the net field will be reduced. This situation is corrected by reversing the field connections. It should be noted that if the direction of rotation of the armature is reversed, the polarity of the generated voltage will be reversed and create the same effect. This condition will only occur with a reversible prime mover such as a dc motor or a polyphase induction motor, since internal combustion engines rotate only in one direction.
3. The field circuit resistance, including any voltage adjusting rheostats, must be less than the critical resistance R_e.
4. The rotational speed of the generator is greater than the critical speed. The critical speed is the minimum rotational speed at which the generator will self-excite. Usually this is not a problem since the generator is designed to run at a definite speed, specified on the nameplate.

Factors Affecting the Terminal Voltage. A number of factors cause the terminal voltage of a generator to decrease as the load increases.

1. An increase in the voltage drop across the armature, interpoles, series, and compensating windings, if present, even though the generated electromotive force remains constant. There will always be a resistance increase as the temperature of the machine rises.
2. A decrease in the generated electromotive force because:
   a. The armature reaction demagnetizing effect reduces the air-gap flux.
   b. The voltage across the shunt field decreases with load, creating a further decrease in the air-gap flux.
   c. The prime mover speed drops with an increase in load.

Shunt Generator

The only difference between the separately excited generator and the shunt generator is that the field circuit is connected across the armature circuit. As result, the voltage across the shunt field is the terminal voltage V_t, which will decrease with load.

From Fig. 5–6(a), and using Ohm's and Kirchhoff's laws, it can be seen that the following relationships exist:

$$I_a = I_L + I_f \tag{5–18}$$

$$I_f = \frac{V_t}{R_f} \tag{5–19}$$

$$E_g = V_t + I_a(R_a + R_{ip}) \tag{5–20}$$

Since the only power input to the generator is the mechanical power input from the prime mover, then

$$\begin{aligned} P_{\text{in}} &= P_{\text{out}} + \text{losses} \\ &= V_t I_L + V_t I_f + I_a^2 R_a + P_{fwc} \\ &= V_t(I_L + I_f) + (I_a R_a) I_a + I_a^2 R_{ip} + P_{fwc} \\ &= (V_t + I_a R_a) I_a + I_a^2 R_{ip} + P_{fwc} \\ &= E_g I_a + I_a^2 R_{ip} + P_{fwc} \\ &= P_d + P_{ip} + P_{fwc} \end{aligned} \tag{5–21}$$

where P_{ip} is the interpole copper loss.

The load characteristic shown in Fig. 5–6(b) is obtained by adjusting the field current using the shunt field rheostat, so that at no load the generated voltage and, hence, the terminal voltage are at rated value. The field current setting then remains unchanged. As the load resistance decreases, the load current increases and the generated and terminal voltages decrease. There are three causes contributing to this voltage drop.

1. The effects of armature reaction, apart from causing a distortion of the air-gap-flux, also cause a reduction in the air-gap flux density and reduce the generated voltage E_g.

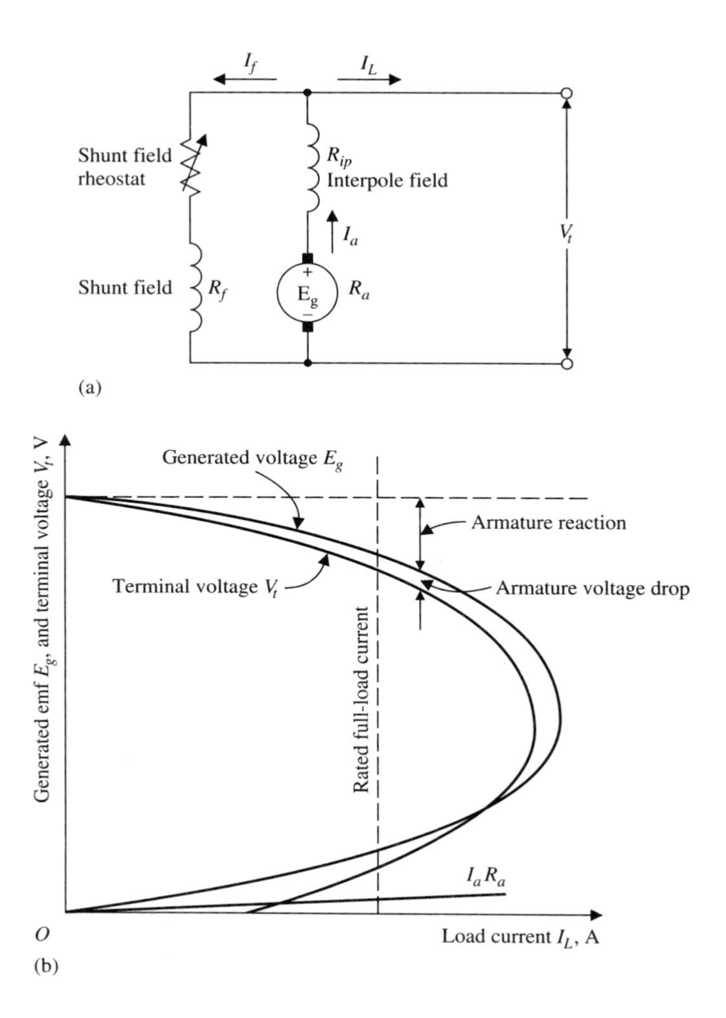

Figure 5-6 Shunt generator. (a) Schematic. (b) Load characteristic.

2. The armature circuit voltage drop, which includes the armature and any fields connected in series with the armature, also reduces the terminal voltage V_t. Recall

$$V_t = E_g + I_a(R_a + R_{ip})$$

3. The reduced terminal voltage, which is also the voltage applied to the shunt field, $V_t = V_f$, causes a decrease in the shunt field current, with a consequent decrease in the generated and terminal voltages.

If the main field poles are unsaturated as the load current increases beyond rated load, the effects of increasing armature reaction, the armature voltage drop, and decreasing field current will all combine to cause the terminal voltage to decrease at a faster rate than the decrease in load resistance. The result is a total collapse of the terminal voltage. At the same time the load current is also decreasing until the generator becomes short-circuited, supplying a short-circuit current *Oa*, which is very much less than the full-load rated current. This current is driven by the residual voltage, which in turn has been weakened by armature reaction.

This situation is peculiar to the shunt generator, since in the separately excited generator the field current is supplied from a separate constant voltage source. In addition, because of the reduced voltage supplied to the shunt field, the rate of decrease of terminal voltage, or voltage regulation, is greater than that of the separately excited generator.

If it is necessary to operate a shunt generator at a constant terminal voltage, then the resistance of the shunt field circuit must be reduced so that the field current can be increased to maintain a constant terminal voltage. This is accomplished by varying the resistance of the shunt field rheostat in series with the shunt field. Shunt generators are used for constant-voltage applications where there are slowly changing load conditions.

➤ EXAMPLE 5-4

A shunt generator has a no-load terminal voltage of 254 V. At rated full load the terminal voltage is 240 V. Calculate: **(a)** the full-load current if the field circuit resistance is 30 Ω and the armature resistance is 0.02 Ω; **(b)** the voltage regulation. Ignore the effects of armature reaction.

SOLUTION

(a)

$$I_f = \frac{V_t}{R_f} = \frac{240\text{ V}}{30\ \Omega} = 8\text{ A}$$

$$E_g = V_t + I_a R_a$$

$$I_a = \frac{E_g - V_t}{R_a} = \frac{254 - 240}{0.02} = 700\text{ A}$$

$$I_L = I_a + I_f = 700 + 8 = 708\text{ A}$$

(b)

$$\%\text{ voltage regulation} = \frac{E_g - V_t}{V_t} \times 100\%$$

$$= \frac{254 - 240}{240} \times 100\% = 5.83\%$$

➤ EXAMPLE 5-5

The magnetization curve of a shunt generator running at a constant speed of 600 r/min (62.83 rad/s) is as follows:

Field current I_f (A)	0.5	1.0	1.5	2.0	2.5	3.0	3.5
Generated voltage E_g (V)	50	100	140	171	191	206	216

Determine: (**a**) the generated voltage when the field circuit resistance is 85.5 Ω; (**b**) the generated voltage at 660 r/min (69.12 rad/s) when the field circuit resistance is 85.5 Ω; (**c**) the critical resistance at 600 r/min (62.83 rad/s).

SOLUTION

1. Draw the open-circuit curve (Fig. 5–7) using the data.
2. Draw the field resistance line for a field circuit resistance of 85.5 Ω (171 V/2 A). Note that only two points are required, one of which is the origin. From the graph, the intersection of the field resistance line and the magnetization curve is 171 V at 600 r/min.
3. The magnetization curve at 660 r/min is obtained by multiplying each ordinate of the 600-r/min curve by 660 r/min/600 r/min = 1.1. The field resistance line of 85.5 Ω intersects this curve at 207.5 V at 660 r/min.
4. To obtain the critical resistance, draw a straight line from the origin, overlaying the straight portion of the 600 r/min magnetization curve. This gives $R_c = 100/1 = 100\ \Omega$.

Series Generator

The schematic of a series generator is shown in Fig. 5–8(a). The armature current also flows through the series field, and as might be expected, the series field wind-

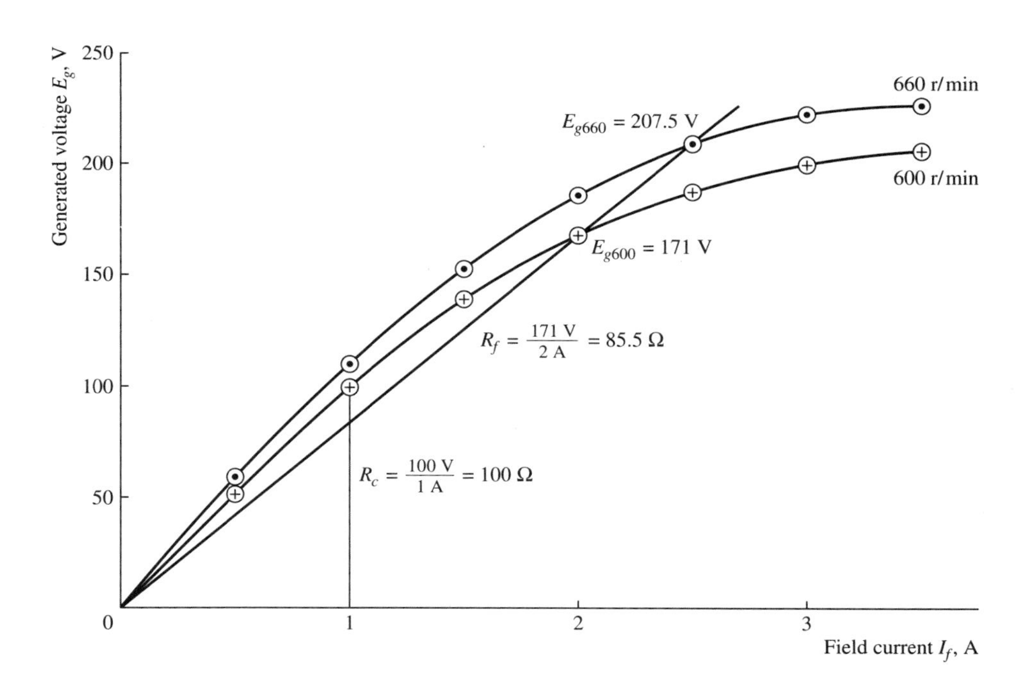

Figure 5-7 Open-circuit curve for Example 5-5.

ings are made up of relatively few turns of large-cross-sectional-area copper wire because of the high currents involved.

The magnetization curve is similar to that of the shunt generator. However, it should be noted that the field current is equal to the armature and load currents. Apart from the small voltage produced by the residual field, the series generator cannot build up voltage unless it is connected to a load since there will be no field

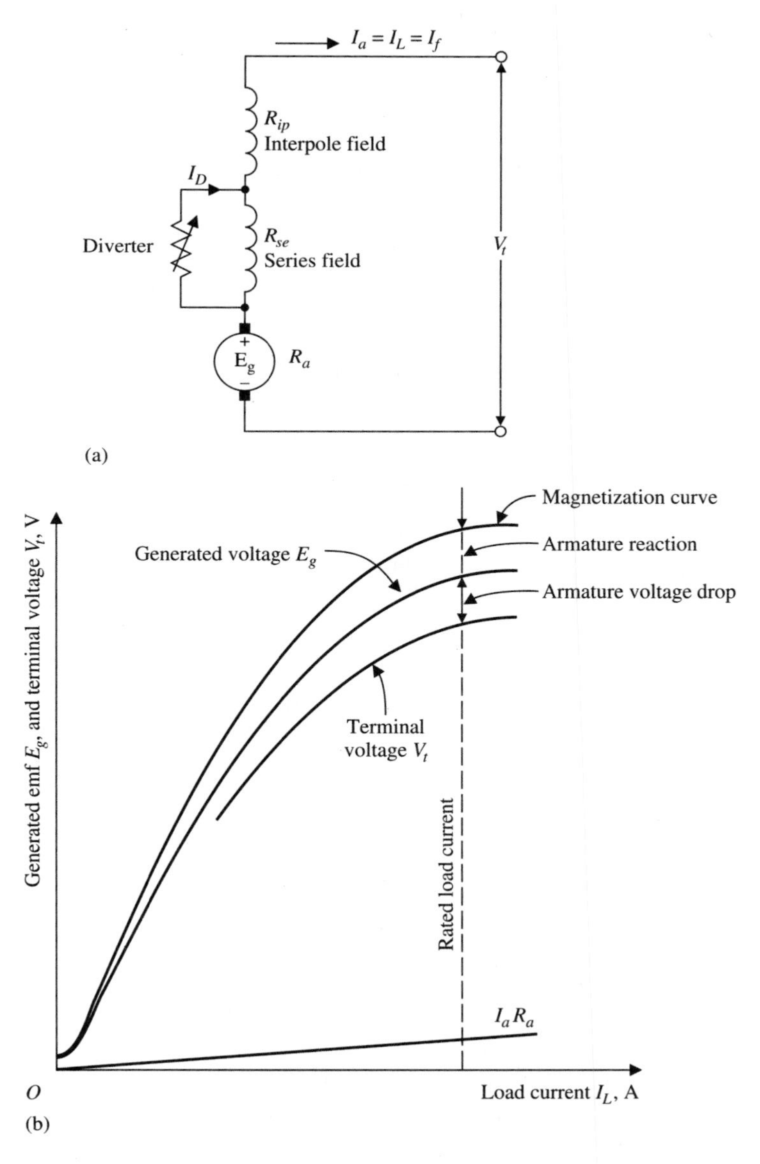

Figure 5-8 Series generator. (a) Schematic. (b) Load characteristic.

current. When connected to a load, it will build up voltage. As was observed with the separately excited and shunt generators, the generated voltage is reduced because of the demagnetizing effects of armature reaction. The terminal voltage reflects the additional drops caused by the resistance of the total armature circuit, that is, the armature, series field, and interpole field resistances. Since the terminal voltage directly depends on the load current until the magnetic circuit is saturated, it can be seen that the series generator is not used in normal service. The terminal voltage is controlled by the diverter resistance in parallel with the series field. The diverter is a low-resistance high-wattage resistance which diverts current from the field. As a result it reduces the field pole flux.

The following relationships apply to the series generator:

$$V_t = E_g - I_a(R_a + R_{se} + R_{ip}) \tag{5–22}$$

$$I_a = I_f = I_L \tag{5–23}$$

Compound Generator

The major reason that the terminal voltage of the shunt generator drops between no load and full load is the demagnetizing effect of armature reaction. However, as we have just seen, the series generator produces an increasing magnetic field as the load current increases. Therefore, by placing a few series winding turns on each main pole, in addition to the shunt field turns, it is possible to neutralize the armature reaction magnetomotive force. Generators with both series and shunt windings on the main poles are called *compound generators,* although the shunt field is still the dominant field.

When the series field magnetomotive force is additive to the shunt field magnetomotive force, the generator is said to be *cumulatively compounded.* There are three degrees of cumulative compounding [Fig. 5–9(c)].

1. Overcompounding occurs when the added series turns cause the terminal voltage to rise as the load current increases, that is, it has a negative voltage regulation.
2. Flat compounding occurs when the added series turns cause the no-load and full-load terminal voltages to be the same, that is, zero voltage regulation. The terminal voltage increases slightly as the load first increases and then decreases so that the full-load voltage is equal to the no-load voltage.
3. Undercompounding occurs when there are only sufficient series turns to reduce the drop in the terminal voltage between no load and full load. The result is a positive voltage regulation, which is slightly better than that of a shunt generator.

Most manufacturers design their machines with excessive series turns. The required degree of compounding is obtained by adjusting the diverter resistance in parallel with the series field. This adjustment is usually carried out when the machine is installed.

When the series field magnetomotive force is subtractive with respect to the shunt field magnetomotive force, the generator is said to be *differentially*

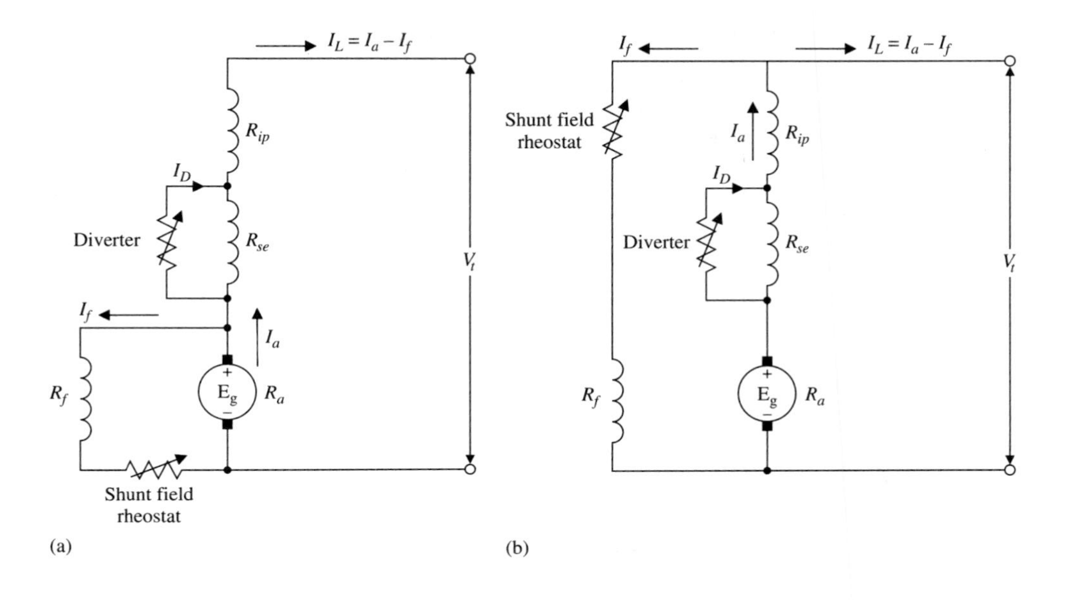

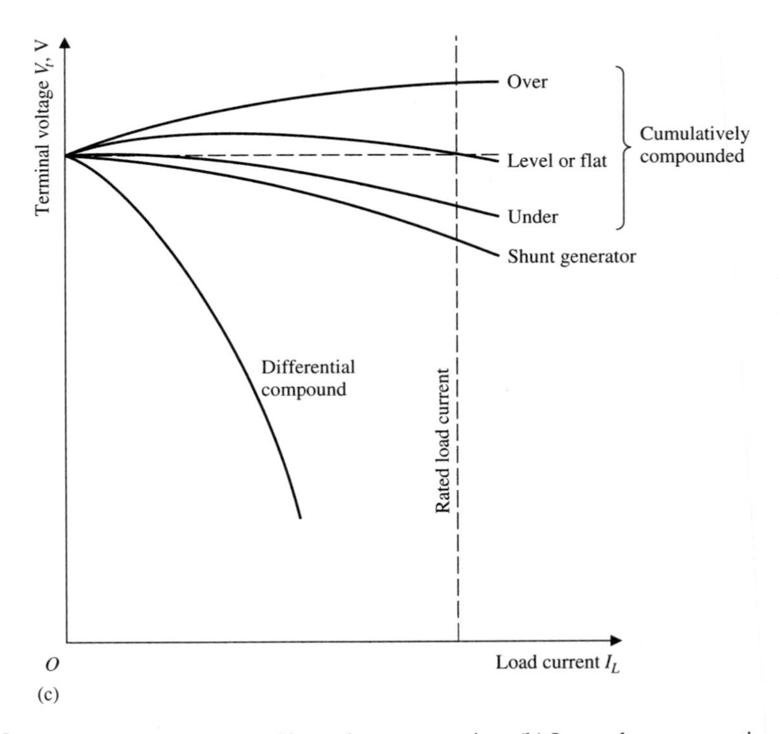

Figure 5-9 Compound generator. (a) Short-shunt connection. (b) Long-shunt connection. (c) Load characteristics.

compounded. A differentially compounded generator has a terminal voltage that rapidly drops with increasing load [Fig. 5–9(c)]. Differentially compounded generators are used in arc welding applications, where it is necessary to reduce the voltage as soon as the arc is struck, or they are used as constant-current generators.

Figure 5–9(a) and (b) shows the *short-shunt connection* and the *long-shunt connection*. The only difference between the two connection arrangements is that in the short-shunt connection the shunt field is in parallel across the armature, whereas in the long-shunt connection it is in parallel with the whole armature circuit, namely, the armature, series, and interpole fields, and as a result the shunt field voltage will be smaller than with the short-shunt connection. The load characteristics of the two connections are essentially similar. The terminal voltage is adjusted by means of the shunt field rheostat.

The following relationships apply to compound generators:

1. *Short-shunt connection:*

$$V_t = E_g - I_a(R_a + R_{se} + R_{ip}) \tag{5–24}$$

$$V_f = E_g - I_a R_a \tag{5–25}$$

$$I_f = \frac{V_f}{R_f} \tag{5–26}$$

$$I_L = I_a - I_f \tag{5–27}$$

2. *Long-shunt connection:*

$$V_t = E_g - I_a(R_a + R_{se} + R_{ip}) \tag{5–28}$$

$$V_f = V_t \tag{5–29}$$

$$I_f = \frac{V_f}{R_f} = \frac{V_f}{R_f} \tag{5–30}$$

$$I_L = I_a - I_f \tag{5–31}$$

➤ EXAMPLE 5-6

A 550-V 150-kW compound generator has a series field resistance of 0.025 Ω, an interpole field resistance of 0.002 Ω, a shunt field resistance of 175 Ω, and an armature resistance of 0.038 Ω. Calculate the generated voltage when the machine is delivering rated power, and is connected: **(a)** short shunt; **(b)** long shunt. Neglect brush drops.

SOLUTION

$$I_L = \frac{150 \text{ kW}}{550 \text{ V}} = 272.73 \text{ A}$$

(**a**) Short shunt: The voltage drop across the series and interpole fields is

$$I_L(R_{se} + R_{ip}) = 272.73(0.025 + 0.002) = 7.36 \text{ V}$$

$$V_f = V_t + I_L(R_{se} + R_{ip}) = 550 + 7.36 = 557.36 \text{ V}$$

$$I_f = \frac{V_f}{R_f} = \frac{557.36 \text{ V}}{175\ \Omega} = 3.18 \text{ A}$$

$$I_a = I_L + I_f = 272.73 + 3.18 = 275.91 \text{ A}$$

$$E_g = V_f + I_a R_a = 557.36 + 275.91 \times 0.038 = 567.84 \text{ V}$$

(**b**) Long shunt:

$$V_t = V_f = 550 \text{ V}$$

$$I_f = \frac{550 \text{ V}}{175\ \Omega} = 3.14 \text{ A}$$

$$I_L = \frac{150 \text{ kW}}{550 \text{ V}} = 272.73 \text{ A}$$

$$I_a = I_L + I_f = 272.73 + 3.14 = 275.87 \text{ A}$$

$$E_g = V_t + I_a(R_a + R_{se} + R_{ip})$$
$$= 550 + 275.87(0.038 + 0.025 + 0.002)$$
$$= 550 + 275.87 \times 0.065 = 567.93 \text{ V}$$

As can be seen from the answers, there is almost no difference in the terminal voltages of the two connections.

5–4 COUNTERTORQUE

Figure 5–10 shows the armature conductors of a one-turn armature coil which is being turned clockwise by an input torque T_i supplied by the prime mover. The direction of current flow in the conductors is determined by Fleming's right-hand rule. The current flow in the conductors creates concentric magnetic fields around each conductor, clockwise around the conductor under the N pole and counterclockwise around the conductor under the S pole. These magnetic fields strengthen and distort the field pole flux upward at the top of the left-hand conductor and downward at the bottom of the right-hand conductor. At the same time they weaken the field pole flux on the opposite side of the conductors. Since the magnetic lines of force exist in a state of tension, this tension causes the lines of force to attempt to straighten. At the same time a mechanical force is applied on the armature conductors, downward on the left-hand side and upward on the right-hand side. These forces produce a countertorque, or developed torque T_d, which is acting in opposition to the input torque T_i. In the case of a generator T_i is greater than T_d. In addition, the friction, windage, and core losses also create a small opposing torque T_{fwc}, which acts in the same direction as T_d. If the armature is rotating at a constant angular velocity ω, then the net rotor torque is zero, that is,

$$T_i = T_d + T_{fwc} \tag{5–32}$$

From Eq. (5–32), assuming that the generator is operating at a constant angular velocity ω, at constant field pole flux density, and that the friction, windage, and core losses are constant, then as the power output of the generator increases, the developed torque $T_d = E_g I_a$ increases, and since T_{fwc} is constant, T_i must increase, that is, the power output of the prime mover must increase. The power balance relationship is

$$\begin{aligned} \omega(T_d + T_{fwc}) &= E_g I_a + P_i \\ &= P_{\text{out}} + P_{cu} + P_{fwc} \end{aligned} \tag{5–33}$$

where P_{out} is the generator output power, P_{cu} the total of all copper losses (armature and fields), and P_{fwc} the power lost to friction, windage, and core losses. These relationships apply to all generators.

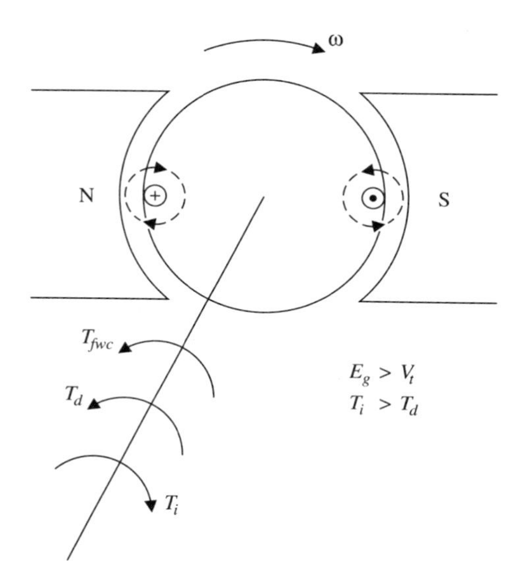

Figure 5-10 Countertorque in dc generator.

QUESTIONS

5–1 Under what conditions can it be said that $E_g = K_g \Phi S$ or $E_g = k_g = k_g \Phi \omega$?

5–2 With respect to the magnetization curve (Fig. 5–1): (**a**) why does the curve not start at the origin; (**b**) why is the curve linear at low to medium voltages; (**c**) why is the curve nonlinear at medium to high voltages?

5–3 What are the effects of magnetic saturation on a generator's output voltage?

5–4 Explain what is meant by separate and self-excitation.

5–5 What factors cause the terminal voltage of a separately excited generator to be less than ideal?

5–6 What is meant by voltage regulation?

5–7 With the aid of a sketch explain the buildup of electromotive force in a self-excited generator.

5–8 Why is the output voltage of a self-excited generator operating on the linear portion of the open-circuit characteristic unstable?

5–9 What is meant by the critical resistance of a self-excited generator?

5–10 What are the four conditions necessary for voltage buildup in a self-excited generator?

5–11 What factors cause the terminal voltage of a generator to decrease with increasing load?

5–12 Why is the voltage regulation of a shunt generator greater than that of a separately excited generator?

5–13 What is meant by brush voltage drop? Why is it expressed as a voltage?

5–14 How is the terminal voltage of a shunt generator normally controlled?

5–15 Why is a series generator rarely used?

5–16 What conditions must be met before a series generator will self-excite?

5–17 How is the output of a series generator controlled?

5–18 What applications would use a series generator?

5–19 Why are dc generators compounded?

5–20 What is meant by cumulative compounding?

5–21 With the aid of a graph explain flat compounding, undercompounding, and overcompounding.

5–22 How is the degree of compounding adjusted?

5–23 What is meant by differential compounding?

5–24 With the aid of sketches explain short- and long-shunt connections.

5–25 What is the effect of short- and long-shunt connections on a generator's performance?

5–26 What is countertorque?

PROBLEMS

5–1 The terminal voltage of a dc generator is 240 V. What will be the terminal voltage if: (**a**) the flux increases 5% with the speed remaining constant; (**b**) the speed is increased 10% with the flux unchanged; (**c**) the flux is increased by 10% and the speed decreased by 15%?

5–2 A separately excited generator has an open-circuit voltage of 250 V at 1,200 r/min (125.66 rad/s). Assuming that the field excitation remains constant, calculate: (**a**) the open-circuit voltage at 900 r/min (94.25 rad/s); (**b**) at what speed it must operate to produce an open-circuit voltage of 275 V.

5–3 A shunt generator supplies a load current of 100 A at 220 V, the efficiency of the generator is 86%, and the windage, friction, and core losses are 1,100 W. If the shunt field resistance is 110 Ω, calculate the armature resistance.

5–4 An eight-pole 1,200-r/min (125.66-rad/s) lap-wound separately excited generator has a flux per pole of 0.03 Wb. There are 400 armature conductors, and the terminal voltage under load is 225 V. The armature copper loss is 750 W, and the brush contact voltage drop is 2.5 V. What is the generator output?

5–5 The magnetization curve of a shunt generator running at 1,200 r/min (125.66 rad/s) is as follows:

Field current (A)	1	2	3	4	5	6	7
Generated voltage (V)	72	140	203	260	301	335	361

Determine: (**a**) the critical resistance; (**b**) the no-load voltage at 900 r/min (94.25 rad/s) if the field resistance is 55 Ω.

5–6 A shunt generator has an open-circuit voltage of 225 V. Under load the terminal voltage is 212 V. Find the load current when the field circuit resistance is 35 Ω and the armature resistance is 0.03 Ω. Neglect armature reaction and brush drop.

5–7 A compound generator is connected in long shunt. The load current is 60 A at 550 V, and the armature, series, and shunt field resistances are 0.045, 0.025, and 250 Ω, respectively. Calculate the generated voltage and armature current. Neglect the brush drop.

5–8 A 50-kW 250-V series generator has an armature resistance of 0.02 Ω and a series field resistance of 0.045 Ω. At rated load calculate: (**a**) the armature current; (**b**) the generated voltage; (**c**) the armature copper loss; (**d**) the field copper loss; (**e**) the efficiency.

5–9 A 50-kW 250-V shunt generator has armature and field resistances of 0.02 and 150 Ω, respectively, and a P_{fwc} of 1,500 W. At full load calculate: (**a**) the load current; (**b**) the field current; (**c**) the armature current; (**d**) the voltage regulation; (**e**) the shunt field copper loss; (**f**) the armature copper loss; (**g**) the efficiency.

5–10 A 50-kW 220-V long-shunt compound generator has an armature resistance of 0.075 Ω, the resistance of the series field is 0.036 Ω, and the shunt field resistance is 110 Ω. If the machine is delivering rated load at rated voltage, find the efficiency.

CHAPTER

6

DC Motors

6–1 INTRODUCTION

In spite of competition from variable-frequency induction motor drives, the dc motor is still the first choice in many applications for the following reasons:

1. It has a wide range of stepless speed control.
2. It can be operated in either the *constant-torque* or the *constant-horsepower* or *constant-power* (kW) modes.
3. It is capable of rapid acceleration, deceleration, and reversal, a requirement that is common in rolling mill operations.
4. When used in conjunction with closed-loop feedback, very precise speed control can be achieved.
5. It is capable of very high short-duration torque output.
6. It can provide regenerative braking in overhauling load applications such as mine hoists and electric traction.

When compared to an equivalent polyphase induction motor, the initial cost and ongoing maintenance costs of the dc motor are considerably greater, but the flexibility and versatility, especially in variable-speed thyristor phase-controlled converter drives, weigh heavily in favor of the dc motor.

6-2 EQUIVALENT CIRCUIT OF THE DC MOTOR

DC machines can be operated as either a motor or a generator, the only difference being the direction of the energy flow. These differences are illustrated in Fig. 6-1. In Fig. 6- 1(a), the generated voltage is

$$E_g = K_g \Phi S \text{ V} \qquad \textbf{(5–2)(E)}$$

or

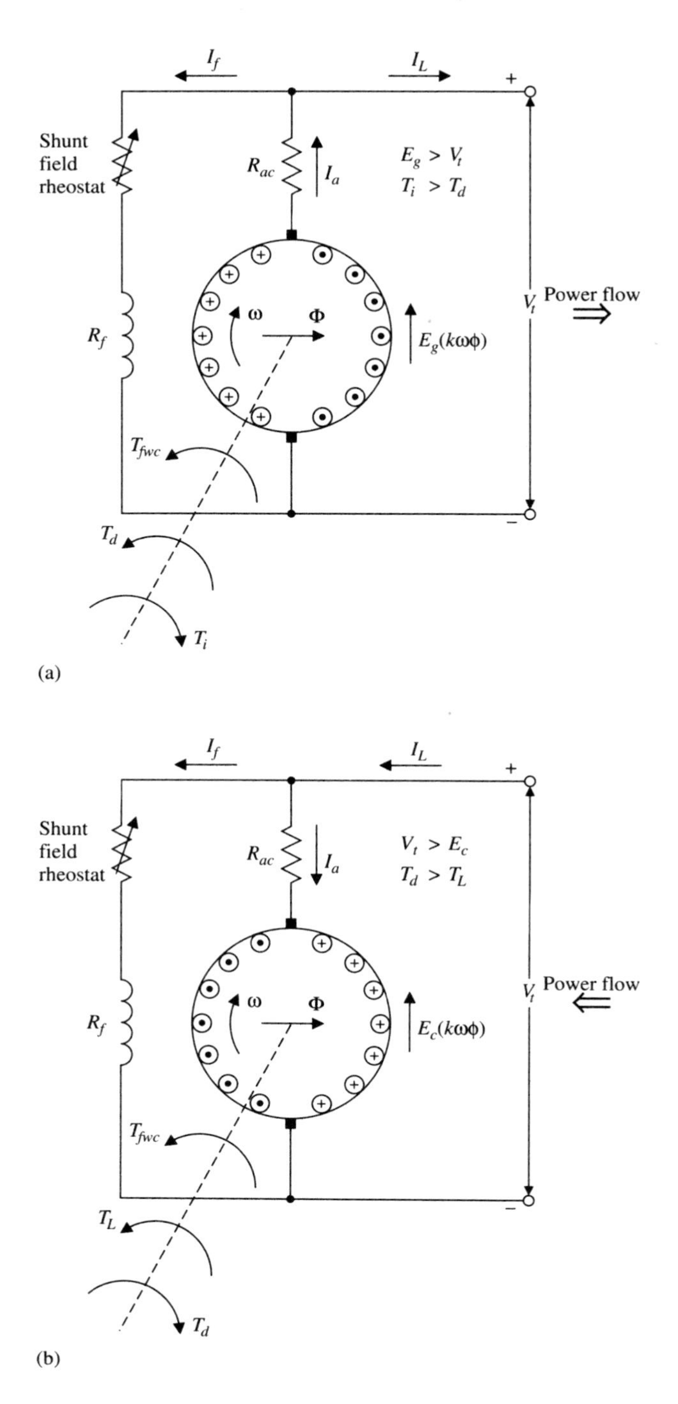

Figure 6-1 Shunt machine. (a) Operating as a generator; (b) Operating as a motor.

$$E_g = k_g \phi \omega \text{ V} \tag{5-2)(SI}$$

where $K_g = (ZP/60\alpha) \times 10^{-8}$ and $k_g = ZP/2\pi\alpha$.

In the case of the motor, although the counter electromotive force E_c is identical to the generated electromotive force E_g in the generator, the symbol E_c is traditionally used. Therefore

$$E_c = K_g \Phi S \text{ V} \tag{6-1)(E}$$

or

$$E_c = k_g \phi \omega \text{ V} \tag{6-1)(SI}$$

The developed torque is

$$T_d = K_g \Phi I_a \text{ lb} \cdot \text{ft} \tag{4-17)(E}$$

or

$$T_d = k_g \phi I_a \text{ N} \cdot \text{m} \tag{4-17)(SI}$$

The results obtained for T_d will be slightly optimistic, since it is assumed that the field pole shoe arcs cover the entire circumferential surface of the armature. In actual practice the field pole faces only cover approximately 70–75% of the armature surface.

From Fig. 6-1(b) the generated electromotive force E_g is greater than the terminal voltage V_t, and the input torque T_i delivered by the prime mover is greater than the countertorque T_d (see Section 5-4). In the case of the motor [Fig. 6-1(b)], the applied terminal voltage V_t is greater than the counter electromotive force E_c, and the developed torque T_d is greater than the opposing load torque T_L. In both the generator and the motor the direction of rotation is the same, and the torque attributable to friction, windage, and magnetic losses T_{fwc} acts to oppose the rotation of the armature. To maintain a constant speed, the developed torque must counterbalance the combined opposing torques, that is,

$$T_d = T_L + T_{fwc} \tag{6-2}$$

and T_d must be greater than $T_L + T_{fwc}$ to achieve acceleration.

From Eq. (4-14),

$$P_d = E_c I_a = T_d \omega \text{ W} \tag{6-3)(SI}$$

or

$$P_d = \frac{2\pi S T_d}{33{,}000} = \frac{E_c I_a}{746} \text{ hp} \tag{6-3)(E}$$

Then

$$\begin{aligned} T_d &= \frac{33{,}000 P_d}{2\pi S} = \frac{33{,}000 E_c I_a}{2\pi S(746)} \\ &= \frac{7.04(K_g \Phi S) I_a}{S} \\ &= 7.04 K_g \Phi I_a \text{ lb} \cdot \text{ft} \end{aligned} \tag{6-4)(E}$$

or

$$T_d = \frac{E_c I_a}{\omega} = \frac{k_g \phi \omega I_a}{\omega} = k_g \phi I_a \ \text{N} \cdot \text{m} \tag{6–4)(SI}$$

In turn,

$$\omega = E_c I_a T_d = \frac{E_c I_a}{k_g \phi I_a} = \frac{E_c}{k_g \phi} \ \text{rad/s} \tag{6–5)(SI}$$

or

$$S = \frac{7.04 E_c I_a}{T_d} = \frac{7.04 E_c I_a}{7.04(K_g \Phi S) I_a}$$

$$= \frac{E_c}{K_g \Phi} \ \text{r/min} \tag{6–5)(E}$$

It should be noted that with generators we are interested mainly in efficiency and voltage regulation. In the case of motors our prime interests are efficiency, torque, and speed regulation. The percentage speed regulation is

$$\% \text{ speed regulation} = \frac{S_{nl} - S_{fl}}{S_{fl}} \times 100\% \tag{6–6)(E}$$

where S_{nl} and S_{fl} are the no-load and full-load rotational speeds (r/min), or

$$\% \text{ speed regulation} = \frac{\omega_{nl} - \omega_{fl}}{\omega_{fl}} \times 100\% \tag{6–6)(SI}$$

where ω_{nl} and ω_{fl} are the low-load and full-load angular velocities (rad/s).

These equations, used in conjunction with Kirchhoff's laws and the magnetization curve, provide all the tools necessary to analyze the steady-state performance of dc motors.

6-3 SEPARATELY EXCITED AND DC SHUNT MOTORS

The equivalent circuits of separately excited and shunt motors are shown in Fig. 6-2. The separately excited motor, just as with the separately excited generator, derives its field excitation from a separate constant-voltage source. The shunt motor field on the other hand is excited from the machines terminals. Provided that the applied terminal voltage is constant, there is no significant difference between the performance of the two types of motors.

Applying Kirchhoff's voltage law to both motors yields

$$V_t = E_c + I_a R_{ac} \tag{6–7}$$

For the shunt motor,

$$I_L = I_a + I_f \tag{6–8}$$

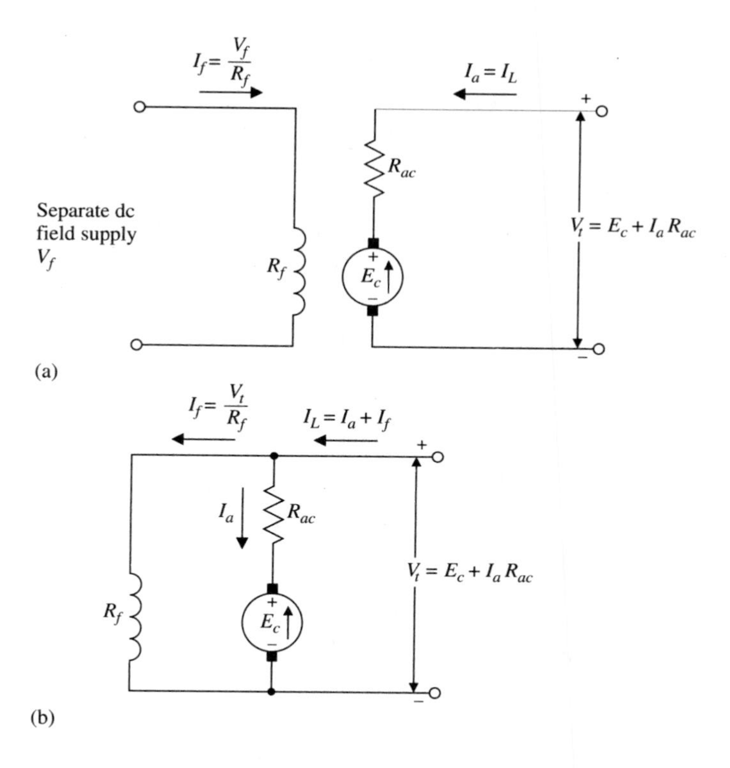

Figure 6-2 Equivalent circuits. (a) Separately excited motor. (b) Self-excited shunt motor.

and

$$I_f = \frac{V_t}{R_f} \tag{6–9}$$

For the separately excited motor,

$$I_f = \frac{V_f}{R_f} \tag{6–10}$$

and

$$I_L = I_a \tag{6–11}$$

In response to an increase in load demand, the load torque T_L will be greater than the developed torque T_d. As a result the armature will start to slow down, and thus the counter electromotive force $E_c = k_g\phi\omega$ will decrease slightly. Since $I_a = (V_t - E_c)/R_{ac}$, the armature current I_a will increase. This increase causes the developed torque $T_d = k_g\phi I_a$ to increase also, and the equilibrium will be restored with $T_d = T_L + T_{fwc}$ at a new lower angular velocity ω.

The relationship between the developed torque T_d and the angular velocity (speed) can be derived as follows:

$$\begin{aligned} V_t &= E_c + I_a + I_a R_{ac} \\ &= k_g \phi \omega + I_a R_{ac} \end{aligned} \tag{6–12}$$

since $T_d = k_g \phi I_a$, then

$$I_a = \frac{T_d}{k_g \phi} \tag{6–13}$$

From Eqs. (6-12) and (6-13),

$$V_t = k_g \phi \omega + \frac{T_d}{k_g \phi} R_{ac} \tag{6–14}$$

Then

$$k_g \phi = V_t - \frac{T_d}{k_g \phi} R_{ac}$$

and therefore

$$\omega = \frac{V_t}{k_g \phi} - \frac{T_d R_{ac}}{(k_g \phi)^2} \tag{6–15}$$

Equation (6-15) represents a straight line with a negative slope and is shown graphically in Fig. 6-3. Constant values are assumed, that is, the terminal voltage and the field pole flux are constant. The effects of armature reaction must be considered. As the load on the motor increases, the armature current increases and the field-weakening effects of armature reaction increase. The result is a net decrease in the field pole flux. From Eq. (6-15) it can be seen that a reduction of the field pole flux will cause an increase in angular velocity (speed) at any given loading, as compared to the angular velocity without armature reaction being considered. It should also be noted that if the motor is fitted with a compensating winding, the field weakening caused by armature reaction will be minimized.

Assuming the field pole flux is constant, then the developed torque T_d is proportional to the armature current I_a. In turn the characteristic curves can be drawn relating angular velocity and torque to armature current (Fig. 6-4).

The developed torque T_d is a straight line through the origin. However, because armature reaction weakens the main field pole flux, there is a slight reduction in the rate of increase of the torque, as can be seen by the departure of curve *Oc* from the straight line. The developed or gross torque must be greater than the opposing torque required to overcome the friction, windage, and magnetic losses. The output or useful torque is the torque applied to the load. The armature current *Oa* represents the current under no-load conditions, which is necessary to overcome T_{fwc}, that is, line *ab*. The speed-torque curve can be obtained by plotting values of angular velocity (speed) against torque at the same armature current.

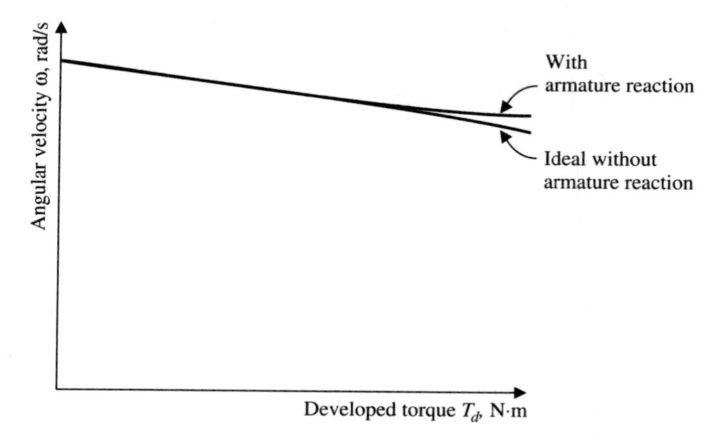

Figure 6-3 Speed verses torque characteristic of separately excited or shunt motor.

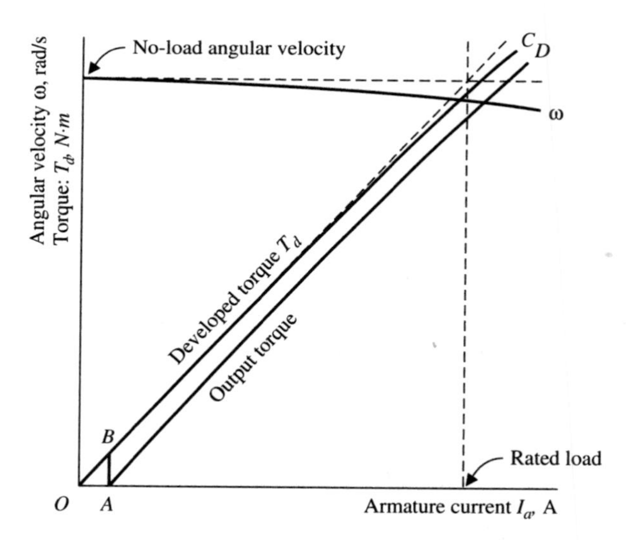

Figure 6-4 Characteristic curves of shunt motor.

➤ EXAMPLE 6-1

A 600-V motor has an armature circuit resistance of 0.14 Ω. At no load, when running at 800 r/min (83.78 rad/s), the armature current is 12 A. The full-load armature current is 225 A. Calculate the full-load speed when the field pole flux is: **(a)** constant; **(b)** reduced to 95% of its original value by armature reaction; **(c)** reduced to 80% of its original value by adjusting the field current.

SOLUTION

At no load,

$$E_{c\,nl} = V_t - I_a R_{ac} = 600 - (12 \times 0.14) = 598.32 \text{ V}$$

At full load,

$$E_{c\,fl} = 600 - (225 \times 0.14) = 568.50 \text{ V}$$

(a) $E_c = K_g \Phi S$. Therefore

$$S = \frac{E_c}{K_g \Phi}$$

Since Φ is constant, S is proportional to E_c. Therefore

$$\frac{S_{fl}}{S_{nl}} = \frac{E_{c\,fl}}{E_{c\,nl}}$$

Then

$$S_{fl} = 800 \times \frac{568.50}{598.32}$$
$$= 760.13 \text{ r/min} = 79.60 \text{ rad/s}$$

(b) When the field pole flux has been reduced to 0.95 Φ,

$$\frac{S_{fl}}{S_{nl}} = \frac{E_{c\,fl}}{E_{c\,nl}} \times \frac{\Phi_{nl}}{\Phi_{fl}}$$

$$S_{fl} = S_{nl} \times \frac{E_{c\,fl}}{E_{c\,nl}} \times \frac{\Phi_{nl}}{0.95\Phi_{nl}}$$
$$= 800 \times \frac{568.60}{598.32} \times \frac{1}{0.95}$$
$$= 800.14 \text{ r/min} = 83.79 \text{ rad/s}$$

(c) When the field pole flux has been reduced to 0.80 Φ,

$$\frac{S_{fl}}{S_{nl}} = 800 \times \frac{568.60}{598.32} \times \frac{1}{0.8}$$
$$= 950.16 \text{ r/min} = 99.50 \text{ rad/s}$$

➤ EXAMPLE 6-2

A 125-V shunt motor has an armature circuit resistance of 0.2Ω and a shunt field resistance of 45 Ω. If the line current is 50 A, calculate: **(a)** E_c; **(b)** P_d.

SOLUTION

(a)

$$I_f = \frac{V_t}{R_f} = \frac{125\ \text{V}}{45\ \Omega} = 2.78\ \text{A}$$

$$I_a = I_L - I_f = 50.00 - 2.78 = 47.22\ \text{A}$$

$$E_c = V_t - I_a R_{ac}$$

$$= 125 - 47.22 \times 0.2 = 115.55\ \text{V}$$

(b)

$$P_d = E_c I_a$$

$$= 115.55 \times 47.22 = 5456.43\ \text{W} = 5.46\ \text{kW}$$

➤ EXAMPLE 6-3

The armature of a dc motor is 14 in long and has a diameter of 12 in. There are 1,000 conductors, and the current in each conductor is 20 A. The pole shoes extend over 80% of the armature surface and the field pole flux density is 0.8 T. Calculate: **(a)** the developed torque T_d; **(b)** the developed power P_d if the armature is turning at 1,000 r/min; **(c)** the output power P_{out} if the rotational losses are 3.25 kW.

SOLUTION

(a) The force exerted by each conductor is

$$F = BIl$$

$$= 0.8\ \text{T} \times 20\ \text{A} \times 14\ \text{in} \times 0.0254\ \text{m/in} = 5.69\ \text{N}$$

The torque developed by each conductor is

$$T = Fd = 5.69\ \text{N} \times 6\ \text{in} \times 0.0254\ \text{m/in} = 0.87\ \text{N}\cdot\text{m}$$

The developed torque is

$$T_d = TZ \times \%\ \text{armature surface covered by pole shoe}$$

$$= 0.87\ \text{N}\cdot\text{m} \times 1{,}000 \times 0.8\ \text{T} = 693.68\ \text{N}\cdot\text{m}$$

(b) The developed power P_d is

$$P_d = T_d \omega$$

$$= 693.68 \times \frac{2\pi \times 1{,}000}{60}$$

$$= 72{,}641.58\ \text{W} = 72.64\ \text{kW}$$

(c) The output power is

$$P_{\text{out}} = P_d - P_{\text{rot}}$$

$$= 72.64 - 3.25 = 69.39\ \text{kW}$$

$$= 69.39\ \text{kW} \times 1.34\ \text{hp/kW} = 93.02\ \text{hp}$$

➤ EXAMPLE 6-4

A 250-V shunt motor runs on no load at 1,750 r/min (183.26 rad/s). The no-load current is 10 A. The armature circuit resistance is 0.25 Ω and the shunt field resistance is 250 Ω. Calculate: **(a)** I_f; **(b)** E_c; **(c)** K_g; **(d)** the speed when the load current is 50 A; **(e)** The speed regulation.

SOLUTION

(a)

$$I_f = \frac{V_t}{R_f} = \frac{250\ V}{250\ \Omega} = 1\ \text{A}$$

$$I_a = I_L - I_f = 10 - 1 = 9\ \text{A}$$

(b)

$$E_c = V_t - I_a R_a$$
$$= 250 - 9 \times 0.25 = 247.75\ \text{V}$$

(c)

$$E_c = k_g \phi \omega = k_g I_f \omega$$

$$k_g = \frac{E_c}{I_f \omega} = \frac{247.75\ \text{V}}{1\ \text{A} \times 183.25\ \text{rad/s}} = 1.35\ \text{V} \cdot \text{s/A} \cdot \text{rad}$$

(d) When $I_L = 50$ A, $I_a = I_L - I_f = 50 - 1 = 49$ A,

$$E_c = V_t - I_a R_a$$
$$= 250 - 49 \times 0.25 = 237.75\ \text{V} = k_g \phi \omega$$

$$\omega = \frac{E_c}{k_g I_f} = \frac{237.75\ \text{V}}{1.35\ \text{V} \cdot \text{s/A} \cdot \text{rad} \times 1\ \text{A}}$$
$$= 176.11\ \text{rad/s}$$

$$\%\ \text{speed regulation} = \frac{\omega_{nl} - \omega_{fl}}{\omega_{fl}} \times 100\%$$

$$= \frac{183.25 - 176.11}{176.11} \times 100\% = 4.05\%$$

6-4 SPEED CONTROL OF SHUNT AND SEPARATELY EXCITED DC MOTORS

From Eq. (6-5), $\omega = E_c / k_g \phi$ or $\omega = (V_t - I_a R_{ac}) / k_g \phi$, three methods of controlling the speed of a separately excited or self-excited shunt motor are indicated: (1) maintaining the voltage applied to the armature constant and then varying the field pole flux; (2) maintaining the field pole flux constant and varying the voltage applied to the armature; (3) a combination of methods (1) and (2). In order to define speed ranges it is necessary to define a reference called the *base speed.* The base speed of a dc motor is defined as the speed produced with rate armature voltage and rated field pole flux applied to the motor.

Field Control

Assuming that the voltage applied to the armature and the load connected to the motor are constant. Then considering a shunt motor, if the resistance of the shunt field circuit is increased, the shunt field current $I_f = V_t/R_f$ decreases. Immediately there is a decrease in the counter electromotive force $E_c = k_g\phi\omega$, which in turn causes the armature current to increase since

$$I_a = \frac{V_t - E_c}{R_{ac}} \tag{6–16}$$

The developed torque $T_d = k_g I_a$ will increase since a relatively small decrease in ϕ will cause a larger percentage change in the armature current. The developed torque T_d will now be greater than $T_L + T_{fwc}$, and the motor accelerates. At the same time, the counter electromotive force E_c increases and the armature current and T_d also decrease until $T_d = T_L + T_{fwc}$ at a new higher speed. The effects of field resistance changes on the output characteristics are shown in Fig. 6-5.

Field resistance changes in shunt motors are usually achieved by adjusting a shunt field rheostat in series with the shunt field. In the case of a separately excited motor, the applied voltage may be decreased, or the resistance of the field circuit can be increased. See Fig. 6-6. Reduction of the field current, and thus the field pole flux, provides speed control above base speed. However, there are several factors that limit the amount that the speed can be increased. The maximum speed is determined by (1) the effects of centrifugal force on the armature and (2) the ability of the motor to complete commutation in the ever decreasingly available time interval. In addition, the effects of armature reaction are to weaken the air-gap flux, which as a result can lead to unstable operation. The effect can be minimized by two or three series

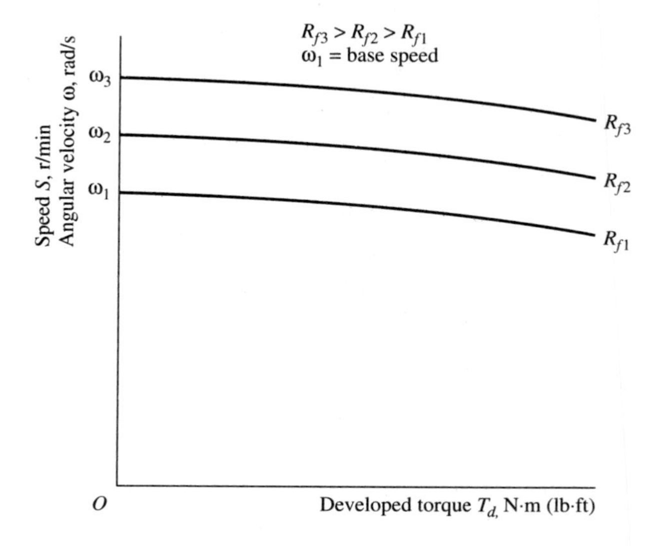

Figure 6-5 Effect of increasing shunt field resistance on rotational speed.

field turns wound on the main field poles. This winding is known as the *stabilizing winding*. Normally the ranges of speed control, that is, the ratio of the maximum safe speed to the base speed, are from 8:1 for small motors to 2:1 for large motors.

At this point the effect of an interruption or an open circuit occurring in the shunt field can be seen as potentially disastrous for two reasons. (1) Because of the high inductance of the shunt field and the rapid rate of change of the field current, the very high induced voltages in the shunt field winding can break down its insulation. (2) Because of the resulting runaway condition this will damage the armature mechanically and electrically. A field loss relay with its operating coil in series with the field, and its contacts in series with the main contactor coil, will prevent this situation by removing power from the motor.

Armature Voltage Control

Varying the voltage applied to the armature and maintaining the field pole flux constant provide speed control between zero and base speed. Achieving variable armature voltage control of a shunt motor is only possible by inserting a high-wattage variable resistance in series with the armature, which wastes energy, and as a result shunt motors rarely use this method of armature voltage control. It is relatively simple to supply the armature circuit of a separately excited motor from a variable-voltage source. In fact, this is the standard method used in dc drives, where the variable voltage V_a is supplied from a thyristor phase-controlled converter [Fig. 6-7(b)].

From Eq. (6-16) it can be seen that increasing the voltage applied to the armature will cause the armature current to increase. As I_a increases, the developed torque $T_d = k_g \phi I_a$ increases and accelerates the armature since T_d is greater than $T_L + T_{fwc}$. At the same time the counter electromotive force $E_c = k_g \phi \omega$ increases, which acts to reduce the armature current, which in turn causes T_d to decrease until it is equal to $T_L + T_{fwc}$ at a new higher rotational speed ω. The effect of varying the applied armature voltage V_a can be seen in Fig. 6-8.

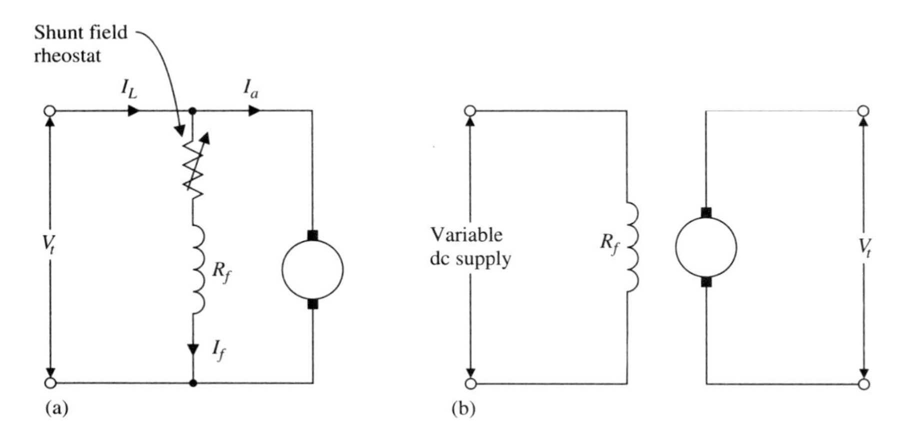

Figure 6-6 Speed control above base speed using field-weakening techniques. (a) Shunt motor. (b) Separately excited motor.

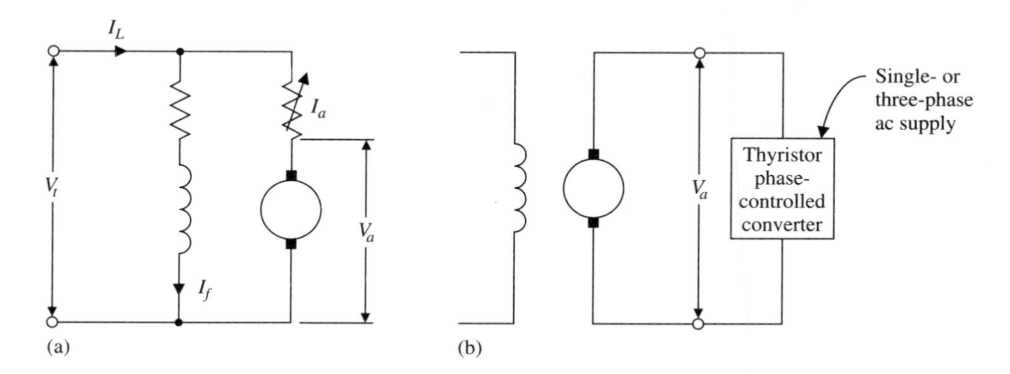

Figure 6-7 Speed control below base speed using variable armature voltage control. (a) Shunt motor using variable series resistance in armature circuit. (b) Separately excited motor supplied from thyristor phase-controlled converter.

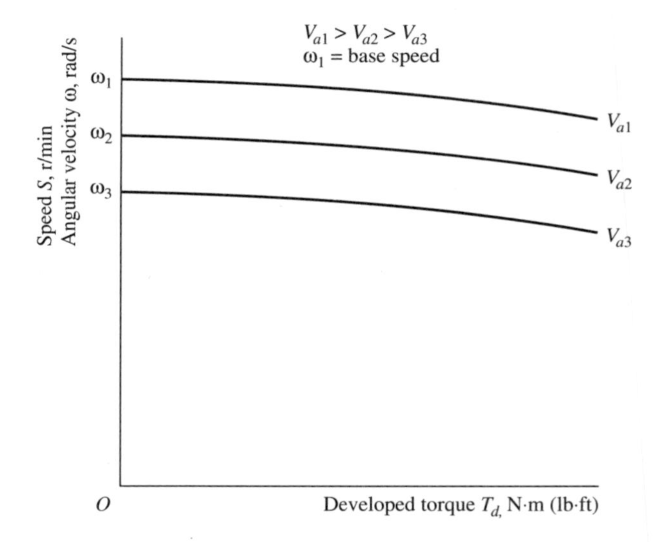

Figure 6-8 Effect of varying the armature voltage on rotational speed.

Variable Armature Voltage and Variable Field Control

Both previous methods of speed control can be combined to give speed control from zero to above base speed, that is, the motor is operated with a variable armature voltage and rated air-gap flux up to base speed, and then with reduced air-gap flux and at rated armature voltage.

These speed control methods imply that the rotational speed is solely determined by V_a and ϕ. However, this is not the complete story. The limiting factor in either case is the maximum permissible armature current which, in turn, determines the developed power P_d and the developed torque T_d. Assuming that the maximum developed power P_d is

$$P_d = E_c I_a = T_d \omega = (V_a - I_a R_{ac}) I_a$$

then

$$I_a^2 - \frac{V_a I_a}{R_{ac}} + \frac{P_d}{R_{ac}} = 0$$

Solving for I_a,

$$I_a = \frac{V_a / R_{ac} \pm \sqrt{(V_a / R_{ac})^2 - 4P_d / R_{ac}}}{2}$$

Then

$$I_a = \frac{V_a}{2R_{ac}} \left(1 - \sqrt{1 - \frac{4R_{ac} P_d}{V_a^2}} \right) \tag{6–17}$$

The plus sign condition is ignored since it would lead to unacceptably high values of I_a.
Analyzing Eq. (6-17) leads to the following conclusions:

1. With V_a constant at rated value, and assuming I_a remains unchanged at rated value, then the developed power P_d remains constant as the field pole flux ϕ is varied. When operating under variable field control, the motor is operating in the *constant power mode*. Then since $P_d = T_d \omega$,

$$\omega_1 T_{d1} = \omega_2 T_{d2}$$

or

$$\frac{\omega_1}{\omega_2} = \frac{T_{d2}}{T_{d1}} \tag{6–18}$$

2. With a constant air-gap flux and a variable armature voltage V_a, and assuming that I_a is constant and does not exceed its rated value,

$$T_d = k_g \phi I_a$$

and since $k_g \phi$ is constant, the developed torque T_d is also constant. When operating with variable armature voltage and rated field pole flux, the motor is operating in the *constant torque mode*. These modes of operation are illustrated in Fig. 6-9.

6-5 PERMANENT-MAGNET MOTORS

Permanent-magnet motors are effectively separately excited dc motors with a constant field excitation. While they are manufactured in all ranges up to 100 hp (74.6 kW), they are most commonly used in low-power battery-supplied tools and appliances. There are a number of advantages in their use:

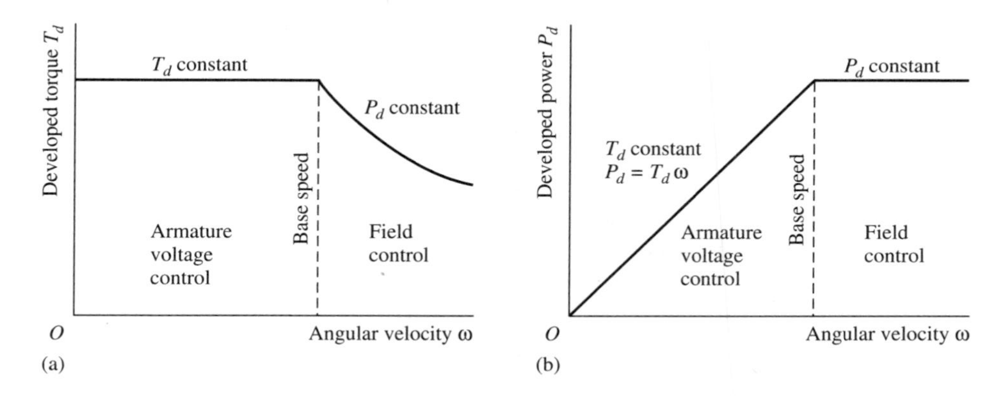

Figure 6-9 Developed power and torque relationships for separately excited motor. (a) Torque versus rotational speed; (b) Developed power versus rotational speed.

1. They have a high efficiency because there is no field loss, and they are cool running.
2. They are lighter and physically smaller than a comparable shunt or separately excited motor.
3. They are more reliable since there is no field to burn out.
4. They have good speed regulation.
5. They have a high dynamic braking capability, which is easily achieved by shunting the armature leads.
6. They are easily controlled with thyristor controls.
7. They are less costly to manufacture, because the air gap is not critical. There are no field windings and the field poles can be produced with large arc pole shoes, thus reducing the number of poles required in a multipole machine.

There are also several disadvantages to their use in higher power applications:

1. The permanent-magnet material may be demagnetized in one or more of the following ways:
   a. Self-demagnetization or aging.
   b. By temperature change. Alnico magnets are relatively stable and there is very little change in the magnetic field strength with temperature increases. Ceramic magnets have a higher coercivity, but show a reduction in magnetic field strength with increasing temperature, about six times that of Alnico.
   c. Demagnetization of the permanent-magnet material caused by armature reaction is not a problem with low-power motors, since it requires armature currents in excess of 10 times full-load rated current, although in some high-power applications this has been a problem. It can be solved in larger motors by using interpoles and current limit control to ensure that excessive load currents are not permitted.

2. Commutation can be a problem since small permanent-magnet motors do not have interpoles or compensating windings, and as a result arcing will occur between the brushes and the commutator. Careful design can minimize commutation problems

Alnico magnets were first developed in the 1930s, and as a result of continuing research into the orientation of the grain structure, their magnetic properties have been much improved. They have the characteristic that, once magnetized, they retain their magnetic properties indefinitely and are little affected by temperature. To combat the demagnetizing effects of armature reaction, the Alnico magnet must have a high enough coercivity to prevent permanent reduction of the magnetic field under peak load current operation. This is achieved by making the magnet long in the direction of magnetization. Alnico permanent-magnet motors produce the highest torque output per watt because they have the highest flux density.

Ceramic or ferrite permanent magnets are now being used more frequently than the Alnico family, because of their high coercivity and high energy product. Another important factor is the lower cost of the basic ingredients, such as iron oxide and barium or strontium. The field poles are compression molded into the desired shapes with the particles oriented to achieve the greatest net magnetic field strength. After molding they are sintered at temperatures in excess of 1,093°C (2,000°F) and magnetized in the motor frame. Because of their high coercivity, the magnet is shorter in the direction of magnetization. However, since the magnets have a lower flux density, the pole face area must be increased to obtain the desired flux. It is interesting to note that ceramic poles are formed with integral shoes, while Alnico magnets usually have a soft or laminated iron pole face. However, in situations where the maximum output for size or weight, or the maximum torque per watt, is the prime requirements, the Alnico permanent-magnet motor is still the preferred choice.

Rare-earth magnetic materials currently being produced usually are alloys of samarium and cobalt, and are characterized by very high energy products. However, to reduce costs, the samarium is being replaced by mischmetal-cobalt and cerium-cobalt, but with the penalty of reduced energy products. The torque and output of a rare-earth cobalt permanent-magnet motor is approximately twice that of an equivalent ceramic permanent-magnet motor with the same power input, as well as being less sensitive to temperature increases. A major disadvantage is that in order to achieve the superior performance, the permanent-magnet motor must be designed specifically for use with these materials. This, combined with the high cost of materials, is restricting the use of these magnets to the low end of the power output range.

The major advantages of using permanent-magnet motors are:

1. They possess very high starting torques.
2. They are easily controlled.
3. They have linear speed-torque characteristics over the entire torque range which, when operated at low speed, permits their substitution for gear motors, thus eliminating the backlash problems associated with gearing.

6-6 SERIES MOTOR

The steady-state equivalent circuit of the dc series motor is shown in Fig. 6-10, where R_{ac} represents the armature circuit resistance and R_{se} the series field resistance.

By Kirchhoff's voltage law,

$$V_t = E_c + I_a(R_{ac} + R_{se}) \quad (6\text{–}19)$$

$$I_a = I_{se} = I_L \quad (6\text{–}20)$$

The speed-torque characteristic is dependent on the fact that the series field flux ϕ_{se} is dependent on the load or armature current, that is, the flux will be a minimum at no load and a maximum at full load. The developed torque T_d is

$$T_d = k_g \phi_{se} I_a \text{ N} \cdot \text{m} \quad (6\text{–}21)$$

Assuming operation on the linear portion of the magnetization curve, ϕ_{se} is directly proportional to I_a. Therefore

$$k_g \phi_{se} = k_f I_a \quad (6\text{–}22)$$

where k_f is a proportionality constant. The developed torque can now be expressed as

$$T_d = k_g \phi_{se} I_a = k_f I_a^2 \text{ N} \cdot \text{m} \quad (6\text{–}23)$$

that is,

$$I_a = \sqrt{\frac{T_d}{k_f}} \quad (6\text{–}24)$$

Therefore the armature current increases as the square root of the developed torque. Recall that for the shunt and separately excited motors the developed torque is directly proportional to the armature current.

From Eqs. (6-22) and (6-23),

$$k_g \phi_{se} = \sqrt{k_f T_d} \quad (6\text{–}25)$$

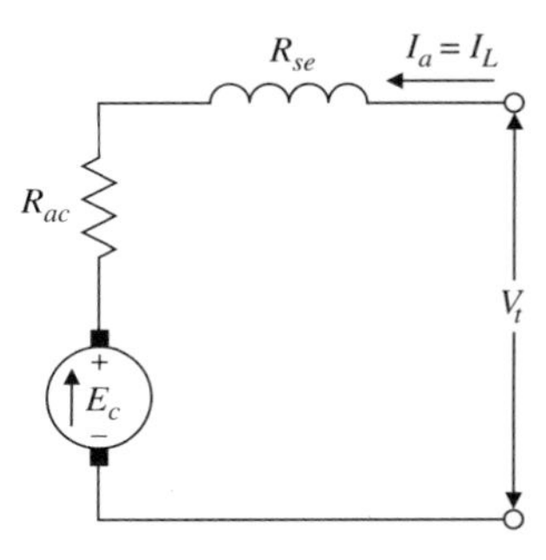

Figure 6-10 Equivalent circuit of series motor.

From Eq. (6-5)(SI),

$$\omega = \frac{E_c}{k_g\phi} = \frac{V_t - I_a(R_{ac} + R_{se})}{k_g\phi} \text{ rad/s} \tag{6–26}$$

Then substituting into Eqs. (6-25) and (6-26) yields

$$\omega = \frac{V_t}{\sqrt{k_f T_d}} - \frac{R_{ac} + R_{se}}{k_f} \text{ rad/s} \tag{6–27}$$

which is the speed-torque relationship of an unsaturated series motor and is illustrated in Fig. 6-11.

Alternatively, in terms of armature current,

$$\omega = \frac{E_c}{k_g\phi} = \frac{V_t - I_a(R_{ac} + R_{se})}{k_g\phi} \text{ rad/s} \tag{6–28}$$

Then substituting for $k_g\phi$ from Eq. (6-22) yields

$$\omega = \frac{V_t}{k_f I_a} - \frac{R_{ac} + R_{se}}{k_f} \text{ rad/s} \tag{6–29}$$

An examination of Eq. (6-29) shows that as I_a approaches zero, the rotational speed ω approaches infinity, that is, the no-load speed is extremely high and is only limited by the friction, windage, and magnetic losses. An obvious precaution is that a series motor must never be permitted to run unloaded, or the motor may be severely damaged. As the load on the motor increases, the rotational speed ω decreases, rapidly at first and then more slowly as the load increases. The developed torque $T_d = k_f I_a^2$ increases parabolically until the knee of the magnetization curve is reached; then it becomes linear. These curves are shown in Fig. 6-12.

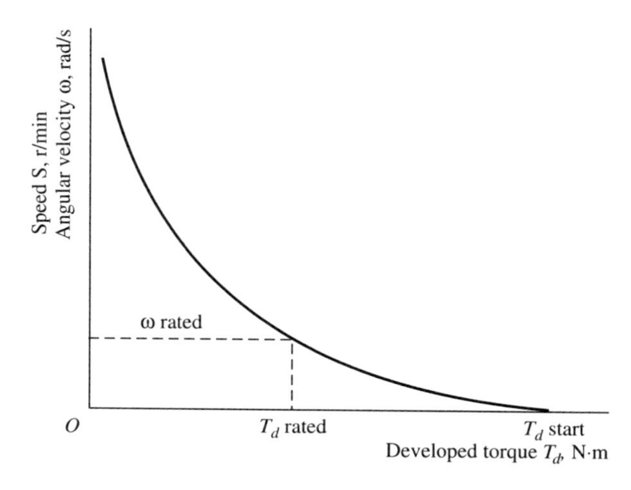

Figure 6-11 Speed-torque characteristic of series motor.

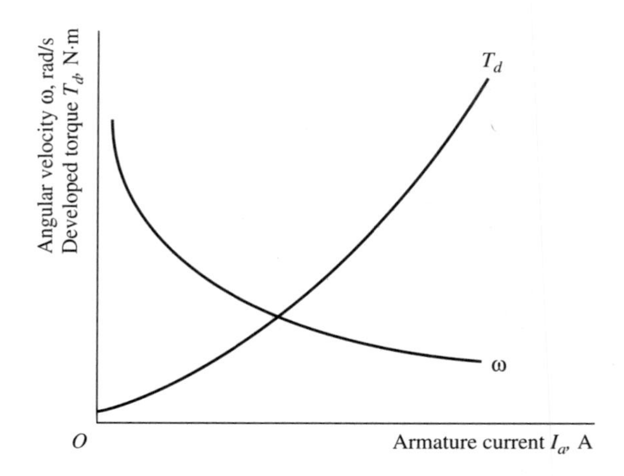

Figure 6-12 Series motor speed and torque characteristics.

Series Motor Speed Control

The speed of a series motor may be changed by:

1. Adding resistance in series with the armature. Increasing the armature circuit resistance reduces the rotational speed.
2. Using a diverter resistance in parallel with the armature. Then as the ohmic value of the diverter resistance R_D is decreased, the motor speed decreases.
3. A combination of 1 and 2.

These methods are illustrated in Fig. 6-13. The major disadvantage of these methods is that they waste a significant amount of energy and must be capable of carrying high currents. The modern approach to controlling the speed of a series motor is to use a variable dc voltage applied to the motor terminals. From the first term of Eq. (6-28) it can be seen that if V_t is increased, the rotational speed ω increases, and vice versa. Speed control of a series motor can easily be obtained by supplying the motor from either a single- or a three-phase thyristor phase-controlled converter, as shown in Fig. 6-14.

Both converters shown in Fig. 6-14 are full converters, that is, they are capable of regeneration. If regeneration is not required, half the SCRs may be replaced with rectifier diodes, which simplifies the control requirements and reduces the initial cost, but will only permit one-quadrant operation. The single-phase configuration is commonly used for motors up to 20 hp (14.92 kW), and up to 100 hp (74.6 kW) for series traction motors. The three-phase configuration is suitable for motors in excess of 20 hp (14.92 kW).

Equation (6-24) shows that the output torque of the series motor is much greater than that of the separately excited and the shunt motors, and as will be seen, it is

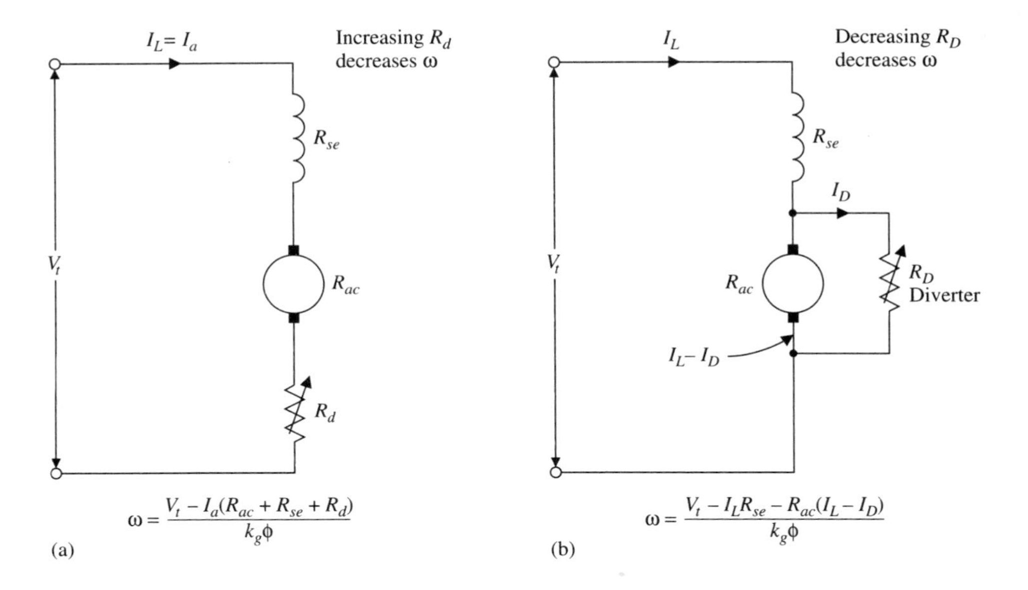

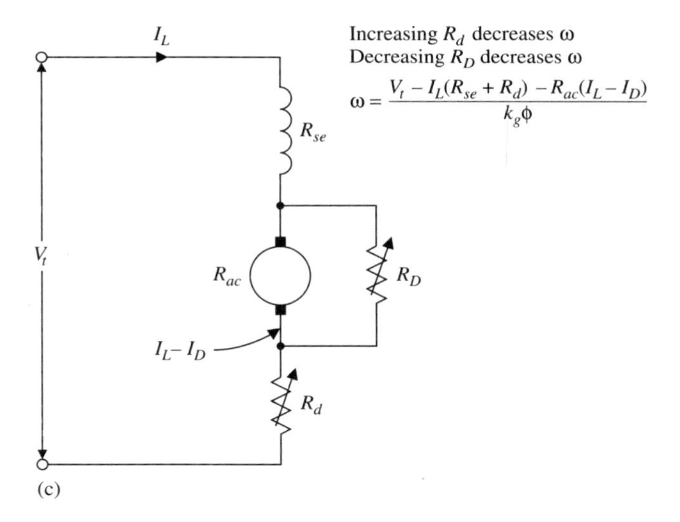

Figure 6-13 Series motor speed control using variable resistors.

much greater than that of the compound motor, which is essentially a shunt motor with a small series field. The high output torque of the series motor immediately suggests that it should be applied where the load demands a high initial torque, such as cranes, hoists, electric vehicles, electric traction, and internal combustion engine cranking motors.

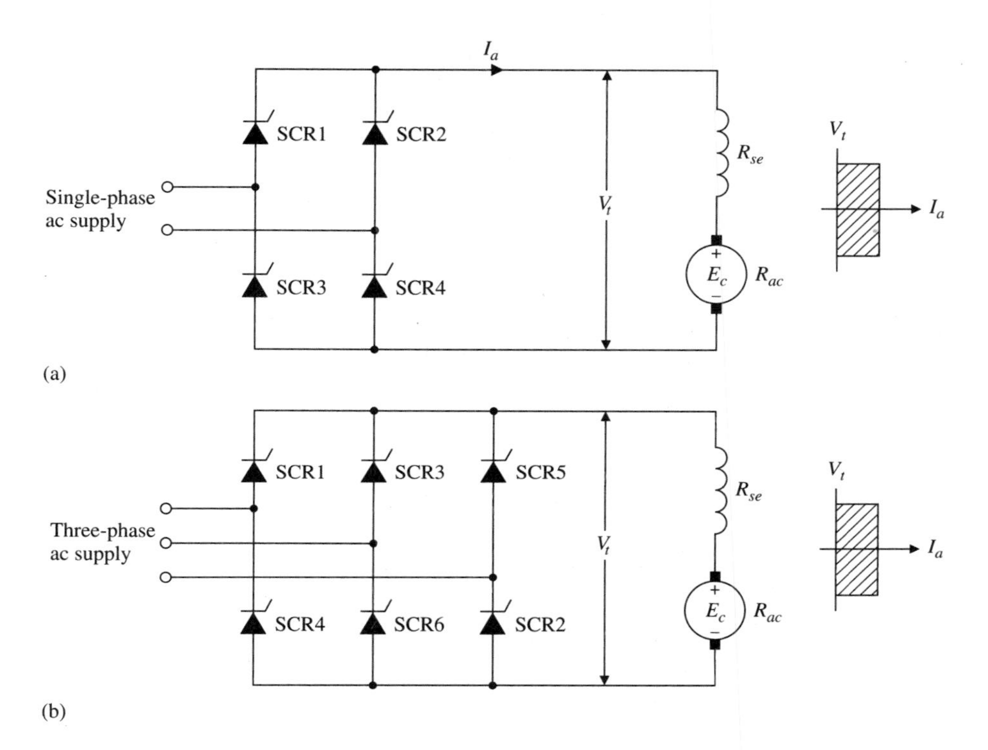

Figure 6-14 Series motor speed control using thyristor phase-controlled converters. (a) Single-phase full converter. (b) Three-phase full converter.

➤ EXAMPLE 6-5

A series motor has combined series field and armature resistance of 0.85 Ω, runs at 1,000 r/min (104.72 rad/s), and draws 20 A from a 250-V source. Calculate: **(a)** the rotational speed if a 3.75 Ω resistor is connected in series with the motor and it draws 20 A from a 250-V source; **(b)** the developed power at both speeds; **(c)** the developed torque at both speeds.

SOLUTION

(a) From Eq. (6-19), the counter electromotive force E_c at 1,000 r/min is

$$E_{c1} = V_t - I_a(R_{ac} + R_{se}) = 250 - 20 \times 0.85$$
$$= 233 \text{ V}$$

Since the terminal voltage and the armature current are unchanged when the additional resistance is in series with the armature, then

$$E_{c2} = 250 - 20(0.85 + 3.75)$$
$$= 158 \text{ V}$$

Now $S_1 = \frac{E_{c1}}{k_g \Phi}$ and $S_2 = \frac{E_{c2}}{k_g \Phi}$ since I_a is the same, $K_g \Phi$ is the same. Then

$$S_2 = \frac{S_1 \times E_{c2}}{E_{c1}} = \frac{1{,}000 \text{ r/min} \times 158 \text{ V}}{233 \text{ V}}$$
$$= 678.11 \text{ r/min}$$

or in SI units,

$$\omega_2 = \frac{\omega_1 \times E_{c2}}{E_{c1}} = \frac{104.72 \text{ rad/s} \times 158 \text{ V}}{233 \text{ V}}$$
$$= 71.01 \text{ rad/s}$$

(b) At 104.72 rad/s

$$P_{d1} = E_{c1} I_a = 233 \text{ V} \times 20 \text{ A} = 4{,}660 \text{ W}$$

At 71.01 rad/s,

$$P_{d2} = E_{c2} I_a = 158 \text{ V} \times 20 \text{ A} = 3{,}160 \text{ W}$$

(c) At 104.72 rad/s,

$$T_d = \frac{P_d}{\omega} = \frac{4{,}660 \text{ W}}{104.72 \text{ rad/s}} = 44.50 \text{ N} \cdot \text{m}$$

At 71.01 rad/s,

$$T_d = \frac{3{,}160 \text{ W}}{71.01 \text{ rad/s}} = 44.50 \text{ N} \cdot \text{m}$$

These answers are consistent since the developed torque T_d is proportional to I_a^2 and $I_a = 20$ A in both cases.

6-7 COMPOUND MOTOR

The equivalent steady-state circuits of the compound motor are shown in Fig. 6-15. Just as with the compound generator, there are two methods of connecting the shunt field: the short shunt [Fig. 6-15(a)] and the long shunt [Fig. 6-15(b)]. Also as was noted for the compound generator, the difference in performance between the two connections is only marginal.

The major factor that affects the performance of the compound motor is the relative direction of the series field flux ϕ_{se} with respect to the direction of the shunt field flux ϕ_{sh}. It should also be remembered that the major portion of the air-gap flux is produced by a shunt field and extends into the saturation region of the magnetization curve. The contribution from the series field as a result will not be proportional to the armature current. The series field flux ϕ_{se} may be additive to the shunt field flux, that is, $\phi_{sh} + \phi_{se}$, in which case the motor is *cumulatively compounded,* or it may be subtractive, that is, $\phi_{sh} - \phi_{se}$ and the motor is *differentially compounded.*

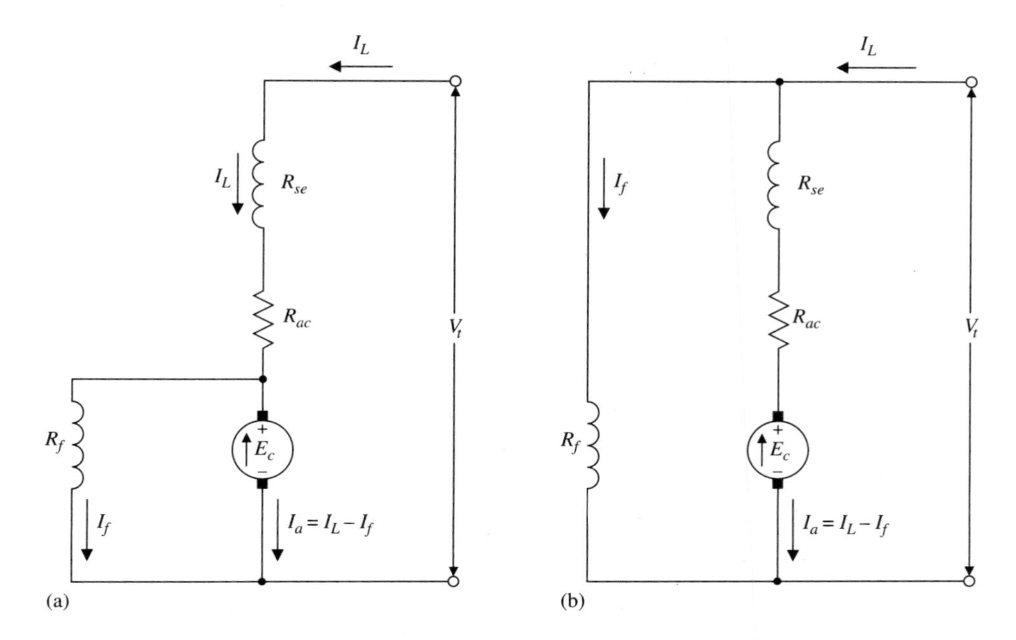

Figure 6-15 Equivalent circuits of compound motor. (a) Short-shunt connection; (b) Long-shunt connection.

The relationships according to Kirchhoff's law for the long-shunt connection are

$$E_c = V_t - I_a(R_{ac} + R_{se}) \text{ V} \tag{6–30}$$

The input power is

$$P_{\text{in}} = V_t I_L \text{ W} \tag{6–31}$$

The shunt field current is

$$I_f = \frac{V_t}{R_f} \text{ A} \tag{6–32}$$

The armature current is

$$I_a = I_L - I_f \tag{6–33}$$

The rotational speed is

$$\omega = \frac{V_t - I_a(R_{ac} + R_{se})}{k_g(\phi_{sh} \pm \phi_{se})} \text{ rad/s} \tag{6–34}$$

Finally the developed torque is

$$T_d = k_g I_a(\phi_{sh} \pm \phi_{se}) \text{ N}\cdot\text{m} \tag{6–35}$$

Here $\pm$ describes the effect of the series field, + when cumulatively compounded, and $-$ when differentially compounded, although a differentially compounded motor will be encountered very rarely.

As can be seen from Eq. (6-34), when considering a cumulatively compounded motor, the numerator will decrease more rapidly than the denominator as the armature current increases, with the result that the cumulatively compounded motor will exhibit a greater drop in rotational speed between no load and full load than would be the case for a comparably rated shunt motor.

For a differentially compounded motor, the numerator of Eq. (6-34) decreases, but not as rapidly as the denominator. As a result, as the armature current increases, the rotational speed also increases. This characteristic makes differentially compounded motors unstable.

By controlling the relative effect of the series field, a cumulatively compounded motor may assume any characteristics between that of a shunt and a series motor. Compound motors are designed by the manufacturer to be excessively compounded. The required degree of compounding is obtained by installing a diverter resistance in parallel with the series field to control the series field magnetomotive force.

The degree of compounding, that is, the effect of the series field, is determined by the nature of the connected load. For example, a drop of 15–20% is desirable in some machining applications to relieve the drive motor under overload conditions, or in applications where there is a large starting torque. In Fig. 6-16 the characteristics of the compound motor are compared to those of the shunt and the series motors.

The methods of controlling the speed of compound motors are almost identical to those used to control shunt motors. Speeds above base speed are achieved by increasing the resistance of the shunt field circuit. Speeds below base speed can be achieved by increasing the resistance in the armature circuit. However, this is not energy efficient. Alternatively speed control below base speed can be achieved by supplying the armature circuit from a variable-voltage source and separately exciting the shunt field circuit.

6-8 DC MOTOR STARTING

With the exception of thyristor phase-controlled dc motors, the dc motor is normally supplied from a constant-voltage source. It is necessary then to provide a number of functions in the control equipment, such as the following:

1. Isolation from the power source.
2. Protection of the motor against short circuits and long-duration overloads.
3. Protection of the motor and source against damage caused by high inrush currents during starting.
4. Providing control of the motor during its operating cycle.

At this point we will concentrate on item 3; the remaining items are dealt with in more depth in Chapter 8.

The resistance of the armature circuit of most dc motors is less than 1 Ω, and the armature current is defined by

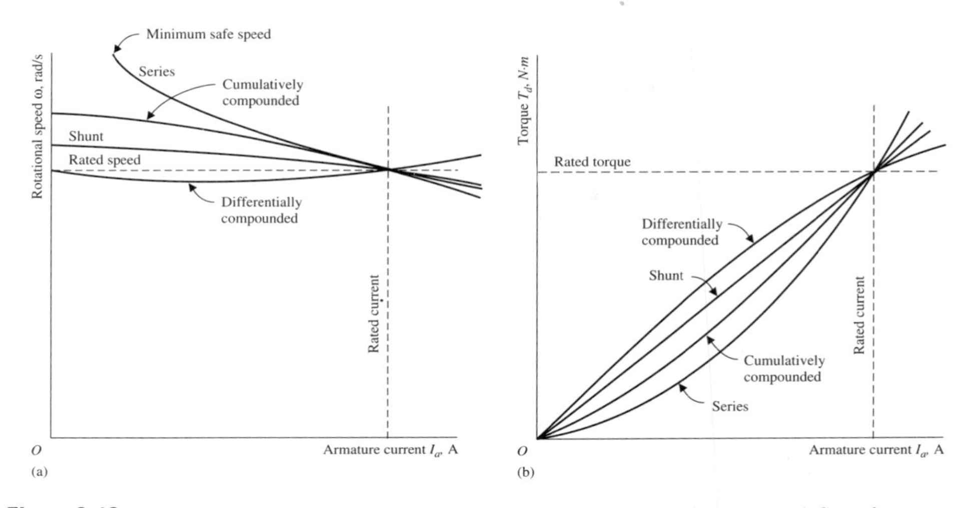

Figure 6-16 Comparison of equally rated shunt, series, and compound motors. (a) Speed versus armature current. (b) Torque versus armature current.

$$I_a = \frac{V_t - E_c}{R_{ac}}$$

Since $E_c = k_g \phi \omega$, obviously at the instant of starting, $E_c = 0$ V. Then

$$I_a = \frac{V_t - 0}{R_{ac}}$$

➤ EXAMPLE 6-6

A 75-hp (55.96-kW) 250-V 280-A shunt motor has an armature resistance of 0.072 Ω. What is the inrush current at the instant of starting?

SOLUTION

$$I_a = \frac{V_t}{R_{ac}} = \frac{250 \text{ V}}{0.072\ \Omega} = 3{,}500 \text{ A}$$

In this example the starting current is 14 times the rated full-load current of the motor. The motor will be severely damaged, even though the current is only present for a short period. The obvious solution is to introduce a resistance into the armature circuit during the starting cycle. This resistance is known as the *accelerating resistance,* and its ohmic value is reduced as the motor armature accelerates, until at rated speed the accelerating resistance is totally removed from the armature circuit. The accelerating resistance is removed in a series of steps, so that the inrush current is limited to an acceptable value, usually between 125 and 175% of rated full-load current to permit acceleration of the connected load without creating disturbances to the voltage source or producing sudden changes in acceleration, which would produce uneven motion in applications such as cranes, hoists, and electric traction equipment.

During the starting period it must be recognized that the motor is operating under transient conditions, and that the inductance of both the field and the armature circuits has a significant effect on current changes. The effect of the inductance is to oppose current changes in their respective circuits, that is, both field and armature currents build up exponentially to their maximum values over a period of time. The time interval is defined in terms of the time constant of the respective circuit. Recall that the time constant τ is defined as the time required for the current to build up to 63% of its final steady-state value, or the time required for the current to decay to 37% of its initial value. The time constant $\tau = L/R$, and the current is assumed to reach a steady-state value in five time constants. When a shunt or compound motor is started, full terminal voltage is immediately applied to the field circuit, and after five time constants, that is, $\tau_f = L_f/R_f$, the field current remains constant at $I_f = V_t/R_f$. In order to simplify our discussion, the effect of the field circuit inductance will not be considered. The armature circuit time constant $\tau_a = L_a/R_{ac}$ is significant and determines the rate of current buildup and decay. Another important factor in machine performance is the moment of inertia J, which is the mechanical equivalent of inductance in the electric circuit, that is, the moment of inertia of the motor and the connected load act to oppose changes in angular velocity. The developed torque $T_d = J\alpha$, where α is the angular acceleration in rad/s^2. Since $\omega = \alpha t$, $E_c = k_g\phi\omega$ will also be affected. The mechanical time constant τ_m can be shown to be equal to $JR_{ac}/(k_g\phi)^2$, and the angular velocity ω of the motor also increases exponentially. While there are two time constants affecting the acceleration of the motor, the armature time constant τ_a and the mechanical time constant τ_m, the latter has the longest duration and has the most effect on the motor's acceleration.

Figure 6-17 shows the variations of armature current and rotational speed versus time as the ohmic value of the accelerating resistance is decreased in a series of steps until the motor is running at rated speed. At time t_0 power is applied to the motor, the armature current I_a builds up exponentially to the predetermined maximum of 1.5 I_a, being limited by the series resistance combination of the whole accelerating resistance and the armature circuit resistance R_{ac}. As the armature accelerates exponentially from 0 to ω_1 rad/s, the counter electromotive force E_c increases exponentially and the armature current decreases exponentially until it reaches I_a. At time t_1 a section of the accelerating resistance is cut out, and once again the armature current increases to $1.5I_a$ and the armature accelerates up to ω_2 at time t_2, at which point the armature current has decreased to I_a once again, and the counter electromotive force has reached a new higher value. At this point the next section of the accelerating resistance is cut out, and the whole process is repeated until all the accelerating resistance has been removed and the motor is running at rated speed. The number of steps of acceleration, and hence the number of resistance steps, is dependent on the output torque of the motor and the inertia of the connected load. The technique of calculating the ohmic value of each step of the accelerating resistance is best demonstrated in the following example.

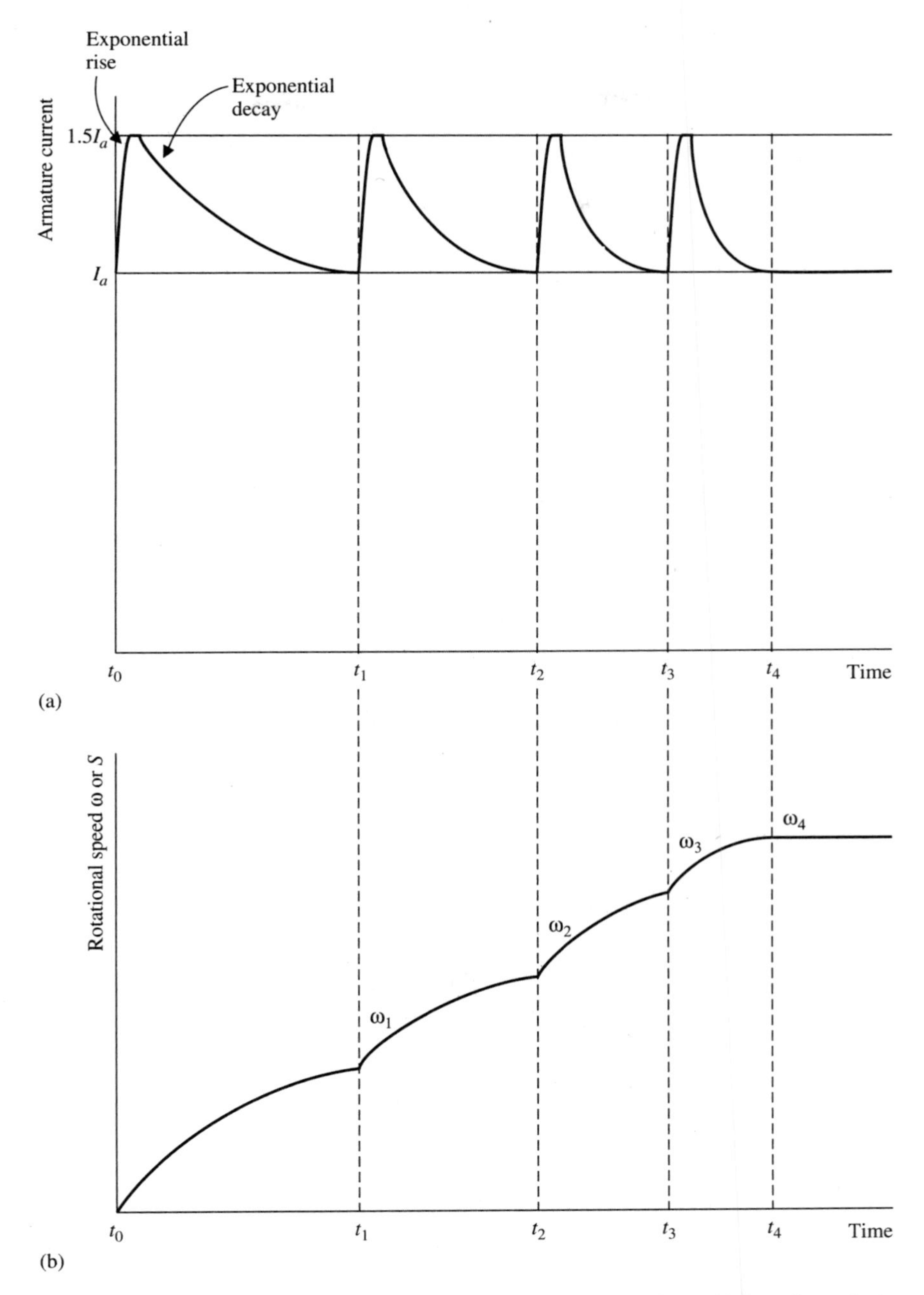

Figure 6-17 Four-step motor starter. (a) Armature current versus time. (b) Angular velocity versus time.

➤ EXAMPLE 6-7

A 200-hp (149.20-kW) 250-V 1,750-r/min (183.26-rad/s) shunt motor draws a full-load current of 650 A. The full-load rated torque is 580 lb·ft (786.36 N·m), and the armature circuit resistance is 0.015 Ω. If the starting inrush current is to be limited to 150% of the rated armature current, calculate: **(a)** the ohmic value of each section of the starting resistance; **(b)** the counter electromotive force at the end of each step; **(c)** the rotational speed at the end of each step; **(d)** the minimum and maximum torques developed during the starting cycle.

SOLUTION

(a)

$$I_a = 650 \text{ A} = \text{ rated current}$$
$$1.5I_a = 1.5 \times 650 = 975 \text{ A} = \text{ peak current}$$

To limit the inrush current to 975 A requires a total resistance in the armature circuit of

$$1.5I_a = \frac{V_t - E_c}{R_{ac} + R_{\text{accel}}}$$

Therefore

$$R_{\text{accel}} = \frac{V_t - E_c}{1.5I_a} - R_{ac}$$

At the instant of starting $\omega = 0$ and $E_c = 0$. Therefore the total accelerating resistance required is

$$R_{\text{accel}} = \frac{250 \text{ V} - 0 \text{ V}}{975 \text{ A}} - 0.015\ \Omega = 0.256 - 0.015 = 0.241\ \Omega$$

(b) To calculate the speed at the end of each step, it is necessary to determine k_g. This can be done at rated speed and current, assuming a constant field flux,

$$V_t = E_c + I_a R_{ac} = k_g \omega + I_a R_{ac}$$
$$250 = k_g (183.26 \text{ rad/s}) + 650 \times 0.015$$
$$k_g = \frac{250 \text{ V} - 9.75 \text{ V}}{183.26 \text{ rad/s}} = 1.311 \text{ V/rad/s}$$

At the end of the first acceleration step $I_a = 650$ A. At time t_1,

$$E_{c1} = V_t - I_a(R_{ac} + R_{\text{accel}})$$
$$= 250 - 650(0.015 + 0.214) = 250 - 166.4 = 83.60 \text{ V} = k_g \omega_1$$

(c)

$$\omega_1 = \frac{83.60 \text{ V}}{1.311 \text{ V/rad/s}} = 63.77 \text{ rad/s} = \frac{60\omega_1}{2\pi} = 608.96 \text{ r/min}$$

At the beginning of the second step,

$$R_{\text{accel}_1} = \frac{V_t - E_{c1}}{1.5I_a} - R_{ac}$$
$$= \frac{250 \text{ V} - 83.60 \text{ V}}{975 \text{ A}} - 0.015\ \Omega = 0.171 - 0.015 = 0.156\ \Omega$$

The ohmic value of the first step of R_{accel} is $R_1 = 0.241 - 0.156 = 0.085\ \Omega$. At time t_2,

$$E_{c2} = V_t - I_a(R_{ac} + R_{\text{accel}_1})$$
$$= 250 - 650(0.015 + 0.156) = 138.85 \text{ V} = k_g\omega_2$$
$$\omega_2 = \frac{138.85 \text{ V}}{1.311 \text{ V/rad/s}}$$
$$= 105.91 \text{ rad/s} = \frac{60\omega_2}{2\pi} = 1{,}011.34 \text{ r/min}$$

At the beginning of the third step,

$$R_{\text{accel}_2} = \frac{V_t - E_{c2}}{1.5I_a} - R_{ac}$$
$$= \frac{250 \text{ V} - 138.85 \text{ V}}{975 \text{ A}} - 0.015\ \Omega = 0.114 - 0.015 = 0.099\ \Omega$$

The ohmic value of the second step of R$_{\text{accel}}$ is $R_2 = 0.156 - 0.099 = 0.057\ \Omega$. At time t_3,

$$E_{c3} = V_t - I_a(R_{ac} + R_{\text{accel}_2})$$
$$= 250 - 650(0.015 + 0.099) = 175.9 \text{ A} = k_g\omega_3$$
$$\omega_3 = \frac{175.9 \text{ V}}{1.311 \text{ V/rad/s}}$$
$$= 134.17 \text{ rad/s} = \frac{60\omega_3}{2\pi} = 1{,}281.25 \text{ r/min}$$

At the beginning of the fourth step,

$$R_{\text{accel}_3} = \frac{V_t - E_{c3}}{1.5I_a} - R_{ac}$$
$$= \frac{250 \text{ V} - 175.9 \text{ V}}{975 \text{ A}} - 0.015\ \Omega = 0.076 - 0.015 = 0.061\ \Omega$$

The ohmic value of the third step of R_{accel} is $R_3 = 0.099 - 0.061 = 0.038\ \Omega$. At time t_4,

$$E_{c4} = V_t - I_a(R_{ac} + R_{\text{accel}_3})$$
$$= 250 - 650(0.015 + 0.061) = 200.60 \text{ V} = k_g\omega_4$$
$$\omega_4 = \frac{200.60 \text{ V}}{1.311 \text{ V/rad/s}}$$
$$= \frac{60\omega_4}{2\pi} = 1{,}461.17 \text{ r/min}$$

At the beginning of the fifth step,

$$R_{\text{accel}_4} = \frac{V_t - E_{c4}}{1.5I_a} - R_{ac}$$

$$= \frac{250 \text{ V} - 200.60 \text{ V}}{975 \text{ A}} - 0.015\ \Omega = 0.051 - 0.015 = 0.036\ \Omega$$

The ohmic value of the fourth step of R_{accel} is $R_4 = 0.061 - 0.036 = 0.025\ \Omega$. At time t_5,

$$E_{c5} = V_t - I_a(R_{ac} + R_{\text{accel}_4})$$

$$= 250 - 650(0.015 + 0.036) = 216.85 \text{ V} = k_g\omega_5$$

$$\omega_5 = \frac{216.85 \text{ V}}{1.311 \text{ V/rad/s}} = 165.41 \text{ rad/s}$$

$$= \frac{60\omega_5}{2\pi} = 1{,}579.53 \text{ r/min}$$

At the beginning of the sixth step,

$$R_{\text{accel}_5} = \frac{V_t - E_{c5}}{1.5I_a} - R_{ac}$$

$$= \frac{250 \text{ V} - 216.85 \text{ V}}{975 \text{ A}} - 0.015\ \Omega = 0.034 - 0.015 = 0.019\ \Omega$$

The ohmic value of the fifth step of R_{accel} is $R_5 = 0.036 - 0.019 = 0.017\ \Omega$. It can be seen that a sixth step of resistance is unnecessary, and as a result, after R_5 is removed, the armature is connected directly across the line and accelerates from 1,579. 53 to 1,750 r/min or, in SI units, from 165.41 to 183.26 rad/s.

(d) Since $T_d = k_g\phi I_a$ and $k_g\phi$ is constant, the minimum torque during the starting period is 580 lb·ft or 786.36 N·m, and the maximum torque is 1.5 × 580 = 870 lb·ft or 1.5 × 786.36 = 1,179.54 N·m.

As can be seen, this type of calculation is very tedious and is best done using a computer or programmable calculator, especially if it is necessary to carry out the computation frequently.

6-9 DC MOTOR REVERSAL

There are many applications where it is necessary to reverse the direction of rotation of a dc motor. This can be accomplished in one of two ways: (1) reversing the current flow in the armature by reversing the voltage supplied to the armature, with the field polarity remaining unchanged, or (2) reversing the current flow in the field circuit, thus reversing the field pole polarity and leaving the armature circuit untouched. It should be noted that only one element is reversed at a time. If both are reversed, the direction of rotation will remain unchanged. The preferred method is to reverse the armature circuit, which has the lower inductance and, as a result, will have less energy stored in its magnetic field, and the current can be reversed more rapidly.

6-10 WARD-LEONARD SYSTEM

The Ward-Leonard system was originally designed to provide highly accurate speed and position control of a separately excited dc motor and is still found in some older mine hoists. The principle of the system is still worth exploring because it clearly demonstrates the concept of four-quadrant operation used in modern dual- or four-quadrant thyristor phase-controlled converters and choppers, which are discussed in Chapter 2.

The advantage of armature voltage control is that the speed may be adjusted from zero to base speed at constant torque. Prior to the introduction of high-power solid-state power supplies the problem was to have a high-power variable-voltage source. The Ward-Leonard system was the solution.

Referring to Fig. 6-18(a), a three-phase squirrel-cage induction motor (SCIM), a wound-rotor induction motor (WRIM), or even a three-phase synchronous motor is mechanically coupled to a separately excited dc generator. The generator output is directly supplied to the armature of a separately excited dc motor. Since the generator operates essentially at constant speed, the output voltage is controlled by varying the generator field current. Below base speed the motor operates with rated field pole flux, with the variable armature voltage being controlled by the generator field. Above base speed the field flux of the motor is weakened and the armature is supplied at rated value from the generator. The direction of rotation of the motor is easily changed by reversing the polarity of the generator field. This arrangement gives smooth stepless speed control in both directions from zero to above base speed. However, the Ward-Leonard system has another significant advantage, namely, it can provide *regenerative braking,* that is, the ability to develop a countertorque and at the same time convert the rotational energy of the motor and its connected load into electric energy, and feed the electric power back to the ac source.

Consider the mine hoist arrangement shown in Fig. 6-18(a), assuming that the cage is being lowered. Initially the generator output voltage V_a is greater than the motor counter electromotive force E_c. As the cage is lowered, the weight of the supporting cable overhanging the winding drum increases and causes the cage to accelerate, which in turn accelerates the motor armature, with the result that E_c increases and the load current I_L drawn from the generator decreases. When $V_a = E_c$, $I_L = 0$. Any further increase in the rotational speed of the motor armature causes E_c to be greater than V_a, and the load current I_L reverses. The motor is now acting as a generator and supplies power to the generator, causing it to act as a motor, which in turn causes it to accelerate the ac drive motor, thus converting it into a generator. The overall effect has been to convert the rotational energy acting on the dc drive motor to dc power, and then return it to the ac source as ac power. The amount of energy fed back, and thus the amount of countertorque produced, is controlled by varying the dc motor field excitation. It should be pointed out that while operating in the regenerative mode, it can only slow down the rate of descent of the cage. It cannot stop and hold the cage at any desired position. This can only be achieved by a conventional braking system.

The disadvantages of the Ward-Leonard system are the initial cost of three large machines in the range of 5,000 hp (3,730 kW) each. Assuming an individual efficiency of 85%, this gives an overall system efficiency of approximately 61%. The modern-day equivalent of the Ward-Leonard system is the four-quadrant or dual converter discussed in Chapter 2, which has no moving parts, does not require any special mounting arrangements, and has an operating efficiency in excess of 90%, as well as possessing a very high reliability coupled with minimum maintenance. Figure 6-18(b) illustrates the operation of the system.

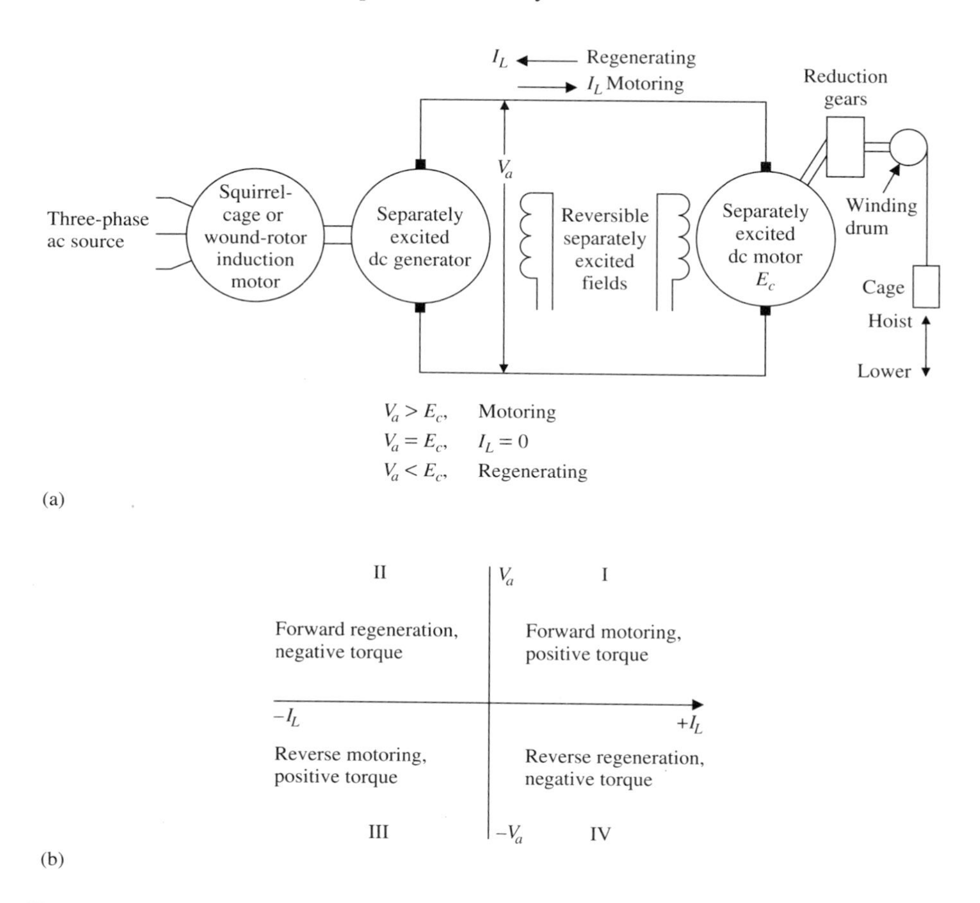

Figure 6-18 Ward-Leonard system applied to mine hoist. (a) Schematic. (b) Four-quadrant operation.

QUESTIONS

6-1 Discuss the reasons why dc motors are still the first choice in some applications.

6-2 Explain what is meant by counter torque.

6-3 Explain why the developed torque in Eq. (4-17)(E) on p. 195 is considered to be optimistic.

6-4 Explain what is meant by speed regulation.

6-5 Under what conditions can the performance of separately excited and shunt motors be considered to be similar?

6-6 Explain the process that takes place when the load torque on a shunt motor increases slightly.

6-7 Explain why a decrease in field pole flux causes a dc motor to increase speed.

6-8 What effect does armature reaction have on the speed of a shunt motor?

6-9 Discuss the speed-torque characteristics of a shunt motor.

6-10 Discuss the reasons why the output torque is less than the developed torque in a dc motor.

6-11 Discuss the methods of obtaining speed control of a shunt motor above and below base speed.

6-12 Discuss the factors that limit the maximum speed when using flux reduction in a dc motor.

6-13 What is the function of a stabilizing winding in a shunt motor?

6-14 What are the effects of an interruption or open circuit in the shunt field circuit of a shunt motor?

6-15 Why is rheostatic speed control of the armature voltage of a separately excited or shunt motor rarely used?

6-16 Even though varying the armature voltage or field pole flux of a dc motor varies the speed, what is the limiting factor and why?

6-17 Explain what is meant by operation in the constant-torque mode.

6-18 Explain what is meant by operation in the constant-power mode.

6-19 What are the advantages of using permanent-magnet motors?

6-20 List six applications using permanent-magnet motors.

6-21 In what ways can a permanent-magnet motor field become demagnetized?

6-22 What are the factors that usually restrict permanent-magnet motors to low power applications?

6-23 Why are Alnico magnets still the most commonly used material in permanent-magnet motors?

6-24 What are the advantages and disadvantages of ceramic magnet motors?

6-25 What are the advantages and disadvantages of using rare-earth magnets in permanent-magnet motors?

6-26 Why can a permanent-magnet motor be substituted for a gear motor?

6-27 With the aid of sketches discuss the methods of controlling the speed of a permanent-magnet motor.

6-28 Why is the developed torque of a series motor so much greater than that of equivalent separately excited and shunt motors?

6-29 What precautions must be exercised when using series motors?

6-30 Discuss the methods of controlling the speed of a series motor.

6-31 Discuss the speed-torque characteristics of a series motor.

6-32 Discuss the types of applications where a series motor would be used.

6-33 Would a series motor be used for both the hoisting and the lowering operations of a crane? If not why not?

6-34 What are the requirements that must be met to reverse the direction of rotation of a dc motor?

6-35 Why does a cumulatively compounded motor have a greater speed regulation than a shunt motor?

6-36 Why is a differentially compounded motor rarely used?

6-37 How can the degree of compounding of a compound motor be varied?

6-38 Why is it necessary to use an accelerating resistance when starting a medium- or large-size dc motor?

6-39 Why are accelerating resistances not required when using a thyristor phase-controlled converter to control a dc motor?

6-40 How do armature inductance and load inertia affect the starting cycle of a dc motor?

6-41 With the aid of a sketch explain the operation of a Ward-Leonard system.

6-42 Explain how a Ward-Leonard system achieves regenerative braking.

6-43 With the aid of a diagram explain four-quadrant operation. Can this be achieved with a mine hoist? If not, give an example where four-quadrant operation can be obtained.

PROBLEMS

6-1 A 250-V 50-hp (37.3-kW) 1,750-r/min (183.26 rad/s) shunt motor, when running at full load, draws a line current of 165 A. The armature circuit resistance is 0.082 Ω, the field resistance is 75 Ω, and the rotational losses are 1,750 W. Calculate: **(a)** the output power; **(b)** the output torque; **(c)** the efficiency; **(d)** the no-load speed; **(e)** the speed regulation.

6-2 A 10-hp (7.46-kW) 250-V shunt motor has an armature circuit resistance of 0.45 Ω and a shunt field resistance of 185 Ω. When running at no load with rated terminal voltage, the speed is 1,250 r/min (130.90 rad/s), and the armature current is 3.1 A. When operating at full load, the line current is 39.2 A. Calculate the full-load speed. Neglect armature reaction.

6-3 Repeat Problem 6-2 assuming that armature reaction has reduced the field pole flux by 5%.

6-4 A 5-hp (3.73-kW) 1,725-r/min (180.64 rad/s) 230-V shunt motor has an armature circuit resistance of 1.25 Ω. At full load the armature current is 20 A. Calculate: **(a)** the counter electromotive force; **(b)** the power developed, in watts and horsepower. Assume a brush contact drop of 2 V.

6-5 At full load the power input into the armature circuit is 10,000 W, and the armature current is 40 A. Given that the armature circuit resistance is 0.45 Ω, what is the output power of the motor in horsepower and kilowatts?

6-6 A 25-hp (18.65-kW) 250-V shunt motor draws a line current of 75 A at full load. The resistances of the armature circuit and the shunt field are 0.21 and 250 Ω, respectively. Assuming a total brush contact drop of 2.5 V, friction, windage, and magnetic losses of 925 W, and a stray-load loss of 1% of the output, calculate the efficiency of the motor.

6-7 A 5-hp (3.73-kW) 1725-r/min (180.64 rad/s) 125-V shunt motor has an efficiency of 82.5% at full load. Calculate: **(a)** the power input; **(b)** the line current; **(c)** the output torque.

6-8 A 250-V shunt motor has an armature circuit resistance of 0.055 Ω and a shunt field resistance of 125 Ω. At no load it runs at 1,200 r/min (125.66 rad/s) and draws a line current of 8.5 A. When loaded the motor draws a line current of 50 A. Calculate: **(a)** the rotational speed; **(b)** the efficiency; **(c)** the friction, windage, and magnetic loss.

6-9 A 250-V shunt motor has an armature circuit resistance of 0.55 Ω and draws an armature current of 50 A at rated load. Calculate the ohmic resistance of each step of the accelerating resistance if the inrush current is to be limited to 175% of the rated armature current.

6-10 A 600-V series motor has an armature circuit resistance of 0.095 Ω and a series field resistance of 0.22 Ω. The motor draws 50 A when running at 1,150 r/min (120.43 rad/s) and develops an output torque T. Calculate the rotational speed when the output torque is $0.8T$.

6-11 A 250-V series motor has an armature resistance of 0.06 Ω and a field resistance of 0.025 Ω and draws a line current of 25 A when running at 850 r/min (89.01 rad/s). Calculate the rotational speed when the line current is 18 A.

6-12 A 250-V series motor runs at 1,100 r/min (115.19 rad/s) and draws a line current of 50 A. The armature circuit resistance is 0.2 Ω. Assuming the motor is operating on the linear portion of the magnetization curve, calculate: **(a)** the armature current when the motor speed is 1,500 r/min (157.08 rad/s); **(b)** the change in developed torque at 1,500 r/min (157.08 rad/s). Neglect armature reaction.

6-13 A 250-V compound motor has an armature circuit resistance of 0.175 Ω and a series field resistance of 0.20 Ω. When drawing an armature current of 105 A, calculate: **(a)** the counter electromotive force; **(b)** the power developed. Neglect brush contact drop.

6-14 A 600-V long-shunt compound motor has an armature circuit resistance of 0.025 Ω, a shunt field resistance of 150 Ω, and a series field resistance of 0.018 Ω. When the counter electromotive force is 545 V, calculate: **(a)** the shunt field current; **(b)** the armature current; **(c)** the line current. Neglect the brush contact drop.

CHAPTER

7

DC Machine Losses, Efficiency, and Testing

7–1 INTRODUCTION

The efficiency of any device or plant is important in the sense that it is the measure of how effectively we are utilizing energy. In the case of dc machines we are converting mechanical energy to electric energy or vice versa, that is, they are energy-converting devices. Any time a machine converts energy from one form to another there are losses. These losses occur within the machine, causing an increase in temperature, and as a result its output rating is reduced while its operating costs are being increased.

7–2 EFFICIENCY RELATIONSHIPS

The efficiency of a rotating machine, transformer, or any other energy conversion device is governed by the law of conservation of energy, that is,

$$P_{\text{in}} = P_{\text{out}} + P_{\text{losses}} \tag{7–1}$$

where P_{in} is the input power, P_{out} the output power, and P_{losses} the total of all losses, such as copper, windage, friction, hysteresis, and eddy current losses.

Obviously the power supplied will always be greater than the power out. The losses usually appear in the form of heat in electric equipment, which in turn contributes to the temperature rise of the machine. Efficiency is usually expressed as a percentage,

$$\eta = \frac{P_{\text{out}}}{P_{\text{in}}} \times 100\% \tag{7–2}$$

This in turn can be expressed as

$$\eta = \frac{P_{in} - P_{losses}}{P_{in}} \times 100\% = 1 - \frac{P_{losses}}{P_{in}} \tag{7-3}$$

or

$$\eta = \frac{P_{out}}{P_{out} + P_{losses}} \times 100\% \tag{7-4}$$

Since in the case of a motor P_{in} is easily measured, Eq. (7-3) is more readily adapted to motors. Similarly, with a generator P_{out} is easily measured, and therefore Eq. (7-4) lends itself more readily to generator applications.

➤ EXAMPLE 7-1

A 100-hp (74.6-kW) 550-V motor draws a line current of 150 A when operated at rated full load. Calculate: **(a)** the losses; **(b)** the efficiency.

SOLUTION

(a)

$$\begin{aligned} P_{in} &= 550 \text{ V} \times 150 \text{ A} = 82.5 \text{ kW} \\ &= P_{out} + P_{losses} \end{aligned}$$

Therefore

$$\begin{aligned} P_{losses} &= P_{in} - P_{out} \\ &= 82.5 - 74.6 = 7.90 \text{ kW} \end{aligned}$$

(b) From Eq. (7-3),

$$\begin{aligned} \eta &= \frac{P_{losses}}{P_{in}} \\ &= 1 - \frac{7.90 \text{ kW}}{82.5 \text{ kW}} = 0.904 = 90.4\% \end{aligned}$$

The efficiency of machines is not constant over the whole operating range. In fact, machines are designed to have their best efficiencies at or near rated load. In addition, smaller machines will have lower efficiencies than a larger capacity machine. For example, a 1-hp (0.746-kW) dc motor would have an efficiency of approximately 75%, while a 100-hp (74.6-kW) motor would be expected to have an efficiency on the order of 87–89%. Also in general, slow-speed machines will have lower efficiencies than high-speed machines, the difference usually being between 3 and 4%.

7-3 DISTRIBUTION AND TYPES OF LOSSES

Normally efficiencies are determined by measurement of the losses, as compared to the direct measurement of input and output. Loss measurement is accurate, economical, and convenient. It has the added benefit that when competing manufacturers' equipment is to be compared, it provides identical methods of measurement so that a fair comparison can be made. Testing standards have been established by the National Electrical Manufacturers Association (NEMA), the Canadian Electrical Manufacturers Association (CEMA), the Institute of Electrical and Electronics Engineers (IEEE), and the American National Standards Institute (ANSI) for North America, and by the International Electrotechnical Commission (IEC) in Europe. Similar organizations exist in other major manufacturing countries in the world.

In dc machines losses occur basically in three areas: (1) electrical or copper losses, (2) rotational or stray power losses, and (3) stray load losses.

Electrical Losses

Electrical or copper losses are mainly those caused by current flow and can be classified as follows:

1. *Shunt field copper losses:*

Shunt field winding	$I_f^2 R_f$
Shunt field rheostat	$I_f^2 R_r$

The shunt field losses usually are reasonably constant and will vary only if there is frequent adjustment of the shunt field rheostat.

2. *Armature circuit copper losses:*

Armature winding	$I_a^2 R_a$
Interpole winding	$I_a^2 R_{ip}$
Series field winding	$I_a^2 R_{se}$
Compensating winding	$I_a^2 R_{\text{com}}$
Brush contact resistance loss	$I_a V_B$

The armature circuit copper losses in general are proportional to the square of the armature current. The preceding listing of course must be modified to suit the specific machine. For example, a shunt machine obviously will not have a series field winding; also it is only in rare instances that a compensating winding will be encountered. The brush contact resistance loss voltage drop V_B can be assumed to be constant at all loads, so the power loss at the brush contact will vary directly with the armature current, that is, it is equal to $I_a V_B$.

Rotational or Stray Power Losses

The rotational or stray power losses are the energy losses attributable to the power required to turn the armature of the machine, that is, they must be subtracted from the power input. Rotational losses have two components, friction and core or iron losses.

The friction losses can be grouped as follows:

1. *Bearing friction loss:* Sleeve bearings are used in large machines and the loss is a fluid friction loss since the bearings are oil lubricated. This loss is independent of the bearing pressure. If the bearing oil is pressure fed or cooled by equipment connected to the shaft, then the power requirement of this equipment must be considered as a loss attributable to the machine. Normally antifriction ball and roller bearings have bearing losses that are low compared to windage losses, so usually there is no attempt to isolate them, and they are included as part of the windage loss.
2. *Brush friction loss:* This is the power wasted in overcoming brush friction. It can be calculated provided the coefficient of friction is known. The coefficient of friction is dependent on the surface speed of the commutator, which is less at high speeds, and on the brush pressure and the brush contact area. It also depends on the type of brush material.
3. *Windage loss:* This loss has two components, the effect of wind friction on the armature surface and the windage loss attributable to the shaft-mounted cooling fan.

All these losses are constant under constant-speed operation, but they will vary directly in proportion to any changes in speed.

Core or iron losses consist of the hysteresis loss $P_h = K_h B^n f$ W, where $n = 1.6$, and the eddy current loss $P_e = K_e B^2 f^2 t^2$ W. The flux in the armature goes through a complete cycle every two pole pitches, namely, $P/2$, that is, frequency of the flux cycles is $(P/2) \times$ (r/s) Hz. The effect of the armature currents is to cause a distortion of the air-gap flux toward the trailing pole tips of dc generators, or toward the leading pole tips of dc motors. This concentration of flux causes the flux density B of the armature teeth to increase with load. Under constant-speed operation the hysteresis losses and even more so the eddy current losses are significant and depend on the flux density, the rotational speed, the mass of the armature iron, and the magnetic quality of the armature laminations. The core or iron losses create a mechanical drag on the armature.

Normally the rotational or stray power losses account for an efficiency loss ranging from 4% for large machines to 11% for small machines. These losses can be measured individually and summed, but normally they are measured taking current and voltage readings at no load. The rotational losses are

$$\begin{aligned} P_{\text{rot}} &= E_g I_a - I_a^2 R_a \\ &\sim E_g I_a \end{aligned} \tag{7–5}$$

since the armature copper loss is small.

Stray Load Losses

Stray load losses, although minor, are caused by the effects of leakage and armature reaction fluxes which produce skin effect losses in the armature conductors and iron losses in the pole faces, armature teeth, and various structural parts of the machine.

These losses are usually assumed to be equal to 1% of the total load in dc generators and ignored in dc motors under 200 hp (149.2 kW).

Figure 7-1 shows the power flow diagrams for a dc generator and motor and provides an insight into the power balance equations.

7-4 CONDITIONS FOR MAXIMUM EFFICIENCY

From Section 7-3 dc machine losses can be thought of as either fixed or variable. The fixed losses, although not constant, remain relatively unchanged with load changes. They include the rotational losses and the shunt field loss. The variable losses are considered to be those that vary with the square of the armature current. Variable losses include all the armature circuit losses, that is, the armature, series, interpole, and compensating field losses as appropriate.

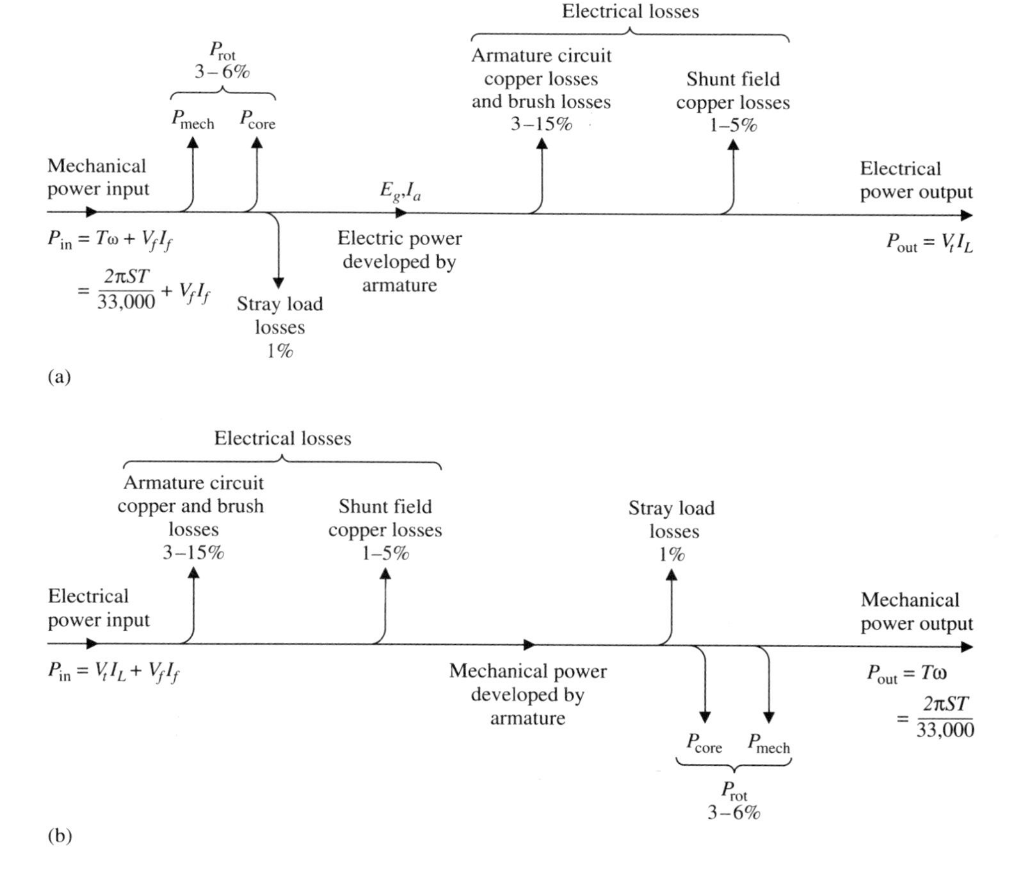

Figure 7-1 Power flow diagrams. (a) DC generator. (b) DC motor.

It can be proven mathematically that maximum efficiency occurs when the fixed losses are equal to the variable losses, that is,

$$P_{\text{rot}} + V_t I_f \sim I_a^2 (R_a + R_{se} + R_{ip} + R_{\text{com}}) \tag{7-6}$$

where P_{rot} is the rotational or stray loss, $V_t I_f$ the shunt field circuit (field and rheostat) loss, R_a the armature resistance, and R_{se}, R_{ip}, and R_{com} are the series, interpole, and compensating winding resistances, respectively.

While Eq. (7-6) is equally applicable to a motor and a generator, some terms are redundant, depending on the application. For example, a series machine obviously will not have a shunt field loss. Similarly not all machines will be fitted with all of the fields in the armature circuit.

Equations (7-3) and (7-4) can be modified in terms of the fixed and variable losses to yield

$$\eta_{\max} = \frac{P_{\text{in}} - 2(\text{variable losses})}{P_{\text{in}}} = \frac{P_{\text{in}} - 2(\text{fixed losses})}{P_{\text{in}}} \tag{7-7}$$

or

$$\eta_{\max} = \frac{P_{\text{out}}}{P_{\text{out}} + 2(\text{variable losses})} = \frac{P_{\text{out}}}{P_{\text{out}} + 2(\text{fixed losses})} \tag{7-8}$$

In turn Eqs. (7-6) and (7-7) can be modified assuming that R_{ac} represents the armature circuit resistance, that is, the armature plus any additional series-connected fields. Then for a motor,

$$\eta_{\max} = \frac{P_{\text{in}} - 2I_a^2 R_{ac}}{P_{\text{in}}} \tag{7-9}$$

and for a generator,

$$\eta_{\max} = \frac{P_{\text{out}}}{P_{\text{out}} + 2I_a^2 R_{ac}} \tag{7-10}$$

The load at which this maximum efficiency occurs is determined by the machine design, but it usually occurs between 50 and 80% of full load (Fig. 7-2).

➤ EXAMPLE 7-2

A 10-kW 220-V 1,750-r/min shunt generator has a shunt field resistance of 110 Ω and an armature resistance of 0.275 Ω. The rotational loss has been measured as 750 W. Assuming a brush contact drop of 2.5 V, calculate the full-load efficiency.

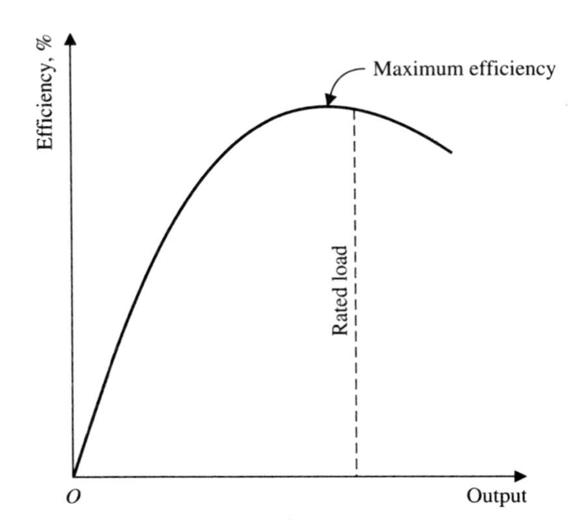

Figure 7-2 Typical efficiency curve.

SOLUTION

$$\text{Full-load current } I_L = \frac{P_{\text{out}}}{V_t} = \frac{10 \text{ kW}}{220 \text{ V}} = 45.45 \text{ A}$$

$$\text{Field current } I_f = \frac{V_t}{R_f} = \frac{220 \text{ V}}{110 \ \Omega} = 2.0 \text{ A}$$

$$\text{Armature current } I_a = I_f + I_L = 2.0 + 45.45 = 47.45 \text{ A}$$

$$\text{Power loss in shunt field } P_{sh} = V_t I_f = 220 \times 2 = 440 \text{ W}$$

$$\text{Armature circuit loss } I_a^2 R_a = 47.45^2 \times 0.275 = 619.16 \text{ W}$$

$$\text{Brush loss } V_B I_a = 2.5 \times 47.45 = 118.63 \text{ W}$$

The total losses are

$$P_{sh} + I_a^2 R_a + V_B I_a + P_{\text{rot}} = 440 + 619.16 + 118.63 + 750 = 1{,}927.79 \text{ W}$$

$$\begin{aligned}\text{Input power } P_{\text{in}} &= P_{\text{out}} + \text{ losses} \\ &= 10{,}000 + 1{,}927.79 = 11{,}927.79 \text{ W}\end{aligned}$$

$$\eta = \frac{10{,}000 \text{ W}}{11{,}927.79 \text{ W}} = 0.8384 = 83.84\%$$

➤ EXAMPLE 7-3

A 10-kW 220-V 1,800-r/min (188.5-rad/s) shunt generator has a shunt field resistance of 220 Ω, an interpole field resistance of 0.015 Ω, and an armature resistance of 0.325 Ω. The rotational loss was measured at 850 W. Calculate the generator efficiency at: **(a)** rated load; **(b)** 25%, 50%, 75%, and 125% of rated load.

SOLUTION

$P_{rot} = 850$ W is constant for all loads and assumes a negligible armature copper loss.

(a) At rated load,

$$I_L = \frac{P_{out}}{V_t} = \frac{10 \text{ kW}}{220 \text{ V}} = 45.45 \text{ A}$$

$$I_f = \frac{V_t}{R_f} = \frac{220 \text{ V}}{220 \ \Omega} = 1.0 \text{ A}$$

$$I_a = I_f + I_L = 1.0 + 45.45 = 46.45 \text{ A}$$

(b) The full-load armature circuit copper loss, that is, armature and interpole field, is

$$I_a^2(R_a + R_{ip}) = 46.45^2(0.325 + 0.015) = 733.58 \text{ W}$$

The shunt field copper loss is

$$V_f I_f = 220 \times 1.0 = 220 \text{ W}$$

$$P_{Cu} = 733.58 + 220 = 953.58 \text{ W}$$

The efficiency at full load is

$$\eta = \frac{P_{out}}{P_{out} + P_{rot} + P_{Cu}}$$

$$= \frac{10,000}{10,000 + 850 + 733.58 + 220} = \frac{10,000}{11,803.58}$$

$$= 0.8472 = 84.72\%$$

The efficiency at any load can be obtained from

$$\eta = \frac{\text{load output}}{\text{load output} + \text{rotational loss} + \text{armature circuit losses} + \text{shunt field copper loss}}$$

Therefore at 25% load,

$$\eta = \frac{10,000 \times 0.25}{10,000 \times 0.25 + 850 + (733.58 \times 0.25^2 + 220)} \times 100\% = 69.14\%$$

At 50% load,

$$\eta = \frac{10,000 \times 0.5}{10,000 \times 0.5 + 850 + (733.58 \times 0.5^2 + 220)} \times 100\% = 79.96\%$$

At 75% load,

$$\eta = \frac{10,000 \times 0.75}{10,000 \times 0.75 + 850 + (733.58 \times 0.75^2 + 220)} \times 100\% = 83.35\%$$

At 125% load,

$$\eta = \frac{10,000 \times 1.25}{10,000 \times 1.25 + 850 + (733.58 \times 1.25^2 + 220)} \times 100\% = 84.94\%$$

7-5 TESTING

Dc machine testing is very specifically defined by IEEE 113 and ANSI C50.4 as well as by the test procedures detailed by the major manufacturers. Testing can be broken down into three areas:

1. Those tests required to determine losses and efficiency.
2. Those tests whose prime purpose is to determine performance under specific operating conditions.
3. Those tests that determine optimum performance and material quality.

Determination of Rotational Losses

The rotational losses of a dc machine include bearing friction, windage, hysteresis, and eddy current losses in the armature and pole faces. Since the mechanical components of the rotational losses are proportional to speed, and the magnetic losses are proportional to the speed squared, it is necessary to test the machine at its designed operating speed. Also the effects of the magnetic fields, that is, flux, must be considered since the hysteresis loss is proportional to B^n where n = 1.6, and the eddy current loss is proportional to B^2. Therefore the field system must be excited during the rotational loss test.

Prior to carrying out the rotational loss test the machine must be at normal operating temperature. The reason for this is that the eddy current losses form the major component of the core losses, and since they depend on the resistance of the core, they will be at their minimum value.

The measurement of the rotational loss is dependent on adjusting the field excitation so that

$$V_t = E_g \pm (V_B + I_a R_{ac}) \tag{7–11}$$

$$E_g = V_t \pm (V_B + I_a R_{ac}) \tag{7–12}$$

where V_t is the terminal voltage, V_B the brush contact drop (2 V normally assumed), and R_{ac} the armature circuit resistance, that is, the armature and all windings in series with the armature. With ±, + denotes a generator, and − a motor. Normally $V_B + I_a R_{ac}$ is calculated for normal rated load.

A simple method of determining the rotational loss of a dc generator or motor is as follows:

1. Run the machine as a motor totally isolated from any load, and adjust the speed to rated value and E_g to the appropriate value determined from Eq. (7-12). If there is a series-excited field, it must be disconnected and excited separately to carry full-load rated current.
2. Measure and record E_g and I_a. Then

$$P_{\text{rot}} \sim E_g I_a$$

➤ EXAMPLE 7-4

A 10-kW 220-V shunt generator is run as a motor at rated speed under no load to determine the rotational loss. Under these conditions the machine draws a current of 2.5 A. The field circuit resistance is 220 Ω and the armature resistance is 0.175 Ω. Calculate the rotational loss at rated load. Neglect brush contact drop.

SOLUTION

From Eq. (7-12), $E_g = V_t + I_a R_a$. Then

$$\text{Rated full-load current } I_L = \frac{P_{\text{out}}}{V_t} = \frac{10 \text{ kW}}{220 \text{ V}} = 45.45 \text{ A}$$

$$\text{Shunt field current } I_f = \frac{V_t}{R_f} = \frac{220 \text{ V}}{220 \ \Omega} = 1.0 \text{ A}$$

Therefore

$$I_a = I_L + I_f = 45.45 + 1.0 = 46.45 \text{ A}$$

Then

$$E_g = 220 + 46.45 \times 0.175 = 228.13 \text{ V}$$

and therefore

$$P_{\text{rot}} = E_g I_a = 228.13 \times 2.5 = 570.32 \text{ W}$$

If it is necessary to identify the individual components of the rotational loss, then the machine must be driven by a calibrated motor, that is, a motor whose torque-speed characteristics and losses have been determined previously.

The various losses, that is, windage and bearing friction, brush friction, and core loss, can be isolated by using the following procedure:

1. With the fields of the machine under test unexcited and the brushes lifted free of the commutator, and then with the calibrated motor driving it at

rated speed, the output of the calibrated motor P_F is equal to the windage and bearing frictional loss of the machine.

2. With the brushes still lifted, the fields are excited to rated values. Then at rated speed the output of the calibrated motor P_{fwc} is equal to the windage, bearing, and core loss of the machine under test. The core loss is $P_{fwc} - P_F$.
3. Once again deenergize the fields and lower the brushes back onto the commutator. Then with the calibrated motor driving the machine under test, the output of the calibrated motor is the windage and bearing friction losses plus the brush friction P_B. The bearing friction will then be the measured output power minus P_F.

This test procedure has permitted us to isolate the various components of the rotational loss. However, it is only undertaken when modifications or design requirements call for this detailed a breakdown.

Armature Circuit Resistance Measurement

The major copper loss in a dc machine occurs in the armature. Therefore it is necessary to be able to measure its resistance accurately. In order to make accurate resistance measurements it is necessary to make direct contact with the commutator surface. If resistance measurement is attempted via the carbon brushes, an error will be introduced because of the relatively high brush to commutator interface resistance. The two basic methods of measuring armature resistance are the voltmeter-ammeter and the bridge megger; if the armature resistance is less than 1 Ω, the Kelvin double-bridge method is preferred.

No matter which method is used, it is essential that the measuring probes be in contact with exactly the right commutator segments. These segments are determined by the way the armature is wound. Recall that a simplex wave-wound armature has two paths in parallel, and a simplex lap winding has as many paths in parallel as there are poles. Therefore for a wave winding the measurement probes should be in contact with segments that are $360°/P$ degrees apart. For example, for a six-pole wave-wound armature the probes should be contacting commutator segments $360°/6 = 60°$ apart. With a lap-wound armature it is necessary to use as many probes as poles, spaced $360°/P$ degrees apart, with alternate probes connected together, that is, probes of the same polarity are connected together. For an eight-pole lap-wound armature there would be eight probes spaced $360°/8 = 45°$ apart.

If the voltmeter-ammeter method is used, the probes are placed as described, with the brushes lifted from the commutator, and then a constant dc voltage is applied. The voltage between probes of opposite polarity is measured as well as the current through the probes. The armature resistance is $R_a = V/I$. Measuring the resistance by this method will not yield the most accurate result. Note that the magnitude and duration of current flow should be controlled so as to minimize errors caused by heating.

When measuring the armature resistance of machines smaller than 1 hp (0.746 kW), a Wheatstone bridge or bridge megger may be used. For machines in excess of 1 hp (0.746 kW) the most accurate method of measuring the armature resistance, because of the low resistances involved (usually less than 1 Ω), is to use the Kelvin double bridge.

Usually the armature and field resistances are measured before a test is commenced and the ambient temperature is also recorded. Then immediately after the test has been completed these resistances are recorded again because of the changes that occur as the winding temperatures increase.

Field Resistance Measurement

Usually the best method of measuring field resistances, that is, shunt, series, interpole, and compensating field, is by the use of a bridge megger. It is usually sufficient to isolate one end of the appropriate winding and measure the resistance across the winding. The measurement should be repeated at the end of the test and the ambient temperature recorded at the beginning of the test.

Insulation Resistance

The *insulation resistance*, that is, the resistance to ground of all windings, should be checked before the machine is run. Normally the preferred method is to use a 500- or 1,000-V megger, and the insulation resistance for integral-horsepower machines should not be less than

$$\text{M}\Omega = \frac{\text{rated voltage of machine}}{1{,}000 + \text{rated hp}} \tag{7–13}$$

In the case of fractional-horsepower machines the insulation resistance should not be less than 1 MΩ.

Effect of Temperature on Resistance

As we know from our studies of electrical fundamentals, the resistance of copper increases linearly as its temperature increases, except for a small portion at −234.5°C. If a straight line is drawn, it intercepts the zero-resistance axis at −234.5°C. This is considered to be the absolute zero for copper. Normally we are interested in temperature values in the range of 20 to 200°C.

From Fig. 7-3, triangles ABC and ADF are similar triangles. Therefore

$$\frac{DF}{BC} = \frac{AD}{AB} = \frac{AD + OD}{AO + OB}$$

or

$$\frac{R_1 + \Delta R}{R_1} = \frac{-234.5°C + T_2}{-234.5°C + T_1}$$

But where T is in °C. But $R_1 + \Delta R = R_2$. Therefore

$$\frac{R_2}{R_1} = \frac{234.5°C + T_2}{234.5°C + T_1}$$

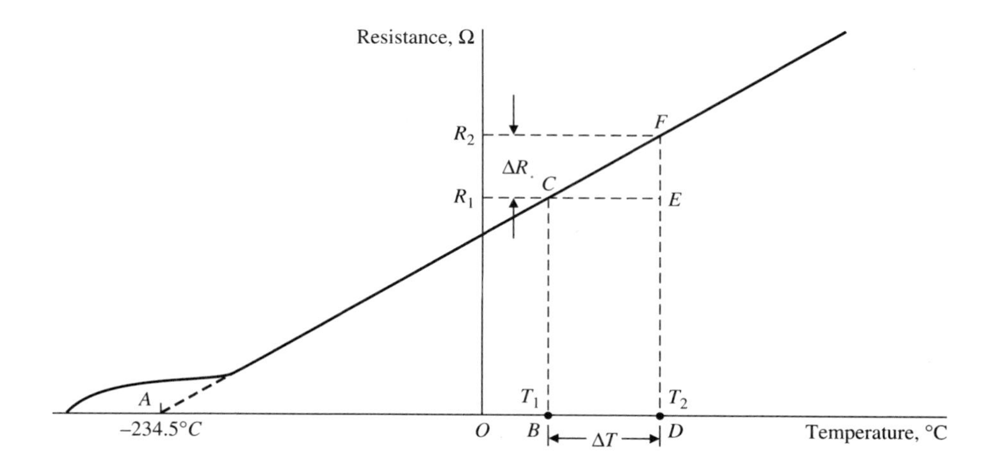

Figure 7-3 Resistance-temperature relationship for copper.

and

$$R_2 = \frac{R_1(234.5°C + T_2)}{234.5°C + T_1}$$

or

$$234.5°C + T_2 = \frac{R_2(234.5°C + T_1)}{R_1}$$

Hence

$$T_2 = \frac{R_2(234.5°C + T_1)}{R_1} - 234.5°C \tag{7–14}$$

As can be seen, Eq. (7-14) provides a simple method of determining the final temperature, and the temperature rise ΔT of the armature and all the field windings. Alternatively, the temperature rise and the final temperature can also be measured by installing thermocouples, which of course will permit continuous temperature readings to be taken, as would be required in a heat run. However, it is not possible to mount thermocouples on any rotating parts unless slip rings are used to provide the interface between stationary and rotating parts.

➤ EXAMPLE 7-5

At the beginning of a dc motor test the temperature of the motor is 20°C, and the measured armature resistance is 0.20 Ω. At the completion of the load test, the armature resistance was found to be 0.22 Ω. Calculate the final temperature of the armature and the increase in temperature.

SOLUTION

From Eq.(7-14),

$$T_2 = \frac{R_2(234.5°C + T_1)}{R_1} - 234.5°C$$

$$= \frac{0.22(234.5°C + 20°C)}{0.2} - 234.5°C = 279.95 - 234.5$$

$$= 45.45°C$$

$$\Delta T = T_2 - T_1 = 45.45 - 20 = 25.45°C$$

7-6 SETTING BRUSH NEUTRAL

Before running a new machine, or after the armature has been rewound or the machine has been disassembled, the position of the brush rocker arm should be checked to ensure that the brushes are positioned on the geometric neutral axis. A simple but effective test consists of applying approximately rated shunt field voltage to the shunt field, and connecting a center-reading voltmeter between the positive and negative brushes. Then slowly making and breaking the field supply, the brush rocker arm position is adjusted until there is little or no deflection of the voltmeter. The brush rocker arm is now locked and its position marked. The accuracy of this method can be improved by removing all the brushes except two that are used to make contact with the commutator. In addition, these brushes should be beveled so that they are making contact with only one commutator segment.

7-7 EFFICIENCY BY LOSS SUMMATION

Generator efficiencies may be determined by measurement of the power output. However, this method requires the use of a calibrated motor or some other prime mover whose output is accurately defined. This method requires wasting a considerable amount of power and is especially difficult to perform with large-kW-output generators. Furthermore it is subject to the introduction of measurement errors, which will make the calculated efficiency grossly inaccurate. This method is known as the *direct method* of efficiency measurement. Conventional efficiency measurement, or efficiency by *loss summation,* is the preferred method of testing generators and motors.

The basics of this method have already been described. Essentially the individual losses are determined and then added together. The tests are as follows:

1. Measurement of armature, shunt, series, interpole, and compensating field resistances as appropriate.
2. Calculation of field copper losses.
3. Calculation of armature copper loss.
4. Determination of rotational losses.
5. Calculation of brush losses.

Stray load losses normally are neglected for machines rated at less than 150 kW (200 hp), and are assumed to be equal to 1% of the total output for machines in excess of 150 kW (200 hp).

The efficiency is then calculated for any load,

$$\eta = \frac{\text{output at that load}}{\text{output at that load} + \text{rotational loss} + \text{armature and field copper loss}} \tag{7–15}$$

7-8 HEAT RUNS

Generators or motors are operated at full load at rated speed for 4 h or until the temperatures stabilize. Motors may be tested using a brake load, although this is usually restricted to motors rated at 5 hp (3.73 kW) or less. Above this rating the motor is used to drive a generator loaded into a resistance bank, or a dynamometer. Similarly generators are driven at rated speed and load by a prime mover. If two identical dc machines are to be tested, a back-to-back test can be used where one machine drives the other as a generator. The generator's output is fed back to the motor so that the only power supplied from the source is for the losses of the two machines.

Initially all temperatures and resistances are measured, including the ambient air temperatures. All intakes, exhausts, bearings, frame, and field windings have temperature-monitoring devices installed in them. The machine is then run at rated load and speed, and voltage, current, speed, and temperature readings are taken and recorded, usually half-hourly. After the temperatures have stabilized, the machine is shut down, and the armature and commutator surfaces, armature end connections, and all winding temperatures are measured. Also at the same time all resistance measurements are taken and recorded, so that the temperature rises may be calculated.

7-9 SWINBURNE TEST

The Swinburne test is used to determine the efficiency of dc shunt and compound machines. The efficiency can be calculated for any load, with the machine running as a motor or generator, from the data for the no-load test when the machine is run as a motor. This test assumes that the rotational losses, that is, friction, windage, and core losses, remain constant at all loads, which is not strictly accurate. As a result the calculated efficiencies are optimistic, because the core losses increase when the machine is loaded because of the effects of armature reaction. Also the eddy current electromotive forces in the armature conductors effectively reduce their cross-sectional area and increase the armature copper losses.

The machine should be run at no load for 10 to 15 min to warm up the bearings. Then the supply voltage V_t, the no-load armature current $I_{a\,nl}$ and the field current I_f are measured. The no-load armature input is now $V_t I_{a\,nl}$ since the no-load armature copper loss is negligible. Then

$$P_{\text{rot}} = V_t I_{a\,nl}$$

The field copper loss is constant at all loads and is $I_f^2R_f$. Therefore

$$\text{Constant losses} = P_{\text{rot}} + I_f^2 R_f \tag{7–16}$$

Then at any load the armature circuit copper losses are $I_a^2R_{ac}$. Therefore the total loss for an armature current I_a is

$$P_{\text{loss}} = P_{\text{rot}} + I_f^2 R_f + I_a^2 R_{ac} \tag{7–17}$$

For a motor the input is $V_t(I_a + I_f)$. Therefore

$$\eta = \frac{\text{input} - \text{losses}}{\text{input}} = \frac{V_t(I_a + I_f) - P_{\text{loss}}}{V_t(I_a + I_f)} \tag{7–18}$$

For a generator the output is $V_t(I_a - I_f)$. Therefore

$$\eta = \frac{\text{output}}{\text{output} + \text{losses}} = \frac{V_t(I_a - I_f)}{V_t(I_a - I_f) + P_{\text{loss}}} \tag{7–19}$$

A further disadvantage of the Swinburne test is that it does not give any indication of operating problems, such as brush sparking and unusual temperature rises. The Hopkinson test, which requires two identical dc machines, permits fully loaded operation without excessive power consumption.

7-10 HOPKINSON TEST

The Hopkinson test, subsequently modified by Kapp, permits heat run testing to be carried out under full-load conditions. The input power required is used to supply the losses of the two similar machines. The connection arrangement for the Kapp modified test is shown in Fig. 7-4.

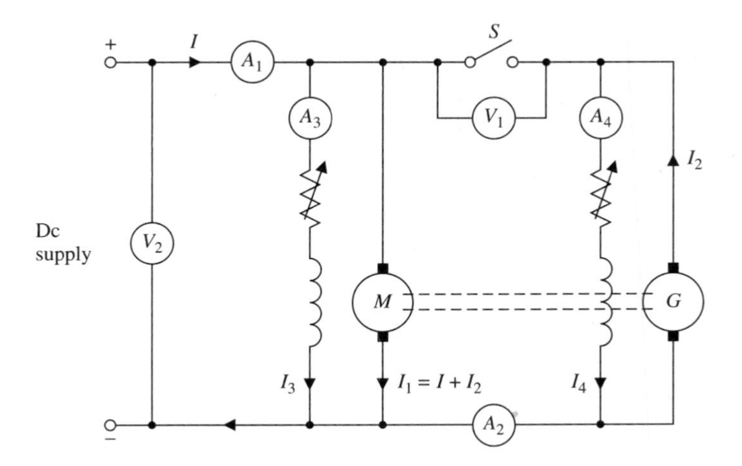

Figure 7-4 Kapp test connections for a pair of shunt machines.

The two machines are mechanically coupled together so that M runs as a motor driving the other as a generator G. With switch S open, motor M is brought up to speed, and the generator's field is adjusted until the voltmeter V across switch S reads zero; S is then closed. At this point the generator G is floating across the supply, that is, it is neither taking nor delivering power. The load supplied by the generator is controlled by increasing its electromotive force, and by decreasing the shunt field circuit resistance, or alternatively, by increasing the field circuit resistance of the motor, or by making both adjustments simultaneously. The output current of the generator I_2 then partially supplies the motor current I_1. The motor current $I_1 = I + I_2$, that is, the current I drawn from the supply, has been reduced by I_2. This arrangement permits efficiency tests and heat runs to be carried out strictly by controlling the shunt field circuit resistances. With the minimum of input power from the dc supply, for example, an input power on the order of 100–125 kW is required to test two 1,000-kW machines. The author has personally seen this test applied to two 7,000-hp (5,222-kW) propulsion motors for an icebreaker.

The efficiency can be obtained from the following:

Motor armature copper loss	$I_1^2 R_{\text{ac } m}$
Generator armature copper loss	$I_2^2 R_{\text{ac } g}$
Total rotational loss per machine	$P_{\text{rot}} = \frac{1}{2}(V_2 I - (I_1^2 R_{\text{ac } m} + I_2^2 R_{\text{ac } g}))$
Motor input power	$P_{\text{in } m} = V_2 I_1 + V_2 I_3$
Motor losses	$P_{\text{loss } m} = P_{\text{rot}} + I_1^2 R_{\text{ac } m} + V_2 I_3$
Motor output	$P_{\text{out } m} = P_{\text{in } m} - P_{\text{loss } m}$
Motor efficiency	$\eta_m = 1 - \dfrac{P_{\text{loss} m}}{P_{\text{in} m}}$
Generator output	$P_{\text{out } g} = V_2(I_2 - I_4)$
Generator losses	$P_{\text{loss } g} = P_{\text{rot}} + I_2^2 R_{ac\ g} + V_2 I_4$
Generator input	$P_{\text{in } g} = P_{\text{out } g} + P_{\text{loss } g}$
Generator efficiency	$\eta_g = 1 - \dfrac{P_{\text{loss} g}}{P_{\text{in} g}}$

This test provides accurate data and is suitable for all dc shunt machines except low-output machines.

7-11 HIGH-POTENTIAL TESTS

As a final test of the insulation of a dc machine, it is normal practice to apply a high voltage between the windings and between windings and ground immediately after a heat run. The test voltage is obtained from a variable ac transformer. The voltage is gradually increased to the specified value and is applied for 1 min. For dc machines above 3 hp (2.24 kW) the test voltage is 1,000 + 2(rated voltage). This is normally a factory test. However, it may also be carried out after installation. If a breakdown occurs, special tests may be required to determine the exact location of the fault.

QUESTIONS

7-1 Why is efficiency important in dc machines?

7-2 Why is Eq. (7-3) the preferred method of determining a motor's efficiency?

7-3 Why is Eq. (7-4) the preferred method of determining a generator's efficiency?

7-4 Why is the efficiency of a dc machine not constant over the whole load range?

7-5 Why do smaller rated dc machines have a lower efficiency than higher rated machines?

7-6 Why are efficiencies normally determined by loss summation, rather than by direct measurement of input and output?

7-7 What are the components of electrical or copper losses in a dc machine?

7-8 Why is the brush contact loss determined from $V_B I_a$?

7-9 What is meant by the term *rotational* or *stray power losses*?

7-10 What are the two components of rotational losses?

7-11 What items form the mechanical losses of a dc machine?

7-12 What are the items that form the core or iron losses of a dc machine?

7-13 Why are the core losses not constant?

7-14 What are stray losses? How are they approximated?

7-15 What conditions must be met to obtain maximum efficiency from a dc machine?

7-16 How are rotational losses determined?

7-17 When performing a rotational loss test, why is it important to run the machine at rated temperature and speed?

7-18 How can the various components of the rotational loss be separated?

7-19 Discuss two methods of measuring armature resistance, and what precautions must be taken to obtain accurate results?

7-20 How is the resistance of a wave-wound armature measured?

7-21 How is the resistance of a lap-wound armature measured?

7-22 What precautions are necessary to measure the individual resistances of the fields of dc machines?

7-23 How is insulation resistance of the armature and fields of dc machines measured?

7-24 How is the brush neutral position of a dc machine checked and adjusted?

7-25 Discuss two methods of determining the temperature rises in the fields and armature of a dc machine. Which of the methods will give the higher reading?

7-26 What is a heat run and why is it performed?

7-27 Discuss the Swinburne test as applied to dc machines. What are its limitations?

7-28 What are the advantages of the Hopkinson test? What are its limitations?

7-29 What is the function of a high-potential test?

PROBLEMS

7-1 A 10-hp (7.46-kW) shunt motor draws a current of 55 A at 150 V when running at rated full load. What is its efficiency?

7-2 A 5-hp (3.73-kW) motor is 87.5% efficient. What are the total losses?

7-3 A 250-kW dc generator has an efficiency of 90.5% at rated full load. Calculate: **(a)** the input power; **(b)** the total losses.

7-4 A 25-hp (18.65-kW) dc motor has an efficiency of 89.25% at rated full load. If the rotational losses are equal to one-third of the total losses, what is the total copper loss?

7-5 A dc motor develops an output torque of 75 lb·ft (101.69 N·m) at 850 r/min (89.01 rad/s). If the power input is 10,250 W, calculate: **(a)** the total loss; **(b)** the efficiency.

7-6 A 100-kW dc generator is 88.75% efficient. Calculate the power output of the prime mover: **(a)** in kW; **(b)** in hp.

7-7 A 10-kW 250-V dc compound generator is operated at rated speed at no load. The rotational losses were found to be 875 W. The armature resistance is 0.375 Ω, the series field resistance is 0.027 Ω, and the shunt field resistance is 115 Ω. Assuming that there is a 2.5-V brush contact drop, calculate the full-load efficiency when it is connected: **(a)** in short shunt; **(b)** in long shunt.

7-8 A separately excited dc generator has a field resistance of 100 Ω, an armature resistance of 0.015 Ω, and a brush contact drop of 2 V. At no load the generated voltage is 275 V, and the rated full-load current is 150 A. The applied field voltage is 250 V, and the rotational loss is 2,750 W. Calculate: **(a)** the rated terminal voltage and power output; **(b)** the full-load efficiency.

7-9 A 25-kW 250-V shunt generator has an armature resistance of 0.25 Ω, a shunt field resistance of 60 Ω, and a rotational loss of 1,200 W. Calculate: **(a)** the efficiency at full load; **(b)** the load at which maximum efficiency occurs; **(c)** the maximum efficiency.

7-10 A 10-hp (7.46-kW) dc motor has an efficiency of 87.8% and operates 300 h per month driving a circulating fan at rated full load. Calculate: **(a)** the losses; **(b)** the cost of operation per month if electrical energy costs 8.5 cents per kilowatthour; **(c)** the cost of the energy loss per month.

7-11 A 20-hp (14.92-kW) 250-V shunt motor draws a line current of 125 A at rated full load. The armature circuit resistance including the interpoles is 0.27 Ω and the shunt field resistance is 215 Ω. The rotational losses are 850 W and the brush contact drop is 2.5 V. Assuming that the stray load losses are 1% of the output, calculate the efficiency.

7-12 A 50-hp (37.3-kW) 1,200-r/min (125.66-rad/s) 250-V shunt motor has an efficiency of 90.5%. Calculate: **(a)** the input power; **(b)** the line current; **(c)** the developed torque if the torque lost to overcoming windage and friction is 5% of the output torque.

7-13 A Hopkinson (Kapp) test is carried out on two identical dc shunt machines with the following results. The supply voltage was 250 V, the motor current was 40 A, the motor field current was 0.7 A, the generator armature current was 33 A, the generator field current was 1.1 A. The armature resistance of each machine was 0.45 Ω. Calculate the efficiency of each machine.

CHAPTER

8

DC Motor Control

8-1 INTRODUCTION

This chapter presents an overview of traditional manual and electromagnetic methods of controlling dc motors. In addition, an introduction is given to the rapidly expanding role of power electronics in the control of dc drives. An overview of power electronics applied to dc motor control was presented in Chapter 2. Obviously it is impossible to provide an in-depth study of traditional and modern dc motor control techniques in one chapter, and it is recommended that the reader study the specialized material available on these subjects.

8-2 MANUAL STARTERS AND CONTROLLERS

Dc motor applications almost always require that some or all of the following features be provided by the controlling system: line isolation, motor protection, controlled acceleration and deceleration, reversing operation, speed control, braking (dynamic or regenerative), sequential operation, and so on. These features are provided by electrical controllers.

A controller is defined by ANSI/NEMA ICS 1 and ANSI/IEEE Std. 100 as a group of devices which serve to govern, in some predetermined manner, the electric power delivered to the apparatus to which it is connected. That is, the function of a controller is to connect the electric power to a motor and control the motor.

A starter is a simple form of a controller whose main function it is to start and accelerate a motor. The starting and speed control principles of dc motors have been studied in Chapter 6, so we shall now concentrate on the specific equipments and methods.

Faceplate Starters

There are two types of faceplate starters that permit the operator to accelerate a dc motor manually up to a predetermined speed: *three-point* and *four-point starters.* They are often used for starting small shunt and compound motors. The function of a faceplate starter is to insert an additional resistance in series with the armature during starting so as to limit the inrush current to an acceptable value and then, by removing increments of the starting or accelerating resistance, to accelerate the motor up to full speed. In the design of the faceplate starter the resistance value is chosen to ensure that the inrush current is limited to the range of 125–200% of the rated full-load current.

The three-point starter is a manual, that is, hand-operated device in which a spring-loaded control arm is moved in a clockwise direction so that a spring-loaded brass contact shoe makes contact with a number of brass contact studs. As the brass contact shoe makes contact with the brass studs, it short-circuits sections of the accelerating resistance that is connected between the studs, and reduces the amount of resistance connected in series with the armature. The result is that the motor accelerates in a series of steps up to full speed. If the control arm is moved too rapidly, arcing will occur at the studs. This can be minimized by increasing the number of studs and, as a result, reducing the current change per step. The last stud is larger than the other studs because it will be carrying full-load current for the entire operating period. The resistance wire is usually a ferro-chrome-nickel alloy to give high resistance while at the same time it can withstand the high temperatures that occur during the starting process.

The control arm has a soft-iron keeper attached to it so that, in the full ON position, it bridges the poles of a holding coil electromagnet which is connected in series with the shunt field. The general arrangement and the electrical connections are shown in Fig. 8-1.

From Fig. 8-1(b) it can be seen that there are two parallel paths, one through the holding coil, the shunt field, and the field rheostat, which is the low-current path, and the second through the accelerating resistance and the armature, which is the high-current path. This arrangement ensures that the shunt field is established before current flows through the armature. The holding coil also acts as an undervoltage relay, because if the line voltage drops, the attractive force of the holding coil electromagnet decreases and the return spring attached to the control arm will cause it to fly back to the OFF position. It should also be noted that if the shunt field rheostat is adjusted to increase the motor speed, the shunt field current will decrease and the holding coil magnetomotive force will also decrease and may even be less than that required to hold the control arm in the ON position.

This deficiency of the three-point starter is corrected in the four-point faceplate starter where, as can be seen from Fig. 8-2, the holding coil is in parallel with the shunt field. As a result it is unaffected by any changes to the shunt field current.

Although the four-point starter eliminates the possibility of the starter tripping except under low-voltage conditions, there is still a problem. In either the three- or the four-point starter, after the motor has been started and the shunt field weakened to increase the motor speed, there is no provision for restoring the shunt field circuit

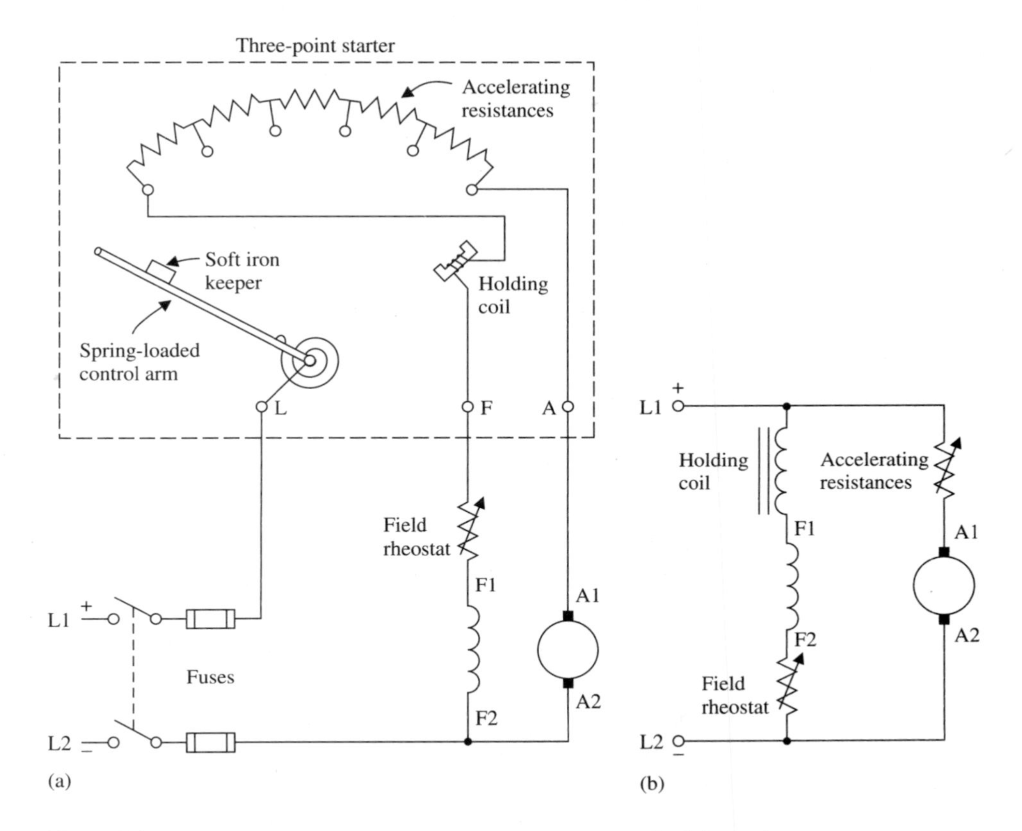

Figure 8-1 Three-point manual faceplate starter. (a) Layout. (b) Schematic.

resistance to its original value, that is, to minimum field circuit resistance, such that for the next start cycle the motor is started with maximum field pole flux and maximum starting torque is developed.

A modified four-point starter incorporating the field rheostat as well as the accelerating resistance has been developed. This starter, known as the *four-point compound starter,* ensures that the motor is always started with maximum shunt field flux. A secondary advantage is that it contains a built-in field rheostat.

Faceplate starters are used in low-horsepower (kW) applications, usually 7.5 hp (5.60 kW) or less. They are limited electrically because of their tendency to arc between the sliding surfaces, which promotes rapid wear of the contacting surfaces. They are not operator-proof and cannot be used in simple control applications such as motor reversal or braking.

An improved and more flexible manual starting arrangement is provided by the *drum controller.* The drum controller consists of a number of stationary contact fingers, insulated from each other and the supporting structure. These fingers are spring loaded and are usually separated by *arc chutes.* In higher current applications they also have *magnetic blowout* protection to extinguish any arcs. The stationary contact

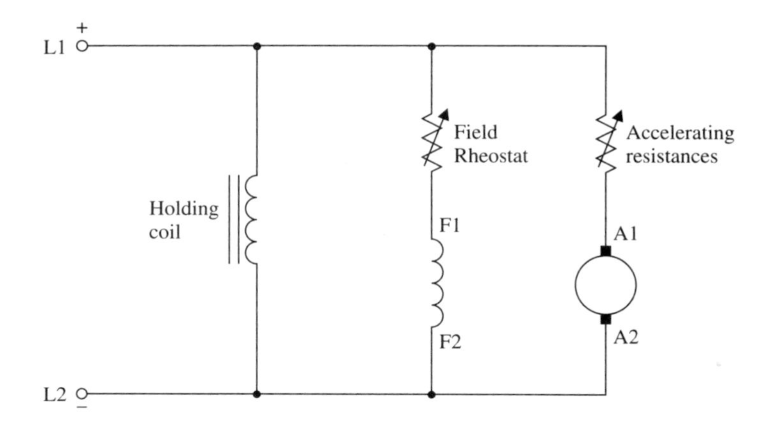

Figure 8-2 Four-point faceplate starter schematic.

fingers make contact with insulated circular copper segments that are moved past the stationary contact fingers. The circular copper segments are mounted on spiders, which in turn are attached to a steel shaft. The steel shaft usually has a handle mounted at one end, which is used to position the moving contacts, the position being determined by a notched wheel with a spring-loaded roller bearing against it. As a result the movable copper segments are moved in a series of well-defined steps. Also since the arc length of the copper segments can be varied easily, as can their horizontal and vertical positions, the whole arrangement permits adjustment to achieve any desired control sequence. Figure 8-3 illustrates the application of a drum controller to the control of a dc compound motor.

The drum controller arrangement in Fig. 8-3 has thermal overloads OL1 and OL2 in series with the incoming lines. Their normally closed (N.C.) auxiliary contacts are in series with the main contractor coil M. Also the undervoltage relay coil UV is in parallel across the line, and its normally closed auxiliary contact is also in series with the main contactor coil. The tapped accelerating resistance provides three steps of acceleration. Figure 8-3(b) shows the position of the copper segments on the rotating drum.

In the OFF position, assuming that the double-pole single-throw (DPST) line-isolating switch is closed, only the undervoltage relay will be energized. At the START position contacts A and B are bridged by the auxiliary segments. As a result, contactor coil M picks up and closes contact M in series with the motor. At the same time auxiliary contact M picks up and seals the circuit of contactor coil M. When the drum controller is moved to position 1, it completes the armature circuit through the full acceleration resistance, and at the same time power is applied to the shunt field. After a suitable interval of time, the drum controller is moved to position 2, the copper segments short-circuit the R1–R2 section of the accelerating resistance, and the motor again accelerates. Once again after a short time interval the drum controller is then moved to position 3, which short-circuits the R1–R2–R3 section of the accelerating resistance, and the motor once again accelerates. Finally when the drum con-

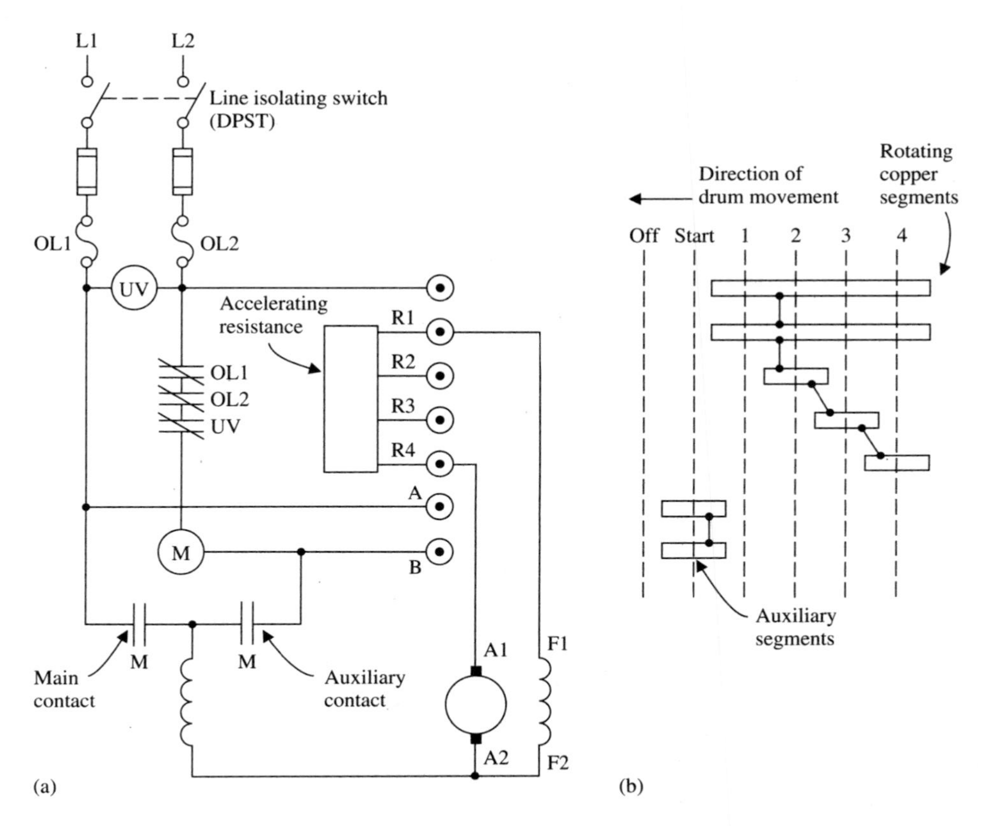

Figure 8-3 Manual nonreversing drum controller with overload and undervoltage protection. (a) Schematic. (b) Contact segment layout.

troller is moved to position 4, the whole accelerating resistance is short-circuited and the motor accelerates to full speed.

If the motor is overloaded, overload relays OL1 and OL2 will be activated and will open their auxiliary contacts in the main contactor coil retaining circuit, thus removing power from the motor. Similarly if an undervoltage occurs, the UV relay will drop out and open the normally closed auxiliary contacts in the coil M circuit. Before power can be restored to the motor, the overloads must be reset, or the line voltage must be restored to cause the UV relay to pick up. It should be noted that if either or both events occur, the drum controller must be returned to the OFF position before the motor can be restarted. This prevents damage to the motor or personnel injury.

Drum controllers are used in many applications, such as crane control. There are a number of advantages for using drum controllers:

1. They are relatively inexpensive as compared to automatic starters.
2. They are compact, mechanically strong, and simple. They can also be easily adapted electrically to meet a wide variety of control requirements.

3. They can be enclosed to meet specific environmental requirements, for example, to be splash-proof, watertight, or explosion-proof.
4. They can be made operator-proof by using time-delay relays to prevent the motor from being accelerated too rapidly.

8-3 AUTOMATIC STARTERS

Most dc motors in modern industrial applications are controlled by automatic starters. There are a number of advantages to using automatic starters, in spite of their increased initial cost:

1. They are not limited in physical size or capacity as are manual starters.
2. They can be controlled locally or remotely.
3. They eliminate operating errors caused by inexperienced or careless operators.
4. They may be integrated into complex manufacturing processes by the use of *programmable logic controllers* (PLCs).

In addition, automatic starters can be operated under open-loop or closed-loop control. Open-loop control means that the operating sequence is totally independent of the motor's performance, while closed-loop control means that the operating sequence depends on the performance of the motor. Definite-time acceleration starters are open-loop systems; current-limit acceleration starters, since they sense the actual motor current and use this information to control the timing of the operating sequence of the starter, are closed-loop systems. The major advantage of the definite-time acceleration starter is its ability to repeat timed operating sequences continuously, especially when teamed up with other machines in a production process.

Definite-Time Acceleration Starters

The timing sequence of definite-time acceleration starters is established by a variety of timing devices, such as mechanical escapement mechanisms, pneumatic dashpot timers, motor-driven timers, and inductive time-delay relays. The disadvantage of this type of starter is that, irrespective of the magnitude of the connected load, the duration of the starting cycle is constant, that is, it does not shorten the duration of the starting cycle under lightly loaded conditions or increase it under heavy loads.

Figure 8-4 shows a definite-time acceleration starter in which the timing sequence is provided by a motor-driven timer causing the movement of a shaft which, in turn, closes a set of contacts. The timing of the complete cycle is adjustable, so that the acceleration rate of the motor may be adjusted. However, this rate will remain constant until readjusted.

Referring to Fig. 8-4, pressing the START push button applies power to contactor coil M provided that the field loss relay FL and the overload relays OL1 and OL2 have not been tripped. The field loss relay is in series with the shunt field, and it will not pick up if the field circuit is open or if there is excessive resistance, that is, if the field rheostat resistance is too high. When contactor M picks up, it closes contact M in the armature circuit and applies power to the armature circuit. At the same time the interlock contacts of contactor M close and provide a sealing path around the START push button. When the sealing contacts M close, the timing relay TR is energized, and the timed to close (T.C.) contacts in series with accelerating relays 1A, 2A, and 3A close in a timed sequence. As 1A, 2A, and 3A close, they short-circuit sections of the accelerating resistance so that when accelerating relay 3A closes, the armature circuit is directly across the line.

Some of the features of this starter are that (1) the STOP-START push-button station can be mounted on or adjacent to the starter, or it can be mounted remotely; and (2) the shunt field has a *thyrite bypass* resistance connected in parallel with it. The thyrite resistance is what is termed a *voltage-dependent resistance,* that is, its resistance decreases rapidly as the voltage across it increases. The thyrite resistor protects the shunt field insulation against voltage breakdown if the field circuit is suddenly opened. It should be noted that voltages on the order of 20,000 V can be induced under this condition.

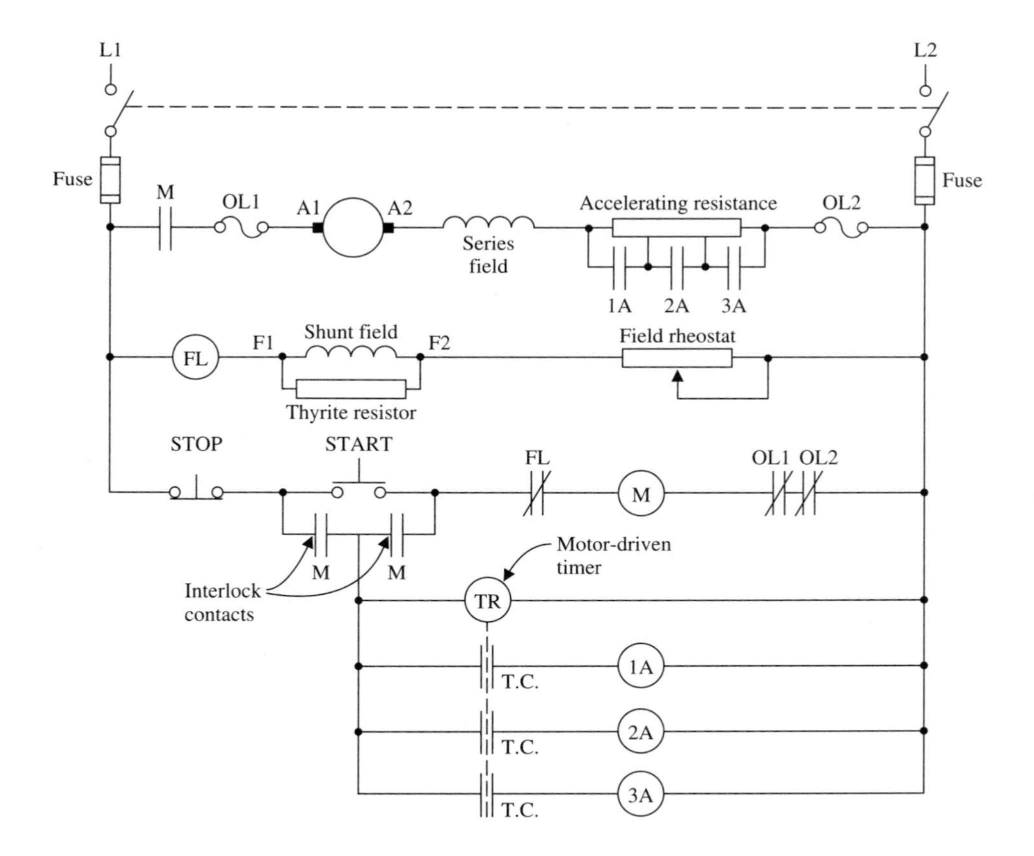

Figure 8-4 Definite-time acceleration dc compound motor starter using motor-driven timer.

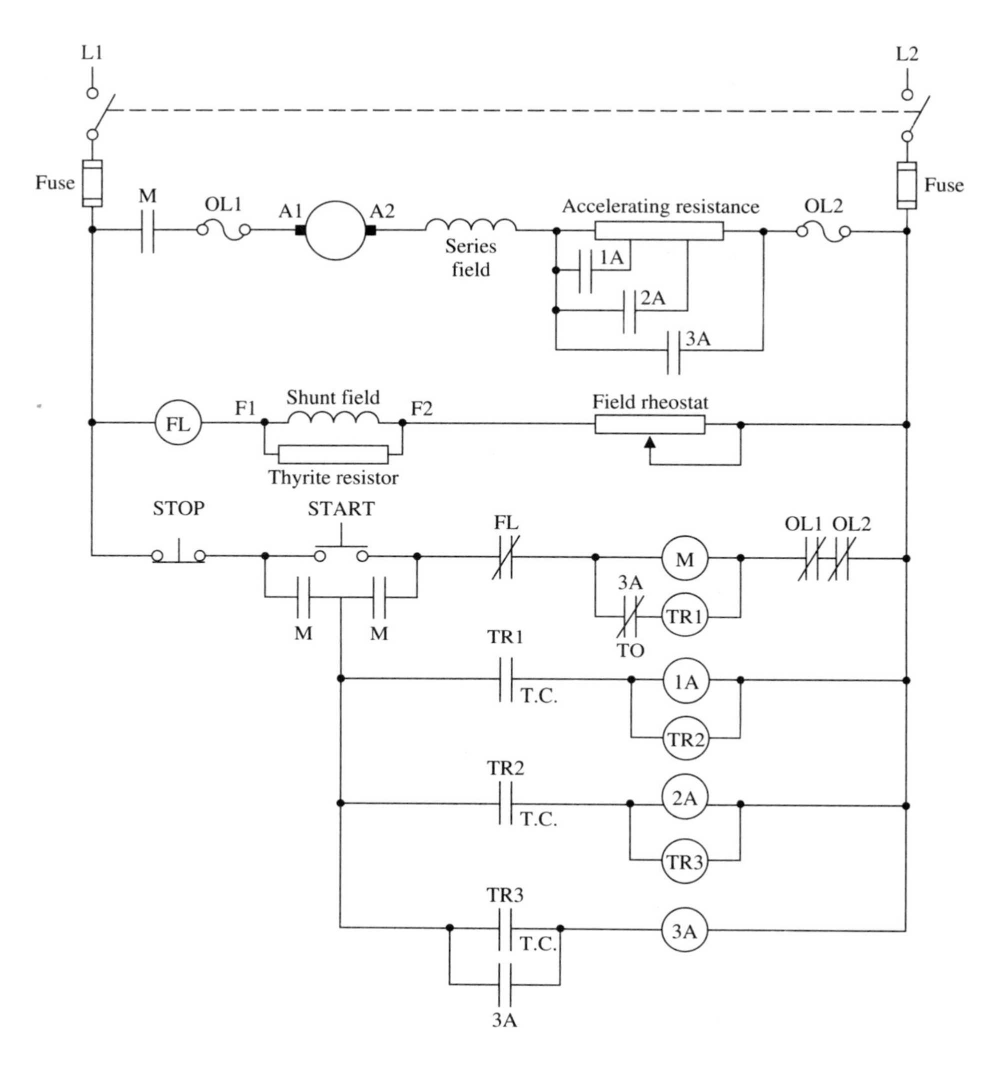

Figure 8-5 Definite-time acceleration dc compound motor starter using pneumatic dashpot timers.

One major disadvantage of this starter is that coils M, 1A, 2A, and 3A are energized whenever the motor is running. This means that all the coils must be designed for continuous duty, and also, the arrangement wastes energy.

The starter shown in Fig. 8-5 uses pneumatic dashpots to obtain the timing control. The pneumatic timer consists of two parts, a solenoid with a set of instantaneous contacts, and the pneumatic unit with two air chambers and a set of timed contacts. The time delay is obtained by controlling the airflow through a variable orifice between the two chambers. The pneumatic dashpot relays are shown as TR1, TR2, and TR3.

The motor is started by pressing the START pushbutton, which permits contactor M to pick up and close contact M in the armature circuit of the compound motor. At the same time timing relay TR1 is energized through normally closed contact 3A, and interlocks M close and provide a sealing path around the START pushbutton and also supply power to the remainder of the control section. After relay TR1 times out, normally open (N.O.) contacts TR1 in series with accelerating relay coil 1A close, causing relay 1A to pick up and close contact 1A, which short-circuits a section of the accelerating resistance, and the motor accelerates. At the same time TR2 is energized, and when it times out, accelerating relay 2A picks up and closes contact 2A, which short-circuits another section of the accelerating resistance, and once again the motor accelerates. At the same time TR3 is energized. When TR3 times out, accelerating relay 3A picks up and short-circuits the last section of the accelerating resistance, and the motor is now running at full speed. At the same time that accelerating relay 3A picks up, it closes the normally open contact in parallel with contact TR3–TC, which seals relay circuit 3A and opens normally closed contact 3A in series with relay TR1. This causes accelerating relays 1A and 2A to drop out, and the only contactors remaining energized are contactors M and 3A.

Current-Limit Acceleration Starters

Current-limit acceleration starters are closed-loop systems, that is, the power flow to the motor is controlled so as to achieve the fastest acceleration possible, irrespective of load variations, without exceeding predefined current limits. The motor acceleration is dependent on current change. There are four basic types of current-limit starters:

1. *Series relay type.* The series relay coils carry the full armature current, that is, the operating coil has only a few turns. In addition, the armature is light and well balanced with a small air gap. As a result the relay has a low inductance and is fast acting. The relay picks up immediately and a high armature current is sensed. When the current decreases, a strong spring opens the contact.

2. *Counter electromotive force type.* This starter is used mainly with low-horsepower (kW) machines, that is, 5 hp (3.73 kW) or less. Voltage-sensitive relays are connected in parallel across the armature, the relays being adjusted to pick up sequentially with the increase in the counter electromotive force as the motor accelerates. This is a speed-limit method of acceleration control.

3. *Holding coil relay type* or *lockout acceleration relay.* A special contactor is used with two operating coils, a *holding coil,* which attempts to hold the contact open, and a *closing coil,* which attempts to close the contact. Both of these coils are in series with the armature circuit and create magnetic fluxes in two independent magnetic circuits so that these opposing magnetic forces act on a spring-loaded pivoted armature that carries the moving contact. The two magnetomotive forces and the magnetic circuits are designed so that, when the armature current is high, the contact is *locked out* by the superior magnetomotive force of the holding coil. As the motor accelerates the armature current decreases, the closing coil magnetomotive force now predominates and the armature moves and closes the contact.

4. *Holding coil and voltage drop type.* This is really a combination of types 2 and 3. The major difference is that the armature current does not pass through the relay coils. As a result smaller coils can be used. This type of starter is most suitable for large-horsepower (kW) motor starting.

The series relay type of current-limit acceleration starter is shown in Fig. 8-6. When the START pushbutton is pressed, the main contactor coil M picks up and closes contact M in series with the armature circuit. The armature starts to rotate with the accelerating resistances R1, R2, and R3, and the accelerating relay 1AR in series with the armature. At the same time the two normally open interlocks M in parallel with the START pushbutton pick up and seal the circuit of contactor coil M. When 1AR picked up with the initial current surge, it opened normally closed interlock 1AR in the contactor coil 1A circuit. As the current in coil 1AR decreases, a point is reached where the armature spring overcomes the relay armature, and the relay contacts open. In so doing interlock 1AR in the contactor coil 1A circuit closes. Contact 1A then closes and short-circuits the R1 section of the accelerating resis-

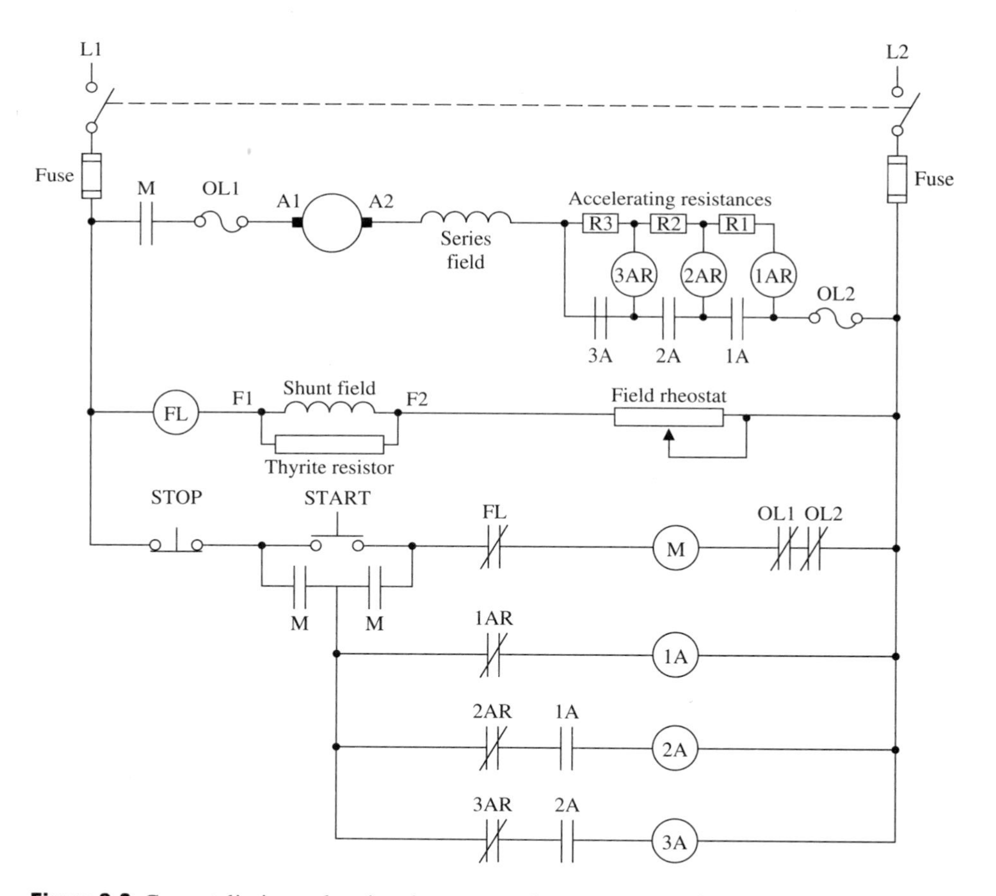

Figure 8-6 Current-limit acceleration dc compound motor starter using series relays.

tance. The armature current increases and flows through series coil 2AR, which then picks up and opens the interlock in the contactor coil 2A circuit. When the armature current decreases as the armature accelerates, relay 2AR drops out and recloses interlock 2AR, and contactor coil 2A picks up and closes contacts 2A, short-circuiting R2. Then series relay 3AR picks up. It opens normally closed interlock 3AR in the contactor coil 3A circuit. When the armature current once again decreases, relay 3AR drops out, contactor 3A picks up, and contact 3A closes and short-circuits the final section of the accelerating resistance R3. At this point the motor is now running at full speed, with the armature directly across the line.

As can be seen, the entire operation of this starter is determined by current changes, that is, the rate of decrease of the armature current.

A current-limit acceleration starter using lockout acceleration relays to start a dc series motor is shown in Fig. 8-7. This starter has three steps of acceleration and requires three lockout relays 1AH–1AC, 2AH–2AC, and 3AH–3AC–3AS, where H designates the holding coil, C the closing coil, and S a shunt coil, which is in parallel with coil M, but is physically wound on the same coil spool as coil 3AC. Included in the contactor coil M circuit are the normally closed overload interlocks OL1 and OL2, and the normally closed overspeed interlock OS. Interlock OS is mechanically coupled to a centrifugal device to detect any overspeeding of the series motor under no- or light-load conditions.

When the START pushbutton is pressed, contactor coil M picks up and closes contact M in series with the armature. At the same time interlocks M in parallel with the START pushbutton close and seal contactor coil M. The motor starts with accelerating resistances R1, R2, and R3 in series. At the same time lockout relays 3AH, 1AH, and 1AC are energized. The holding coils of 3AH and 1AH prevent contacts 1A and 3A from closing. As the armature current decreases, the magnetomotive force of coil 1AH decreases until the magnetomotive force of coil 1AC causes it to close. At the same time contact 1A closes, short- circuiting R1. The armature current is now passing through coils 2AH and 2AC. The current surge through 2AH prevents 2AC from closing. As the motor accelerates, the current through 2AC decreases, and coil 2AC overpowers coil 2AH, closing contact 2A, which short-circuits the R2 section of the accelerating resistance. The armature current now flows through 3AH–3AC, 2AC, and 1AC. Once again as the armature current decreases, coil 3AC picks up and closes contact 3A. At this point the magnetomotive force produced by coil 3AS in parallel with coil M maintains contact 3A closed, and the remainder of the lockout relays drop out, leaving the armature and series field connected across the line.

The foregoing discussion of definite-time and current-limit acceleration starters is by no means meant to be exhaustive, but is intended to provide an understanding of their operation.

8-4 DC MOTOR REVERSAL TECHNIQUES

Dc motors can be reversed by either reversing the main pole flux or reversing the direction of current flow in the armature. It should be noted that if both the field

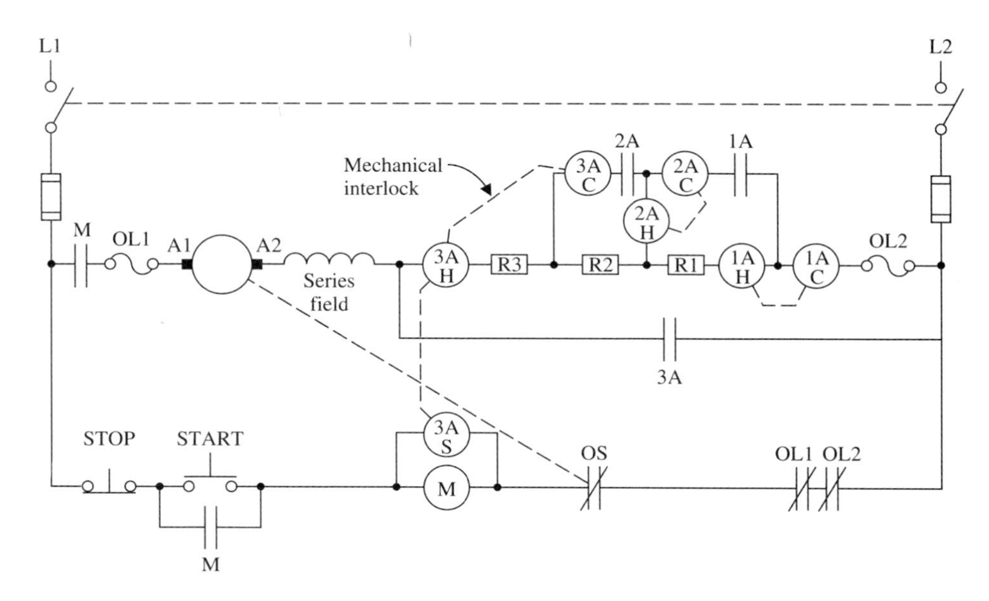

Figure 8-7 Current-limit acceleration dc series motor starter using holding or lockout acceleration relays with overspeed protection.

pole flux and the armature current are reversed at the same time, the direction of rotation will remain unchanged. The preferred method is to reverse the direction of the armature current. There are two main reasons for reversing the armature current in spite of the high currents that may be present:

1. The inductance of the field is very much greater than that of the armature. As a result, when the field circuit is opened, a very high induced voltage is produced ($e = Ldi/dt$), which causes heavy arcing at the switching contacts and possibly a voltage breakdown of the field insulation.
2. Interrupting the field current will create a weak field, which may result in dangerous overspeeding, high armature currents, and unstable operation.

Shunt, series, and compound motors are reversed by either reversing the armature current in the armature circuit, that is, the armature, interpoles, and compensating windings if fitted, or reversing the direction of the main field pole flux, that is, in all shunt and series field windings, with the direction of current in the armature circuit remaining unchanged. Obviously permanent-magnet motors can only be reversed by reversing the armature circuit. In actual practice, prior to reversing the armature current, it is reduced to zero or an acceptable value.

Figure 8-8 shows the basic techniques of achieving dc motor reversal. Figure 8-8(c) illustrates the reversal of a series motor by reversing its field. This is quite acceptable since the inductance of the series field is relatively small as compared to that of a shunt field.

Figure 8-9 shows these concepts applied to an automatic reversing controller for a compound motor. The distinctive features of the control circuit are that (1) the forward F and reverse R contactors are interlocked mechanically and electrically, and (2) a timing relay TR picks up and opens the normally closed contact as soon as the motor is started in either direction, and both the forward and the reverse push buttons are inoperative until some predetermined interval after the STOP push button has been pressed. This arrangement permits the motor to slow down before it is restarted. Then when it is reversed or restarted, it will have the full accelerating resistance in series with the armature.

The controller operates in the following way. Pressing the FOR push button will energize the F contactor coil through the normally closed R contacts. The F contacts in the armature circuit then close, and the F interlock seals the FOR push button contacts. At the same time the normally closed contact in the R contactor coil circuit is

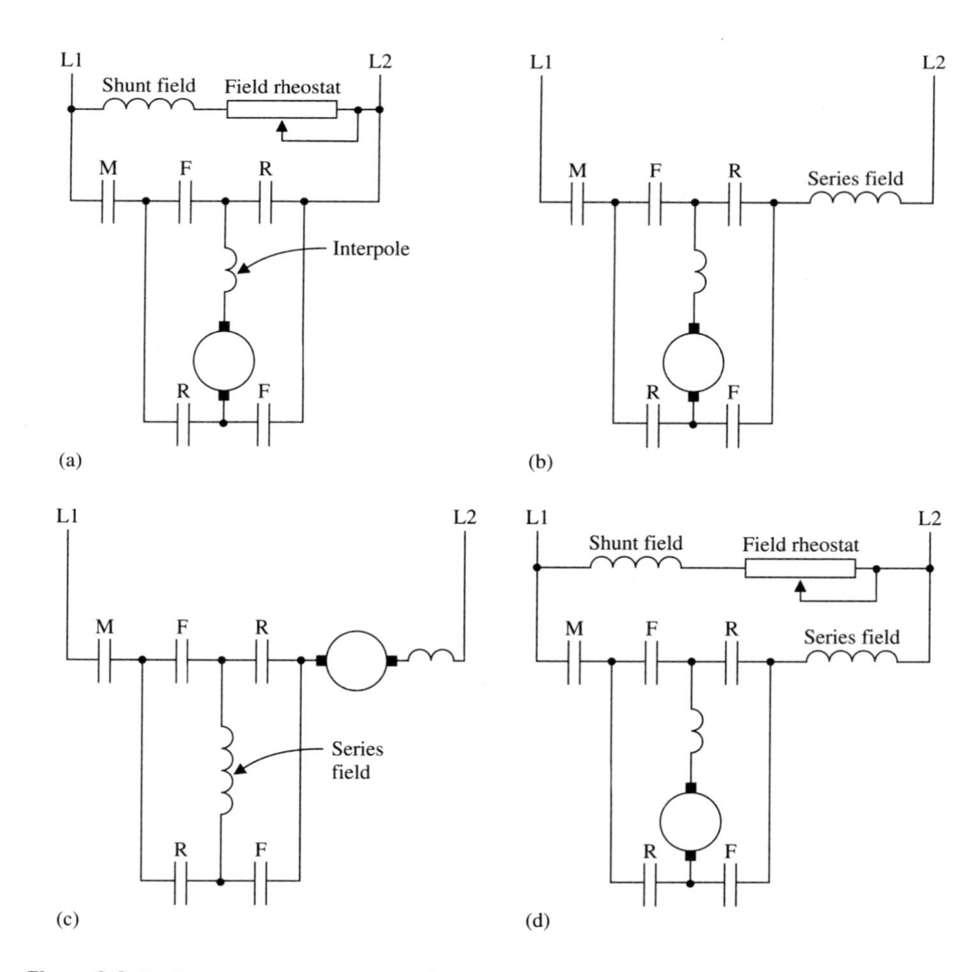

Figure 8-8 Basic dc motor reversal techniques. F–forward; R–reverse. (a) Shunt motor armature circuit reversal. (b) Series motor armature circuit reversal. (c) Series motor field reversal. (d) Compound motor armature reversal.

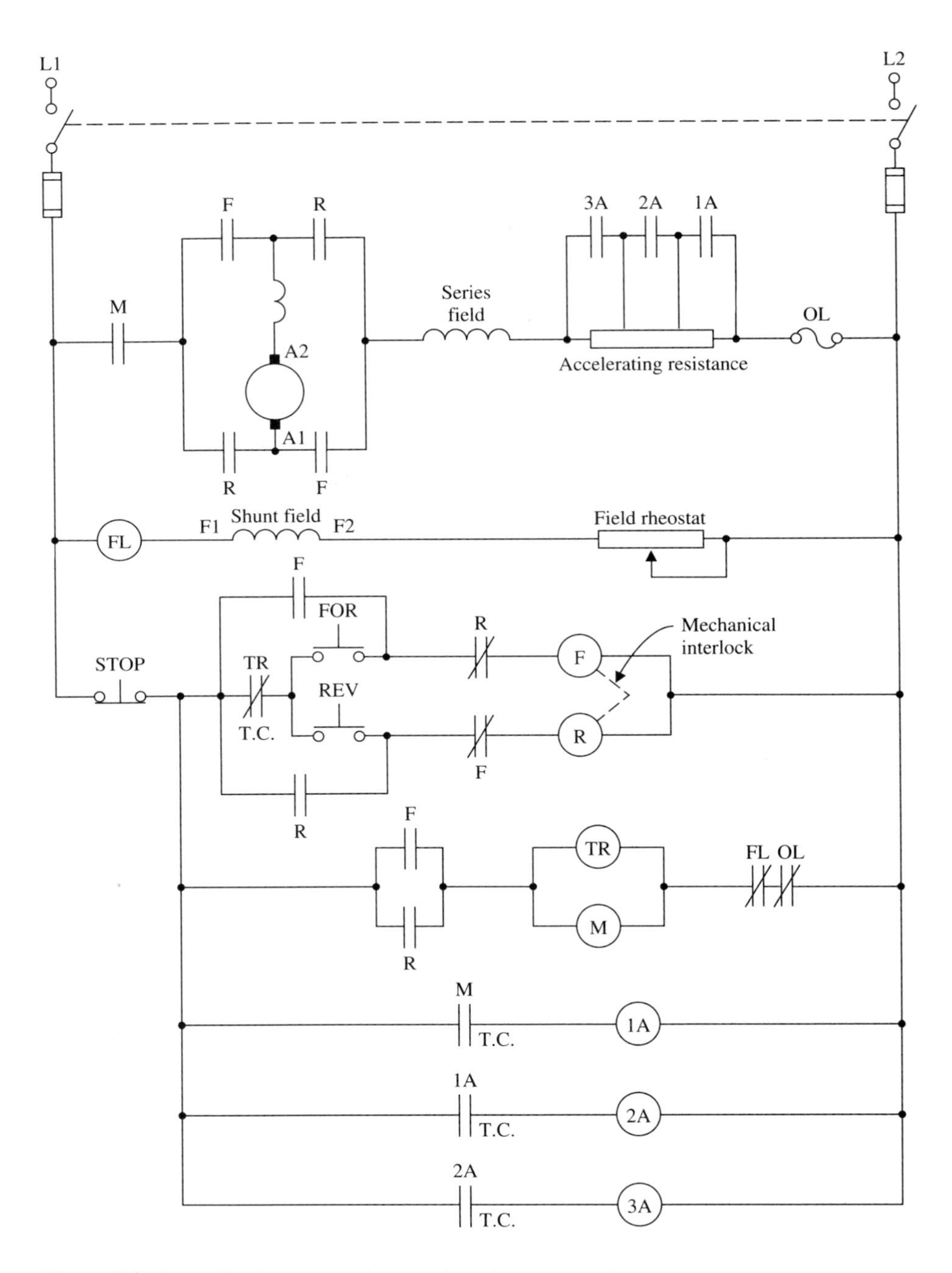

Figure 8-9 Controller for automatic reversing of a compound motor.

opened (electrical interlock), and the F contact in the contactor coil M circuit closes. As soon as contactor coil M is energized, contact M in the armature circuit closes and the armature starts rotating with all the accelerating resistances in series with it across the line. Simultaneously relay TR energizes and opens the normally closed contacts in series with the FOR and REV pushbuttons. When interlock M in the contactor 1A circuit times out, coil 1A is energized and closes contact 1A, short-circuiting a section of the accelerating resistance. It also initiates the timing sequence for interlock 1A in the contactor coil 2A circuit. When interlock 1A times out, coil 2A is energized and contact 2A short-circuits another section of the accelerating resistance, and similarly for the accelerating relay 3A circuit. To reverse the motor, the STOP pushbutton must be pressed since the motor cannot be reversed until contacts TR close after timing out. Then pressing the REV pushbutton will cause the motor to start in the opposite direction and accelerate to full speed through the normal sequence of accelerating relays.

It is advisable in many cases to either reduce the motor speed or stop it completely before attempting to reverse the motor. Small motors will coast to a stop in a short time interval if the power is removed, while large motors, especially if connected to high-inertia loads, will coast for several minutes, or if there is an overhauling load, they may even accelerate after the power is removed. To shorten the time interval before going to the next step in the operating cycle, it is necessary to brake the machine. Braking can take many forms, ranging from completely reversing the power applied to the armature, known as *plugging,* to dynamic and regenerative braking.

8-5 PLUGGING

Plugging is an electrical braking technique in which a strong countertorque is developed by reversing the power applied to the armature, which, in turn, slows down the motor armature rapidly. However, if it is not removed at the point of reversal, the motor will start turning in the opposite direction. At the instant that the reversed power is applied, the generated voltage E_g and the terminal voltage V_t are approximately equal and additive. Then by Ohm's law $I_a = (V_t + E_g)/R_{ac}$, which because of the low resistance of the armature circuit will give rise to a very high armature current. Normally the armature current is not permitted to exceed 150–200% of the rated full-load current. The current is limited to this range by inserting a *plugging resistance* in series with the armature circuit during plugging. At the point where the armature will come to rest, a centrifugal switch, called a zero-speed switch ZSS, will initiate the removal of power from the armature circuit. The plugging technique is illustrated in Fig. 8-10.

The distinctive feature of this circuit is the zero-speed switch ZSS, which is connected directly to the motor shaft. The normally open contact of the ZSS closes, immediately the armature starts rotating. When the START pushbutton is depressed, the main contactor coil M is energized, contacts M in the armature circuit close, and interlocks M in the coil 1A circuit and in the ZSS close. At the same time the undervoltage relay UV is energized, closes the sealing contact around the START contacts,

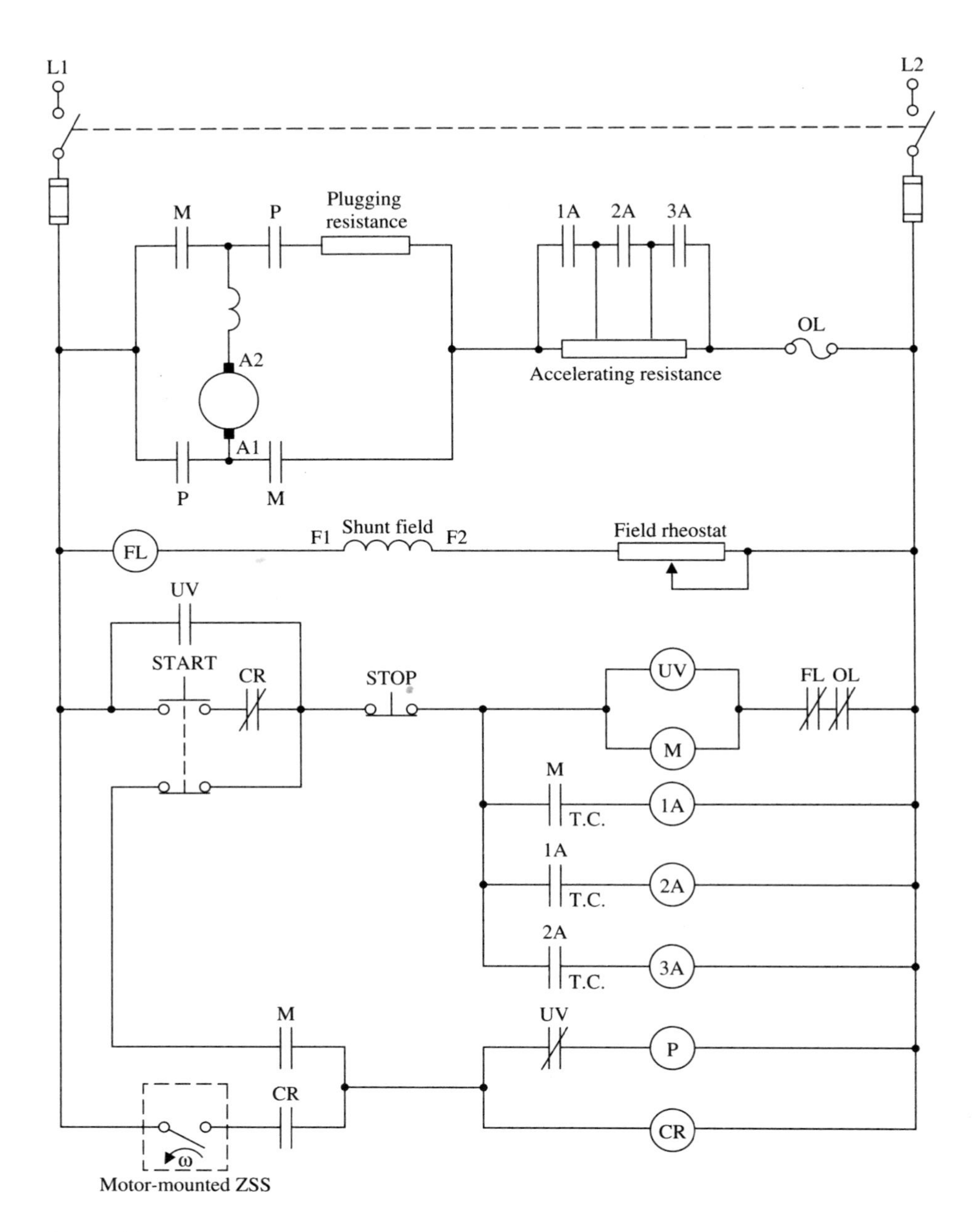

Figure 8-10 Plugging applied to definite-time nonreversing acceleration starter for a dc shunt motor.

and opens the normally closed contact in series with the plugging coil P. Simultaneously control relay CR picks up and opens the normally closed interlock in series with the START pushbutton and closes the interlock in series with the ZSS. The motor accelerates normally, and the armature circuit will be across the line after contact 3A closes.

When the STOP pushbutton is depressed, coils M and UV are deenergized, the sealing circuit around the START pushbutton is opened, and the UV interlock in series with the plugging coil P closes. Simultaneously contacts M in the armature circuit open and all the accelerating relays drop out. As coil P is energized through the ZSS and interlock CR, it closes contacts P in the armature circuit. This causes the armature current to reverse, its magnitude being limited by the series combination of the plugging and accelerating resistances. The armature will then decelerate rapidly until it stops. At this point the ZSS opens and deenergizes both the plugging relay P and the control relay CR, which in turn causes contacts P in the armature circuit to open, and interlock CR in series with the START pushbutton recloses.

8-6 ELECTROMAGNETIC BRAKING

A dc motor can be stopped by removing the power, in which case the inertia of the motor and load will cause the armature to decelerate slowly to a standstill over a period of time. This time interval may be unacceptable in many applications. Also there are some instances where the load is overhauling, that is, the load drives the motor. A good illustration is the cage of a mine hoist. There are two methods of braking: electromagnetic and mechanical. It should be noted that electromagnetic braking will cause rapid deceleration, but it will not stop and hold a load in a given position. To hold the load, it must be stopped by a mechanical brake. Electromagnetic braking techniques depend on the production of a countertorque, that is, if the direction of the field pole flux remains unchanged and the armature current is reversed, a countertorque will be produced which opposes the armature rotation. There are two common methods of producing electromagnetic braking: dynamic and regenerative braking.

Dynamic Braking

Dynamic braking depends on the production of the counter electromotive force or speed voltage that is always present when the armature conductors cut the field pole flux. The counter electromagnetic force opposes the applied terminal voltage and limits the amount of current drawn from the power source. If the armature circuit is disconnected from the power source and a low-resistance high-wattage resistor is connected across the armature terminals with the main field remaining energized, then the load inertia will act as a prime mover and drive the motor so that it now acts as a generator. Since the polarity of the generated voltage is opposite to that of the original terminal voltage, the direction of the armature current is reversed. The reversed current flow will produce a torque which acts in the opposite direction to the armature rotation. As a result the angular velocity of the armature decreases rapidly, as will also the amplitude of the generated voltage. At standstill the generated voltage is zero, and therefore there is no current flow, and also no torque is produced. The rate of braking is determined by the ohmic value of the dynamic braking resistance. Because the braking torque is proportional to the armature current, a low-ohmic-value resistor will stop the armature and connected load in the shortest possible time. Although dynamic braking is effective, it is not energy efficient since energy output from the motor into the dynamic braking resistance is dissipated as heat. Dynamic braking can be easily implemented in separately excited, shunt, permanent-magnet, and lightly compounded motors. It can also be applied to series motors, but it then requires all four leads to be brought out so that the armature can be connected across the field. However, it is not too effective because the series field flux is decreasing as the armature slows down. The principle of dynamic braking applied to a shunt motor is shown in Fig. 8-11, where the dynamic braking resistance is shunted across the armature terminals by either a dynamic braking contactor or a thyristor.

In Fig. 8-11 (a), when contact M opens, thus removing the applied power from the armature, the dynamic braking contact DB closes and places the dynamic braking resistance across the motor armature terminals. Otherwise, as shown in Fig. 8-11(b), the SCR is substituted for contact DB and is turned on by applying a gate pulse.

These arrangements can also be used with a permanent-magnet motor. This motor has the advantage that in the event of a power failure it can still be stopped using dynamic braking, which is impossible with a shunt or lightly compounded motor.

Regenerative Braking

With both friction braking and dynamic braking the kinetic energy stored in the rotating mass is expended as heat during the braking process. Regenerative braking is a form of dynamic braking, but the stored energy is returned to the power source as electric energy. This situation can only occur when the counter electromotive force is greater than the terminal voltage, that is, under overhauling load conditions. Normally in the motoring mode the load torque opposes the armature rotation, but in an overhauling load situation, the load torque is acting in the same direction as the armature rotation. As a result the armature is accelerated, with the result that the counter electromotive force E_c increases. When E_c is greater than V_t, the machine now acts as a generator, and the power flow is from the motor acting as a generator to the power source. The armature current produces a braking torque which is proportional to $E_c - V_t$, that is, the braking torque is controlled by the magnitude of E_c, which in turn can be controlled by adjusting the motor field current.

Regenerative braking, just as dynamic braking, cannot stop the rotation of the armature completely. However, in situations where the load is accelerating the armature, such as an electric train descending a grade, regenerative braking prevents excessive speed buildup and is simple to control as well as being energy efficient. The final stopping of the armature must still be achieved by mechanical braking.

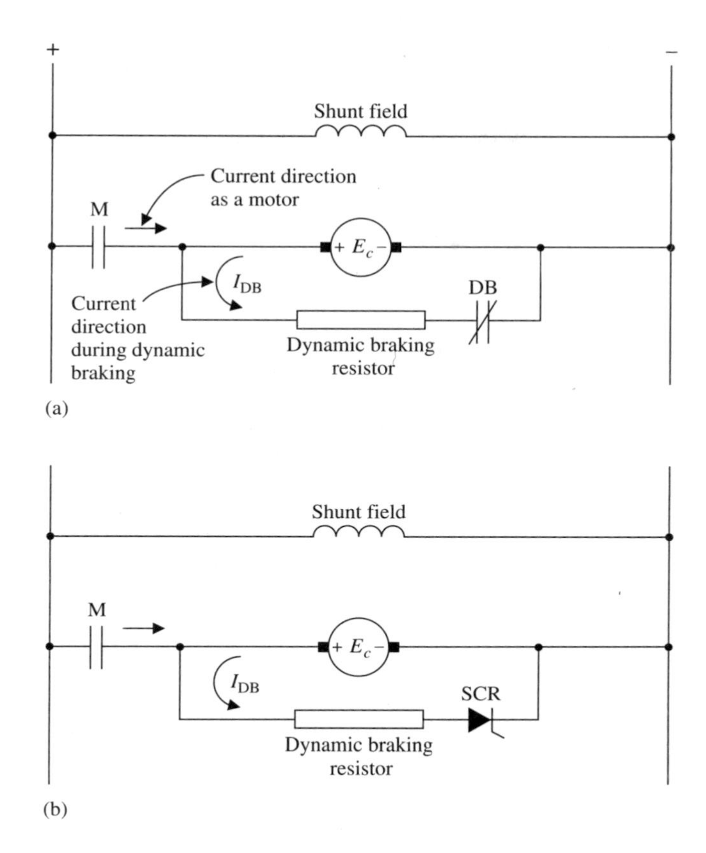

Figure 8-11 Dynamic braking of shunt motor. (a) With dynamic braking contactor. (b) With SCR.

Friction Brakes

Various types of friction brakes are used with dc and ac machines to prevent motor rotation when power is not applied. Friction brakes can be an integral part of the motor, that is, built onto the nondrive end of a double-shaft motor, or they can be mounted separately from the motor, applying the braking force to a braking pulley mounted on the load shaft or, as is the case with some mine hoists, as a disk brake system mounted on the end of the winding drum.

Brakes that are actuated mechanically or hydraulically are not normally directly connected to the motor controller. Magnetically operated friction brakes are electrically controlled from the motor controller, that is, the braking torque is applied by means of a spring, and removed by energizing an electromagnet to overcome the force of the spring.

There are two methods of providing excitation for the electromagnet: (1) series brakes, which are designed to carry the armature current, release when the current builds up to approximately 40% of rated motor current, and reset when the current drops to approximately 10% of rated current; and (2) shunt brakes, which derive their power from the controller and release the brake with approximately 80% of rated voltage. Because of the high currents involved, series brakes are usually rated for $^1/_2$- and 1-h duty cycles. But because of the relatively small number of coil turns their inductance is low, and as a result they are fast acting. The shunt brake with a large number of turns has a large time constant and thus is not fast acting. But because of the lower excitation currents, they are usually rated for 1 h or continuous duty cycles. In all cases they apply a braking force in the event of a power failure.

8-7 JOGGING

Jogging or *inching* is defined by NEMA as "the quickly repeated closing of a circuit in order to start a motor from rest for accomplishing *small* movements of the driven machine." Jogging is manually controlled by a machine operator to accurately position tools, equipment, machines, cranes, and so on, in small incremental steps. By its very nature it is obvious that it is not desirable to permit the motor to accelerate to normal speeds during jogging. As a result jogging momentarily applies power to the main contactor coil, but does not permit the accelerating contacts to pick up. Therefore the speed of rotation of the armature is limited by the full value of the accelerating resistance.

8-8 SOLID-STATE SPEED CONTROL

In Chapter 6 we determined that the torque developed by a dc motor is directly proportional to the armature current I_a and the flux per pole, that is,

$$T_d = K_g \Phi I_a \text{ ft} \cdot \text{lb} \qquad \textbf{(4–16)(E)}$$

or, in SI units,

$$T_d = k_g \phi I_a \text{ N} \cdot \text{m} \qquad \textbf{(4–16)(SI)}$$

The horsepower (kW) output is proportional to both torque and speed. Therefore

$$\text{hp} = KTS \qquad \textbf{(8–1)(E)}$$

or

$$\text{kW} = kT\omega \qquad \textbf{(8–1)(SI)}$$

Equations (4-16) and (8-1) show that if the armature current I_a remains constant, then:

1. By varying the applied armature voltage V_a, with the field pole flux remaining constant, the output torque will remain constant, and the output horsepower (kW) will then be directly proportional to the speed. This is known as the *constant-torque mode.*

2. By applying nameplate rated voltage to the armature, and weakening the field pole flux, the output horsepower (kW) is constant and the output torque decreases. This is known as the *constant-horsepower (kW) mode.*

Figure 8-12 illustrates these modes of operation. It can be seen that when the field pole flux is maintained constant at its rated value, variation of armature voltage will produce rated current at any speed between zero and base speed. It can also be seen that the output horsepower (kW) of the motor is directly proportional to speed, and that rated torque is produced at all speeds up to base speed. Armature voltage control drives are constant-torque drives and are used in such applications as general machinery, hoists, conveyers, and printing presses. Constant-torque drives represent about 90% of all dc drives. It should also be noted that in the dc shunt motor the field and armature are in parallel. As a result it is not possible to maintain a constant field and at the same time vary the armature voltage to obtain speed control. Consequently dc drive motors are almost exclusively separately excited motors.

Constant-horsepower (kW) drives have an output torque that varies inversely with speed. Typical applications are metal-cutting machines that operate over a wide speed range, electric traction motors, and cranes.

When operating above base speed with field weakening techniques, the range of speed control is dependent on the type of motor, that is, whether it is designed for fixed-speed operation or adjustable-speed operation. If designed for fixed-speed operation, the speed may be varied over a 1.5:1 speed range, and for adjustable-

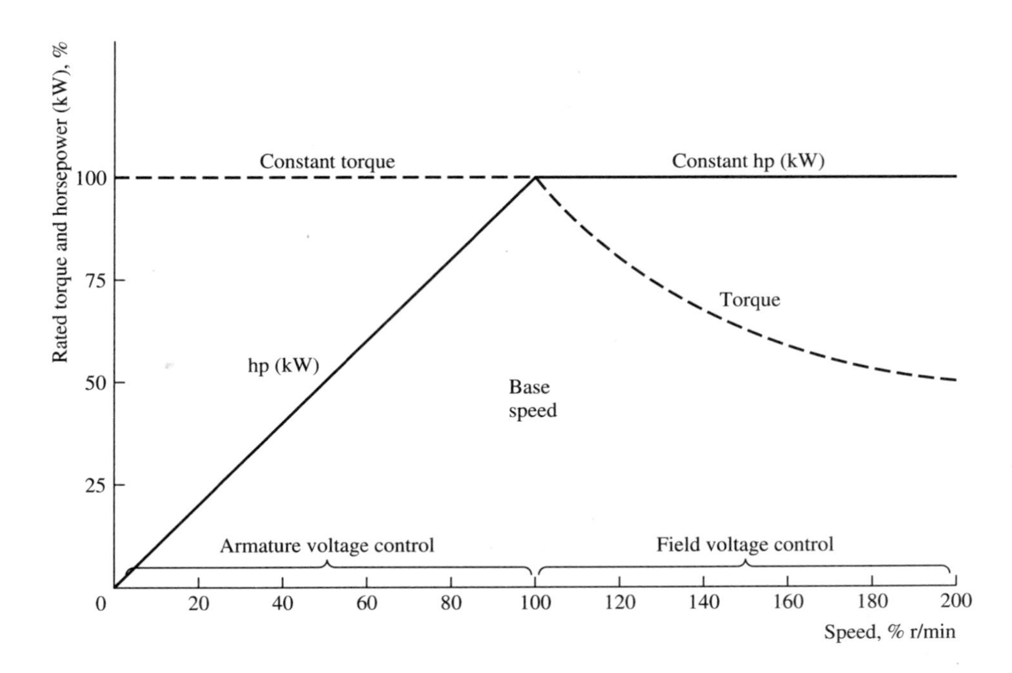

Figure 8-12 Constant-torque and constant-horsepower (kW) curves.

speed motors a 6:1 speed range can be obtained. The speed range is limited by mechanical factors such as centrifugal forces, and electrically by the current that can be commutated because the effects of armature reaction increase as the field pole flux is decreased, because the time available to complete commutation is decreased, and also because the speed regulation worsens. High-speed motors require special frames and stabilizing windings, that is, a few series turns wound on the main field poles which prevent the flux from decreasing below an established minimum and therefore prevent overspeeding. All of these factors increase the cost of the motor.

Closed-Loop Phase-Controlled Systems

The block diagram in Fig. 8-13 illustrates the principle of closed-loop control applied to an armature voltage-controlled dc motor.

There are two requirements that must be met by the control system: (1) it must limit the inrush current during starting to a predefined maximum and (2) it must maintain the output speed within an acceptable deviation for any designated speed setting.

Referring to Fig. 8-13, the operator designates the desired speed, usually known as the *set speed,* by varying the setting of a potentiometer. The set speed signal is usually a dc voltage varying from 0 to 10 V. This signal is applied to the *ramp generator,* whose function it is to control the rate of advance of the gate firing pulses applied to the thyristors in the thyristor phase-controlled converter. The ramp generator generates an output dc voltage signal representing the desired speed. An acceleration ramp circuit determines the rate at which this output voltage is achieved. The output signal from the ramp generator is compared at the summing junction against a suitably attenuated dc signal from the tachogenerator TG connected to the motor

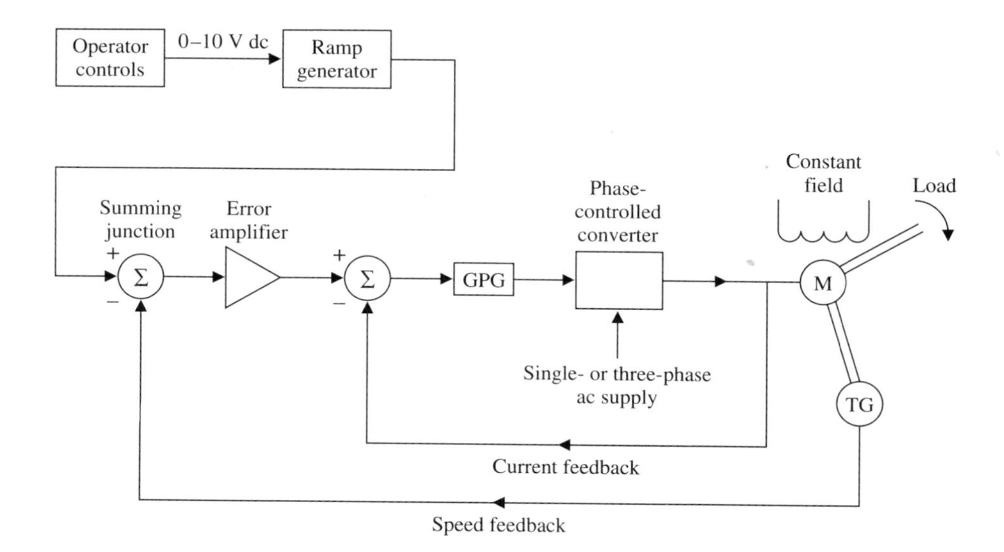

Figure 8-13 Block diagram of a phase-controlled dc motor speed control.

shaft, which represents the actual motor speed. The difference between the two voltages is known as the *speed error signal.* If the ramp generators output voltage is, for example, +5 V, and the output of the TG is −4 V, there is a positive error signal. It causes the gate firing points of the thyristors in the phase-controlled converter to be phased forward, that is, the firing delay angle is reduced and the output voltage of the converter is increased, which in turn causes the motor to accelerate. As the motor accelerates, the output of the TG increases, and the speed error signal decreases toward zero. If the motor accelerates beyond the designated speed, the speed error signal becomes negative, and the firing point of the thyristors is retarded, thus reducing the output voltage of the converter. It should be noted that the motor speed never remains constant at the set speed, but continually oscillates about that point. The amount of oscillation is a measure of the quality of the control system. The error signal is applied to an *error amplifier,* whose function it is to increase the amplitude of the signal. The output of the error amplifier is compared at a summing junction with a dc voltage representing the motor armature current. The output signal from the summing junction also modifies the firing delay angle, that is, if the current is in excess of a predetermined maximum, it will cause the firing delay angle to be retarded, thus reducing the armature voltage and current. This *current limit* feature ensures that the inrush current to the armature is within safe limits, and has completely eliminated the need for the accelerating resistances that are required by conventional automatic starters. It is also interesting to note that the whole system is acting as a current-limit acceleration starter.

The error signal produced by the second summing junction is applied to the *gate pulse generators* GPG, one for each thyristor. The GPGs usually use NAND or CMOS logic and are responsible for producing an output pulse, which when applied to the *gate pulse transformers* or *optocouplers,* will produce the actual gate firing signal applied to the thyristor. Usually incorporated into the GPG section is an electronic overload, which will disconnect the converter from the ac supply under heavy overload or stalled armature conditions.

The TG has a linear output characteristic, that is, its voltage is directly proportional to the motor speed. The TG's feedback signal performs three important functions: (1) it improves speed regulation; (2) it reduces speed drift; and (3) it ensures speed reproducibility.

Closed-loop feedback control systems are error actuated, that is, an error must be present before the system can respond. As a result the actual motor speed will not remain constant, but will vary slightly about the set speed. The amount of the variation is the measure of the speed-regulating capability of the drive control system. Armature voltage feedback senses the counter electromotive force as the speed signal and usually produces a speed-regulating capability of ±2%, whereas an ac or dc tachogenerator speed sensor will provide a speed regulation of ±1%. Speed regulation capabilities of ±0.001% can be provided by using *phase-locked-loop* (PLL) techniques.

Typical speed-torque curves for a constant-torque drive are shown in Fig. 8-14 for a number of different speed settings. These are a family of parallel curves, one for each speed setting. The minimum operating speed of the motor is determined by

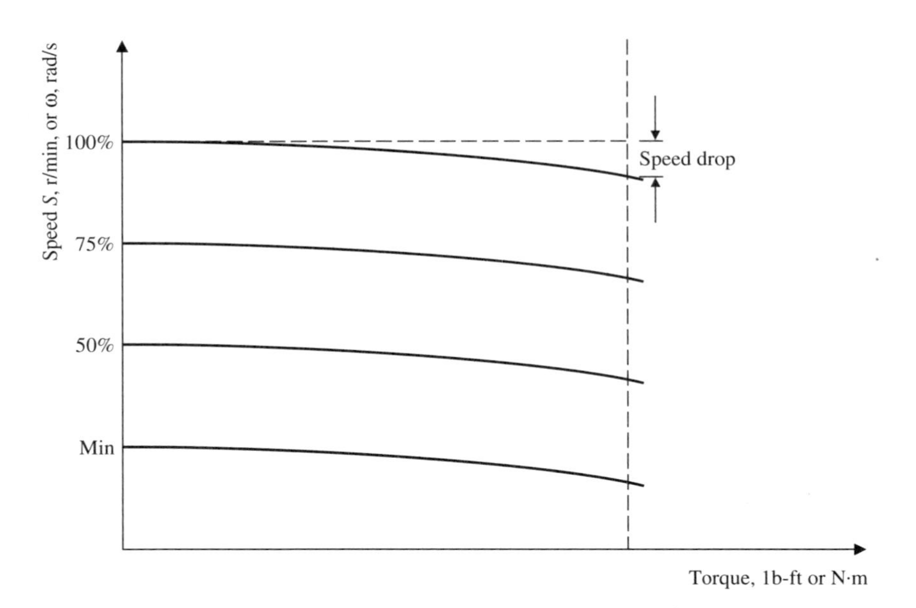

Figure 8-14 Typical speed-torque curves of variable-speed drive.

its ability to dissipate the heat resulting from the copper and iron losses. A number of drive motors are fitted with an externally powered ventilation fan that is independent of the motor rotation. The maximum torque is limited by the current limit circuit since it is not a motor characteristic. It should also be noted that the speed regulation, which is the ability of the drive motor to maintain its speed under varying load conditions, is usually defined in terms of the base speed, and as a result when the speed setting is reduced, the drop in speed between no load and full load remains constant, but the actual speed regulation increases.

In constant-torque applications, that is, constant-field variable armature voltage control, the field excitation is usually obtained from an uncontrolled diode rectifier connected across the ac supply lines. In large-horsepower (kW) machine applications a separate controlled rectifier is the preferred method of providing the field supply.

Commercial SCR adjustable-speed drives are available as off-the-shelf items in single- or three-phase, fully or half-controlled configurations. They feature, usually as standard items, instantaneous electronic overload protection, phase loss and phase sequence protection, isolated control circuit power supplies, semiconductor and control circuit fuse protection, *RC* snubber networks, and so on. Optional controls that can be added to the basic configuration include dynamic braking, jogging, plugging, reversing, motor thermal protection, diagnostics, field loss protection, process control followers, and speed and load indicators among others.

Variable-speed drives are available in the following ranges: one-quadrant units 1–1,000 kW (1.34 to 1,340 hp), with custom-built units available up to 10,000 kW

(13,400 hp); two-quadrant units 18–830 kW (24.12–1,112.20 hp), with custom-built units up to 10,000 kW (13,400 hp); and four-quadrant units 18–416 kW (24.12–557.44 hp), with custom-built units up to 10,000 kW (13,400 hp).

In the past it was common practice to use three-phase half-controlled converters because of lower initial costs. However, it has been found that the worsened form factor of the mean output dc voltage has created commutation as well as motor overheating problems. As a result the drive motor has to be derated. On the other hand if a three-phase full-converter is used, practically all dc motors can be used without derating or commutation problems. The three-phase full-converter has the added advantage of being able to operate in the synchronous inversion mode, and can be effectively used in an overhauling load situation to provide regenerative braking.

If regenerative braking is not possible, dynamic braking can be used. Usually this is achieved by placing a dynamic braking resistance across the armature terminals when the converter is disconnected by pressing the STOP pushbutton.

There are three commonly used methods of reversing dc drive motors: (1) reversing the armature current, which is the preferred method; (2) reversing the field pole flux; and (3) using a four-quadrant converter. The latter method is illustrated in Fig. 8-15.

Four-Quadrant or Dual Converters

This method, which completely eliminates the use of contactors, is best explained by means of Fig. 8-15. Assume that converter 1 is driving the motor in the forward direction and converter 2 is blocked, that is, no gate signals are being supplied to the thyristors. Then to reverse the motor, the gate firing pulses being supplied to the thyristors in converter 1 are retarded, which causes the firing delay angle to be increased, thus decreasing the output voltage of converter 1. This also causes the armature current to decrease. Using an ammeter shunt in the armature circuit enables the control electronics to detect the point at which the armature current becomes zero. At this point the gate signals are removed from converter 1 and applied to converter 2. As the gate signals are advanced, the motor now accelerates up to the desired speed in the opposite direction.

This configuration is used in many applications, such as mine hoists and reversing rolling mill drives, in which forward and reverse motor operation is required. By simultaneous firing angle control of both converters very rapid reversals of speed and torque can be achieved. Converters operated in this manner are known as dual-converters, and the armature current can flow in either direction in the armature circuit. The ideal control technique is to adjust the firing delay angles of both controllers to produce the same output voltage, that is, $v_{d\text{pos}} + v_{dneg} = 0$, $v_{d\text{pos}}$ and $v_{d\text{neg}}$ being the instantaneous voltages of the positive and negative converters, and $V_{d\text{pos}} + V_{d\text{neg}} = 0$, $V_{d\text{pos}}$ and $V_{d\text{neg}}$ being the average voltages of the positive and negative converters. This condition cannot be met because the ac ripple components of the output dc voltages do not occur simultaneously. As a result, even if the firing delay angles are varied simultaneously, the gating signals to the negative converter are suppressed when the armature current is positive, and similarly, those of the positive converter

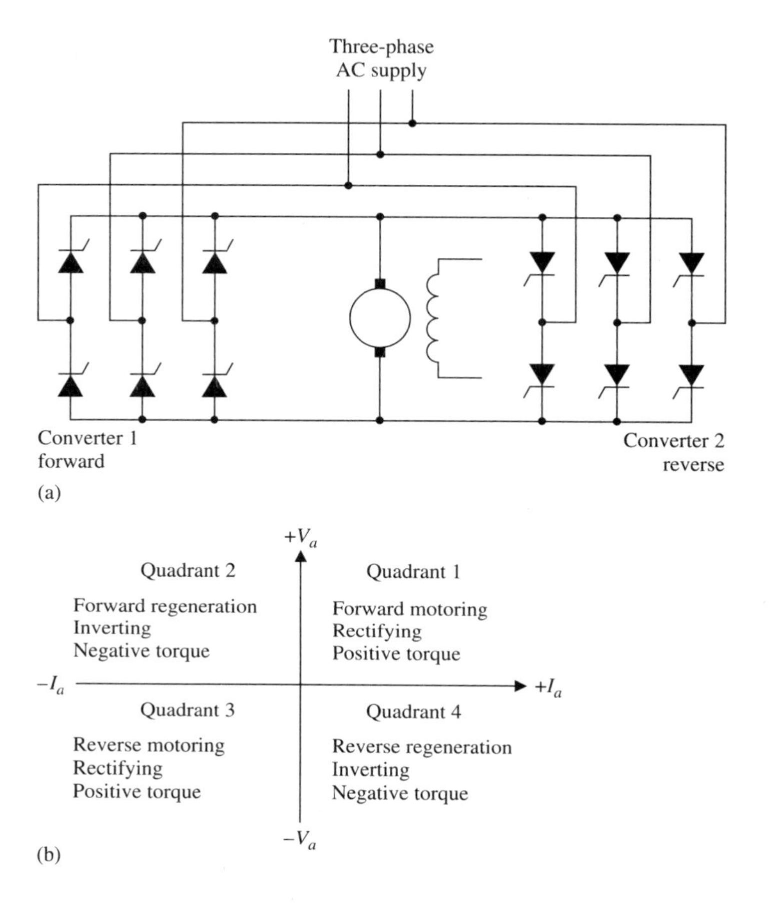

Figure 8-15 Reversible operation of dc drive using a four-quadrant converter. (a) Schematic. (b) Converter V–I plane.

are blanked when the armature current is negative. As a result only one converter at a time is supplying power and the other converter appears as a high impedance. Effectively the transition from motoring to regeneration, that is, quadrant 1 to 2 or quadrant 3 to 4, or vice versa, takes place in three steps: (1) The current in the rectifying converter is reduced to zero by firing angle control. (2) Its gating pulses are suppressed and the gating pulses to the other converter are activated. (3) Its firing delay angle is reduced to increase the current from zero. The polarity of the voltage at the armature terminals has remained unchanged. When operated in this manner, the converters are operating in the *circulating current-free mode.*

Operation in the circulating current-free mode is illustrated in Fig. 8-16. The actual speed voltage signal suitably attenuated to match the desired speed voltage is obtained from a TG mechanically connected to the drive motor. This signal is compared against the desired speed signal. The resulting error signal in turn is compared against a voltage signal representing the actual motor current. Provided the drive is not in current limit, the resulting error signal is supplied to the steering logic. The function of the steering logic is to determine which converter will be in conduction. This is determined by (1) the signal representing the desired direction and (2) the current sensor, a shunt or dc transformer (DCCT), in the armature circuit that senses current direction and magnitude, whose function it is to inhibit the changeover of converters until the armature current becomes zero. If all these conditions have been met, then a reference signal will be applied to the appropriate gate pulse generators, which will then apply the correctly sequenced firing pulses to the thyristors, with the correct firing delay angle.

Phase-Locked-Loop Speed Control

In industrial processes there are many applications where the inherent speed or the open-loop regulation of the motor is unacceptable. In many other applications closed-loop feedback techniques are used to obtain fast response and accurate speed control. So far we have looked briefly at two methods of sensing variations in the speed of the drive motor: armature counter electromotive force sensing and tachogenerator speed sensing. Neither of these analog systems is completely satisfactory where precise speed control combined with a fast response to load disturbances is a major requirement. The inability of analog feedback control systems to overcome these problems may be eliminated by using a digital phase-locked-loop (PLL) control system.

The basic concept of a PLL system is illustrated in Fig. 8-17. In its simplest form the PLL consists of a *phase detector* or *comparator*, a *loop filter*, and a *voltage-controlled oscillator* (VCO). As can be seen from Fig. 8-17(a), two pulse trains are applied to the phase detector, f_i being the reference pulse train and f the output frequency of the VCO. The phase detector, after comparing these two signals, produces

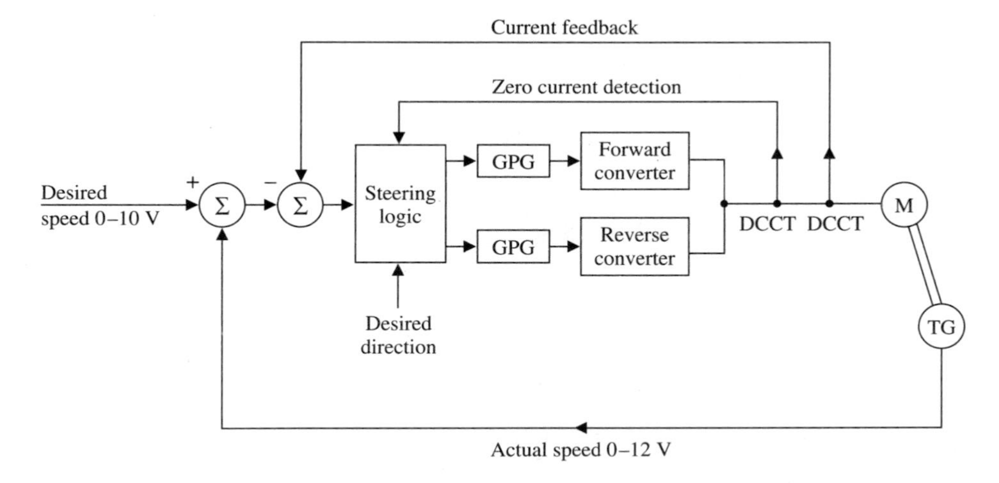

Figure 8-16 Block diagram of closed-loop control system controlling a dual-converter dc drive.

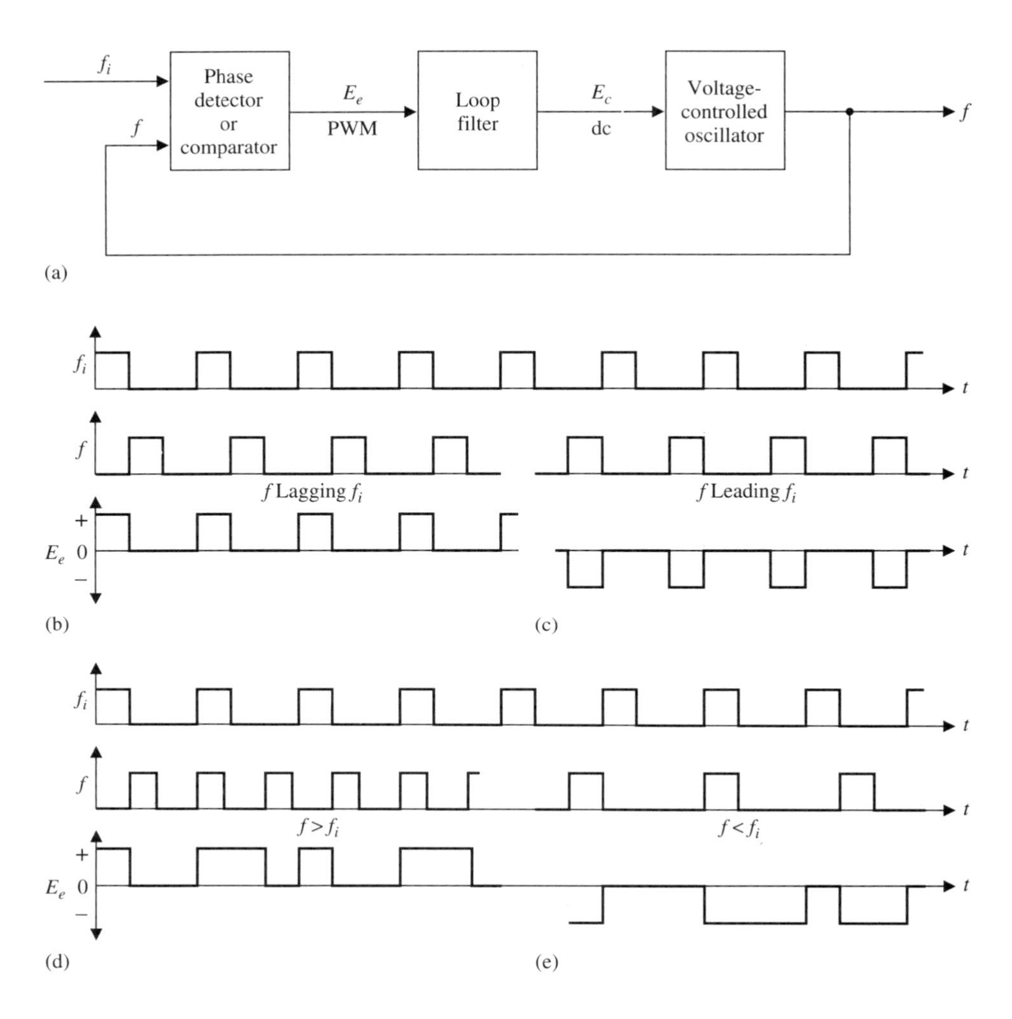

Figure 8-17 Principle of phase-locked-loop control. (a) Block diagram of basic PLL. (b) f lagging f_i. (c) f leading f_i. (d) $f > f_i$. (e) $f < f_i$.

a pulse-width-modulated output voltage E_e whose pulse width is proportional to the phase difference between f_i and f. When f_i and f have exactly the same frequency and phase $E_e = 0$, for all other conditions the phase detector produces an error voltage E_e. If f is of the same frequency as f_i but lags it, E_e consists of a series of positive pulses; if f is of the same frequency but leads f_i, E_e consists of a series of negative pulses. Similarly, if the frequency of f is greater than f_i, the output E_e is a positive pulse-width-modulated voltage, but if the frequency of f is less than f_i, E_e is a negative pulse-width-modulated output. Since the output E_e representing the phase is either a negative or a positive pulse train, it is necessary to convert it to a variable dc output voltage E_c. This is the function of the loop filter. In turn E_c is supplied to the

VCO. As a result the VCO will vary its output frequency f in response to the input voltage E_c. The system operates so as to produce a zero error, that is, $f = f_i$. The system is then said to be *phase locked.*

In Fig. 8-18 the concept is applied to the control of a three-phase six-pulse bridge converter driving a dc motor. It should be noted that the thyristor phase-controlled converter, the motor, and the mechanically coupled shaft encoder combine together functionally to form the VCO. The desired speed is represented by a digital pulse train f_i. This is compared in the phase detector with the digital pulse train f representing the actual speed signal generated by the shaft encoder. The error signal output E_e is converted by the loop filter to the dc version of the error signal, which in turn controls the firing delay angle of the thyristor phase-controlled converter and hence its output voltage V_a. The response of a PLL system is very rapid, and any motor speed changes are detected almost immediately and rapidly corrected. The PLL control can be applied to any size machine, but is most commonly encountered in paper and textile plants, where it is exceedingly important to maintain accurately the speed of succeeding motors in the production process.

Chopper or DC-to-DC Control

Choppers are used in applications where a constant-potential dc voltage source is available. Such applications include streetcars, trolley buses, electric trains, and subway cars, which are supplied from overhead lines or conducting third-rail systems, or battery-operated equipment such as forklift trucks, automatic guided vehicles, golf carts, and small delivery vehicles. The main advantages of chopper control are as follows:

1. There is reduced weight because of the elimination of the traditional rheostatic control systems.
2. The rectified voltage is usually provided by a diode rectifier and, as a result, inefficient motor-generator sets have been eliminated.
3. They appear as a high-power-factor load to the ac source.
4. They may be operated as four-quadrant systems.
5. They are currently being applied to motors up to 3,000 hp (2,240 kW). Recalling from Section 2-6, the output voltage V_{do} is

$$V_{do} = \frac{t_{\text{on}}}{t_{\text{on}} + t_{\text{off}}} V_d = \frac{t_{\text{on}}}{\text{T}} V_d \qquad (2\text{–}20)$$

It can be seen that as t_{on} approaches zero, V_{do} approaches V_d, and as t_{on} approaches the periodic time T, V_{do} approaches infinity.

Normally the chopping frequency is limited by the thyristor and the forced commutation circuitry to the range of 200–400 Hz. The chopper frequency may be raised to approximately 1 kHz by using GTOs. However, the frequency at which the chopper is operated is determined mainly by the limitations of the chopper.

The basic problem in a chopper controlling a dc motor is the maximum current that can be commutated by the thyristor, As the motor's power output increases, the worst-case situation occurs when the motor is first started, that is, locked-armature

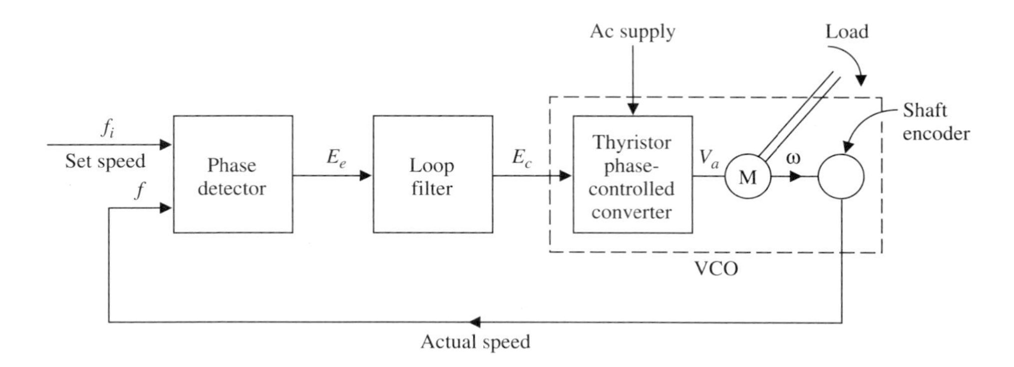

Figure 8-18 Phase-locked-loop dc drive system.

conditions. This current can be calculated, or it can be obtained from a locked-armature test, as can also the rate of change of the armature current. These data are used to design the circuit so that commutation is achieved before the load-carrying thyristor is subjected to the maximum armature current. Commutation is assured by using a current feedback to sense the buildup of armature current and provide a current limit control, as well as controlling the acceleration and deceleration rates.

The basic principle of chopper control of a series traction motor supplied from a constant-potential dc source is shown in Fig. 8-19(a). The chopper is operated in the pulse-width-modulated mode, that is, t_{on} and t_{off} variable and the periodic time T constant. The chopper supplies pulses of current from the supply to the motor. During t_{off} the motor current freewheels through diode D2. The motor current is thus made up of two components, the chopper current drawn from the supply and the freewheel current supplied by the motor. The supply current as a result contains harmonics that cause voltage fluctuations, distortion of the supply, harmonic heating, and in electric train applications, possible interference with communication and speed-control signals. These effects can be minimized by introducing an *LC* filter, namely L1–C1, where the capacitor supplies ripple current to the chopper so that only the average current is drawn from the supply, and the inductor reduces the size of the required capacitance and at the same time provides transient isolation between the load and the supply during any short circuits.

The operation of the chopper is as follows. Applying a gate signal to SCR2 will cause capacitor C2 to precharge to approximately the source voltage with the left-hand plate positive. Power is applied to the series motor by turning on SCR1. At the same time the charge on capacitor C2 reverses so that the right-hand plate is now positive. Turn-off is initiated by applying a gate pulse to SCR2, which applies a reverse bias to SCR1, causing it to commutate off. Simultaneously the polarity of the charge on capacitor C2 has reversed, ready for the next cycle of operation. Alternatively, toggling SCRs 1 and 2 on and off controls the voltage level applied to the motor. Resistor R, which is a high-ohmic-value resistor, ensures that any capacitor leakage current is shunted through SCR2 and D1. As can be seen from Fig. 8-19(b), the motor current rises and falls exponentially, rising exponentially when

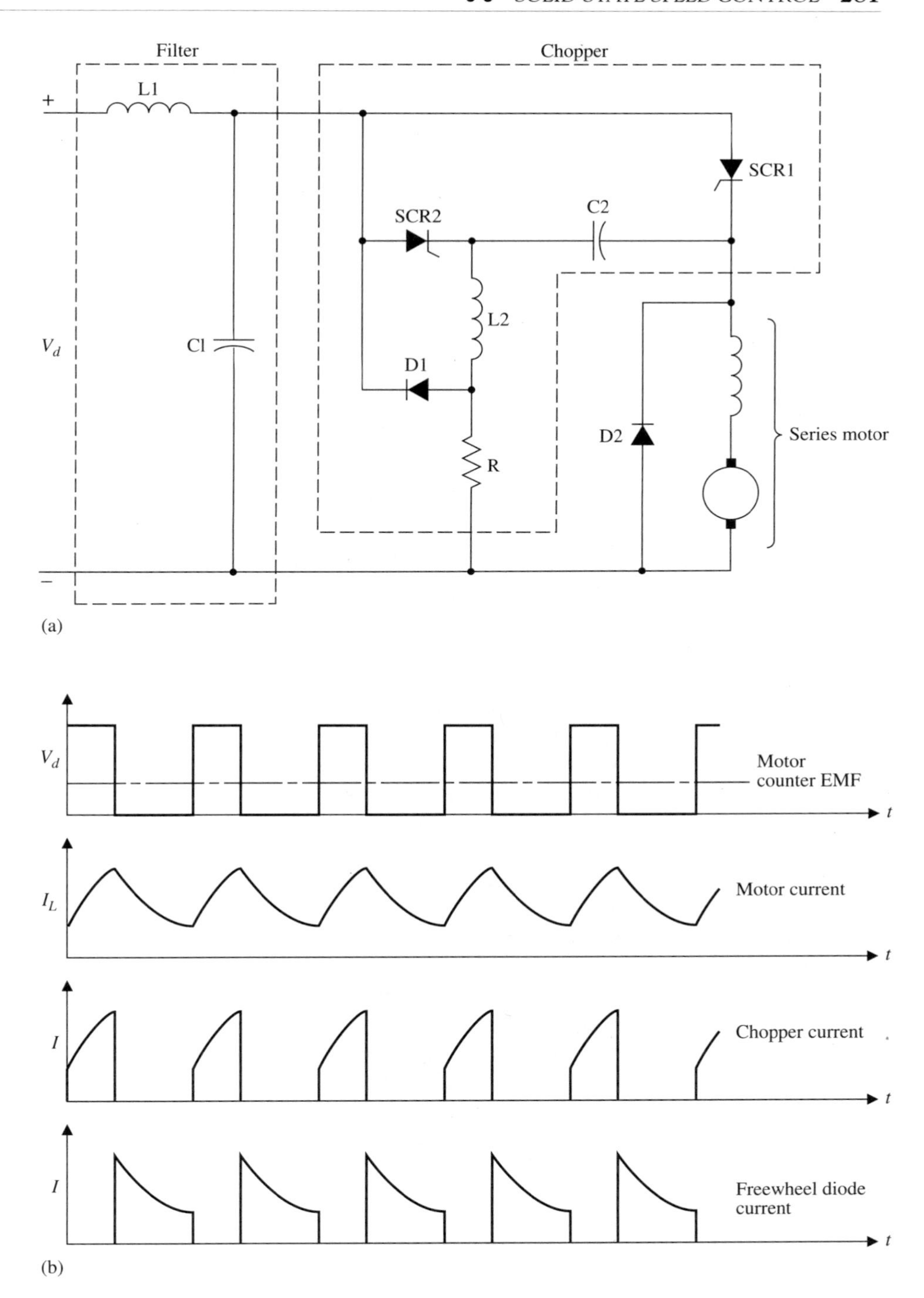

Figure 8-19 Chopper control of series traction motor. (a) Schematic. (b) Voltage and current waveforms.

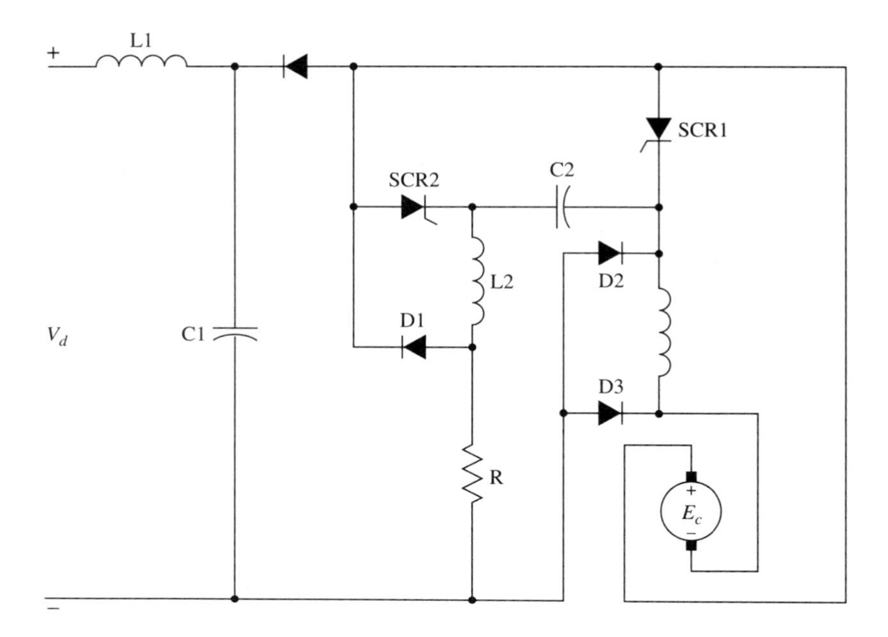

Figure 8-20 Regenerative braking applied to circuit of Fig. 8-19.

SCR1 is conducting, and decaying exponentially through the freewheel diode D2 when SCR2 is commutated off. At the point of starting the ratio of t_{on}/t_{off} will be low and determined by the current limit. As the motor accelerates, the ratio t_{on}/t_{off} automatically increases until the desired speed is achieved.

Dynamic braking can be achieved by connecting, through a suitable arrangement of contactors, the armature to the dynamic braking resistance and at the same time ensuring that the field remains energized. Regenerative braking can also be accomplished by arranging the armature connections as shown in Fig. 8-20.

When SCR1 is turned on, the current build up exponentially in the motor circuit, and energy is stored in the magnetic fields. Current sensing in the armature circuit determines the point at which SCR2 is turned on to initiate the turn-off of SCR1. If the armature voltage E_c is less than the source voltage V_d, the collapsing current returning the stored magnetic energy flows through D2 to the source. After the current has decayed sufficiently, SCR1 is again fired and the cycle is repeated. If the motor speed is high enough and E_c is greater than V_d, D3 will conduct, but with the penalty that the current through the series field will decrease with a subsequent reduction in E_c.

QUESTIONS

8-1 What is the function of a starter?

8-2 Discuss the relative merits of three- and four-point manual faceplate starters.

8-3 Why are automatic starters preferred to manual starters?

8-4 What are the disadvantages of manual faceplate starters?

8-5 Why is a manually operated drum controller preferred to a manual faceplate starter?

8-6 What are the advantages of using drum controllers?

8-7 What are the advantages of using automatic starters?

8-8 What is meant by a definite-time acceleration starter? What are its advantages and disadvantages?

8-9 What is meant by a current-limit acceleration starter? What are its advantages and disadvantages?

8-10 With the aid of a sketch explain the operation of a definite-time acceleration starter using pneumatic dashpot timing relays.

8-11 What is the function of a thyrite resistor connected across the shunt field of a motor?

8-12 Why is a current-limit acceleration starter considered to be a closed-loop control system?

8-13 What are the four types of current-limit starters?

8-14 Explain the operation of a series relay type starter.

8-15 Explain the operation of a counter electromotive force type starter.

8-16 Explain the operation of a lockout relay type starter.

8-17 Explain the operation of a holding coil type starter.

8-18 What requirements must be met in order to reverse a dc motor?

8-19 Why is armature current reversal preferred to field reversal in shunt and compound motors?

8-20 How is a permanent-magnet motor reversed?

8-21 What precautions must be observed before reversing a large dc motor?

8-22 What is plugging? What precautions must be taken before plugging a motor?

8-23 What is meant by electromagnetic braking?

8-24 What is dynamic braking?

8-25 Explain what is meant by regenerative braking.

8-26 Under what conditions can regenerative braking be used?

8-27 What is meant by mechanical braking?

8-28 When must mechanical braking be used?

8-29 What is jogging? Where would jogging be used?

8-30 Explain what is meant by constant-torque mode.

8-31 What is meant by base speed?

8-32 Explain what is meant by constant-power mode operation.

8-33 In what applications would a constant-torque drive be used?

8-34 In what applications would a constant-power drive be used?

8-35 What factors limit the speed control range of a dc motor?

8-36 What is the function of a stabilizing winding in a shunt motor?

8-37 Explain with the aid of a block diagram the operation of a closed-loop thyristor phase-controlled dc motor drive.

8-38 What is the function of a ramp generator?

8-39 What is the function of an accelerating ramp?

8-40 What is the purpose of current limiting?

8-41 What is the purpose of the tachogenerator?

8-42 Explain the operation of a four-quadrant or dual converter in the circulating current free mode.

8-43 With the aid of a diagram and waveforms explain the operation of a PLL control system.

8-44 Explain with the aid of a block diagram the operation of a PLL-controlled SCR drive.

8-45 What are the benefits of using PLL? What applications would be best suited to PLL control?

8-46 Explain the principles of chopper control.

8-47 Explain the operation of a series motor chopper control.

8-48 What are the advantages of chopper-controlled drives?

8-49 Explain the operation of a regenerative chopper.

CHAPTER

9

DC Machine Selection

9-1 INTRODUCTION

The selection of a dc motor for a specific application is far more complex than would be the case in selecting an ac motor. The effective selection of a dc motor involves careful coordination between the nature of the power source, the motor, and the nature of the load. A number of factors must be considered when dc machines are being selected or specified. The principal factors are:

1. Rated horsepower or kilowatt output.
2. Load characteristics.
3. Speed rating.
4. Frame size.
5. Ambient temperature.
6. Temperature rise.
7. Duty cycle.
8. Voltage rating.
9. Enclosure.
10. Maintenance.

9-2 RATED HORSEPOWER OR KILOWATT RATING

The rated horsepower or kilowatt rating defines the continuous power output rating guaranteed by the machine manufacturer at rated speed, full-load current and voltage. Adherence to the nameplate conditions means that the machine should not experience excessive temperatures or overheat. Implicit is the fact that temporary overloads may be encountered intermittently, but the machine should not be subjected to long-duration overloading. With the exception of the United States, all generator and motor ratings are expressed in kilowatts. Final agreement has not yet been reached on international standardized power ratings, frame sizes, and dimensions.

9-3 LOAD CHARACTERISTICS

When selecting dc generators, the choice depends on the nature of the load. For example, shunt generators are selected for constant-voltage applications where the load is close to the generator, such as battery charging, or in older installations, as the excitation source for waterwheel ac generators. The cumulatively compounded generator, however, is the most commonly used dc generator. It is used where a constant dc voltage source is required, as for dc rolling mill drives. An example would be a synchronous-motor-driven generator set, where the synchronous motor, operating as a synchronous condenser, is used to control the plant power factor to prevent penalty payments to the utility company because of low plant power factors, but at the same time can be used to drive dc generators supplying power to the rolling mill drive motors. The series generator is used very rarely since the output voltage increases with the load current.

In the case of dc motors, the nature of the load is a major factor in determining which type of motor should be selected. For example, in electric traction applications usually the high starting torque and low running torque requirements are best satisfied by choosing a series motor. A heavy-duty crane, on the other hand, requires a series motor for the hoist cycle, but for the lowering cycle the motor must be reconnected as a shunt motor to prevent a runaway condition. Shunt motors are best suited for constant-speed applications such as machine tools. Separately excited motors are used extensively in conjunction with thyristor phase-controlled converters and choppers or dc-to-dc converters for variable-speed dc drives.

9-4 SPEED RATING

A number of factors determine the speed range of dc motors.

1. Commutation ability is not unlimited, but is affected by a factor known as *divergence,* which is caused by the interpole field strength. It can be either too great at the maximum operating speed or not adequate when operating at base speed if the interpole field strength has been adjusted for maximum speed. The effect is to place a ceiling on the maximum speed.

2. Mechanical stress is the effect of centrifugal forces acting on the commutator and the armature winding end turns. In small machines centrifugal forces can be largely overcome; however, dynamic balance, critical speed, and bearing limitations impose limits on the speed range. In medium- to large-size machines commutator stresses are the major problem, and usually the peripheral speed of the commutator is limited to 8,500 ft/min (2,590 m/min), although in special applications such as traction motors peripheral speeds of up to 10,000 ft/min (3,048 m/min) are acceptable. Banding on the armature core is usually designed to accept peripheral speeds of up to 12,000 ft/min (3,658 m/min). Fast-response machine armatures, because of their small armature diameters, rarely approach these peripheral limits.

3. There are limits on flashover. Shunt motors operating at high speeds as a result of shunt field weakening are limited in the amount of armature current that can be safely commutated. This limitation is a direct result of the effects of armature reaction being more pronounced with a weakened shunt field. The results are a significant increase in brush arcing and instability. Since torque is proportional to the product of armature current and field pole flux, a decrease in flux will lead to an increase in armature current for the same torque. As a result the output must be decreased at higher speeds in order that the armature current is restricted to levels that minimize arcing. The effects of armature reaction during weak field operation can be reduced by adding a few series turns on the main field poles. The resulting series field helps to neutralize the armature reaction effects. This winding is known as the stabilizing winding.

Motors are classified by NEMA in terms of their speed-load characteristics as follows:

1. *Constant-speed motors*: A motor whose speed changes from no load to full load by less than 20%. Shunt and compound motors belong to this class.
2. *Varying-speed motors*: A motor whose speed change from no load to full load is greater than 20%. Series motors belong to this class.
3. *Adjustable-speed motors*: A motor whose speed can be adjusted over a wide range, but at any particular speed setting the variation of speed between no load and full load is less than 20%.
4. *Adjustable varying-speed motor*: A motor whose speed can be varied over a wide range, but whose speed variation between no load and full load will be greater than 20% for a particular speed setting. Series and some compound motors fit into this class.
5. *Reversing motor*: A motor that can be reversed at any time under any load condition. This classification applies to all dc motors.

9-5 FRAME SIZE

Frame sizes have been standardized by NEMA, and a series of frame numbers have been assigned. Each frame number defines a physical motor size, with fixed mounting-hole dimensions and locations as well as standard shaft sizes and locations. The same frame size is quite often common to a number of differently rated machines. For example, for a General Electric drip-proof, 60°C temperature-rise motor, frame size 326A is used for six different output and speed ratings, namely, 30 hp, 2,500 r/min; 25 hp, 1,750 r/min; 15 hp, 1,150 r/min; 10 hp, 850 r/min; 7.5 hp, 650 r/min; and 5 hp, 500 r/min (see Table 9-1).

Table 9-1 Sample motor ratings and frame sizes.

hp	Rotational Speed (r/min)								
	3,500	2,500	1,750	1,150	850	650	500	400	300
$\frac{1}{2}$					187A	215A	216A	216A	
$\frac{3}{4}$				187A	215A	216A	218A	254A	
1			186A	187A	216A	218A	254A	256A	
$1\frac{1}{2}$	186A	186A	187A	215A	218A	254A	256A	284A	
2	186A	187A	215A	216A	254A	256A	284A	286A	324A
3	187A	215A	216A	218A	256A	284A	286A	324A	
5	216A	216A	218A	256A	286A	324A	326A		
$7\frac{1}{2}$	218A	218A	256A	286A	324A	326A			
10	256A	256A	284A	286A	326A				
15	284A	284A	286A	326A					
20	286A	286A	324A						
25		324A	326A						
30		326A							

The NEMA ratings have permitted a reasonable degree of standardization in frame sizes. However, dc machines sometimes have longer frames in order to accommodate the commutator and brush gear. International standardization is slowly being agreed on, with all dimensions being metric.

9-6 AMBIENT TEMPERATURE

The term *ambient temperature* is defined by IEEE as the temperature of the medium used for cooling, either directly or indirectly, which is to be subtracted from the measured temperature of the machine to determine the temperature rise under specified test conditions.

Specific cases are defined as follows:

1. For self-ventilating equipment the ambient temperature is the average temperature of the air in the immediate neighborhood of the apparatus.
2. For air- or gas-cooled machines with forced ventilation or secondary water cooling, the ambient temperature is taken as that of the in-going air or cooling gas.

Most dc machines are air cooled. As a result the temperature of the air in the immediate vicinity of the machine determines the effectiveness of the cooling, that is, it is the reference point against which all temperature rises in the windings, and so on, are compared. The winding temperatures are determined by the copper and iron losses, and by the ambient temperature. Most electric machines are designed for an ambient temperature of 40°C. If the ambient temperature is greater than 40°C, then we have three choices: (1) select a machine with a higher power rating, (2) use a higher temperature insulation such as class H, or (3) use forced cooling.

9-7 TEMPERATURE RISE

Normally dc machines are designed to operate with a temperature rise of 40°C above ambient temperature. As was discussed in Section 3-3, insulation materials are assigned specific temperature ceilings which, if exceeded, will result in deterioration of the insulating material. For example, class H insulating materials are permitted to have a temperature rise of 140°C above an ambient temperature of 40°C, with a maximum hot-spot temperature of 180°C. Operating at temperatures in excess of the prescribed limits for each insulation class will shorten the life expectancy of the insulating material. A rule of thumb shows that for every 12°C increase in the motor temperature above the recommended hot-spot temperature, the designed life expectancy of the insulation material will be cut in half. On the other hand, every 12°C reduction in the operating temperature below the rated temperature will double the insulation life expectancy.

With self-ventilated machines it is important to realize that the volume of air flowing through the machine varies directly with the speed of the machine. For speeds less than base speed, that is, if a machine is to be operated for significant periods at medium to low speeds, consideration must be given to using a separate blower to drive cooling air through the machine.

9-8 DUTY CYCLE

The type of duty cycle has a very important effect on the selection of a dc machine. The application may vary from one where the load remains relatively constant over long periods, to one where the load varies cyclically over a wide range of loads, such as a crane or a skip hoist motor. From our previous studies it can be appreciated that the average heating of the machine is affected by a number of factors, such as speed, type of enclosure, and nature of the load variations. When estimating the output rating of a motor under relatively constant loads, it is assumed that the temperature rise of the insulation material varies as the square of the output, an assumption that emphasizes the armature copper losses, but at the same time does not recognize the core losses. This method of determining a motor rating in terms of cyclic load is known as the *rms horsepower* or *rms power rating* and is obtained from

$$\text{rms hp} = \sqrt{\frac{\sum (hp)^2 \times \text{time}}{\text{running time} + \text{standstill time}/k}} \tag{9–1)(E}$$

or

$$\text{rms power} = \sqrt{\frac{\sum \text{kW}^2 \times \text{time}}{\text{running time} + \text{standstill time}/k}} \tag{9–1)(SI}$$

where k is a constant that introduces a factor compensating for the lack of a cooling medium at standstill. (k is assumed to be 4 for an open motor.)

The complete cycle time must be less than the time required for the motor to come up to operating temperature if it is running continuously. The rms horsepower (kW) method, although crude, is used frequently, and while an rms rating is obtained, the result must be tempered with caution since the resulting rating must be rounded off to permit the selection of a commercially available machine.

➤ EXAMPLE 9-1

A motor is connected to a load that demands various power outputs over a 1-h period. The motor operates at 45 hp (33.75 kW) for 2 min, 20 hp (14.92 kW) for 15 min, 17 hp (12.69 kW) for 10 min, and is off for a period of 15 min. What is the rms horsepower and kilowatt power output of the required motor?

SOLUTION

$$\text{rms kW} = \sqrt{\frac{(33.75\ \text{kW})^2 \times 2 + (14.92\ \text{kW})^2 \times 15 + (12.69\ \text{kW})^2 \times 10}{2 + 15 + 10 + 15/4}}$$

$$= \sqrt{\frac{7{,}227.58}{30.75}} = \sqrt{235.04}$$

$$= 15.33\ \text{kW or } 20.45\ \text{hp}$$

The logical choice would be the next larger NEMA rated standard motor, which would be a 25-hp (18.6-kW) motor. The motor would then operate at 33.75 kW/18.6 kW = 181% of rated load for 2 min, 14.92 kW/18.6 kW = 80.2% of rated load for 15 min, and 12.69 kW/18.6 kW = 68.2% of rated load for 10 min.

The highest power output has a duration of 2 min. However, NEMA standard industrial motors are capable of 150% of full-load rated current for 1 min at all speeds, although in our example it is unlikely that the motor temperature will exceed its designed limits.

Other factors that affect the motor temperature are repeated starts with a short interval between starts, plug braking, and plug reversal. It should also be noted that the manufacturers' application engineers are always available to assist in the correct motor selection for a specific application.

Machines are designated as suitable for operation in one of four duty cycles. NEMA defines these duty cycles as follows:

1. *Continuous duty*: Machines that operate at relatively constant loads over long time intervals.
2. *Periodic duty*: Machines that operate periodically at relatively constant loads over a long time interval.
3. *Intermittent duty*: Machines that operate at irregular loads, including long intervals of rest, over a long time period.
4. *Varying duty*: Machines that operate with widely varying loads and varying operating periods without rest over a long time period.

Some duty cycles have load peaks which can exceed the capability of a continuous-duty machine to supply the load peak without overheating. To meet this requirement, short-time rated motors are available which have a high torque capability without exceeding the short-time temperature limits. Short-time temperature limits vary considerably with the motor size, and usually for continuously rated motors they are 5 min at 150% overload, 15 min at 145% overload, 30 min at 140% overload, and 60 min at 135% overload.

Service factors, that is, the multiplier by which the normal output rating may be increased under the conditions specified, of usually between 1.15 to 1.3 are available in limited ranges of ratings.

9-9 VOLTAGE RATINGS

Standard voltage ratings have been adopted by NEMA. For dc generators the standard voltages are 125, 250, 275, and 600 V, and for dc motors the standard voltages are 120, 240, and 550 V.

The differences between generator and motor voltages are accounted for by the line-voltage drop between the generator and the motor. Industrial systems usually operate with voltages of between 120 and 125 V or 240 and 250 V. With the increasing use of thyristor phase-controlled converters as the voltage source, dc drive motors now are operating with a variety of armature and field voltages.

9-10 ENCLOSURES

Enclosures for dc machines were dealt with at some length in Section 3-7. A few remarks are in order at this point. Obviously the cost of a given power output machine will increase as compared to that of an open machine as the degree of protection afforded by the enclosure increases.

In the absence of ventilating air, as is the case for a totally enclosed machine, the power output must also be derated to ensure that field and armature winding temperatures do not exceed those specified for their insulation class.

9-11 MAINTENANCE

Preventive maintenance requires well thought-out planning and appropriate action to identify and correct operating and environmental conditions before the situation develops into a catastrophic repair action. Experience has shown that a well thought-out planned maintenance system which is effectively monitored will reduce repair costs and increase production time, and as a result will reduce lost service time or lost production.

Unfortunately the maintenance requirements of dc machines are the most demanding, and hence the most costly, as compared to equivalently rated polyphase induction motors. The major maintenance problems occur with the brushes and the commutator and require regular inspection to ensure their best performance. Typical brush and commutator problems are:

1. Short brushes, where the brush shunt or rivet is making contact with the commutator and causes short circuits between commutator segments as well as grooving of the commutator surface.
2. Loose brush shunts force the current through the brush-holder pressure finger or the brush spring, resulting in the spring losing its properties.
3. Brush pressure should be on the order of 3 to 4 lb per square inch of brush contact area. Excessive brush pressure causes heat and excessive wear; insufficient brush pressure permits the brush to jump from the commutator surface if it is rough; and unequal brush pressure causes unequal current sharing between the brushes.
4. Commutator surface has lost its patina, and as a result brush arcing will worsen progressively.
5. Cleanliness. The brush rigging and supports can acquire an accumulation of carbon dust from the brushes and oil from the surrounding atmosphere. These in turn can lead to flashovers to ground as well as causing the brushes to stick in the brush holders.

QUESTIONS

9-1 What is the significance of the horsepower or kilowatt rating of a dc machine?

9-2 Why is the characteristic of the load an important criterion in selecting a dc machine?

9-3 What are the factors that determine the speed range of a dc machine?

9-4 Discuss the NEMA classification of dc motors in terms of their speed-load characteristics.

9-5 What is meant by frame size?

9-6 What is meant by ambient temperature?

9-7 What is meant by nameplate stating "40°C temperature rise"?

9-8 What factors determine the temperature rise above ambient temperature?

9-9 What are the effects of operating a dc machine with an excessive temperature rise?

9-10 What effects will running a dc machine at reduced speeds have on the insulation life? How can this problem be solved?

9-11 What effect does an ambient temperature greater than that specified on the nameplate have on the machine's performance?

9-12 What is meant by duty cycle?

9-13 What is meant by rms horsepower (kW) rating?

9-14 What are the four types of duty cycles?

9-15 Why would a continuous-duty machine be physically larger than an intermittent-duty machine?

9-16 What is meant by the term *service factor*?

9-17 What is meant by a short-time rated motor?

9-18 How is the output rating of a motor affected by: (**a**) ambient temperature; (**b**) type of enclosure; (**c**) duty cycle?

9-19 Why are the voltage ratings of dc motors less than those of dc generators?

9-20 What are the functions and benefits of preventive maintenance?

9-21 Why are dc machines more costly to maintain than their ac counterparts?

9-22 Discuss the major causes and problems leading to the breakdown of dc machines.

CHAPTER

10

AC Machine Principles

10-1 INTRODUCTION

Nearly all electric energy is generated by three-phase ac generators or alternators that convert mechanical energy to three-phase electric energy. Three-phase and single-phase ac motors convert electric energy to mechanical energy. There are two major types of polyphase machines: synchronous machines and induction machines. Synchronous machines can be operated as either generators or motors and require dc excitation of their field systems. Induction machines are mainly used as motors and depend on induction to produce a field current in the rotor circuit.

The load-carrying or armature windings of polyphase synchronous and induction machines are always located in the *stator*, and the field windings are located in the rotor. In the case of synchronous generators the rotation of the dc energized rotor induces three-phase alternating voltages in the stator windings. This is generator action. When a three-phase alternating voltage is applied to the stator of a polyphase synchronous or induction motor, stator currents will flow which produce a rotating magnetic field which reacts with the rotor field to produce rotation of the rotor, or motor action. In the study of polyphase machines it is essential to have a clear understanding of motor and generator action to understand the performance, operation, and control of polyphase machines.

10-2 STATOR CONSTRUCTION

The stator or stationary component of polyphase machines consists of a structure supporting the stator laminations. These laminations form part of the magnetic circuit and are punched out from thin sheets of silicon steel, usually 0.014 in (0.35 mm), either in segments for large-diameter machines, or in one piece for smaller frame-size machines. The laminations are insulated from each other with either organic or inorganic insulating material to prevent interlaminar current flow caused by eddy currents. The conductors, usually copper, are random wound for small machines and form wound from rectangular cross-section copper for larger machines. They are placed in axial slots in the immediate vicinity of the air gap, and are formed into coils to form the windings that carry the stator currents. The coils are externally connected and insulated from each other as well as from the stator iron, and are held in position by nonmagnetic wedges.

The use of stationary stators has become standard in polyphase ac machines for a number of reasons:

1. The larger peripheral area permits the use of higher voltages and currents, which would not be practical if the armature or load-carrying windings were located in the rotor.
2. Since the armature windings are not rotating, direct connection to the external circuits can be made without the complications resulting from stationary to rotary interfaces, that is, slip rings and the associated brush gear.
3. Since the power handled by the stator circuit is much greater than that of the rotor circuit, the cooling of the stator is more easily achieved with a stationary stator, and in the case of large-power-output alternators, cooling is greatly enhanced by using hollow water-cooled conductors, which can increase output by as much as 500%.

10-3 STATOR WINDINGS

Most polyphase machines, mainly for economic reasons, use a double-layer multiturn coil winding, that is, there are two separate coil slides in each stator slot. The use of double-layer coils with open slots enables form-wound coils to be used as well as obtaining a lower leakage reactance and, in the case of synchronous generators, an improved voltage waveform. Former-wound coils are diamond shaped and are connected in either lap or wave (Fig. 10-1).

Normally coils are connected together to form an open winding, that is, a winding in which there is a continuous path through the conductors of each phase, terminating with two free ends. If there are three such series-connected windings displaced physically 120° from each other, then the stator has been wound for a three-phase winding.

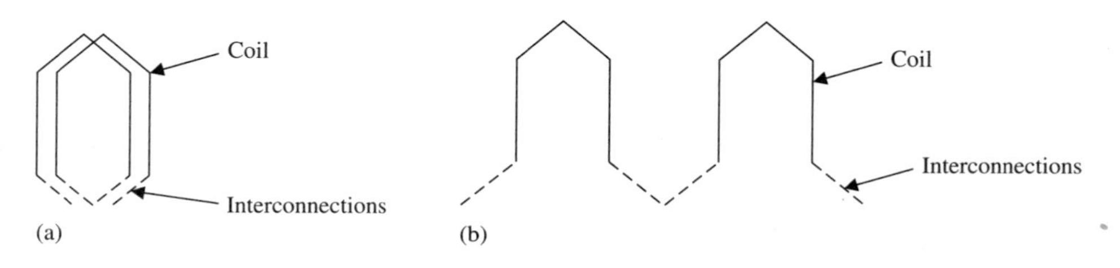

Figure 10-1 Polyphase stator winding. (a) Lap. (b) Wave.

There are two types of windings: *concentrated* and *distributed.* Concentrated windings have all the conductors of any one phase concentrated in one slot per pole. This type of winding requires only a few stator slots, the remainder of the stator iron being used to distribute the magnetic flux. In other words, the stator periphery is very poorly utilized. Distributed windings have the conductors spread between several slots so that the whole winding utilizes the whole stator periphery. In addition, distributed windings reduce the effects of armature reaction and reactance and distribute the heating effects of the armature copper loss more evenly throughout the stator. Also in the case of synchronous generators, the distributed winding gives a better voltage waveform than can be achieved using concentrated windings.

Single-layer or half-coil windings, which have one coil per phase per pole pair, are sometimes used in small-frame polyphase induction motors. More commonly the double-layer or full-coil winding, which has as many coils per phase as there are poles, is used in larger frame polyphase machines and synchronous generators.

Single-layer or half-coil windings are usually full pitch, that is, the coil sides are spaced 180° electrical apart. Double-layer or full-coil windings are usually *fractional pitch* or *chorded*; that is, the coil sides are spaced less than 180° electrical apart. The advantages of using fractional-pitch coils are:

1. It reduces the amount of copper required in the overhang or end connections of the winding.
2. It reduces the flux harmonics produced by load currents flowing in the stator windings.
3. It reduces the harmonics present in the voltage waveform without significantly reducing the amplitude of the fundamental voltage waveform.

These three factors show why nearly all synchronous generator windings are two-layer fractional pitch. In the case of synchronous generators, when using fractional-pitch coils, it is necessary to apply a *pitch factor* k_p when calculating the induced voltages. Since the coil sides are not 180° apart and as a result the voltages generated in the two coil sides are out of phase, in a full-pitch coil these voltages are in phase. In other words, the pitch factor of a full-pitch winding is unity.

Multicircuit windings are used whenever it is necessary to reduce the current loading in a circuit. But care must be taken to ensure that the individual circuits do not become independently loaded, or magnetic unbalance and mechanical stresses will occur.

10-4 INDUCED EMF EQUATION

Referring to Section 1-8, the average electromotive force induced in a single-turn coil was

$$E_{\mathrm{AV}} = \frac{\Phi}{t} \text{ V/turn} \qquad \textbf{(1-50)(SI)}$$

or

$$E_{AV} = \frac{\Phi}{t} \times 10^{-8} \text{ V/turn} \qquad \textbf{(1–50)(E)}$$

Since the time t for the coil side to travel one quarter of a revolution is $\frac{1}{4}s$, where s is the number of revolutions per second, then substituting for t,

$$E_{AV} = 4s\Phi \times 10^{-8} \text{ V/turn} \qquad \textbf{(1–51)(E)}$$

or

$$E_{AV} = \frac{4\omega\Phi}{2\pi} \text{ V/turn} \qquad \textbf{(1–51)(SI)}$$

$$= 0.637\omega\Phi \text{ V/turn} \qquad \textbf{(1–52)(SI)}$$

Therefore for an N-turn coil Eq. (1-51) becomes

$$E_{AV} = 4sN\Phi \times 10^{-8} \text{ V/coil} \qquad \textbf{(1–53)(E)}$$

and

$$E_{AV} = \frac{4\omega\Phi N}{2\pi}$$

$$= 0.637\omega\Phi N \text{ V/coil} \qquad \textbf{(1–53)(SI)}$$

Since the induced voltage in a coil passes through one complete cycle as a pair of poles pass by it, a relation between frequency, rotational speed, and number of poles can be developed,

$$f = \frac{sP}{2} = \frac{SP}{120} \text{ Hz} \qquad \textbf{(10–1)(E)}$$

or

$$f = \frac{\omega P}{4\pi} \qquad \textbf{(10–1)(SI)}$$

where s is the number of revolutions per second, S the number of revolutions per minute, ω is in rad/s, and P the number of poles.

As can be seen for a given frequency, the rotational speed S or ω is inversely proportional to the number of pole pairs, that is,

$$S = \frac{120f}{P} \text{ r/min} \qquad \textbf{(10–2)(E)}$$

or

$$\omega = \frac{4\pi f}{P} \text{ rad/s} \qquad \textbf{(10–2)(SI)}$$

where f is the frequency in hertz, s the revolutions per second, S the revolutions per minute, and P the number of poles.

Since 1 cycle occurs in $1/f$ s, the time taken to complete $\frac{1}{4}$ of a cycle is $\frac{1}{4}f$. Therefore substituting $\frac{1}{4}f$ for t in Eq. (1-53)(E) yields

$$E_{\text{AV}} = 4\Phi N f \times 10^{-8} \text{ V/coil} \tag{10–3)(E}$$

and since $\omega = 2\pi f$ rad/s, then Eq.(1-53)(SI) becomes

$$E_{\text{AV}} = 4\Phi N f \text{ V/coil} \tag{10–3)(SI}$$

To be more meaningful, the voltage generated per phase should be used instead of the voltage per coil. Therefore the total turns per phase is Nn, where N is the number of turns per coil and n the number of coils per phase. Therefore Eqs. (10-3) can be modified to give

$$E_{\text{AVph}} = 4\Phi N n f \times 10^{-8} \text{ V/phase} \tag{10–4)(E}$$

or

$$E_{\text{AVph}} = 4\Phi N n f \text{ V/phase} \tag{10–4)(SI}$$

However, we normally deal with rms or effective voltages. Since rms value = average value × form factor, where the form factor is

$$\text{Form factor} = \frac{\text{rms value}}{\text{average value}} \tag{10–5}$$

then for a sine wave,

$$\text{Form factor} = \frac{0.707 \text{ maximum value}}{0.637 \text{ maximum value}} = 1.11$$

Then the rms induced voltage per phase $E_{\text{ph}} = 1.11E_{\text{AV ph}}$, that is,

$$E_{\text{ph}} = 4.44\Phi N n f \times 10^{-8} \text{ V} \tag{10–6)(E}$$

or

$$E_{\text{ph}} = 4.44\Phi N n f \text{ V} \tag{10–6)(SI}$$

Equations (10-6) are valid for full-pitch coils.

10-5 COIL PITCH AND PITCH FACTOR

The coil sides of full-pitch coils are one pole pitch or 180° electrical apart and the generated voltage is the phasor difference of the induced voltages in the two coil sides. These voltages are in phase for full-pitch coils. In order to improve the induced voltage waveform, the span of the coils in a double-layer winding is often made less than a pole pitch, that is, the coil sides are spaced less than 180° electrical apart. These coils are called *fractional-pitch coils*. As a result there is now a phase difference between the induced voltages in the two coil sides, and the resultant voltage is less than that of a full-pitch coil. The factor by which the voltage generated in a fractional-pitch coil is less than that generated in a full-pitch coil is called the *pitch factor*. The pitch factor is

$$k_p = \sin\frac{p^\circ}{2} \tag{10–7}$$

where k_p is the pitch factor ($k_p \le 1$), and p° the span of the coil in degrees electrical. $p^\circ = 180^\circ$ for a full-pitch coil, that is, $k_p = 1$.

➤ EXAMPLE 10-1

Calculate the pitch factor for the following winding combinations: **(a)** 54 slots, six poles, coil span 1:8; **(b)** 72 slots, four poles, coil span 1:15.

SOLUTION

(a)

$$\text{Full-pitch coil span} = \frac{54 \text{ slots}}{6 \text{ poles}} = 9 \text{ slots/pole}$$

$$p^\circ = \frac{\text{fractional-pitch coil span}}{\text{full-pitch coil span}} \times 180^\circ$$

$$= \frac{7}{9} \times 180^\circ = 140^\circ$$

$$k_p = \sin\frac{p^\circ}{2} = \sin\frac{140^\circ}{2} = \sin 70^\circ = 0.94$$

(b)

$$\text{Full-pitch coil span} = \frac{72 \text{ slots}}{4 \text{ poles}} = 18 \text{ slots/pole}$$

$$p^\circ = \frac{14}{18} \times 180^\circ = 140^\circ$$

$$k_p = \sin\frac{p^\circ}{2} = \sin\frac{140^\circ}{2} = \sin 70^\circ = 0.94$$

10-6 DISTRIBUTION FACTOR

Distributing the phase winding in a number of stator slots in each pole pitch has the effect of reducing the magnitude of the resultant voltage because of the phase displacement of the induced voltages in the coil sides in each slot. For example, if there are three slots per phase, that is, nine slots per pole for a three-phase stator, the phase displacement α between adjacent slots is $180^\circ/9 = 20^\circ$. The resultant of the two phasors is displaced by 20° and, as can be seen from Fig. 10-2, is less than their arithmetic sum.

The ratio of the resultant voltage of a winding distributed in n slots to the voltage obtained from a concentrated winding, that is, a winding in which all the coil side voltages are in phase, is called the *distribution factor* k_d. It is given by

$$k_d = \frac{\text{resultant voltage of conductors in } n \text{ slots}}{n\,(\text{voltage of conductors in one slot})}$$

$$= \frac{E_r}{nE_c} \tag{10–8}$$

Assuming that there are three slots per pole per phase, that is, $n = 3$, then

$$\alpha = \frac{180°}{3n} = 20°$$

The voltages induced in the slots are represented by ac, ce, and eg, and are mutually displaced by 20° (Fig. 10-2). Then

$$E_{c1} = 2ab \sin 10°$$
$$E_r = 2aX = 20a \sin 30°$$

and the distribution factor is

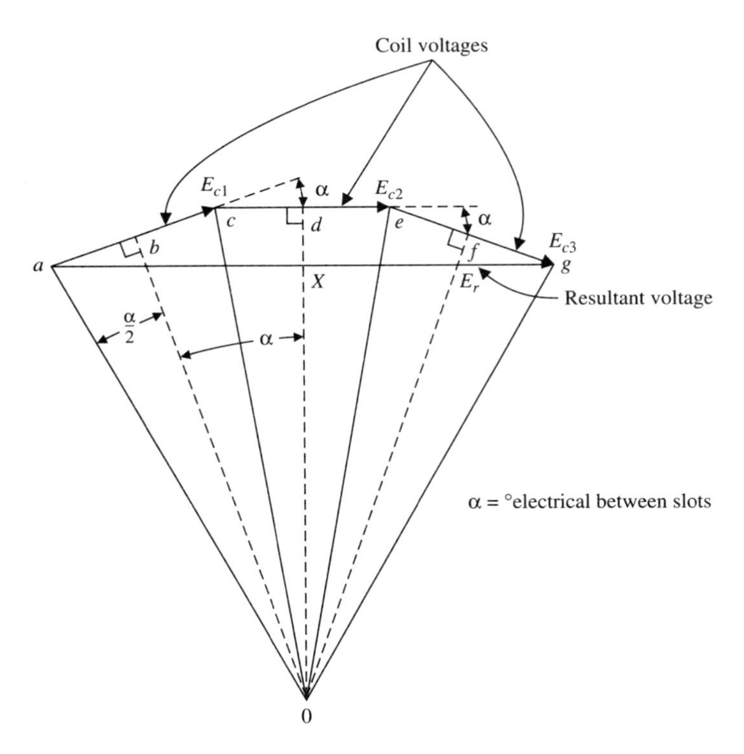

Figure 10-2 Determination of distribution factor.

$$k_d = \frac{E_r}{3E_c} = \frac{\sin 30°}{3 \sin 10°} = 0.96$$

From this the general form can be deduced,

$$k_d = \frac{\sin (n\alpha/2)}{n \sin (\alpha/2)} \qquad \textbf{(10–9)}$$

where k_d is the distribution factor, n the number of slots per phase per pole, and α the number of electrical degrees between adjacent slots.

➤ EXAMPLE 10-2

Calculate the distribution factor k_d for a 48-slot four-pole three-phase stator.

SOLUTION

First we must determine α ,

$$\text{Slots/pole} = \frac{48 \text{ slots}}{4 \text{ poles}} = 12 \text{ slots/pole}$$

$$\alpha = \frac{180°}{12 \text{ slots/pole}} = 15°$$

$$n = \frac{48 \text{ slots}}{4 \text{ poles} \times 3 \text{ phases}} = 4 \text{ slots/phase/pole}$$

$$k_d = \frac{\sin (4 \times 15°/2)}{4 \sin (15°/2)} = \frac{\sin 30°}{4 \sin 7.5°} = 0.96$$

The generated voltage represented by Eqs. (10-4) must be corrected for the reduced voltages resulting from using fractional-pitch coils and distributed windings. Equations (10-4) become

$$E_{\text{ph}} = 4.44\Phi N n f k_p k_d \times 10^{-8} \text{ V/phase} \qquad \textbf{(10–10)(E)}$$

or

$$E_{\text{ph}} = 4.44\phi N n f k_p k_d \text{ V/phase} \qquad \textbf{(10–10)(SI)}$$

➤ EXAMPLE 10-3

A three-phase eight-pole 60-Hz star-connected synchronous generator has 96 slots with 10 turns per coil. The coil span is 10 slots. If the air-gap flux per pole is 0.050 Wb, calculate: **(a)** the rms phase voltage; **(b)** the rms line voltage; **(c)** the rotational speed.

SOLUTION

$$E_{ph} = 4.44\Phi Nnfk_pk_d \text{ V/phase}$$

$$\text{Full-pitch coil span} = \frac{96 \text{ slots}}{8 \text{ poles}} = 12 \text{ slots/pole}$$

$$p^\circ = \frac{\text{fractional-pitch coil span}}{\text{full-pitch coil span}} \times 180^\circ$$

$$= \frac{10}{12} \times 180^\circ = 150^\circ$$

$$k_p = \sin \frac{p^\circ}{2} = \sin 75^\circ = 0.97$$

$$\alpha = \frac{180^\circ}{12 \text{ slots/pole}} = 15^\circ$$

$$n = \frac{96 \text{ slots}}{8 \text{ poles} \times 3 \text{ phases}} = 4 \text{ slots/phase/pole}$$

$$k_d = \frac{\sin(4 \times 15^\circ/2)}{4 \sin(15^\circ/2)} = \frac{\sin 30^\circ}{4 \sin 7.5^\circ} = 0.96$$

Since there are 96 slots, the stator contains 96 coils. Therefore

$$\text{Number of coils per phase } n = \frac{96 \text{ coils}}{3 \text{ phases}} = 32 \text{ coils/phase}$$

$$\text{Turns per coil } N = 10$$

$$\text{Turns per phase } Nn = 10 \times 32 = 320 \text{ turns}$$

(a)

$$E_{ph} = 4.44\Phi Nnfk_pk_d \text{ V/phase}$$
$$= 4.44 \times 0.05\text{Wb} \times 320 \text{ turns/phase} \times 60\text{Hz} \times 0.97 \times 0.96$$
$$= 3{,}969.15 \text{ V/phase}$$

(b) Since the stator winding is star connected, the terminal or line-to-line voltage is

$$E_{line} = \sqrt{3}E_{ph} = \sqrt{3} \times 3{,}969.15 \text{ V/phase}$$
$$= 6{,}874.76 \text{ V}$$

(c)

$$f = \frac{SP}{120}$$

Therefore

$$S = \frac{120f}{P} = \frac{120 \times 60\text{Hz}}{8 \text{ poles}} = 900 \text{ r/min} = 94.25 \text{ rad/s}$$

10-7 HARMONICS

Implicit in the development of the generated voltage relationship is the assumption that the voltage waveform is sinusoidal. Unfortunately this is not correct because the generated voltage waveform contains harmonics. It is necessary to eliminate or reduce harmonics because (1) they increase the eddy current and hysteresis losses and thus decrease the overall efficiency of the machine, and (2) nearly all equipment that is connected to an electric system is designed assuming that the source voltage is free of harmonics.

Harmonics appear in the generated waveform because (1) the air-gap flux is nonsinusoidal, and (2) there is a cyclic variation of the air-gap reluctance caused by the stator teeth. The air-gap flux can be modified by modifying the arrangement of the rotor coils in the case of a turboalternator rotor, or by modifying the radius of the pole shoe arc faces for salient-pole rotors. However, these methods will cause a con-

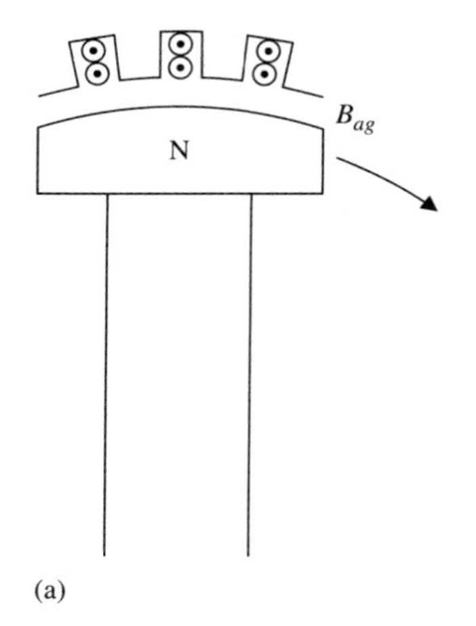

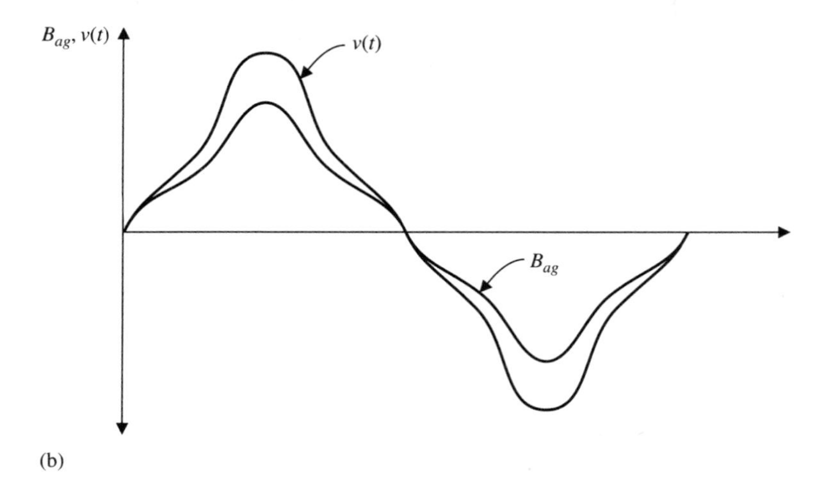

Figure 10-3 Flux and voltage waveforms. (a) Modified air gap. (b) Symmetrical flux density and voltage waveforms.

centration of the air-gap flux at the center line of the pole face and will be markedly lower away from the center of the pole face. Since the air-gap flux is symmetrical about the centerline of the rotor poles [Fig. 10-3(b)], the voltage $v(t)$ generated in the stator coils is also symmetrical about the centerline of the rotor poles. As a result there will be no even harmonics present in the phase voltage waveform. However, all the odd harmonics, that is, the third, fifth, seventh, ninth, and so on, will be present with varying amplitudes, being at their greatest at the lower frequencies and decreasing as the frequency increases. It is necessary in the design of the synchronous generator to take corrective action to reduce or, if possible, eliminate the ninth and lower harmonics.

Depending on whether the stator is star or delta connected will eliminate some harmonics from the line voltages. Star connecting the phase windings will eliminate the third and triplen harmonics from the line or terminal voltage, even though third harmonics are present in the phase voltages as long as no neutral point path is provided. In delta-connected windings the third-harmonic components are additive and cause a third-harmonic current to circulate in the delta-connected phase windings. This current is limited by the machine's internal impedance and is not present in the line voltages. However, the fifth, seventh, eleventh, thirteenth, etc., harmonics are present in the line voltage, but since the amplitude of the harmonic components decreases with increasing frequency, the major harmonics of interest are the fifth and seventh. These harmonics can be reduced by attempting to produce a sinusoidal flux in the air gap between stator and rotor, but additional improvement is obtained by using fractional-pitch windings that produce a pitch factor of zero for that harmonic. For example, a 4/5 pitch coil eliminates the fifth harmonic and a 5/6 pitch coil reduces both the fifth and the seventh harmonics. By using distributed fractional-pitch windings the voltage waveform can be made to approach that of the air-gap flux. An additional problem occurs because of the cyclic variation of both the reluctance and the air-gap flux caused by the stator teeth. This introduces a regular harmonic component into the voltage waveform. This type of harmonic is called the *tooth* or *slot harmonic*.

Slot harmonics appear as odd high-number components, such as the twenty-third and the twenty-fifth, which for a 60-Hz machine have a frequency of 1,380 and 1,500 Hz, respectively. The harmonics are particularly objectionable because:

1. They introduce a harmonic component into the output voltage waveform of an alternator.
2. They cause an increase in the eddy current and hysteresis losses in the stator teeth.
3. The combination of slot harmonics in the stator voltage and rotor slot harmonics in polyphase induction motors creates parasitic torques which degrade motor performance.
4. They also increase the noise level and vibration of polyphase induction motors.

The effects of slot harmonics can be reduced by (1) using fractional-pitch coils and (2), and more commonly, axially skewing the rotor slots of polyphase induction motors by an amount equal to the pitch of the stator slot.

10-8 ROTATING MAGNETIC FIELDS

The rotation of a magnetically polarized rotor inside a three-phase wound stator induces three-phase voltages in the stator windings. Conversely the application of three-phase currents equally displaced in time, that is, mutually displaced by 120° to a three-phase stator winding whose phase windings are equally displaced in space by 120°, will create a constant-amplitude magnetic field rotating at synchronous speed ($S_s = 120f/P$ r/min or $\omega_s = 4\pi f/P$ rad/s).

Figure 10-4(a) shows the three-phase currents mutually displaced by 120°. Figure 10-4(b) illustrates the time relationship between instantaneous currents, and Fig. 10-4(c) the phase displacement of two-pole three-phase stator windings, the starts of the stator windings being designated by S and the finishes by F.

Figure 10-4(d), (e), (f), and (g) shows snapshots of the components of the magnetic field at times t_1, t_2, t_3 and t_4. as indicated in Fig. 10-4(b).

At time t_1, i_a is at its maximum positive value, and i_b and i_c are at 50% of their maximum negative values as represented by flux phasors ϕ_A, ϕ_B, and ϕ_C in Fig. 10-4(d). These flux phasors, when added vectorially, give the resultant flux ϕ_R, which is 1.5 times that of ϕ_A, or the maximum flux produced by any phase, and is oriented horizontally with its direction from right to left.

At time t_2, i_c is at its maximum negative value, and i_a and i_b are at 50% of their maximum positive values as represented by flux phasors ϕ_A, ϕ_B, and ϕ_C in Fig. 10-4(e). These flux phasors, when added vectorially, give the resultant flux ϕ_R, which is 1.5 times the maximum flux produced by any phase. Also it should be noted that the resultant flux has rotated 60° clockwise in the time interval t_2–t_1 = 60° electrical. At the time t_3, i_b is at its maximum positive value, and i_a and i_c are at 50% of their maximum negative values as represented by flux phasors ϕ_A, ϕ_B, and ϕ_C in Fig. 10-4(f). These flux phasors, when added vectorially, give the resultant flux ϕ_R, which is 1.5 times the maximum flux produced by any phase, and once again it has moved 60° clockwise during the time interval t_3–t_2 = 60° electrical.

At time t_4, i_a is at its maximum negative value, and i_b and i_c are at 50% of their maximum positive values, as represented by flux phasors ϕ_A, ϕ_B, and ϕ_C in Fig. 10-4(g). Once again these phasors are added vectorially and the resultant flux ϕ_R is 1.5 times the maximum flux produced by any phase, and once again it has moved 60° clockwise during the time interval t_4–t_3 = 60° electrical.

As can be seen for the two-pole stator being considered, the resultant flux has rotated through 180° during the time interval corresponding to half a cycle of the three-phase ac source. Expressed in another way, the rotating field of a two-pole stator will rotate through 360° for each cycle of the three-phase ac source. For a four-pole stator the rotating magnetic field will turn through 180° for each complete cycle. The rotating magnetic field rotates at synchronous speed S_s or ω_s, which is inversely proportional to the number of pole pairs and is expressed by Eqs. (10-2).

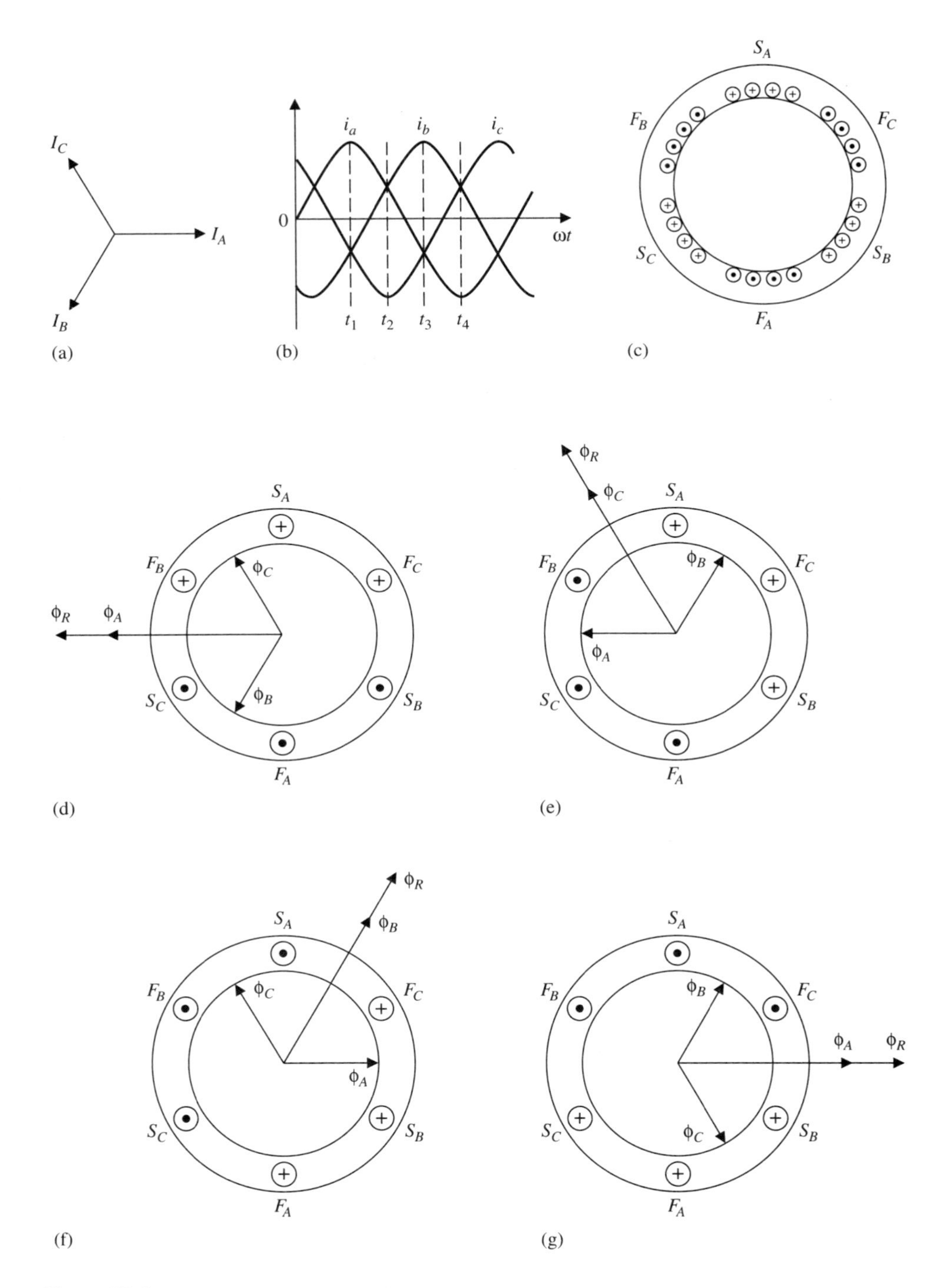

Figure 10-4 Graphic representation of constant amplitude synchronous rotating magnetic field. (a) Three-phase current phasor diagram. (b) Current waveforms. (c) Stator windings displaced 120°. Rotating magnetic field at: (d) Time t_1. (e) Time t_2. (f) Time t_3. (g) Time t_4.

➤ EXAMPLE 10-4

Calculate the synchronous speed of the rotating magnetic field of a stator having eight poles per phase, when supplied from a 60-Hz source.

SOLUTION

$$S_s = \frac{120f}{P} = \frac{120 \times 60\text{Hz}}{8 \text{ poles}} = 900 \text{ r/min}$$

or

$$\omega_s = \frac{4\pi f}{P} = \frac{4\pi \times 60\text{Hz}}{8 \text{ poles}} = 94.25 \text{ rad/s}$$

It should be noted that the direction of rotation of the rotating magnetic field is reversed by reversing the stator connections of any two phases, which in turn reverses the phase sequence of the stator currents.

QUESTIONS

10-1 What are the major differences between a synchronous machine and an induction machine?

10-2 Discuss the construction of a polyphase machine stator.

10-3 Why are the load-carrying windings of polyphase machines located in the stator?

10-4 What are the advantages of using double-layer multiturn coil stator windings?

10-5 What is the difference between a concentrated winding and a distributed winding?

10-6 What are the advantages of using concentrated windings?

10-7 What are the advantages of using distributed windings?

10-8 What are the advantages of using fractional-pitch or chorded coil windings?

10-9 What is meant by pitch factor, and why is it necessary when calculating the induced phase voltage?

10-10 What are multicircuit windings? When would they be used?

10-11 What is meant by the term *form factor*?

10-12 What is meant by the term *distribution factor*?

10-13 What are harmonics? Why is it desirable to reduce or eliminate harmonics?

10-14 Why are there no even harmonics in the phase voltage waveform of an alternator?

10-15 Why are there no third harmonics present in the line voltages of star- and delta-connected alternators?

10-16 What harmonics can appear in the line voltage waveforms of an alternator?

10-17 How are the fifth, seventh, ninth, eleventh, thirteenth, etc., harmonics reduced in the line voltage waveforms of an alternator?

10-18 What are tooth harmonics? How are they produced?

10-19 How can tooth or slot harmonics be reduced?

10-20 What are the effects of tooth harmonics on the performance of polyphase induction motors?

10-21 With the aid of sketches and waveforms explain the production of a polyphase rotating magnetic field.

PROBLEMS

10-1 **(a)** Calculate the number of poles required by a synchronous generator driven by a diesel engine at 720 r/min (75.40 rad/s) to produce a 60-Hz output, **(b)** Develop a table showing the rotor speeds, E and SI, for 4, 6, 8, 10, 12, 14, and 16 poles to produce a 60-Hz output.

10-2 A 1,000-kVA 2,300-V delta-connected alternator is reconnected in star. Calculate **(a)** the line voltages; **(b)** the line currents; **(c)** kVA.

10-3 An eight-pole three-phase stator has 120 slots. Calculate: **(a)** the number of coils per phase; **(b)** the number of coils per pole phase group.

10-4 A three-phase six-pole synchronous generator has a total of 90 slots and a coil span of 13 slots. Calculate: **(a)** the pitch factor; **(b)** the distribution factor.

10-5 An eight-pole three-phase stator has a double-layer winding placed in 48 slots. Determine the distribution factor.

10-6 A four-pole 60-Hz three-phase delta-connected 48-slot synchronous generator is wound with a fractional-pitch double-layer winding. The pitch of the stator coils is 5/6, and there are 15 turns per coil. The field pole flux is 0.055 Wb. Calculate: **(a)** the pitch factor; **(b)** the distribution factor; **(c)** the line voltage.

CHAPTER

11

Synchronous Generators

11-1 INTRODUCTION

Three-phase synchronous generators or alternators convert mechanical power to electric energy. Synchronous generators driven at constant speed by a prime mover, such as a steam, gas, or hydraulic turbine or a diesel or gas engine, are used by electric-power utilities to generate three-phase power at a constant frequency, 60 Hz in North America and 50 Hz in Europe and Asia. Prime movers can be classified as high speed, where the rotor is turned at either 1,800 or 3,600 r/min (188.5 or 376.99 rad/s) by steam- or gas-driven turbines, or low speed, usually less than 600 r/min (62.8 rad/s), by hydraulic turbines. Approximately 85% of the electric energy generated in the United States is by steam-driven turboalternators in sizes ranging from 200 to 1,500 MVA, with stator terminal voltages less than 25 kV. The remaining 15% is mainly produced by hydroelectric generators in ratings ranging from 50–600 MVA.

11-2 CONSTRUCTION—CYLINDRICAL AND SALIENT-POLE MACHINES

Polyphase synchronous generators are classified as having either a *cylindrical* or *salient-pole* rotor. The cylindrical rotor is driven at high speeds by a steam or gas turbine and contains two or four poles. These rotors are characterized by being relatively long in axial length and small in diameter; a ratio of 10 to 1 is typical. The rotor is made small in diameter to minimize the effects of centrifugal forces on the rotor conductors. A typical cylindrical two-pole rotor for a turboalternator is shown is Fig. 11-1. The rotor coils are former wound and inserted in rotor slots that have been machined out of a single steel forging, with the coils held in position by slot wedges in the slots, and the overhanging coil ends are restrained against tangential forces by retaining rings. Four-pole cylindrical rotors are two to three times heavier than a two-pole rotor of the same rating, making it difficult to produce a four-pole rotor from one forging. Brown Boveri fabricate their four-pole rotors from six forgings. The three center sections are located on a central shaft, which is then heated, and the two end sections are screwed onto the shaft. After the shaft is carefully cooled, the three center sections are firmly held in position by compression. Then the axial slots for the rotor conductors are milled ready for the installation of the rotor conductors. Cylindrical rotors have axial and radial passages drilled in them to permit the flow of cooling gases, air, or hydrogen through the rotor. Considerable attention has been given to cooling the rotor directly by water in high-output alternators, typically 1,000 MVA and above.

Hydroelectric generator rotors, since they operate at slower speeds, are characterized by short axial length and large diameter, usually a ratio of 1:5. This type of rotor construction is called a *salient-pole* rotor. Salient-pole rotors are normally assembled in situ and consist of the rotor shaft, to which a spider and a rim are attached, an arrangement similar to a wagon wheel. The rim is built up from thin steel laminations sandwiched between end plates and keyed to the spider. The cores of the rotor poles are built up from steel laminations and bolted together. These poles usually have a single or double dovetail, which permits their being mounted in matching slots around the rim of the rotor, where they are held in position by tapered keys. The rotor coils are usually preformed, mounted on the poles, and suitably insulated from the pole core and rotor rim (Fig. 11-2). In addition, a squirrel-cage winding is usually imbedded in slots in the pole faces. This winding will not have any currents induced in it when the rotor is turning at synchronous speed. However, under the sudden load change conditions the rotor may speed up or slow down momentarily, and as a result voltages are induced in the squirrel-cage windings, which in turn produce rotor currents. These rotor currents produce magnetic fields which react with the rotating stator magnetic field to create generating or motoring torques, which dampen out the rotor oscillations. These pole face windings are called *damper* or *amortisseur windings*.

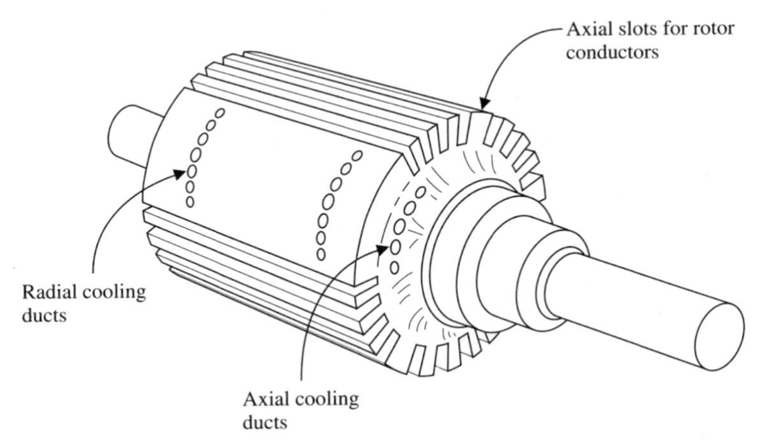

Figure 11-1 Two-pole cylindrical rotor for a 3,600 rpm (376.99 rad/s) synchronous generator.

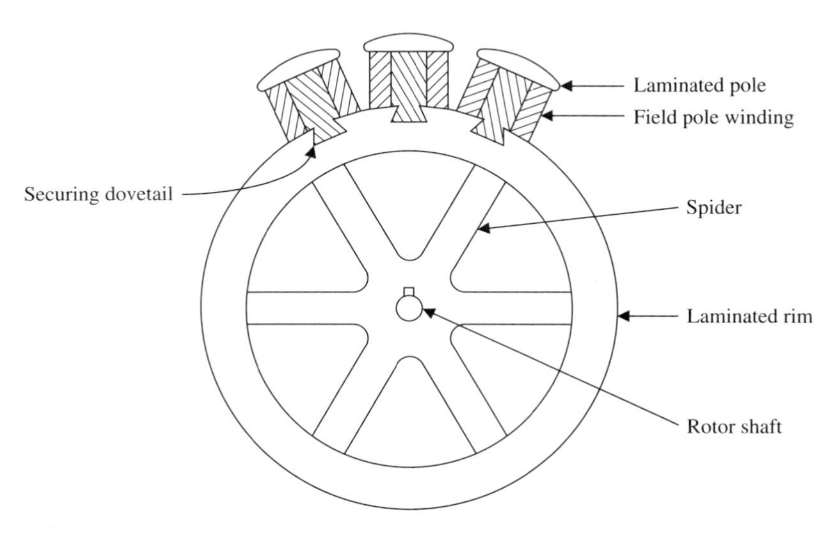

Figure 11-2 Salient-pole rotor construction.

Rotors must be designed to withstand overspeeding of the prime mover. Usually cylindrical rotors are subjected to a maximum of 20% overspeed. However, hydraulic turbines can produce overspeeds as great as 125% when load is suddenly removed from the alternator. As a point of interest it should be noted that the prime mover and the alternator rotor have a *critical speed*, which is the natural frequency of the rotating system in response to residual unbalanced forces. The critical speed is affected by the shaft supports, that is, bearings, and by internal and external damping. Normal practice is that the operating speed should be at least ± 20% of the critical speed. Waterwheel generators usually operate at speeds less than the critical speed. Since two- and four-pole alternators operate above the critical speed, it is essential that these rotating systems be dynamically balanced to ensure that torsional forces are minimized as the rotating parts pass through the critical speed during either start-up or shutdown.

Although the physical dimensions of stators for cylindrical and salient-pole machines differ significantly in axial length and diameter, their construction is basically the same. The stator is built up of thin prepunched laminations as either segments or rings, depending on the diameter. The joints in successive layers are staggered to minimize any increase in the reluctance of the magnetic circuit. At regular intervals I-shaped spacers are placed between the laminations to provide radial ventilation passages. The stator core is clamped between pressure plates to prevent vibration of the laminations. The stator windings are inserted in the stator slots and wedged in position by slot wedges. The coil end connections are securely fastened to coil support rings to prevent any movement under short-circuit conditions. The stator core is supported by a stator frame or yoke fabricated from structural steel. This structure also provides accurate positioning of the rotor as well as forming part of the closed-loop ventilating system.

High power output alternators always use oil-lubricated babbitt bearings, which are either self-lubricating or pressure fed. High-speed horizontal alternators use journal bearings which must be rigid enough to prevent the introduction of any movement to the rotor, especially at the critical speed. Waterwheel alternators have their rotors mounted vertically, and the combined load of the generator rotor, turbine rotor, and water passing through the turbine is supported by a Kingsbury thrust bearing immersed in a water-cooled oil bath. The rotor shaft is constrained in position by self-lubricating guide bearings. Two types of bearing arrangements are used with waterwheel generators: (1) the suspended unit, with a thrust and guide bearing above the rotor and a guide bearing below the rotor, and (2) the umbrella unit, with a thrust and guide bearing below the rotor.

Cooling

Turboalternators are usually cooled by air, hydrogen, or water, and can be grouped into the following categories:

1. *Air cooled:* Usually restricted to units less than 30 MW. Air is circulated in a closed-loop system through ducts provided in the rotor and stator and is cooled by external water-cooled heat exchangers.
2. *Conventional hydrogen cooled:* Usually encountered in units rated between 30 and 200 MW. The air is evacuated and hydrogen under pressure is substituted as the direct cooling medium in exactly the same manner as for the air-cooled type.
3. *Hydrogen inner cooled:* Usually applied to units between 200 and 1,200 MW. Hydrogen is circulated through ventilating tubes between the conductors of the stator windings. The rotor windings also are cooled by ventilating tubes placed between the rotor conductors. Hydrogen under pressures as high as 75 lb/in^2 (52,733 kg/m^2) is pumped through ventilating ducts in a closed-loop system with heat exchangers.
4. *Water cooled:* Usually applied to units between 700 and 1,250 MW. Water is passed through hollow copper conductors, which form part of the stator winding. The closed- loop system consists of pumps, heat exchangers, filters, and demineralizers, which maintain the conductivity of the water in the stator windings. The rotor is usually cooled in the same manner as the inner cooled system. However, some units also use water-cooled rotors.

Waterwheel generators are usually cooled by the passage of cooling air through radial slots in the stator laminations and ducted through water-cooled heat exchangers back to the rotor pit. Fan blades on the rotor periphery create the airflow.

Hydrogen under pressure is used in turboalternators because of its superior heat transfer capability and nonoxidizing effect on the winding insulation, as well as its superior fire-retardation properties.

Rotor Excitation

The dc power required to energize the rotor windings is usually supplied by a static excitation system. This method takes three-phase ac power from the generator terminals, transforms it to a suitable voltage level, rectifies it via a thyristor phase-controlled converter, and supplies the dc current via slip rings to the rotor windings. An automatic voltage regulator controls the terminal voltage and the reactive loading of the generator.

Figure 11-3 shows the block diagram of a typical static excitation system for controlling the output terminal voltage of a three-phase alternator. Initially during start-up the residual magnetic field is not great enough to permit the output voltage of the alternator to build up. As the alternator is being brought up to speed, the rotating field system is energized directly from the power station battery-supplied dc bus bars. When the generator voltage has built up to approximately 40% of the rated terminal voltage, the automatic bus transfer switch switches over to the thyristor phase-controlled converter. At this point power for the thyristor converter is obtained from the three-phase step-down transformer connected to the output terminals of the generator. The function of the automatic voltage regulator is to produce gate firing signals which are applied to the gates of the thyristors. The firing signals are derived by comparing a dc voltage representing the desired generator terminal voltage against low-level dc signals representing the actual generator terminal voltage. The resulting error signal advances or retards the firing signal applied to the thyristors in the converter, which in turn controls the air-gap flux of the generator and therefore maintains the terminal voltage within prescribed limits.

This system has the disadvantage that the power supplied to the rotor must be supplied via slip rings from a floor-mounted static excitation system. An alternative system, which totally eliminates the slip rings, is shown in Fig. 11-4.

The brushless excitation system illustrated in Fig. 11-4 has a permanent-magnet

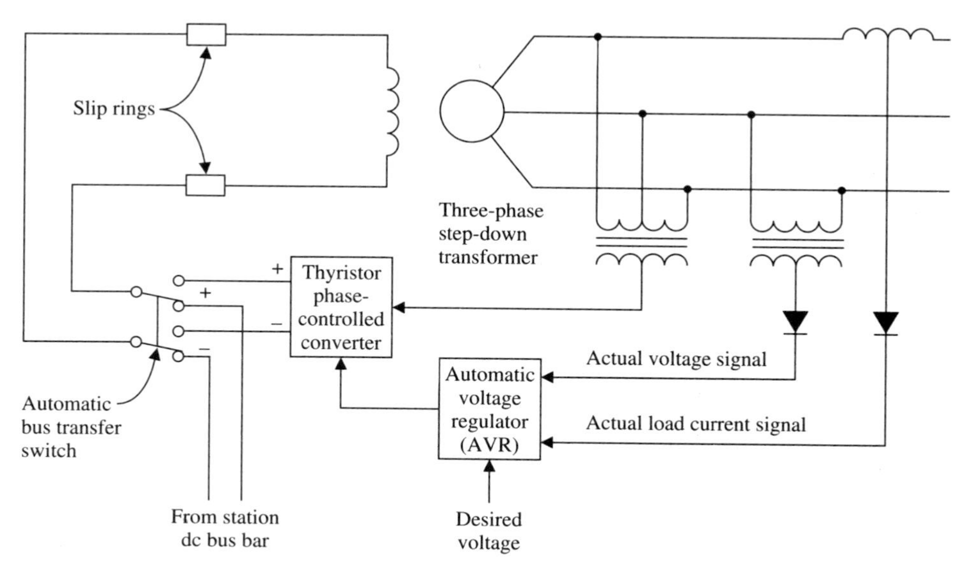

Figure 11-3 Block diagram of static excitation system for a synchronous generator.

pilot exciter, a three-phase main exciter winding, and a three-phase diode bridge, all mounted on the rotor shaft. The stationary pilot exciter stator supplies three-phase power to the three-phase phase-controlled converter. The dc output of the converter is controlled by the automatic voltage regulator, and is supplied to a stationary field surrounding the three-phase rotor of the main exciter. In turn the output of the main exciter is supplied to the rotating three-phase diode bridge. The generator output voltage is regulated by varying the firing delay signals to the thyristors of the phase-controlled converter supplying the main exciter field.

Static excitation systems have completely replaced the dc shunt pilot and main generators, which were mounted on the rotor shaft of older alternators. The main disadvantages of the dc rotary exciters were the maintenance problems always associated with dc machines, and their slow response caused by the relatively high time constants of their field windings. Static excitation systems on the other hand have a fast response, low maintenance requirements, as well as a much higher efficiency. It should be noted that the power requirements of the dc excitation system usually are about 0.28% of the total power output of the alternator, that is, a 1,000-MW alterna-

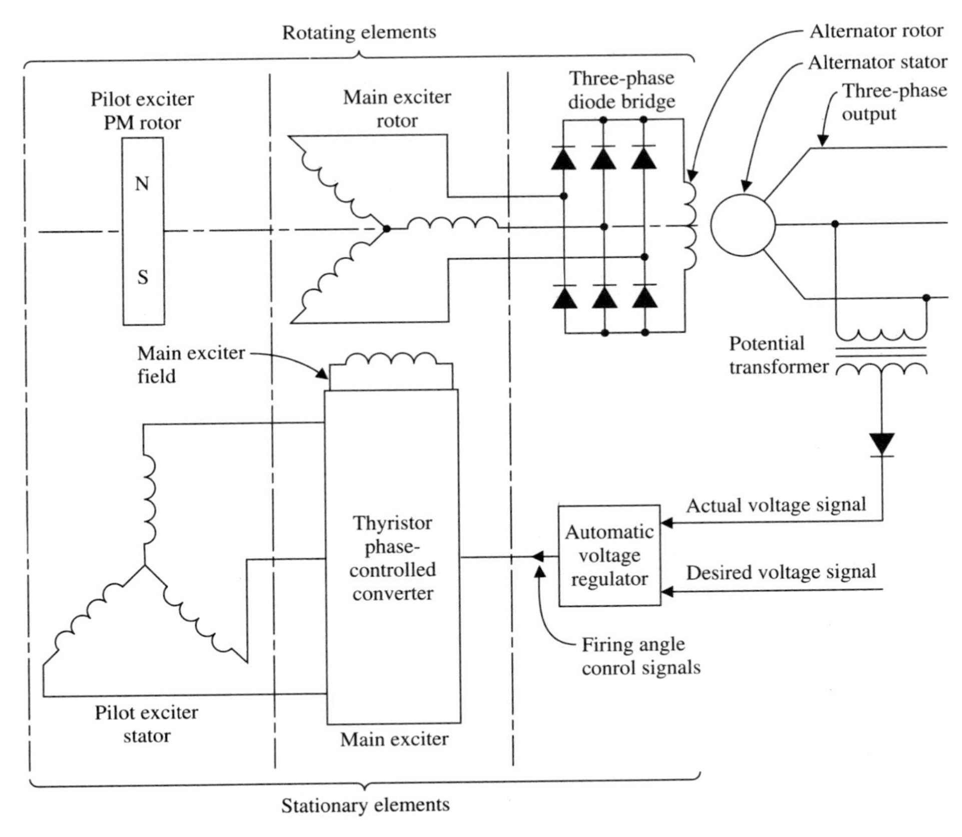

Figure 11-4 Brushless excitation system.

tor will require 2.8 MW for the field excitation, or for a 600-V system 4,666 A must be supplied to the rotor windings.

11-3 RELATIONSHIP BETWEEN SPEED, NUMBER OF POLES, AND FREQUENCY

Public utilities generate three-phase electric power in North and South America at 60 Hz, and in Europe and Asia at 50 Hz. To maintain a constant-frequency output the synchronous generator must operate at a constant speed called the *synchronous speed*. As has been discussed previously, there is a relationship between the mechanical rotational speed of the rotating magnetic field pole system, the number of pole pairs on the rotor, and the output frequency of the synchronous generator, namely,

$$f = \frac{SP}{120} \text{ Hz} \qquad \textbf{(11–1)(E)}$$

or

$$f = \frac{\omega P}{4\pi} \text{ Hz} \qquad \textbf{(11–1)(SI)}$$

where f is the generated frequency in hertz, S or ω the synchronous speed (r/min or rad/s), and P the number of field poles.

➤ EXAMPLE 11-1

A six-pole alternator rotates at 1,200 r/min (125.66 rad/s). Calculate: **(a)** the output frequency; **(b)** the number of field poles if an alternator rotates at 720 r/min (75.40 rad/s) to produce a 60-Hz output.

SOLUTION

(a)

$$f = \frac{SP}{120} = \frac{1{,}200 \text{ r/min} \times 6 \text{ poles}}{120} = 60 \text{ Hz}$$

or

$$f = \frac{\omega P}{4\pi} = \frac{125.66 \text{ rad/s} \times 6 \text{ poles}}{4\pi} = 60 \text{ Hz}$$

(b)

$$P = \frac{120 f}{S} = \frac{120 \times 60 \text{ Hz}}{720 \text{ r/min}} = 10 \text{ poles}$$

or

$$P = \frac{4\pi f}{\omega} = \frac{4\pi \times 60 \text{ Hz}}{75.4 \text{ rad/s}} = 10 \text{ poles}$$

Table 11-1 shows the relationship between the number of rotor poles and angular velocity for 60- and 50-Hz synchronous generators.

11-4 ARMATURE REACTION

In dc machines armature reaction was shown to to be the effect of the fluxes produced by the load-carrying armature conductors upon the distribution of the main field pole flux in the air gaps. A similar situation exists in synchronous generators, where the three-phase stator currents produce fluxes which have two effects: (1) they modify the magnitude and distribution of the main field pole fluxes, called *armature reaction*; and (2) they produce leakage fluxes surrounding the stator conductors which link with the stator coils, and as a result make the stator coils inductive, that is, they produce a *leakage reactance*.

The effect of armature reaction is best illustrated by considering the combined effect of the stator currents in the three-phase stator. The resulting armature reaction flux ϕ_{AR} waveform is assumed to be sinusoidal, is rotating at synchronous speed, and is fixed in position relative to the rotor field poles for a specific power factor load.

It can be shown that when supplying a unity power factor load, that is, a pure resistive load, the armature reaction flux is at a maximum midway between the rotor poles and changes polarity at the midpoint between the rotor poles [Fig. 11-5(a)]. From Fig. 11-5(a), when supplying a unity power factor load the armature reaction flux lags the field pole flux by 90° as a result the two fluxes are additive on the left-hand side of the rotor pole and subtractive on the right-hand side of the rotor pole. The net result is a weakening of the air-gap flux on the leading edge of the pole and a strengthening of the air-gap flux on the trailing edge of the rotor pole, that is, the air-gap flux has been distorted, with some reduction of the total air-gap flux because of a decrease in the permeability of the magnetic path under the saturated portion of the field pole. This effect is very similar to the condition that exists in a dc machine with the brushes set on the geometric neutral axis.

Table 11-1 Relationship between poles and angular velocities for 60 and 50 Hz synchronous generators.

	Angular Velocity			
Poles	r/min		rad/s	
	60 Hz	50 Hz	60 Hz	50 Hz
2	3,600	3,000	376.99	314.16
4	1,800	1,500	188.50	157.08
6	1,200	1,000	125.66	104.72
8	900	750	94.25	78.54
10	720	600	75.40	62.83
12	600	500	62.83	52.36
24	300	250	31.42	26.18
48	150	125	15.71	13.09
96	75	62.5	7.05	6.54

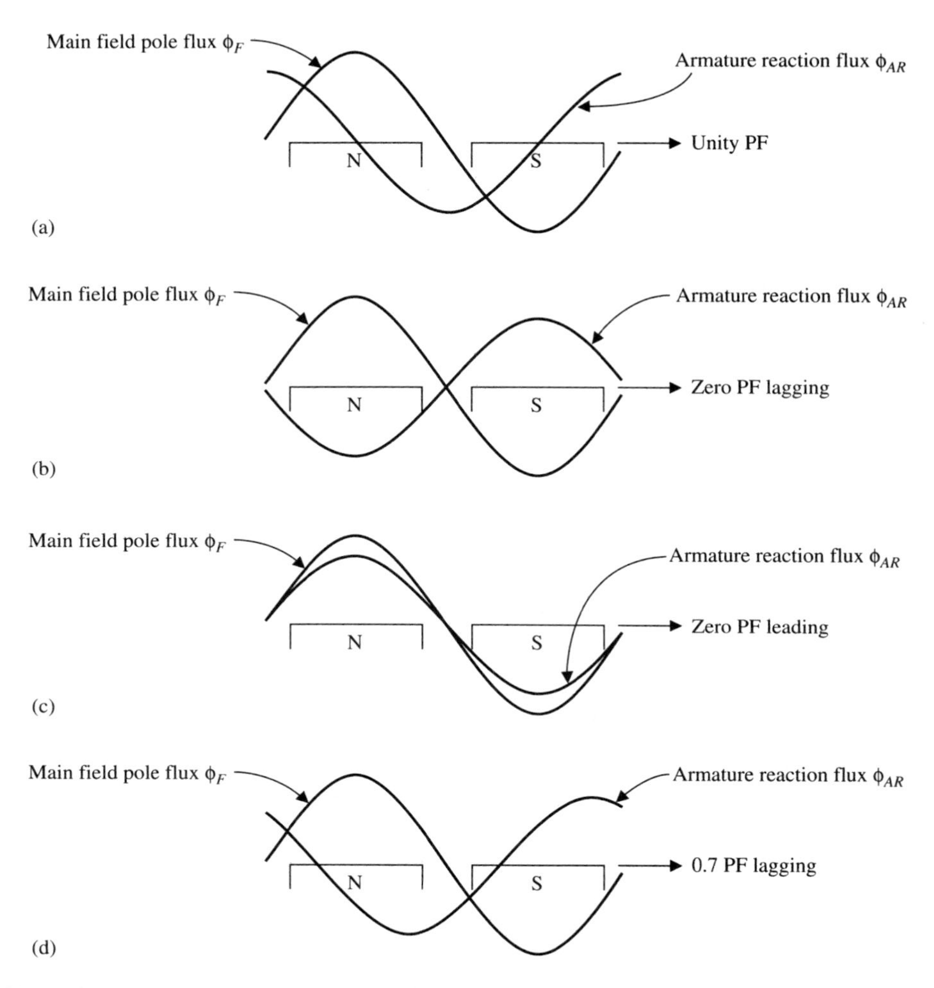

Figure 11-5 Armature reaction and effects of varying power factors. (a) Unity power factor. (b) Zero power factor lagging. (c) Zero power factor leading. (d) 0.7 power factor lagging.

Figure 11-5(b) represents the condition that exists when the alternator is supplying a purely inductive load, that is, a zero power factor lagging load. The armature reaction flux waveform is now 180° out of phase with the field pole flux, that is, the armature reaction flux is lagging by 90° with respect to the unity power factor condition. As can be seen, the armature reaction flux opposes the main field pole flux, and as a result there is a net reduction of the total air-gap flux. Similarly when supplying a zero power factor leading load or a purely capacitive load, the armature reaction flux is in phase with the main field pole flux, and there is a net increase in the total air-gap flux [Fig. 11-5(c)]. However, in actual practice the alternator load is usually slightly lagging. A typical 0.7 power factor lagging condition is represented in Figure 11-5(d). As can be seen, the air-gap flux is a combination of distortion or cross-magnetization and demagnetization.

To summarize, the amplitude of the armature reaction flux is proportional to the magnitude of the load current. Its position relative to the field poles is determined by the load power factor. At unity power factor the air-gap flux is distorted with a slight reduction of the flux density under the trailing pole tips caused by saturation of the magnetic paths. For intermediate leading or lagging power factor loads the air-gap flux is both distorted and increased (decreased).

11-5 ARMATURE LEAKAGE REACTANCE

In addition to creating changes in the air-gap flux, currents flowing in the stator conductors produce magnetic fluxes which do not cross the air gap (Fig. 11-6). These fluxes are confined to the paths of least magnetic reluctance, that is, where the stator conductors are surrounded on three sides by the stator laminations, the flux density between the stator teeth being a minimum because of the reluctance of the air path. The effect of the magnetic flux linking with the stator conductors is to make the stator windings inductive, with the result that the stator leakage reactance is much greater than the stator resistance.

11-6 EQUIVALENT CIRCUITS

In Chapter 10 the rms generated voltage per phase was found to be

$$E_{\text{ph}} = 4.44\Phi Nnfk_pk_d \times 10^{-8} \text{ V/phase} \qquad \textbf{(10–10)(E)}$$

or

$$E_{\text{ph}} = 4.44\phi Nnfk_pk_d \text{ V/phase} \qquad \textbf{(10–10)(SI)}$$

The generated voltage per phase is dependent on the flux per pole ϕ and the frequency or rotational speed of the rotor. Equations (10-10) can be rewritten in terms of flux and rotational speed,

$$E_{\text{ph}} = K\Phi S \text{ V/phase} \qquad \textbf{(11–2)(E)}$$

or

$$E_{\text{ph}} = k\phi\omega \text{ V/phase} \qquad \textbf{(11–2)(SI)}$$

where

$$K = \frac{4.44Nnfk_pk_dP \times 10^{-8}}{120} \qquad \textbf{(11–3)(E)}$$

and

$$k = \frac{4.44Nnfk_pk_dP}{4\pi} \qquad \textbf{(11–3)(SI)}$$

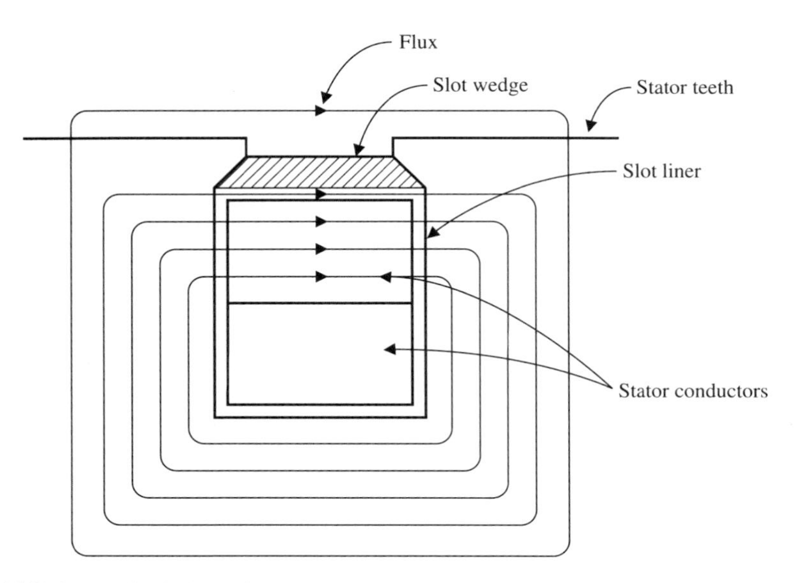

Figure 11-6 Stator slot leakage flux.

K and k being constants that represent the characteristics of the machine.

Equations (11-2) show that the generated voltage per phase is directly proportional to the flux per pole. However, the flux pole ϕ is determined by the rotor field current I_f and the magnetic properties of the rotor iron. Figure 11-7(a) shows the relationship between ϕ and I_{f9} and Fig. 11-7(b) shows the relationship between the generated voltage per phase E_{ph} and the rotor field current I_f at constant speed or frequency.

The generated voltage per phase E_{ph} is the voltage produced by one phase of the synchronous generator. However, the voltage is not the voltage present at the output terminals of one phase of the synchronous generator V_{ph}. A number of factors account for the difference between generated and terminal voltages:

1. The effect of armature reaction on the air-gap flux
2. In the case of salient pole machines the effect of a nonuniform air gap caused by the shape of the rotor pole face
3. The resistance of the stator or armature coils
4. The self-inductance of the stator or armature coils

The effect of armature reaction, which is the most significant, is best illustrated by means of phasor diagrams. Referring to Fig. 11-8, the emf phasor E_g represents the actual electromotive force (Fig. 11-8) and the emf phasor E_t represents the electromotive force that would be induced at no load, that is, when there is no stator or armature current and therefore no armature reaction. Expressed another way, E_t is the electromotive force produced by the rotor field ϕ_f, and E_g is that produced by the

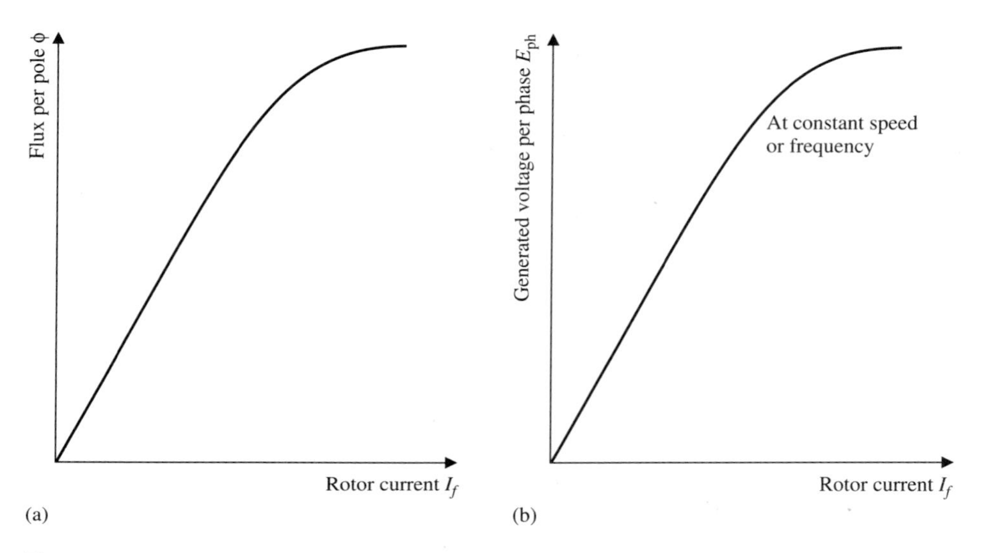

Figure 11-7 Magnetization curve or open-circuit characteristic. (a) Flux per pole versus rotor current. (b) Generated voltage per phase versus rotor current.

resultant field ϕ_R, where ϕ_R is the phasor sum of ϕ_f and ϕ_{AR}. The actual electromotive force E_g can be seen to be the phasor sum of E_t and a fictional voltage E_{AR}, which is proportional to the stator current I_a, that is, E_g is the phasor sum of E_t and E_{AR}.

The fictional voltage E_{AR} is always in phase quadrature with I_a and proportional to it, that is, it can be considered as representing the electromotive force induced in an inductive reactance. This means that the effect of armature reaction can be represented by assuming that the stator windings have an inductive reactance $X_a = E_{AR}/I_a$. It should be noted that this reactance does not exist, but the introduction of the fictional reactance X_a is very useful in the analysis of the performance of an alternator.

A qualitative understanding of the effects of armature reaction at varying power factors can be obtained by considering the three power factor conditions represented in Fig. 11-8. Figure 11-8(a) represents the unity power factor condition. The rotor pole flux ϕ_f produces the no-load voltage E_t, where E_t lags ϕ_f by 90° electrical, since an induced voltage lags 90° electrical behind the flux producing it. The armature reaction flux ϕ_{AR} is in phase with the load current I_a and induces the fictional voltage E_{AR} lagging it by 90° electrical. The voltage E_g is the phasor sum of E_{AR} and E_t. Similarly the resultant flux ϕ_R is the phasor sum of the field pole flux ϕ_f and the armature reaction flux ϕ_{AR}. These phasors confirm the fact that at unity power factor the resultant air-gap field is distorted. Figure 11-8(b) represents a zero power factor lagging condition. Once again E_{AR} lags ϕ_{AR} by 90° electrical. However, it can be seen that since the field pole flux ϕ_f and the armature reaction flux ϕ_{AR} are 180° out of phase with each other, the resultant flux ϕ_R has been reduced, and as a result E_g has also been reduced. This confirms that the effect of a zero power factor lagging load is to reduce the air-gap flux and thus the generated voltage E_g. Similarly Fig. 11-8(c)

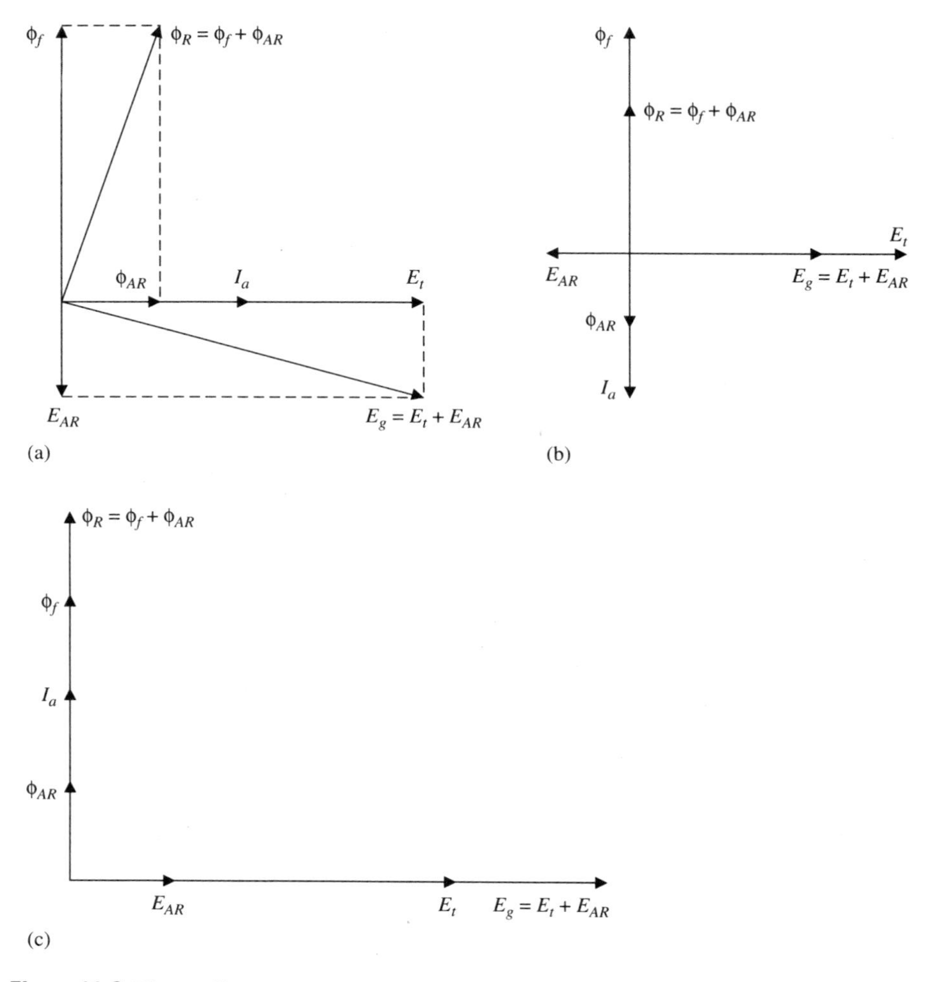

Figure 11-8 Phasor diagrams showing effects of armature reaction at various power factors on terminal voltage of a synchronous generator. (a) Unity power factor. (b) Zero power factor lagging. (c) Zero power factor leading.

shows that the effect of a zero power factor leading load is to cause an increase in the resultant air-gap flux ϕ_R with a subsequent increase of E_g.

So far the effects of stator resistance R_a and stator leakage reactance X_l have been neglected. The equivalent circuit diagram in Fig. 11-9(a) includes the fictional inductive reactance X_a representing the effects of armature reaction, X_l representing the stator leakage reactance, and R_a the stator resistance separated from the ideal synchronous generator producing an electromotive force E_t induced entirely by the field pole flux ϕ_f. The electromotive force E_t is reduced by the inductive voltage drop I_aX_a, which accounts for the effect of armature reaction. The net electromotive force E_g is the phasor sum of E_t and I_aX_a. In turn the terminal voltage V_t is the phasor sum

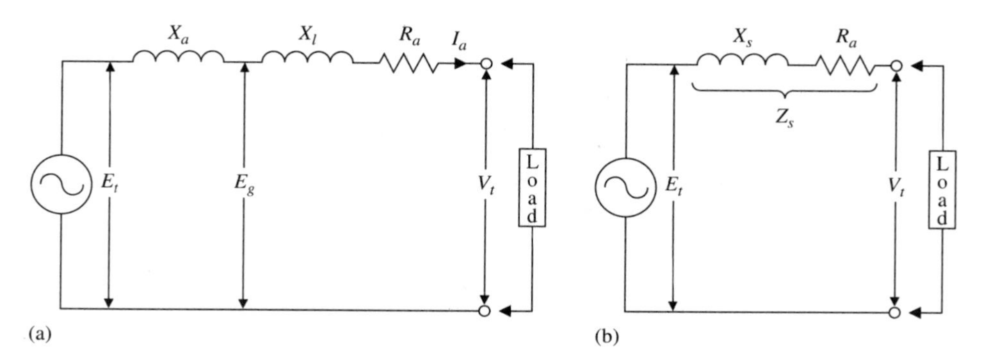

Figure 11-9 Synchronous generator equivalent circuits.

of E_g and the resistive and leakage reactance voltage drops I_aX_a and I_aX_l respectively. This may be expressed as

$$\begin{aligned} V_t &= E_t - jI_aX_a - jI_aX_1 - I_aR_a \\ &= E_t - I_a\left[R_a - j(X_a + X_1)\right] \text{ V/phase} \end{aligned} \tag{11–4}$$

This can be simplified by letting

$$X_s = X_a + X_1 \ \Omega \tag{11–5}$$

where X_s is the *synchronous reactance*. The equivalent circuit diagram can be simplified as shown in Fig. 11-9(b), where

$$Z_s = R_a + jX_s \tag{11–6}$$

Z_s being the *synchronous impedance*. Equation (11-4) can be rewritten as

$$V_t = E_t - I_a(R_a + jX_s) \text{ V/phase} \tag{11–7}$$

where I_aX_s is the synchronous reactance voltage drop and I_aR_a the resistive voltage drop. Equation (11-7) can be expressed as

$$V_t = E_t - I_aZ_s \text{ V/phase} \tag{11–8}$$

The voltage regulation of a synchronous generator can be obtained from phasor diagrams developed from these equivalent circuits.

11-7 VOLTAGE REGULATION OF A SYNCHRONOUS GENERATOR

Synchronous generators usually supply lagging power factor loads. However, to be complete, unity and leading power factor loads must be considered. Since synchronous generators usually operate at a constant terminal voltage, it is necessary to be able to determine the terminal voltage and thus the voltage regulation of the machine under all load and power factor conditions. This can be obtained from the equivalent circuit diagram.

Unity Power Factor Loads

Figure 11-10(a) shows the relationship of the voltages in a synchronous generator operating under unity power factor load conditions. The terminal voltage per phase V_t is drawn as the reference phasor, and the phase current I_a is in phase with V_t. The voltage drop across the stator resistance I_aR_a is in phase with the current I_a, and therefore in phase with V_t. The synchronous reactance voltage drop I_aX_s by definition leads the current by 90°. The voltage drop across the synchronous impedance I_aZ_s is the phasor sum of I_aR_a and I_aX_s. The generated voltage E_t is then the phasor sum of V_t and I_aZ_s. The generated voltage per phase E_t is

$$E_t = \sqrt{(V_t + I_aR_a)^2 + (I_aX_s)^2} \text{ V/phase} \tag{11–9}$$

or, expressed in complex form,

$$E_t = V_t + I_a(R_a + jX_s) \text{ V/phase} \tag{11–10}$$

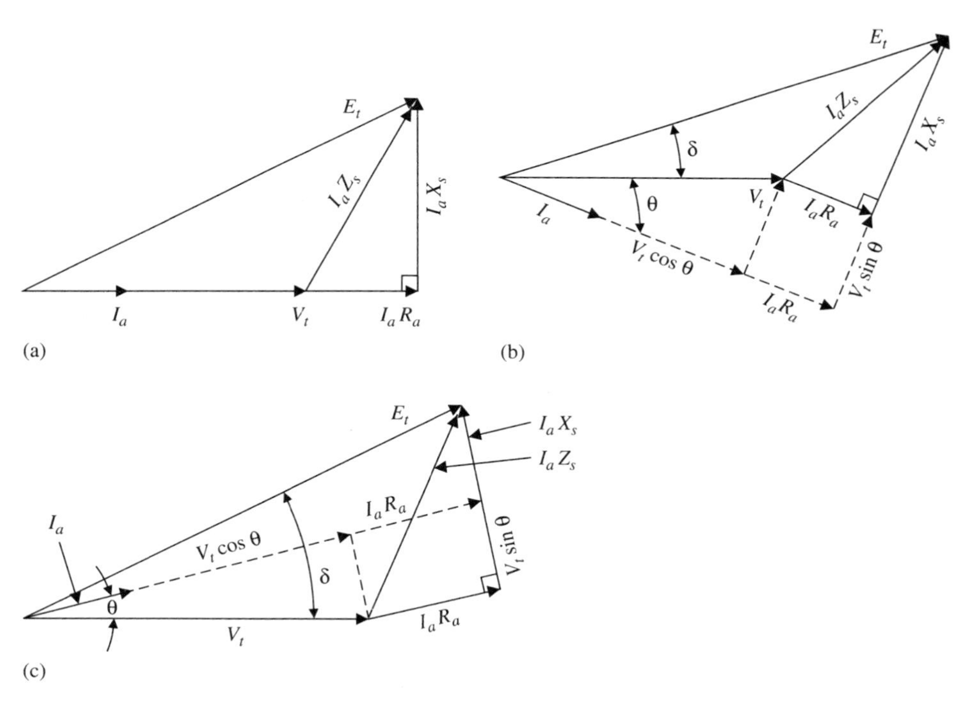

Figure 11-10 Synchronous generator voltage relationships. (a) Unity power factor. (b) Lagging power factor. (c) Leading power factor.

The latter is the preferred form since most engineering and scientific calculators permit an easy solution of Eq. (11-10).

It should be noted that Fig. 11-10 is not drawn to scale, and that the resistive voltage drop I_aR_a is usually much smaller than the synchronous reactance voltage drop I_aX_s, and in large multicircuit synchronous generators the I_aR_a drop is sometimes ignored.

Lagging Power Factor Loads

Normally synchronous generators supply lagging power factor loads. These loads are lagging because of the lagging magnetizing current requirements of transformers, polyphase and single-phase induction motors, and fluorescent lighting.

As can be seen from Fig. 11-10(b), the load current I_a lags the terminal voltage reference phasor V_t by the power factor angle θ. The stator resistance voltage drop I_ar_a is in phase with I_a, and the synchronous reactance voltage drop I_aX_s leads I_a by 90°. The generated voltage per phase E_t is then the phasor sum of V_t and the synchronous impedance voltage drop I_aZ_s. The generated voltage per phase E_t is then, by Pythagoras' theorem,

$$E_t = \sqrt{(V_t \cos\theta + I_aR_a)^2 + (V_t \sin\theta + I_aX_s)^2} \text{ V/phase} \tag{11-11}$$

or, expressed in complex form,

$$E_t = (V_t \cos\theta + I_aR_a) + j(V_t \sin\theta + I_aX_s) \text{ V/phase} \tag{11-12}$$

Perusal of Fig. 11-10(b) shows that as the load power factor worsens, the phase angle θ becomes larger, and as a result E_t must increase to maintain a constant terminal voltage V_t. At the extreme condition, that is, $\theta \cong 90°$, the synchronous impedance voltage drop I_aZ_s will be in phase with V_t, that is, the generated voltage E_t will be a maximum. The angle δ between E_t and V_t is called the *torque angle* or *power angle* of the alternator. This angle, which is load dependent, is a measure of the air-gap power developed in the alternator.

Leading Power Factor Loads

Figure 11-10(c) shows the relationships that exist when the alternator is supplying a leading power factor or capacitive load. As can be seen, the load current I_a leads the reference voltage phasor V_t by the angle θ. As before, the stator resistance voltage drop I_aR_a is added vectorially to V_t, and the synchronous reactance voltage drop I_aX_s leads I_a by 90°. The generated voltage per phase E_t is the phasor sum of V_t and the synchronous impedance voltage drop I_aZ_s,

$$E_t = \sqrt{(V_t \cos\theta + I_aR_a)^2 + (V_t \sin\theta - I_aX_s)^2} \text{ V/phase} \tag{11-13}$$

or, in complex form,

$$E_t = (V_t \cos\theta + I_aR_a) + j(V_t \sin\theta - I_aX_s) \text{ V/phase} \tag{11-14}$$

As can be seen from Eqs. (11-13) and (11-14) and Fig. 11-10(c), as I_a becomes more leading, a point is reached where $E_t = V_t$, that is, zero voltage regulation. If there is any further increase in the power angle θ, then the rotor excitation must be reduced so as to maintain the terminal voltage constant.

Equations (11-9), (11-11), and (11-13) can be consolidated to become

$$E_t = \sqrt{(V_t \cos\theta + I_a R_a)^2 + (V_t \sin\theta \pm I_a X_s)^2} \text{ V/phase} \tag{11–15}$$

or

$$E_t = (V_t \cos\theta + I_a R_a) + j(V_t \sin\theta \pm I_a X_s) \text{ V/phase} \tag{11–16}$$

Note that + is used when dealing with lagging power factors, − when dealing with leading power factors.

When θ = 0°, the unity power factor case, then cos θ = 1 and sin θ = 0, and Eqs. (11-15) or (11-16) are identical to Eqs. (11-9) and (11-10), respectively.

At this point a brief recap of three-phase voltage and current relationships is in order, since all generators are either star (wye) or delta connected. Recall for a star-connected winding

$$V_l = \sqrt{3} V_{\text{ph}} \tag{11–17}$$

and

$$I_l = I_{\text{ph}} \tag{11–18}$$

and for a delta-connected winding,

$$V_l = V_{\text{ph}} \tag{11–19}$$

and

$$I_l = \sqrt{3} I_{\text{ph}} \tag{11–20}$$

where V_l and I_l are line values and V_{ph} and I_{ph} are phase values.

The power supplied per phase is

$$P_L = V_t I_a \cos\theta \text{ W/phase} \tag{11–21}$$

and the total power supplied to a three-phase balanced load is

$$P_L = 3 V_t I_a \cos\theta \text{ W} \tag{11–22}$$

Normally only line values of current and voltage are monitored in a generating station. As a result Eq. (11-22) can be modified since $I_l = I_{\text{ph}} = I_a$ and $V_t = V_l/\sqrt{3}$ for a star-connected stator winding, so that

$$P_L = \frac{3V_l}{\sqrt{3}} I_l \cos\theta = \sqrt{3} V_l I_l \cos\theta \qquad (11\text{–}23)$$

For a delta-connected stator winding, $I_{ph} = I_l/\sqrt{3}$ and $V_l = V_p$, so that

$$P_l = 3V_l \frac{I_l}{\sqrt{3}} \cos\theta = \sqrt{3} V_l I_l \cos\theta \qquad (11\text{–}24)$$

As can be seen from Eqs. (11-23) and (11-24) the power supplied to a balanced three-phase star- or delta-connected load is identical.

11-8 VOLTAGE REGULATION

The change in the terminal voltage that occurs between no load and full load of a synchronous generator when operated at constant speed and field current is called *voltage regulation*. It is defined as

$$\%\ \text{voltage regulation} = \frac{V_{nl} - V_{fl}}{V_{fl}} \times 100\% \qquad (11\text{–}25)$$

where V_{nl} is the no-load terminal voltage and V_{fl} the full-load terminal voltage.

A synchronous generator supplying a lagging power factor load will have a fairly large positive voltage regulation; at unity power factor the voltage regulation will be slightly positive; and for leading power factor loads the regulation will be negative.

As mentioned earlier, it is normal practice to operate synchronous generators at a constant terminal voltage. Therefore, when supplying lagging power factor loads it is necessary to increase the rotor flux as the power factor worsens. Conversely, as the power factor approaches unity and becomes leading, it is necessary to reduce the rotor flux. These variations of rotor flux are achieved by varying the current supplied to the rotor circuit. In modern systems using static exciters, the variation of rotor current is achieved by varying the firing delay angle of the thyristor phase-controlled converter in response to the control signals developed by the closed-loop voltage control system.

The variation of terminal voltage with field current for various load conditions is illustrated in Fig. 11-11. Figure 11-11(a) shows the changes in terminal voltage with varying power factor loads when the excitation current is maintained constant at the value required to give rated voltage on open circuit. Figure 11-11(b) shows the required change of excitation current to maintain a constant terminal voltage. Both figures clearly show the effects of the load power factor, that is, with lagging power factors the effect of armature reaction is to reduce the air-gap flux and decrease the terminal voltage; and with leading power factors armature reaction increases the air-gap flux and the terminal voltage.

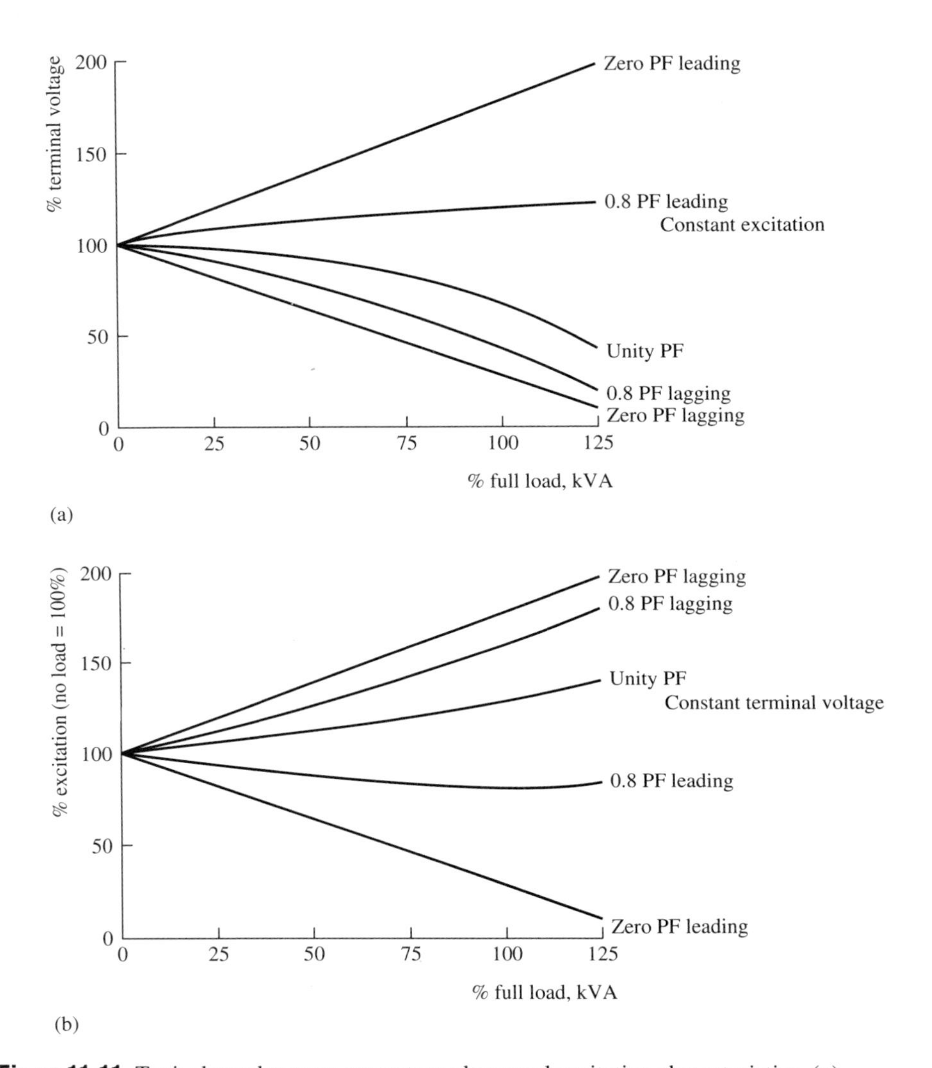

Figure 11-11 Typical synchronous generator voltage and excitation characteristics. (a) Terminal voltage variation with constant excitation. (b) Excitation variation to maintain constant terminal voltage.

➤ EXAMPLE 11-2

A 1,000-kVA 4,600-V three-phase star-connected alternator has a stator winding resistance of 1.75 Ω/phase and a synchronous reactance of 20 Ω/phase. Calculate the full-load generated voltage per phase at: **(a)** unity power factor; **(b)** 0.8 power factor lagging; **(c)** 0.85 power factor leading.

SOLUTION

$$V_{\text{ph}} = \frac{V_l}{\sqrt{3}} = \frac{4{,}600\text{V}}{\sqrt{3}} = 2{,}655.81 \text{ V}$$

$$\text{I}_{\text{ph}} = \frac{\text{kVA} \times 1{,}000}{3V_{\text{ph}}} = \frac{1{,}000 \times 1{,}000}{3 \times 2{,}655.81} = 125.51\text{A} = I_a$$

$$I_a R_a = 125.51\text{A} \times 1.75\Omega = 219.64\text{V}$$

$$I_a X_s = 125.51\text{A} \times 20\Omega = 2{,}510.20\text{V}$$

(a) At unity power factor, from Eq. (11-16),

$$\begin{aligned} E_t &= (V_t \cos\theta + I_a R_a) + j(V_t \sin\theta + I_a X_s) \\ &= (2{,}655.81 + 219.64) + j2{,}510.20 = 2{,}875.45 + j2{,}510.20 \\ &= 3{,}816.98 \text{ V/phase} \end{aligned}$$

(b) At 0.8 power factor lagging, from Eq. (11-16),

$$\begin{aligned} E_t &= (2{,}655.81 \times 0.8 + 219.64) + j(2{,}655.81 \times 0.6 + 2{,}510.20) \\ &= 2{,}344.29 + j4{,}103.69 = 4{,}726.09 \text{ V/phase} \end{aligned}$$

(c) At 0.85 power factor leading, from Eq. (11-16),

$$\begin{aligned} E_t &= (2{,}655.81 \times 0.85 + 219.64) + j(2{,}655.81 \times 0.53 - 2{,}510.20) \\ &= 2{,}477.08 - j1{,}102.62 = 2{,}711.40 \text{ V/phase} \end{aligned}$$

➤ EXAMPLE 11-3

Using the data of Example 11-2, calculate the voltage regulation at each power factor.

SOLUTION

(a) At unity power factor,

$$\begin{aligned} \%\text{ regulation} &= \frac{V_{nl} - V_{fl}}{V_{fl}} \times 100\% \\ &= \frac{3{,}816.98 - 2{,}655.81}{2{,}655.81} \times 100\% = 43.72\% \end{aligned}$$

(b) At 0.8 power factor lagging,

$$\%\text{ regulation} = \frac{4{,}726.09 - 2{,}655.81}{2{,}655.81} \times 100\% = 77.95\%$$

(c) At 0.85 power factor leading,

$$\%\text{ regulation} = \frac{2{,}711.40 - 2{,}655.81}{2{,}655.81} \times 100\% = 2.09\%$$

11-9 DETERMINATION OF SYNCHRONOUS GENERATOR PARAMETERS

The parameters of the equivalent circuit of a synchronous generator are found by conducting three simple tests:

1. Stator or armature resistance test
2. Open-circuit test
3. Short-circuit test

These tests permit the measurement of the stator resistance R_a, the synchronous impedance Z_s, and the calculation of the synchronous reactance X_s, all on a per-phase basis. From these results the voltage regulation can be calculated, since it is not practical to obtain the voltage regulation under full-load operating conditions.

Determination of Stator Resistance

The dc phase resistance R_{dc} of three-phase stator windings may be measured by applying a dc voltage between a pair of stator terminals and measuring the applied voltage and the current passing through the windings, then applying Ohm's law to calculate the dc resistance. Alternatively a bridge-megger may be used to obtain a direct readout of the resistance (Fig. 11-12).

The dc resistance per phase R_{dc} of a star-connected winding is

$$R_{dc} = \frac{V}{A} \times \frac{1}{2} = \frac{V}{2A} \ \Omega/\text{phase} \tag{11–26}$$

and for a delta winding,

$$R_{dc} = \frac{3V}{2A} \ \Omega/\text{phase} \tag{11–27}$$

The resistance of any conductor is greater when used in an ac application. At low frequencies the increase in resistance is small, but as the frequency of the ac source increases, there is a marked increase in the resistance of the conductor. The ac resistance is called the *effective resistance* and is greater than the dc resistance because of the effect of eddy currents and hysteresis losses, dielectric losses in the insulation, radiation losses, and skin effect. These losses are all frequency dependent. Therefore, for accurate results the effective resistance at the operating frequency must be known. The skin effect decreases the effective cross-sectional area of the conductor because the current-produced magnetic fields induce voltages at the center of the conductor, which oppose the current flow.

As was discussed in Chapter 10, harmonics are generated in the individual coils of the stator windings. These high-frequency currents, because of skin effect, are crowded to the outer perimeter of the conductor. The dc resistance of the stator must be increased by a factor to account for the increased resistance in the ac applications. For 60-Hz machines the factor is usually taken to be 1.5 times the dc resistance, that is,

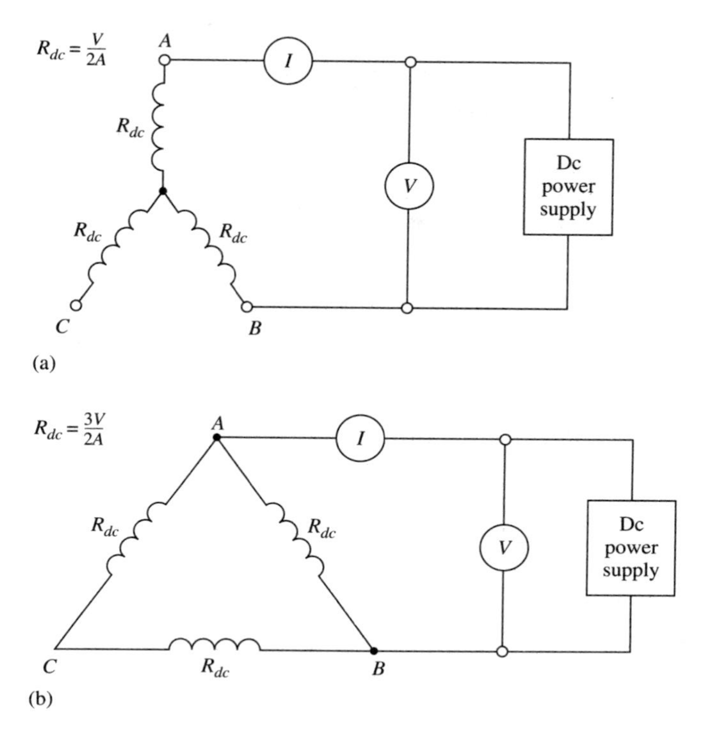

Figure 11-12 Determination of phase resistance of three-phase windings. (a) Star connected. (b) Delta connected.

$$R_a = 1.5 R_{dc}\ \Omega \tag{11-28}$$

It should be noted that especially in high-output machines, the effective resistance R_a is only a fraction of an ohm. When using the voltmeter-ammeter method of determining the dc resistance, currents as close as practical to the actual phase currents should be used to obtain accurate values of R_a since an inaccurate value will introduce errors into calculations, especially efficiency calculations.

Open-Circuit Characteristic

The open-circuit characteristic (OCC) is the relationship between the terminal voltage V_{oc} under open-circuit conditions and the rotor current I_f or rotor magnetomotive force when the synchronous generator is driven at sychronous speed. The test is conducted over a range of values of the open-circuit voltage V_{oc} up to usually 130% of rated terminal voltage, with corresponding readings of V_{oc} and I_f being recorded and plotted. The connections and the resulting graph are shown in Fig. 11-13.

If input mechanical power is also measured during this test, then the no-load rotational losses are obtained. These losses consist of the friction (bearing and slip-ring brush friction), windage, and core losses. The friction and windage losses are constant at synchronous speed, and are obtained by measuring the mechanical power input with the rotor unexcited. The core losses are the difference in mechanical power inputs with the rotor excited and not excited.

Short-Circuit Characteristic

The short-circuit characteristic (SCC) is the relationship between the armature current I_a and the rotor current I_f or rotor magnetomotive force when the synchro-

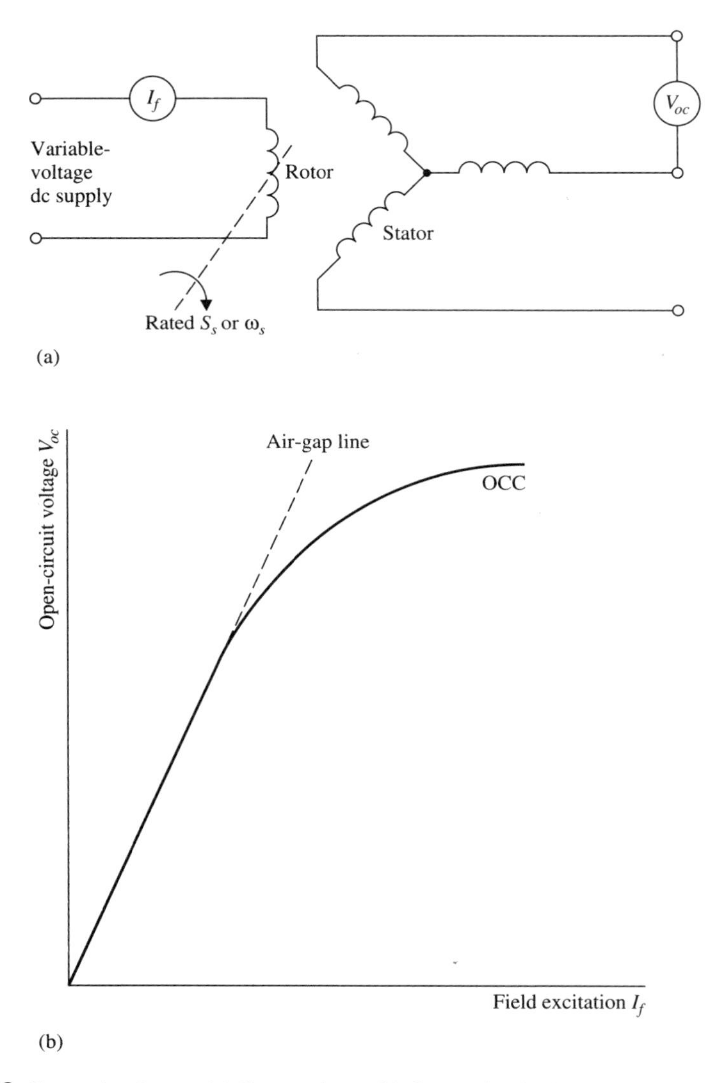

Figure 11-13 Open-circuit test. (a) Connections. (b) Open-circuit characteristic.

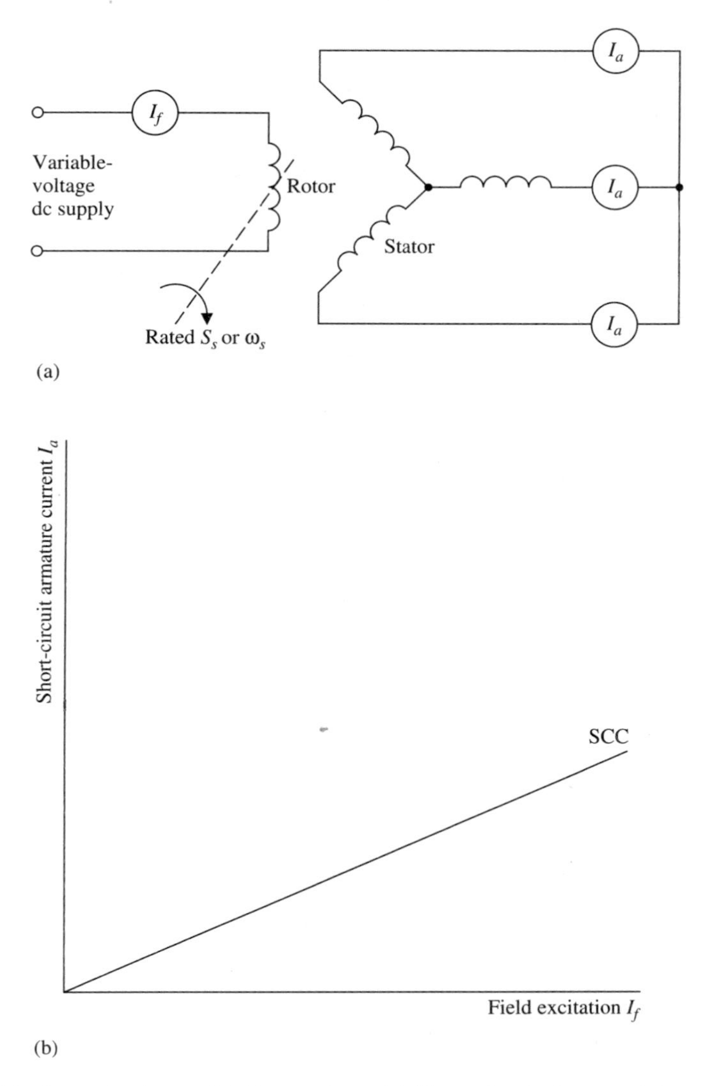

Figure 11-14 Short-circuit test. (a) Connections. (b) Short-circuit characteristic.

nous generator is being driven at synchronous speed with a symmetrical short circuit placed across its terminals. Normally readings of armature current I_a and rotor current I_f are recorded and plotted for a range of armature currents up to 150% of rated armature current. This plot is normally a straight line through the origin (Fig. 11-14).

By plotting both OCC and SCC with common abscissa, field excitation, as shown in Fig. 11-15, it is possible to obtain the synchronous impedance per phase.

In Fig. 11-15 for a field current OA, the unsaturated synchronous impedance is AD/AB, and the saturated impedance is AC/AB. However, in actual practice the synchronous impedance is obtained by assuming that the generator is unsaturated and has the air-gap line passing through the rated voltage point of the OCC. Then the synchronous impedance is the ratio of $A'C'/A'B'$, that is,

$$Z_s = \frac{V_{oc}}{I_{a(\text{sc})}} \tag{11-29}$$

with I_f constant. In turn the synchronous reactance X_s is

$$X_s = \sqrt{Z_s^2 - R_a^2}\ \Omega \tag{11-30}$$

The *short-circuit ratio* (SCR) is by definition the ratio of the field excitation current required to produce rated terminal voltage on an open circuit to that required to produce rated armature current on a short circuit, that is, from Fig. 11-15,

$$\text{SCR} = \frac{OA}{OA'} \tag{11-31}$$

The short-circuit ratio provides information with respect to the effects of arma-

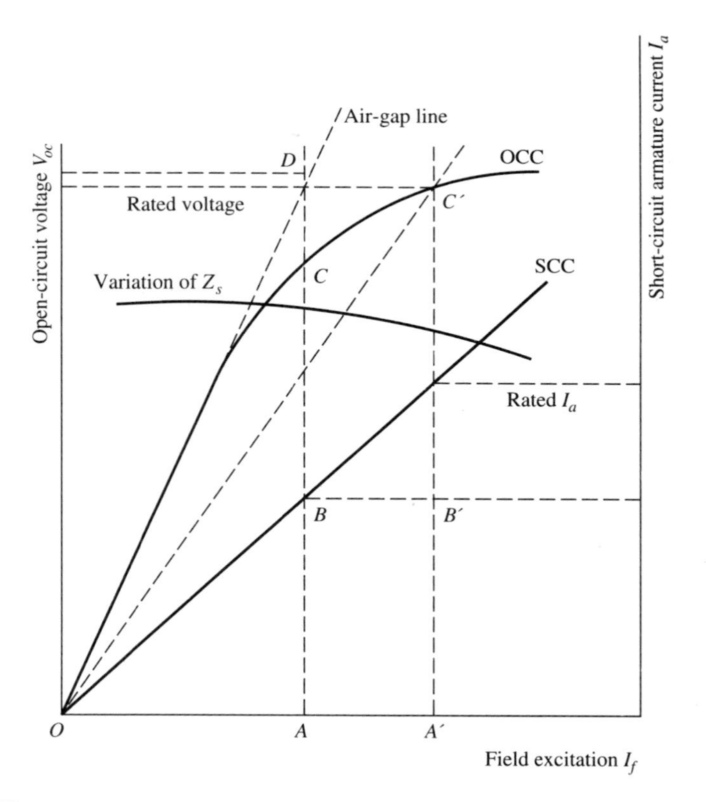

Figure 11-15 Open- and short-circuit characteristics on a per-phase basis.

ture reaction. For example, a low value of SCR indicates a large armature reaction, and as a result the alternator is sensitive to load variations. On the other hand, a large SCR indicates a small armature reaction, and as a result the alternator is far less sensitive to load variations.

These tests permit the performance of large alternators to be tested without the expenditure of the energy that would be necessary if actual full-load testing were carried out. It should also be noted that most hydroelectric alternators are only fully assembled at their final location, and that most turboalternators have been assembled and tested in the manufacturer's plant.

➤ EXAMPLE 11-4

A 500-kVA 2,330-V three-phase star-connected alternator yielded the following test results:

dc resistance test $V_{dc} = 20.0$ V, $I_{dc} = 30.0$ A
Open-circuit test $I_f = 10.0$ A, $V_l = 810$ V
Short-circuit test $I_f = 10.0$A, $I_a = 125.5$ A = rated full-load current

Calculate the regulation at rated full load with a 0.85 lagging power factor. F

SOLUTION

$$V_{oc} = \frac{V_l}{\sqrt{3}} = \frac{810 \text{ V}}{\sqrt{3}} = 467.65 \text{ V/phase}$$

$$Z_s = \frac{V_{oc}}{I_{a(sc)}} = \frac{467.65 \text{ V}}{125.5 \text{ V}} = 3.73 \text{ Ω/phase}$$

$$R_a = 1.5 \times \frac{V_{dc}}{2I_{dc}} = 1.5 \times \frac{20 \text{ V}}{2 \times 30 \text{ A}} = 0.50 \text{ Ω/phase}$$

$$X_s = \sqrt{Z_s^2 - R_{a2}} = \sqrt{3.73^2 - 0.5^2} = 3.70 \text{ Ω/phase}$$

$$I_a R_a = 125.5 \text{ A} \times 0.5 \text{ Ω} = 62.75 \text{ V}$$

$$I_a X_s = 125.5 \text{ A} \times 3.70 \text{ Ω} = 464.35 \text{ V}$$

$$V_t = \frac{V_l}{\sqrt{3}} = \frac{2,300 \text{ V}}{\sqrt{3}} = 1,327.91 \text{ V/phase}$$

The generated voltage per phase is, from Eq. (11-16),

$$\begin{aligned} E_t &= (V_t\cos\theta + I_a R_a) + j(V_t\sin\theta + I_a X_s) \\ &= (1,327.91 \times 0.85 + 62.75) + j(1,337.91 \times 0.53 + 464.35) \\ &= 1,191.47 + j1,168.14 = 1,668.58 \text{ V/phase} \end{aligned}$$

The voltage regulation per phase is

$$\% \text{ regulation} = \frac{V_{nl} - V_{fl}}{V_{fl}} \times 100\%$$

$$= \frac{1{,}668.58 - 1{,}327.91}{1{,}327.91} \times 100\% = 25.65\%$$

A study of Fig. 11-15 and Eq. (11-16) shows that the synchronous impedance is the ratio of the open-circuit characteristic to the short-circuit characteristic. When the OCC is asymptotic to the air-gap line, the synchronous impedance is constant. As the knee of the OCC is approached and passed, the magnitude of the synchronous impedance decreases. For the calculation of the synchronous impedance it was assumed that the generator was unsaturated, and as a result our calculations will give a higher, or pessimistic, value. However, it must also be recognized that the stator resistance is small and in large machines can be ignored when compared to the synchronous reactance, that is, $Z_s \cong X_s$. Therefore under short-circuit conditions the armature current will lag the generated voltage by almost 90°. As a result the effect of the armature reaction flux is to be almost totally in opposition to the field pole flux. This method of calculating the synchronous impedance leads to too high a value of Z_s and thus X_s. This can be compensated for by applying a correction factor, usually about 0.75.

Under short-circuit conditions only the stator resistance limits the buildup of current. Since there is a time lag of several cycles before the demagnetizing effect of armature reaction has any appreciable effect on the air-gap flux, during this time interval the short-circuit current can build up to as much as 10 to 15 times the rated full-load current. When armature reaction has reduced the air-gap flux, the peak short-circuit current is reduced to a steady value.

The overload protection devices of synchronous generators such as circuit breakers require several cycles of the supply frequency before they can interrupt the circuit and prevent damage to the generator. Although the generator has an inherent capability of reducing short-circuit currents, damage may occur to the circuit breakers, station bus bars, and most importantly the generator's stator windings. Most electrical utilities use current-limit reactors external to but in series with each phase of the alternator windings. The function of these reactors is to limit the short-circuit current to an acceptable value, usually about twice the rated full-load current, by introducing an inductive reactance voltage drop of about 20% of rated terminal voltage. The series current-limiting reactors, which consist of a number of turns of heavily insulated cable held in position in a concrete form, permit the operation of the circuit breakers without any damage occurring to the equipment. As can be seen, armature reaction under these circumstances is beneficial, and in fact is deliberately designed into the generator by permitting a high ratio of synchronous reactance to stator resistance to exist, so that sustained short-circuit currents are limited to the order of 200% of rated full-load current.

11-10 LOSSES AND EFFICIENCY

It is not practical to test high power output synchronous generators at rated loads and power factors because of the power consumption that is required. As a result the individual losses are summed and added to the power output to determine the input power and efficiency. These losses include the following:

1. *Friction and windage losses* P_{fw}: These losses are proportional to speed, and since an alternator operates at synchronous speed, these losses are constant and are obtained from the no-load or open-circuit test by measuring the input power required to drive the alternator with the rotor field unexcited.

2. *Core loss* P_c: These losses consist of the eddy current and hysteresis losses with the assumption that the air-gap flux density is constant. The core losses are proportional to speed, and since the alternator operates at constant speed, they are constant. The core loss is obtained from the open-circuit test by measuring the input required to drive the alternator at synchronous speed with the rotor excited. The core loss is the difference in power input required to drive the alternator at no load with and without the rotor poles being excited.

3. *Copper losses:* These consist of the armature copper loss $P_a = I_a^2 R_a$ and the field copper loss $P_f = V_f I_f$. Since the effective armature resistance R_a and the armature current I_a are per-phase quantities, the total copper loss is the number of phases, normally three, times the phase copper loss. The dc resistance must also be corrected for temperature, that is, ambient temperature 25°C plus the observed temperature rise, or to the appropriate temperature for the insulation class being used, for example, 115°C for class F.

4. *Stray load losses* P_{stray}: The stray load losses are mainly caused by eddy currents in the armature conductors and distortions of the magnetic flux caused by load currents. These losses are almost impossible to calculate. However, tests can provide data to enable a graph of stray load losses versus load current to be plotted. The simplest correction is to use always the effective ac resistance instead of the dc resistance of the windings.

5. *Excitation losses:* Normally the losses of the excitation system are not included in the determination of the alternator efficiency. Even though the power supplied to the rotor may be large, it usually accounts for less than 1% of the total power output of the alternator; but even this can still account for 10-11 MW for a 1,500-MW alternator.

The overall efficiency η is calculated from

$$\eta = \frac{\text{kVA} \times \text{power factor}}{\text{kVA} \times \text{power factor} + \sum \text{losses}} \times 100\% \tag{11–32}$$

where η is the percent efficiency, kVA the load on the alternator, and the power factor is the power factor of the connected load.

Efficiency can also be expressed as

$$\eta = \frac{P_{\text{out}}}{P_{\text{in}}} \times 100\% = \frac{P_{\text{out}}}{P_{\text{out}} + \sum \text{losses}} \times 100\% \tag{11–33}$$

➤ EXAMPLE 11-5

A 1,000-kVA 2,300-V star-connected alternator operates at rated kVA at a power factor of 0.85 lagging. The synchronous reactance is 1.2 Ω/phase, and the effective armature resistance per phase is 0.12 Ω. When running at synchronous speed, the friction and windage losses were found to be 23.75 kW, and the core loss was 19.5 kW. The rotor field was supplied with 14.5 A at 600 V dc from a static exciter. The stray power losses are 1.1 kW. Calculate the efficiency of the alternator at rated load.

SOLUTION

$$P_{\text{out}} = 1{,}000 \text{ kVA} \times 0.85 = 850 \text{ kW}$$

$$I_a = I_l = \frac{1{,}000 \text{ kVA} \times 1{,}000}{\sqrt{3} \times 2{,}300 \text{ V}} = 251.02 \text{ A}$$

Losses	kW
Friction and windage	23.75
Core	19.5
Field copper loss $V_f I_f = 600 \times 14.5$	8.7
Armature copper loss $= 3 \times 251.02^2 \times 0.12$	22.68
Stray losses	1.1
Total	75.73

$$\eta = \frac{1{,}000 \text{ kVA} \times 0.85}{1{,}000 \text{ kVA} \times 0.85 + 75.73 \text{ kW}} = \frac{850 \text{ kW}}{925.73 \text{ kW}} = 91.82\%$$

The maximum efficiency that can be achieved by a synchronous generator can be shown to occur when the fixed losses are equal to the variable losses. The fixed losses consist of the friction and windage, core, and field copper losses. The variable losses are made up of the armature copper loss, which is proportional to I_a^2.

The point at which maximum efficiency occurs can be controlled in the design process, and will usually be around 90% of the rated output of the alternator since alternators are rarely operated continuously at 100% output.

11-11 PARALLEL OPERATION

Most hydroelectric and thermal power stations contain more than one alternator. These alternators are connected either singly or in groups in parallel to the power station bus bars, and from there to the high-voltage transmission system and the distribution system to the customer. The major reasons for operating alternators in parallel are:

1. The power demand exceeds the power output capability of one or more alternators.
2. From the economic point of view alternators should be operated at maximum efficiency. This is achieved by shutting down partially loaded alternators until all alternators in the power station are operating efficiently.
3. Planned or emergency maintenance can be carried out on one or more alternators with the load being carried by the remaining operating units.
4. Permits future expansion by permitting the installation of additional alternators as the connected load increases.
5. Permits the transfer of power anywhere in an interconnected system. A typical example is the high-voltage transmission system interconnecting the Provinces of Ontario, Manitoba, and Quebec and the states of Minnesota, Wisconsin, Illinois, Michigan, Ohio, and New York. This system permits the efficient transfer of electric energy anywhere in the system.

Before alternators can be paralleled, the incoming alternator must be synchronized to the system, that is, each alternator must be synchronized to the power station bus bars. In turn each power station must be synchronized to the interconnecting system.

There are certain requirements that must be met before an alternator can be paralleled to the bus bars:

1. The terminal voltage of the incoming alternator must be equal to the bus bar voltage.
2. The frequency of the incoming alternator must be equal to the bus bar frequency. *Note:* A 3,600-r/min two-pole 60-Hz alternator can be paralleled with a 1,200-r/min six-pole 60-Hz alternator.
3. The phase sequence of the incoming three-phase alternator must be the same as the phase sequence of the bus bars. This is normally checked prior to commissioning an alternator or after major repairs.
4. The incoming alternator voltages are in phase with the bus bar voltages.
5. The speed control governors of the prime movers and the automatic voltage regulators of the alternators all have similar drooping characteristics.

Failure to follow these requirements during paralleling most probably will lead to excessive currents circulating between alternators through the interconnecting bus bars, and may cause serious damage because of excessive heating of the stator windings.

Synchronization

Our discussions can be simplified by first considering the parallel operation of single-phase alternators. From Fig. 11-16(a), alternators B and C are connected in parallel to the single-phase bus bar. Their terminal voltages V_B and V_C are *in phase* relative to the bus bars, and in *phase opposition* in the local circuit *abcd* between the alternators. If alternator A is brought up to nearly the correct speed, that is, its frequency is slightly less than that of the bus bar, with its terminal voltage V_A adjusted to be equal to V_B and V_C, as measured by the voltmeter across the paralleling circuit breaker contacts, then since the frequencies are not the same, the phase relationship of V_A with respect to V_{bus} will vary at a rate that is equal to the frequency difference or beat frequency [Fig. 11-16(b)]. This results in the voltage across the synchronizing circuit breaker varying in amplitude from zero to twice the normal operating voltage. The incoming alternator may be synchronized after its speed is adjusted until the difference is small. Then when the voltmeter reads zero, the circuit breaker can be closed, and the machine has been synchronized.

The procedure for synchronizing three-phase alternators is basically the same. First the incoming alternator is brought up to the correct speed by remotely adjusting the governor of the prime mover. Then the terminal voltage is adjusted to be equal to the bus bar voltage by adjusting the set voltage of the automatic voltage regulator. Further adjustment of the prime mover governor may be necessary to minimize the frequency difference between the incoming alternator and the bus bars. Then when the voltage difference across the circuit breaker is zero, the circuit breaker is closed. The correct point of synchronization is determined by synchronizing lamps or by a synchroscope.

Dark-Lamp Synchronization

Figure 11-17(a) shows the connection arrangement for the dark-lamp synchronization method paralleling the incoming alternator to the bus bars. As the incoming alternator is brought up to speed, and its terminal voltage is adjusted to be equal to the bus bar voltage, the synchronizing lamps see the difference voltage across the open paralleling switch. If the phase sequence is correct, they will go bright and dark together at the difference frequency between the alternator and the bus bars. The paralleling switch is closed when the voltage difference across the contacts of the same phase of the paralleling switch is zero, that is, the lamps are all dark. There are two major disadvantages to this method: (1) There is no appreciable light emitted from the lamps, even though there may be 50% of the rated voltage applied to the lamps. (2) Since the correct point synchronization is at the middle of the dark period, it is very difficult for the operator to judge the exact point of synchronization. As a result this method is rarely used.

Bright-Lamp Synchronization

Figure 11-17(b) shows the connections for bright-lamp synchronization. This method overcomes the disadvantages of the dark-lamp method since it is relatively easy to judge when the lamps are at their brightest. It should also be noted that both the dark-lamp and the bright-lamp methods provide an indication of phase sequence, that is, if the phase sequence is correct, the lamps will all go bright and dark together; if the phase sequence is incorrect, the lamps will be bright and dark individually.

Two Bright- One Dark-Lamp Synchronization

Figure 11-17(c) shows the connections for this method. The correct point of synchro-

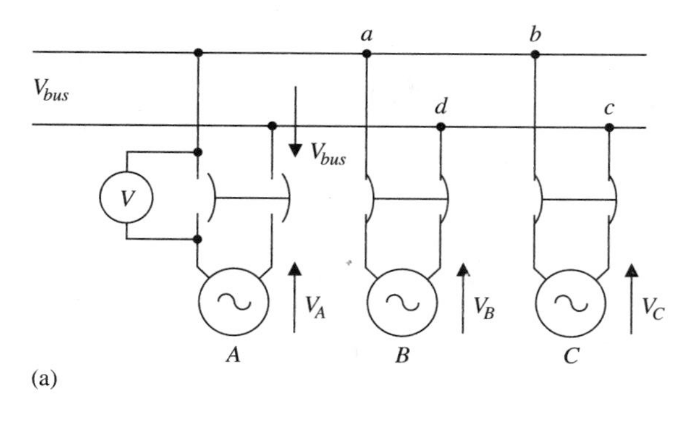

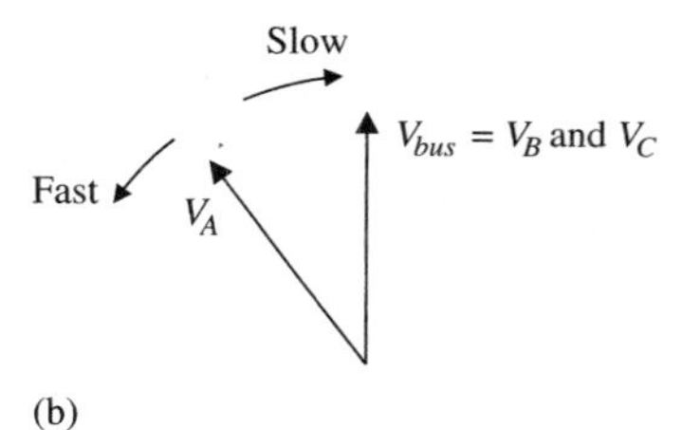

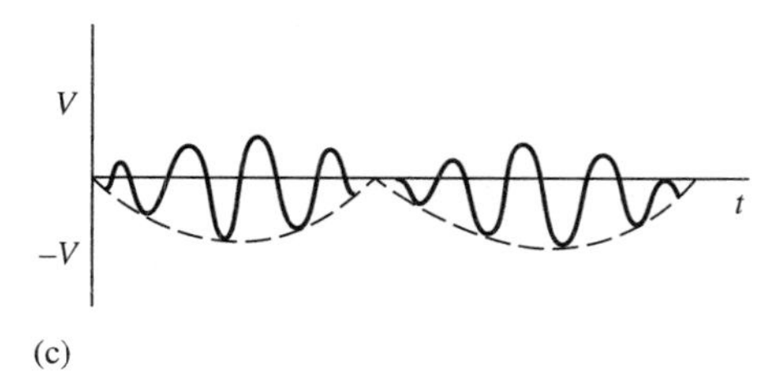

Figure 11-16 Parallel operation of single-phase alternators. (a) Schematic. (b) Phasor representation of frequency difference. (c) Voltage waveform of frequency difference across synchronizing circuit breaker.

nization is when one lamp is dark, one is increasing in brightness, and the other is decreasing in brightness. The paralleling switch is closed when these two lamps are equally bright.

While these methods are effective, modern practice is to use a *synchroscope*. The synchroscope provides a physical indication of the correct point of synchronization. It also shows by the needle movement whether the incoming machine frequency is greater or less than the bus bar frequency, that is, whether the incoming machine is running fast or slow. Normally because of operator reaction times and the operating times of remotely controlled paralleling circuit breakers, the switch is closed slightly before the marked synchronizing point, with the incoming alternator running slightly fast. This ensures that the alternator will immediately generate and supply power to the bus bars. In modern power stations the whole synchronizing procedure is automated to eliminate human error.

11-12 SYNCHRONIZING TORQUE

When alternators are paralleled, they run in synchronism with each other, that is, their rotors do not vary their relative angular positions with respect to each other by more than a few degrees. This locking together of the rotors is a direct result of synchronizing torque.

As was mentioned previously, although the terminal voltages of the alternators are in phase with respect to the external circuit, that is, the bus bars, their terminal voltages are in phase opposition with respect to the load circuits between the machines. To simplify our discussion we will consider single-phase machines. Figure 11-18(a) shows the phase relationship in the local circuit when alternators A and B are correctly paralleled. If alternator B slows down momentarily, then V_B lags V_A and a resultant voltage V_R is produced, which in turn causes a synchronizing current I_s to flow in the load circuit between the machines, where

$$I_s = \frac{V_R}{Z} \tag{11-34}$$

Z being the synchronous impedance of the phase windings of both alternators plus the impedance of the connections between the machines. In addition, since the synchronous reactance of the alternators is much greater than their resistances, the synchronizing current will lag V_R by $\theta \cong 90°$.

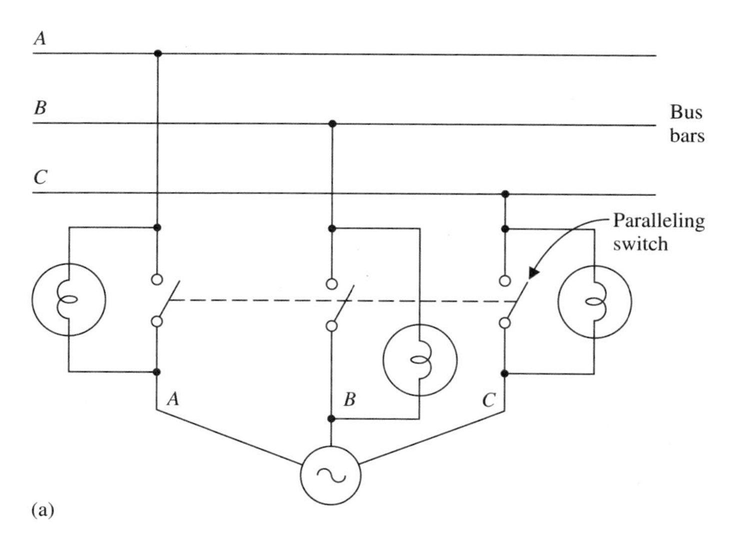

Figure 11-17 (a) Synchronizing lamp circuits. Dark-lamp synchronizing.

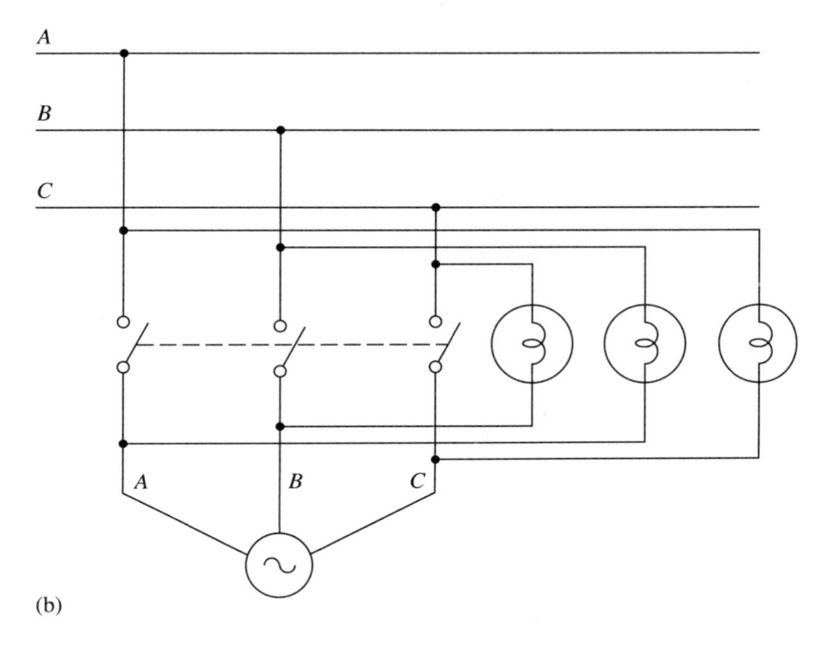

Figure 11-17 (b) Synchronizing lamp circuits. Bright-lamp synchronizing.

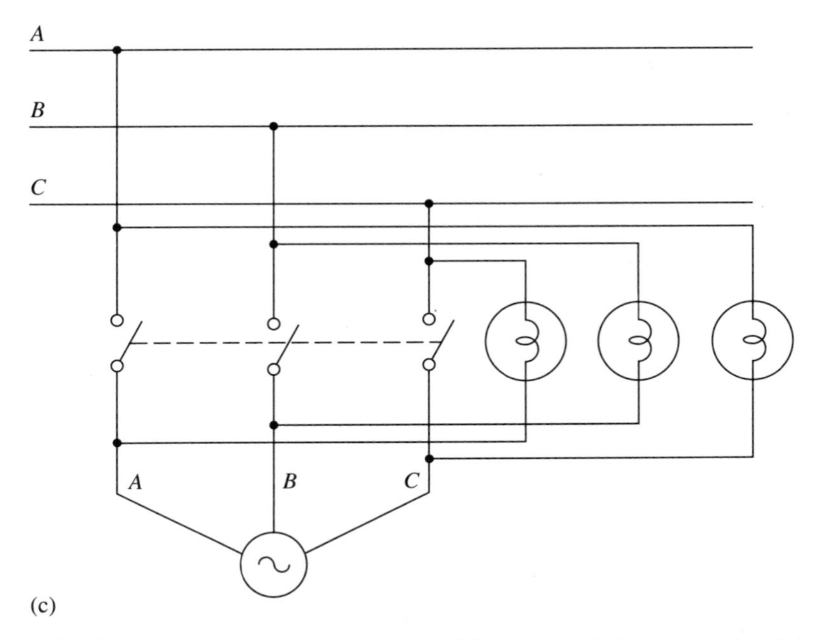

Figure 11-17 (c) Synchronizing lamp circuits. Two bright- and one dark-lamp synchronizing.

As can be seen from Fig. 11-18(b), when alternator B slows momentarily with respect to alternator A, I_s is almost in phase with V_A, that is, alternator A is acting as a generator and a *retarding torque* is acting on the rotor. Similarly, I_s is almost 180° out of phase with alternator B. As a result I_s produces motor current in alternator B, which creates an *accelerating torque*, speeding up its rotor momentarily. Figure 11-18(c) illustrates the production of synchronizing torque when alternator B speeds up with respect to alternator A. Under this condition alternator B acts as a generator and the resulting retarding torque acts to slow down its rotor. At the same time I_s produces an accelerating torque in alternator A.

As can be seen from the preceding discussion, whenever there is a departure of the alternator rotors from being exactly synchronized, a circulating current is developed between the alternators. This in turn creates a synchronizing torque, which accelerates the slower alternator rotor and retards the faster alternator rotor. The synchronizing torque is proportional to the sine of the angle of displacement (sin δ) and becomes zero when $\delta = 0°$. It should be noted that the production of synchronizing torque depends entirely on the synchronous reactance of the alternator. If the alternator had zero reactance and only resistance, the synchronizing current would be in phase with the resultant voltage V_R and approximately at right angles to V_A or V_B, that is, it is a *wattless current* and will not produce any synchronizing torque.

11-13 LOAD AND POWER FACTOR ADJUSTMENT—PARALLEL OPERATION

With the use of interconnecting systems between power stations the total connected generating power output of all the paralleled alternators is far greater than the power output of any one generator. As a result any change in the operating state of a single alternator on the system is negligible. The alternator is then considered to be connected to an *infinite bus bar system*, that is, the frequency, voltage, and phase relationships of the infinite bus bar are unaffected by changes occurring in a single alternator. For example, if an alternator's rotor tends to lag, then synchronizing power is supplied from the bus bar to accelerate it back into synchronism. But the effect on the remaining alternators paralleled to the system is negligible.

Figure 11-19(a) represents a three-phase alternator connected to a lagging power factor load. Then assuming that the armature resistance is very small, which is normally the case, the output power developed P_d is

$$P_d = 3V_t I_a \cos\theta \text{ W} \tag{11–35}$$

or it can be expressed as

$$P_d = 3E_t I_a \cos\phi \text{ W} \tag{11–36}$$

where $\phi = \delta + \theta$ degrees electrical. From Fig. 11-19(a), side *ab* is

$$ab = E_t \sin\delta = I_a X_s \cos\theta \tag{11–37}$$

Therefore

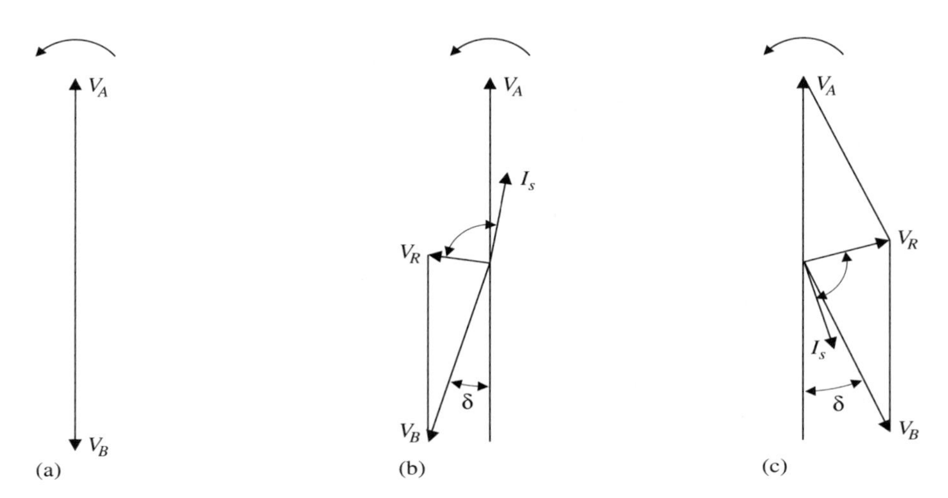

Figure 11-18 Production of synchronizing torque. (a) V_A and V_B synchronized. (b) V_B lagging V_A. (c) V_B leading V_A.

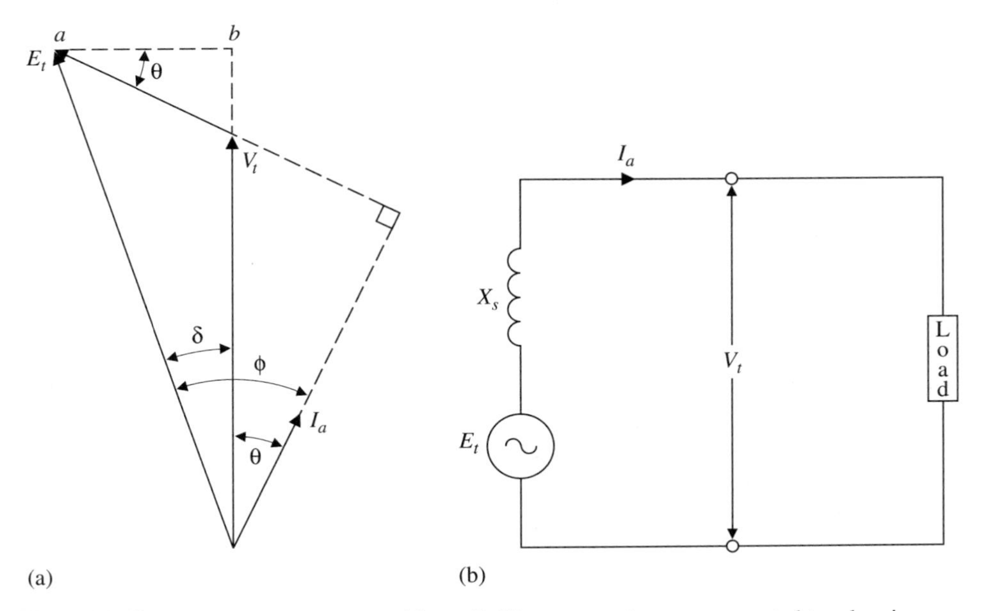

Figure 11-19 Synchronous generator with negligible stator resistance connected to a lagging power factor load. (a) Phasor diagram. (b) Equivalent circuit.

$$I_a \cos \theta = \frac{E_t \sin \delta}{X_s} \tag{11–38}$$

Then substituting Eq. (11-38) into Eq. (11-35) yields

$$P_d = \frac{3E_t V_t \sin \delta}{X_s} \text{ W} \tag{11–39}$$

and the developed power P_d will be a maximum when the torque or power angle is $\delta = 90°$. Equation (11-39) is only valid when R_a is negligible.

Figure 11-20 represents a three-phase alternator connected to an infinite bus bar, that is, a system operating at constant frequency and voltage. The effect of the reactance X_l between the alternator terminals and the infinite bus bars must now be considered. This external reactance includes the impedance of the step-up transformer, the circuit breakers, and the line impedances up to the point where the connection is made to the interconnecting high-voltage system.

Equation (11-39) can be modified to give the power delivered to the infinite bus bar as follows:

$$P_{\text{del}} = \frac{V_t V_B \sin \Delta}{X_{\text{ext}}} \text{ W} \tag{11–40}$$

where P_{del} is the power delivered to the infinite bus bar, X_{ext} the reactance from the station bus bar to the infinite bus bar, V_B the infinite bus bar voltage, considered to be the reference voltage, and Δ the phase angle between V_t and V_B. When power is being delivered from the alternator to the infinite bus bars, V_t leads V_B by Δ electrical degrees.

At the instant of synchronization with an infinite bus bar, the induced voltage E_t is equal to and in phase with the bus bar voltage V_B [Fig. 11-21(a)]. Under these conditions there is no voltage drop or potential difference across the synchronous reactance X_s, and the load current I_a is zero. The alternator is "floating" across the

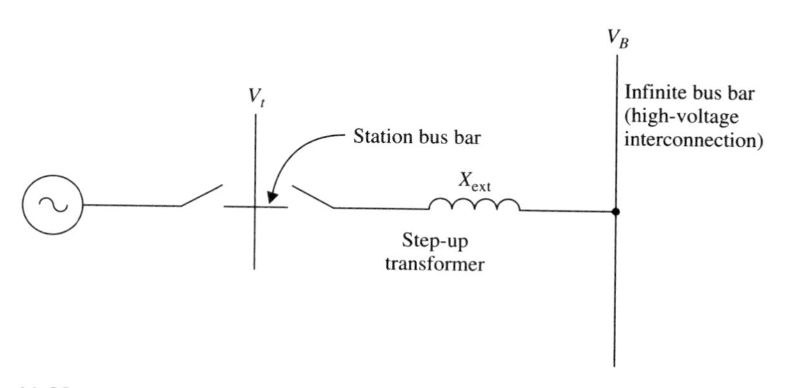

Figure 11-20 Synchronous generator connected to infinite bus bar.

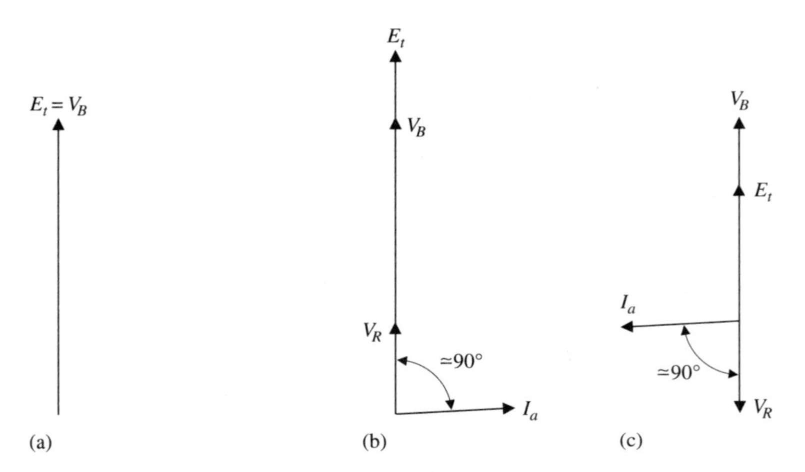

Figure 11-21 Effect of excitation changes on power factor. (a) Floating. (b) Overexcited. (c) Underexcited.

system, that is, it is neither delivering nor receiving electric power.

If the rotor excitation current of the alternator is increased, the induced voltage E_t is also increased and a resultant voltage V_R in phase with E_t is produced, where $V_R = E_t - V_B$. However, since the stator resistance is very small compared to the synchronous reactance X_s, a current $I_a = V_R/X_s$ will flow. This current will lag the induced voltage E_t and V_R by approximately 90°, and is almost a pure *wattless current.* Expressed another way, the infinite bus bar load is seen as an inductive reactance [Fig. 11-21(b)], and the alternator supplies reactive power to the bus bar.

If the rotor excitation current is decreased, E_t will be less than V_B, and $V_R = E_t - V_B$, and I_a will lead V_B, [Fig. 11-21(c)]. The infinite bus bar load is seen as a capacitive reactance, that is, the alternator will draw reactive power from the infinite bus bar system.

The effect of excitation changes is exaggerated because an increase in excitation current will be counteracted by the increase in the demagnetizing effect of armature reaction caused by the lagging current. On the other hand a decrease in excitation current will result in an increased air-gap flux produced by the magnetizing effect of armature reaction when a leading current is flowing.

The preceding explanations show that the *true power* output (kW or MW) of an alternator can only be varied by changing the mechanical power output of the prime mover. The results of changing the excitation are to change the *volt-amperes reactive* or *apparent power*, that is, the kVAR or MVAR output of the alternator. This can be illustrated by means of Fig. 11-22.

Figure 11-22(a) shows the effect of increasing the prime mover output so that the rotor is advanced, that is, supplying power to the infinite bus bar without an increase in excitation current. Fig. 11-22(b) shows the effect of increasing the excitation current without a further increase in the mechanical power input. As can be seen, E_t is increased, and V_R is increased, and the angle by which I_a lags V_B is increased, that is, the power factor is more lagging. Fig. 11-22(c) shows the effect of decreasing the excitation current without changing the mechanical power input to the alternator. As can be seen, both V_R and I_a have advanced with respect to V_B, that is, the alternator is supplying a leading current. These effects are shown in the V curves of Figure

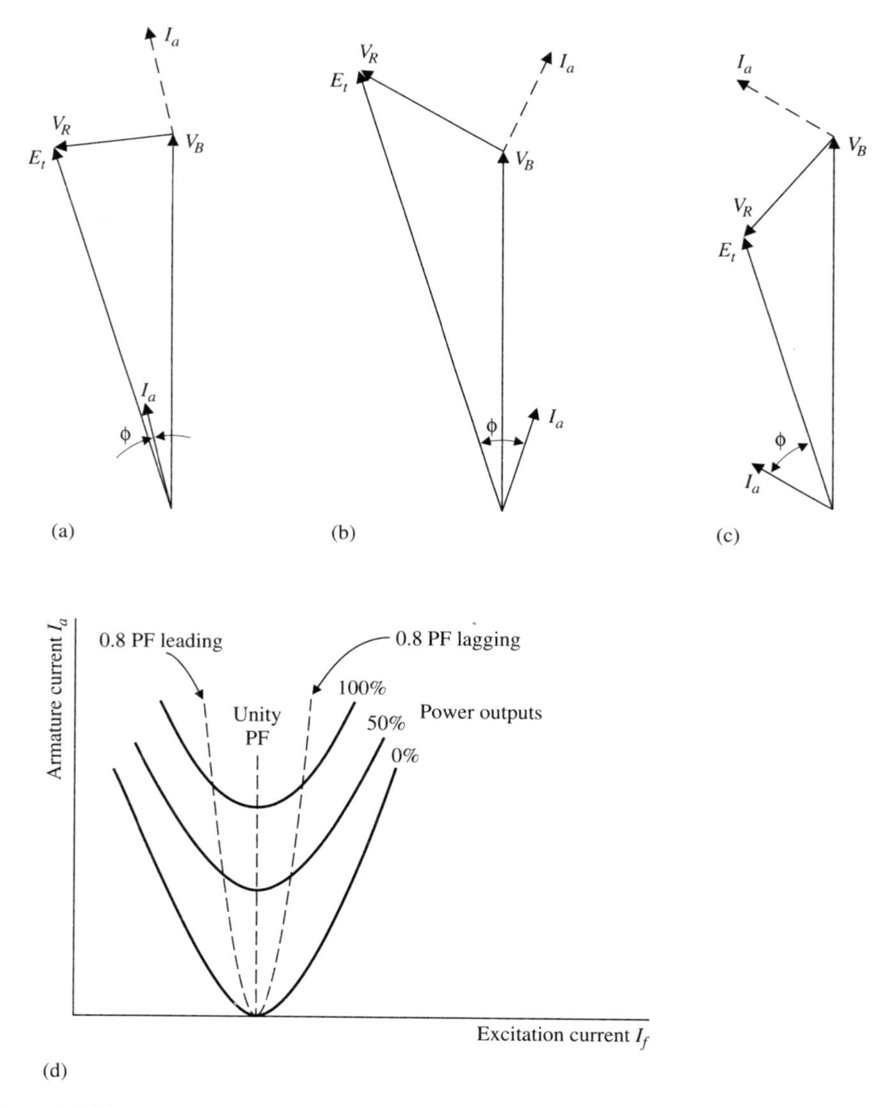

Figure 11-22 Combined effect of torque and excitation power and power factor of alternator connected to infinite bus bars. (a) Increased torque. (b) Increased excitation. (c) Decreased excitation. (d) V curves.

11-22(d). To summarize, the power output of the alternator is solely controlled by the angular movement of the rotor, that is, by increasing or decreasing the torque output of the prime mover. The power factor is solely controlled by the rotor excitation, becoming more lagging as the excitation current is increased, or more leading as the excitation current is reduced.

11-14 SYNCHRONOUS GENERATOR RATINGS

The limitations of synchronous generators are defined by their ratings, which are usually specified on their nameplates and in the manufacturer's literature. The most important ratings are voltage, frequency, speed (angular velocity), apparent power (kVA or MVA), power factor, rotor excitation current, and service factor.

Voltage, Frequency, and Speed Ratings

Synchronous generators are usually manufactured in the following NEMA voltage ratings: 120, 240, 480, 600, 2,400, 4,160, 4,330, 6,990, 11,500, 13,800, and 23,000 V. The most common voltages used in large power stations are 13,800 and 23,000 V, with the higher voltage ratings being associated with high-output machines.

The standard output frequency of generators in use in North and South America is 60 Hz; 50 Hz is the standard in Europe and Asia. In addition, 400 Hz is commonly used in aircraft and control system applications. However, the power output of these machines is confined to the low end of the power spectrum. Since the relationship between frequency and poles defines the rotational speed, synchronous generators operate at specific and constant speeds (see Table 11-1).

The output terminal voltage is proportional to the air-gap flux and the angular velocity, and since the angular velocity is constant, the output voltage is determined by the maximum allowable rotor excitation current. In addition, the output voltage is also limited by the breakdown voltage rating of the stator insulation.

Apparent Power and Power Factor

Alternators are rated in terms of their apparent power output S, where

$$S = \sqrt{3}V_l I_l \text{ VA} \tag{11–41}$$

Since the terminal voltage is known, the maximum allowable armature current is also known. Also since the armature copper losses are proportional to I_a^2, they are dependent on the phase angle or power factor of the connected load. As a result alternators are rated in terms of their kVA or MVA outputs, usually at 0.85 power factor lagging. This means that as the system power factor worsens, the kVA or MVA output must be reduced to ensure that the armature current does not exceed design limits.

QUESTIONS

11-1 Discuss the characteristics and construction of cylindrical and salient-pole rotors.

11-2 Explain the function and operation of amortisseur windings.

11-3 What is meant by the term *critical speed?*

11-4 Discuss the types of cooling used with turboalternators.

11-5 What are the advantages of hydrogen cooling?

11-6 With the aid of a sketch explain the operation of a static excitation system for a synchronous generator.

11-7 With the aid of a sketch explain the operation of a brushless excitation system for a synchronous generator.

11-8 What is meant by armature reaction in a synchronous generator?

11-9 With the aid of sketches and waveforms explain the effect of leading and lagging power factor loads on the air-gap flux of a synchronous generator.

11-10 What is leakage reactance? What is its effect on a synchronous generator?

11-11 What is synchronous reactance? What factors are involved in the synchronous reactance voltage drop?

11-12 What tests are required to determine the parameters of the equivalent circuit of a synchronous generator?

11-13 Discuss with the aid of a graph the effect of varying power factors on the terminal voltage of a synchronous generator.

11-14 Discuss how the effective ac resistance per phase of a synchronous generator is obtained.

11-15 What is meant by the power or torque angle?

11-16 Discuss with the aid of a schematic the connections and procedure for conducting an open-circuit test on an alternator.

11-17 How can the no-load losses be found from the open-circuit test?

11-18 Discuss with the aid of a schematic the connections and procedure for conducting a short-circuit test on an alternator.

11-19 What is the significance of the short-circuit ratio? What information is obtained from the SCR?

11-20 Why is the synchronous impedance not constant over the operating range of a synchronous generator? What value of synchronous impedance should be used?

11-21 Discuss the behavior of a synchronous generator under short-circuit conditions. What is the function of a current-limiting reactor?

11-22 Why are synchronous generators paralleled?

11-23 What conditions must be met before synchronous generators are paralleled?

11-24 Discuss the steps that must be taken to parallel a synchronous generator to a power station bus bar.

11-25 Discuss with the aid of sketches the relative merits of: **(a)** dark-lamp; **(b)** bright-lamp; **(c)** two bright- one dark-lamp synchronization.

11-26 What is synchronizing torque and how is it produced?

11-27 With the aid of phasor diagrams explain the effect of increasing the input mechanical power to an alternator connected to an infinite bus bar system.

11-28 With the aid of phasor diagrams explain the effect of varying the rotor excitation of an alternator connected to an infinite bus bar system.

PROBLEMS

11-1 A 60-Hz hydroelectric alternator has a rated speed of 80 r/min. Calculate the number of rotor poles.

11-2 A 10,000-kVA 11,500-V three-phase star-connected alternator is operated at full load. Calculate: **(a)** the phase voltage; **(b)** the phase current.

11-3 The terminal voltage of an alternator rises from 11,500 V at full load to 14,789 V at no load. Calculate the percentage voltage regulation.

11-4 A three-phase star-connected synchronous generator supplies a unity power factor load at 4,160 V. If the resistance voltage drop per phase is 9.5 V/phase and the synchronous reactance voltage drop is 115 V/phase, calculate the percent voltage regulation.

11-5 A 1,500-kVA 6,600-V three-phase star-connected alternator has an effective resistance of 1.75 Ω/phase and a synchronous reactance of 18.0 Ω/phase. Calculate the full-load generated voltage per phase at: **(a)** unity power factor; **(b)** 0.85 power factor lagging.

11-6 Calculate the voltage regulation of the alternator in Problem 11-5 at full load at: **(a)** 0.8 power factor lagging; **(b)** 0.75 power factor leading.

11-7 A 1,000-kVA 13,800-V three-phase star-connected alternator has an effective armature resistance of 2 Ω/phase. The friction and windage losses are 9.75 kW, the core loss is 14.5 kW, and the rotor power input is 5.25 kW. Calculate the full-load efficiency at a 0.85 power factor lagging.

11-8 Repeat Problem 11-7 at 25%, 50%, 75%, and 125% of rated full load.

11-9 A 2,000-kVA 2,300-V three-phase star-connected alternator has an effective armature resistance of 0.2 Ω/phase. The results of tests are:
Short-circuit test: field current = 12 A; line current = rated current
Open-circuit test: field current = 12 A; line voltage = 975 V

Calculate: **(a)** the synchronous impedance per phase; **(b)** the synchronous reactance per phase; **(c)** the voltage regulation at 0.75 power factor lagging.

CHAPTER

12

Synchronous Motors

12-1 INTRODUCTION

Polyphase synchronous generators can be operated as synchronous motors or vice versa. The rotor of the synchronous motor, as its name implies, rotates in synchronism with the rotating magnetic field; that is, they operate at a constant speed determined by the ac source frequency and the number of rotor poles. Although operation at constant speed is an important factor, synchronous motors are selected because:

1. They may be used to provide system power factor correction, as well as providing output torque to the connected load at the same time.
2. When operated at unity power factor, they are more efficient than polyphase induction motors of equivalent voltage and horsepower (kW) output ratings.
3. They are especially suited to low-speed high-power-output applications below 514 r/min (53.83 rad/s) and high-speed 514–1,200-r/min (53.83–127.76-rad/s) high-horsepower (kW), typically 700–10,000-hp (522.20–7,460-kW), applications.

Normally synchronous motors are available in ratings from 150 to 20,000 hp (112 to 14,920 kW), and are used in applications where a constant-speed output is desirable. In addition, because of their improved efficiencies and power factor characteristics, they permit making substantial reductions in utility bills. However, the control and protection requirements are more complex than those required for polyphase induction motors. To this must be added the requirement for dc excitation of the rotor. These factors contribute to increased maintenance costs for the synchronous motor. In addition, the synchronous motor has increased no-load losses because of the rotor excitation requirements, and as a result the synchronous motor is best suited for constant-load applications such as blowers, compressors, centrifugal pumps, and cement kilns. Figure 12-1 illustrates the typical power output and speed ranges of synchronous and induction motors.

12-2 CONSTRUCTION

The synchronous motor is identical in construction to the salient-pole alternator, except that the rotor shaft is normally horizontal. Depending on the diameter, the stator laminations are punched out either in one piece or in segments. The three-phase stator windings are connected either star or delta, in some cases with the windings designed for dual-voltage operation. The stator frame also provides support for the brush rigging and brush supports. Most commonly synchronous motors are purchased in either open- or drip-proof configurations. However, totally enclosed fan-cooled motors are rarely encountered.

The rotor consists of a spider to which laminated steel field poles are attached with prewound dc field pole windings. The spider construction is determined by the designed rotor speed. In low-speed machines the spider is either cast iron or cast

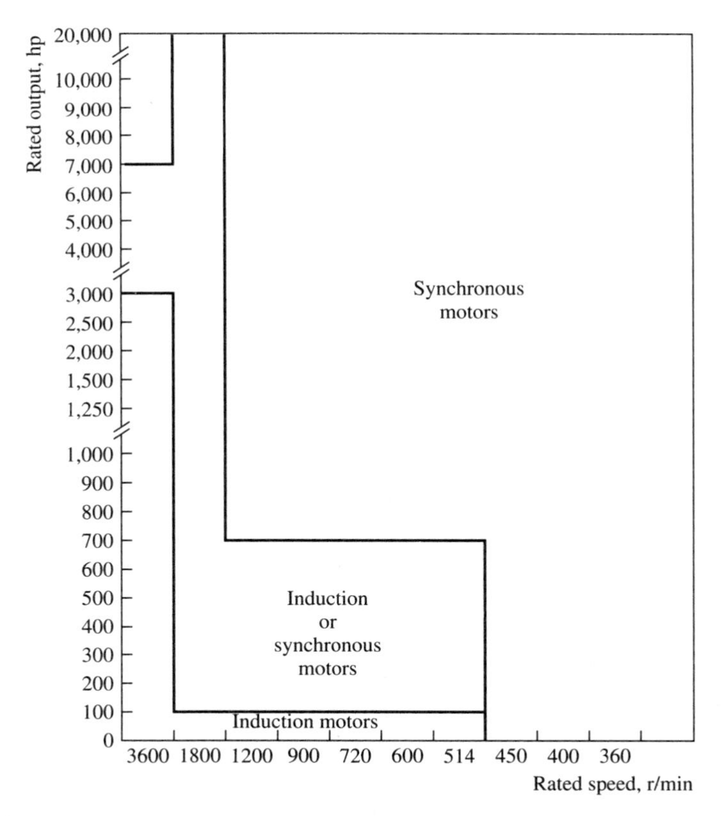

Figure 12-1 Graph showing application ranges of synchronous and induction motors.

steel with a split hub. This type of spider can be supplied without a shaft and is directly attached to the load and supported by its bearings. High-speed rotor spiders are made up from a number of stamped laminations riveted together. The faces of the salient poles are slotted axially to carry the *starting* or *amortisseur windings*, which consist of copper conductors placed in the slots with their ends short-circuited together. The individual pole face windings are in turn short-circuited together by a short-circuiting ring to form a squirrel-cage winding similar to that of an induction motor. The main purpose of this winding is to act as a start winding since the synchronous motor is not self-starting. A secondary purpose is to damp out load-induced shaft oscillations. It must be remembered that this squirrel-cage winding is not as efficient as the rotor of a polyphase induction motor.

The rotor poles are excited from an external dc source or from a directly connected dc exciter mounted on the rotor shaft. However, modern practice is to use brushless excitation similar to that previously described for synchronous generators.

12-3 PRINCIPLE OF OPERATION

The synchronous motor is a *doubly excited* machine; that is, three-phase ac power is supplied to the stator windings to produce a constant-amplitude rotating magnetic field ϕ_R, and dc power is supplied to the rotor circuit to produce the rotor field ϕ_{rot}. The basic principle of operation can most easily be explained by considering a two-pole synchronous motor. The rotating magnetic field is represented by the U-shaped permanent magnet shown in Fig. 12-2(a). This magnet is free to rotate about the axis shown. The rotor field ϕ_{rot} is represented by a second permanent magnet, also free to rotate about the same axis. In the position shown, the rotor field is aligned with the stator field [see Fig. 12-2(b)]. Assume now that the U-shaped magnet is rotated through a small angle in either direction with the rotor restrained. Then the flux phasors are no longer in alignment and a magnetic torque T_m is created, which acts in such a direction as to restore alignment of the two magnetic fields [see Fig. 12-2(c)]. The larger the angle up to 90° between the flux phasors, the greater the restoring effect of the magnetic torque. After the U-shaped magnet has turned through 180°, the fluxes are antiphase and the restoring torque is zero. As the U-shaped magnet continues its rotation, the restoring torque increases, but in the opposite direction until it reaches a maximum after the U-shaped magnet has rotated through 270°, and it decreases to zero when the fluxes are once again in alignment. The torque acting on the rotor is alternating at the supply-line frequency. If the rotor is now free to rotate, rotation will be opposed by the inertial force T_J and frictional forces T_B acting on the rotor, so that it will not rotate in either direction.

When power is applied to the stator of a polyphase synchronous motor, the rotating magnetic field is revolving at synchronous speed ω_s rad/s. It is obvious that the rotor must then be accelerated by external means to be in synchronism with the rotating field, so that it turns through one pole pitch during each half-cycle of the supply frequency; that is, the stator pole reverses polarity as the next rotor pole of opposite polarity is aligned with it, with the result that a unidirectional torque is applied to the rotor and the rotor field poles are locked in synchronism with the rotating magnetic

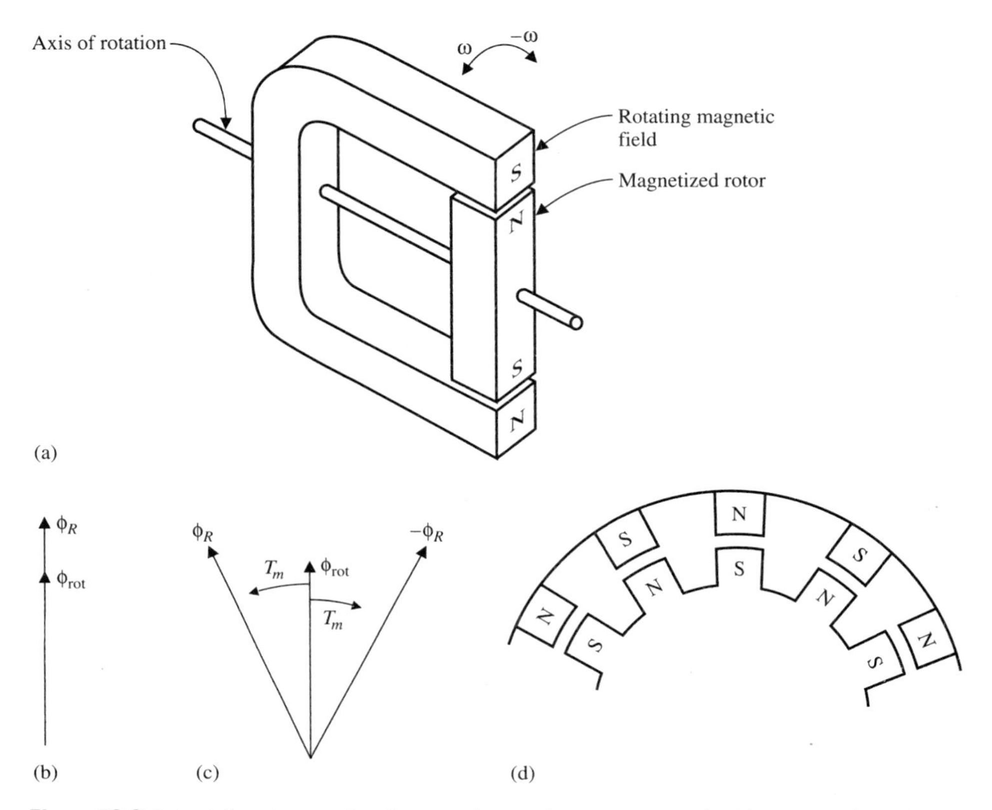

Figure 12-2 Principle of torque development by synchronous motor. (a) Representation of rotating magnetic field by U-shaped magnet. (b) Flux phasors when aligned. (c) Flux phasors showing production of magnetic torque. (d) Alternative concept of torque production.

field. This condition can be best visualized by imagining the rotating stator field to be made up of the same number of salient poles as there are salient poles on the rotor, with both sets of poles locked together by a radial magnetic force [Fig 12-3 (a)].

Under no-load conditions, assuming no frictional or inertial loads are present, the axes of the stator and rotor poles coincide. As mechanical loading on the rotor shaft increases, there will be a momentary drop in the rotor speed, and the rotor poles will fall back slightly from the stator poles, but they will still turn at synchronous speed. The angle δ, or torque angle, will increase progressively as the mechanical load on the rotor increases. The developed torque T_d is dependent on the torque angle δ, and T_d must be great enough to overcome the load torque T_L and the combined windage and friction load. Magnetic attraction is keeping the rotor and stator poles locked in synchronism with the rotating magnetic field.

As the mechanical loading on the rotor shaft increases, the torque angle δ increases, which in turn is accompanied by an increase in the developed torque T_d.

However, there is a limit, which is reached when the load torque T_L exceeds the maximum developed torque T_d. At this point, known as the *pull-out torque*, the rotor pulls out of synchronism with the rotating magnetic field, and the rotor stops rotating. The rated pull-out torque is defined as the maximum torque that the motor can supply for 1 min, and is expressed as a percentage of full-load torque with rated rotor current and rated stator current and supply frequency. The maximum torque that is developed is determined by the magnetic field strength of the rotor and stator poles. The field strength of the rotor poles is determined by the excitation current in the rotor field windings and because of the nonuniform air gap also on the rotor position. The stator pole strength is determined by the stator current, which in turn is dependent on the ac line voltage. Pull-out torques range between 150 and 200% of rated full-load torques. Under rapid load fluctuations the transient or instantaneous pull-out torque can exceed the maximum continuous pull-out torque by as much as 20%, which prevents the motor from stalling under severe short-duration load fluctuations.

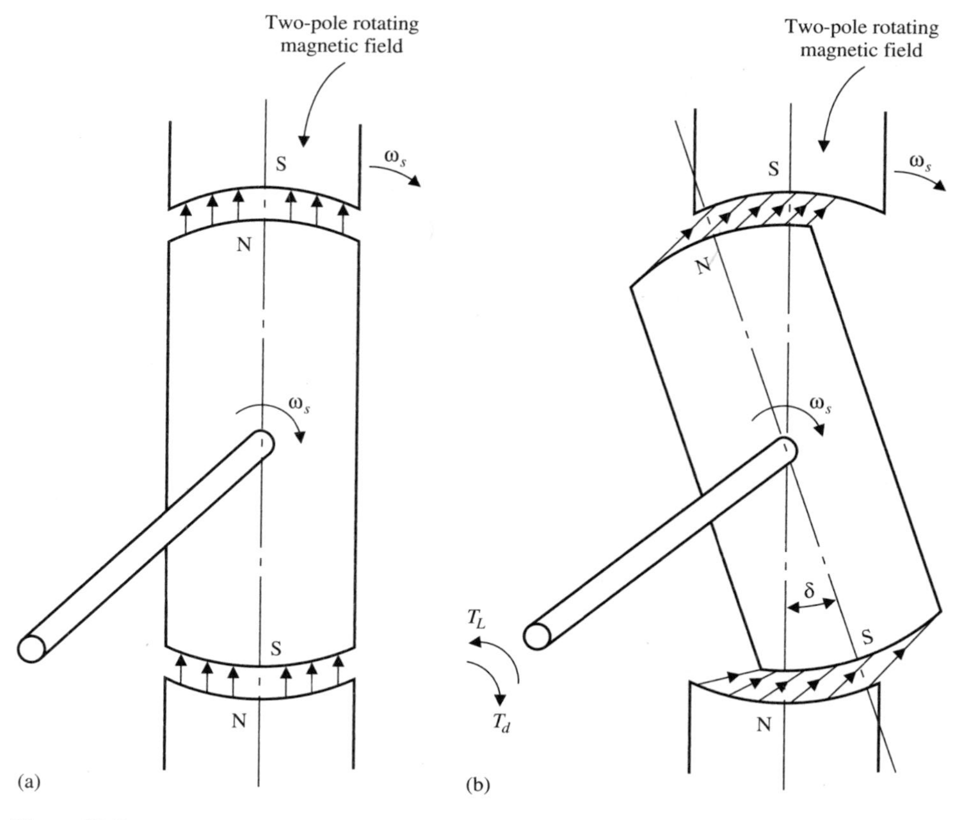

Figure 12-3 Torque development of two-pole synchronous motor. (a) No load. (b) Full load.

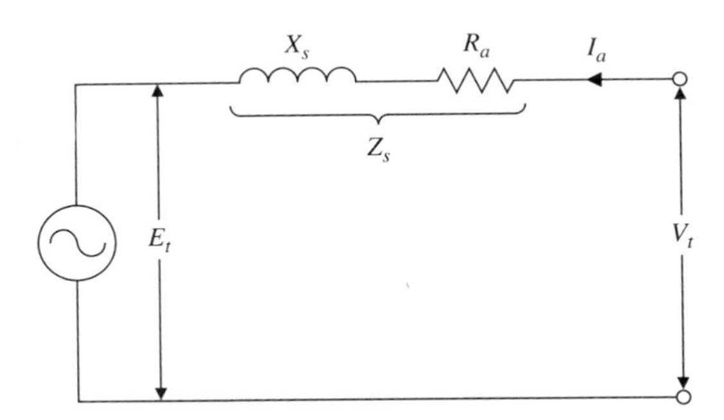

Figure 12-4 Synchronous-motor equivalent circuit.

12-4 EQUIVALENT CIRCUIT

The equivalent circuit of a synchronous motor is identical to that of the synchronous generator, with the exception that the direction of the current I_a is into the armature windings.

The per-phase equivalent circuit of a synchronous motor shown in Fig. 12-4 is a valuable tool in analyzing the motor's performance, provided all the phase components are balanced, and

$$E_t = V_t - I_a(R_a + jX_s) \text{ V/phase} \tag{12-1}$$

from which

$$I_a = \frac{V_t - E_t}{R_a + jX_s} = \frac{V_R}{Z_s} \text{ A} \tag{12-2}$$

where I_a is the armature current per phase drawn by the motor, V_t the applied terminal voltage per phase, E_t the generated voltage per phase, Z_s the synchronous impedance per phase, R_a the effective ac resistance per phase, and X_s the synchronous reactance per phase.

12-5 TORQUE PRODUCTION

Equation (12-2) shows that the armature current I_a is limited by the counter electromotive force and the synchronous impedance Z_s in a manner similar to the limitation of armature current in a dc motor by the counter electromotive force E_c and the armature resistance R_a. However, there is one major difference, namely, when a dc shunt motor is mechanically loaded, the increase in load torque produces a drop in the angular velocity of the shaft, which, since E_c is proportional to flux and angular velocity, causes E_c to decrease and the armature current to increase. From this it must be noted that the counter electromotive force in a dc motor must always be less than the applied terminal voltage if torque is to be developed.

The synchronous motor by definition is a constant-speed motor, and the generated voltage E_t, which is proportional to flux and angular velocity, is only varied by varying the dc excitation of the rotor field. Assuming constant field excitation, that is, E_t is constant, then I_a can only increase or decrease in response to varying load demands by varying the phase relationship of E_t with respect to V_t, that is, V_R is equal to the phasor difference of E_t and V_t.

Consider the situation where a synchronous generator is synchronized to the bus bar and the generated voltage and the terminal voltage are equal and 180° out of phase with respect to each other. Then the resultant voltage $V_R = 0$ [Fig. 12-5(a)], and the stator current $I_a = 0$.

If the prime mover is disconnected from the synchronous machine, the rotor of the machine now acting as a synchronous motor will slow momentarily and then continue rotating at synchronous speed, but the centerlines of the rotor poles will lag behind the centerlines of the poles of the revolving stator magnetic field. This angle δ is called the torque angle. As a result the generated voltage E_t, which is produced by the rotor field, is also displaced by δ [Fig. 12-5(b)], and the resultant voltage V_R will produce an armature current I_a, which lags it by approximately 90°, since the synchronous reactance X_s is much greater than the effective armature resistance R_a. If

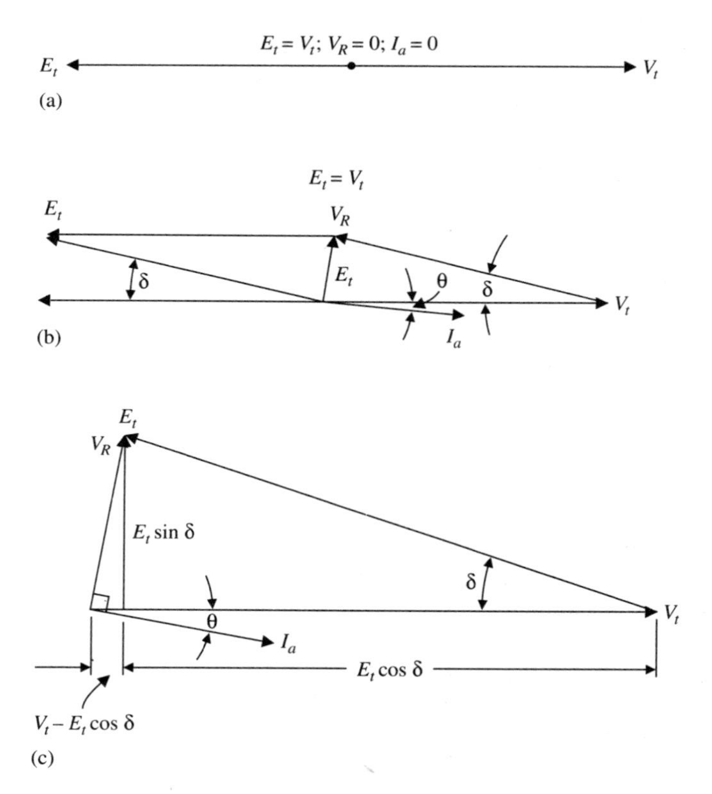

Figure 12-5 Torque development by synchronous motor. (a) Synchronous motor driven by prime mover at synchronous speed. (b) Operating under no-load conditions. (c) Phasor diagram showing relationship between V_t, E_t, and V_R.

the synchronizing power produced is insufficient to overcome the load torque, that is, the windage and friction torques, the rotor will fall back with respect to the rotating magnetic field until both the resultant voltage V_R and the armature current I_a produce sufficient synchronizing power to counterbalance the load, friction, and windage countertorques; that is, the rotor will rotate at synchronous speed with the rotor displaced by the torque angle δ. The torque angle is increasing or decreasing with increasing or decreasing load countertorques. The amount of synchronizing power developed per phase is

$$P_d = V_t I_a \cos\theta \text{ V/phase} \tag{12–3}$$

where P_d is the synchronizing power developed per phase, V_t the applied terminal voltage per phase, and I_a the armature current per phase.

The torque angle δ between the stator poles and the rotor poles is measured in electrical degrees, and the actual mechanical displacement angle β is

$$\beta = \frac{2\delta}{P} \text{ ° mechanical} \tag{12–4}$$

where P is the number of rotor poles and δ degrees electrical. As can be seen from Eq. (12-4), ß is equal to δ (for $P = 2$) or less than δ.

From Fig. 12-5(c) the resultant voltage per phase V_R can be calculated from

$$V_R = (V_t - E_t \cos\delta) + j(E_t \sin\delta) \text{ V/phase} \tag{12–5}$$

➤ EXAMPLE 12-1

A 24-pole, 500-hp (373-kW) 4,160-V 60-Hz three-phase star-connected synchronous motor is operated at no load with the generated voltage per phase adjusted to be equal to the applied stator phase voltage. At no load the rotor poles lag the stator poles by 0.75° mechanical, the synchronous reactance per phase is 22 Ω, and the effective armature resistance per phase is 2 Ω. Calculate: **(a)** the torque angle δ; **(b)** the resultant voltage V_R per phase; **(c)** the phase current; **(d)** the total developed synchronizing power; **(e)** the total armature copper loss; **(f)** the total shaft power output in horsepower and kW.

SOLUTION

(a) From Eq. (12-4),

$$\beta = \frac{2\delta}{P}$$

Therefore

$$\delta = \frac{\beta P}{2} = \frac{0.75° \times 24 \text{ poles}}{2} = 9°$$

(b)

$$V_t = \frac{V_l}{\sqrt{3}} = \frac{4{,}160\,\text{V}}{\sqrt{3}} = 2{,}401.78\,\text{V} = E_t$$

From Eq. (12-5),

$$\begin{aligned} V_R &= (V_t - E_t\cos\delta) + j(E_t\sin\delta) \\ &= (2{,}401.78 - 2{,}401.78\cos 9°) + j(2{,}401.78\sin 9°) \\ &= 29.57 + j375.72 = 376.88\angle 85.50° \text{ V/phase} \end{aligned}$$

(c) The synchronous impedance per phase Z_s is

$$Z_s = R_a + jX_s = 2 + j22 = 22.09\angle 84.81° \ \Omega\text{/phase}$$

Therefore from Eq. (12-2),

$$\begin{aligned} I_a &= \frac{V_R}{Z_s} = \frac{376.88\angle 85.50° \text{ V/phase}}{22.09\angle 84.81° \ \Omega\text{/phase}} \\ &= 17.06\angle 0.69° \text{A/phase} \end{aligned}$$

(d) From Eq. (12-3), the synchronizing power developed per phase is

$$\begin{aligned} P_d &= V_t I_a \cos\theta \\ &= 2{,}401.78 \times 17.06 \times \cos 0.69° = 40{,}971.40 \text{ W/phase} \end{aligned}$$

The total developed power P_T is

$$\begin{aligned} P_T &= 3P_d = 3 \times 40{,}971.40 \text{ W} \\ &= 122{,}914.19 \text{ W} = 122.91 \text{ kW} \end{aligned}$$

(e) The total armature copper loss is

$$P_{Cu} = 3I_a^2 R_a = 1{,}746.26 \text{ W} = 1.75 \text{ kW}$$

(f) The total shaft power output is

$$\begin{aligned} \text{hp} &= \frac{P_T - P_{Cu}}{0.746 \text{ kW/hp}} = \frac{122.91 \text{ kW} - 1.75 \text{ kW}}{0.746 \text{ kW/hp}} \\ &= 162.40 \text{ hp} = 121.15 \text{ kW} \end{aligned}$$

It should be noted that since the armature current I_a is nearly in phase with V_t, the power absorbed by the motor is almost entirely converted into mechanical power output, except for that required to supply the copper and iron losses.

Since synchronous machines can be operated as generators or motors, the relationships for developed power for the generator, Eq. (11-39), are also valid for the synchronous motor; that is,

$$P_d = \frac{3E_tV_t \sin\delta}{X_s} \text{ kW} \tag{12-6}$$

Since

$$P_d = T_d\omega \text{ kW} \tag{12-7}$$

then

$$T_d = \frac{3E_tV_t \sin\delta}{\omega X_s} \text{ N} \cdot \text{m} \tag{12-8}$$

From Eqs. (12-6) and (12-8) it can be seen that maximum power and torque occur when $\delta = 90°$, that is, $\sin\delta = 1$, assuming that V_t and X_s are constant. However, the magnitude of the maximum torque varies with the rotor excitation current since E_t is proportional to the rotor excitation current. Although the losses have been neglected, the torque calculated by Eq. (12-8) is reasonably accurate.

It should be noted that salient-pole-rotor synchronous motors actually develop two torques: the electromagnetic torque and the reluctance torque. The result of the combination of these two torques is to shift the point at which the maximum torque is developed to $\pm\delta_{max} \sim 70°$. This is illustrated in Fig. 12-6. For stable operation the torque is maintained in the range of $-\delta_{max} \le \delta \le +\delta_{max}$.

12-6 EFFECT OF EXCITATION CHANGES ON A SYNCHRONOUS MOTOR

We saw that changes in the rotor excitation current of an alternator operating in parallel with an infinite bus bar produced a change in the power factor. It would seem reasonable that changes in the rotor excitation of a synchronous motor would produce changes in the power factor.

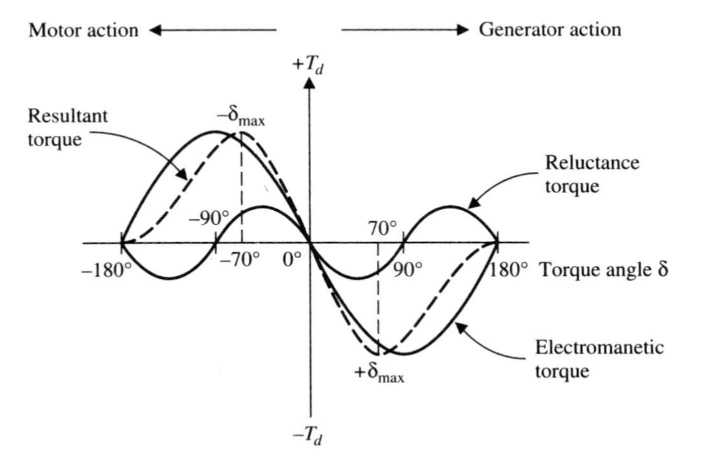

Figure 12-6 Torque versus torque angle for salient-pole synchronous motors.

Figure 12-7 shows the phasor relationships of a synchronous motor supplying a constant-kW load from a constant-voltage source V_t. Initially, as is shown in Fig. 12-7(a), the armature current I_a lags V_t by θ. The angle between V_R and I_a, which is approximately 90°, is determined by the synchronous reactance X_s and the effective armature resistance $R_a(\theta = \tan^{-1} X_s/R_a)$ and remains constant. It can be seen in Fig. 12-7(a) that E_t is less than V_t. In Fig. 12-7(b) the excitation current has been increased, thus increasing E_t, which has resulted in an advance of the resultant voltage phasor V_R, with a corresponding advance of the armature current phasor I_a until it is now in phase with V_t, that is, the motor is operating at unity power factor. Further increases in the field current increase the generated voltage E_t, and further advance the resultant voltage phasor V_R and the current phasor I_a. It can be seen from Fig. 12-7(c) that the increase in E_t has resulted in I_a leading V_t, that is, the motor is acting as a capacitive load.

Summarizing, when the motor is underexcited, that is, E_t is small, I_a lags V_t and the motor is acting as an inductive load and consumes lagging reactive power Q. Increasing the field excitation current causes the phase angle θ to decrease to 0°, that is, unity power factor, where the motor is acting as a resistive load. Then as the excitation current is further increased, the motor is overexcited, the armature current becomes leading, and the motor acts as a capacitive load and supplies leading reactive power to the system. It should also be noted from Fig. 12-7 that as the excitation increases, the torque angle δ decreases, that is, the maximum torque that can be developed before the synchronous motor pulls out of step with the rotating magnetic field increases as the field current is increased.

The relationship between the excitation current and the armature current at various power levels is shown in Fig. 12-8. These curves are called *V curves*. For each V curve the minimum armature current I_a occurs at unity power factor. For excitation currents less than the value that produces minimum armature current, the armature current I_a lags the applied terminal voltage V_t, and the motor consumes reactive power Q. When the excitation current is greater than that necessary to produce minimum armature current, the armature current leads the applied voltage, and the motor supplies reactive power to the supply system, that is, by controlling the field excitation the amount of reactive power consumed or supplied to the system is readily controlled.

12-7 SYNCHRONOUS CAPACITOR OR CONDENSER

A synchronous capacitor or condenser is a synchronous motor operated under no-load conditions, that is, the only power drawn from the ac source is that required to overcome rotational, copper, and iron losses. The function of the synchronous capacitor is (1) to absorb or deliver reactive kVARs from or to a three-phase ac system and, as a result, to reduce supply system voltage fluctuations caused by rapidly fluctuating loads such as ac furnaces; and (2) to improve the system power factor caused mainly by lagging power factor loads such as polyphase induction motors, transformers, and fluorescent lighting. Synchronous capacitors are custom built usually in ratings of between 20 and 200 MVAR. Waterwheel generators are also operated as

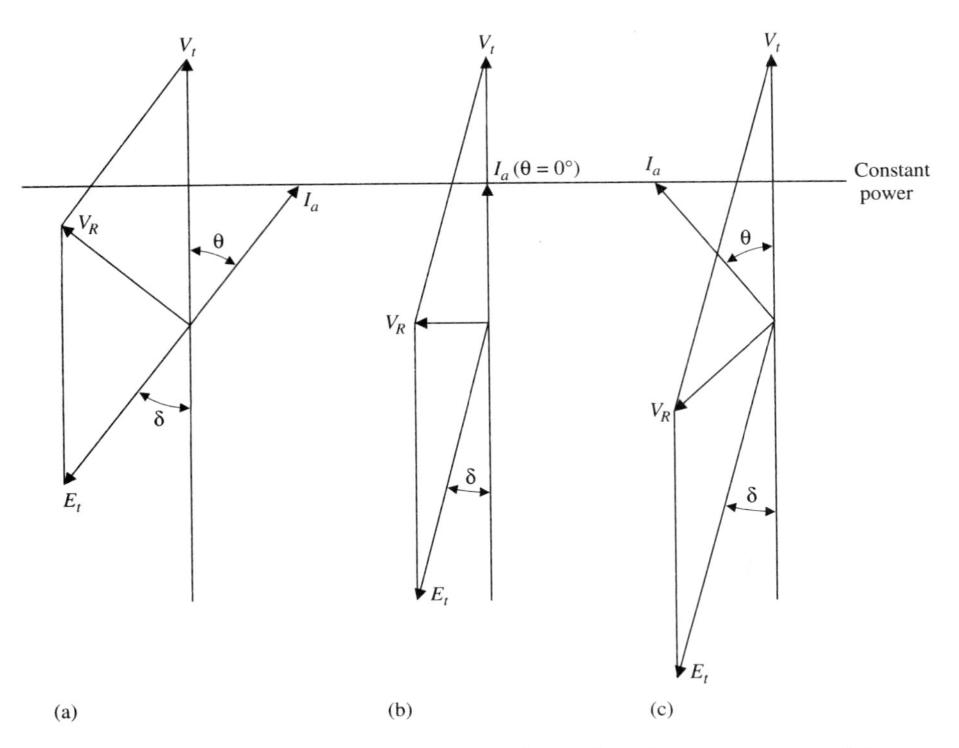

Figure 12-7 Effect of excitation changes on power factor of synchronous motor supplying a constant-kW load. (a) Lagging power factor, underexcited. (b) Unity power factor. (c) Leading power factor, overexcited.

synchronous capacitors during off-peak load periods to provide power factor improvement and regulate transmission-line voltages.

Commercial and industrial loads usually have lagging power factors because of the magnetizing currents required by fluorescent lighting, transformers, and induction motors. Low power factors affect the system in three ways:

1. The circuits are more inductive than resistive and the reactive current components cause larger voltage drops than the resistive current components. This results in poor voltage regulation, which in turn requires additional equipment to improve the voltage regulation.
2. For a given true power load (kW or MW) a low power factor causes increased current flow, which in turn causes increased I^2R losses, as well as effectively derating protective components such as relays, fuses, and wiring. Improving the power factor by injecting leading kVARs increases the system capacity and loading without increased costs in the distribution system.

3. The increased current resulting from a low power factor causes the losses to increase as the current squared, and also to be inversely proportional to the square of the power factor.

As can be seen, it is cost-effective to use power factor improvement techniques, the overexcited synchronous motor being one method. The amount of power factor correction that can be obtained can be determined from the V curve. Normally VAR compensation in industrial applications is obtained by using a combination of synchronous and polyphase induction motors to drive the connected loads, or by static VAR compensation techniques.

➤ EXAMPLE 12-2

An industrial plant has an average load of 25,750 kW at 0.7 power factor lagging. A 5,000-hp (3,730-kW) polyphase induction motor having an efficiency of 91.5% and a power factor of 0.85 lagging breaks down and is replaced by an equivalent synchronous motor with the same output and efficiency. If the synchronous motor is operated at 0.7 power factor leading, determine the new plant power factor.

SOLUTION

$$\text{True power } P \text{ kW} = \text{total kVA } S \times \cos\phi$$
$$\text{Reactive power } Q \text{ kVAR} = \text{total kVA } S \times \cos\phi$$

Therefore

$$\text{Total kVA } S = \sqrt{P^2 + Q^2} \tag{12–9}$$

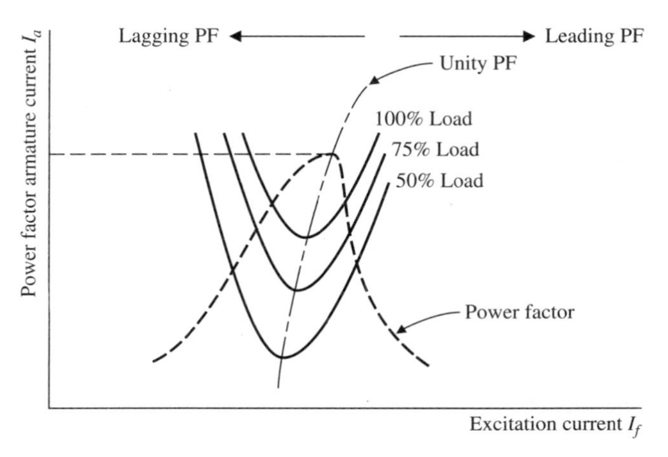

Figure 12-8 Synchronous-motor V curves.

Initial conditions:

$$\text{Plant kVA} = \frac{\text{true power } P}{\text{power factor}} = \frac{25,700 \text{ kW}}{0.7}$$
$$= 36,714 \text{ kVA}$$
$$\phi = \cos^{-1} 0.7 = 45.57°$$
$$\text{Plant kVAR} = \sqrt{S^2 - Q^2}$$
$$= \sqrt{36,714^2 - 25,700^2}$$
$$= 26,218.84 \text{ kVAR lagging}$$

or

$$\text{Plant kVAR} = \text{plant kVA} \times \sin \phi$$
$$= 36,714 \text{ kVA} \sin 45.57°$$
$$= 26,218 \text{ kVAR lagging}$$

For the polyphase induction motor,

$$\text{True power input} = \frac{5,000 \text{ hp} \times 0.746 \text{ kW/hp}}{0.915} = 4,076.5 \text{kW}$$
$$\text{kVA input} = \frac{4,076.5 \text{ kW}}{0.85} = 4,795.89 \text{ kVA}$$
$$\text{kVAR} = \sqrt{4,795.89^2 - 4,076.5^2} = 2,526.4 \text{ kVAR lagging}$$

For the synchronous motor,

$$\text{True power input} = 4,076.5 \text{ kW}$$
$$\text{kVA input} = \frac{4,076.5 \text{ kW}}{0.7} = 5,823.57 \text{ kVA}$$
$$\text{Leading kVAR} = \sqrt{5,823.57^2 - 4,076.5^2} = 4,158.86 \text{ kVAR leading}$$

When the polyphase induction motor is taken out of service, 2,526.4 kVAR lagging is removed from the overall plant kVAR. When the synchronous motor is installed, a further 4,158.86 kVAR is neutralized by the leading kVARs of the synchronous motor. Therefore the final plant parameters are:

$$\text{Final plant kVAR} = 26,218.84 - 2,526.4 - 4,158.86 - 19,533.58 \text{ kVAR lagging}$$
$$\text{Final plant kVA} = \sqrt{25,750^2 + 19,533.58^2} = 32,320 \text{ kVA}$$
$$\text{Final plant power factor} = \frac{25,750 \text{ kW}}{32,320 \text{ kVA}} = 0.80 \text{ lagging}$$

that is, the replacement of the polyphase induction motor by the overexcited synchronous motor has caused an improvement in the overall plant power factor of from 0.7 to 0.8 lagging.

12-8 EFFICIENCY

The procedure for testing synchronous motors is almost identical to that used to test synchronous generators. Since it is impractical to load large synchronous motors, their efficiencies must be determined by summing the losses, that is, the efficiency is

$$\eta = \frac{P_{out}}{P_{out} + \sum \text{losses}} \times 100\% \qquad (12\text{–}10)$$

where P_{out} is the output shaft power in watts and is equal to

$$P_{out} = \omega T_{out} \text{ W} \qquad (12\text{–}11)$$

The input power P_{in} is

$$P_{in} = \sqrt{3} V_l I_l \cos\theta \text{ W} \qquad (12\text{–}12)$$

Equation (12-10) can be rewritten as

$$\eta = \frac{\omega T_{out}}{\omega T_{out} + 3I_a^2 R_a + P_f + P_{fwc}} \times 100\% \qquad (12\text{–}13)$$

where

$$P_{fwc} = P_{in} - 3I_a^2 R_a \text{ W} \qquad (12\text{–}14)$$

P_{fwc} being the windage, friction, and core loss at rated speed. P_{in} and I_a are measured at no load with rated field current I_f applied to the rotor, and R_a is the effective ac resistance per phase, as previously described in Chapter 11. The P_{fwc} loss remains constant at rated speed.

The rotor power input is

$$P_f = V_f I_f \text{ W} \qquad (12\text{–}15)$$

and the desired air-gap power is

$$P_d = P_{in} - 3I_a^2 R_a \text{ W} \qquad (12\text{–}16)$$

In turn the output shaft power is

$$P_{out} = P_d - P_{fwc} \text{ W} \qquad (12\text{–}17)$$

➤ EXAMPLE 12-3

A 100-hp (74.6-kW) 2,300-V 60-Hz eight-pole star-connected synchronous motor has a synchronous impedance of $1.2 + j21.25$ Ω/phase and draws a stator current of 22.75 A/phase when operating at full load at a 0.9 power factor leading. P_{fwc} = 6,500 W. Calculate: **(a)** the input power; **(b)** the output power; **(c)** the efficiency; **(d)** the output shaft horsepower and torque in N · m and lb · ft.

SOLUTION

(a)

$$P_{in} = \sqrt{3}V_l I_l \cos\theta \text{ W}$$
$$= \sqrt{3} \times 2,300 \text{ V} \times 22.75 \text{ A} \times 0.9 = 81,566.6 \text{ W}$$

(b) The armature copper loss is

$$3I_a^2 R_a = 3 \times 22.75^2 \text{ A} \times 1.2\ \Omega = 1,836.23 \text{ W}$$

The developed air-gap power is, from Eq. (12-16),

$$P_d = P_{in} - 3I_a^2 R_a \text{ W}$$
$$= 81,566.66 - 1,863.23 = 79,703.38 \text{ W}$$

Then the output shaft power is, from Eq. (12-17),

$$P_{out} = P_d - P_{fwc} \text{ W}$$
$$= 79,703.38 - 6,500 = 73,203.38 \text{ W}$$

(c) The efficiency is, from Eq. (12-13),

$$\eta = \frac{\omega T_{out}}{\omega T_{out} + 3I_a^2 R_a + P_f + P_{fwc}} \times 100\%$$
$$= \frac{73,203.38 \text{ W}}{73,203.38 \text{ W} + 1,863.23 \text{ W} + 6,500 \text{ W}} \times 100\% = 89.75\%$$

(d) The output shaft horsepower is

$$\text{hp} = \frac{P_{out}}{746 \text{ W/hp}} = \frac{73,203.38 \text{ W}}{746 \text{ W/hp}} = 98.13 \text{ hp}$$

and the output torque is, from Eq. (12-11),

$$T_{out} = \frac{P_{out}}{\omega} \text{ N} \cdot \text{m}$$
$$= \frac{73,203.38 \text{ W}}{2\pi \times 900 / 60\text{r/ps}} = \frac{73,203.38 \text{ W}}{94.25 \text{ rad/s}} = 776.71 \text{ N} \cdot \text{m}$$

or

$$T_{out} = \frac{33,000 \times \text{hp}}{2\pi S} \text{ lb} \cdot \text{ft}$$
$$= \frac{33,000 \times 98.13 \text{ hp}}{2\pi \times 900 \text{ r/min}} = 572.65 \text{ lb} \cdot \text{ft}$$

12-9 SYNCHRONOUS MOTOR CHARACTERISTICS

Synchronous motors are used to drive constant-speed loads and are usually supplied from a stiff source, that is, an infinite bus bar. As a result the applied terminal volt-

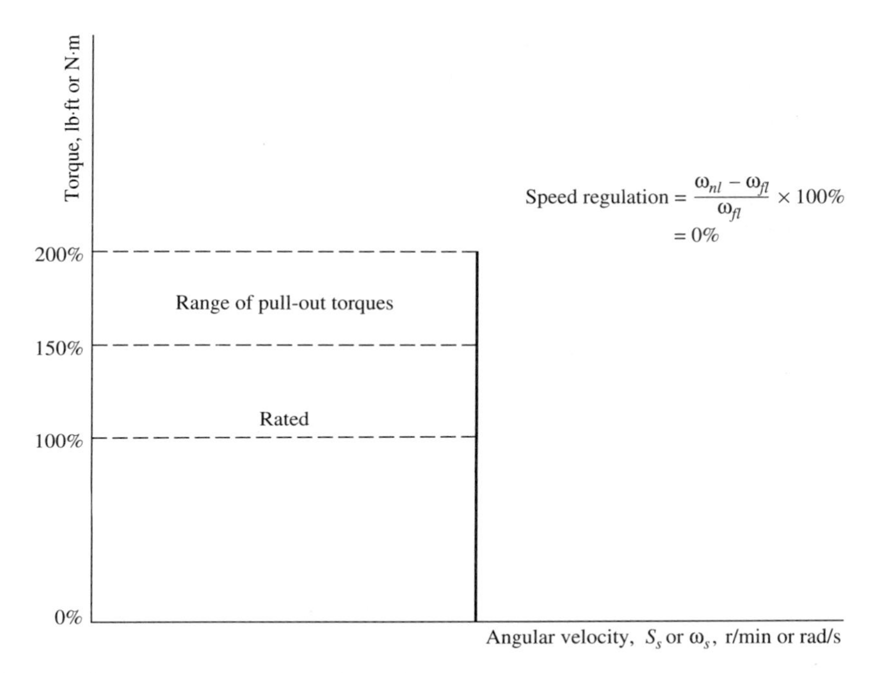

Figure 12-9 Torque-speed characteristic of synchronous motor.

age can be assumed to be constant, irrespective of the load demand on the motor. Also the ac source frequency is basically constant. Therefore once the rotor has been brought up to synchronous speed, its speed will remain constant for all load torques up to the pull-out torque. When the rotor pulls out of synchronism, the rotating magnetic field is continually overtaking the energized rotor poles. As a result the induced torque in the rotor is reversed each time the stator field laps the rotor field. This condition creates alternating torque surges, which can cause severe motor vibration. A study of Eq. (12-8) shows that the maximum pull-out torque can be increased by maintaining a high induced electromotive force in the stator windings, which is achieved by maintaining a high rotor excitation. This increase in pull-out torque is used to obtain more stable operation when driving equipment that produces load torque pulsations, such as low-speed reciprocating pumps and air compressors. Also load torque pulsations are rapid short-period oscillations which are opposed by a pulsating torque output from the synchronous motor. These torque pulsations also create stator and line current pulsations. These pulsations can be reduced by the addition of rotor inertia, which permits a fraction of the load torque peaks to be supplied from the stored energy, thus reducing the amplitude of the current fluctuations, and minimizes the possibility of the rotor pulling out of synchronism. Figure 12-9 shows the torque-speed characteristics of a synchronous motor after it is running at synchronous speed.

As can be seen, the synchronous motor has a flat torque-speed characteristic for all loads from no load to the maximum or pull-out torque. General-purpose motors have pull-out torques that are defined by NEMA. For synchronous motors between 500 and 1,800 r/min (52.36 and 188.50 rad/s) the pull-out torques are 150% of rated full-load torque for motors operating at unity power factor and 175% for motors operating at 0.8 power factor leading. Motors operating at 450 r/min (47.12 rad/s) or less have pull-out torques ranging from 150% of rated full-load torque at unity power factor to 200% at 0.8 power factor leading.

12-10 ROTOR EXCITATION

The function of a rotor excitation system is to provide a controlled dc current to the rotor field pole windings when operating as a synchronous motor. It also must be able to control the point at which the rotor is energized during the starting cycle so that the rotor is pulled into synchronism with the rotating magnetic field. The dc current can be obtained from:

1. Rotating dc exciters directly or indirectly connected to the shaft of the synchronous motor
2. Motor-generator sets operating independently of the synchronous motor
3. Separate dc bus bars
4. A static excitation system
5. A brushless excitation system

Rotating DC Exciters

These consist of dc shunt or compound generators directly connected to the motor shaft in the case of high-speed synchronous motors, or belt or chain driven from the motor shaft for low-speed synchronous motors. The dc excitation current applied to the rotor field is controlled by manual or automatic rheostatic control of the shunt field current. This arrangement has been very commonly used in the past.

Motor-Generator Sets

Polyphase induction-motor-driven dc generator sets also have been used in the past for low-speed synchronous motors, especially where space in the immediate vicinity of the motor is at a premium. Once again the dc excitation current is controlled by rheostatic control of the shunt field of the dc generator.

DC Supply Bus Bars

This system depends on the availability of a separately supplied dc bus bar system. Variation of the rotor excitation current is achieved by either a series rheostat or a tapped resistance external to the rotor slip rings.

Static Excitation

Static excitation systems using silicon diode rectifiers mounted external to the synchronous motor supply dc current via the slip rings to the rotor are being used increasingly in modern installations. The static exciter is characterized by having a high efficiency and power factor combined with low maintenance requirements, unlike the systems described previously. A block diagram of a typical system is shown in Fig. 12-10.

Brushless Excitation

The major disadvantage of the previously described systems is that they all require slip rings to supply the dc excitation current to the rotor field from the dc source. Brushless excitation techniques permit the elimination of the brush–slip ring interface by using an alternator-rectifier assembly mounted on the rotor shaft. The block diagram of Fig. 12-11 shows a brushless excitation system applied to a synchronous motor.

The stationary dc field winding of the alternator is supplied from a variable ac supply which has been rectified by a single-phase full-wave diode rectifier. The alternator stator and the three-phase diode bridge are mounted on the rotor shaft of the synchronous motor. The output of the three-phase diode bridge is directly connected to the rotor field pole windings. The function of the field discharge resistance is to provide a shunt path during starting to prevent the high induced voltages from damaging the field winding insulation. Also during normal operation, if the synchronous motor pulls out of synchronism, the field winding must be short-circuited to prevent large torque and current pulsations. A disadvantage of brushless excitation systems is that the dc excitation current cannot be monitored directly. If closed-loop operation is desired, it is necessary to monitor the field input of the on-board alternator. Brushless excitation is not suitable if the synchronous motor is to be dynamically braked or inched.

12-11 SYNCHRONOUS-MOTOR STARTING TECHNIQUES

Synchronous motors are not inherently self-starting. In Section 12-3 we showed that if the rotor was at standstill, the rotating magnetic field sweeping past the energized rotor field poles produced alternating torques that, over a complete cycle, averaged out to zero. These alternating torques cause motor vibration and overheat the windings. The synchronous motor can be brought up to synchronous speed using one of the following methods:

1. Use an external-drive motor to bring the rotor of the synchronous motor up to synchronous speed and parallel the machine onto the bus bars as a generator. Then deenergize or disconnect the external drive motor. The machine will draw power from the bus bars and run as a synchronous motor.
2. Use the amortisseur or damping windings connected as a squirrel-cage rotor to bring the rotor up to slip speed prior to energizing the rotor windings. It is common practice with this method to relieve the motor load during the starting cycle. For example, if driving an air compressor, the air compressor is vented to the atmosphere during the starting cycle.
3. With the rotor energized, apply a variable-frequency three-phase voltage to the stator, and with the rotor energized, as the frequency is increased from zero up to rated frequency, the rotor will accelerate up to rated speed.

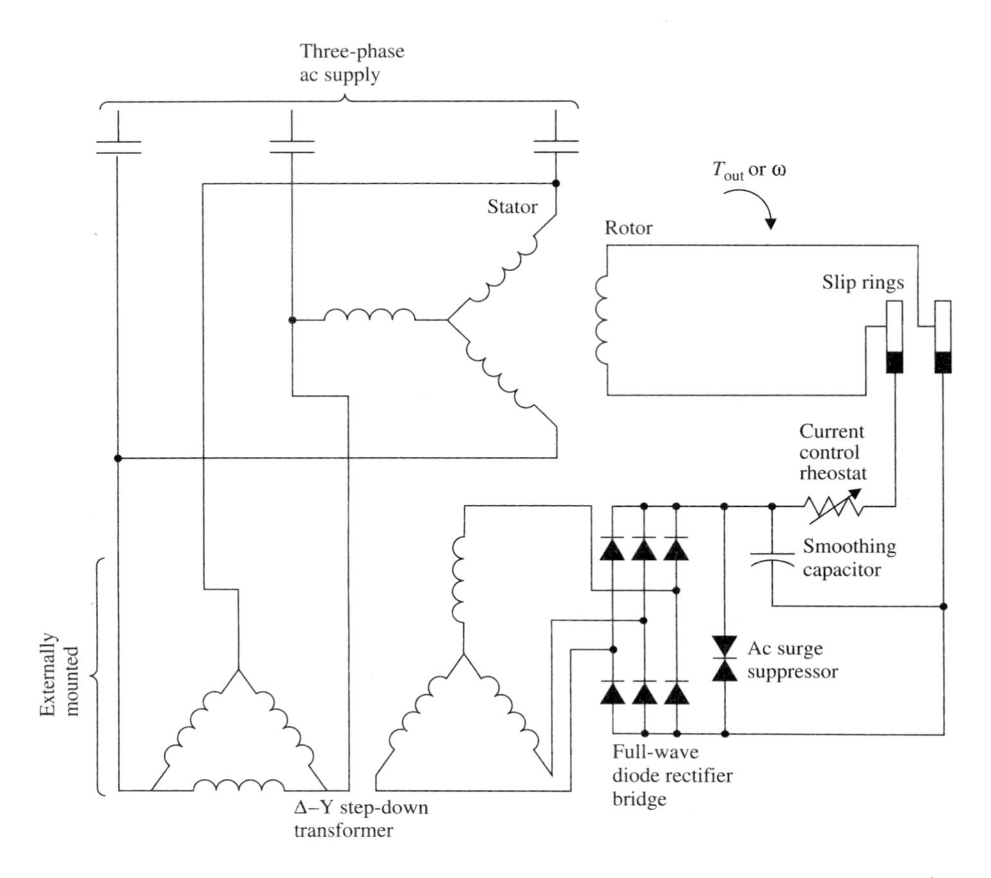

Figure 12-10 Block diagram of static excitation control applied to synchronous motor.

Starting Using an External-Drive Motor

The unloaded synchronous motor is brought up to synchronous speed by means of an external drive or pony motor. The rotor windings are deenergized and short-circuited during this process. The synchronous motor is paralleled onto the bus bars operating as a generator. The pony motor is then disengaged, and the synchronous machine is now operating as a motor and ready to take load. The pony motor can be the dc exciter operating as a motor, or a squirrel-cage induction motor. However, in the case of the polyphase induction motor care must be taken to ensure that the stator windings have at least one pair of poles less than the synchronous motor to compensate for the fact that the induction motor operates at less than synchronous speed. It should be noted that since the load is disengaged or relieved during the starting cycle, the pony motor rating needs only be great enough to overcome the inertia and friction of the synchronous motor. As a result the pony motor needs only be rated at 5 or 10% of the synchronous motor. This method of starting is not in common use.

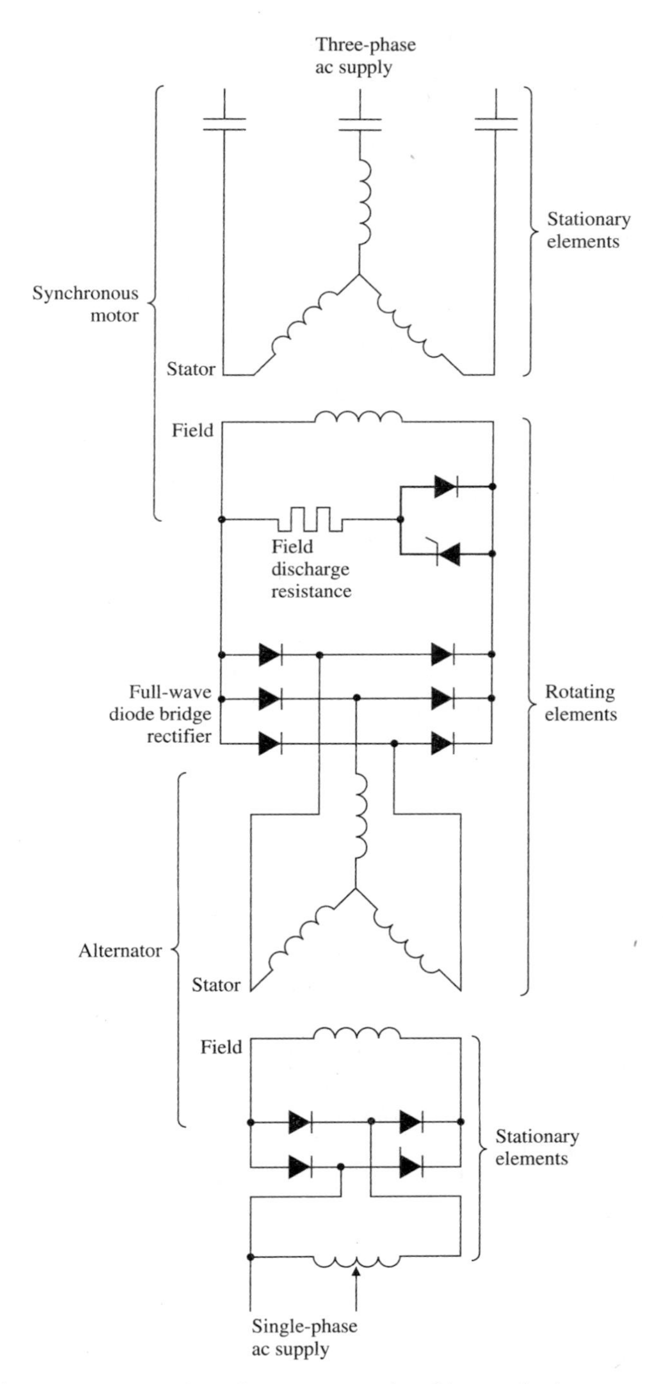

Figure 12-11 Block diagram of synchronous-motor brushless excitation system.

Starting Using the Amortisseur or Damping Windings

The most common method of starting synchronous motors is as a polyphase squirrel-cage induction motor using the amortisseur or damping windings. These windings are not capable of carrying the rated synchronous motor load, but are capable of starting the motor as a polyphase induction motor. With large synchronous motors it may be necessary to limit the inrush current during starting by a reduced-voltage starter. Reduced-voltage starters, since they are also common to polyphase induction motors, will be dealt with in detail in Chapter 14. During the starting process the rotor field windings are short-circuited to prevent insulation damage from high induced voltages. A side benefit is that the resulting current flow creates a magnetic field that assists the damping winding. Under these conditions the rotor accelerates up to approximately 95% of synchronous speed, that is, the motor is running with a low slip as a polyphase induction motor. If the rotor windings are energized at this point, the rotor poles will be polarized with alternate N and S poles. If at the instant of excitation the rotor poles are aligned or nearly aligned with opposite-polarity stator poles, there will be a strong attractive force, which will pull the rotor poles into alignment and the rotor will rotate in synchronism with the rotating stator field. The rotor is said to have pulled in. The starting current of the squirrel-cage rotor winding can be improved by using high-resistance bars in the pole face windings. However, this increases the slip. If the rotor winding is opened momentarily and then short-circuited again for an instant before applying the dc excitation, it will cause the rotor to surge forward in the direction of rotation so that it will lock into synchronism with the rotating magnetic field and then run as a synchronous motor.

The pull-in torque of a synchronous motor is defined as the maximum constant load that the motor will pull into synchronism at rated voltage and frequency when field excitation is applied. It is expressed as a percentage of the rated full-load torque of the motor. Synchronous-motor pull-in torques have been standardized by NEMA as 100% for high-speed general-purpose motors in the 1 to 200 hp (0.746 to 149.2 kW) range, 60% for large high-speed motors from 250 to 1,250 hp (186.5 to 932.50 kW), and 30% for all low-speed synchronous motors.

It should be noted that if the rotor is energized so that the rotor poles are opposite stator poles of the same polarity, a repulsive force will be produced which will attempt to rotate the rotor through 180° electrical. This will produce a violent shock as the rotor attempts to slip one pole pitch. At the same time since the rotor slows down, an overload condition exists, which will cause the protective devices to trip. This situation can be avoided by using a *polarized-field frequency relay* (PFR). This relay senses the point where the revolving stator field is about to overtake a rotor pole of the correct polarity and then initiates the application of the dc field excitation.

The characteristic curves of speed versus torque during the starting process for synchronous motors using damping windings can be modified by controlling the resistance of the squirrel-cage windings. The greater the resistance, the greater the starting torque, but at the expense of a smaller torque contribution from the amortisseur windings at pull-in with a consequent reduction in the pull-in torque because of the greater slip. Typical curves are shown in Fig. 12-12.

As can be seen from Fig. 12-12, the load on the synchronous motor must be relieved for the rotor to accelerate to the normal 5% slip so that it can pull in when the rotor is excited.

The effects of inertia Wk^2 of the motor and load are very important. A high-inertia load will damp out rotor oscillations at the point of synchronization and prevent the rotor from locking into synchronism with the stator field.

So far it has been assumed that the synchronous motor has full voltage applied to the stator during the starting cycle. However, in high-speed motors larger than 500 hp (373 kW), the inrush starting currents are in the range of 350–450% of rated full-load current, and for high-speed motors of less than 500 hp (373 kW) the inrush currents are in the range of 450–600%. Depending on the requirements of the power-supply authorities, these inrush currents may exceed their specifications. As a result the synchronous motor must be started using reduced-voltage starting techniques. The most common methods of obtaining reduced voltages are by using autotransformer or reactor starters.

Starting Using a Variable-Frequency Inverter

A very effective method of starting synchronous motors is by means of a variable-frequency source, either a variable-frequency inverter or a cycloconverter. These methods have the advantage that they can also be used to control the synchronous motor under normal operating conditions such as in a fiber-spinning mill, where close speed tracking is required. Specific examples of open- and closed-loop control including starting are discussed in Section 12-15.

12-12 THREE-PHASE SYNCHRONOUS RELUCTANCE MOTORS

With the strong emphasis on automation in manufacturing plants there has been a corresponding demand for simple and efficient motors that operate at constant speed between no load and full load, even though there may be voltage fluctuations. Three-phase synchronous reluctance motors are available in integral-horsepower ratings in NEMA frame sizes 140, 180, and 210 with two-, four-, or six-pole rotors. The major difference between these motors and the general-purpose polyphase induction motor lies entirely in their rotor construction. The rotor is formed from specially shaped rotor laminations, which are designed to shape the magnetic flux paths in the rotor (Fig. 12-13).

After the rotor laminations have been stacked, the aluminum conductors, end rings, and ventilating fan blades are die cast in exactly the same way as is used in the construction of polyphase squirrel-cage induction motor rotors. These rotors are similar to salient-pole rotors with amortisseur windings, except that there are no rotor field windings. The synchronous reluctance motor starts and accelerates up to slip speed as a polyphase induction motor. At the same time the rotor field is focused along the low-reluctance paths formed by the aluminum flux barriers, effectively producing N and S poles at the rotor periphery. Once the rotor has accelerated to slip

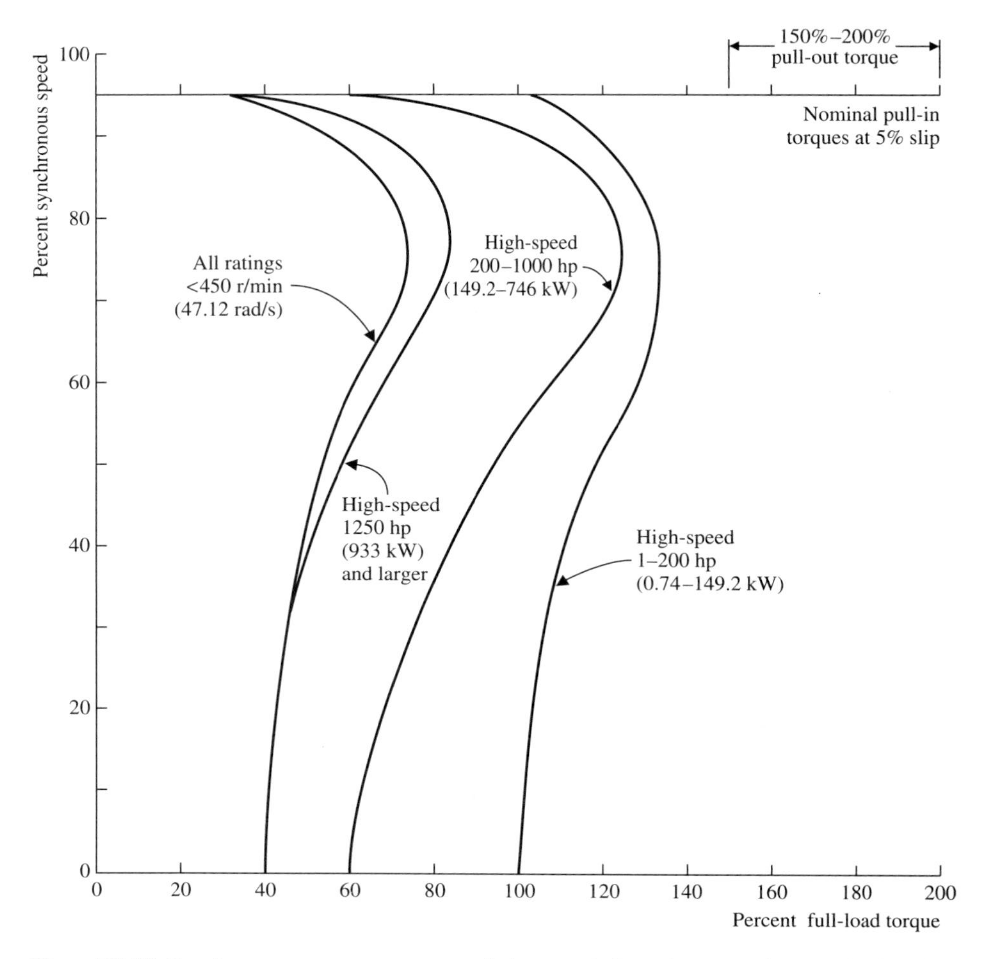

Figure 12-12 Synchronous-motor torque-speed characteristics using amortisseur windings for starting.

speed, usually around 95% of synchronous speed, and as long as the load torque and load inertia are within the design limits, the rotor will pull into synchronism with the stator field and it will run at constant speed, provided the load torque does not exceed the pull-out torque. If the pull-out torque is exceeded, the motor will operate as a squirrel-cage induction motor, although it is normal to design the motor so that it will only operate in this manner under transient load conditions.

Synchronous reluctance motors are most commonly used in multimotor applications with a variable-frequency inverter supplying power to as many as 90 or 100 motors operating in synchronism in an industrial process over a range of predetermined speeds.

These motors are commercially available in horsepower ratings from 1 to 50 hp (0.746 to 37.30 kW) in two-, four-, and six-pole configurations. Typical efficiencies range from 75% for a 1-hp (0.746-kW) motor to 91% for a 50-hp (37.30-kW) motor. Specially designed motors can attain angular velocities of 12,000 r/min (1,256.64 rad/s) at 200 Hz for a two-pole motor, or as low as 200 r/min (20.94 rad/s) for a six-pole motor. These motors operate with low lagging power factors, typically 0.65 to 0.75. Also they can be braked by plugging or using regenerative braking as a synchronous generator up to pull-out torque, and as an induction generator above this point.

12-13 THREE-PHASE PERMANENT-MAGNET SYNCHRONOUS MOTORS

Integral-horsepower three-phase permanent-magnet synchronous motors were first commercially available with the introduction of stable ceramic permanent magnets. However, the rotors were fragile, and as a result the motor speed was restricted. In the 1970s rare-earth permanent magnets were introduced. These magnets with their high energy content and low temperature coefficient have permitted the fabrication of permanent-magnet rotors so that with variable-frequency inverters rotor speeds in excess of 10,000 r/min (1,047.20 rad/s) can now be achieved.

The performance of the permanent-magnet excited synchronous motor is comparable to that of the synchronous reluctance motor. It has a pull-out torque of approximately 150% of rated full-load torque. The rotor contains a squirrel-cage winding as well as the permanent magnets, and as a result it has the starting characteristics of a polyphase induction motor. As was the case with the synchronous reluctance motor, if the pull-out torque is exceeded, the motor will pull out of synchronism and then run as an induction motor or stall. It also has the additional advantage that the motor can be braked dynamically by disconnecting it from the power source and short-circuiting the stator leads together. An additional advantage in favor of its use is the improvement in both efficiency and operating power factor as compared to the synchronous reluctance motor.

Permanent-magnet excited synchronous motors are used in applications where precise speed control is essential, such as in synthetic-fiber drawing machines. They are also relatively maintenance free since there are no brushes and slip rings.

12-14 THREE-PHASE SYNCHRONOUS INDUCTION MOTORS

The three-phase synchronous induction motor has a three-phase wound rotor, and the motor is started as a three-phase wound-rotor induction motor. The rotor winding is connected to an external variable three-phase starting resistance via slip rings. As the external resistance is decreased, the rotor accelerates. At the point where the external starting resistance is short-circuited, the rotor winding is reconnected by the starting switch [Fig. 12-14(b)] and supplied with dc current from a shaft-mounted dc exciter.

As can be seen from Fig. 12-14(b), the rotor is dc energized when operating as a synchronous motor, and since the output of the dc exciter is controlled by the resistance in the shunt field circuit, the rotor excitation and thus the power factor can be readily controlled.

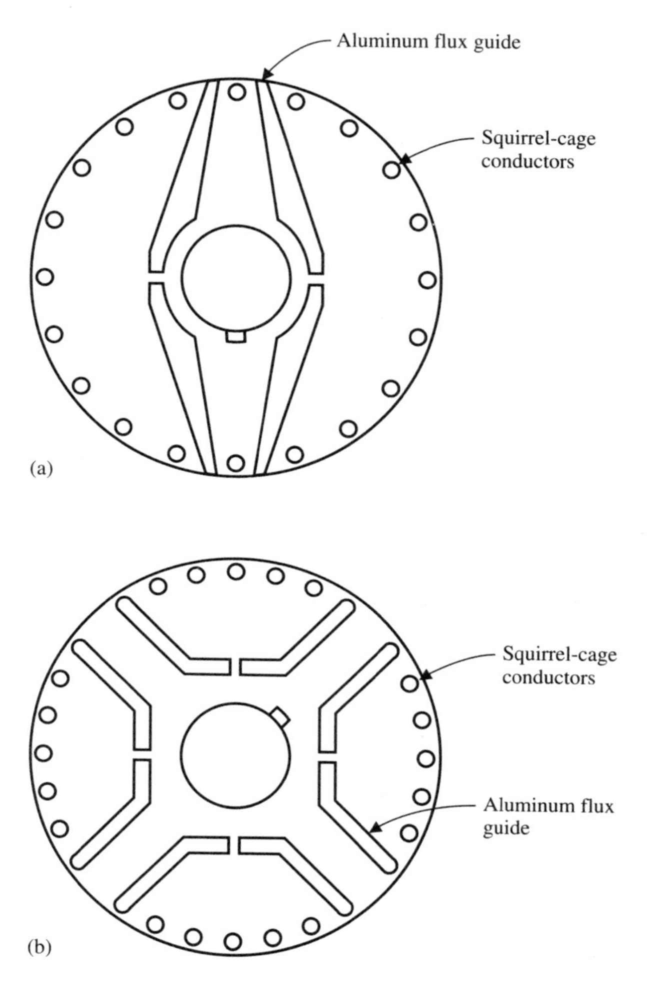

Figure 12-13 Synchronous reluctance motor rotor laminations. (a) Two-pole. (b) Four-pole.

The synchronous induction motor has a relatively high starting torque, typically 400% of rated full-load torque, with a pull-in torque of approximately 125% and a pull-out torque about 200% of rated full-load torque when operating as a synchronous motor. If the load torque exceeds the pull-out torque, the motor will continue operating as a wound-rotor induction motor. The polyphase synchronous induction motor is manufactured in low-power-output ratings up to 50 hp (37.30 kW). The main disadvantages to its use are low efficiency, size, and weight. However, these are offset by the ease of starting, high starting torques which can be varied by modifying the rotor resistance, and low maintenance costs.

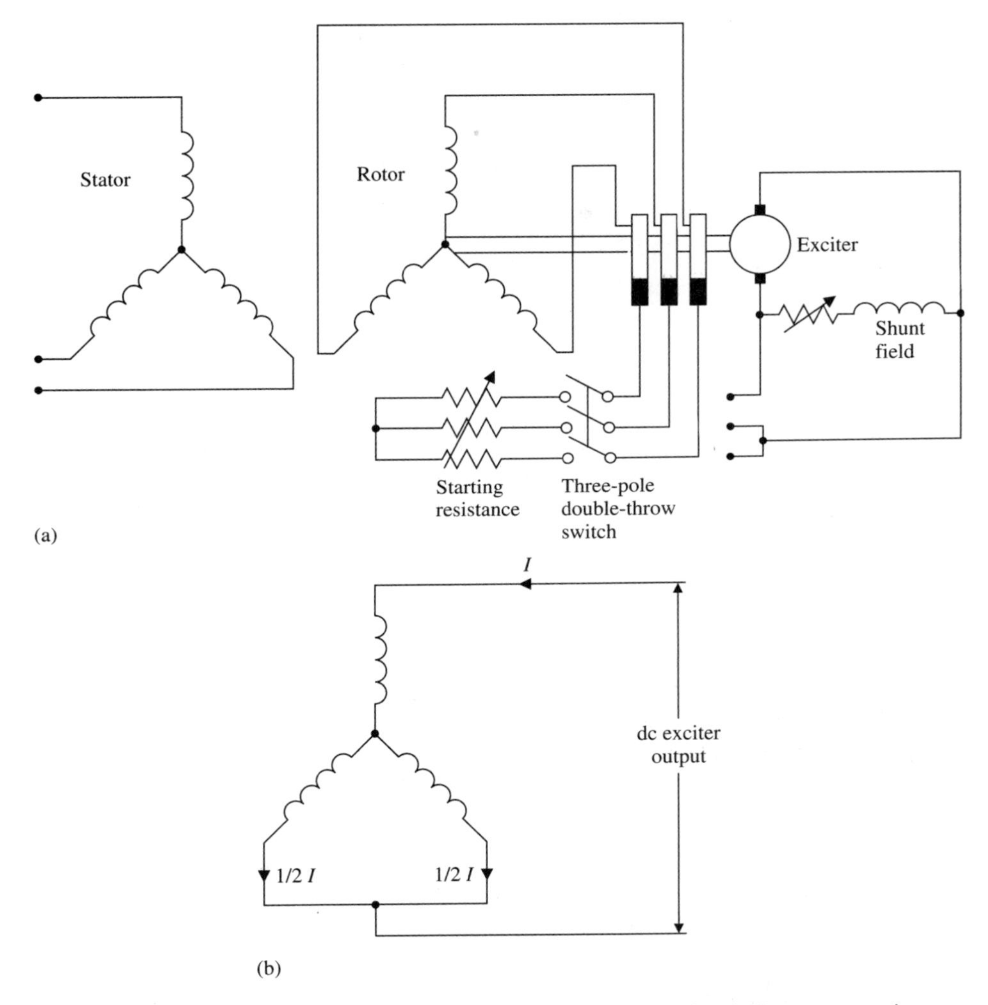

Figure 12-14 Polyphase synchronous induction motor. (a) Schematic. (b) Rotor connections as synchronous motor.

12-15 SOLID-STATE CONTROL OF POLYPHASE SYNCHRONOUS MOTORS

With the introduction of the silicon-controlled rectifier (SCR) in 1957 it was economically feasible to develop variable-frequency converters for polyphase induction and synchronous motors. This has resulted in an ever-increasing use of ac drives in industrial applications where precise speed and position control is required.

Polyphase synchronous motors are increasingly being supplied from solid-state variable-frequency inverters with a frequency range usually from 0 to 150 or 200 Hz, or from cycloconverters with a frequency range from 0 to one-third of the supply frequency. There are two basic control techniques that are in common use: open-loop and closed-loop. In the case of open-loop control the output frequency, and therefore the rotational speed, is determined by a precisely controlled oscillator in the inverter, that is, there is no feedback from the motor being controlled. The closed- loop control requires a speed feedback signal derived from either a tachogenerator or a rotor position sensor which is mounted on the rotor shaft.

Variable-Frequency Control

The most commonly used variable-frequency system is the dc link converter, which can be used to control polyphase induction and synchronous motors. The operating characteristics of the motor are retained over the operating frequency range of the inverter. At rated stator voltage and frequency the air-gap flux will be constant. However, if the stator terminal voltage is maintained at its rated value, then as the frequency is decreased, the stator current increases since the impedance of the stator winding has decreased. As a result the air-gap flux increases and the stator iron is saturated. The air-gap flux must remain constant if a constant torque output is to be produced. This can be accomplished if the applied stator terminal voltage is reduced as the frequency is reduced, that is, V_t/f is constant. In most variable-frequency drives the constant Volts/Hz ratio is maintained from zero to the rated motor frequency, and then the stator voltage is maintained constant as the frequency is increased. As a result the motor operates in the constant-torque mode up to rated frequency, usually 60 Hz, and then in the constant-power mode for higher frequencies. Modern dc link converters are programmed to operate in a constant Volts/Hz mode to produce a constant-torque output up to rated frequency by controlling the amplitude of the inverter output voltage by pulse-width modulation (PWM). The torque-speed characteristic of a synchronous motor operating from a variable-frequency source is shown in Fig. 12-15.

Open-Loop Control

A typical open-loop variable-frequency dc link converter system using pulse-width modulation for output voltage control is illustrated in Fig. 12-16. The three-phase input is rectified by a diode bridge, filtered by the *LC* network, and supplied to the inverter as a constant dc voltage. A voltage signal representing the desired speed is applied simultaneously to the voltage-controlled oscillator (VCO) and the PWM control. The output signals from both these units are applied to the control logic unit. The function of the control logic unit is to generate the correct sequence of gate firing pulses, which are then applied to the inverter switching devices. The output of the three-phase variable-frequency variable-voltage inverter is then supplied to the stator of the synchronous motor. It is usual practice in solid-state variable-frequency inverters to use a ramp circuit to control the rate of acceleration, that is, the rate of frequency increase from zero to the desired speed. This arrangement, which is adjusted during the installation of the drive, enables the synchronous motor to be started up and run up to the desired speed as a synchronous motor provided that the rotor poles are energized, instead of using the damping windings to bring it up to slip speed as an induction motor. It also has the added benefit that the current limit con-

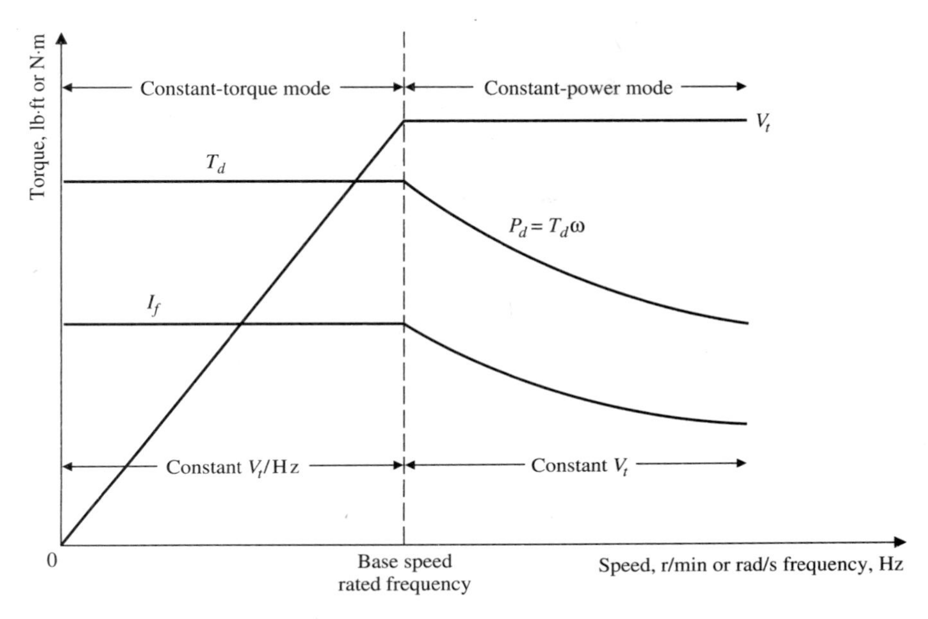

Figure 12-15 Synchronous motor torque-speed characteristics when operating from variable-frequency source.

trol prevents excessive currents from being present during the starting cycle. Permanent-magnet and synchronous reluctance motors are also controlled in a similar manner.

Provided that the kVA output of the inverter is adequate, this method of control can be extended to drive a number of synchronous motors in parallel from the same inverter. This is termed *multimotor* operation. This method of control is in common use in textile mills, where it is used to drive multiple permanent-magnet synchronous or synchronous reluctance motors under closely regulated speed conditions.

Closed-Loop Control—Self-Synchronous Control

In open-loop control the rotor speed is controlled by the frequency of the inverter, which is dependent on an independent oscillator, and the motor operates as a true synchronous motor. This method is sometimes called *imposed frequency control.* Synchronous motors can be operated in the *self-synchronous mode*, where the synchronous motor operates as a dc commutatorless or electronically commutated motor. In this mode the inverter thyristor firing pulses are derived from a shaft-mounted rotor position sensor. To fully understand this concept it is necessary to view the dc motor from a slightly different perspective, namely, to think of the dc motor as a synchronous motor with a stationary field pole system, and a rotating multiphase armature which is supplied with ac power from a mechanical inverter, that is, the brushes and commutator from a dc source. This concept can be trans-

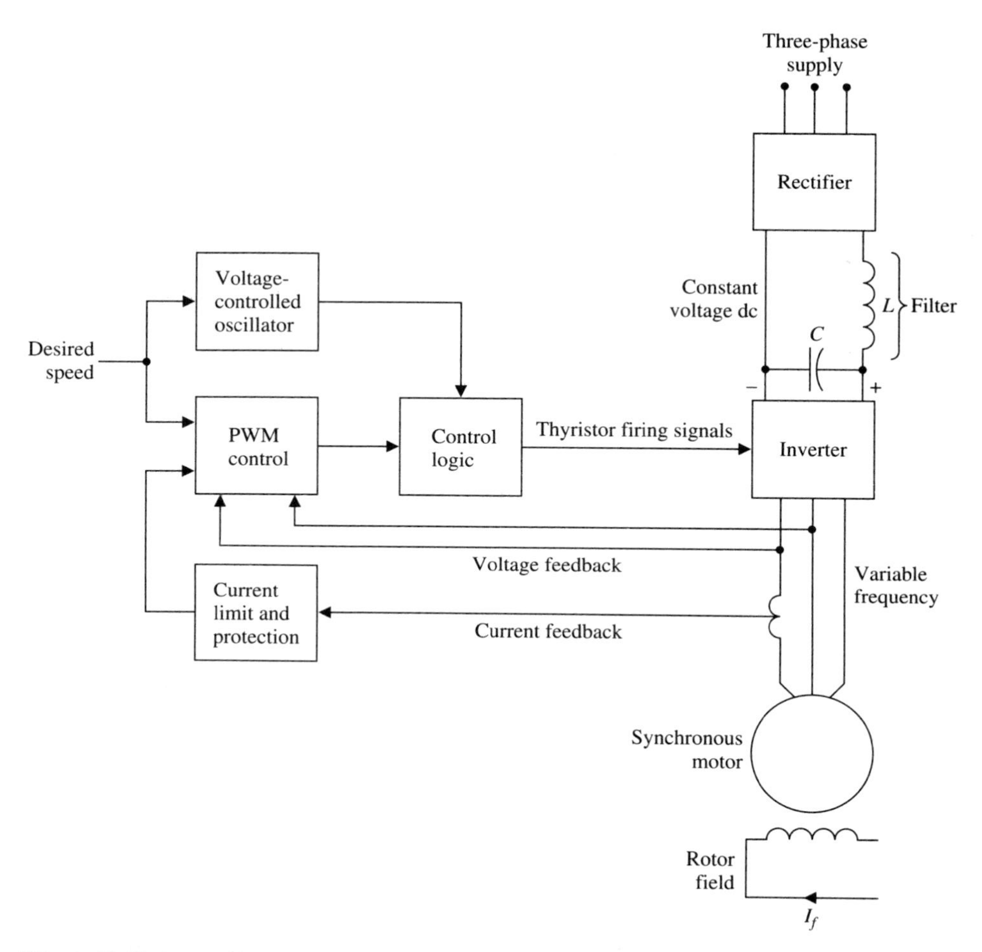

Figure 12-16 Block diagram of variable-frequency dc link converter with pulse-width modulation controlling a synchronous motor.

ferred to the synchronous motor, where the field system is rotating and the multiphase armature is stationary. It is supplied from an inverter whose switching devices are controlled from shaft-mounted rotor position sensors, which is the electronic equivalent of commutator action in the dc motor.

When operating in the self-synchronous mode, the output frequency of the inverter is controlled by the rotor speed, and not vice versa. An added advantage of this arrangement is that the synchronous motor cannot pull out of synchronism, since any drop in rotor speed caused by an increase of load torque will cause a corresponding decrease in the output frequency of the inverter. The output torque and speed of the synchronous motor are controlled by the dc current supplied to the inverter or static commutator.

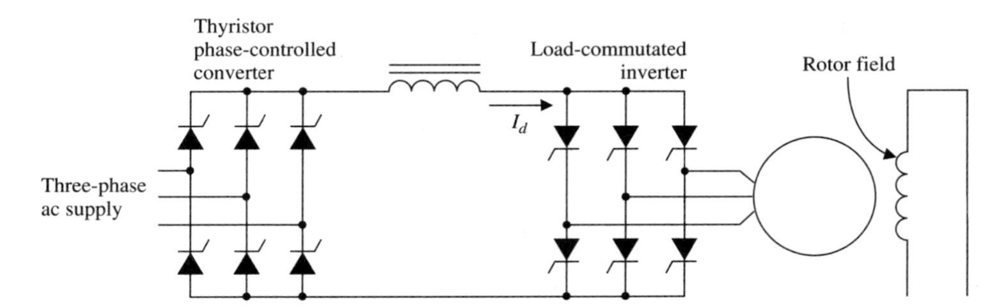

Figure 12-17 Self-synchronous controlled synchronous motor using a current source load-commutated inverter.

Figure 12-17 shows the most common method of achieving self-synchronous control of a synchronous motor. The current source load commutated inverter is supplied from a thyristor phase-controlled converter via a dc link inductor. When the synchronous motor is driving the connected load, the thyristor phase-controlled converter operates in the rectifying mode and supplies dc power to the load-commutated inverter (LCI). The gating signals for the LCI are derived from the rotor shaft position sensors. In turn the switching devices produce a six-step three-phase output which is correctly phased to achieve natural commutation or turn-off of the outgoing switching devices. For correct commutation the motor must be overexcited, that is, it is operating with a leading power factor.

The current supplied to the inverter is controlled by a current regulator, which in turn is slaved to the speed regulator. At low speeds the inverter cannot be naturally commutated since the motor terminal voltages are not great enough. As a result at speeds less than 5% of rated speed commutation is achieved by momentarily reducing the dc link current to zero by means of the current controller, so that the next step in the switching sequence of the inverter switching devices can be made.

A major advantage of self-synchronous control is that the arrangement is capable of four-quadrant operation, that is, the inverter section acts as a rectifier and the thyristor phase-controlled converter operates in the synchronous inversion mode when the synchronous motor is being driven by the load, that is, regeneration takes place. Also by changing the firing sequence of the inverter switching devices the motor can be reversed.

Load-commutated inverter synchronous motor drives are usually found in high-horsepower (kW) applications such as pumps and compressors, gas-turbine starting, and pumped-storage hydroelectric power stations. The system can also act as a soft-start solid-state starter for high-power-output machines, where it brings the motor up to speed in three steps. During the initial starting step the inverter is force commutated. Then, when the motor has accelerated sufficiently so that natural commutation can take place, the motor is ramped up to rated voltage and frequency.

At this point the motor is synchronized to the ac lines and operates as a normal synchronous motor.

Cycloconverter Drives

Cycloconverters are used to drive polyphase induction and synchronous motors in high-horsepower, usually above 7,500-hp (5,595-kW), low-speed applications, such as gearless mills. The cycloconverter output frequency is usually no greater than one-third of the ac source frequency and is easily adjusted from 0 to 20 Hz when operating from a 60-Hz source. In this mode the cycloconverter is line commutated, that is, it does not require forced commutation as is the case with an inverter. The thyristor gate firing pulses are derived from rotor shaft position sensors, and the synchronous motor is operated at unity power factor by independent control to the rotor field excitation.

The majority of cycloconverters are used in one-quadrant applications. However, they can also be used in four-quadrant reversing drives such as rolling mills and mine hoists. The schematic of a four-quadrant cycloconverter drive is shown in Fig. 12-18.

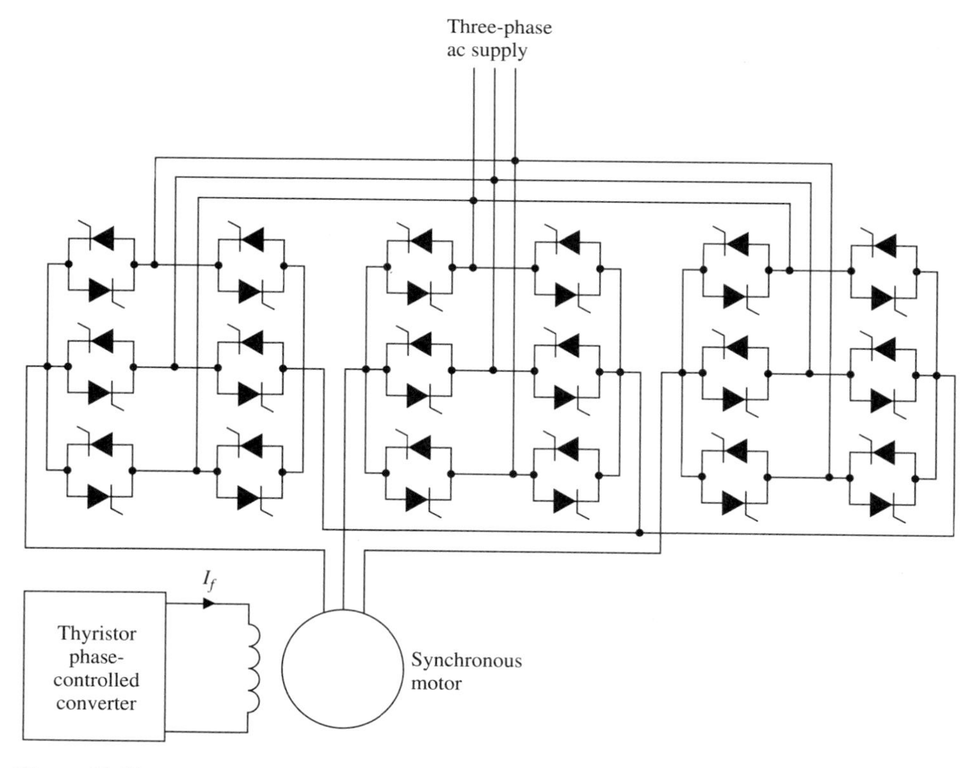

Figure 12-18 Three-phase four-quadrant cycloconverter supplying a synchronous motor.

The converter consists of three reversing thyristor phase-controlled converters (one per phase), connected as bridge converters. By sinusoidally modulating the firing delay angle signals to the thyristors in the converters at the desired output frequency, and by varying the amount of firing angle variation, the output three-phase voltage can be controlled in both amplitude and frequency. The output torque of the motor is controlled using *field-oriented control* or *transvector control*. This is a closed-loop control technique which is dependent on the fact that maximum output torque occurs when the magnetic flux and the stator or armature current are mutually displaced by 90°. As a result by determining the magnetic flux amplitude and direction at any instant, the cycloconverter can be controlled to ensure that the stator current is always at right angles to the magnetic flux. The major advantage of field-oriented control is that the synchronous motor has the same operating characteristics as a dc shunt motor, and does not exhibit load angle oscillations resulting from sudden load torque variations. Nor is it susceptible to pull-out when excessive load is applied.

A number of factors heavily favor a cycloconverter-supplied synchronous motor drive as compared to a comparable thyristor phase-controlled dc motor drive:

1. The construction of the polyphase synchronous motor is simpler and less costly than the equivalent dc motor. However, cycloconverter drives are more expensive than thyristor phase-controlled dc drives in output ratings up to 2,500 hp (1,865 kW); above this rating the cycloconverter drive is cheaper.
2. Synchronous motors can be manufactured to meet all load ratings and are more efficient than dc motors and therefore require less cooling.
3. Cycloconverter-supplied synchronous motors can produce high-output torques at very low speeds down to standstill without the commutation problems that exist with the dc motor.
4. The synchronous motor weighs less than an equivalent dc motor, has a lower moment of inertia, and is easier and less costly to repair.

QUESTIONS

12-1 What factors favor the selection of polyphase synchronous motors over polyphase induction motors?

12-2 Discuss the construction of a synchronous motor.

12-3 What are the functions of amortisseur windings? Describe their construction.

12-4 Explain the principle of operation of a synchronous motor.

12-5 Discuss the production of torque by a synchronous motor.

12-6 What is meant by the term *pull-out torque*?

12-7 What is meant by the terms *torque angle* and *displacement angle*?

12-8 Explain with the aid of phasor diagrams the effect of rotor excitation changes on the power factor.

12-9 Explain the significance of synchronous-motor V curves.

12-10 What is a synchronous condenser?

12-11 What are the effects of low power factor on an ac distribution system?

12-12 How can the pull-out torque of a synchronous motor be increased?

12-13 What are the effects on the performance of a synchronous motor when the load torque exceeds the pull-out torque?

12-14 List five methods of providing excitation to the rotor of a synchronous motor.

12-15 With the aid of a sketch explain the operation of a static excitation system for a synchronous motor.

12-16 Explain the operation of a brushless excitation system for a synchronous motor. What are its advantages and disadvantages?

12-17 Why are synchronous motors not self-starting?

12-18 What methods are used to start synchronous motors?

12-19 Describe how a synchronous motor can be started using the damping windings. How can the starting torque be increased? What precautions must be taken to protect the rotor circuit?

12-20 What is the function of a polarized field frequency relay?

12-21 With the aid of a sketch describe the construction, operation, and applications of a synchronous reluctance motor.

12-22 Describe the construction and operation of a three-phase permanent-magnet synchronous motor.

12-23 With the aid of a sketch describe the construction, operation, and applications of a synchronous-induction motor.

12-24 With variable-frequency control, why is it necessary to maintain a constant Volts/Hz ratio when operating a motor at less than rated frequency?

12-25 With the aid of a block diagram explain the operation of a variable-frequency dc link converter supplying a synchronous motor.

12-26 Describe the concept of self-synchronous motor control using a load-commutated inverter.

12-27 With the aid of a schematic discuss the use of a cycloconverter synchronous motor drive.

PROBLEMS

12-1 A polyphase synchronous motor is supplied from a 60-Hz source. It is used to drive a 400-Hz alternator. Calculate: **(a)** the number of poles in each machine; **(b)** the angular velocity at which the machines operate.

12-2 A 100-hp (74.6-kW) 550-V 1,200-r/min (125.66-rad/s) star-connected synchronous motor has an effective armature resistance of 0.25 Ω/phase and a synchronous reactance of 2.5 Ω/phase. Assuming an efficiency of 92% and a 0.85 power factor leading at rated load, calculate: **(a)** the induced phase voltage; **(b)** the power angle; **(c)** the mechanical power developed.

12-3 A 100-hp (74.6-kW) 550-V three-phase star-connected synchronous motor has an effective resistance and synchronous reactance of 0.03 Ω and 0.3 Ω, respectively. At rated load and 0.82 power factor leading and lagging, calculate: **(a)** the induced voltage per phase; **(b)** the total mechanical power developed. Assume that the motor is 91% efficient.

12-4 A 11,000-V three-phase star-connected synchronous motor draws a line current of 75 A. The effective armature resistance and synchronous reactance per phase are 1.25 Ω and 25.0 Ω, respectively. Calculate: **(a)** the total power supplied to the motor; **(b)** the induced voltage per phase for a power factor of 0.80 leading; and **(c)** 0.80 lagging.

12-5 A 2,300-V three-phase star-connected synchronous motor has negligible armature resistance and a synchronous reactance per phase of 2.5 Ω. The input power to the motor is 850 kW at rated line voltage, and the induced line voltage is 2,450 V. Calculate: **(a)** the torque angle; **(b)** the line current; **(c)** the power factor.

12-6 An industrial plant has the following connected loads: (1) a 2,500-hp (1,865-kW) polyphase induction motor; (2) a 125-kW heating load; (3) a 45-kW lighting load; (4) a 500-hp (373-kW) synchronous motor. The synchronous motor is 92% efficient and operates at 0.82 power factor leading; the induction motor is 90% efficient and operates at 0.76 power factor lagging. Calculate the system power factor: **(a)** when the synchronous motor is not running; **(b)** when the synchronous motor is running.

CHAPTER

13

Transformers

13-1 INTRODUCTION

The static transformer, so named because there are no moving parts, uses the principle of *mutual induction* between two or more magnetically coupled coils. Although the static transformer is not an energy-conversion device, it is used in energy-conversion systems.

Transformers are used to:

1. Step up ac voltages. This property is used extensively in electric-power systems to step up voltages when transmitting electric energy in order to achieve minimum I^2R losses and minimum conductor size. It should be noted that for a given amount of power transmitted, the cross-sectional area of the conductor varies inversely as the square of the voltage, that is, doubling the ac voltage requires a conductor that is one-fourth the original cross-sectional area.
2. Step down ac voltages. Electric power is distributed to users in a variety of voltages so as to meet their individual requirements.
3. Provide electrical isolation between systems.
4. Provide impedance matching to obtain maximum power transfer, a prime consideration in electronics applications.
5. Provide reduced ac voltages and currents for use with relay protection, metering, instrumentation, and control systems.
6. Provide phase conversion. Three-to-two-phase conversion and vice versa is used for servomotors and three-to-six-, twelve-, and twenty-four-phase conversion of ac voltages is used in solid-state power electronics applications.

13-2 IDEAL SINGLE-PHASE TRANSFORMER

An ideal transformer has no losses, that is, core and copper losses, no leakage fluxes, and has a magnetic core with infinite permeability and infinite resistivity. In its simplest form it consists of two windings, a primary winding of N_1 turns and a secondary winding of N_2 turns, as shown in Fig. 13-1. When the primary winding is connected to an ac voltage source v_1, an alternating flux ϕ is produced in the magnetic core. The magnitude of the flux is dependent on the voltage, the frequency, and the number of primary turns. Since the alternating flux is confined to the magnetic core, it will link the turns of the secondary winding and induce a voltage e_2 whose magnitude is dependent on the number of turns in the secondary winding. At the same time the flux links the primary winding and induces a primary induced voltage e_1, which opposes the applied primary voltage v_1.

Since by our assumptions neither the primary nor the secondary windings have any resistance, and there is no leakage flux,

$$v_1 = e_1 = N_1 \frac{d\phi}{dt} \qquad (13\text{–}1)$$

where v_1 is the applied primary voltage, e_1 the induced primary voltage, ϕ the core flux, and N_1 the number of primary turns.

Since there is no leakage flux, the entire core flux ϕ links the N_2 turns of the secondary winding and induces the secondary voltage e_2, where

$$v_2 = e_2 = N_2 \frac{d\phi}{dt} \qquad (13\text{–}2)$$

Since there is no load connected to the secondary winding, the secondary terminal voltage v_2 is equal to e_2.

Taking the ratio of Eq. (13-1) to Eq. (13-2) we obtain

$$\frac{v_1}{v_2} = \frac{e_1}{e_2} = \frac{N_1}{N_2} \qquad (13\text{–}3)$$

This means that in an ideal transformer the voltage ratio v_1/v_2 is equal to the turns ratio N_1/N_2.

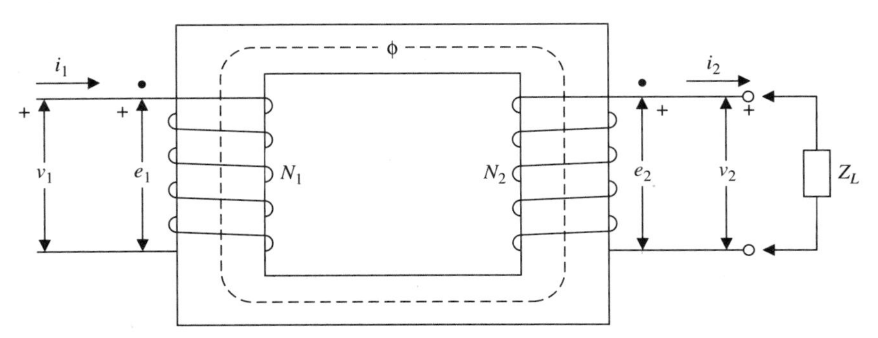

Figure 13-1 Ideal transformer.

It should be noted in Fig. 13-1 that the upper ends of the primary and secondary windings are marked with dots. These dots are called *polarity marks* and indicate that the voltages have the same polarities at the same instant of time, when the primary current enters at the dot and the secondary current leaves at the dot.

If the secondary winding is connected to a load, than an instantaneous secondary current i_2 will flow in the direction shown in Fig. 13-1. Neglecting losses, the instantaneous power input is equal to the instantaneous power output, that is,

$$v_1 i_1 = v_2 i_2 \tag{13–4}$$

Therefore combining Eqs. (13-3) and (13-4) we obtain

$$\frac{i_1}{i_2} = \frac{N_2}{N_1} \tag{13–5}$$

Alternatively, using the right-hand grasp rule, the primary magnetomotive force $N_1 i_1$ produces a core flux that is acting clockwise around the core, whereas the secondary magnetomotive force $N_2 i_2$ is producing a flux that is acting counterclockwise around the core. In our assumptions we defined the core as having an infinite permeability and resistivity. Therefore to produce a finite core flux requires that the primary and secondary magnetomotive forces be equal and opposite, that is,

$$N_1 i_1 = N_2 i_2 \tag{13–6}$$

or

$$\frac{i_1}{i_2} = \frac{N_2}{N_1} \tag{13–7}$$

This means that in the ideal transformer the current ratio i_1/i_2 is the inverse of the voltage ratio v_1/v_2. The transformation ratio α is

$$\alpha = \frac{v_1}{v_2} = \frac{e_1}{e_2} = \frac{i_2}{i_1} = \frac{N_1}{N_2} \tag{13–8}$$

A transformer is classified as a step-up transformer if N_1/N_2 is less than 1, that is, v_2 is greater than v_1. A step-down transformer has N_1/N_2 greater than 1, that is, v_2 is less than v_1.

Assuming that the core flux is sinusoidal,

$$\phi = \phi_{max} \sin \omega t \tag{13–9}$$

Then Eq. (13-1) becomes

$$\begin{aligned} e_1 &= \frac{N_1 d(\phi_{max} \sin \omega t)}{dt} \\ &= \omega N_1 \phi_{max} \cos \omega t \end{aligned}$$

Also since $\omega = 2\pi f$, where f is the supply frequency,

$$e_1 = 2\pi N_1 f \cos \omega t$$
$$= 2\pi N_1 f (\sin \omega t + 90°)$$

The rms value of the primary induced electromotive force is

$$E_1 = \frac{2\pi}{\sqrt{2}} N_1 f \phi_{max}$$
$$= 4.44 N_1 f \phi_{max} \tag{13-10}$$

This means that the rms primary induced electromotive force E_1 is proportional to the number of primary turns N_1, the maximum flux, and the frequency, and that the voltage phasor E_1 leads the flux phasor by 90°.

In terms of rms values Eq. (13-8) can be rewritten as

$$\frac{V_1}{V_2} = \frac{E_1}{E_2} = \frac{I_2}{I_1} = \frac{N_1}{N_2} \tag{13-11}$$

Also from Eq. (13-10), since the primary winding is assumed to have zero resistance, the induced or counter electromotive force is equal to the applied voltage. Then with a sinusoidal voltage V_1 applied to the primary, a sinusoidal flux whose maximum value is ϕ_{max} is produced, where

$$\phi_{max} = \frac{V_1}{4.44 f N_1} \tag{13-12}$$

From Fig. 13-1, when the load impedance Z_L is connected to the transformer secondary, the secondary voltage is

$$V_2 = I_2 Z_L \tag{13-13}$$

Then from Eqs. (13-3) and (13-5) it follows that

$$\frac{V_1}{I_1} = \left(\frac{N_1}{N_2}\right)^2 \frac{V_2}{I_2} \tag{13-14}$$

However, the load impedance Z_L is

$$Z_L = \frac{V_2}{I_2} \tag{13-15}$$

Therefore

$$\frac{V_1}{I_1} = \left(\frac{N_1}{N_2}\right)^2 Z_L \tag{13-16}$$

This means that the load impedance can be replaced by an equivalent impedance Z_L' in the primary circuit of the transformer. Therefore

$$Z'_L = \left(\frac{N_1}{N_2}\right)^2 Z_2 = \alpha^2 Z_2 \qquad \textbf{(13–17)}$$

➤ EXAMPLE 13-1

A single-phase transformer has a primary winding with 1,500 turns and a secondary winding with 80 turns. If the primary winding is connected to a 2,300-V 60-Hz supply, calculate: **(a)** the secondary voltage; **(b)** the maximum value of the core flux. Neglect the primary resistance.

SOLUTION

(a) From Eq. (13-11),

$$V_2 = \frac{V_1 N_2}{N_1}$$

$$= \frac{2{,}300\text{ V} \times 80\text{ turns}}{1{,}500\text{ turns}} = 122.67\text{ V}$$

(b) From Eq. (13-12),

$$\phi_{max} = \frac{2{,}300\text{ V}}{4.44 \times 60\text{ Hz} \times 1{,}500\text{ turns}} = 0.0058\text{ Wb}$$

Equation (13-12) represents a sinusoidal flux. However, since the *B–H* curves of transformer steels are nonlinear, the waveform of the magnetizing current I_ϕ will be different from the flux waveform. The waveform of the magnetizing current can be constructed graphically, as shown in Fig. 13-2.

As can be seen from Fig. 13-2, the exciting current I_ϕ is nonsinusoidal, and if analyzed using Fourier analysis, it is found to consist of a fundamental phasor and a series of odd harmonics. Concentrating on the fundamental phasor, it can be resolved into two phasors at right angles to each other: (1) I_c in phase with the

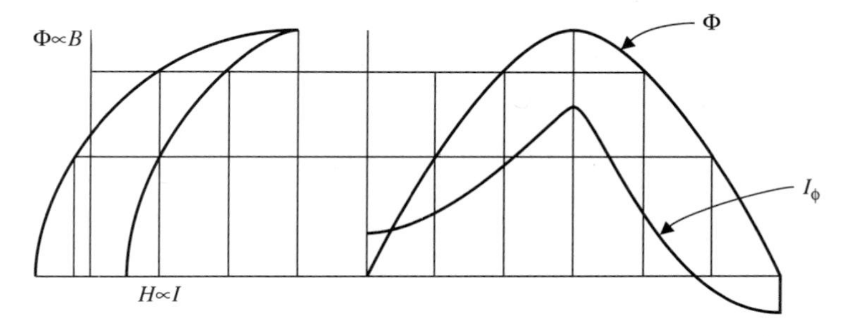

Figure 13-2 Graphic determination of exciting current from sinusoidal flux waveform.

primary induced electromotive force E_I, and (2) I_m in phase with the core flux ϕ. The in-phase component I_c is the *wattful* component that supplies the power absorbed by the hysteresis and eddy current losses in the magnetic core. This component is called the *core-loss* component. The component I_m is called the *magnetizing current,* and since it lags the applied voltage by 90°, it is called the *wattless* component. The wattless component also contains all the odd harmonics, the major one being the third, with an amplitude approximately 40% of I_ϕ. Normally, since the exciting current is only approximately 5% of the rated full-load current, we ignore the effects of the harmonics and consider the exciting current as being represented by an equivalent sine wave, with the same rms value and frequency producing the same average power as the actual exciting current.

Figure 13-3 is the phasor diagram of a transformer, with the secondary open-circuited, from which it can be seen that the core loss P_c is

$$P_c = V_1 I_\phi \cos\theta_c \tag{13–18}$$

where P_c is the sum of the hysteresis and eddy current losses and appears as heat generated in the magnetic core.

➤ EXAMPLE 13-2

A single-phase 2,300/230-V 500-kVA 60-Hz distribution transformer is tested with the secondary open-circuited. The following test results were obtained: $V_I = 2{,}300$ V, $I_\phi = 10.5$ A, and $P_c = 2{,}300$ W. Calculate: **(a)** the power factor; **(b)** the core-loss current; **(c)** the magnetizing current.

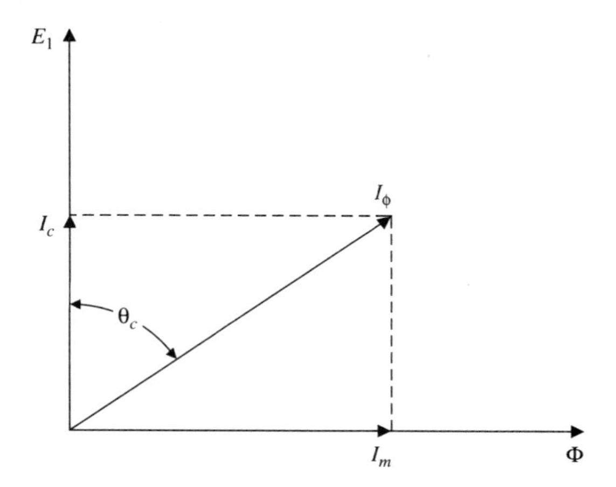

Figure 13-3 No-load phasor diagram.

SOLUTION

(a)

$$\text{Power factor} = \cos\theta_c = \frac{P_c}{V_1 I_\phi}$$

$$= \frac{2{,}300 \text{ W}}{2{,}300 \text{ V} \times 10.5 \text{ A}} = 0.0952$$

$$\theta_c = 84.53°$$

(b)

$$I_c = I_\phi \cos\theta_c$$

$$= 10.5 \text{ A} \times 0.0952 = 1.0 \text{ A}$$

(c)

$$I_m = I_\phi \sin\theta_c$$

$$= 10.5 \text{ A} \times 0.9955 = 10.45 \text{ A}$$

13-3 PRACTICAL SINGLE-PHASE TRANSFORMER

The transformer concepts stated so far have assumed that the transformer is ideal. The practical transformer has windings that have both resistance and leakage inductance. Under load conditions, both the primary and the secondary windings develop magnetmotive forces producing leakage fluxes ϕ_{l1} and ϕ_{l2}, which link only with the turns of the winding producing them, and are external to the core. Figure 13-4 represents a core-type transformer with the mutual and leakage flux paths shown.

The effect of the primary and secondary leakage fluxes is to produce primary and secondary leakage inductances. In turn, each leakage inductance introduces the leakage reactances X_{l1} into the primary winding and X_{l2} into the secondary winding.

When load is applied to the secondary, a secondary current I_2 flows, and in accordance with Lenz's law, the secondary magnetomotive force $N_2 I_2$ links with the core

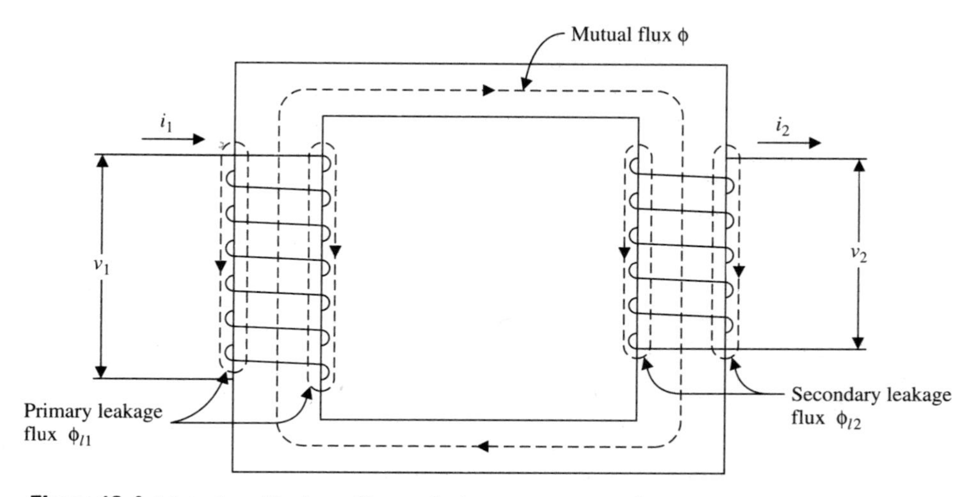

Figure 13-4 Mutual and leakage flux paths in core-type transformer.

and produces a flux which opposes the original core flux produced by the primary magnetomotive force N_1I_1. There is a momentary reduction in the mutual flux ϕ, which reduces the primary induced electromotive force E_1 and in turn allows the primary current to increase. Since the primary resistance is small, only a slight reduction in E_1 is necessary for the primary current to increase to counter the demagnetizing effect of the secondary load current and restore the mutual flux to its original level.

In the practical transformer, since the reluctance of the magnetic circuit is small, the self-inductance of the primary winding will be large compared to the resistance. When load is applied to the secondary of the practical transformer, the primary current I_1 has two functions to perform. First it must supply the no-load current I_ϕ, and second it must supply a component I_2 to counteract the demagnetizing effect of the secondary load current I_2. Figure 13-5(a) is the equivalent circuit of the primary of a loaded transformer.

The applied primary voltage V_1 is the phasor sum of the primary circuit voltage drops,

$$V_1 = E_1 - I_1(R_1 + j\omega X_{l1}) \tag{13–19}$$

the no-load current I_ϕ is the phasor sum

$$I_\phi = I_c + I_m \tag{13–20}$$

and the primary current I_1 is the phasor sum

$$I_1 = I_\phi + I_2' \tag{13–21}$$

The no-load current I_ϕ can be resolved into two components at right angles to each other, as was seen earlier. The effect of the no-load current can be represented by the shunt circuit consisting of a noninductive resistance R_c in parallel with a pure inductance X_m, where $E_1^2/R_c = P_c$, the core loss, and the magnetizing current $I_m = E_1/X_m$.

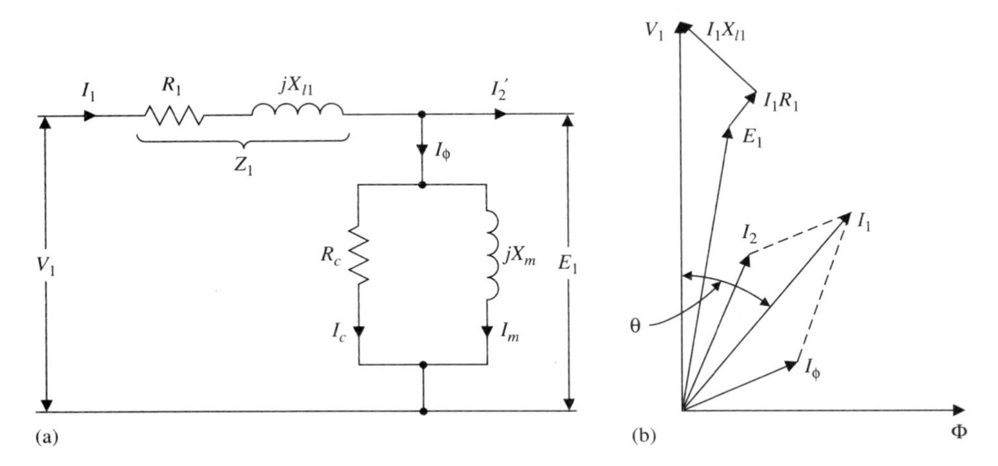

Figure 13-5 (a) Equivalent circuit of loaded transformer primary. (b) Phasor diagram.

The practical transformer can be represented as illustrated in Fig. 13-6(a), and the corresponding phasor diagrams are shown in Fig. 13-6(b), (c), and (d).

In the phasor diagram for a resistive load [Fig. 13-6(b)] the induced voltages E_1 and E_2 are in phase with each other, since they are both induced simultaneously by the mutual flux Φ. The secondary phase angle θ_2, between the secondary current I_2 and the secondary terminal voltage V_2, is solely determined by the nature of the connected load. The secondary induced voltage E_2 must overcome the secondary impedance voltage drop I_2Z_2. The secondary impedance voltage drop has two components, the secondary resistive voltage drop I_2R_2 in phase with the current and the secondary leakage reactance voltage drop I_2X_{l2}, which leads I_2 by 90°. The secondary terminal voltage V_2 is

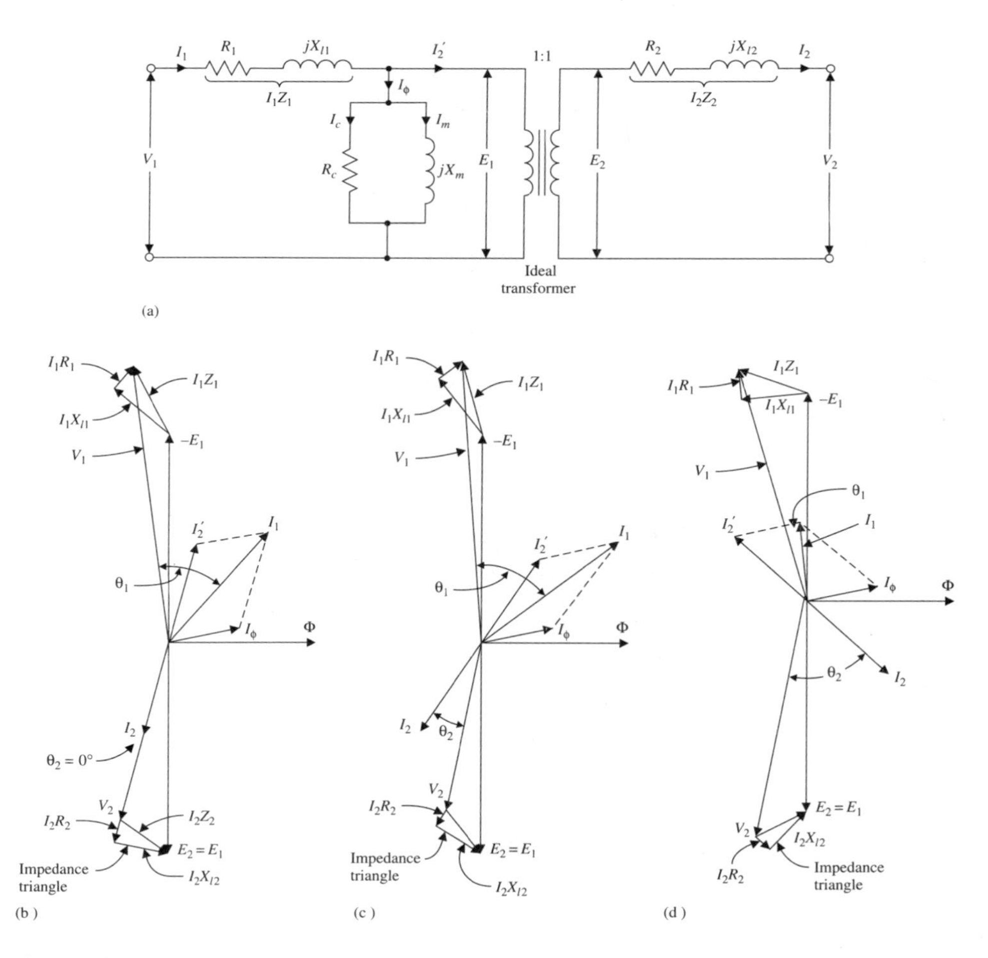

Figure 13-6 Loaded practical transformer. (a) Schematic. Phasor diagrams. (b) Resistive load. (c) Inductive load. (d) Capacitive load.

$$V_2 = E_2 - I_2(R_2 + jX_{l2})$$
$$= E_2 - I_2 Z_2 \quad (13\text{–}22)$$

Since the load is resistive, V_2 and I_2 are in phase and $\theta_2 = 0°$.

In order to neutralize the demagnetizing effect of the secondary magnetomotive force N_2I_2, a counteracting primary magnetomotive force N_1I_2' is required. This is provided by I_2'. The primary line current I_1 is the phasor sum of I_ϕ and I_2'. The primary induced voltage $-E_1$ is the component of the primary voltage required to overcome the voltage induced by the mutual flux Φ, and leads Φ by 90°. The primary impedance voltage drop I_1Z_1 must be added to $-E_1$ to obtain the applied primary terminal voltage V_1. The primary impedance voltage drop I_{Z1} consists of the primary resistance voltage drop I_1R_1 in phase with I_1, and the primary leakage reactance voltage drop I_1X_{l1}, which leads I_1 by 90°. The primary phase angle θ_1 is the angle between V_1 and I_1. As can be seen, in spite of the in-phase relationship of the secondary load, the current I_1 lags V_1 by θ_1. This is attributable to the leakage reactances of both the primary and the secondary windings, as well as the magnetizing current.

Similar phasor diagrams for inductive and capacitive secondary loads are shown in Fig. 13-6(c) and (d). Two points of interest should be noted: (1) for a given power factor load, the primary phase angle θ_1 decreases as the secondary load current increases, and (2) in the case of a capacitive load, that is, I_2 leads V_2 by θ_2, there is a rise in the secondary terminal voltage V_2.

13-4 SINGLE-PHASE EQUIVALENT CIRCUITS

The inclusion of the ideal transformer tends to complicate the analysis of a transformer's performance. The ideal transformer [Fig. 13-6(a)] can be eliminated by referring all quantities to one side or the other of the transformer, that is, secondary components are referred to the primary, or primary components are referred to the secondary.

We saw earlier that the load impedance could be reflected across the transformer to the primary [Eq. (13-17)]. Applying similar techniques, the resistances and reactances can be reflected across the transformer, and at the same time the ideal transformer of Fig. 13-6(a) is eliminated. The secondary resistance and reactance can be reflected to the primary as follows:

$$R_{e1} = R_1 + \alpha^2 R_2 \ \Omega \quad (13\text{–}23)$$

$$X_{e1} = X_{l1} + \alpha^2 X_{l2} \ \Omega \quad (13\text{–}24)$$

$$Z_{e1} = Z_1 + \alpha^2 Z_2 \ \Omega \quad (13\text{–}25)$$

and the secondary terminal voltage reflected to the primary is

$$V_2' = \alpha V_2 \ \text{V} \quad (13\text{–}26)$$

The corresponding relationships for referring the primary quantities to the secondary are

$$R_{e2} = R_2 + \frac{R_1}{\alpha^2}\ \Omega \tag{13–27}$$

$$X_{e2} = X_{l2} + \frac{X_{l1}}{\alpha^2}\ \Omega \tag{13–28}$$

$$Z_{e2} = Z_2 + \frac{Z_1}{\alpha^2}\ \Omega \tag{13–29}$$

and the primary terminal voltage referred to the secondary is

$$V_1'' = \frac{V_1}{\alpha^2}\ V \tag{13–30}$$

The developed transformer equivalent circuits are shown in Fig. 13-7.

➤ EXAMPLE 13-3

A 150-kVA 2,400/240-V 60-Hz single-phase transformer has the following resistances and reactances: $R_1 = 0.225\ \Omega$, $X_{l1} = 0.525\ \Omega$, $R_2 = 0.002{,}20\ \Omega$, and $X_{l2} =$ $0.044{,}5\ \Omega$. Calculate the transformer equivalent values: **(a)** referred to the primary; **(b)** referred to the secondary.

SOLUTION

(a) The transformation ratio is

$$\frac{V_1}{V_2} = \frac{2{,}400\ \text{V}}{240\ \text{V}} = 10 = \alpha$$

From Eq. (13-23),

$$R_{e1} = R_1 + \alpha^2 R_2 = 0.225\ \Omega + 10^2 \times 0.0022\ \Omega$$
$$= 0.225\ \Omega + 0.220\ \Omega = 0.445\ \Omega$$

From Eq. (13-24),

$$X_{e1} = X_{l1} + \alpha^2 X_{l2} = 0.525\ \Omega + 10^2 \times 0.00445\ \Omega$$
$$= 0.525\ \Omega + 0.445\ \Omega = 0.970\ \Omega$$

(b) From Eq. (13-27),

$$R_{e2} = R_2 + \frac{R_1}{\alpha^2} = 0.00220\ \Omega + \frac{0.225}{10^2}\ \Omega$$
$$= 0.00220\ \Omega + 0.00225\ \Omega = 0.00445\ \Omega$$

From Eq. (13-28),

$$X_{e2} = X_{l2} + \frac{X_{l1}}{\alpha^2} = 0.00445\ \Omega + \frac{0.525\ \Omega}{10^2}\ \Omega$$
$$= 0.00445\ \Omega + 0.00525\ \Omega = 0.00970\ \Omega$$

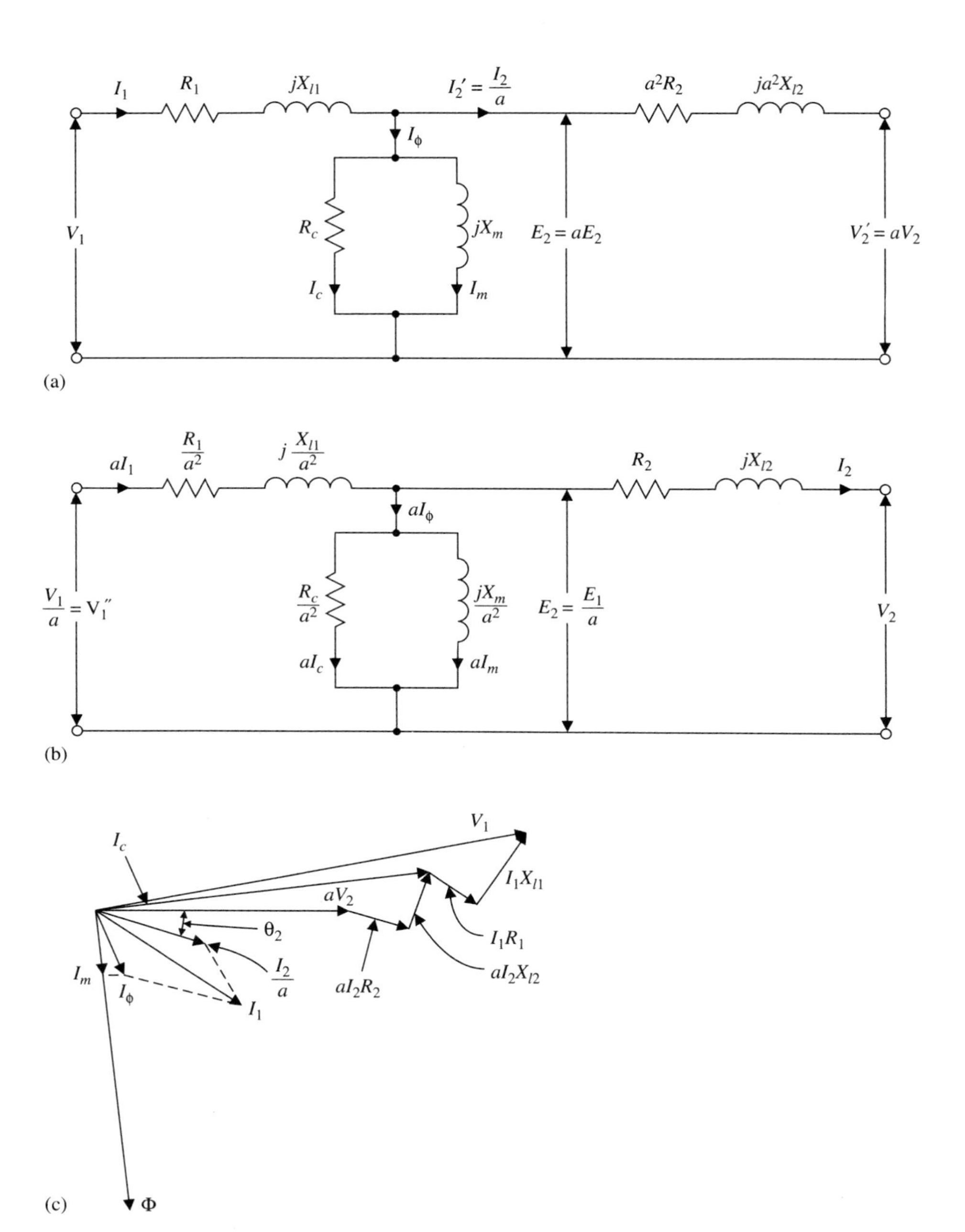

Figure 13-7 Transformer equivalent circuits. (a) Exact equivalent circuit, secondary referred to primary. (b) Exact equivalent circuit, primary referred to secondary. (c) Phasor diagram for (a).

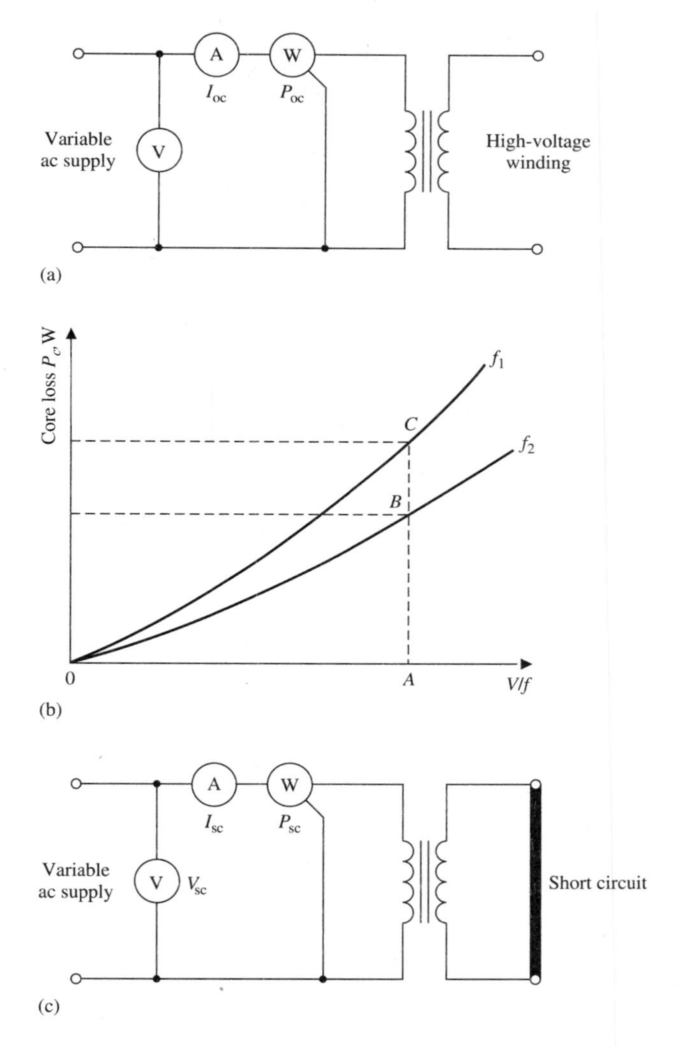

Figure 13-10 Determination of equivalent-circuit parameters. (a) Open-circuit test connections. (b) Separation of iron losses. (c) Short-circuit test connections.

Where P_{c1} and P_{c2} are the total iron losses at frequencies f_1 and f_2 and A and B are constants.

The constants A and B can be calculated from these equations, and the hysteresis and eddy current losses at this V/f ratio can be calculated for any frequency. Similarly, if a number of values of V/f are considered, the individual components of the core loss can be determined at any V/f.

Short-Circuit or Impedance and Copper-Loss Tests

The short-circuit test enables us to determine the copper losses and impedances. The terminals of the low-voltage winding are short-circuited by a suitably sized conductor. Then with the instrumentation setup shown in Fig. 13-10(c), if a variable ac voltage V_{sc} at rated frequency is applied to the high-voltage terminals, a voltage of approximately 3–15% of rated terminal voltage will cause rated current to flow in the high-voltage winding. Normally the applied voltage is varied to obtain a series of short-circuit currents ranging from zero to 125% of the rated winding current. The input current supplies the reflected current from the short-circuited low-voltage winding and the no-load excitation current. The input power P_{sc} supplies the copper, iron, and dielectric losses. However, the iron and dielectric losses are small compared to the copper loss and are usually neglected. At rated current,

$$Z_{sc} = \frac{V_{sc}}{I_{sc}} \tag{13–38}$$

$$R_{e1} = R_1 + \alpha^2 R_2 = \frac{P_{sc}}{I_{sc}^2} \tag{13–39}$$

$$X_{e1} = X_{l1} + \alpha^2 X_{l2} = \sqrt{Z_{sc}^2 - R_{e1}^2} \tag{13–40}$$

➤ EXAMPLE 13-6

The results of open- and short-circuit tests carried out on a 230/115-V 60-Hz single-phase transformer are:

	Primary	Secondary
Open circuit	Open	115 V, 6.5 A, 192 W
Short circuit	17.5 V, 43.5 A, 234 W	Short-circuited

Calculate the parameters of the approximate equivalent circuit of Fig. 13-8(a).

SOLUTION

From the short-circuit test,

$$R_{e1} = \frac{P_{sc}}{I_{sc}^2} = \frac{234 \text{ W}}{43.5^2} = 0.124\ \Omega$$

$$Z_{sc} = \frac{V_{sc}}{I_{sc}} = \frac{17.5 \text{ V}}{43.5 \text{ A}} = 0.402\ \Omega$$

$$X_{e1} = \sqrt{Z_{sc}^2 - R_{e1}^2} = \sqrt{0.402^2 - 0.124^2} = 0.37\ \Omega$$

The open-circuit test was conducted with the primary open. Therefore the results of the open-circuit test must be referred to the primary. Then

$$V_1 = 115 \text{ V} \times \frac{230 \text{ V}}{115 \text{ V}} = 230 \text{ V}$$

$$I_1 = I_\phi = 6.5 \text{ A} \times \frac{115 \text{ V}}{230 \text{ V}} = 3.25 \text{ A}$$

$$P_c = P_{\text{oc}} = 192 \text{ W}$$

$$R_c = \frac{V_1^2 \text{ V}}{192 \text{ V}} = 275.52 \ \Omega$$

$$I_c = \frac{P_{\text{oc}}}{V_1} = \frac{192 \text{ W}}{230 \text{ V}} = 0.835 \text{ A}$$

$$I_m = \sqrt{I_{\text{oc}}^2 - I_c^2} = \sqrt{6.5^2 - 0.835^2} = 6.446 \text{ A}$$

$$X_m = \frac{V_1}{I_m} = \frac{230 \text{ V}}{6.446 \text{ A}} = 35.68 \ \Omega$$

These values can now be substituted into the equivalent circuit.

13-6 VOLTAGE REGULATION AND EFFICIENCY

We have assumed that changes in the mutual flux and secondary terminal voltage under load conditions are small and can be neglected. However, it is necessary to study the effect of loads at varying power factors on the secondary terminal voltage. Normally it is assumed that the primary terminal voltage remains constant, that is, the transformer as a result should deliver rated output power at rated secondary terminal voltage. Obviously the secondary terminal voltage at no load will differ from the secondary terminal voltage under load, because there will not be an impedance voltage drop in the secondary winding. The change in the secondary terminal voltage under load conditions is called the *voltage regulation* of the transformer. The *percentage regulation* of a transformer is defined as

$$\% \text{ regulation} = \frac{V_{2nl} - V_{2fl}}{V_{2fl}} \times 100\% \tag{13–41}$$

The change in the secondary terminal voltage with increasing load and varying power factors is a direct result of the effect of the impedances of both windings. The primary resistance and the leakage reactance cause a decrease in E_1 with load increases. In turn there is a corresponding decrease in the mutual flux, which reduces the secondary induced electromotive force E_2. The secondary terminal voltage is then further reduced because of the secondary resistance and leakage reactance voltage drops. The foregoing is true for unity power factor (UPF) and lagging loads.

To calculate the voltage regulation, the transformer impedances can be referred to the primary or secondary. However, since most calculations involve loads connected to the secondary terminals, it is more convenient to refer all impedances to the secondary. Referring to Fig. 13-11(a), with the approximate transformer equivalent circuit referred to the secondary, the excitation branch can be ignored without introducing an unacceptable error. As a result only the effects of the series impedances, that is, the primary impedance referred to the secondary, the secondary impedance, and the load impedance, need be considered. The easiest method of understanding the effects of the impedances with varying power factors is by means of phasor diagrams.

Figure 13-11(b) is the phasor diagram of a transformer supplying a resistive or unity power factor load. As can be seen, the secondary current I_2 is in phase with the secondary voltage V_2. As a result the referred resistive voltage drop I_2R_{e2} is in phase with and added to V_2. The referred reactance voltage drop I_2X_{e2} leads I_2R_{e2} by 90°, and I_2Z_{e2} is the phasor sum of I_2X_{e2} and I_2R_{e2}. The relationship between the primary input voltage V_1 and the secondary output voltage V_2 is given by

$$\frac{V_1}{\alpha} = \sqrt{(V_2 + I_2R_{e2})^2 + (I_2X_{e2})^2} \tag{13–42}$$

In the case of a lagging power factor load, the secondary current I_2 lags the secondary voltage V_2 by the angle θ_2 [Fig. 13-11(c)]. The referred resistance voltage drop I_2R_{e2} is in phase with I_2 and is added to V_2. Similarly, the referred reactance voltage drop I_2X_{e2} leads I_2 by 90°. The referred primary voltage V_1/α is then

$$\frac{V_1}{\alpha} = \sqrt{(V_2 \cos \theta_2 + I_2R_{e2})^2 + (V_2 \sin \theta_2 + I_2X_{e2})^2} \tag{13–43}$$

Similarly with a leading power factor load [Fig. 13-11(d)], that is, the secondary load current I_2 leads the secondary voltage V_2 by the angle θ_2. The referred primary voltage V_1/α is then

$$\frac{V_1}{\alpha} = \sqrt{(V_2 \cos \theta_2 + I_2R_{e2})^2 + (V_2 \sin \theta_2 - I_2X_{e2})^2} \tag{13–44}$$

Equations (13-42), (13-43), and (13-44) can be consolidated into

$$\frac{V_1}{\alpha} = \sqrt{(V_2 \cos \theta_2 + I_2R_{e2})^2 + (V_2 \sin \theta_2 \pm I_2X_{e2})^2} \tag{13–45}$$

and expressed in complex form,

$$\frac{V_1}{\alpha} = (V_2 \cos \theta_2 + I_2R_{e2}) + j(V_2 \sin \theta_2 \pm I_2X_{e2}) \tag{13–46}$$

where the last term is positive for unity and lagging power factor loads and negative for leading power factor loads.

From Fig. 13-11(b), (c), and (d) it can be seen that for a lagging power factor load V_1/α is greater than V_2, that is, the voltage regulation of the transformer is greater than zero. In the case of a unity power factor load, V_1/α is still greater than V_2. However, the voltage regulation is still greater than zero, but less than that with a lagging power factor load. When the transformer is supplying a capacitive or leading power factor load, the secondary voltage V_2 can be greater than V_1/α, that is, the voltage regulation is less than zero, or negative.

Equation (13-41) can be modified in terms of Eqs. (13-42), (13-43), (13-44), and

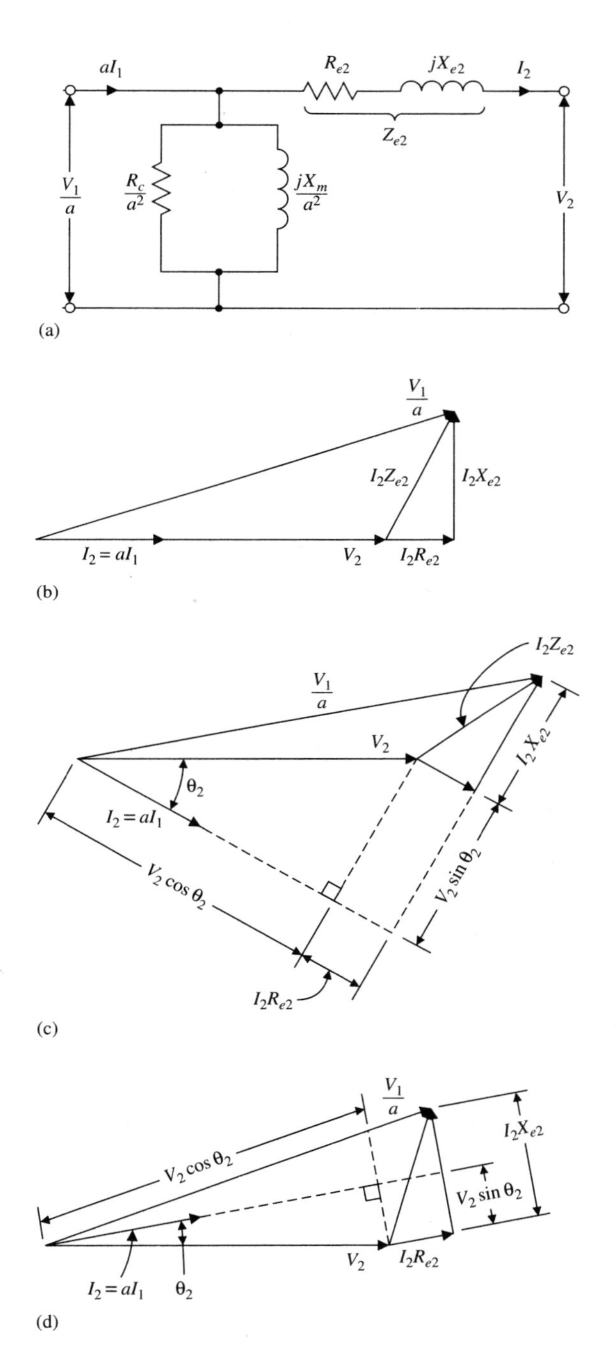

Figure 13-11 Transformer secondary relationships. (a) Approximate equivalent circuit referred to the secondary. Phasor diagrams with operation at: (b) Unity power factor. (c) Lagging power factor. (d) Leading power factor.

(13-45) to become

$$\% \text{ regulation} = \frac{V_1/\alpha - V_2}{V_2} \times 100\% \quad \textbf{(13–47)}$$

where V_2 is the measured secondary voltage, α the transformation ratio, and V_1 the calculated primary voltage required to produce V_2.

Transformers are also compared on the basis of their efficiencies. Efficiency is given by

$$\eta = \frac{P_{\text{out}}}{P_{\text{in}}} \times 100\%$$

$$= \frac{P_{\text{out}}}{P_{\text{out}} + P_{\text{losses}}} \times 100\%$$

As can be seen from our work so far with equivalent circuits, and the determination of the equivalent circuit parameters, that is, iron and copper losses, by the open- and short-circuit tests, transformer efficiency calculations are easily performed.

The efficiency of a transformer at any given load and power factor is

$$\eta = \frac{V_2 I_2 \cos \theta_2}{P_{\text{Cu}} + P_c + V_2 I_2 \cos \theta_2} \times 100\% \quad \textbf{(13–48)}$$

where P_{Cu} is the copper loss incurred in the series resistance of the equivalent circuit, that is, $I_2^2 R_{e2}$, and P_c is the core loss measured in the open-circuit test.

➤ EXAMPLE 13-7

The results of open- and short-circuit tests on a 100-kVA 11,000/2,200-V 60-Hz single-phase transformer are:

	Primary	**Secondary**
Open circuit	Open	2,200 V, 1.5 A, 800 W
Short circuit	600 V, 10.0 A, 1,000 W	Short-circuited

Determine: **(a)** R_{e1} and X_{e1}; **(b)** R_{e2} and X_{e2}; **(c)** the percentage regulation at 0.75 power factor leading, unity power factor, and 0.85 power factor lagging.

SOLUTION

$$\alpha = \frac{11{,}000 \text{ V}}{2{,}200 \text{ V}} = 5$$

(a)

$$Z_{e1} = \frac{V_{sc}}{I_{sc}} = \frac{600\ \text{V}}{10\ A} = 60\ \Omega$$

$$R_{e1} = \frac{P_{sc}}{I_{sc}^2} = \frac{1,200\ \text{W}}{10^2} = 12\ \Omega$$

$$X_{e1} = \sqrt{Z_{e1}^2 - R_{e1}^2} = \sqrt{60^2 - 12^2} = 58.79\ \Omega$$

(b)

$$R_{e2} = \frac{R_{e1}}{\alpha^2} = \frac{12\ \Omega}{5^2} = 0.48\ \Omega$$

$$X_{e2} = \frac{X_{e1}}{\alpha^2} = \frac{58.79}{5^2} = 2.35\ \Omega$$

(c)

$$I_2 = \frac{100\,\text{kVA}}{2,200\,\text{V}} = 45.45\,\text{A}$$

For 0.75 power factor leading, arccos 0.75 = 41.41° and sin 41.41° = 0.66. From Eq. (13-46),

$$\frac{V_1}{\alpha} = (2,200 \times 0.75 + 45.45 \times 0.48) + j(2,200 \times 0.66 - 45.45 \times 2.35)$$
$$= 1,671.82 + j1,345.19 = 2,145.82 \quad 38.82°\ \text{V}$$

From Eq. (13-47),

$$\%\ \text{regulation} = \frac{2,145.82\ \text{V} - 2,200\ \text{V}}{2,200\ \text{V}} \times 100\% = -2.46\%$$

For unity power factor, arccos 1.0 = 0° and sin 0° = 0. From Eq. (13-46),

$$\frac{V_1}{\alpha} = (2,200 \times 1.0 + 45.45 \times 0.48) + j(2,200 \times 0 + 45.45 \times 2.35)$$
$$= 2,221.82 + j106.81 = 2,224.38\angle 2.75°\ V$$

From Eq. (13-47),

$$\%\ \text{regulation} = \frac{2,224.38\ \text{V} - 2,200\ \text{V}}{2,200\ \text{V}} \times 100\% = 1.11\%$$

For 0.85 power factor lagging, arccos 0.85 = 31.79° and sin 31.79° = 0.53. From Eq. (13-46),

$$\frac{V_1}{\alpha} = (2,200 \times 0.85 + 45.45 \times 0.48) + j(2,200 \times 0.53 + 45.45 \times 2.35)$$
$$= 1,891.82 + j1,272.81 = 2,280.14\angle 33.93°\ V$$

From Eq. (13-47),

$$\%\ \text{regulation} = \frac{2,280.14\ \text{V} - 2,200\ \text{V}}{2,200\ \text{V}} \times 100\% = 3.64\%$$

Identical results will be obtained if all quantities are referred to the primary. The results are confirmed by the phasor diagrams of Fig. 13-11. Normally the requirement is to maintain the secondary voltage constant irrespective of load and power factor variations. This can be achieved by the use of tapped windings. Taps are brought out from the primary or the secondary, or from both windings, and their function is to change the transformation ratio. The need for changing the ratio depends on the service requirements of the transformer. Large power transformers usually have a relatively constant primary voltage, whereas the secondary voltage may vary significantly with varying load conditions. As a result there is a frequent need to change the transformation ratio by means of tap changing. Distribution transformers usually have a relatively constant load, and as a result only require occasional changes to the transformation ratio. There are two types of tap changers: off-load and on-load. Off-load tap changers require the power to be removed prior to making any changes, and are commonly used with distribution transformers. On-load tap changers permit the voltage to be maintained at a desired level without removal or interruption of the load. On-load tap changers are used where frequent ratio changes are required because of the nature of the load, and where there is interconnection of subsystems into a large system. Tap changing can be automated by using a contact-making voltmeter or a primary voltage relay to sense voltage changes. However, if there are frequent short-duration load spikes, time delay can be introduced to reduce unnecessary operation.

Whenever tap changing is taking place, care must be taken to ensure that the transformer rating is not exceeded, that is, that excessive currents do not occur, since they will contribute to increased heating and, probably, deterioration of the insulating materials, as well as shortened equipment life expectancy. This results in the requirement that the permitted kVA rating be reduced unless full-capacity taps are used. The nameplate should specify if the taps are not full-capacity taps.

➤ EXAMPLE 13-8

From the data of Example 13-7 calculate the transformer efficiency at 0.8 power factor lagging for: **(a)** 25%; **(b)** 50%; **(c)** 75%; **(d)** 100%; **(e)** 125% of rated full load.

SOLUTION

The core loss P_{oc} is constant at all loads. The copper loss is proportional to I^2.

(a) At 25% of rated full load,

$$P_c = P_{oc} = 800 \text{ W}$$

$$P_{Cu} = 0.25^2 P_{sc} = 0.25^2 \times 1{,}000 \text{ W} = 62.5 \text{ W}$$

$$\eta = \frac{0.25 \times 2{,}200 \times 45.45 \times 0.8}{62.5 + 800 + (0.25 \times 2{,}200 \times 45.45 \times 0.8)} \times 100\%$$

$$= \frac{19{,}998}{20{,}860.5} \times 100\% = 95.87\%$$

(b) At 50% of rated full load,

$$P_{Cu} = 0.5^2 P_{sc} = 0.5^2 \times 1{,}000 \text{ W} = 250 \text{ W}$$

$$\eta = \frac{0.5 \times 2{,}200 \times 45.45 \times 0.8}{250 + 800 + (0.5 \times 2{,}200 \times 45.45 \times 0.8)} \times 100\%$$

$$= \frac{39{,}996}{41{,}046} \times 100\% = 97.44\%$$

(c) At 75% of rated full load,

$$P_{Cu} = 0.75^2 P_{sc} = 0.75^2 \times 1{,}000 \text{ W} = 562.5 \text{ W}$$

$$\eta = \frac{0.75 \times 2{,}200 \times 45.45 \times 0.8}{562.5 + 800 + (0.75 \times 2{,}200 \times 45.45 \times 0.8)} \times 100\%$$

$$= \frac{59{,}994}{61{,}356.5} \times 100\% = 97.78\%$$

(d) At 100% of rated full load,

$$P_{Cu} = P_{sc} = 1{,}000 \text{ W}$$

$$\eta = \frac{2{,}200 \times 45.45 \times 0.8}{1{,}000 + 800 + (2{,}200 \times 45.45 \times 0.8)} \times 100\%$$

$$= \frac{79{,}992}{81{,}792} \times 100\% = 97.8\%$$

(e) At 125% of rated full load,

$$P_{Cu} = 1.25^2 P_{sc} = 1.25^2 \times 1{,}000 \text{ W} = 1{,}562.5 \text{ W}$$

$$\eta = \frac{1.25 \times 2{,}200 \times 45.45 \times 0.8}{1{,}562.5 + 800 + (1.25 \times 2{,}200 \times 45.45 \times 0.8)} \times 100\%$$

$$= \frac{99{,}990}{102{,}352.5} \times 100\% = 97.69\%$$

A study of Eq. (13-48) shows that the transformer efficiency is dependent on the power factor. The highest efficiency occurs when supplying a unity power factor load, that is, $\cos\theta = 1$. As the load becomes more reactive, the efficiency decreases. This emphasizes the reason why alternators and transformers are rated in terms of voltage and current, that is, kVA or MVA, not in terms of their kW or MW ratings. The effect of the power factor on the efficiency of the transformer, which we have been studying in Examples 13-7 and 13-8, is shown in Fig. 13-12. The data were obtained by running the BASIC program given in Appendix B.

Just as with dc machines, the maximum efficiency of a transformer occurs when the fixed losses are equal to the variable losses, that is,

$$I_2^2 R_{e2} = P_c \tag{13–49}$$

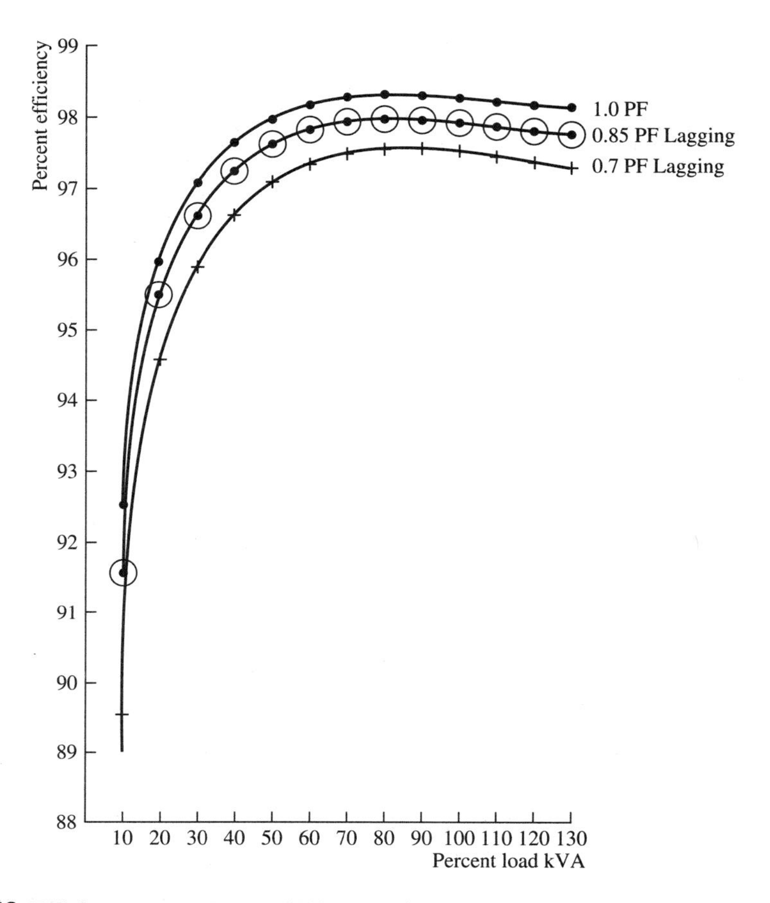

Figure 13-12 Efficiency versus percent kVA at varying power factors.

Therefore the secondary current at which maximum efficiency occurs is

$$I_2 = \sqrt{\frac{P_c}{R_{e2}}} \tag{13–50}$$

➤ EXAMPLE 13-9

A 100-kVA 11,000/220-V 60-Hz single-phase transformer has a core loss Pc 5 800 W and Re2 5 0.48 V. Calculate the secondary current for maximum efficiency.

SOLUTION

From Eq. (13-50),

$$I_2 = \sqrt{\frac{P_c}{R_{e2}}} = \sqrt{\frac{800\text{ W}}{0.48\ \Omega}} = 40.82\ A$$

From Fig. 13-12 it can be seen that the maximum efficiency occurs at 89% of rated kVA output. The maximum efficiency at any power factor will occur at the same load, and the highest efficiency is obtained when the transformer is supplying a unity power factor load. The high efficiencies that we have seen in our calculations result from the fact that a transformer is a static device, that is, there are no moving parts and as a result there are no rotational or stray losses.

The transformers that have been considered so far are all power transformers, that is, they operate at a substantially constant load. The aim of the designer is to ensure that the maximum efficiency occurs at the normal operating load. However, as can be seen from Fig. 13-12, there is very little variation in the efficiency from 50 to 130% of rated kVA load.

Distribution transformers on the other hand are subjected to wide load variations over a 24-h period. As a result they are rated in terms of their *all-day efficiency,* which is computed on the basis of energy output versus energy input over a 24-h period,

$$\text{All-day efficiency} = \frac{\sum W_{out}}{\sum W_{int}} \times 100\% \tag{13–51}$$

where ΣW_{out} is the sum of the energy demands supplied by the transformer over a 24-h period, and ΣW_{int} is the sum of the constant core energy loss, the variable copper loss, and the primary energy input over the same period.

➤ EXAMPLE 13-10

A 10-kVA 2,400/240/120-V 60-Hz single-phase distribution transformer has a full-load copper loss of 125 W and a core loss of 63 W. It supplies the following loads over a 24-h period:

No load for 2.5 h

15% rated load at 0.7 power factor lagging for 3 h

40% rated load at 0.75 power factor lagging for 4.5 h

75% rated load at 0.8 power factor lagging for 5 h

100% rated load at 0.85 power factor lagging for 6 h

110% rated load at unity power factor for 3 h

Calculate: **(a)** the total energy loss over the 24-h period; **(b)** the total energy output over the 24-h period; **(c)** the all-day efficiency.

SOLUTION

(a)

$$\text{Core energy loss} = P_c t = \frac{63\ \text{W} \times 24\ \text{h}}{10^3\ \text{W/kW}}$$
$$= 1.51\ \text{kWh}$$

The copper loss calculations can be simplified by using the ratio of load kVA to rated kVA. The full-load copper loss is P_{sc}. Then

$$\frac{P_{\text{Cu(load)}}}{P_{\text{sc}}} = \frac{I_1^2 R_{e1}}{I_2^2 R_{e2}} = \left(\frac{V_1}{V_1} \times \frac{I_1}{I_2}\right)^2 = \left(\frac{\text{kVA}_{\text{load}}}{\text{kVA}_{\text{rated}}}\right)^2 \quad (13\text{–}52)$$

Then

$$\text{Output} = \%\text{kVA} \times t \text{ h} \times \text{power factor} = \text{kWh}$$

$$\text{Copper loss} = \left(\frac{\%\text{kVA}}{\text{kVA}_{\text{rated}}}\right)^2 \times P_{sc} \times t \text{ h} = \text{kWh}$$

$$\text{Core loss} = P_c \times 24 \text{ h}$$

The results of these calculations are as follows:

kVA	t (h)	Power Factor	Output (kWh)	Copper Loss (kWh)	Core Loss (kWh)
0	2.5		0	0	
1.5	3.0	0.7	3.15	0.008	
4.0	4.5	0.75	13.50	0.090	
7.5	5.0	0.8	30.00	0.352	
10.0	6.0	0.85	51.00	0.750	
11.0	3.0	1.0	33.00	0.454	
		Totals	130.65	1.654	1.51

$$\textit{All-day efficiency} = \frac{130.65}{130.65 + 1.654 + 1.51} \times 100\% = 97.64\%$$

13-7 TESTING

A number of tests are conducted on transformers. They are defined by ANSI C57 Series of Standards and in accordance with the procedures specified in ANSI test code C57.12.90a. It is only intended to describe here briefly the more important tests, in addition to the open-and short-circuit tests already described.

Phasing Out

While a transformer is out of its tank or during manufacture, it is very easy to trace all its winding connections. But when it is totally enclosed and with no markings, the problem is a little more complex. The primary and secondary connections may be identified using a small dc source and a voltmeter in the following manner. Apply the dc voltage across a pair of supposedly primary leads, and connect the voltmeter across the suspected secondary leads. If the dc supply is made and broken and at the same time a momentary deflection is observed on the voltmeter, then the primary and secondary of one phase have been identified. The process is then repeated for the remaining phases. Some difficulty may be experienced when testing three-phase internally connected windings such as a zigzag or interconnected star winding.

Polarity Testing

The term *polarity* refers to the voltage phasor relationship of the external transformer leads, with both the high- and the low-voltage leads being viewed in the same order, that is, from left to right or vice versa, when facing the same side of the transformer. Polarity is totally independent of the way in which the windings are installed. The polarity can be changed by interchanging the two external leads of the same winding. The ANSI standardization rules specify that the external high-voltage leads are to be marked H_1, H_2, H_3, H_4, H_5, . . . , and the low-voltage leads are to be marked X_1, X_2, X_3, X_4, X_5, . . . ,with the lowest and highest numbers representing the full winding and the intermediate numbers representing taps.

The primary and secondary voltages are induced by the same core flux, and they must have the same direction in the coil turn [Fig. 13-13]. The direction of the induced voltages at the external terminals is dependent on the direction in which the coils were wound. Referring to Fig. 13-13(a), the induced voltages between H_1 and H_2 and X_1 and X_2 are in the same direction; this is termed *subtractive polarity*. In Fig. 13-13(b) the induced voltages between H_1 and H_2 and X_1 and X_2 are in opposite directions; this is termed *additive polarity.*

The polarity of a transformer can be determined by a simple test. Connect terminals H_1 and X_1 together and connect a voltmeter with a suitable range between terminals H_2 and X_2. Then apply a low ac voltage between H_1 and H_2 and take readings of the applied voltage and the voltage between H_2 and X_2. If the latter voltage reading is less than the applied voltage, the polarity is subtractive; if it is greater, then the polarity is additive. This method is satisfactory for transformers with a transformation ratio less than 30:1.

Transformers up to 500 kVA and 34.5 kV normally have additive polarities. Large high-voltage power transformers are normally subtractive because of the voltage difference between adjacent external leads H_1X_1 or H_2X_2. As a result if there is an accidental short circuit between them, the voltage difference would be the sum of the

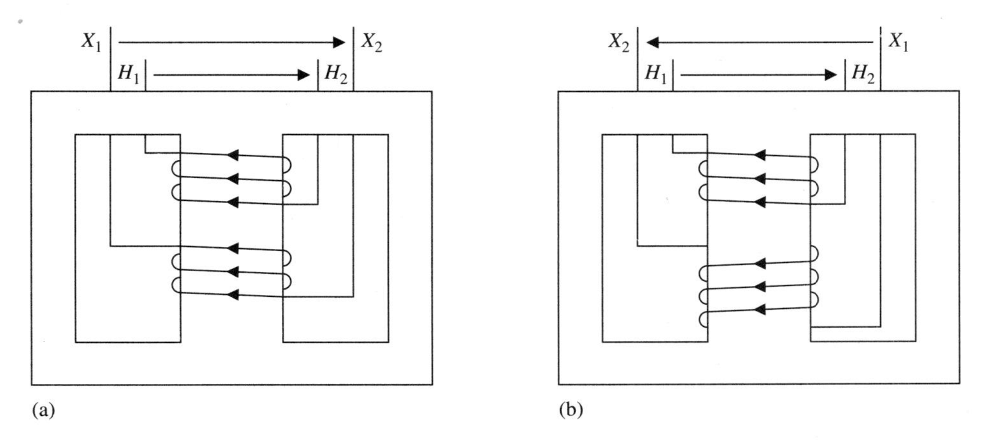

Figure 13-13 Transformer polarities. (a) Subtractive. (b) Additive.

voltages if the transformer has additive polarity, and the difference if the polarity is subtractive. Under operating conditions, the potential stress between adjacent high- and low-voltage connections is one-half the sum of the high and low voltages for additive polarity, and one-half the difference for subtractive polarity.

Back-to-Back or Sumpner Test

When conducting heat runs on large power transformers, it is not efficient to dissipate the output to a load bank. The Back-to-Back or Sumpner test permits transformers to be loaded without drawing a large amount of energy from the ac supply (Fig. 13-14). Two identical transformers are required, with their primaries connected in parallel across the main supply. The secondaries are connected in series opposing and are supplied from a variable ac auxiliary supply source, which can supply full-load current at a reduced voltage. When there is no power being supplied from the auxiliary source, wattmeter W1 measures the core loss of both transformers. As power is supplied from the auxiliary source, an increasing current I_2 flows in the secondary windings. Since the secondary windings are connected in series opposing, the power supplied from the auxiliary source supplies the copper losses. Wattmeter W2 measures the copper loss. For any load current I_2 the total loss for each transformer is $(W_1 + W_2)/2$, and as a result the efficiency can be calculated.

Under heat run conditions the winding hot spot, cooling medium, and ambient temperatures are all monitored and recorded periodically, usually every half-hour.

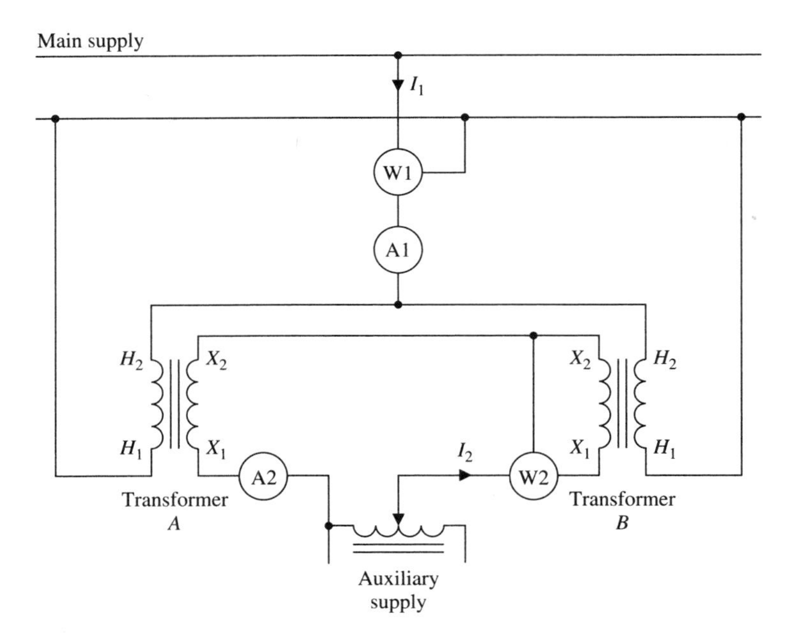

Figure 13-14 Back-to-back or Sumpner test connections.

Dielectric Tests

Dielectric tests are made to check the standards of workmanship and insulation. Several tests are defined by ANSI test code C57.12.90 which include the following:

1. High potential or applied potential dielectric test. A high potential, usually twice rated plus 1,000 V at 60 Hz, is applied for 1 min between all terminals (connected together) of the winding being tested, and all other windings and ground. The test voltage between the windings and ground is measured by means of a calibrated-sphere spark gap.
2. Induced voltage dielectric test. This is normally the final dielectric test. It is usually carried out by applying twice the normal voltage of the winding being tested, taking care that the voltage of any winding to ground does not exceed safe values. The exciting current is limited by using supply frequencies of between 120 and 500 Hz. The duration of the test is usually 1 min for 120 Hz or less, and 7,200 cycles for frequencies greater than 120 Hz.
3. Whenever specified, radio-influence and audible noise tests are carried out in accordance with NEMA Publ. TR1.

13-8 SINGLE-PHASE TRANSFORMER CONNECTIONS

So far we have only discussed transformers with two windings: a primary and a secondary. However, many transformers have one or more primary and secondary windings, which may be connected in various parallel and series arrangements to obtain a variety of voltage combinations. We will consider a common combination of windings and voltages used in distribution transformers. A typical nameplate would read 4,600 V–2,300 V/ 460 V–230 V. This transformer would have two primary windings rated at 2,300 V and two secondary windings rated at 230 V. This arrangement of windings permits four possible combinations:

1. 4,600/460 V, series-series connection [Fig. 13-15(a)]
2. 2,300/460 V, parallel-series connection [Fig. 13-15(b)]
3. 4,600/230 V, series-parallel connection [Fig. 13-15(c)]
4. 2,300/230 V, parallel-parallel connection [Fig. 13-15(d)]

As can be seen from the polarity markings, the odd-numbered terminals have the same instantaneous polarity. When making parallel connections, the coils with the same instantaneous polarity and voltage are connected in parallel, that is, the even-numbered terminals are connected together, and the odd-numbered terminals are connected together on the primary or secondary side of the transformer. When making series connections, the primary or secondary windings of opposite polarity are connected together, for example, H_2 to H_3 or X_2 to X_3. The respective windings will then be connected in *series-aiding,* that is, the induced voltages cancel each other.

The four possible combinations shown in Fig. 13-15 provide four voltage and current combinations, but only three transformation ratios: 4,600 V/460 V = 10:1; 4,600 V/230 V = 20:1; and 2,300 V/460 V = 5:1.

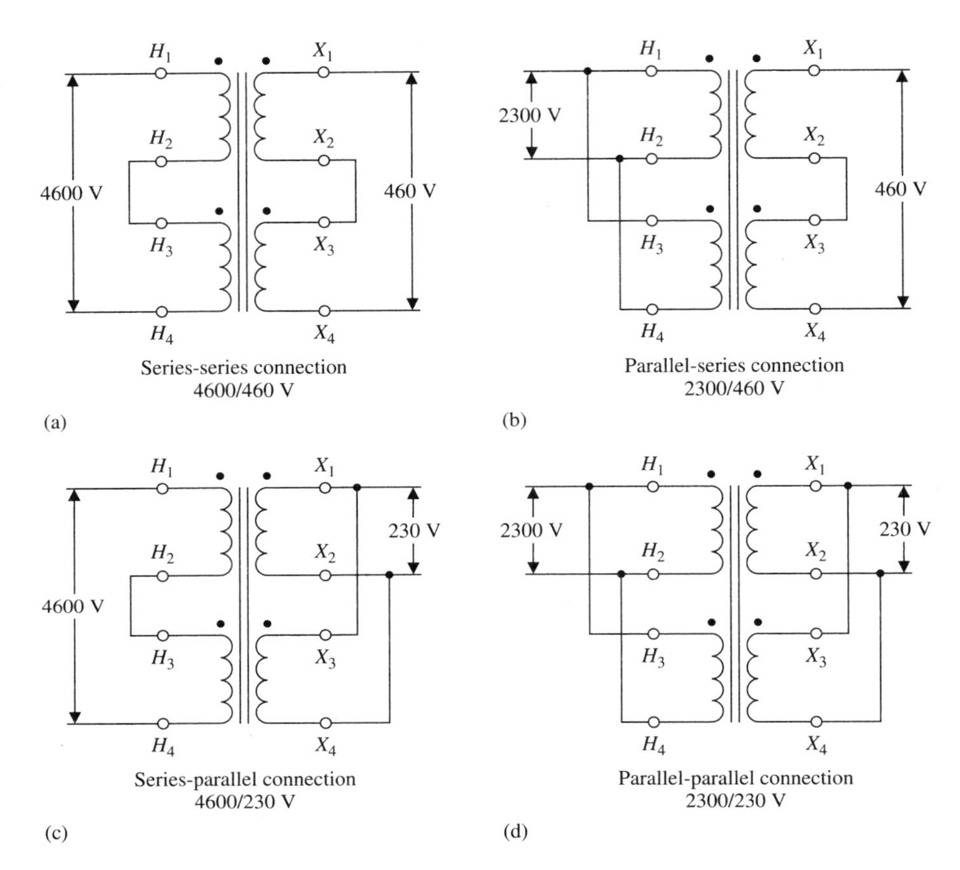

Figure 13-15 Typical distribution transformer with equal-voltage windings connected in series and parallel.

It is important to note that only equal-voltage coils are connected in parallel since their induced voltages oppose each other. If coils with unequal voltages are paralleled, because of the low equivalent impedance of the windings, large circulating currents will be present.

13-9 AUTOTRANSFORMER

So far our discussion of transformers has been limited to transformers with separate primary and secondary windings, where energy transfer has been entirely by transformer action, that is, by the magnetic circuit, with total electrical isolation between the windings. The autotransformer is a transformer where the primary and the secondary have a section of winding that is common to both (Fig. 13-16). There are a number of advantages to this arrangement, as compared to a conventional transformer of the same rating:

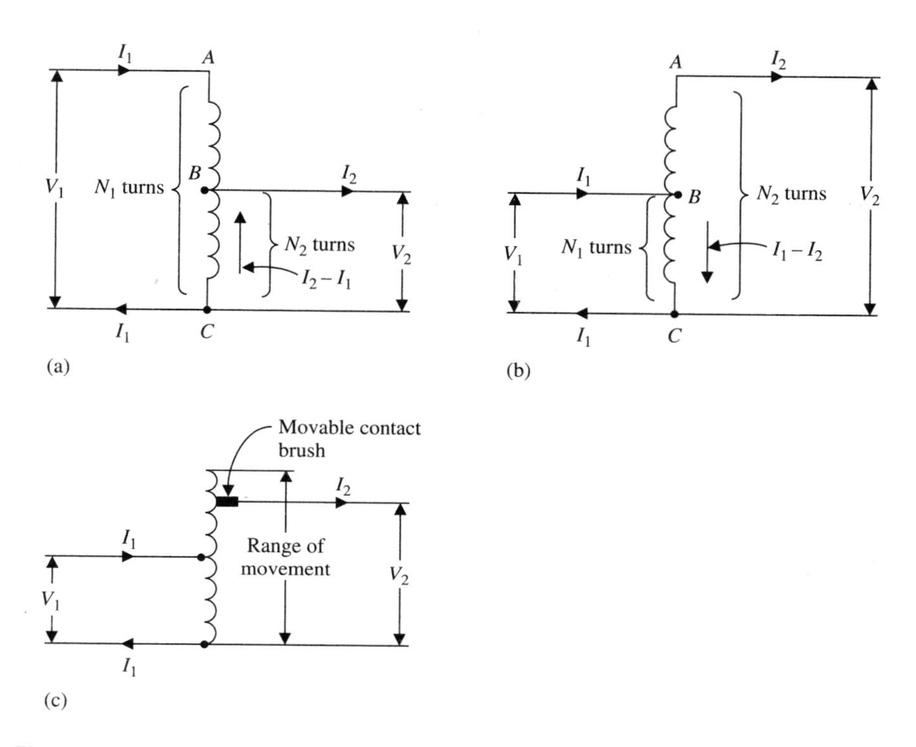

Figure 13-16 Autotransformer. (a) Step-down connection. (b) Step-up connection. (c) Variable or variac connection.

1. Lower initial cost
2. Higher efficiency
3. Improved voltage regulation
4. Physically smaller and lighter
5. Requiring smaller excitation current because of the reduced core volume.

These advantages are due to the fact that a significant proportion of the output kVA is transferred by conduction, and only a small proportion is transferred by transformer action. The conventional transformer transfers all the energy by transformer action.

Neglecting losses and power factor, $V_1I_1 = V_2I_2$. Assuming that the difference between V_1 and V_2 is small, then the difference between I_1 and I_2 will also be small. The current flow in the section of the winding between B and C is much less than the load current, that is, $I_2 - I_1$ for a step-down autotransformer and $I_1 - I_2$ for a step-up autotransformer. As a result it can be seen that the conductor cross-sectional area between B and C can be reduced, with a subsequent reduction in weight and cost. In comparison the normal two-winding transformer must have windings capable of carrying the respective primary and secondary currents.

Assuming that the weight of the copper in a winding is proportional to NI, then if the current density and the turn length remain constant, the total weight of copper in a conventional transformer is

$$W_T \propto (N_1 I_1 + N_2 I_2)$$

In the step-down autotransformer the weight of copper in section AB is proportional to $(N_1 - N_2)I_1$, and the weight of copper in section BC is proportional to $N_2(I_2 - I_1)$. As a result the total weight of copper in the autotransformer is

$$\begin{aligned} W_{AT} &\propto (N_1 - N_2)I_1 + N_2(I_2 - I_1) \\ &\propto (N_1 - 2N_2)I_1 + N_2 I_2 \end{aligned}$$

Therefore

$$\begin{aligned} \frac{W_{AT}}{W_T} &= \frac{(N_1 - 2N_2)I_1 + N_2 I_2}{N_1 I_1 + N_2 I_2} \\ &= \frac{N_1 / N_2 - 2 + I_2 / I_1}{N_1 / N_2 + I_2 / I_1} \end{aligned}$$

But the transformation ratio $\alpha = N_1/N_2 = I_2/I_1$. Therefore

$$\frac{W_{AT}}{W_T} = \frac{\alpha - 2 + \alpha}{\alpha + \alpha} = \frac{2\alpha - 2}{2\alpha} = \frac{\alpha - 1}{\alpha} \tag{13–53}$$

This result can be best illustrated as follows:

1. If $\alpha = 1.25$, the copper saving is 80%.
2. If $\alpha = 2.5$, the copper saving is 40%.
3. If $\alpha = 5$, the copper saving is 20%.

These results show that for a small increase or decrease of voltage there is a considerable reduction in the copper required, and since the current in section BC is less, the copper loss is less. As a result the autotransformer will have a higher efficiency. The gains made in reducing the amount of copper in the windings rapidly decrease as the transformation ratio increases, and in fact very little saving is achieved for transformation ratios greater than 3.

Referring to Fig. 13-16(a) for a step-down transformer, the transformation ratio is

$$\alpha = \frac{N_{AC}}{N_{BC}} = \frac{V_1}{V_2} = \frac{I_2}{I_1} \tag{13–54}$$

where N_{AC} and N_{BC} are the number of primary and secondary turns, respectively.

The power supplied to the load P_L is

$$P_L = V_2 I_2$$

Now

$$I_2 = I_1 + (I_2 - I_1)$$

Therefore

$$P_L = V_2[I_1 + (I_2 - I_1)]$$
$$= V_2 I_1 + V_2(I_2 - I_1) \tag{13–55}$$

Equation (13-55) shows that the load is supplied in two ways, namely,

$$P_{\text{con}} = V_2 I_1 \tag{13–56}$$

where P_{con} is the conducted power from the source through winding AB to the load, and

$$P_{\text{tr}} = V_2(I_2 - I_1) \tag{13–57}$$

where P_{tr} is the power transformed to the load by winding BC, that is, power is conducted to the load through the winding that is not common between source and load, and the balance of the power is transformed by the winding section common to source and load. In order to determine the ratio of power conducted to power transformed,

$$\frac{P_{\text{con}}}{P_L} = \frac{V_2 I_1}{V_2 I_2} = \frac{I_1}{I_2} = \frac{1}{\alpha} \tag{13–58}$$

and

$$\frac{P_{\text{tr}}}{P_L} = \frac{V_2(I_2 - I_1)}{V_2 I_2} = \frac{I_2 - I_1}{I_2} = \frac{\alpha - 1}{\alpha} \tag{13–59}$$

from which

$$P_{\text{con}} = \frac{P_L}{\alpha} \tag{13–60}$$

and

$$P_{\text{tr}} = \frac{P_L(\alpha - 1)}{\alpha} \tag{13–61}$$

where α is greater than 1 for a step-down autotransformer and less than 1 for a step-up autotransformer. From Fig. 13-16(b) the power supplied to the load is

$$P_L = V_1 I_1 = V_1[I_2 + (I_1 - I_2)]$$
$$= V_1 I_2 + V_1(I_1 - I_2) \tag{13–62}$$

Once again the power is supplied to the load in two ways,

$$P_{\text{con}} = V_1 I_2 \tag{13–63}$$

and

$$P_{\text{tr}} = V_1(I_1 - I_2) \tag{13–64}$$

where P_{con} is the power conducted to the load directly from the source and P_{tr} is the power transformed by the common winding section BC. The relative amounts of power conducted and transformed are

$$\frac{P_{\text{con}}}{P_L} = \frac{V_1 I_2}{V_1 I_1} = \frac{I_2}{I_1} = \alpha \quad \textbf{(13–65)}$$

that is,

$$P_{\text{con}} = \alpha P_L \quad \textbf{(13–66)}$$

and

$$P_{\text{tr}} P_{\text{L}} = \frac{V_1(I_1 - I_2)}{V_1 I_1} = \frac{I_1 - I_2}{I_1} = 1 - \alpha \quad \textbf{(13–67)}$$

that is,

$$P_{\text{tr}} = P_{\text{L}}(1 - \alpha) \quad \textbf{(13–68)}$$

➤ EXAMPLE 13-11

A 10-kVA 2,400/240-V 60-Hz single-phase transformer is reconnected to step down a voltage from 2,640 to 2,400 V. Calculate: **(a)** the kVA rating as an autotransformer; **(b)** P_{con}; **(c)** P_{tr}. The transformer is connected as shown in Fig. 13-16(a), where the 240-V winding is section *AB*.

SOLUTION

The rated current I_1 in the 240-V winding (section *AB*) is

$$I_1 = \frac{10 \text{ kVA}}{240 \text{ V}} = 41.67 \text{ A}$$

The rated current in the 2,400-V winding (section *BC*) is

$$I_2 - I_1 = \frac{10 \text{ kVA}}{2,400 \text{ V}} = 4.17 \text{ A}$$

The transformation ratio is

$$\alpha = \frac{2,640 \text{ V}}{2,400 \text{ V}} = 1.1$$

Since

$$I_2 = (I_2 - I_1) + I_1 = 4.17 + 41.67 = 45.84 \text{ A}$$

the power supplied to the load is

$$P_L = V_2 I_2 = 2,400 \text{ V} \times 45.84 \text{ A} = 110 \text{ kVA}$$

Therefore

$$P_{\text{con}} = \frac{P_L}{\alpha} = \frac{110 \text{ kVA}}{1.1} = 100 \text{ kVA}$$

and

$$P_{tr} = P_L \frac{\alpha - 1}{\alpha} = 110 \text{ kVA} \frac{1.1 - 1}{1.1} = 10 \text{ kVA}$$

➤ EXAMPLE 13-12

Repeat Example 13-11 with the transformer reconnected as in Fig. 13-16(b) to step up the voltage from 2,400 to 2,640 V.

SOLUTION

The rated current I_2 in the 240-V winding (section AB) is

$$I_2 = \frac{10 \text{ kVA}}{240 \text{ V}} = 41.67 \text{ A}$$

The power supplied to the load is

$$P_L = V_2 I_2 = 2{,}640\ V \times 41.67 \text{ A} = 110 \text{ kVA}$$

The transformation ratio is

$$\alpha = \frac{2{,}400 \text{ V}}{2{,}640 \text{ V}} = 0.91$$

Therefore

$$P_{con} = \alpha P_L = 0.91 \times 110 \text{ kVA} = 100 \text{ kVA}$$

and

$$P_{tr} = P_L(1 - \alpha) = 110 \text{ kVA}(1 - 0.91) = 10 \text{ kVA}$$

From these two examples it can be seen that the autotransformer has a greater power-handling capability than the equivalent two-winding transformer, namely, $\alpha/(\alpha - 1)$ times greater in the case of a step-down transformer and $1(1 - \alpha)$ times greater for a step-up transformer, that is, the 10-kVA two-winding transformer supplies 110 kVA as an autotransformer, or 11 times its rated output. This is because only 10 kVA is transformed and 100 kVA is conducted directly to the load.

A major disadvantage of the autotransformer is that it is possible under fault conditions to have high voltages applied to the low-voltage side through the common section of the windings. This could be extremely dangerous if the voltage difference between the two sides were large. Because of this possibility, autotransformers used in power transmission systems are used in grounded neutral systems.

The major uses of autotransformers are usually confined to reduced-voltage starters for polyphase ac motors, power factor correction circuits, high-voltage discharge lighting, in large kVA ratings to raise a generator terminal voltage to the transmission-line voltage, or between two systems operating at different voltage levels.

A special form of the autotransformer is very commonly used in laboratories. This type of autotransformer is a variable-ratio transformer commonly called a *variac.* The variac consists of a toroidal coil wound over a rectangular-cross-section annular magnetic core. The output connection is made by a carbon brush which bears against a cleaned section of the winding [Fig. 13-16(c)]. This arrangement gives an output voltage range from zero to usually 120% of the primary input voltage.

13-10 CORE TYPES AND CONSTRUCTION

The laminations of large power transformers are usually cut by numerically controlled (NC) guillotines from rolls, up to 48 in (1220 mm) wide, of cold-rolled grain-oriented silicon steel. The cuts are angles (Fig. 13-17) so as to retain the magnetic properties imparted by the rolling process. After cutting, the laminations are usually annealed at 1,472°F (800°C) to restore the crystalline structure that was disturbed by the cutting process.

In order to minimize eddy current losses in the core, laminations 0.014 in (0.36 mm) thick are used for transformers operating at power-line frequencies.

There are two types of core construction commonly used in large power transformers: the core type and the shell type (Fig. 13-17). The core type has the longer mean magnetic circuit length and a shorter mean turn length, while the shell type has a shorter mean magnetic circuit length and a longer mean turn length. For the same service conditions, that is, voltage, frequency, and kVA ratings, the core type has a smaller core and a larger number of turns than the comparable shell type. The core type with a relatively larger winding space is better suited to high-voltage applications, while the shell type is better suited to transforming lower voltages. In addition, the effects of leakage flux are also lessened with the shell-type core.

Normally the core-type transformer uses *cylindrical* coil windings, which consist of a multiturn, multilayer winding wound over an insulated cylinder, the low-voltage winding cylinder being closest to the core. Spaces are provided between core and

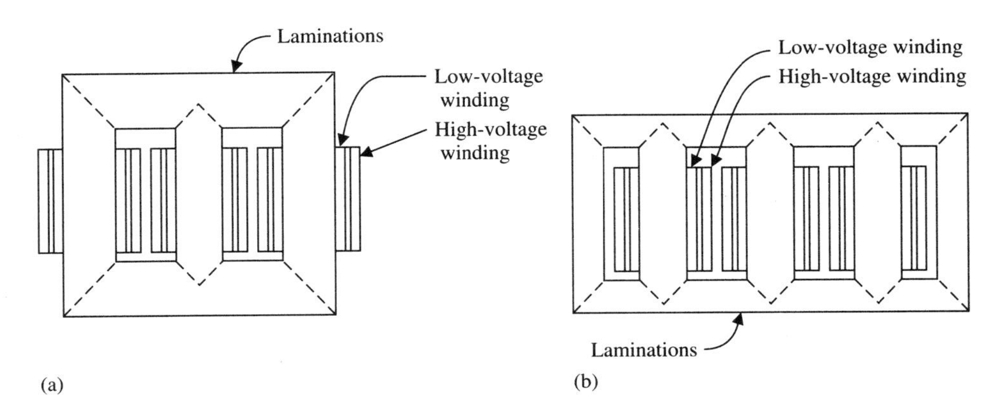

Figure 13-17 Core construction of three-phase transformers. (a) Core type. (b) Shell type.

winding cylinders and between the winding cylinders to provide passages for the cooling medium (air, oil, or silicone fluid). The shell-type transformer normally uses *pancake* coil windings, which are single-layer multiturn coils, with the high- and low-voltage pancake coils being stacked over the inner legs of the transformer core. The coils are separated by spacers to provide passage for the cooling medium. The coils are usually wound using insulated rectangular-cross-section copper wire.

The core and winding structures are rigidly braced to prevent movement and noise. Noise is caused by *magnetostriction,* which causes dimensional changes in the laminations and is dependent on both the flux density and the power-line frequency. Tremendous advances have been made in noise reduction by minimizing resonance in the core and tank.

Usually the core and the windings of large power transformers are enclosed in large steel tanks, which are filled with the insulating medium. Insulating media in common use are liquids or gases. For many years polychlorinated biphenyls (PCBs) or askarels were used as both the insulating and the cooling media in transformers and capacitors. However, their use has now been banned. Silicone fluids, although initially more costly, are chemically inert and thermally stable, and transformers have only required minor changes to permit the use of silicone fluids instead of the askarels. Advances have also been made in the use of fluorogases such as sulfur hexaflouride SF_6. These gases have high dielectric strength and heat-transfer capabilities, both of which increase with pressures up to 3 atmo. Nitrogen and air-insulated transformers are commonly used for voltage levels under 34.5 kV. Air-insulated transformers are normally open to the atmosphere, and as a result must be protected from the weather. In contaminated atmospheres the transformer is totally enclosed and filled with nitrogen at atmospheric pressure.

Removal of the heat losses, that is, iron and copper losses, is important since this affects the temperature rise and thus the rating of the transformer. The method of cooling is designated in the United States by the following abbreviations:

1. *AA:* Self-cooled, dry type
2. *AA/FA:* Self-cooled with forced-air cooling as the transformer is loaded, dry type
3. *OA:* Oil-immersed, self-cooled
4. *OA/FA:* Oil-immersed, self-cooled with forced-air cooling as the transformer is loaded
5. *OA/FA/FA:* Oil-immersed, self-cooled with two stages of forced-air cooling as the transformer is loaded
6. *FOA:* Oil-immersed, forced-oil, forced-air cooling (*Note:* The oil pumps and cooling fans must be operational when the transformer is loaded.)
7. *OA/FA/FOA*: Oil-immersed, self-cooled with forced-air and forced-oil cooling under load conditions
8. *OA/FOA/FOA*: Oil-immersed, self-cooled with two stages of forced-air–forced-oil cooling (This system permits the required cooling to be introduced as the transformer load varies.)
9. *FOW:* Oil-immersed, forced-oil, forced-water cooling

Similar designations have been established by the Canadian Standards Association (CSA):

1. *ONS:* Oil-immersed, natural circulation, self-cooled
2. *ONW:* Oil-immersed, natural circulation, water cooled
3. *ONP:* Oil-immersed, natural circulation, forced-air cooled
4. *OFW:* Oil-immersed, forced-oil, water cooled
5. *OFP:* Oil-immersed, forced-oil, forced-air cooled
6. *ONPP:* Oil-immersed, natural circulation with two stages of forced-air cooling

Oil-immersed transformers are usually limited to either a 55°C or a 65°C temperature rise to prevent oil deterioration. In the case of dry-type transformers temperature rises as great as 180°C are permissible, depending on the type of insulation material used.

Self-cooled dry-type transformers are available in sizes up to 3,750 kVA and voltages up to 15 kV. Self-cooled, forced-air-cooled transformers are available in ranges from 300 to 15,000 kVA and voltages up to 15 kV. Oil-immersed transformers range from the self-cooled single-phase pole-type distribution transformer up to 1,300 MVA, 24.5 kV/345 kV and higher voltages.

To compensate for internal impedance drops and line-voltage drops, transformers are designed with tapped windings, which permit changes to be made to the transformation ratio so as to maintain the secondary voltage within prescribed limits. These tap changers may be manual, that is, off-load tap changers, or automatic, that is, on-load tap changers.

13-11 THREE-PHASE TRANSFORMER CONNECTIONS

All power generation, and nearly all transmission (except for high-voltage dc transmission systems) and distribution systems are three-phase ac systems. In these systems there are frequent voltage-level changes, for example, from the generator voltage to the transmission-line voltages for long-distance transmission, then from transmission-line voltage levels to distribution voltage levels. These voltage level changes are made using three-phase transformers—either a three-phase transformer or a bank of three single-phase transformers. The three-phase transformer is cheaper, occupies less space, is lighter, and is slightly more efficient than three single-phase transformers. However, the latter combination does provide some benefits: it is cheaper to carry a spare unit for repair action, and it is possible to transform voltages using the open-delta or vee connection.

In addition, it is possible to produce *phase shifts* between the primary and secondary voltages, as well as obtaining *phase changes,* such as three-to-six phase or three-to-two phase or vice versa. If three-phase transformers are formed from three single-phase transformers, the transformers should have the same kVA and voltage ratings as well as the same polarities.

The primaries and secondaries of any three-phase transformer arrangement can be connected in star (wye) or delta, which gives rise to four possible connection combinations:

1. Star-star
2. Delta-delta
3. Delta-star
4. Star-delta

Star-Star Connection

The star-star or wye-wye connection is shown in Fig. 13-18. If the transformation ratio of each phase is the same, the same ratio will exist between the line voltages on each side, that is,

$$\frac{V_{AB}}{V_{ab}} = \frac{V_{BC}}{V_{bc}} = \frac{V_{CA}}{V_{ca}} = \alpha \tag{13–69}$$

The star-star connection is most suitable for high-voltage applications since the phase voltage is 57.5% of the line voltage, that is, the winding may be insulated for the lower voltage level. The arrangement is satisfactory provided the secondary load is balanced. If the secondary load is unbalanced, the electrical neutral will be displaced, that is, it becomes a *floating neutral,* and the line-to-neutral voltages are unequal. The floating neutral problem can be eliminated by grounding the primary-side neutral point.

When the primary neutral is grounded, under balanced load conditions the neutral

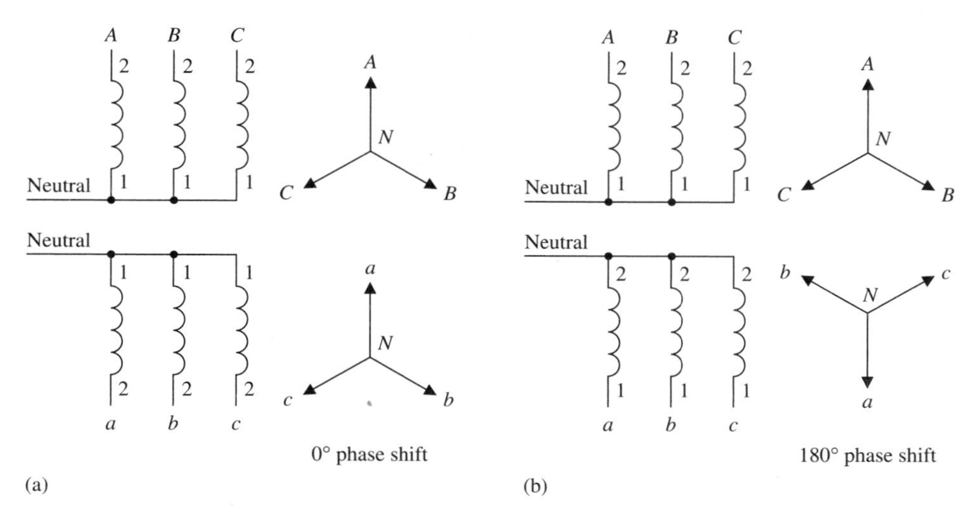

Figure 13-18 Star-star or wye-wye connection. (a) Schematic 0° phase shift. (b) Schematic 180° phase shift.

will carry the third harmonic current, and the primary and secondary phase voltages are sinusoidal. If there is not a primary ground, the primary and secondary phase voltages will contain pronounced third-harmonic voltages, and the neutral point voltage will oscillate at three times the line frequency. However, the line voltages are sinusoidal. This problem can be eliminated by introducing a third winding, called the *tertiary* winding, connected in delta, which provides a path for the third-harmonic currents that are required to produce sinusoidal primary and secondary phase voltages. The tertiary winding can also be used to supply power at an intermediate voltage as well as assisting in distributing the load more equally between phases under balanced load conditions. The tertiary winding is usually rated at about 33% of the transformer rating. From Fig. 13-18 it can be seen that the incoming and outgoing line voltages are either in phase (0° phase shift) or phase shifted by 180°.

➤ EXAMPLE 13-13

A 75-kVA 230-V three-phase load is supplied from a 6,600-V three-phase supply using a star-star-connected transformer bank made up of three equally rated single-phase transformers. What are the voltage, current, and kVA ratings of the single-phase transformers?

SOLUTION

On the primary side,

$$\text{Line voltage } V_{lp} = 6{,}600\ V$$

$$\text{Line current } I_{lp} = \text{phase current } I_{pp} = \frac{\text{kVA}}{\sqrt{3}} = \frac{75\times10^3\ \text{VA}}{\sqrt{3}\times 6{,}600\ \text{V}} = 6.56\ \text{A}$$

$$\text{Phase voltage } V_{pp} = \frac{V_{lp}}{\sqrt{3}} = \frac{6{,}600\ \text{V}}{\sqrt{3}} = 3{,}810.51\ \text{V}$$

On the secondary side,

$$\text{Line voltage } V_{ls} = 230\ \text{V}$$

$$\text{Phase voltage } V_{ps} = \frac{V_{ls}}{\sqrt{3}} = \frac{230\ \text{V}}{\sqrt{3}} = 132.79\ \text{V}$$

$$\text{Line current } I_{ls} = \text{phase current } I_{ps} = \frac{\text{kVA}}{\sqrt{3}\text{V}_{ls}} = \frac{75\ \text{kVA}}{\sqrt{3}\times 230\ \text{V}} = 188.27\ \text{A}$$

$$\begin{aligned}\text{kVA rating of each transformer} &= V_{pp}I_{pp} = V_{ps}I_{ps}\\ &= 3{,}810.51\ \text{V}\times 6.56\ \text{A}\\ &= 132.79\ \text{V}\times 188.27\ \text{A}\\ &= 24.99\ \text{kVA} \cong 25\ \text{kVA}\end{aligned}$$

It should be noted that the star-connected secondary can provide a dual-voltage output, namely, the line-to-line voltage, in this case 230 V, and the line-to-neutral or phase voltage of 132.79 V. This arrangement is very useful since it permits three-phase loads to be supplied at the line voltage, and single-phase loads to be supplied to the line-to-neutral voltage simultaneously. Obviously care must be exercised to ensure that the single-phase loads are balanced.

Delta-Delta Connection

As can be seen from Fig. 13-19, the primary and secondary phase windings are at line potential and must both be insulated for the full line voltage. The relationship between the primary and secondary line voltages is

$$\frac{V_{AB}}{V_{ab}} = \frac{V_{BC}}{V_{bc}} = \frac{V_{CA}}{V_{ca}} = \alpha \tag{13–70}$$

Neglecting third harmonics, the line currents are $\sqrt{3}$ times the phase currents, or the phase currents are 57.74% of the line currents. The primary and secondary windings will have a greater number of turns than are required for the star-star connection, but the conductor cross-sectional area is 57.74% of that required for the star-connected winding. The delta-delta connection is best suited for large low- to medium-voltage transformer banks using single-phase transformers, but is rarely encountered in a three-phase transformer.

Large unbalanced line currents pose no problem, and an additional benefit is that the arrangement can be operated with one phase removed if a fault occurs in that

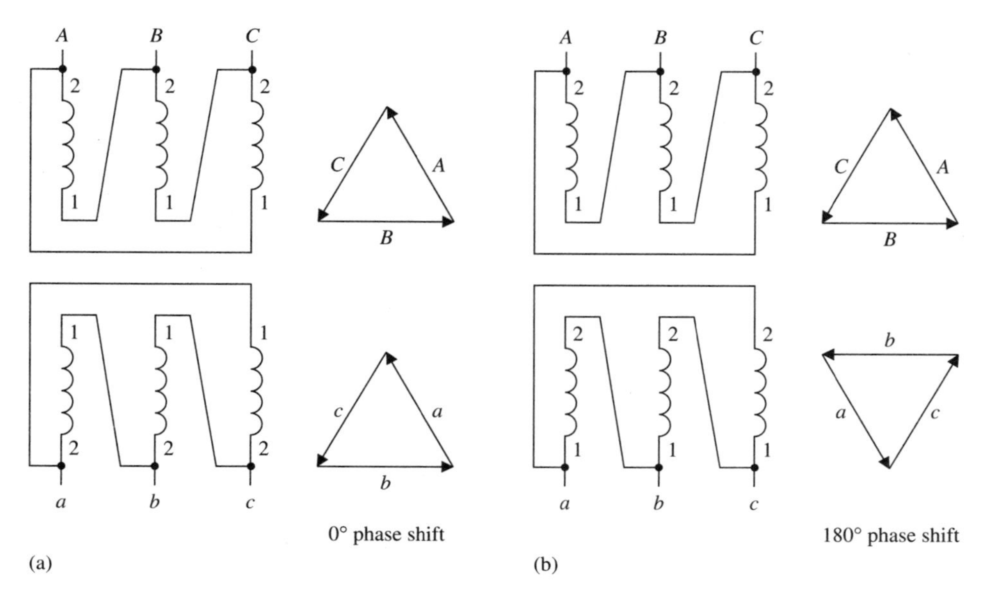

Figure 13-19 Delta-delta connection. (a) Schematic 0° phase shift. (b) Schematic 180° phase shift.

transformer, and yet still provide three-phase power. This connection is known as the *open-delta* or *vee* connection, and can supply up to 57.74% of the original load capability of the delta-delta connection.

It was noted earlier that third-harmonic components of the no-load current are present when the transformer is excited from a sinusoidal source. These third-harmonic components can have a significant effect on the operation of three-phase transformer banks. Assuming that the phase currents in a delta-connected transformer winding each contain a third harmonic, since the fundamental components are 120° out of phase with respect to each other, then the third-harmonic components must be in phase. As a result, the third-harmonic components will circulate around the delta connection, and the line currents will be free of any third-harmonic components.

As can be seen from Fig. 13-19, the secondary voltages can be in phase or 180° out of phase with the primary voltages, depending on the way the secondary windings are connected. A disadvantage of the delta-delta connection is the absence of a neutral point, and as a result there is no path for unbalanced currents caused by unbalanced loading. Also unlike the star-star or delta-star connections, dual-voltage outputs cannot be obtained.

➤ EXAMPLE 13-14

Three single-phase transformers are connected in delta-delta, and are used to step down a line voltage of 110 kV to 66 kV to supply an industrial plant drawing 50 MW at a 0.80 lagging power factor. Calculate: **(a)** the high-voltage-side line current; **(b)** the low-voltage-side line current; **(c)** the primary phase currents; **(d)** the secondary phase currents.

SOLUTION

Recall $P = \sqrt{3}\, V_l I_l \cos\theta$. Therefore

$$I_l = \frac{P}{\sqrt{3} V_l \cos\theta}$$

(a)

$$\text{HV line current} = \frac{50 \times 10^6 \text{ W}}{\sqrt{3} \times 110 \times 10^3 \text{ V} \times 0.80} = 328.04 \text{ A}$$

(b)

$$\text{LV line current} = \frac{50 \times 10^6 \text{ V}}{\sqrt{3} \times 66 \times 10^3 \text{ V} \times 0.80} = 546.73 \text{ A}$$

(c)

$$\text{Primary phase current } I_{p1} = \frac{I_{l1}}{\sqrt{3}} = \frac{328.04 \text{ A}}{\sqrt{3}} = 189.39 \text{ A}$$

(d)

$$\text{Secondary phase current } I_{p2} = \frac{I_{l2}}{\sqrt{3}} = \frac{546.73 \text{ A}}{\sqrt{3}} = 315.65 \text{ A}$$

It should be noted that the way the load is connected is unimportant, as long as the magnitude of the load is known. In actual practice, the loads may be combinations of three-phase and single-phase loads operating at a variety of voltages.

Delta-Star or Wye Connection

The delta-star or delta-wye connection is shown in Fig. 13-20. These connection arrangements are commonly used to step up alternator voltages to the transmission-line voltage. Another common application is in distribution service, where a four-wire dual-voltage system is required. This arrangement takes advantage of the fact that the transformer secondary voltage is $1/\sqrt{3}$ of the line voltage, with the result that the secondary winding insulation needs are reduced.

The primary phase voltage is equal to the primary line-to-line voltage, while the secondary line voltages are $\sqrt{3}$ times the secondary phase voltages, that is,

$$\frac{V_{AB}}{V_{ab}} = \frac{V_{p\,(\text{primary})}}{\sqrt{3}V_{p\,(\text{secondary})}} = \frac{\alpha}{\sqrt{3}} \tag{13–71}$$

The delta-star connection can be operated with the neutral point of the star-connected secondary grounded to provide a four-wire dual-voltage output, for example, three-phase 208 V between lines and 120 V between lines and neutral. Phase shifts of $-30°$ and $+30°$ between primary and secondary voltages are obtained by the way the delta primary is connected. The delta-connected primary also ensures that the

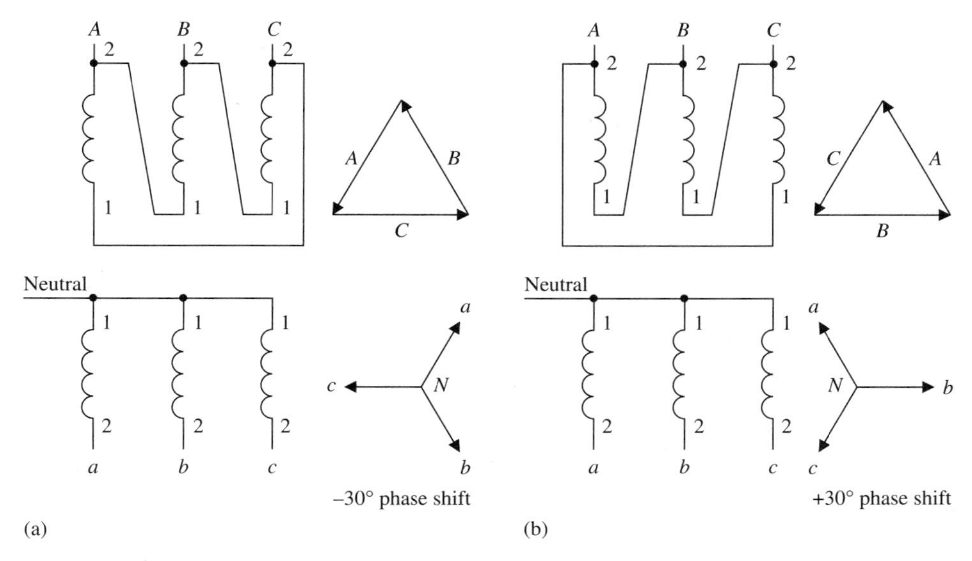

Figure 13-20 Delta-star or delta-wye connection. (a) Schematic –30° phase shift. (b) Schematic +30° phase shift.

secondary line voltages are free of third harmonics, as well as stabilizing the system when unbalanced loads are present at the secondary terminals.

Star-Delta Connection

The star-delta or wye-delta connection is shown in Fig. 13-21. The primary phase voltage is $1/\sqrt{3}$ of the line voltage, and the secondary phase and line voltages are equal.

The relationship between the primary and secondary line voltages is

$$\frac{V_{AB}}{V_{ab}} = \sqrt{3}\alpha \tag{13–72}$$

A major application of star-delta-connected transformers is to step down transmission-level voltages to distribution levels at the load end of the transmission line. There is a phase shift between the primary and secondary line voltages of either $-30°$ or $+30°$, depending on the way the delta-connected secondaries are hooked up, although it is common practice in the United States for the secondary line voltage to lag the primary line voltage. The presence of the delta winding prevents third harmonics from appearing in the line voltages, as well as being more stable under unbalanced load conditions, since the out-of-balance current circulates in the delta connection.

Before any of the four types of transformer connections described are made using three single-phase transformers, the following conditions must be met:

1. The transformation ratios must be the same.
2. The voltage and current ratings must be the same.
3. The equivalent resistances, reactances, and impedances must be the same.

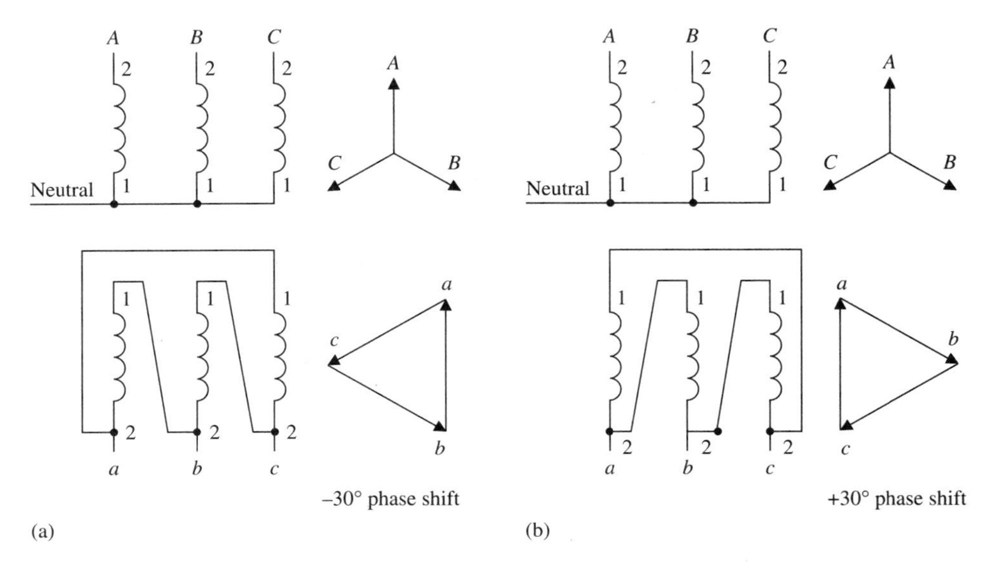

Figure 13-21 Star-delta or wye-delta connection. (a) Schematic –30° phase shift. (b) Schematic +30° phase shift.

4. The polarities must be the same.
5. Transformers must be of the same type of construction, that is, core or shell type.

Open-Delta or Vee-Vee Connection

The open-delta or vee-vee connection requires two identical transformers to provide three-phase transformation (Fig. 13-22).

The open-delta connection can be used to supply a three-phase load which is expected to increase over a period of years. As a result the initial cost is limited to the purchase of two identical single-phase transformers, and later, as the load increases, the third transformer can be added to form a delta-delta arrangement. Also, in the event of damage to one phase of a delta-delta transformer bank formed from three single-phase transformers, the defective unit can be removed for repair.

The total power-handling capability of an open-delta arrangement is 57.74% of the delta-delta-connected transformer bank. As can be seen from Fig. 13-22, each phase of the open-delta arrangement is carrying the line current, not the phase current, as would be the case with a delta-delta-connected transformer bank. Then

$$\frac{\text{Power carried by each transformer}}{\text{Total three-phase power}} = \frac{V_l I_l \cos\theta}{\sqrt{3} V_l I_l \cos\theta} = \frac{1}{\sqrt{3}} = 57.74\% \qquad \textbf{(13-73)}$$

where V_l and I_l are the line voltage and the line current, respectively.

From Eq. (13-73) it can be seen that the load on the transformers must be reduced. Otherwise the two transformers would be operating at 173.19% of their rated output. If originally only two equally rated transformers are operating in open

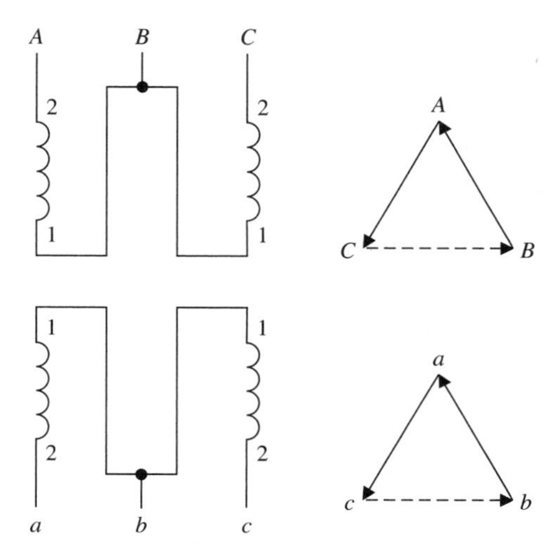

Figure 13-22 Open-delta or vee-vee connection.

delta, for the additional cost of a third transformer and reconnecting the transformer bank in delta-delta, the combination can now handle a load of 173.19% of the open-delta arrangement. The open-delta arrangement is frequently used with autotransformers. This connection arrangement is mainly used to provide reduced-voltage starting of polyphase induction motors.

13-12 PARALLEL OPERATION

Parallel operation of two or more transformers to supply a common load is particularly important in distribution systems when, because of load growth, a transformer has insufficient capacity to satisfy the load demand. To operate satisfactorily in parallel it is necessary that the transformers satisfy the following requirements:

1. The transformation ratios are the same.
2. The polarities are the same for single-phase transformers, and in the case of three-phase transformers, the phase shift between the primary and secondary voltages in the same phase must be the same.
3. The phase sequence is the same for three-phase transformers.
4. The percentage impedance and the ratio of the percentage resistance and percentage reactance are the same.

The requirement that the phase shift between primary and secondary voltages be the same immediately establishes combinations of transformers that can be paralleled. These are:

1. Star-star to star-star, because there is a 0° phase shift between primary and secondary voltages.
2. Star-delta to star-delta, because the 30° phase shift between primary and secondary voltages is common.
3. Delta-star to delta-star. The same as for 2.
4. Delta-delta to delta-delta, because there is a 0° phase shift between primary and secondary voltages.
5. Delta-delta to star-star, because there is a 0° phase shift between primary and secondary voltages, but the phase voltages will be different.
6. Star-star to delta-delta. The same as 5.
7. Star-delta to delta-star, because there is a 30° phase shift between primary and secondary voltages; however, the line voltages must be the same.

When these conditions are met, three-phase transformers can be successfully paralleled, with the load being shared between them in proportion to their kVA ratings, that is, provided all requirements are met, a 500-kVA and a 1,000-kVA transformer can be paralleled. It should be noted that only three-phase transformers with the same phase shift can be paralleled.

13-13 MULTIWINDING TRANSFORMERS

Multiwinding or multicircuit transformers are frequently used where three or more circuits operate at different voltage levels. This arrangement is cheaper than supply-

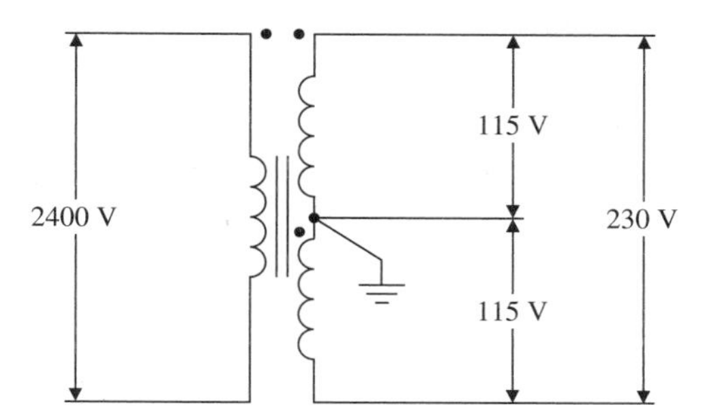

Figure 13-23 Distribution-type multicircuit transformer.

ing the same circuits from conventional two-winding transformers. A common example is the pole- or vault-type transformer operating at 4,160/240/120 V, illustrated in Fig. 13-23.

As has been mentioned previously, large three-phase power transformers interconnecting transmission systems that operate at different voltage levels frequently include a delta-connected tertiary winding to provide intermediate-level voltages to a distribution system. The tertiary windings can also be used with capacitor banks to improve the system power factor, or they can be used to provide a path for third-harmonic components of the exciting current.

13-14 PHASE-CHANGING CONNECTIONS

Frequently a requirement exists where it is necessary to change the number of phases, for example, three to two phases or vice versa, or from three phases to six, twelve, or even twenty-four phases, which are used to reduce the ripple content in the output of solid-state power converters. Because of its high efficiency and the absence of moving parts, the transformer is the ideal conversion device.

Scott Connection

The Scott connection provides a relatively simple method of converting a three-phase input to a two-phase output or vice versa. This system requires two single-phase transformers, one with a center-tapped primary and the other with a primary with an 86.6% tap. These transformers are called the *main* (M) and *teaser* (T) transformers, respectively. The secondaries of the main and teaser transformers are identical. The Scott connection is shown in Fig. 13-24 together with the associated phasor diagrams.

Two-phase power is required for the operation of two-phase servomotors. The Scott connection provides a simple method of obtaining the two-phase power. It is also possible to convert from two-phase to three-phase using the same connection. However, the three-phase voltages will not be mutually displaced by 120°, and two of the three phase voltages will be approximately 12% greater than the third.

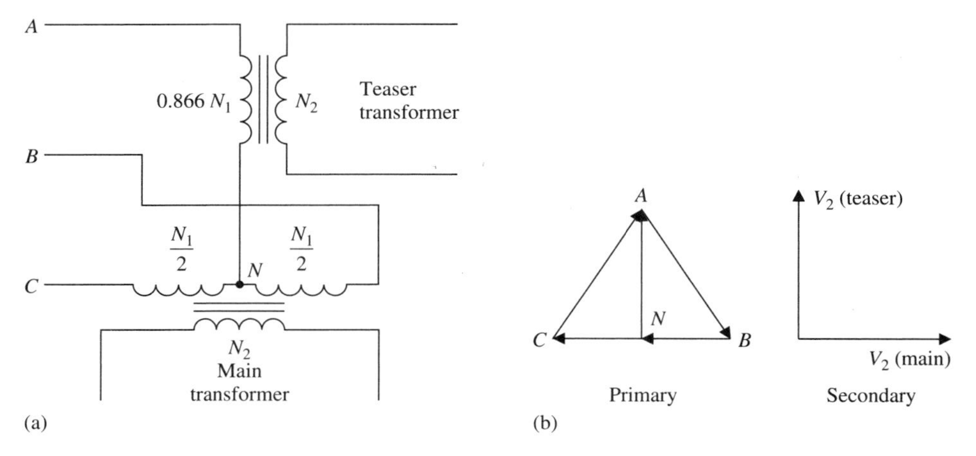

Figure 13-24 Scott connection. (a) Schematic. (b) Phasor diagram.

Three-Phase to Six-Phase Connections

The major use today of three-to-six-phase transformer connections is in conjunction with solid-state rectification and thyristor phase-controlled power converters. The main advantages for using six-phase and even twelve- and twenty-four-phase converters are:

1. The amplitude of the ac ripple component in the output dc voltage is reduced.
2. Lower order harmonics are eliminated.
3. A combination of 1 and 2 reduces the complexity of any filtering networks that may be required.
4. A higher mean output dc voltage is obtained as the number of phases increases.
5. Transformers are utilized more efficiently because the ratio of the dc output power to the transformer kVA rating increases as the number of phases increases.

There are three three-to-six-phase transformer connection arrangements in common use.

1. *Diametrical connection.* This connection arrangement is the one most commonly used to obtain three-to-six-phase conversion (Fig. 13-25). It consists of three identical transformers with center-tapped secondaries. The primary may be connected in star or delta. Delta is the preferred connection because of the third harmonics introduced by the star-star-connected secondary.

The six secondary leads a_1, c_4, b_1, a_4, c_1, and b_4 normally are connected to the anodes of the SCRs in a six-pulse midpoint converter, and the neutral serves as the return for the negative side of the dc load (see Fig. 2-15).

The distinctive feature of this connection is that it produces a true six-phase output, with six equal phase voltages mutually displaced by 60°, as can be seen from Fig. 13-25(b).

2. *Double-delta connection.* The double-delta connection [Fig. 13-26(a)] consists of three single-phase transformers, each with two identical but electrically separate secondary windings. The primary is normally connected in delta, which provides the opportunity, if necessary, for the primary to be operated in open delta.

As can be seen from Fig. 13-26(a), the two sets of secondary windings are each connected in delta, but one set of delta connections is reversed with respect to the other so that the voltages of each delta connection are 180° displaced with respect to each other. This results in the formation of two independent deltas. The six-phase output is obtained by connecting the load to terminals 1 through 6. It should be noted that a six-phase output is only obtained when the transformer secondary is connected to a load.

3. *Double-star connection.* The double-star connection shown in Fig. 13-27(a) is formed from three single-phase transformers, each with two separate secondary windings which are connected in star so that voltages a_2, b_2, c_2 and a_3, b_3, c_3 are 180° out of phase with respect to each other. If the two neutrals N_1 and N_2 are connected together, a true six-phase output is produced. However, if the neutrals are joined together and a six-phase load is connected to terminals a_2, c_3, b_2, a_3, c_2, and b_3, a six-phase output is obtained, but only when power is being supplied to the connected load.

The neutrals can also be connected by means of an interphase reactor whose center tap provides a path for the negative side of the dc load (see Fig. 2-15) when used with a six-pulse midpoint converter formed from two three-pulse midpoint converters.

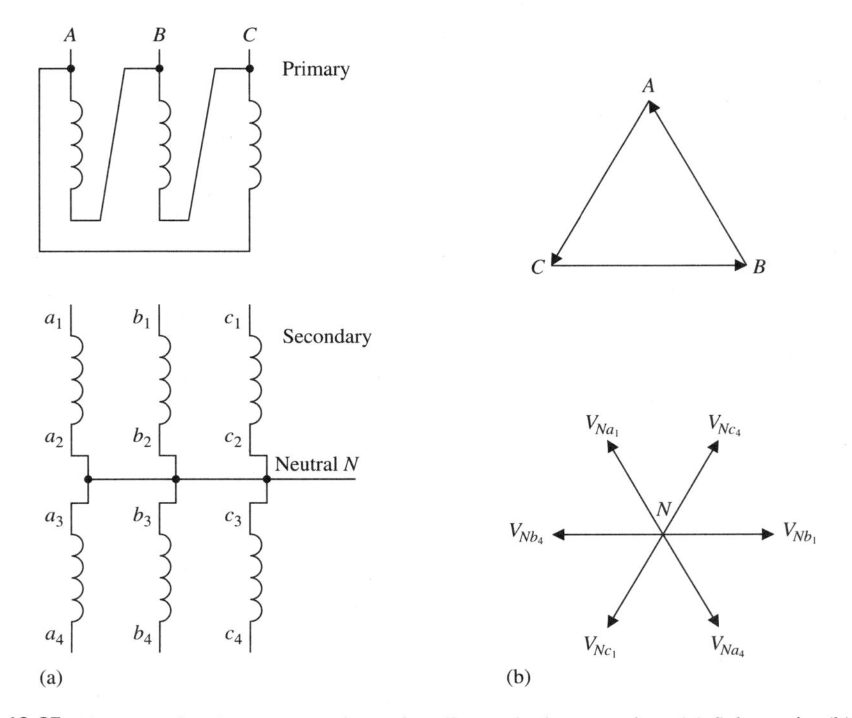

Figure 13-25 Three-to-six-phase conversion using diametrical connection. (a) Schematic. (b) Primary and secondary phasor diagrams.

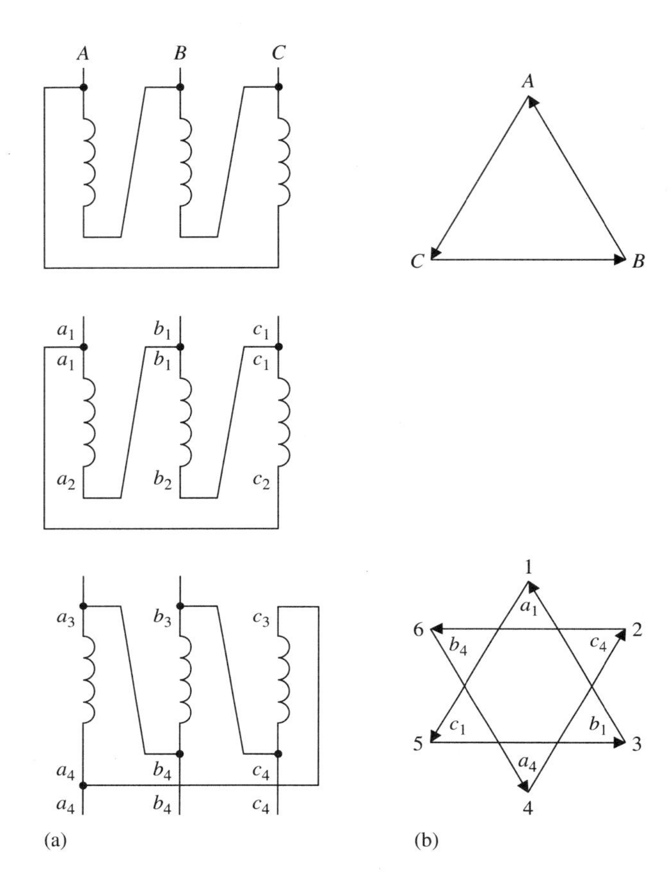

Figure 13-26 Three-to-six-phase conversion using double-delta connection. (a) Schematic. (b) Primary and secondary phasor diagrams.

13-15 INSTRUMENT TRANSFORMERS

There are two types of instrument transformers: potential transformer and current transformer. Their major function is to step down a voltage or a current to a value where the output can be used to operate instrumentation such as voltmeters, ammeters, or wattmeters, as well as protective relay equipment. A second, but extremely important, function is to provide electrical isolation between high-voltage circuits and measurement circuits.

Instrument transformers have a low VA rating, which is sufficient to supply the connected measuring device. As would be expected with instrumentation applications, precision is extremely important. Two sources of error are the transformation error and the phase-angle error.

Potential Transformers

The principle of a potential transformer is identical to that of a power transformer, except that they usually have ratings in the range of 100–500 VA. The low-voltage winding output is usually rated at 115 V. The load or *burden* consists of the potential coils of the connected metering or relay equipment.

The primary winding is connected in parallel across the circuit being measured and must be suitably insulated. Potential transformers are used to measure power-line voltages up to 765 kV. It is common practice when measuring potentials greater than 25 kV to use transformers designed for outdoor service. These transformers have one or more porcelain bushings. For voltage levels of 25 kV or less an outdoor or indoor type transformer may be used.

In order to obtain a constant primary-voltage to secondary-terminal-voltage ratio, potential transformers are designed for minimum resistance and leakage reactance, and the core is operated with a lower flux density than would be used in a power transformer. As a result the core is physically larger than would be the case if it was designed in the same manner as an equivalently rated power transformer. Since the potential transformer is designed for a specific burden, the transformation ratio can be accurately specified. Phase-angle error is only important when power measurements are being made.

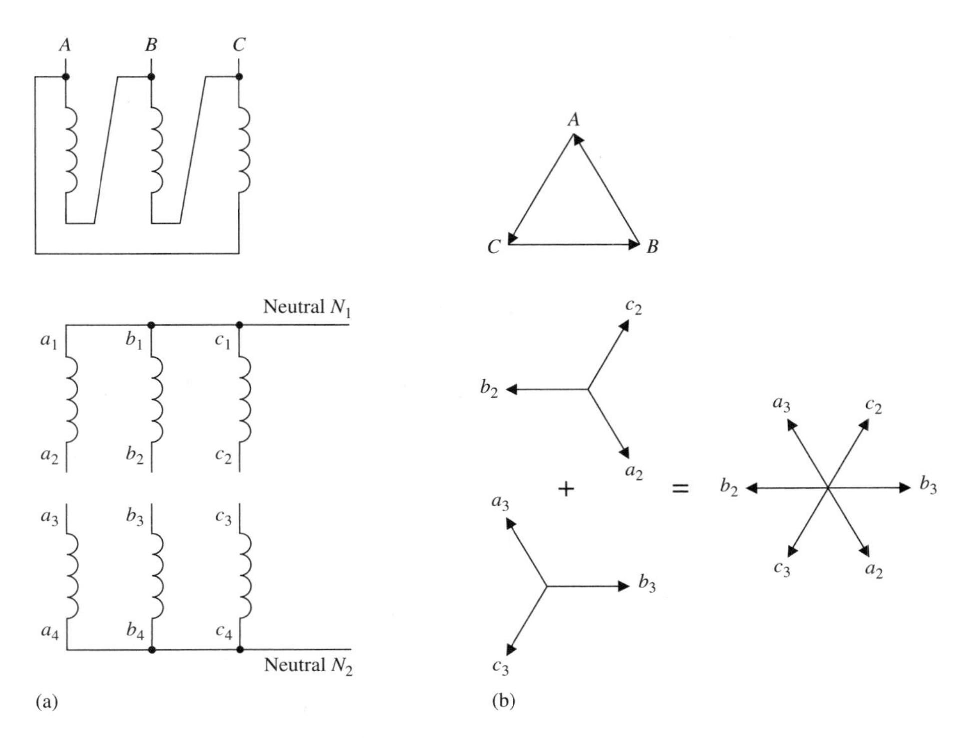

Figure 13-27 Three-to-six-phase conversion using double-star connection. (a) Schematic. (b) Primary and secondary phasor diagrams.

For very-high-voltage measurement, usually in excess of 100 kV, capacitor potential transformers are used. This type of transformer consists of a capacitive voltage divider network made up of a number of capacitors connected in series, with a normal potential transformer connected across a portion of the voltage divider.

Current Transformers

Current transformers have a primary winding of usually one or two turns connected in series with the circuit being measured. The secondary winding consists of many turns of insulated wire wound around a magnetic core surrounding and suitably insulated from the primary. The secondary of the current transformer is designed to supply 5 A when rated primary current is flowing. For example, if the maximum value of the primary current is 1,000 A, then the current ratio between primary and secondary is 1,000:5, or 200:1.

To achieve a constant current ratio, the magnetic core is made up of high-quality laminations and operated at relatively low flux densities. When primary current is flowing, it is producing the core flux. If a load is not connected across the secondary terminals, there will not be a secondary counter magnetomotive force acting in opposition to the primary magnetomotive force. In this condition the entire primary current is acting as the magnetizing current I_m, with the result that a very high core flux density is produced. In turn this flux linking with the multiturn secondary winding will induce an extremely high voltage, which will appear at the secondary terminals as well as a large impedance voltage drop across the primary. This high voltage may be sufficient to kill or severely injure any personnel coming in contact with the secondary terminals, or seriously damage the secondary winding insulation. This gives rise to the cardinal rule: always short-circuit the terminals of a current transformer if no load is connected to the terminals.

13-16 TRANSFORMER RATINGS

Transformer nameplates provide information on the important parameters which affect a transformer's usage as well as its performance. The major items are:

1. *Voltage rating*: This is the standard rated circuit voltage at which the transformer is designed to operate. These voltages are open-circuit values. The voltage rating defines (1) a safe value to prevent insulation breakdown, and (2) the core loss since an increase in voltage produces a nonlinear increase in core flux density, with a consequent increase in the magnetization current and core loss. A corollary is if the transformer is operated at a reduced frequency, for example, if a 60-Hz transformer is operated at 50 Hz, then the applied voltage must be reduced to 50/60 = 83% of its original value to maintain the same core flux.

2. *Kilovolt-ampere rating*: This rating combined with the voltage rating establishes permissible current levels in the transformer. This rating is important because if it is exceeded, excessive copper losses will occur, and the resulting temperature rise of the windings may drastically reduce the insulation life expectancy. If the transformer is operated at a reduced frequency, not only the voltage rating, but also the kVA rating must be reduced. Preferred ratings for distribution and power transformers have been established. The ratings for single-phase transformers have been set so that when the transformers are combined for three-phase operation, they will have the same kVA rating as the equivalent three-phase transformer.

3. *Basic impulse insulation level (BIL)*: Transformers are subjected at various time to momentary overvoltages that are caused by lightning surges, switching transients, and system faults. The transformer insulation must be able to handle three voltage levels: the normal operating voltage, the maximum fault condition voltage, and the impulse voltage. It is obviously impossible to design the insulation to meet each requirement. Instead a decision is based on the rated terminal voltage and the voltage level of the impulse test voltage, and then the BIL for the transformer is selected to be 15–20% greater than the anticipated overvoltage.

4. *Temperature rise*: The permissible temperature rise in °C above ambient is determined by the type of insulation materials used in the transformer construction, that is, class A, B, F, or H. Class H permits a 150°C temperature rise above a 40°C ambient, with a hot-spot temperature of 190°C. Class H insulation is commonly used in dry-type transformers, although they are operated with temperature rises not greater than 115°C to increase their operating life expectancy.

The nameplate also defines cooling medium, capacity, total weight, tap changer (if applicable), physical terminal arrangement, a complete schematic showing all internal connections, and impedance.

QUESTIONS

13-1 What is the significance of polarity markings?

13-2 What effect does the application, or sudden change, of secondary load have on the mutual flux of a transformer?

13-3 What is leakage flux? What are its effects on the performance of a transformer?

13-4 With the aid of phasor diagrams discuss the effects of varying power factor loads on the secondary terminal voltage of a transformer.

13-5 What is meant by a transformer equivalent circuit?

13-6 Why is the magnetization current of a transformer not sinusoidal?

13-7 What is meant by voltage regulation of a transformer? What are the effects of varying power factors on voltage regulation?

13-8 When performing the short-circuit test, why is the low-voltage side normally short-circuited?

13-9 When performing the open-circuit test, why is the high-voltage side normally open?

13-10 Why is it necessary to perform the open-circuit test at rated voltage and frequency?

13-11 What are the conditions for maximum efficiency of a transformer?

13-12 What is meant by all-day efficiency?

13-13 Why is the all-day efficiency calculated on an energy basis?

13-14 What is the function of a tap changer?

13-15 What is meant by off-load tap changing?

13-16 What is meant by on-load tap changing?

13-17 What conditions must exist before tap changing takes place?

13-18 Why is the efficiency dependent on the power factor?

13-19 What is meant by the term *phasing out a transformer*?

13-20 What is meant by polarity?

13-21 What is meant by additive and subtractive polarity?

13-22 With the aid of a schematic explain how the efficiency of two large identical transformers can be determined using the Sumpner test.

13-23 What is the purpose of a dielectric test?

13-24 What is the purpose of a high potential test?

13-25 What is the function of an induced voltage test?

13-26 With the aid of sketches show the four possible voltage combinations that can be obtained from a transformer with two separate and identical high-voltage and low-voltage windings.

13-27 Explain with the aid of a sketch the operation of an autotransformer.

13-28 Why, for the same transformation ratio, is an autotransformer lighter and more efficient than an equivalent two-winding transformer?

13-29 What are the advantages of using an autotransformer?

13-30 Why are autotransformers not used more than conventional transformers?

13-31 In what applications are autotransformers most commonly used?

13-32 Discuss the advantages and disadvantages of core- and shell-type transformers.

13-33 Discuss the advantages and disadvantages of using cylindrical and pancake transformer windings.

13-34 What is meant by magnetostriction?

13-35 What are the cooling media most commonly used in medium and large power transformers?

13-36 What are the advantages of using SF_6 as a cooling medium in a large power transformer?

13-37 Where is nitrogen used as a cooling medium?

13-38 Briefly discuss the various methods of cooling transformers.

13-39 What are the advantages of using three equally rated single-phase transformers instead of a single three-phase transformer?

13-40 What conditions must be met before single-phase transformers can be connected together in a three-phase combination?

13-41 With the aid of a schematic and phasor diagrams, show the relationship between line and phase voltages and line and phase currents on both sides of a star-star-connected transformer bank.

13-42 Repeat Question 13-41 for a star-delta transformer bank.

13-43 Repeat Question 13-41 for a delta-star transformer bank.

13-44 Repeat Question 13-41 for a delta-delta transformer bank.

13-45 What are the possible phase shifts between the primary and secondary line voltages of: **(a)** a star-star; **(b)** a star-delta; **(c)** a delta-star; **(d)** a delta-delta connected transformer bank?

13-46 Discuss typical applications where: **(a)** a star-star; **(b)** a star-delta; **(c)** a delta-star; **(d)** a delta-delta-connected transformer bank would be used.

13-47 What is the purpose of a tertiary winding?

13-48 What are the advantages of the vee-vee connection?

13-49 What are the advantages of a delta-star-connected transformer?

13-50 What conditions must be met before two three-phase transformer banks can be paralleled?

13-51 Discuss the seven possible three-phase transformer combinations that can be operated in parallel.

13-52 What three-phase transformer combinations cannot be paralleled? Why?

13-53 Why is it desirable to be able to change the number of phases when using solid-state converters?

13-54 With the aid of sketches, discuss how three-to-two-phase and two-to-three-phase conversion is possible using a Scott connection.

13-55 Where would two-phase power be used?

13-56 What are the advantages of being able to convert three-phase power to six- or twelve-phase power when using solid-state rectifiers or thyristor phase-controlled converters?

13-57 Explain with the aid of sketches and phasor diagrams the conversion of three-phase power to six-phase power by means of the diametrical connection.

13-58 Repeat Question 13-57 for the double-delta connection.

13-59 Repeat Question 13-57 for the double-star connection.

13-60 What are the functions of instrument transformers?

13-61 Discuss the characteristics of a potential transformer.

13-62 What type of potential transformer would be used to measure 765 kV? Describe the principle of operation.

13-63 What is the burden of a potential transformer?

13-64 A potential transformer must have a constant voltage ratio; how is this achieved?

13-65 A current transformer must have a constant current ratio, how is this achieved?

13-66 Explain exactly what happens if the secondary of an energized current transformer is opened.

13-67 What is the burden of a current transformer?

13-68 What is the significance of the voltage rating of a power transformer?

13-69 What is the effect of an increase in the primary terminal voltage of a power transformer?

13-70 What precautions must be taken if a 60-Hz transformer is connected to a 50-Hz system?

13-71 Why are transformers rated in kVA instead of kW?

13-72 What is meant by the BIL rating of a power transformer?

13-1 The voltage per turn of a 13,800/550-V 60-Hz transformer is 7.29 V. Calculate: **(a)** the number of primary turns; **(b)** the number of secondary turns; **(c)** the maximum value of the core flux.

13-2 A 2,400/240-V 60-Hz transformer has a primary with 1,500 turns. Calculate: **(a)** the number of secondary turns; **(b)** the volts per turn; **(c)** if the primary current is 20.83 A, what is the kVA rating of the transformer?

13-3 A single-phase 2,400/240-V 60-Hz transformer is tested with the secondary open-circuited. The following readings were obtained: $V_1 = 2400$ V, $I_\phi = 12.5$ A, $P_c = 2150$ W. Calculate: **(a)** the no-load power factor; **(b)** the no-load primary phase angle θ_c; **(c)** I_ϕ; **(d)** I_m.

13-4 A transformer primary has 350 turns and is supplied from a 230-V 60-Hz source. What is the maximum value of the core flux?

13-5 Calculate: **(a)** the primary current; **(b)** the secondary current of a 50-kVA 2,400/240-V 60-Hz transformer when it is operating at rated kVA.

13-6 The maximum core flux of a 60-Hz transformer with 1,521 primary turns and 53 secondary turns is 0.0326 Wb. Calculate: **(a)** the primary induced voltage; **(b)** the secondary induced voltage.

13-7 A single-phase 33,000/6,600-V 60-Hz transformer is designed to operate at a maximum flux density of 1.2 T. There are 1,500 primary turns. Calculate: **(a)** the number of secondary turns; **(b)** the core cross-sectional area.

13-8 A 10-Ω resistive load is connected across the 120-V terminals of a step-down transformer. If the primary current is 2.5 A, what is the primary voltage?

13-9 A 20-kVA 2,400/240-V 60-Hz transformer has an equivalent impedance of 12.5 Ω referred to the high-voltage side. If the high-voltage side is short-circuited, what voltage must be applied to the low-voltage terminals to obtain full-load current?

13-10 A 150-kVA 2,400/240-V 60-Hz transformer yielded the following test results:

	Primary	Secondary
Open circuit	Open	240 V, 17.25 A, 610 W
Short circuit	64.5 V, 62.5 A, 1,650 W	Short-circuited

Calculate: **(a)** R_{e1}; **(b)** Z_{e1}; **(c)** X_{e1}; **(d)** R_c; **(e)** X_m; **(f)** I_ϕ; **(g)** I_c; **(h)** I_m.

13-11 A 250-kVA 4,160/440-V 60-Hz transformer yielded the following results:

	Primary	Secondary
Open circuit	Open	440 V, 4.85 A, 550 W
Short circuit	102 V, 60 A, 650 W	Short-circuited

Calculate: **(a)** R_{e1}; **(b)** Z_{e1}; **(c)** X_{e1}; **(d)** R_c; **(e)** I_ϕ; **(f)** I_c; **(g)** I_m; **(h)** X_m; **(i)** % voltage regulation at 0.85 power factor lagging at full-load.

13-12 Using the data from Problem 13-11, calculate: **(a)** % voltage regulation at 0.8 power factor lagging at 75% of rated load; **(b)** the efficiency at 75% of rated output.

13-13 The total iron losses of a transformer operating at a constant flux density are 130 W at 60 Hz, and 110 W at 50 Hz. Calculate: **(a)** the hysteresis loss; **(b)** the eddy current loss at 60 Hz.

13-14 A 50-kVA 4,600/240-V 60-Hz single-phase transformer with the primary open takes 400 W when rated voltage is applied to the secondary. With the secondary short-circuited a voltage of 92 V applied to the primary causes full-load current to flow, and the copper loss is measured to be 950 W. Calculate the transformer efficiency at 0.75 power factor lagging at **(a)** 50% rated load; **(b)** 100% rated load; **(c)** 125% rated load.

13-15 From the data of Problem 13-10 calculate the load at which maximum efficiency occurs when supplying a load at 0.8 power factor lagging.

13-16 A 150-kVA 4,160/240-V 60-Hz distribution transformer has the following 24-h load cycle:

10% rated load at 0.9 PF lagging for 2 h
75% rated load at 0.85 PF lagging for 6 h
100% rated load at 0.88 PF lagging for 7 h
110% rated load at 0.95 PF lagging for 3 h
80% rated load at 0.86 PF lagging for 6 h

The core loss is 750 W, and the full-load copper loss is 1,950 W. Calculate the all-day efficiency.

13-17 A 2,400/2,300-V autotransformer supplies a 100-kVA load at 0.8 power factor lagging. Calculate: **(a)** the secondary current; **(b)** the primary current; **(c)** the kVA rating of the transformer.

13-18 A 10-kVA 2,400/240 V distribution transformer is used as an autotransformer to step up the voltage to 2,640 V. **(a)** Draw a connection diagram showing all connections. **(b)** If 10 kVA is transformed, what is the kVA supplied to the load?

13-19 A step-down autotransformer, 4,160/2,300 V, supplies a load of 40 kW at

0.85 power factor lagging. Calculate: **(a)** the conducted power; **(b)** the transformed power.

13-20 Determine the primary and secondary line voltages and currents, assuming a balanced load, of three single-phase 1,500-kVA 13,400/69,000-V transformers when they are connected: **(a)** delta-star; **(b)** star-delta; **(c)** star-star; **(d)** delta-delta.

13-21 Three single-phase transformers supply a three-phase balanced load of 118 kW at 0.85 power factor lagging at 550 V, from a 4,160-V three-phase source. Calculate: the required kVA and voltage ratings of each transformer when they are connected in **(a)** star-delta; **(b)** delta-star; **(c)** open-delta.

13-22 Three 50-kVA transformers are connected in delta-delta. Calculate: **(a)** the kVA load that can be supplied; **(b)** the load that can be supplied if one phase fails.

13-23 Determine the primary and secondary line currents and voltages for a balanced load applied to three 1,000-kVA single-phase transformers that are connected in **(a)** star-star; **(b)** delta-delta; **(c)** star-delta; **(d)** delta-star.

CHAPTER

14

Polyphase Induction Motors

14-1 INTRODUCTION

Initially electric power was produced by dc generators and distributed at 110–220 V dc. It was not until 1885–1886, when Nikola Tesla invented the induction motor, that the ac synchronous generator became available. Up to that time dc motors were the only means of converting electric energy to a mechanical power output. However, since Tesla obtained his patent, in 1888, the polyphase induction motor in sizes above 5 hp (3.73 kW) has rapidly become the preferred choice in industrial and commercial applications, with the squirrel-cage induction motor (SCIM) accounting for approximately 90% of the power consumed by three-phase motors. The squirrel-cage induction motor is now the major polyphase motor because of its low cost, high reliability, and high efficiency over a wide range of power outputs. When operated from a constant-frequency source, the squirrel-cage induction motor has a low speed regulation, that is, it operates at a relatively constant speed between no load and full load. If two-speed operation is required, a 2:1 speed reduction can be obtained by the consequent pole connection of the stator windings. If speeds other than 2:1 are required, the stator must be wound with two electrically isolated stator windings, bringing about an increase in initial cost and size and a reduction in overall efficiency. For continuous speed control over a wide speed range the stator is usually supplied from a solid-state variable-frequency inverter, which increases the initial cost of the drive. If continuous control at speeds less than synchronous speed is required, this can be obtained by using eddy current or magnetic particle clutches.

There are several penalties involved when using polyphase induction motors:

1. The inrush starting current is high. It can be as high as 8.5 times full-load rated current.

2. They operate at lagging power factors, which become very low when the motor is loaded lightly.
3. They require special techniques to obtain speed control.

14-2 CONSTRUCTION

The stators of polyphase synchronous motors—squirrel-cage and wound-rotor induction motors—are electrically identical. The stator coils, which in the smaller NEMA frame sizes are wound randomly from round insulated magnet wire and in the larger frame sizes are form-wound using rectangular insulated magnet wire, are inserted in slots in the stator iron, with ground insulation between the coils and the stator and phase insulation between the coils. Modern insulation materials are usually synthetic to provide a higher thermal capability as well as improved insulation characteristics. The stator coils are grouped together according to the number of phases and the number of poles per phase. The minimum number of poles in a three-phase stator is equal to three times the number of poles per phase.

There are two types of rotors used with polyphase induction motors: the squirrel-cage rotor and the wound rotor. The squirrel-cage rotor consists of a number of rotor conductors or bars inserted into axial slots around the periphery of the rotor laminations, with both ends of the rotor conductors short-circuited together. The rotor slots are often slightly skewed to reduce noise and magnetic pulsations caused by variations in the air-gap reluctance. The electric circuit of squirrel-cage rotors may be formed by pressure injection of molten aluminum into the rotor slots in a special mold. This process forms the rotor conductors, the short-circuiting end connections, and usually the cooling fan blades in one operation. Alternatively, in larger power output machines, copper bars of rectangular cross section are inserted into the rotor slots. After insertion the rotor bars are crimped along the exposed edge to prevent movement, and the ends of the rotor conductors are then brazed or welded to conducting end rings. By careful choice of the rotor conductor material and cross section the overall rotor resistance can be controlled accurately.

The wound rotor, as the name suggests, consists of a rotor with a three-phase star- or wye-connected winding wound into the axial rotor slots, with the three phase ends brought out to three insulated slip rings. An adjustable or tapped resistance is then connected to the slip rings via carbon brushes. The function of the external resistance is twofold: (1) to control the magnitude of the stator inrush currents during the starting cycle, and (2) to provide speed control. The external resistances are categorized by NEMA as starting, intermittent, or continuous duty. In addition, continuous-duty resistances are further classified either as fan duty, for use in variable-torque applications, where the rotor current is approximately proportional to the rotor speed, which is the case for fans, centrifugal pumps, and centrifugal compressors, or as machine duty, for use in constant-torque applications, where the rotor current is approximately constant over the speed range.

Both the stator and the rotor magnetic cores are formed from thin punched steel laminations. Depending on the machine size, these laminations are punched out either in one piece or in segments, complete with conductor slots and all required openings. The laminations are stacked and clamped together, with openings and slots matched. In larger machines radial ventilation passages are provided to permit the free passage of cooling air through the stator, and in the rotor both axial and radial cooling passages are provided.

Antifriction ball and sometimes roller bearings are used in integral-horsepower machines and, with the use of lithium-based grease, may be operated at elevated temperatures. In horizontally mounted motors of higher power output it is common practice to use sleeve bearings, especially at 1,800 or 3,600 r/min (188.50 or 376.99 rad/s). A properly installed sleeve bearing with clean oil will give many years of service. Vertically mounted motors may use ball or occasionally roller bearings, but more commonly a combined thrust journal antifriction bearing is used. In large power output vertically mounted motors, Kingsbury thrust bearings and babbitted guide bearings are used. The lubricating oil for sleeve and Kingsbury bearings is usually supplied under pressure and cooled by an oil-to-water heat exchanger. Figure 14-1 shows a cutaway view of a three-phase squirrel-cage induction motor.

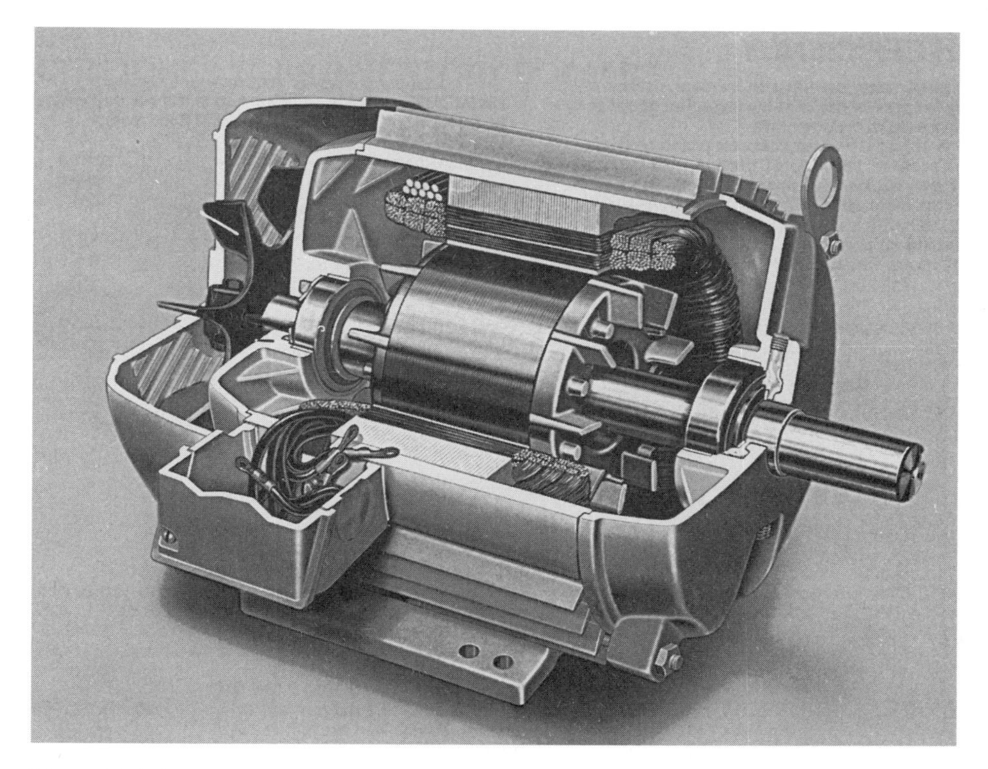

Figure 14-1 Cutaway view of three-phase squirrel-cage induction motor. (*Courtesy of U.S. Electrical Motors*)

14-3 PRINCIPLE OF OPERATION

The three-phase stator or armature windings are distributed in slots in the stator. When a three-phase source is connected to the stator windings, a three-phase rotating magnetic field is produced (see Section 10-8). This field is a constant-amplitude field, which for a two-pole stator winding completes one revolution per cycle of the applied terminal voltage. If the stator is wound for four poles per phase, the rotating magnetic field will turn through 180° per cycle, or one revolution every two cycles. From this we can conclude that the angular velocity of the rotating magnetic field is inversely proportional to the number of pole pairs formed by the stator windings. When a two-pole stator winding is connected to a 60-Hz supply, the magnetic field rotates through 60 complete revolutions in 1 s; if the source frequency is reduced to 30 Hz, the rotating magnetic field completes 30 revolutions in 1 s. As a result we find that the rotational speed of the magnetic field is directly proportional to the supply frequency. This reasoning can be consolidated into an expression for the synchronous speed of the rotating magnetic field as follows:

$$S_S = \frac{120f}{P} \text{ r/min} \qquad \textbf{(14–1)(E)}$$

or

$$\omega_S = \frac{4\pi f}{P} \text{ rad/s} \qquad \textbf{(14–1)(SI)}$$

where S_S and ω_S are the synchronous speeds in r/min and rad/s, respectively, and P is the number of stator poles per phase.

The operation of the polyphase induction motor depends on Faraday's law of electromagnetic induction and the Lorentz force acting on a conductor (see Sections 1–5 and 1–9). The development of an induced torque can be best understood by considering the effect of a moving magnetic field cutting across the rungs of an aluminum ladder [Fig. 14-2(a)], the rungs being short-circuited by sides *A* and *B* of the ladder. When the permanent magnet with flux density B Wb/m^2 is moved to the right at velocity v m/s over the stationary ladder rungs, it induces a voltage in the rungs.

The voltage induced in each rung is

$$e = Blv \text{ V} \qquad \textbf{(14–2)(SI)}$$

or

$$e = Blv \times 10^{-8} \qquad \textbf{(14–2)(E)}$$

where v is the linear velocity in m/s, of the rungs with respect to the magnetic field B in Wb/m^2 and l is the length of the conducting rung, in meters.

This induced voltage (Faraday's law) produces a current I in the conducting rung, which flows through the rung, splits, flows back through the short-circuits formed by the ladder sides, and returns through the adjacent rungs. The magnetic fields created by the currents flowing through the conducting rungs react with the permanent-

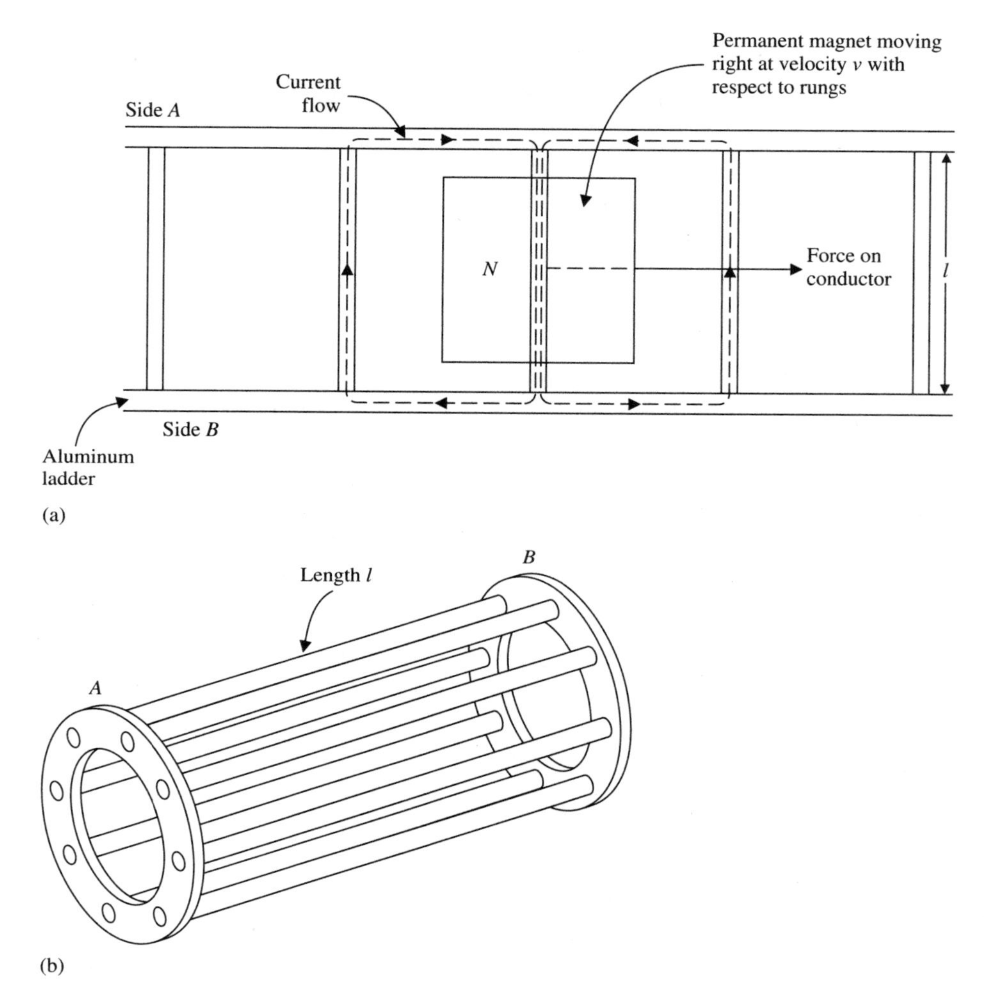

Figure 14-2 Concept of torque production by induction motor. (a) Permanent magnet across rungs of an aluminum ladder. (b) Ladder formed into squirrel-cage rotor.

magnet field. The reaction of these fields creates a mechanical force (Lorentz force), which attempts to move the ladder to the right, that is, the ladder attempts to move in the same direction as the permanent-magnet field. However, as the ladder accelerates, the induced voltage and current, and thus the accelerating force, will decrease since the relative rate of cutting the lines of force is decreasing, until at the point where the velocities of the permanent magnet and the ladder are the same, the induced voltage, current, and accelerating force are zero.

From this concept it is relatively easy to transfer the analogy to an actual squirrel-cage rotor. If the aluminum ladder is bent into a circle and the ends of the two sides A and B are joined so that they form the short-circuiting end rings, we have now formed the electric circuit of a squirrel-cage rotor, as illustrated in Fig. 14-2(b).

In the practical polyphase squirrel-cage induction motor the rotating magnetic field produced by the polyphase stator windings sweeps past the rotor conductors, and in turn these rotor currents develop the same number of rotor poles as there are stator poles. The rotor fluxes react with the stator pole fluxes, thus producing a torque which causes the rotor to turn in the same direction as the rotating magnetic field. As long as the rotor speed is less or greater than the synchronous speed of the rotating magnetic field, there is relative motion between rotor and stator fields and a voltage will be induced in the rotor conductors. If the rotor turns at synchronous speed, there is no motion of the rotating magnetic field relative to the rotor, and no voltages will be induced in the rotor conductors. As a result the induction motor must always operate at less than synchronous speed when driving a load. For this reason the induction motor is often called an *asynchronous* motor. It can also operate as an *induction generator* if it is driven above synchronous speed. It should be noted that power is transferred to the rotor from the stator by induction and not by conduction, and in fact the polyphase induction motor can be thought of as a rotating transformer.

14-4 SLIP, ROTOR SPEED, FREQUENCY, AND VOLTAGE

Since the voltages induced in the rotor circuit are dependent on the rotor speed relative to the synchronous speed of the rotating field, and in turn the performance of the polyphase induction motor is dependent on the induced rotor voltages and currents, it is logical to talk in terms of the relative speed. Since the rotor speed must always be less than the synchronous speed of the rotating field, the difference in the two speeds, that is, the speed at which the rotor slips with respect to the rotating field, is called the *slip* and is expressed as a decimal fraction or percentage of the synchronous speed,

$$s = \frac{S_S - S_R}{S_S} \times 100\% \qquad \textbf{(14–3)(E)}$$

or

$$s = \frac{\omega_S - \omega_R}{\omega_S} \times 100\% \qquad \textbf{(14–3)(SI)}$$

where S_S and ω_S are the synchronous speeds in r/min and rad/s, respectively, and S_R and ω_R are the rotor speeds in r/min and rad/s, respectively.

It should be noted that at standstill $s = 1$, and if the rotor is turning at synchronous speed, $s = 0$. The induction motor is acting entirely as a transformer when $s = 1$.

The actual rotor speed can be expressed in terms of slip and synchronous speed by solving Eqs. (14-3) to yield

$$S_R = (1 - s)S_S \qquad \textbf{(14–4)(E)}$$

or

$$\omega_R = (1 - s)\omega_S \tag{(14–4)(SI)}$$

➤ EXAMPLE 14-1

A 230-V 10-hp (7.46-kW) six-pole 60-Hz star-connected induction motor has a full-load slip of 4.5%. Calculate: **(a)** the synchronous speed in r/min and rad/s; **(b)** the rotor speed in r/min and rad/s.

SOLUTION

(a) The synchronous speed of the motor is

$$S_S = \frac{120f}{P} = \frac{120 \times 60\ \text{Hz}}{6\ \text{poles}} = 1,200\ \text{r/min}$$

or in SI units,

$$\omega_S = \frac{4\pi f}{P} = \frac{4\pi \times 60\ \text{Hz}}{6\ \text{poles}} = 125.66\ \text{rad/s}$$

(b) The rotor speed at rated load is

$$\begin{aligned} S_R &= (1 - s)S_S \\ &= (1 - 0.045) \times 1,200\ \text{r/min} = 0.955 \times 1,200\ \text{r/min} \\ &= 1,146\ \text{r/min} \end{aligned}$$

or in SI units,

$$\begin{aligned} \omega_R &= (1 - s)\omega_S \\ &= 0.955 \times 125.66\ \text{rad/s} = 120.01\ \text{rad/s} \end{aligned}$$

Since the induction motor is functionally a rotating transformer, at standstill $s = 1$. The frequency of the induced rotor voltages and currents is equal to the source frequency. For example, with a 60-Hz supply the frequency of the rotor voltages and currents is 60-Hz. As the rotor starts turning, the relative rate of flux linkages between the rotating field and the rotor conductors decreases until, when $s = 0$, the frequency of the rotor voltages and currents is 0 Hz. The rotor frequency at any slip is

$$f_R = sf \tag{(14–5)}$$

This relationship can be expressed in other forms, such as

$$f_R = \frac{S_S - S_R}{S_S} f \tag{(14–6)}$$

However, since $S_S = 120f/P$ [Eq. (14-1)(E)], then

$$f_R = (S_S - S_R)\frac{P}{120f}f$$

$$= \frac{P}{120}(S_S - S_R)\ \text{Hz} \qquad \textbf{(14–7)(E)}$$

or in SI units,

$$f_R = (\omega_S - \omega_R)\frac{P}{4\pi f}f$$

$$= \frac{P}{4\pi}(\omega_S - \omega_R)\ \text{Hz} \qquad \textbf{(14–7)(SI)}$$

Similarly when $s = 1$, that is, at standstill, the induced rotor electromotive force E_R is at a maximum since the rotating magnetic field is cutting the rotor conductors at its maximum rate. The induced rotor voltage is also dependent on the number of stator turns, the applied stator voltage, and the number of rotor turns. As the rotating field sweeps past the stator turns, an electromotive force is induced in the stator winding, which is nearly equal and opposite in phase to the applied stator voltage. At the same time the field is sweeping past the rotor windings and induces a rotor electromotive force, which is in phase with the induced stator electromotive force. The stator and rotor induced electromotive forces are related by the turns ratio on a per-phase basis between the stator and rotor turns. As the rotor accelerates, the rate at which the rotating field is linked by the rotor conductors decreases. This relationship can be expressed in terms of the standstill or *blocked-rotor* voltage E_{BR} as

$$E_R = sE_{BR} \qquad \textbf{(14–8)}$$

where E_R is the induced rotor voltage at slip s and E_{BR} the blocked-rotor voltage.

➤ EXAMPLE 14-2

A 208-V six-pole 60-Hz wound-rotor induction motor has a star-connected stator and rotor winding. The rotor winding contains 45% as many turns as the stator winding. If the rotor speed is 1,150 r/min (120.43 rad/s), calculate: **(a)** the slip; **(b)** the blocked-rotor voltage per phase; **(c)** the induced rotor voltage per phase; **(d)** the rotor voltage between slip rings; **(e)** the frequency of the rotor currents.

SOLUTION

(a) The synchronous speed is

$$S_S = \frac{120f}{P} = \frac{120 \times 60\ \text{Hz}}{6\ \text{poles}} = 1,200\ \text{r/min}$$

or in SI units,

$$\omega_S = \frac{4\pi f}{P} = \frac{4\pi \times 60\ \text{Hz}}{6\ \text{poles}} = 125.66\ \text{rad/s}$$

Therefore

$$s = \frac{S_S - S_R}{S_S} \times 100\%$$
$$= \frac{1,200 - 1,150}{1,200} \times 100\%$$
$$= 4.17\% = 0.0417 \text{ (decimal slip)}$$

or in SI units,

$$s = \frac{\omega_S - \omega_R}{\omega_S} \times 100\%$$
$$= \frac{125.66 - 120.43}{125.66} \times 100\%$$
$$= 4.17\% = 0.0417 \text{(decimal slip)}$$

(b) The blocked-rotor voltage per phase is

$$E_{BR} = 45\% \times \frac{208 \text{ V}}{\sqrt{3}} = 54.04 \text{ V/phase}$$

(c) The induced rotor voltage per phase is

$$E_R = sE_{BR} = 0.0417 \times 54.04 = 2.25 \text{ V/phase}$$

(d) Since the rotor is star connected, the rotor voltage between the slip rings is

$$V_{\text{slip ring}} = \sqrt{3}E_R$$
$$= \sqrt{3} \times 2.25 \text{ V/phase} = 3.90 \text{ V}$$

(e) The frequency of the rotor electromotive force at 1,150 r/min (120.43 rad/s) is

$$f_R = sf = 0.0417 \times 60 \text{ Hz} = 2.5 \text{ Hz}$$

14-5 EQUIVALENT CIRCUITS

Since the polyphase induction motor can be thought of as a rotating transformer, the same rationale as used on developing transformer equivalent circuits can be applied to the polyphase induction motor.

Rotor Equivalent Circuit

The polyphase induction motor is equivalent to a polyphase transformer operating with variable-frequency voltages and currents in a short-circuited secondary winding. As a result it is possible to develop a per-phase equivalent circuit representing the rotor. The rotor circuit has both resistance and leakage reactance. However, the leakage reactance will be a maximum when the rotor is stationary since it is dependent on the frequency of the rotor currents. Therefore the rotor reactance is

$$X_R = \omega_R L_R = 2\pi f L_R$$

From Eq. (14-5),

$$f_R = sf$$

Therefore

$$\begin{aligned} X_R &= 2\pi s f L_R \\ &= sX_{BR} \end{aligned} \tag{14–9}$$

where X_{BR} is the blocked-rotor reactance and L_R the rotor inductance. The rotor equivalent circuit is shown in Fig. 14-3.

From Fig. 14-3 the rotor current is

$$I_R = \frac{sE_{BR}}{\sqrt{R_R^2 + (sX_{BR})^2}} \tag{14–10}$$

or in complex form,

$$I_R = \frac{sE_{BR}}{R_R + jsX_{BR}} \tag{14–11}$$

Dividing both the numerator and the denominator of Eqs. (14-10) and (14-11) by the slip s yields

$$I_R = \frac{E_{BR}}{\sqrt{(R_R / s)^2 + X_{BR}}} \tag{14–12}$$

or in complex form,

$$I_R = \frac{E_{BR}}{R_R / s + jX_{BR}} \tag{14–13}$$

The term R_R/s can be split into two components,

$$\begin{aligned} \frac{R_R}{s} &= \frac{R_R}{s} + R_R - R_R \\ &= R_R + R_R\left(\frac{1}{s} - 1\right) = R_R + R_R\left(\frac{1-s}{s}\right) \end{aligned} \tag{14–14}$$

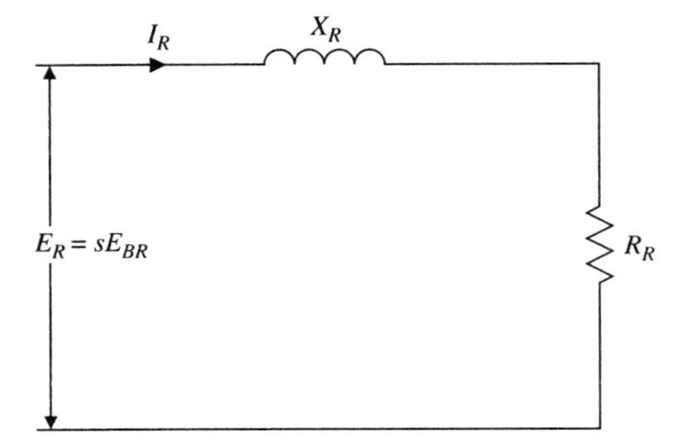

Figure 14-3 Rotor equivalent circuit of polyphase induction motor on a per-phase basis.

Equations (14-12) and (14-13) define the rotor current in terms of the blocked-rotor voltage E_{BR}. Equation (14-14) shows that the equivalent rotor resistance R_R/s consists of two components, the rotor resistance R_R and a dynamic component $R_R[(1 - s)/s]$. The latter represents the mechanical load being driven by the induction motor. Figure 14-4 represents the rotor circuit when modified to represent Eqs. (14-12), (14-13), and (14-14).

Stator Equivalent Circuit

On a per-phase basis the stator equivalent circuit is identical to that of a transformer, as illustrated in Fig. 14-5. Figure 14-5(a) shows the exact equivalent circuit, where R_S and X_S represent the stator resistance and the leakage reactance on a per-phase basis, and the shunt branch of R_c in parallel with X_m represents the iron losses and magnetizing reactance. The core loss represented by R_c is constant when operating at constant frequency, and can be determined from a no-load test and included as a constant loss in efficiency calculations. The magnetizing current I_m on the other hand is approximately 30–50% of the rated current. It is this large when compared to that of a transformer because of the air gap between the stator and rotor. As a result the magnetizing reactance is very significant. However, only a small inaccuracy is introduced by moving it to be in parallel across the input terminals.

Complete Equivalent Circuit

To simplify the analysis of the induction motor, the stator and rotor equivalent circuits should be combined to produce the complete equivalent circuit on a per-phase basis. This requires that the rotor equivalent circuit be referred to the stator, which is accomplished in exactly the same way as in the case of a transformer. Figures 14-4(b) and 14-5(b) can be combined by referring the rotor parameters to the stator, provided E_{BR} is equal to E_S. If the turns ratio between stator and rotor windings is α, then

$$E_S = E'_{BR} = \alpha\, E_{BR} \qquad (14\text{–}15)$$

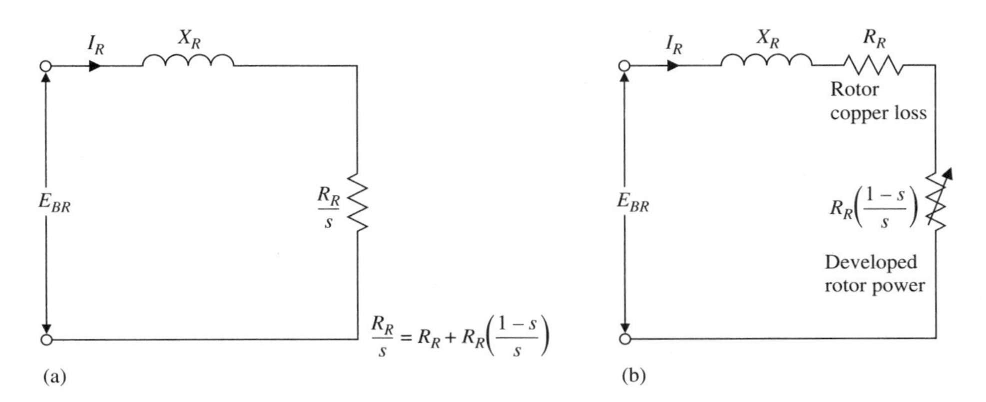

Figure 14-4 Equivalent rotor circuit on a per-phase basis with rotor copper loss and developed power. (a) Combined. (b) Separated into component parts.

where E_S is the induced stator voltage and E'_{BR} the blocked-rotor voltage referred to the stator. The referred rotor current I'_R is

$$I'_R = \frac{I_R}{\alpha} \tag{14–16}$$

the referred rotor resistance R'_R is

$$R'_R = \alpha^2 R_R \tag{14–17}$$

and the referred rotor reactance X'_R is

$$X'_R = \alpha^2 X_R \tag{14–18}$$

The complete equivalent circuit is shown in Fig. 14-6.

Let us take a few moments to study some of the implications that can be observed from the equivalent circuit diagram. When the induction motor is operating at no load or lightly loaded, the slip s is very small. Therefore $R'_R(1 - s)/s$ will be very large. Then I'_R will be small when compared to the magnetizing current I_m. As a

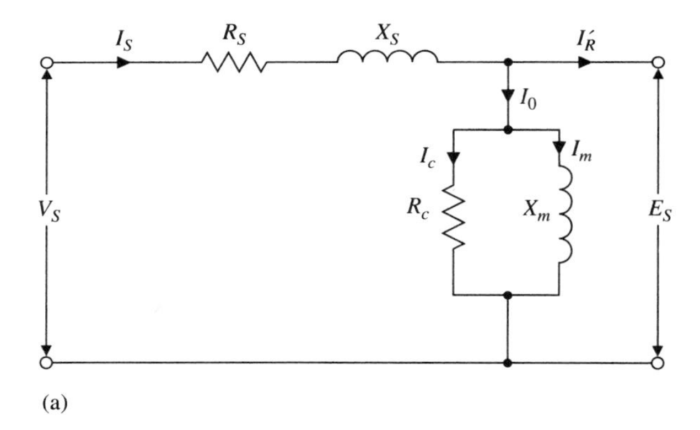

(a)

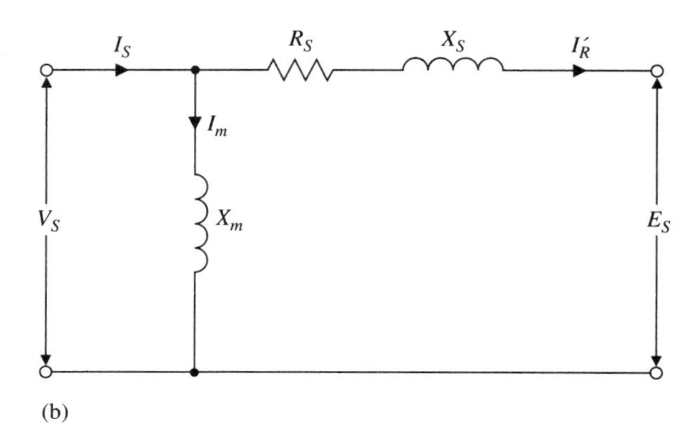

(b)

Figure 14-5 Equivalent stator circuit on a per-phase basis. (a) Exact. (b) Approximate.

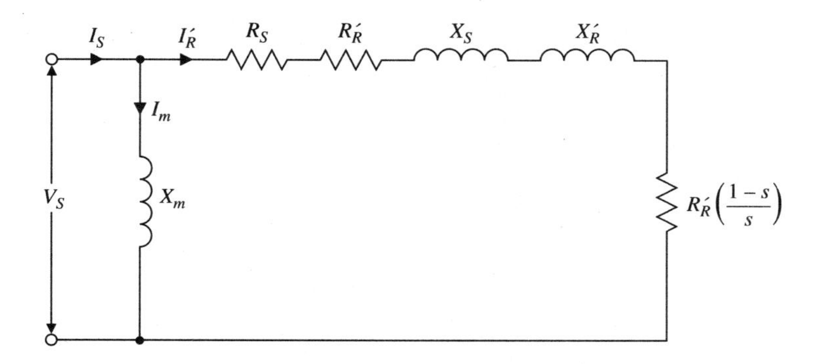

Figure 14-6 Approximate equivalent circuit of induction motor on a per-phase basis.

result, when viewed from the input terminals, the circuit is predominantly inductive because of the magnetizing reactance X_m. The end result is that under lightly loaded conditions the motor operates with a strongly lagging power factor. As the mechanical loading on the induction motor increases, s increases, and the $R'_R(1 - s)/s$ term decreases; thus the circuit as seen from the input terminals has become more resistive and the power factor has become less lagging.

14-6 POWER BALANCE EQUATIONS

The analysis of the induction motor can now be developed from the approximate equivalent circuit diagram for an m-phase machine. The input power is

$$P_{in} = mV_S I_S \cos \phi_S \text{ W} \tag{14-19}$$

where ϕ_S is the phase angle between V_S and I_S. The stator copper loss SCL is

$$\text{SCL} = mI_S^2 R_S \text{ W} \tag{14-20}$$

The power transferred across the air gap to the rotor, namely, the rotor power input RPI, is

$$\text{RPI} = mI'^2_R \frac{R'_R}{s} \text{ W} \tag{14-21}$$

The rotor copper loss RCL is then

$$\text{RCL} = mI'^2_R R'_R \text{ W} \tag{14-22}$$

the gross developed output power P_g, is obtained by subtracting the rotor copper loss RCL from the rotor power input RPI, that is, Eq. (14-21) − Eq. (14-22), to yield

$$\text{RPD} = P_g = mI'^2_R R'_R \left(\frac{1-s}{s}\right) = \text{RPI}(1-s) \text{ W} \tag{14-23}$$

The actual developed output power P_d is the gross developed power P_g minus the fixed losses, that is, friction, windage, and core losses P_{fwc},

$$P_d = P_g - P_{fwc} \text{ W} \tag{14–24}$$

Now

$$P_g = T_g\omega_R = mI_R'^2 R_R' \left(\frac{1-s}{s}\right) \text{W}$$

But from Eq. (14-4)(E) and (SI),

$$S_R = (1-s)S_S \qquad \textbf{(14–4)(E)}$$

or

$$\omega_R = (1-s)\omega_S \qquad \textbf{(14–4)(SI)}$$

Therefore

$$T_g\omega_R = mI_R'^2 R_R' \left(\frac{1-s}{s}\right)$$

$$T_g(1-s)\omega_S = mI_R'^2 R_R' \left(\frac{1-s}{s}\right)$$

and hence

$$T_g = \frac{mI_R'^2 R_R'(1-s)/s}{(1-s)\omega_S}$$

$$= \frac{mI_R'^2 R_R'/s}{\omega_S} \text{ N·m} \qquad \textbf{(14–25)(SI)}$$

or

$$T_g \frac{2\pi S_R}{60} = mI_R'^2 R_R' \left(\frac{1-s}{s}\right)$$

$$T_g \frac{2\pi(1-s)S_S}{60} = mI_R'^2 R_R' \left(\frac{1-s}{s}\right)$$

and hence

$$T_g = \frac{mI_R'^2 R_R'(1-s)/s}{2\pi(1-s)S_S/60}$$

$$= \frac{mI_R'^2 R_R'/s}{2\pi S_S/60} \text{ lb · ft} \qquad \textbf{(14–25)(E)}$$

Now $2\pi S_S/60$ or ω_S is constant for a specific induction motor supplied from a constant-frequency source. It follows that $(2\pi S_S/60)T_g$, which is the gross power transferred across the air gap, must be a measure of the gross torque T_g. Therefore

$$T_g = mI_R'^2 R_R' \left(\frac{1-s}{s}\right) \text{W} \tag{14–26}$$

The developed torque is

$$T_d = \frac{\text{RPD} - P_{fwc}}{\omega_R} \text{ N · m} \tag{14–27}$$

From Fig. 14-6 the rotor current referred to the stator is

$$I_R' = \frac{V_S}{(R_S + R_R'/s) + j(X_S + X_R')} \text{ A} \tag{14–28}$$

Also from Fig. 14-6 the magnetizing current I_m is

$$I_m = \frac{V_S}{jX_m} \text{ A} \tag{14–29}$$

and the line current drawn from the source I_S is the phasor sum of

$$I_S = I_m + I_R' \tag{14–30}$$

The efficiency of the motor is

$$\eta = \frac{P_d}{P_{\text{in}}} \tag{14–31}$$

➤ EXAMPLE 14-3

A three-phase 208-V 60-Hz four-pole star-connected induction motor has the following circuit parameters on a per-phase basis referred to the stator: $R_S = 0.15\ \Omega$; $X_S = 0.38\ \Omega$; $R_R' = 0.17\ \Omega$; $X_R' = 0.38\ \Omega$; $X_m = 16\ \Omega$.

The friction, windage, and core losses are assumed constant at 275 W. For a slip of 3.2%, calculate: **(a)** the line current; **(b)** the power factor; **(c)** the output hp and kW; **(d)** the developed torque; **(e)** the efficiency.

SOLUTION

(a) The phase voltage V_S is

$$V_S = \frac{208\text{ V}}{\sqrt{3}} = 120.09\text{ V}$$

Therefore

$$I_R' = \frac{120.09\angle 0°\text{ V}}{(0.15 + 0.17/0.032) + j(0.38 + 0.38)}$$

$$= \frac{120.09\angle 0°\text{ V}}{5.46 + j0.76} = \frac{120.09\angle 0°\text{ V}}{5.513\angle 7.92°}$$

$$= 21.78\angle -7.92°\text{ A} = 21.57 - j3.00\text{ A}$$

$$I_m = \frac{120.09\angle 0°\text{ V}}{j16} = -7.51\angle 0°\text{ V A} = 0 - j7.51\text{ A}$$

Therefore I_S is the phasor sum of $I_m + I_R'$,

$$\begin{aligned} I_S &= (0 - j7.51) + (21.57 - j3.00) \\ &= 21.57 - j10.51 = 23.99 \angle -25.98^\circ \text{ A} \end{aligned}$$

(b) The power factor is $\cos(-25.98^\circ) = 0.90$ lagging.

(c) The synchronous speed is

$$S_S = \frac{120f}{P} = \frac{120 \times 60 \text{ Hz}}{4 \text{ poles}} = 1,800 \text{ r/min}$$

or in SI units,

$$\omega_S = \frac{4\pi f}{P} = \frac{4\pi \times 60 \text{ Hz}}{4 \text{ poles}} = 188.50 \text{ rad/s}$$

Therefore the rotor speed is

$$S_R = (1 - 0.032) \times 1,800 = 1,742 \text{ r/min}$$

or in SI units,

$$\omega_R = (1 - 0.032) \times 188.50 = 182.47 \text{ rad/s}$$

The rotor power input is

$$\text{RPI} = \frac{mI_R'^2 R_R'}{s} = \frac{3 \times 21.78^2 \times 0.17}{0.032} = 7,560.25 \text{ W}$$

The rotor power developed is, from Eq. (14-23),

$$\begin{aligned} \text{RPD} = P_g &= mI_R'^2 R_R' \left(\frac{1-s}{s}\right) = \text{RPI}(1-s) \\ &= 7,560.25(1 - 0.032) = 7,318.22 \text{ W} \end{aligned}$$

The output power developed P_d is, from Eq. (14-24),

$$P_d = P_g - P_{fwc} = 7,318.22 - 275 = 7,043.32 \text{ W} = 7.04 \text{ kW}$$

or in E units,

$$P_d = \frac{7,043.32 \text{ W}}{746 \text{ W/hp}} = 9.44 \text{ hp}$$

(d) Since $T_d = P_d/\omega_R$, then

$$T_d = \frac{7,034.32 \text{ W}}{182.47 \text{ rad/s}} = 38.60 \text{ N} \cdot \text{m}$$

or in E units,

$$T_d = 38.60 \text{ N} \cdot \text{m} \times 0.738 \text{ lb} \cdot \text{ft} = 28.49 \text{ lb} \cdot \text{ft}$$

(e) The total losses are

$$\text{Friction, windage, and core loss} = 275 \text{ W}$$
$$\text{Rotor copper loss RCL} = s\text{RPI} = 241.93 \text{ W}$$
$$\text{Stator copper loss} = 3 \times 23.99^2 \times 0.15 = 258.89 \text{ W}$$
$$\text{Total} = 775.91 \text{ W}$$

$$\eta = \frac{\text{RPI}}{\text{RPI} + \sum \text{losses}} = \frac{7{,}560.25 \text{ W}}{7{,}560.25 \text{ W} + 775.91 \text{ W}} \times 100\% = 90.69\%$$

Figure 14-6 represents the approximate circuit where the magnetizing reactance X_m has been moved so that it is shunted across the input terminals. This arrangement introduces a small error into the efficiency calculation. However, if the efficiency is calculated from $\eta = P_d/P_{in}$, then

$$\begin{aligned} P_{in} &= \sqrt{3} I_S V_1 \cos\theta \\ &= \sqrt{3} \times 23.99 \times 208 \times \cos -25.98° \\ &= 7{,}769.41 \text{ W} \end{aligned}$$

Therefore

$$\eta = \frac{P_d}{P_{in}} = \frac{7{,}043.32 \text{ W}}{7{,}769.41 \text{ W}} \times 100\% = 90.65\%$$

A comparison of the efficiencies obtained by the two methods shows that only a small error has been introduced.

14-7 TORQUE-SPEED CURVE

A typical induction motor torque-speed curve is shown in Fig. 14-7. A study of this figure shows a number of interesting points. For example, there are three different torques: the starting torque; the breakdown torque, and the full-load or rated torque. Also there are two speeds shown: the no-load speed and the full-load speed.

The starting torque or blocked-rotor torque is the torque developed at standstill ($s = 1$). The rotor current at standstill can be calculated by substituting $s = 1$ in Eq. (14-28) to yield

$$I'_{Rst} = \frac{V_S}{(R_S + R'_R) + j(X_S + X'_R)} \tag{14–32}$$

where I'_{Rst} is the blocked-rotor or starting current.

The rotor power input at standstill is $\text{RPI}_{st} = 3 I'^{2}_{Rst} R'_R$, and from this the starting torque can be calculated.

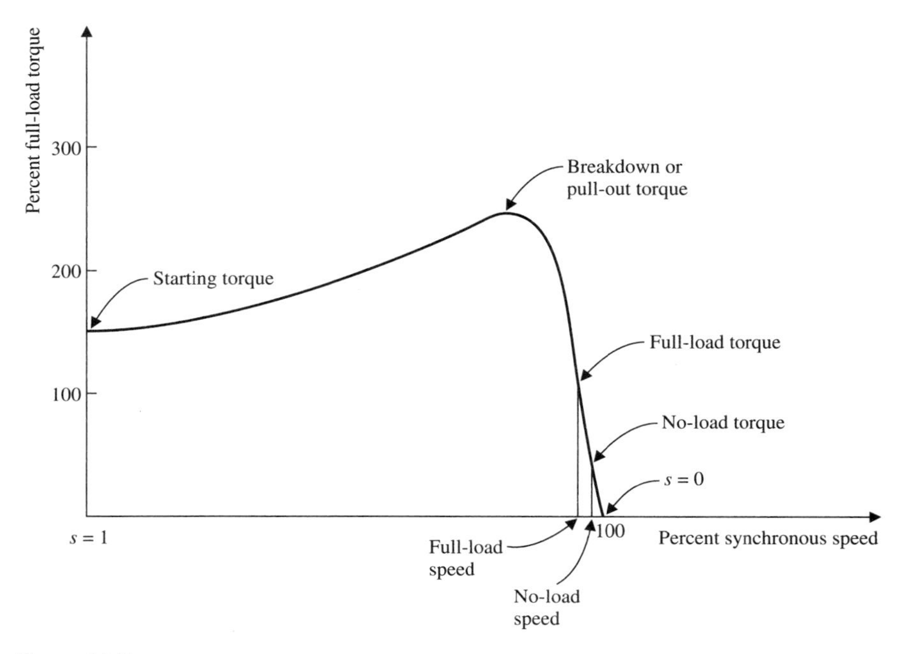

Figure 14-7 Torque-speed curve of polyphase squirrel- cage induction motor.

➤ EXAMPLE 14-4

Calculate the starting torque for the motor in Example 14-3.

SOLUTION

$$I_{R'\text{st}} = \frac{120.09\angle 0°\,\text{V}}{(0.15+0.17)+j(0.38+0.38)} = \frac{120.09\angle 0°\,\text{V}}{0.32+j0.76}$$

$$= \frac{120.09\angle 0°\,\text{V}}{0.82\angle 67.17°} = 145.63\angle -67.17°\,\text{A}$$

$$\text{RPI}_{\text{st}} = 3\times 145.63^2 \times 0.17 = 10{,}816.13\,\text{W}$$

$$T_{\text{st}} = \frac{\text{RPD}_{\text{st}}}{\omega_R} = \frac{\text{RPI}_{\text{st}}(1-s)}{(1-s)\omega_S}$$

$$= \frac{\text{RPI}_{\text{st}}}{\omega_S} = \frac{10{,}816.13\,\text{W}}{188.50\,\text{rad/s}}$$

$$= 57.38 + \text{N}\cdot\text{m} = 42.35\,\text{lb}\cdot\text{ft}$$

It should be noted that I'_{Rst} is 6.69 times greater than the normal full-load current and that the starting torque is 149% of the full-load torque.

Once the motor has accelerated up to operating speed, it will operate at the slip appropriate to drive the connected load. In normal operation the motor will be operating between the slip and speed corresponding to no load and the slip and speed corresponding to full load. However, the connected load can increase; at the same time the speed decreases and the slip increases. The motor can continue in this manner until the breakdown or pull-out torque is reached. At this point the motor cannot produce any more torque. It then stalls if any extra load is applied. It can be shown by the Maximum Power Transfer Theorem that the maximum torque occurs when R_R'/s equals the impedance looking toward the source. The slip at maximum torque is

$$s_{\max t} = \frac{R_R'}{R_S + j(X_S + X_R')} \tag{14–33}$$

If the equivalent reactance is

$$X_e = X_S + X_R' \tag{14–34}$$

then $X_e >> R_S$ and Eq. (14-32) can be approximated by

$$s_{\max t} = \frac{R_R'}{X_e} \tag{14–35}$$

If Eq. (14-35) is substituted into Eq. (14-28), then the rotor current at maximum torque is

$$I'_{R \max t} = \frac{V_S}{\sqrt{2}X_e} \tag{14–36}$$

and therefore the rotor power input at maximum torque is

$$\begin{aligned} \text{RPI}_{\max t} &= 3I_{R \max t}'\frac{R_R}{s_{\max t}} \\ &= \frac{3V_S^2}{2X_e} \end{aligned} \tag{14–37}$$

The rotor power developed at maximum torque is

$$\text{RPD}_{\max t} = \text{RPI}_{\max t}(1 - s_{\max t}) \tag{14–38}$$

and in turn the power developed at maximum torque is

$$P_{d \max t} = \text{RPD}_{\max t} - P_{fwc} \tag{14–39}$$

Then the maximum torque is

$$T_{\max t} = \frac{P_{d \max t}}{\omega_R} \tag{14–40}$$

It is interesting to note from Eq. (14-33) that the slip $s_{\max t}$ is a direct function of the rotor resistance. But from Eq. (14-40) the maximum torque $T_{\max t}$ is independent of the rotor resistance.

➤ EXAMPLE 14-5

Calculate: **(a)** the slip at which maximum torque occurs; **(b)** the maximum torque for the motor in Example 14-3.

SOLUTION

(a)

$$s_{\max t} = \frac{R'_R}{X_e} = \frac{0.17}{0.76} = 0.22$$

$$\omega_R = (1 - s)\omega_S = (1 - 0.22) \times 188.50 = 146.34 \text{ rad/s}$$

(b)

$$\text{RPI}_{\max t} = \frac{3V_S^2}{2X_e} = \frac{3 \times 120.09^2}{2 \times 0.78} = 27{,}733.86\,\text{W}$$

$$\text{RPD}_{\max t} = \text{RPI}_{\max t}(1 - s_{\max t}) = 27{,}733.86(1 - 0.22) = 21{,}632.41\,\text{W}$$

$$T_{\max t} = \frac{P_{d\max t}}{\omega_R} = \frac{21{,}632.41\text{ W} - 275\text{ W}}{146.34\text{ rad/s}} = 145.94\,\text{N}\cdot\text{m} = 107.71\,\text{lb}\cdot\text{ft}$$

In summary a number of important points can be made about the operation of polyphase squirrel-cage induction motors from Figs. 14-7 and 14-8. They are:

1. At synchronous speed ($s = 0$) no torque is produced by the motor.
2. Since the rotor reactance is a minimum at normal operating speeds, the rotor resistance is the predominant component of the rotor impedance. As a result the rotor current, the rotor field, and the induced torque are directly proportional to the slip. Expressed another way, over the normal operating range the torque-speed or torque-slip curve is linear between no load and full load.
3. Each induction motor design has a finite breakdown or pull-out torque, usually between 200 and 300% of the full-load rated torque of the motor. The slip or rotor speed at which the breakdown torque occurs is determined by the rotor resistance.
4. The starting torque ($s = 1$) is usually slightly larger than the full-load rated torque of the motor.
5. If the connected load accelerates the rotor beyond the synchronous speed, the direction of the induced torque is reversed and the induction motor

becomes an induction generator, converting the input mechanical energy to an electric energy output.

6. The rotor will be braked rapidly if its direction of rotation is opposite to that of the stator field. This property is used in plug braking, where the motor is decelerated rapidly by reversing any two stator phases, using a zero-speed switch to detect the actual point where the rotor has stopped, and then immediately removing power from the stator (Fig. 14-8).

14-8 DETERMINATION OF EQUIVALENT-CIRCUIT PARAMETERS

The data required to be able to analyze the performance of polyphase induction motors are obtained from the no load and the blocked-rotor tests and the measurement of the dc phase resistance. The data obtained by these tests enable the equivalent circuit to be used to predict the performance of the induction motor under different operating conditions. The "Test Procedure for Polyphase Induction Motors and Generators," ANSI/IEEE Std. 112, defines the precise conditions for conducting the tests.

No-Load Test

The no-load test is designed to determine the rotational losses and the magnetization reactance. The motor and instrumentation are connected as shown in Fig. 14-9.

The motor is run at no load at rated frequency, and with balanced three-phase voltages applied until the power input becomes constant. When testing wound-rotor induction motors, the slip ring brushes should be short-circuited. When running at no load at rated voltage the power input P_{nl} supplies the friction, windage, and core losses as well as the no-load stator copper loss. The stator copper loss at the operating temperature, when subtracted from the no-load power input, gives the combined

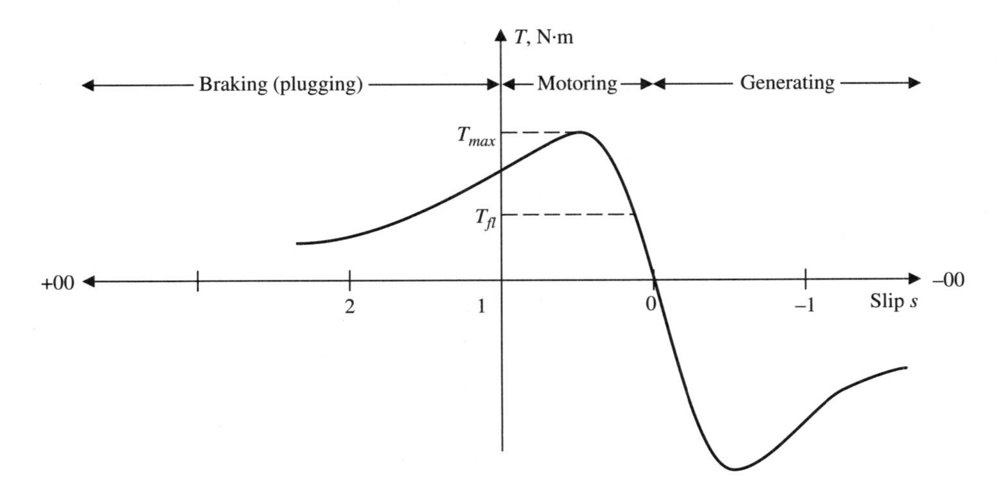

Figure 14-8 Extended-range torque-slip curve.

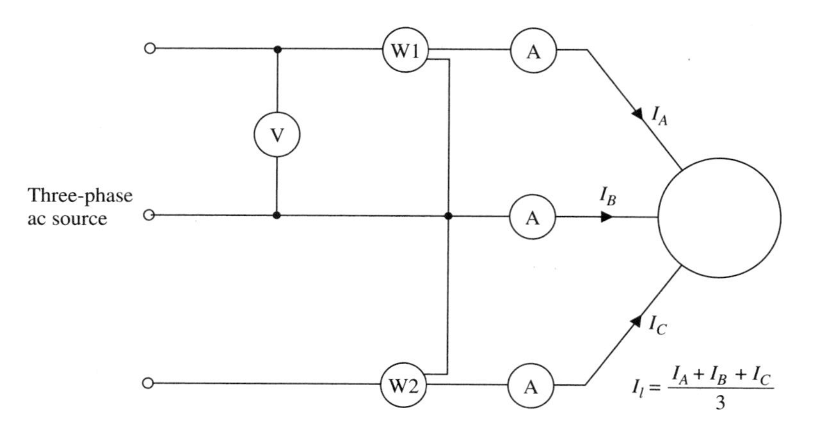

Figure 14-9 Polyphase induction motor no-load test connections.

friction, windage, and core loss P_{fwc}. It should be noted that under no-load conditions the slip s is very small. As a result the rotor can be considered as an open circuit since $R_R'(1 - s)/s >> R_R'$, and it is also much larger than X_R'. Therefore the no-load test can be represented by the equivalent circuit shown in Fig. 14-10.

The rotor copper losses can be ignored since the referred rotor current I_R' is very small. The stator copper loss SCL per phase is

$$\text{SCL} = I_{nl}^2 R_S \tag{14-41}$$

Therefore the no-load input power is

$$\begin{aligned} P_{nl} &= 3\text{SCL} + P_{fwc} + P_{\text{stray}} \\ &= 3I_{nl}^2 R_S + P_{\text{rot}} \text{ W} \end{aligned} \tag{14-42}$$

where

$$P_{\text{rot}} = P_{fwc} + P_{\text{stray}} \text{ W} \tag{14-43}$$

$$= P_{nl} - 3I_{nl}^2 R_S \text{ W} \tag{14-44}$$

From Fig. 14-10 the no-load reactance X_{nl} is the sum of the stator leakage reactance X_S and the magnetizing reactance X_m,

$$\begin{aligned} X_{nl} &= \sqrt{\left(\frac{V_{nl}}{I_{nl}}\right)^2 - \left(\frac{P_{nl}}{mI_{nl}}\right)^2} \ \Omega \\ &= X_S + X_m \ \Omega \end{aligned} \tag{14-45}$$

The no-load impedance per phase of a star-connected stator is

$$Z_{nl} = \frac{V_{nl}}{\sqrt{3}I_{nl}} \ \Omega\,/\,\text{phase} \tag{14–46}$$

where V_{nl} is the line-to-line applied terminal voltage at no load. In turn, the no-load resistance per phase is

$$R_{nl} = \frac{P_{nl}}{3I_{nl}^2} \ \Omega\,/\,\text{phase} \tag{14–47}$$

where P_{nl} is the three-phase power input. Consequently the no-load reactance per phase is

$$X_{nl} = \sqrt{Z_{nl}^2 - R_{nl}^2} = X_S + X_m \Omega\,/\,\text{phase} \tag{14–48}$$

Under no-load conditions the motor operates with a power factor of approximately 0.1. Therefore $X_{nl} \cong Z_{nl}$.

Blocked-Rotor Test

As implied by the name, the rotor is prevented from turning, that is, ω_R and $S_R = 0$ and $s = 1$. As we saw from Example 14-4, the blocked-rotor or starting current can easily be six or seven times the normal full-load current. Therefore to obtain rated stator current, a balanced reduced three-phase voltage must be applied to the stator terminals. The input power, current, and voltage readings must be taken quickly to prevent overheating. In addition the winding temperatures before and after the test should be noted to ensure that errors are not introduced by temperature-induced resistance changes. It should also be noted that under blocked-rotor conditions the stator current can vary because of the rotor position. Also since the applied stator voltage is approximately 10–20% of its normal rated value, the air-gap flux will also be reduced, with the result that the magnetization reactance X_m will appear to be larger than normal. As a result it is acceptable to ignore the no-load current I_{nl}. The blocked-rotor equivalent circuit is shown in Fig. 14.11.

When rated stator current per phase is flowing, the blocked-rotor input power to the motor is

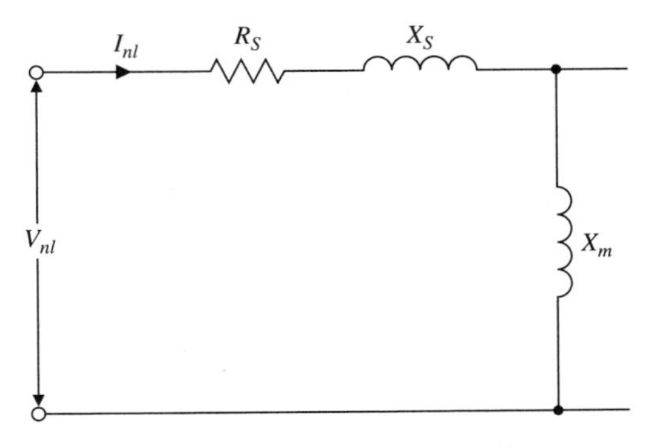

Figure 14-10 No-load equivalent circuit on a per-phase basis.

$$P_{BR} = mI_{BR}^2(R_S + R_R') \text{ W} \tag{14–49}$$

and

$$X_{BR} = X_S + X_R' = \sqrt{\left(\frac{V_{BR}}{I_{BR}}\right)^2 - \left(\frac{P_{BR}}{mI_{BR}^2}\right)^2} \ \Omega \tag{14–50}$$

The blocked-rotor impedance is

$$Z_{BR} = R_{BR} + jX_{BR} \ \Omega \tag{14–51}$$

where R_{BR} is the blocked-rotor resistance, which is equal to

$$R_{BR} = R_S + R_R' \tag{14–52}$$

The rotor resistance referred to the stator is therefore

$$R_R' = R_{BR} - R_S \tag{14–53}$$

where R_S is the stator resistance per phase and is determined by measuring the resistance of the stator windings.

The blocked-rotor reactance referred to the stator is

$$X_{BR} = X_S + X_R' \ \ \Omega \tag{14–54}$$

However, there is no simple method of separating out the component reactances other than as a result of experience. An empirical method has been developed by ANSI C52.1 and is shown in Table 14-1.

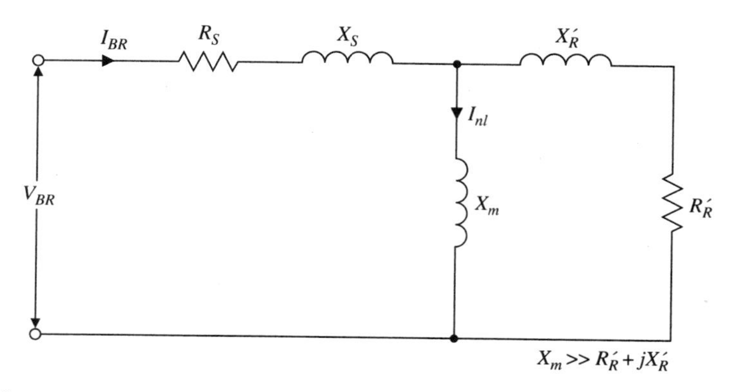

Figure 14-11 Blocked-rotor equivalent circuit on a per-phase basis.

Table 14-1 Distribution of Leakage Reactances in Polyphase Induction Motors.

Motor Class	X_S	X_R'
Class A, normal starting torque and starting current	0.5	0.5
Class B, normal starting torque and low starting current	0.4	0.6
Class C, high starting torque and low starting current	0.3	0.7
Class D, high starting torque and high running slip	0.5	0.5
Wound rotor	0.5	0.5

The magnetizing reactance can be determined from the value obtained in the no-load test [Eqs. (14-45) or (14-48)], from which

$$X_m = X_{nl} - X_S \ \Omega \tag{14-55}$$

Normally, however, X_{BR} is not broken down, since the sum of the reactances $X_S + X_R'$ is used in the calculations.

The technique of measuring the phase resistance of the stator has been described in Section 11-9.

➤ EXAMPLE 14-6

The following test data were obtained on a class A 50-hp (37.30-kW) four-pole 460-V 60-Hz delta-connected squirrel-cage induction motor.

	Line Voltage	Line Current	Total Power
Blocked rotor	144 V	85.3 A	7.09 kW
No load	460 V	25.3 A	1.55 kW

The stator resistance is 0.53 Ω/phase. Determine the parameters of the complete equivalent circuit of the motor.

SOLUTION

From Eq. (14-44) and the no-load test, the rotational losses are

$$\begin{aligned} P_{\text{rot}} &= P_{nl} - 3I_{nl}^2 R_S \\ &= 1,550 - 3 \times (25.3 / \sqrt{3})^2 \times 0.53 = 1,210.75 \text{ W} \end{aligned}$$

The no-load input impedance is

$$Z_{nl} = \frac{V_{nl}}{I_{nl} / \sqrt{3}} = \frac{460 \text{ V}}{25.3 / \sqrt{3}} = 31.48 \ \Omega / \text{phase}$$

The no-load resistance per phase is

$$R_{nl} = \frac{1,550/3}{(25.3/\sqrt{3})^2} = \frac{516.67}{14.61^2} = 2.42\ \Omega$$

From Eq. (14-48) the no-load reactance per phase is

$$X_{nl} = X_S + X_m = \sqrt{Z_{nl}^2 - R_{nl}^2}$$
$$= \sqrt{31.48^2 - 2.42^2} = 31.39\ \Omega / \text{phase}$$

From the blocked-rotor test [Eq. (14-49)],

$$P_{BR} = mI_{BR}^2(R_S + R_R')\ \text{W}$$

Then on a per-phase basis,

$$\frac{7,090\ \text{W}}{3} = \left(\frac{85.3}{\sqrt{3}}\right)^2 (0.53 + R_R')$$
$$2,363.33\ \text{W} = 2,425.36(0.53 + R_R')$$
$$= 1,285.44 + 2,425.36R_R'$$
$$1,077.89 = 2,425.36R_R'$$
$$R_R' = \frac{1,077.89\ \text{W}}{2,425.36} = 0.44\ \Omega / \text{phase}$$

The blocked-rotor input impedance is

$$Z_{BR} = \frac{144\ \text{V}}{85.3/\sqrt{3}} = \frac{144\ \text{V}}{49.25} = 2.92\ \Omega / \text{phase}$$

Then from Eq. (14-50),

$$X_S + X_R' = \sqrt{2.92^2 - 0.97^2} = 2.75\ \Omega / \text{phase}$$

From Table 14-1 for a class A motor, $X_S = X_R' = 1.38\ \Omega$. Therefore from Eq. (14-55), $X_m = X_{nl} - X_S = 31.39 - 1.38 = 30.01\ \Omega/\text{phase}$.

14-9 WOUND-ROTOR MOTOR CHARACTERISTICS

Wound-rotor induction motors (WRIMs) have a rotor that contains a star-connected three-phase winding with the ends of the winding brought out to three insulated slip rings mounted on the rotor shaft. Stationary carbon brushes bearing against the rotating slip rings are connected to external resistances. Wound-rotor motors are normally started with a relatively high value of external resistance connected in series with the rotor winding. As the motor accelerates, the external resistance is reduced in a number of steps until, at the point where the motor is operating at rated speed, the external resistance is short-circuited across the slip rings. In high-power wound-rotor induction motors liquid rheostats are used as the external resistance. The use of the external resistance permits the resistance of the rotor circuit to be changed without any change in the rotor reactance taking place. Normally the relationship between the rotor reactance and the actual rotor resistance at standstill (s = 1) is X_{BR} = 3 to 5 R_{BR}. From Eq. (14-35) the slip at which maximum torque occurs is directly related to the resistance of the rotor circuit. The addition of external resistance in the rotor circuit enables the starting torque to be made equal to the maximum torque. This property is extremely valuable when high starting torques are required to accelerate high-inertia loads. An additional advantage is that during the starting cycle the I^2R losses in the rotor circuit are largely taking place in the external resistances. As a result the heat losses in the rotor are reduced during the starting cycle.

Figure 14-12 shows the effect of different values of rotor circuit resistance on the torque-speed characteristics. Perusal of the curves shows that as the rotor circuit resistance is increased, the slip at which maximum torque occurs increases (compare curves 1, 2, and 3). By making the rotor circuit resistance equal to the blocked-rotor reactance, $R_R + R_{ext} = X_{BR}$, maximum torque can be made to occur at standstill, that is, the starting torque equals the maximum torque. It should also be noted that increasing the rotor circuit resistance beyond this point will result in a reduced starting torque (see curve 4).

Apart from being able to control the acceleration of high-inertia high-torque loads, the added resistance also reduces the inrush stator current during the starting cycle. Also over the normal operating range, increases in the rotor circuit resistance increase the speed regulation of the motor, that is, by the use of an adjustable resistance external to the rotor the rotor speed can be controlled. However, this method of speed control is not without its problems, since the additional resistance introduces a

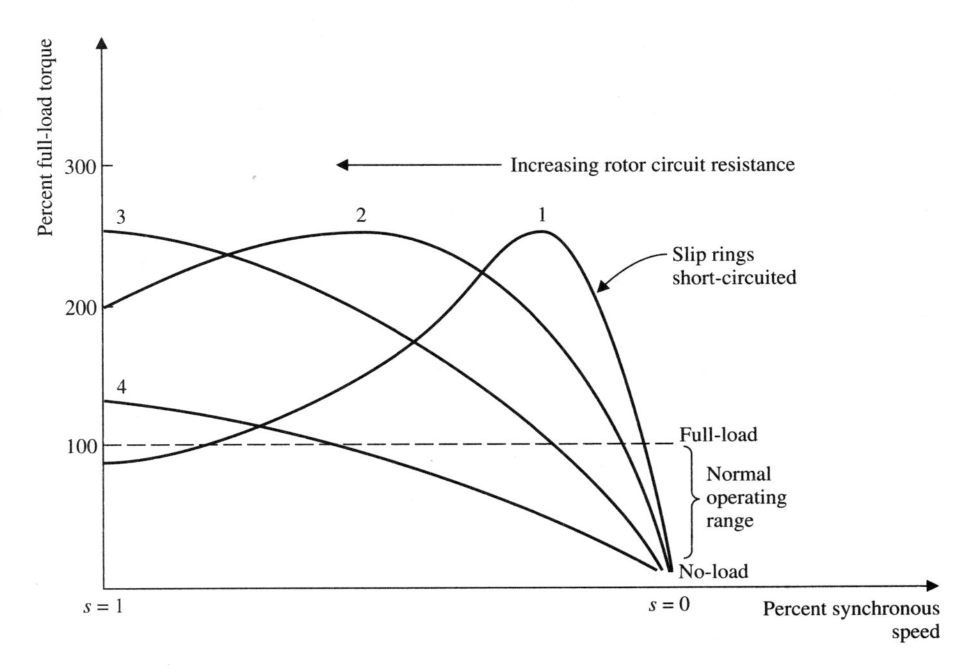

Figure 14-12 Torque-speed curves for typical wound-rotor induction motor.

significant drop in the overall efficiency of the motor. It is normal practice to restrict rheostatic speed control to 70% of rated speed for self-ventilated motors used in constant-torque applications, or to 50% of rated speed with the motor operating at 40% of its rated output.

14-10 MODIFICATION OF TORQUE-SPEED CHARACTERISTICS

From our review of the wound-rotor induction motor we observed that the starting torque could be made equal to the maximum torque by adjusting the rotor circuit resistance to be equal to X_{BR}. A squirrel-cage induction motor with a high rotor resistance would have a high starting torque, but at normal operating load it would have a high slip. From Eq. (14-23), that is, $P_g = \text{RPI}(1 - s)$, it can be seen that the greater the slip, the smaller will be the amount of the rotor power input that is converted into useful mechanical power output, with a consequent reduction in the motor efficiency. Conversely a low-resistance rotor will have a low starting torque and a high inrush starting current, but it will have a low slip and a high efficiency.

The designer's dilemma is to produce a high starting torque, that is, a motor with a high-resistance rotor that will, at normal operating speeds, have a low resistance and slip. The solution to the problem lies in the shape of the rotor slots. When power is first applied to the motor, the frequency of the rotor currents is equal to the source frequency, and the rotor resistance is the effective ac resistance. When running at normal speed, the frequency of the rotor currents is usually on the order of 3 or 4 Hz, and the rotor resistance is approximately equal to the dc resistance. At start it is also desirable that the starting current be low. This can be achieved by having a high rotor reactance. But the maximum torque requires that the rotor reactance be low. A high rotor reactance can be obtained by using semienclosed rotor slots with the minimum possible air gap between rotor and stator. This minimizes the magnetizing current, improves the power factor, and decreases the reluctance of the leakage flux paths.

14-11 DEEP-BAR AND DOUBLE-CAGE ROTORS

The rotor leakage reactance by definition is the reactance that is produced when a portion of the rotor flux does not link with the stator windings. The magnitude of the rotor leakage reactance X_R is in general determined by the relative distance between the active portion of the rotor conductor and the stator. A squirrel-cage rotor, with its rotor conductors close to the rotor's surface, will have a lower leakage reactance than a motor whose conductors are buried deeper in the rotor.

Deep-bar rotor conductors are formed from deep narrow bars (Fig. 14-13). The leakage reactance of the lower portion of the conductor is greater than that of the upper portion since it is linked by a larger portion of the leakage flux. This variation in leakage reactance will result in a higher current density in the lower leakage reactance portion, and a lower current density in the high leakage reactance portion. The uneven current distribution over the cross section of the rotor conductor has effectively increased the resistance of the bar. At standstill the effect is maximized because the frequency of the rotor currents is a maximum. However, at minimum slip

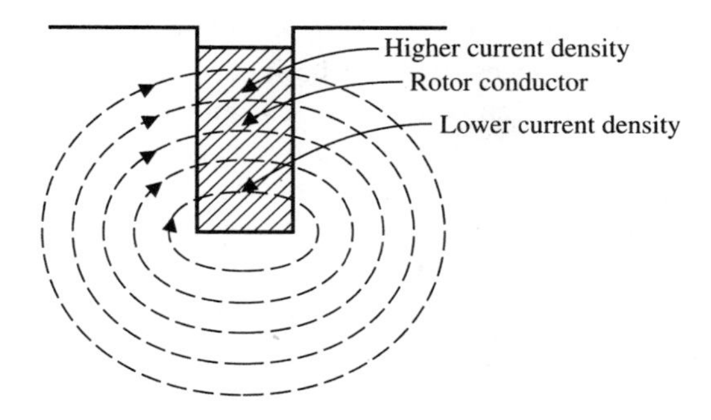

Figure 14-13 Deep-bar rotor conductor.

the rotor frequency is a minimum and the rotor resistance is basically equal to the dc resistance. Careful design of the rotor slot and the rotor conductor permits the effective resistance at standstill to be several times greater than the resistance at normal operating speeds.

An alternative method of obtaining a high starting torque with a low starting current is to use a rotor with two separate concentric squirrel-cage windings. This type of rotor is called a *double-cage* rotor. A typical conductor arrangement is shown in Fig. 14-14.

The outer cage consists of a high-resistance low-reactance winding, and the inner cage is a low-resistance high-reactance winding. The reactance of the inner cage is increased by narrowing the rotor slots between the two windings. When power is first applied to the stator, the frequency of the currents in both windings is equal to the source frequency. As a result the reactance, and therefore the impedance, of the inner cage is greater than the reactance of the outer cage. The rotor current distribution is determined by the relative impedance of each winding. Therefore at the instant of starting the high-resistance low-reactance outer cage will carry the greater portion of the rotor current, and the motor will have the starting characteristics associated with a high-resistance rotor. As the motor accelerates, the frequency of the rotor currents will decrease until, at normal operating speed, the reactances will have

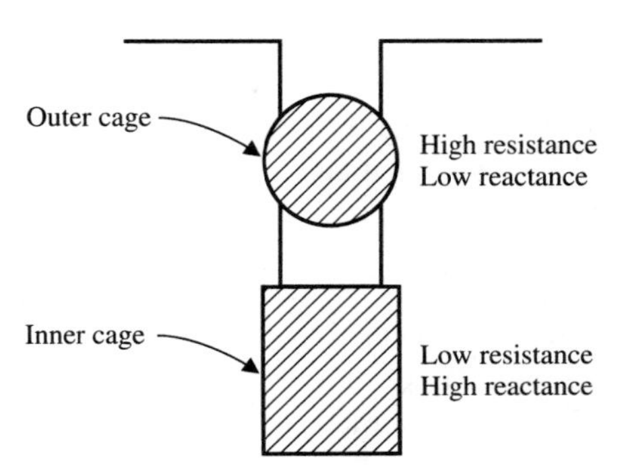

Figure 14-14 Double-cage rotor construction.

decreased significantly. The division of current between the two cages is determined by their resistances, that is, the inner cage low-resistance winding will be carrying the greater portion of the rotor current. It can be seen that by controlling the resistance of the conductors in the two cages, and by controlling the shape of the rotor slots, the double-cage induction motor can be made to have a high starting torque with a low to moderate starting current, and at rated load it operates with a low slip, typically 0.1, and moderate efficiency.

Double-cage induction motors are relatively expensive, but their manufacturing cost is less than for an equivalently rated wound-rotor motor. They combine the best features of the wound-rotor motor and are usually used in high-acceleration high-impact load applications such as punch presses.

14-12 NEMA CLASSIFICATIONS

Polyphase squirrel-cage induction motors up to approximately 500 hp (373 kW) have been classified by the National Electrical Manufacturers Association (NEMA), the Canadian Electrical Manufacturers Association (CEMA), and the International Electrotechnical Commission (IEC) into four classes based on rotor design, shape, dimensions, and resistance of the rotor conductors. These classifications are:

1. *Design class A*: These motors are designed for full-voltage starting and have a relatively low rotor resistance. This results in a low starting torque with a high starting current, typically five to eight times full-load current, and a low slip, between 0.005 and 0.01 at rated load, with a high operating efficiency. In sizes above 7.5 hp (5.60 kW) it is usual practice to use reduced-voltage starting in order to reduce the inrush current and prevent line-voltage dips. The starting torque varies from being equal to the rated torque for larger motors to as great as 200% of rated torque for smaller power output motors. The breakdown or pull-out torque is in the range of 200–300% of rated full-load torque, and usually occurs at slips less than 0.2. Class A motors are especially suited for high-speed applications where rapid acceleration is required to drive a low-load torque load at high efficiencies. Design class A motors are mainly used in special applications.
2. *Design class B*: These motors are designed for full-voltage starting. They produce a starting torque of about 170% and a breakdown torque of about 200% of rated full-load torque, with a slightly lower starting current than a class A motor, and operate with a slip less than 0.05 at rated full load. Design class B motors are considered to be general-purpose motors and are used in applications where a lower starting current and relatively high efficiency are the main concerns. Typical applications are uniform loads such as fans, pumps, and centrifugal compressors.
3. *Design class C*: These motors are designed for full-voltage starting. They have a high starting torque, up to 250% of rated full load, a low starting current, and a slip less than 0.05 at full load, and they operate with moderate efficiency. Typical applications include uniform high-starting-torque

loads such as conveyers and reciprocating compressors. This class of motor uses a double-cage rotor and is initially more expensive to purchase than the other motor design classes.

4. *Design class D*: These motors are designed for full-voltage starting. They have a high starting torque, up to 300% of rated full-load torque, and a low starting current, but operate with slips of about 0.1 at rated load. This in turn reduces the operating efficiency at full load. Class D motors are used in high-inertia high-intermittent-torque applications, such as punch presses or shears. The motor usually drives a flywheel, which is used to supply energy to the load from its stored kinetic energy.

Typical torque-speed characteristics for NEMA design motors are shown in Fig. 14-15.

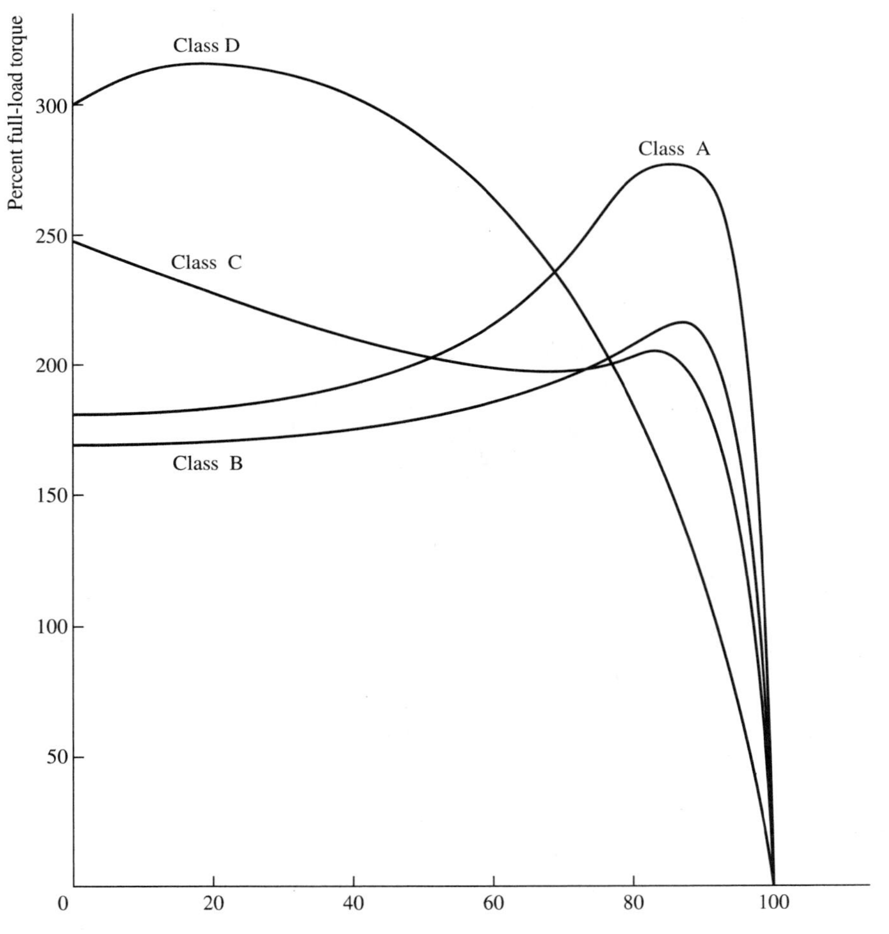

Figure 14-15 Typical torque-speed characteristics for NEMA class polyphase squirrel-cage induction motors.

14-13 POLYPHASE INDUCTION MOTOR OPERATING CURVES

The dominance of the polyphase induction motor in industrial and commercial applications, apart from its low cost and high reliability, is clearly shown in the operating curves presented in Fig. 14-16.

The curves show that the rotor speed is essentially constant over the normal load range, and that the motor operates at an acceptable power factor from approximately 50% of rated load. The overall efficiency curve is almost constant down to 50% of rated load, and the line current is almost linear over the whole range of operation.

14-14 POLYPHASE INDUCTION MOTOR STARTING METHODS

Starting polyphase induction motors is comparatively simple when compared to the methods required for dc motor and synchronous motor starters. There are three methods of starting squirrel-cage induction motors: (1) full-voltage, (2) reduced-voltage, and rarely (3) part-winding starting. These starting methods can be achieved using manual or automatic starters. However, manual starters are usually restricted to polyphase motors less than 7.5 hp (5.50 kW) at 440 V.

Full-Voltage Starting

Full-voltage starting subjects the motor and the connected load to severe shock because of the high starting torque. In addition, if the motor rating is high enough, the supply system is subjected to line-voltage dips which may affect voltage-sensitive systems. Normally the factors which determine whether or not full-voltage starting is employed are:

1. Power output rating and design class of the motor
2. Actual application
3. Capacity of the distribution system and the supply system
4. Specific rules defined by the public-utility authority

Magnetic starters are the most commonly used device and can be operated locally or remotely. A magnetic starter usually consists of a magnetic contactor and overload relays. The inclusion of a branch disconnect switch or a fusible disconnect switch results in a combination starter. Irrespective of the line voltage of the supply system, the control circuit is usually supplied with 115 V single phase.

Figure 14-17 shows a combination starter with a fusible disconnect switch and thermal overloads OL1 and OL2. When the disconnect switch is closed, power is available to the control circuit. Pressing the normally open START push button causes contactor coil M to pick up. This in turn closes contacts M in series with the motor and auxiliary contact M in parallel with the START push button, and seals contactor coil M. In the event of a small persistent overload, overload relays OL1 and OL2 will cause the contacts in the control circuit to open, thus deenergizing coil M and removing power from the motor. Similarly, pressing the STOP push button momentarily will cause coil M to deenergize.

Full-Voltage Reversing Starters. Polyphase induction motors are easily reversed by reversing any two phases. This is normally accomplished in a magnetic starter by using two magnetic contactors with mechanical and electrical interlocks to prevent both contactors from being closed at the same time, which would apply a short circuit across the power lines. A typical full-voltage reversing starter is shown in Fig. 14-18.

Pressing the FOR push button will energize the contactor coil F through the normally closed STOP push-button contacts and the normally closed interlock R and contacts OL1 and OL2. At the same time contactor coil F is sealed on by auxiliary contact F in parallel with the FOR push button, and contacts F close in the power

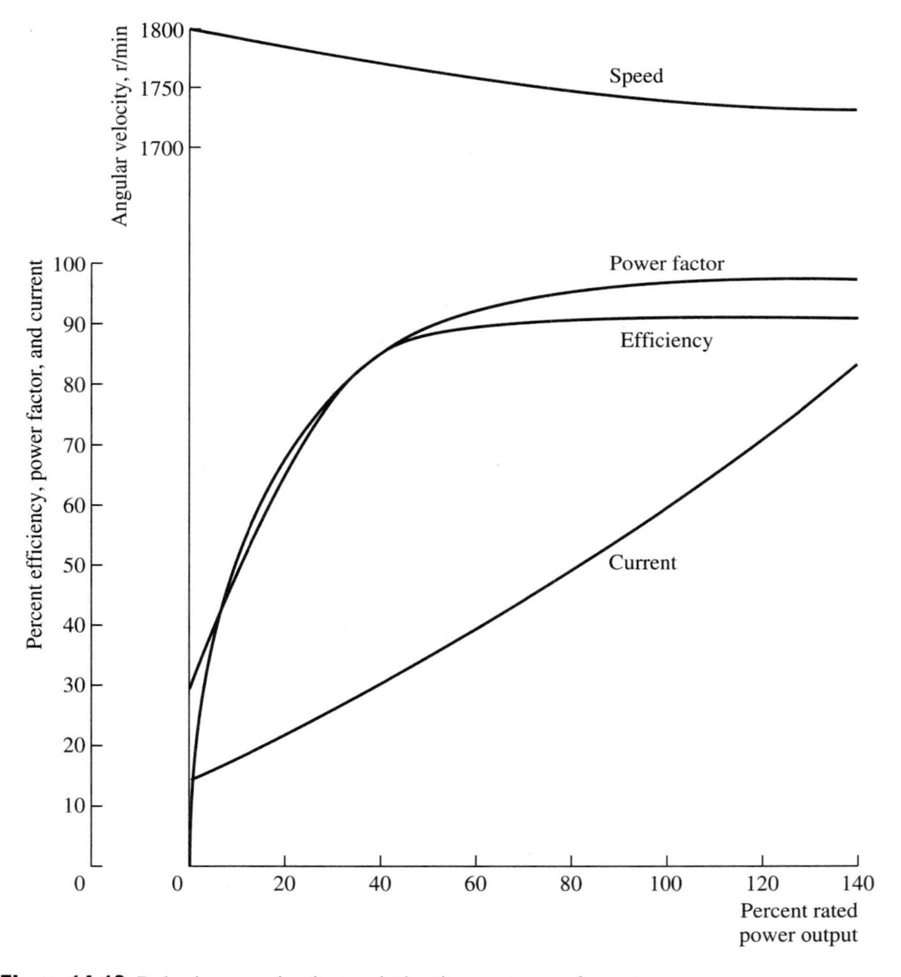

Figure 14-16 Polyphase squirrel-cage induction motor performance curves.

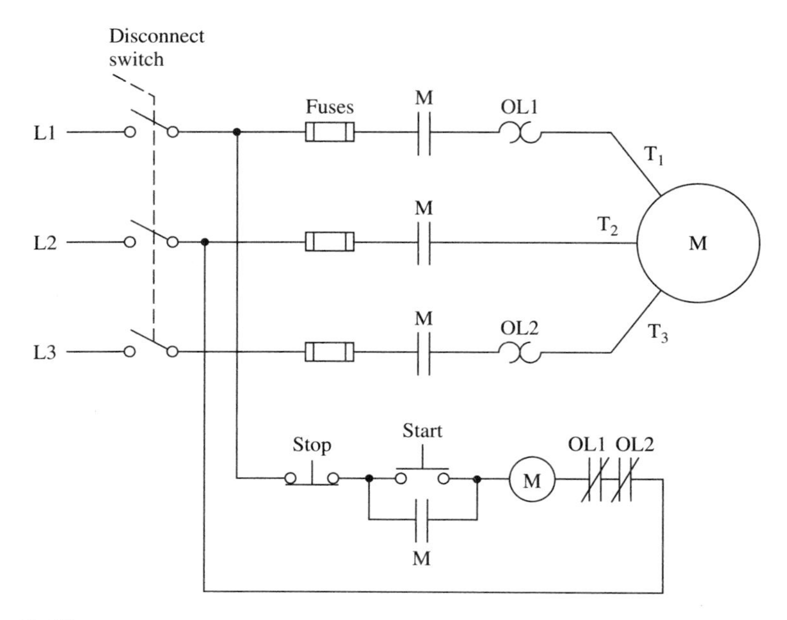

Figure 14-17 Full-voltage combination starter with low-voltage control.

circuit to the motor and simultaneously open interlock F in the contactor coil R circuit, thus preventing contactor coil R from picking up.

Pressing the REV push button deenergizes contactor coil F and opens contacts F in the motor circuit. Contactor coil R picks up through the normally closed FOR contacts and the normally closed interlock F, closing contacts R in the motor power circuit and opening interlock R in the contactor coil F circuit, and seals the circuit by closing the auxiliary contact in parallel with the REV push button. In addition to the electrical interlocks, mechanical interlocks shown by the dashed lines physically prevent closing more than one contactor at the same time.

Reduced-Voltage Starting

There are four main methods of providing reduced-voltage starting for polyphase squirrel-cage induction motors: (1) line resistance; (2) line reactor; (3) autotransformer; and (4) star-delta connection.

Line-Resistance Reduced-Voltage Starters. The line-resistance reduced-voltage starter is the most commonly used starter. It is illustrated in Fig. 14-19. When the START push button is pressed, the START contactor coil S picks up and closes contacts S in the motor power circuit, applying power through the line resistances to the motor and at the same time sealing the contactor coil S circuit. Simultaneously timer T in the run contactor coil R circuit begins to time out and, after a suitable time interval, closes and energizes contactor coil R, which in turn closes contacts R in the power circuit, short-circuiting the line resistances and applying full voltage to the motor.

The line-resistance starter provides a smooth start and acceleration without an interruption of power to the motor. This is known as a *closed-transition starter.* As the motor accelerates, the line current decreases, and at the same time the motor terminal voltage increases. This type of starter, especially in the smaller ratings, is the least expensive and is simple. A high breakdown torque is produced by the motor, and the motor torque increases as the motor accelerates.

Line-Reactor Reduced-Voltage Starters. The main reasons for choosing a line-reactor starter are that a higher starting torque and reduced starting losses are obtained. However, because of increased cost they are mainly used in high-voltage applications. A line-reactor starter is basically identical to that shown in Fig. 14-19, except that the line resistances are replaced by reactors.

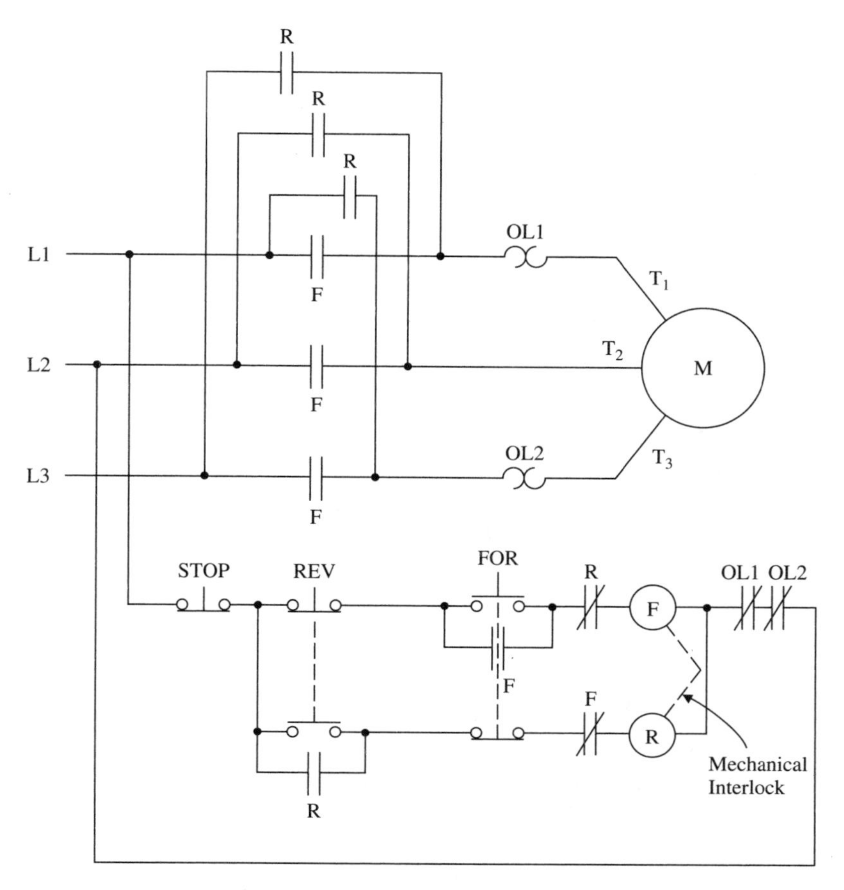

Figure 14-18 Full-voltage reversing starter with mechanical and electrical interlocks.

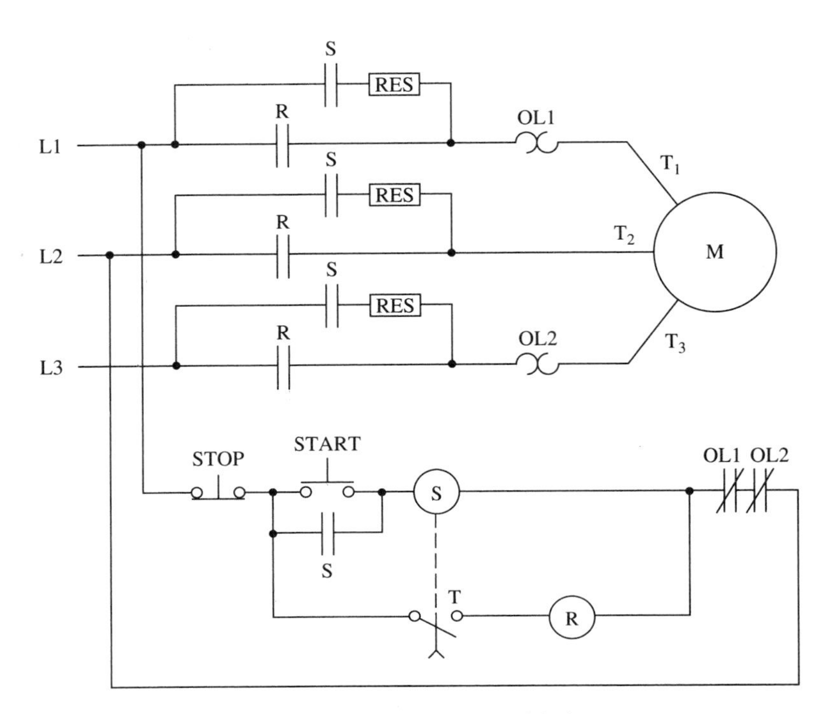

Figure 14-19 Line-resistance reduced-voltage starter with timer.

Autotransformer Reduced-Voltage Starters. If the prime concern is the reduction of the inrush current, then an autotransformer starter should be used, since the line current is the same fraction of the motor current as the ratio of secondary to primary voltages.

Usually a three-phase autotransformer starter or *compensator* is formed from two single-phase autotransformers connected in open-delta or vee-vee, although a conventional three-phase arrangement is used when it is essential that the motor develop maximum starting torque. There are two types of autotransformer starters: *open transition* and *closed transition*. The major disadvantage of the open-circuit transition type is that in the interval when the motor is isolated from the power source during the changeover from the start contacts to the run contacts, voltages are induced in the stator windings. The phase relationship of these voltages may be such as to cause a high current surge as the run contacts are closed. Closed-transition or Korndorfer starters ensure that during the transition from start to run the motor is not isolated from the power source. This prevents a current surge.

Figure 14-20 illustrates a typical open-transition autotransformer starter. When the START push button is pressed, start contactor coil S is energized through the normally closed timing relay contact T and the normally closed interlock R, and all contacts S in the power circuit close, applying a reduced voltage to the motor. At the same time timing relay T is energized, seals the contactor coil S circuit, and initiates the timing sequence on the normally closed contact T in the contactor coil S circuit and the normally open contact T in the contactor coil R circuit. At the end of the timing cycle the normally closed contacts open and the normally open contacts close. Contactor S deenergizes and opens all contacts S, and contactor coil R picks up and closes all contacts R. The motor then has full voltage applied to its terminals.

Star-Delta Reduced-Voltage Starters. A reduced-voltage starter that is in common use in Europe is the star-delta starter. This starter requires that both ends of each phase winding be brought out to the controller. The starter, which does not require the use of line resistances, line reactors, or autotransformers, depends on the stator winding being connected in delta during normal operation. During the starting cycle,

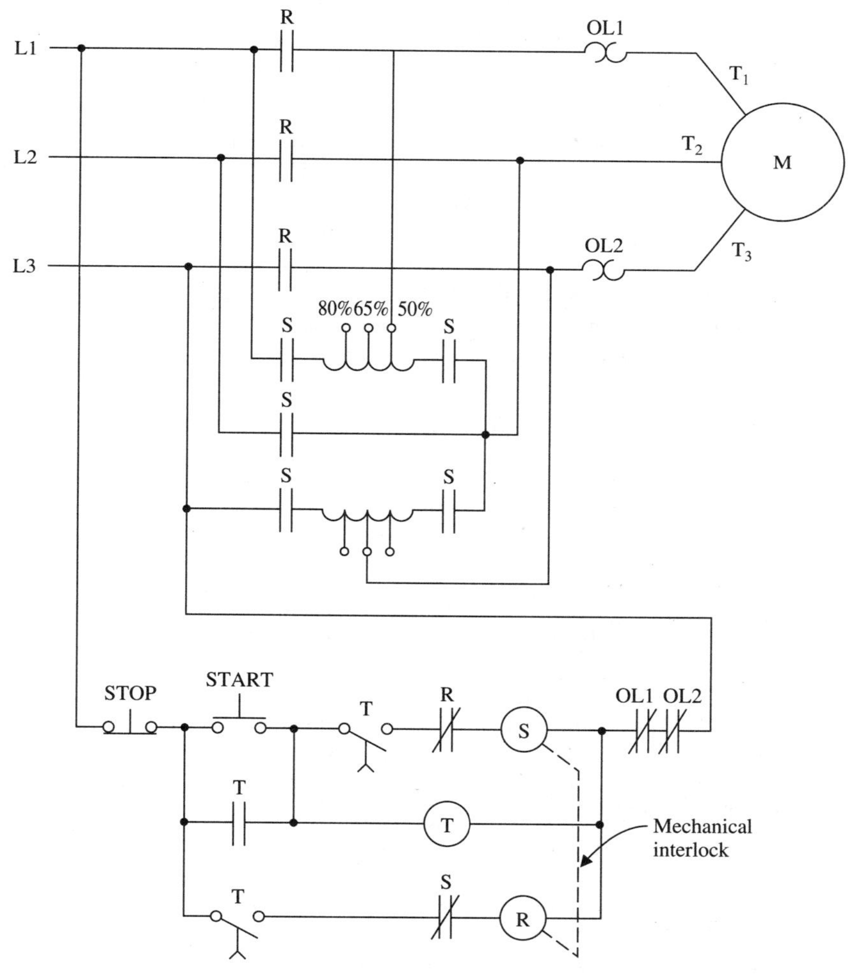

Figure 14-20 Open-transition autotransformer reduced-voltage starter.

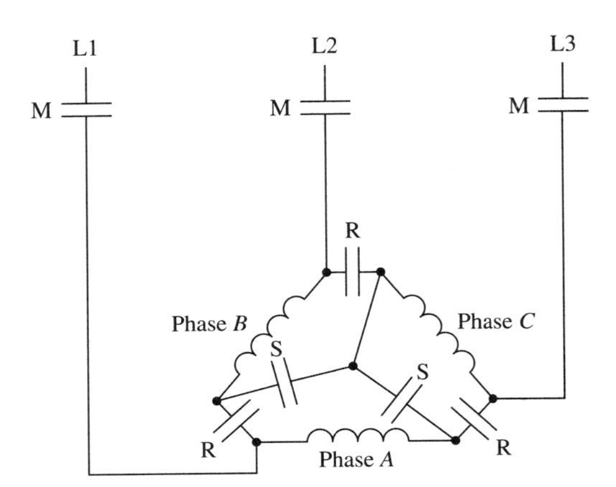

Figure 14-21 Open-transistion star-delta reduced-voltage starter.

the starter externally connects the stator windings in star, that is, the phase voltage is $V_l/\sqrt{3}$, or 58% of the line voltage. When the motor has accelerated, the windings are reconnected in delta. The star-delta starter reduces both the inrush starting current and the starting torque. The power circuit of a star-delta starter is shown in Fig. 14-21.

In this circuit, when the starting cycle is initiated, contacts S close and connect the stator windings in star. After a suitable time delay contacts S open and contacts R close, reconnecting the stator windings in delta.

14-15 WOUND-ROTOR MOTOR STARTING

Normally wound-rotor induction motors are started by using an across-the-line starter to supply the stator windings, and resistances in the rotor circuit to control the inrush current. A typical three-step wound-rotor induction motor starter is shown in Fig. 14-22.

When the START push button is pressed, contactor coil M picks up through the normally closed contacts 2A, OL1, and OL2. Contacts M close in the power circuit and apply full voltage to the stator windings. Simultaneously interlock M seals the START push button, and the normally open timing contacts 1T commence their timing cycle. When timer 1T times out, the contacts close and accelerating contactor 1A picks up and short-circuits one section of the rotor resistance. The operation of contactor 1A initiates the timing for timer 2T. After a suitable time interval contacts 2T close, energizing accelerating contactor 2A, which applies a short circuit across the rotor slip rings. The motor is now operating as a squirrel-cage motor.

14-16 ELECTRIC BRAKING

Squirrel-cage induction motors can be quickly stopped by reversing the motor, so that at the instant that the rotor shaft stops turning, power is removed. Zero-speed switches are used to sense the instant that the shaft stops turning. This technique is called *plug stopping*. Although plug stopping is simple and inexpensive, it is possible in some applications that the plugging torque may be too great. In some situations it is necessary to insert series resistances, called *plugging resistances,* to limit the torque to an acceptable value. Plugging is a satisfactory method of braking for small and medium-size motors, but if attempted with high-horsepower (kW) motors, unacceptable line disturbances will be produced. An acceptable alternative is to use dc dynamic braking.

Just as with dynamic braking of dc motors, the induction motor functions as a generator during the braking cycle. This is accomplished by isolating the stator from the three-phase source, connecting two of the stator leads together, and applying dc power between these leads and the remaining stator lead. The amplitude of the applied dc voltage is controlled to limit the current to the order of two or three times the normal full-load current. This arrangement creates as many stator poles as there were originally in the rotating magnetic field. As the rotor cuts this magnetic field, voltages and currents are induced in the rotor conductors, and the mechanical energy of rotation is converted to an I^2R loss in the rotor. Unlike dc dynamic braking, the braking torque increases as the motor slows down. This result is caused by the fact that at the beginning of the braking period the rotor leakage reactance is a maximum, and the rotor current has a low lagging power factor. Since this current has a large reactive component, it will reduce the effect of the dc field produced by the stator,

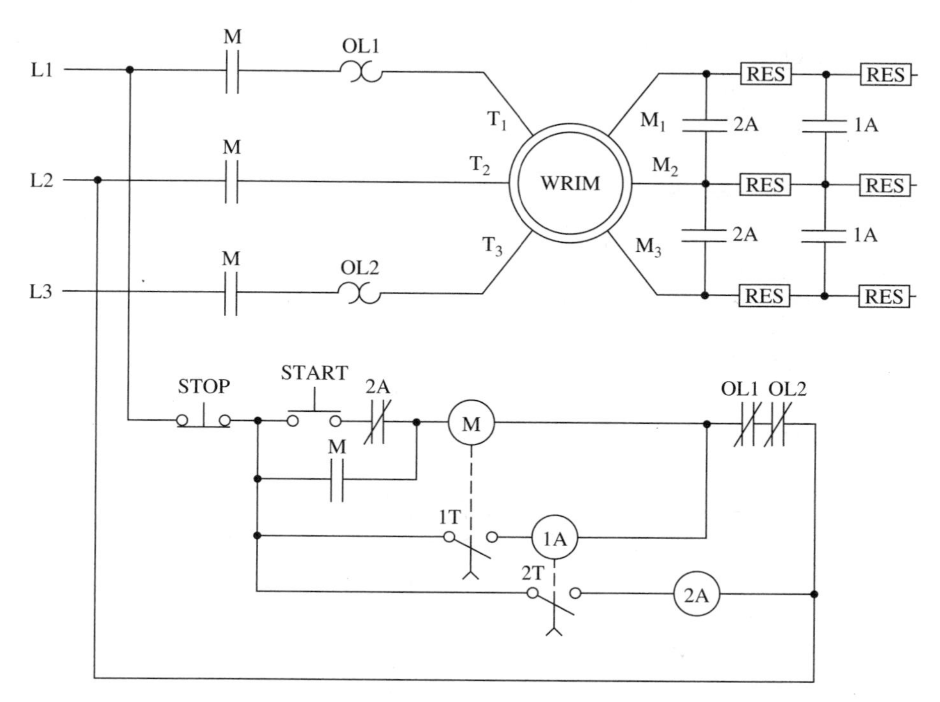

Figure 14-22 Wound-rotor induction motor starter with three steps of acceleration.

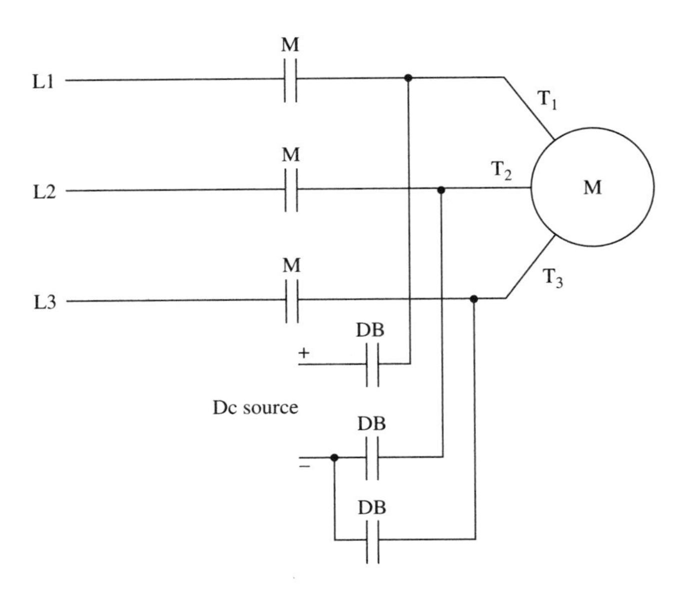

Figure 14-23 Dc dynamic braking of squirrel-cage induction motor.

which in turn reduces the air-gap flux and the braking torque. However, as the rotor slows down, the rotor leakage reactance decreases and the rotor power factor improves. As a result the air-gap flux increases, and the braking torque also increases as the rotor slows down. The basic connections are shown in Fig. 14-23. This method is equally applicable to wound-rotor induction motors.

14-17 SPEED-CONTROL TECHNIQUES

As was discussed in Section 14-13, the squirrel-cage induction motor is essentially a constant-speed motor over its normal operating range. Unlike the dc motor, it is not readily capable of speed control without special techniques, such as pole changing or varying the frequency.

Squirrel-cage induction motors have many advantages over dc motors. They are lighter (20–40% lighter than an equivalently rated dc motor), they are less expensive to procure, and over their planned life they are significantly cheaper to maintain.

Equations (14-1) show that the synchronous speed is inversely proportional to the number of stator poles per phase and directly proportional to the frequency of the applied voltage. The most commonly used methods of controlling the speed of squirrel-cage induction motors are: (1) changing the number of stator poles per phase, or (2) varying the frequency of the voltage applied to the stator.

Pole-Changing Methods

By suitable design of the stator windings it is possible to change the number of poles per phase by either manual or automatic switching. This technique is especially suited to the squirrel-cage induction motor since the rotor will automatically have the same number of induced poles as the stator. Squirrel-cage motors, both polyphase and single phase, which are designed to have the number of poles in their stator windings changed by switching, are called *multispeed motors*. Multispeed motors are available in two- or four-speed configurations using single- or two-winding stators, respectively.

Synchronous and wound-rotor induction motors cannot have their speed changed by this method since the number of rotor poles is fixed. Speed control by pole changing is therefore limited to squirrel-cage induction motors. Single-winding two-speed motors are normally restricted to speed changes in a 2:1 ratio, for example, four-pole 1,800 r/min (188.50 rad/s) to 900 r/min (94.25 rad/s). This type of control is called the *consequent-pole* method.

Consequent-Pole Connection. The consequent-pole method of speed control depends on the external reconnection of the stator windings. Figure 14-24 illustrates the concept of the consequent-pole connection. The stator coils are connected so that they produce alternate N and S poles. This is the high-speed connection. The low-speed connection is obtained by external switching so that the stator coils produce all N or S poles. As a result an equal number of opposite-polarity poles is induced in the stator iron between the stator poles. The consequent-pole connection suffers from the disadvantage that speed changes are always in a 2:1 ratio and intermediate speeds are not available.

Two-Winding Connection. A wider range of speeds can be obtained by using two separate stator windings, which when combined with the consequent-pole connection will give four separate speeds. A typical two-winding motor could have a four-pole and a six-pole winding. This combination would then produce synchronous speeds of 1,800, 1,200, 900, and 600 r/min (188.50, 125.66, 94.25, and 62.83 rad/s). Once again, however, intermediate speeds cannot be obtained. Another major disadvantage to this design is that the stator must be larger in diameter to provide stator slots deep enough to accommodate the two windings. As a result the motor is physically larger and more expensive for the same power output, as well as having a poorer speed regulation, resulting from the increased reactances of each winding, and a lower efficiency as compared to the single-winding stator squirrel-cage induction motor.

Pole-Amplitude Modulation. The 2:1 speed ratio obtained from the consequent-pole connection is not suitable for many applications. Ratios such as 6:8, 8:10, 10:12, etc., are better suited for applications such as centrifugal pumps and fans, where the load torque is proportional to the speed squared. In addition there are a number of applications requiring speed ratios of 4:1, 5:1, and so on. These speed ratios can be obtained by using a pole-changing technique called *pole-amplitude modulation* (PAM), which at the present time is mainly restricted to larger power output squirrel-cage induction motors. Pole-amplitude modulation permits a single-winding

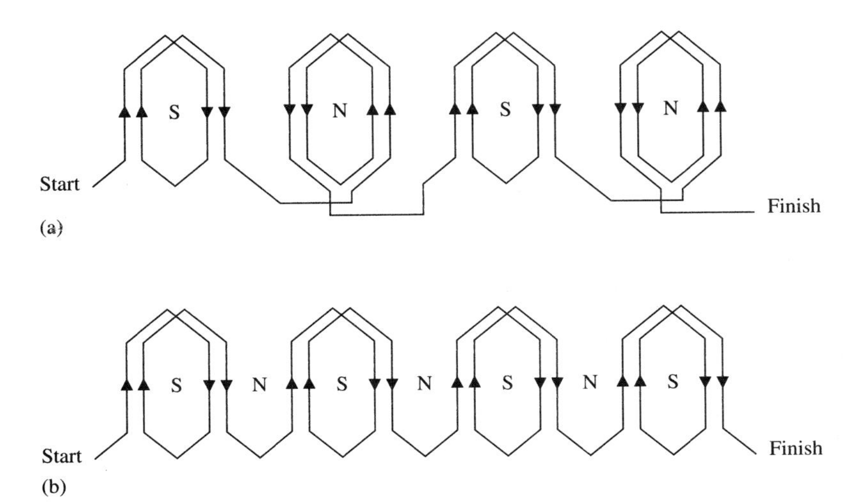

Figure 14-24 Consequent-pole connections. (a) Four-pole high speed. (b) Eight-pole low speed.

squirrel-cage induction motor to operate equally efficiently at either of two speeds. The principle pole-amplitude modulation depends on the fact that when a sinusoidal magnetomotive force with a given number of poles P, which is spatially distributed in the stator, is multiplied by another spatially distributed sinusoidal magnetomotive force with N poles, the resultant magnetomotive force will consist of two separate magnetomotive forces, one with $P + N$ poles and the other with $P - N$ poles. One of these forces must be eliminated. The concept of pulse-amplitude modulation is illustrated in Fig. 14-25.

The original winding produces a stator field with $P = 8$ poles [Fig. 14-25(a)]. The modulating wave $N = 2$ is shown in Fig. 14-25(b) and the resultant stator field $P - N = 8 - 2 = 6$ poles in Fig. 14-25(c). It should be noted that there is also a $P + N = 8 + 2 = 10$-pole stator field present. The effect of the second field is reduced by chording and distributing the winding. A well-designed pole-amplitude-modulated winding will produce two speeds, with balanced phase currents, a smooth torque, and a minimum of harmonics.

Frequency Control

The speed of polyphase induction motors can also be varied by varying the frequency of the applied voltage. By definition the base speed of an induction motor is the operating speed when the motor is supplied with rated voltage at rated frequency. When operating at speeds less than the base speed, with a constant terminal voltage, as the frequency is decreased, the phase resistance becomes the major component of the phase impedance. Since the impedance has decreased because the reactances have decreased, the stator currents increase and cause saturation of the air-gap flux, and the output torque is no longer constant.

If the terminal voltage is maintained constant and the frequency is increased above that required for base speed, then the reactance becomes the major component of the stator impedance, and both the air-gap flux and the torque decrease.

Voltage and Frequency Control

It has been shown that the developed torque is proportional to V/f. The air-gap flux must be constant at all frequencies in order to produce a constant torque output. As a result the V/Hz ratio must also be constant to produce a constant air-gap flux. Then to reduce the speed of the motor, it is not only necessary to reduce the applied frequency, but the applied terminal voltage must also be reduced in the same ratio.

At low frequencies, that is, approximately 10 Hz, the phase resistance is the major component of the phase impedance. As a result there will be a decrease in the air-gap flux at low frequencies. If this condition is not acceptable, the constant V/Hz ratio must be modified to restore the air-gap flux, even through though there is a possibility that the stator iron may be saturated.

The breakdown or maximum torque is constant for a given machine at all frequencies less than that for base speed when operating under a constant V/Hz control.

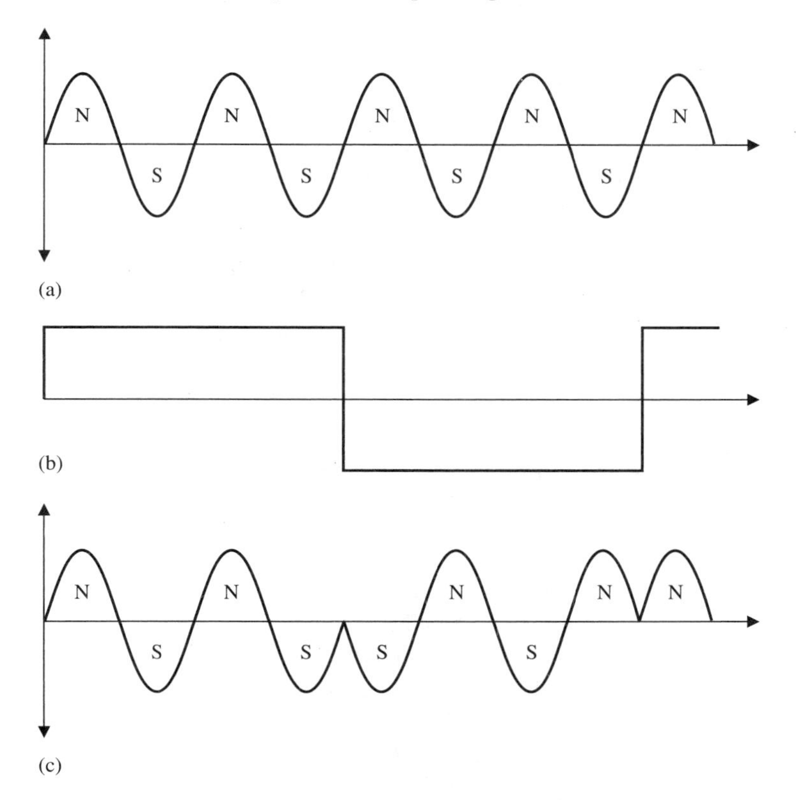

Figure 14-25 Principle of pole-amplitude modulation. (a) Original eight-pole stator field. (b) Modulating wave. (c) Six-pole modulated stator field.

It is proportional to the air-gap flux squared and inversely proportional to the rotor leakage reactance. When a squirrel-cage or a wound-rotor induction motor is operated with a constant *V*/Hz ratio, it will result in higher starting torques, and with the same full-load slip, the torque is greater at the higher frequencies [Fig. 14-26(b)]. In addition the horsepower (kW) output and the efficiency are also greater at the higher frequencies. Failure to maintain the constant *V*/Hz ratio will affect the constant-torque output by the square of the flux density, or will cause the stator current to increase and overheat the motor.

Most variable-frequency ac drives maintain the constant *V*/Hz ratio up to the nameplate voltage and frequency of the motor. Then the stator voltage is maintained at its nameplate value and the frequency is increased. As a result the motor operates in the constant-torque mode up to rated frequency, either 60 or 50 Hz, as appropriate, and then it operates in a constant-horsepower (constant-kW) mode above nameplate frequency [Fig. 14-27].

Modern dc link converters are programmed to provide a constant *V*/Hz ratio for constant-torque mode operation. This is achieved by controlling the amplitude of the three-phase output ac voltage using pulse-width modulation (PWM). The basic control requirements of a variable-frequency dc link converter are shown in block diagram form in Fig. 14-28.

A dc voltage signal representing the desired speed is applied to both the voltage-controlled oscillator (VCO) and the pulse-width-modulation (PWM) control to establish the constant *V*/Hz ratio. The ring counter in conjunction with the control logic distributes the gate firing pulses to the inverter switching devices (power transistors, SCRs, or GTOs). During each half-cycle of the output ac voltage the PWM control will ensure that the switching devices are switched on and off to achieve voltage control within the inverter. The voltage feedback loop permits control of the output voltage of the inverter both in the constant-torque mode (it maintains a constant *V*/Hz ratio) and in the constant-power mode (it maintains a constant output voltage). Simultaneously the input line current to the motor is sensed to ensure that the preset current limits are not exceeded. In the case of an excessive overload it will shut down the converter. Because of the different dc potentials that exist at the control terminals of the switching devices, the gating signals are isolated from the devices by either gate pulse transformers or optocouplers, the latter being the preferred device.

Because polyphase induction motors always draw a lagging load current, this current is not immediately transferred to the incoming switching device after commutation. As a result feedback diodes are connected in inverse parallel with the switching devices to provide a path around them.

Since polyphase induction motors are essentially constant-speed machines, it is normal to operate variable-frequency inverters as open-loop systems. However, if precise speed control is required, the output speed is monitored and a voltage signal representing the actual speed is supplied to the voltage-controlled oscillator (VCO) so that the frequency of the inverter output may be modified. Constant-speed outputs can also be obtained by using either a synchronous-reluctance or a synchronous motor.

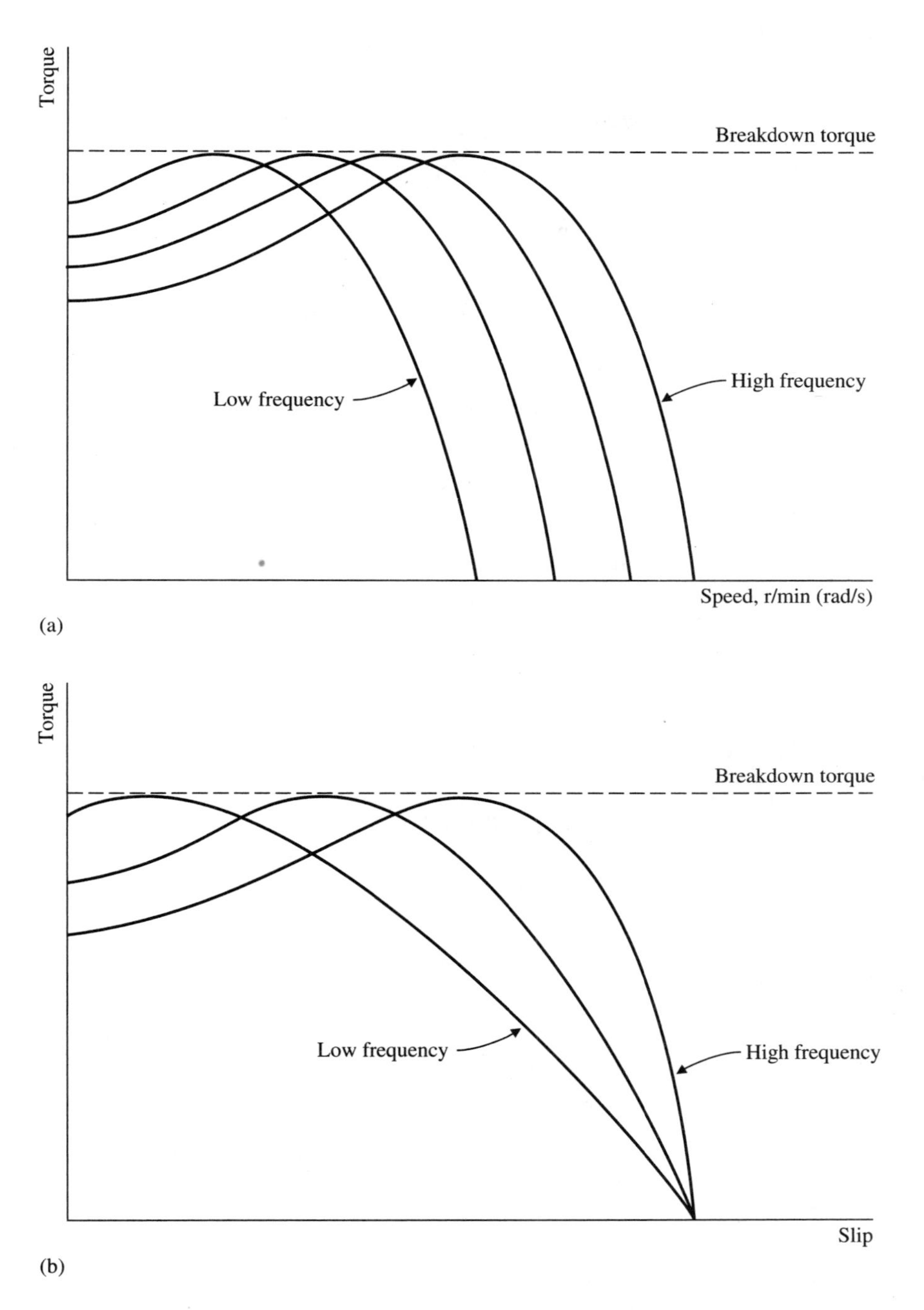

Figure 14-26 Torque-speed and torque-slip curves for polyphase squirrel-cage induction motor operating under constant *V*/Hz variable-frequency control. (a) Torque-speed curves. (b) Torque-slip curves.

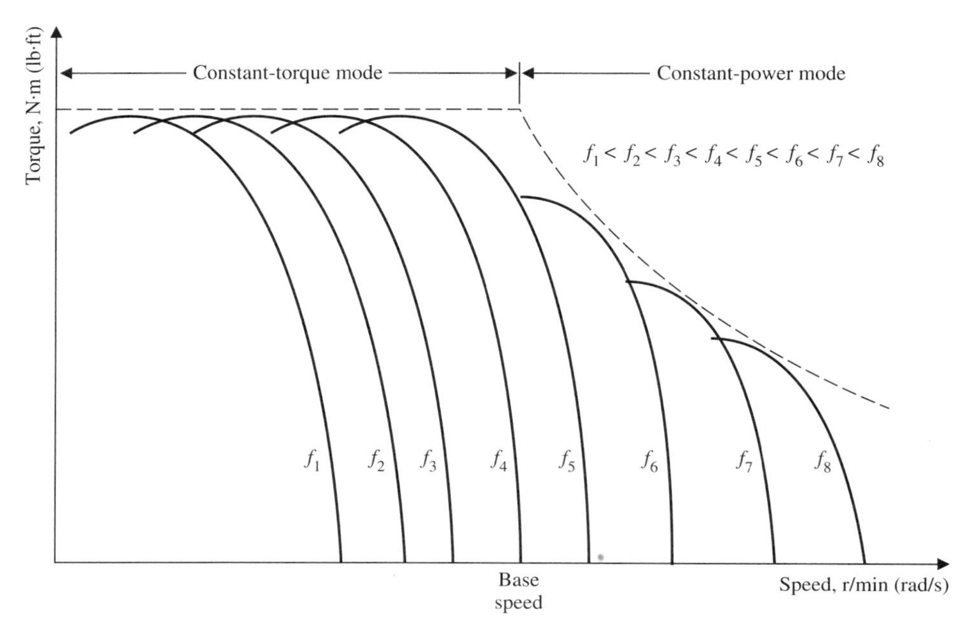

Figure 14-27 Torque-speed curves for variable-frequency operation in constant-torque and constant-power (hp) modes.

In the lower rated inverters it is common practice to use transistorized inverters, which simplifies the control requirements by eliminating the need for forced-commutation circuitry required by inverters using SCRs. Higher kVA rated inverters are currently using gate turn-off thyristors (GTOs), which also eliminate the requirement for forced-commutation circuits.

If regenerative braking is required, the dc rectifier must be replaced by a full converter which can operate in the synchronous inversion mode to permit power to be returned to the ac source. Since the current flow is reversed when the inverter is returning power, a condition that exists when the frequency is rapidly lowered or under overhauling load conditions, it will be necessary to reverse the converter output leads. This is not very satisfactory, and as a result it is normal practice to connect a second full converter in inverse parallel with the first converter and use it as an inverter during regeneration to form a dual-converter configuration (Fig. 14-29). The converter acting as an inverter during regeneration is prevented from operating until there is a reversal of current. It can be appreciated that this configuration increases the control complexity and the total cost, and is only used if dynamic braking is not feasible.

Dynamic braking is accomplished by connecting a discharge resistance across the inverter input as the dc voltage rises during regeneration. This can be done by using an overvoltage relay or a voltage-detection circuit energizing an auxiliary SCR.

The output of the dc rectifier is filtered by a capacitor connected across the output. If provision for regeneration or dynamic braking is not made, then under regenerative conditions the capacitor will become overcharged and the input voltage to the inverter will rise, with a subsequent increase in the iron and copper losses of the machine, which in turn will indirectly provide a substitute for dynamic braking.

The major advantages of dc link converters are:

1. The steady-state speed accuracy is excellent, typically 0.05% over a 10:1 speed range.
2. They have a rapid dynamic response.
3. The full-load power factor as seen from the ac source is usually about 0.95 lagging.
4. The overall efficiency of the converter and motor is 85% or better.
5. Installation costs are low.
6. Reduced-voltage starting is unnecessary, since the frequency is reduced to the order of 5 Hz and then increased to the desired value. This permits motor starting without high starting currents.

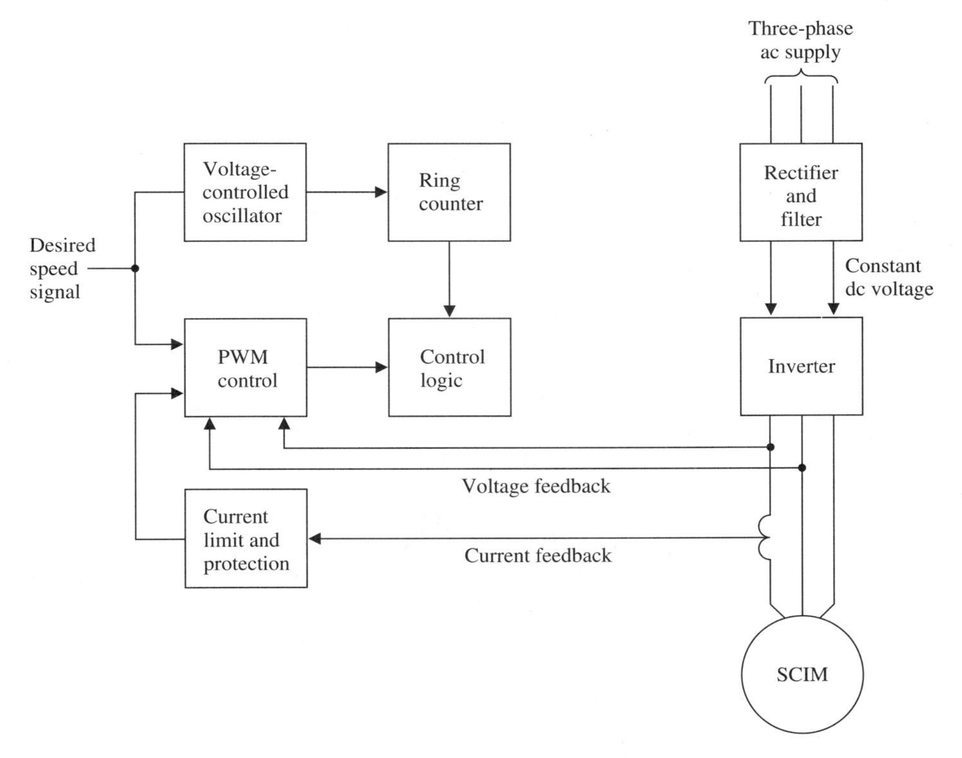

Figure 14-28 Block diagram of variable-frequency pulse-wave-modulated (PWM) dc link converter driving a squirrel-cage induction motor.

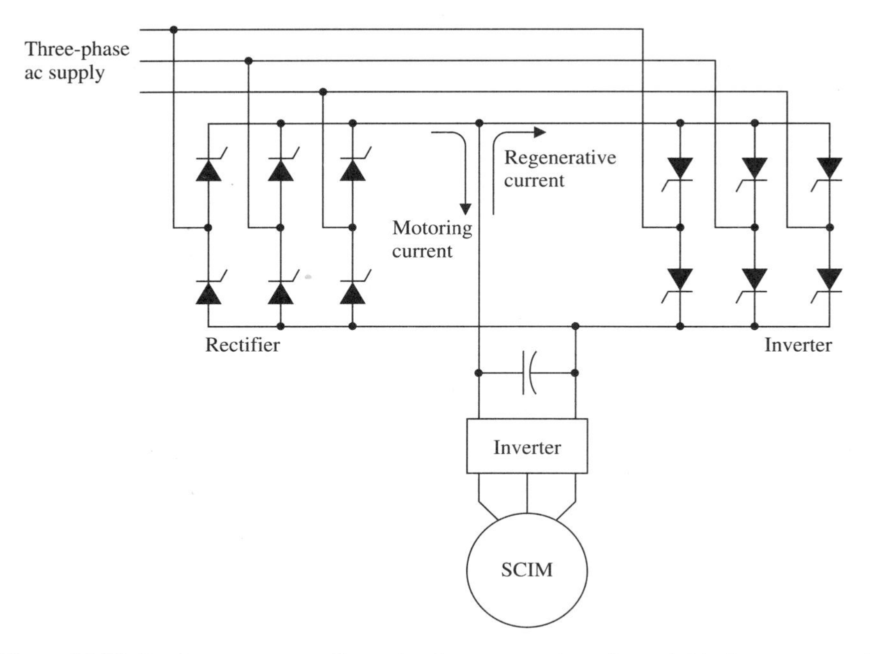

Figure 14-29 Dual-converter configuration for regeneration of a variable-frequency ac drive.

7. Multimotor control is easily achieved from one inverter, thus permitting synchronized control of a conveyer line or, in the case of electric traction, simultaneous control of a number of drive motors.

Slip Power Recovery—Wound-Rotor Motors

The most common method of obtaining variable-speed control of a polyphase wound-rotor induction motor was by the use of an adjustable external resistance connected to the rotor circuit to obtain changes in the speed-torque characteristics of the drive. This arrangement was particularly suited to applications requiring a high starting torque with a low inrush current. Typical examples are elevators, forced draft fans, printing presses, cranes, and conveyer systems. The major disadvantage of rotor resistance control is that the energy in the rotor is dissipated in the controlling resistance, which is directly proportional to the ohmic value for a given load. As a result the motor efficiency drops significantly as the speed is reduced. In general, with this method the speed is never reduced below 50% of the synchronous speed because the overall efficiency will be less than 50%.

In this age of increasing energy costs it only makes good sense to convert the traditional rheostatic method of speed control by solid-state means to a regenerative control system. This type of system is known as *slip power recovery*. A typical system is illustrated in Fig. 14-30. Slip power recovery systems have a high overall efficiency. Up to 98% of the previously wasted energy is recovered, thus improving the overall motor efficiency by close to 25%.

The motor stator is energized from the three-phase supply when contacts M close. The rotor voltage $V_R = sV_{BR}$ appears across the rotor slip rings. It should be noted that V_R decreases as the slip s decreases. For successful speed control it is necessary to develop motor torque over the entire range of speed control. It should also be noted that the developed torque T_d is proportional to the rotor current I_R. When power is applied to the stator, the resulting rotor voltage V_R is supplied to the uncontrolled diode bridge. The resulting dc voltage after filtering by the smoothing reactor is then applied to the phase-controlled thyristor converter. This converter operates in the synchronous inversion mode with a firing delay angle range of $\pi/2 \leq \alpha \leq \pi$. The rotor speed is controlled by adjusting the rotor current and, hence, the developed torque as a function of the load. The rotor current in turn is determined by the firing angle control of the thyristor converter, so that at standstill the firing delay angle is a maximum, which corresponds to the condition that the maximum inverter voltage be greater than the rectified dc link voltage. Advancing the firing point in response to a speed error signal will cause the dc link voltage to drop below the rectified rotor voltage. As a result the rotor current will begin to flow, thus producing torque, and the rotor accelerates. The output of the full converter is applied via the voltage-matching transformer to the motor stator terminals, thus returning the energy removed from the rotor to the source. The voltage-matching transformer is used to match the voltages at any given speed in order to obtain the best power factor.

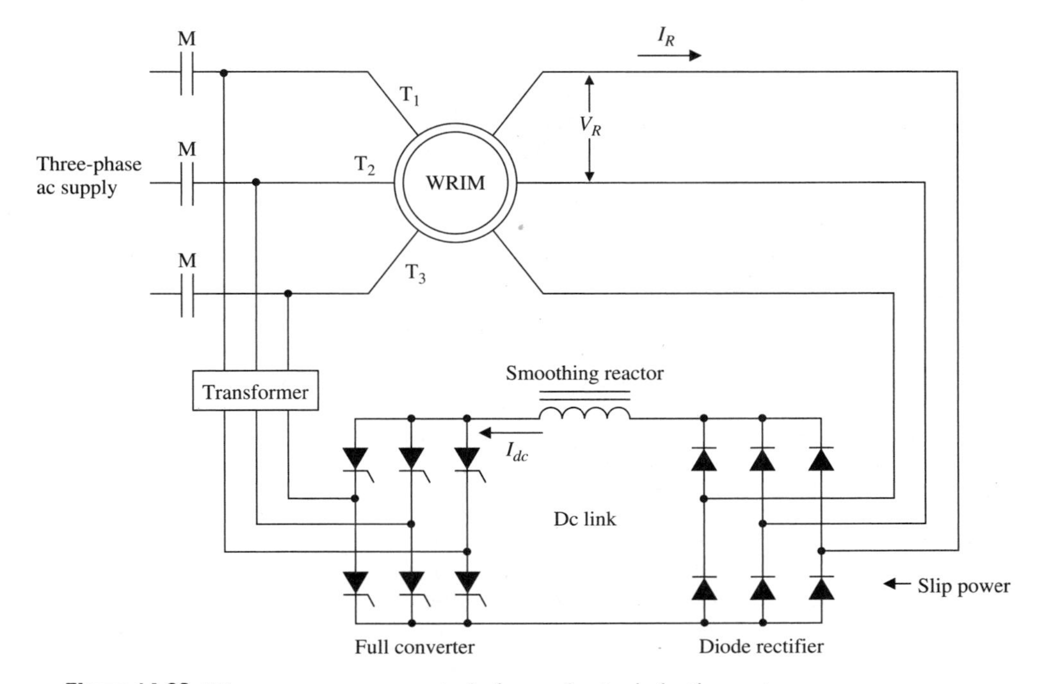

Figure 14-30 Slip power recovery control of wound-rotor induction motor.

To operate as a closed-loop system with variable-speed and -torque capabilities requires that input signals representing the desired speed and torque be compared with the actual speed signal derived from a tachogenerator coupled to the motor shaft, and with the actual torque signal derived as voltage drop across a resistor in the dc link. The error signals are then used to control the inverter firing angle, so that as the firing angle decreases, the motor speed drops, or vice versa.

Drives of this type are called *static Kramer drives* and are used to control high-power-output wound-rotor induction motors driving pumps or forced-draft fans.

Cycloconverter Drives

The basic theory of cycloconverters has been discussed in Chapter 2. The cycloconverter converts three-phase ac power at one frequency to three-phase ac power at a lower frequency, usually less than one-third the input frequency, in one step without an intermediate power conversion, as is the case with a dc link converter. The variable-voltage variable-frequency power output can be directly applied to the stator of a polyphase ac motor, or in the case of a wound-rotor induction motor it can be applied to the rotor circuit. The latter system is called *static Scherbius drive.*

Since the output voltage waveform is reconstituted from segments of the input ac voltages, it will contain complex harmonics. These harmonics will normally be attenuated by the motor inductance. However, harmonic currents will also be reflected to the ac source. Since the cycloconverter uses phase shift control, it will always present a lagging power factor to the ac source, irrespective of the load power factor. A major advantage of a cycloconverter drive is that it is capable of regeneration. Cycloconverter drives are found mainly in low-speed high-horsepower (kW) applications such as cement mills, ore mills, rolling mills, and propulsion systems with speeds as low as 10 r/min (1.05 rad/s) up to 600 r/min (62.85 rad/s) and power output ratings ranging from 1,000 to 40,000 kW (1,340 to 53,600 hp).

AC Voltage Control

The torque developed by polyphase squirrel-cage induction motors is proportional to the square of the applied voltage. As the stator voltage is reduced, the air-gap flux and the torque are also reduced, with the result that the rotor speed is reduced. The disadvantage of this method of speed control is that the range of control is small. Also stator voltage control is not suitable for constant-torque applications, which mainly limits stator voltage control to applications demanding a low starting torque and a narrow range of speed control at a low slip.

Reduced stator voltages may be obtained by: (1) variable-voltage variable-frequency inverters, (2) pulse-width modulation inverters, and (3) ac voltage controllers. The ac voltage controller (see Chapter 2) is the most commonly used method. In low-power applications TRIACs are used, while higher power applications use phase-controlled inverse-parallel-connected SCR converters. These systems suffer from harmonics in both the load and the line sides and have a worsened input power factor resulting from the reactive power demand of the converters and the added harmonics.

A typical system is illustrated in Fig. 14-31. The actual speed signal is produced by the tachogenerator coupled to the motor shaft and is compared to the desired speed signal. The output of the error amplifier is used to control the firing delay angle of the inverse-parallel-connected SCRs in series with the motor stator. If there is an increase in load torque, the resulting speed decrease is detected by the tachogenerator. The error signal will then cause the firing delay to be increased, which in turn will increase the stator voltage and speed, and vice versa. This circuit can be modified to give a reduced-voltage starting capability by the introduction of a ramp generator, which will control the rate of acceleration of the motor to the desired speed. Other options can also be supplied, such as regeneration, zero-voltage switching, reversing, current limit, and protective and diagnostic features. The major disadvantage of reduced stator voltage speed-control techniques is that they are wasteful of energy.

Eddy Current Clutch Variable-Speed Drives

Eddy current clutches are used to provide a variable-speed output from a constant-speed squirrel-cage induction motor, and are available with soft-start capability over

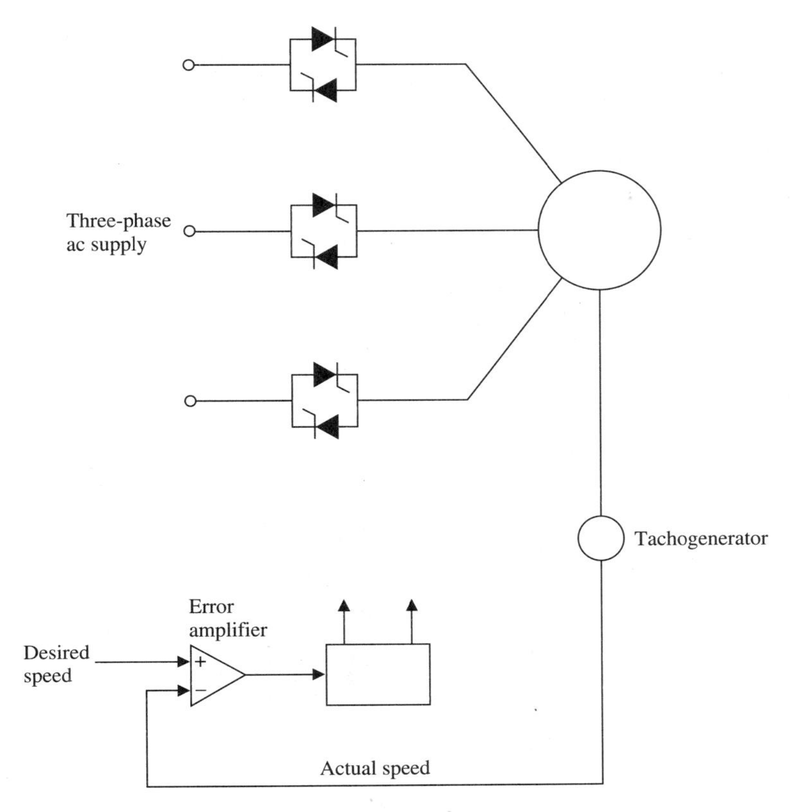

Figure 14-31 Closed-loop stator voltage speed-control system.

a speed range of 34:1 with $\frac{1}{2}\%$ speed regulation in power outputs from $\frac{1}{4}$ to 4,000 hp (0.19 to 2.984 kW).

The eddy current coupling consists of three basic parts: the driving member or drum assembly, the driven member or rotor assembly, and a magnetic member which is the field coil assembly. These parts are shown in Fig. 14-32.

The high-strength iron drum is connected to the shaft of the constant-speed motor in the motor section. This drum rotates at the same speed as the motor shaft. The drum which is the driving member has its inner surface surrounding and partially enclosing the field coil and rotor, or driven member. The variable-speed output shaft is attached to the rotor assembly and rotates at speeds determined by the controller. The cylindrical portion of the rotor lies between the drum and the field coil assembly and is divided into magnetically isolated sections by a nonmagnetic ring. Poles cast into each rotor section project alternately across the outer surface of the rotor. As the field coil is energized, the rotor sections are polarized with alternate N and S poles. The field coil assembly is a stationary toroidal coil mounted concentrically within the rotor. When the field coil is excited, magnetic lines of force flow from it through the N poles of the rotor, across the air gap into the drum, and through the drum across the air gap, and return via the S pole of the rotor to the field coil. As the drum rotates, the magnetic lines of force are cut by the drum, and eddy currents are induced in the drum. These eddy currents produce a second magnetic field. The interaction of this field with that produced by the field coil causes the rotor to rotate in the same direction as the drum.

When there is no current flow in the field coil, the rotor and drum are free to rotate independently of each other, that is, the drum will rotate at the same speed as the motor and the output shaft will be stationary. When current flows in the field coil, the output shaft will start turning and pick up speed. As long as the field coil remains excited, the output shaft will accelerate until it is rotating at a speed slightly less than the drum, since there must be a speed difference between rotor and drum to

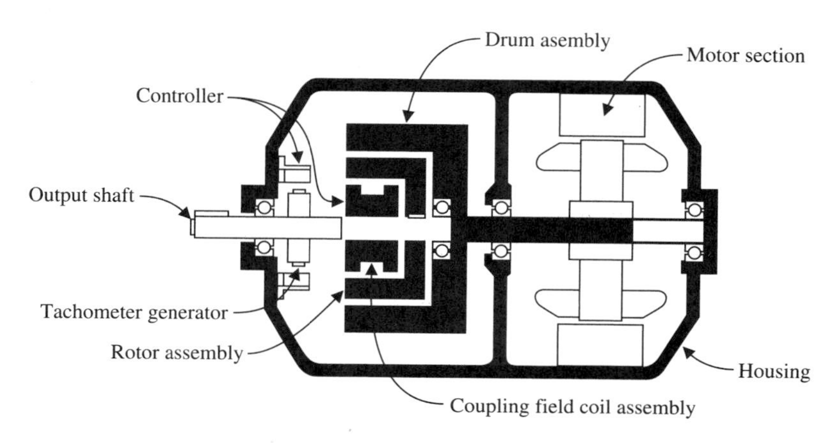

Figure 14-32 Sectional view of eddy-current drive. *(Courtesy of Eaton Corporation)*

produce the eddy current field. The controller automatically varies the field excitation in response to an error signal derived by comparing the desired speed signal to the actual speed signal produced by the tachometer generator. An Eaton Dynamatic eddy current drive is shown in Fig. 14-33.

Eddy current drives have a number of features:

1. There is adjustable speed control for constant torque or variable torque over the full speed control range, so that they may be used for constant- or variable-load applications.
2. Maintenance is low since power is transferred by the magnetic field.
3. They are ideally suited to the acceleration of high-inertia loads and intermittent overloads up to 250%.
4. There are no starting current surges during start-stop operation.
5. With the use of a tachometer generator feedback they can maintain speed within 0.5% for load changes from no load to full load.

Typical applications of eddy current drives are blowers, compressors, conveyers, cranes, dredges, elevators, and winders. They have the advantage that they are easy to maintain, are rugged, and provide an excellent substitute for static inverters in the speed range below the base speed of the drive motor.

14-18 ENERGY-EFFICIENT MOTORS

Approximately 58% of the total electric power generated in the United States is used to supply electric-motor-driven equipment in manufacturing, transportation, and electrical utilities. The most commonly used motor is the polyphase induction motor, which in 5-hp (3.73-kW) or greater ratings accounts for 54% of the total generated power. In addition, approximately 27% of the total power consumption is accounted for by motors ranging between 5 and 125 hp (3.73 and 83.25 kW). From these figures it is obvious that the use of energy-efficient motors has the potential to produce significant dollar savings in reduced electric power bills.

Factors Affecting Motor Losses

Since the most commonly used motor in industry is the polyphase squirrel-cage induction motor, we will confine our studies to this type of motor. The function of a polyphase induction motor is to convert input electric energy efficiently to output mechanical energy. The difference between these two quantities represents the losses or power consumed by the motor. The major loss components of a polyphase induction motor are:

1. Stator copper loss $I_s^2R_s$
2. Rotor copper loss $I_R^2R_R$
3. Iron losses
4. Windage and friction losses
5. Stray load losses

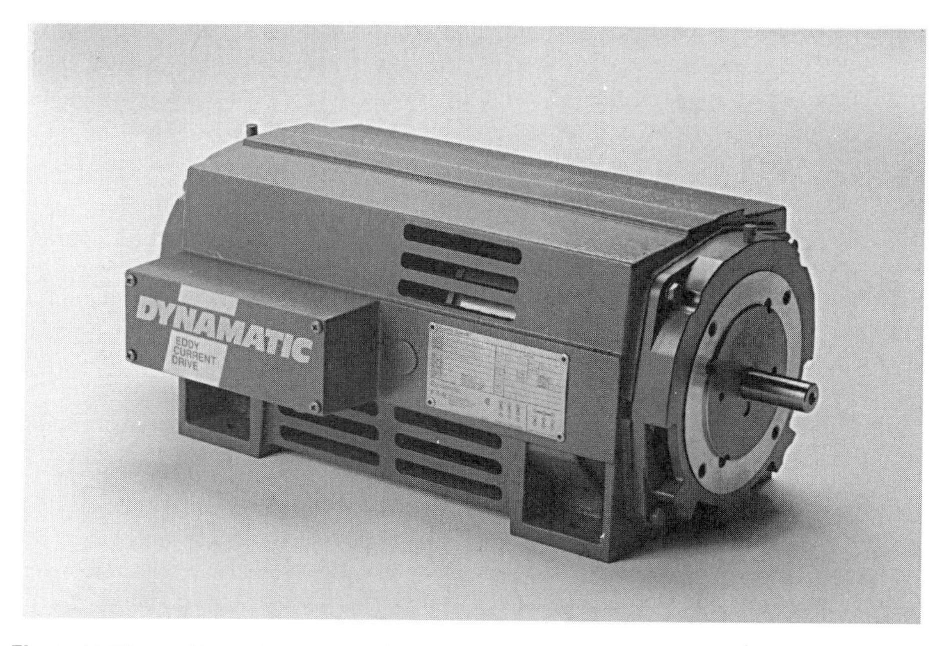

Figure 14-33 An Eaton Dynamatic eddy-current drive. *(Courtesy of Eaton Corporation)*

The stator copper loss is directly controlled by the resistance of the stator windings. There are two methods of reducing the stator resistance: (1) decreasing the specific resistance of the conductor material, which is not very practical, and (2) increasing the cross-sectional area of the stator conductors. The latter is the preferred method. However, it increases the cost and physical weight of the motor.

The rotor conductor system in the lower horsepower (kW) squirrel-cage induction motors is usually an injected aluminum conductor system together with integral conductor short-circuiting rings and fan blades. However, in the case of the polyphase induction motor the starting torque is proportional to the rotor resistance. Minimum acceptable starting torques are defined by NEMA. The most commonly used energy-efficient motor is the NEMA class B motor, which has a limited range by which the rotor resistance can be varied without exceeding the maximum permissible starting current and at the same time ensuring that the minimum values of breakdown, starting, and pull-up torques are achieved. These constraints severely limit the amount by which the rotor resistance can be reduced. As a result the rotor copper loss cannot be reduced significantly.

The greatest proportion of the iron losses occurs in the laminated steel stator core. These losses, which appear in the form of heat, are caused by hysteresis and eddy currents. Since the hysteresis loss is proportional to (flux density)^n, where n is the Steinmetz constant and is usually taken as 1.6, a figure closer to 2 is more consistent with the improvements that have been taking place in steel-making technology. The eddy current loss is proportional to (flux density)^2. The most effective measure to reduce these losses is to reduce the flux density. The most common method of reducing the flux density is to increase the axial length of the stator core. It should be noted as a result that most energy-efficient motors contain more stator copper and iron than does the normal standard motor, which in turn permits a greater heat transfer to the surrounding air, thus reducing the motor temperature.

Additional methods of reducing the stator iron losses are the use of grain-oriented silicon steel, thinner laminations, which reduce the eddy current losses, and annealing the steel. The rotor iron losses are negligible at normal operating speeds since the frequency of the rotor currents is on the order of 4–6 Hz.

The windage and friction losses are rotational losses and must be supplied from the input electric power. In energy-efficient motors the cooling requirement is reduced because of the reduced iron and copper losses. The amount of the energy component attributable to cooling is dependent on the means. For example, a smaller volume of cooling air is required for a drip-proof motor as compared to an explosion-proof motor. This is due to the fact that in the drip-proof motor the rotor fan blades create the necessary circulation of cooling air, while in the case of the explosion-proof motor there is no interchange of air between the inside and outside of the motor. As a result the heat being dissipated by the motor must be transferred by convection to the outside of the motor casing, where it is removed by high-velocity air flowing over the motor casing. It should be noted, however, that the size of the cooling fans in energy-efficient motors is smaller, implying smaller windage losses than those associated with conventional motors.

The friction loss is due entirely to the bearing friction loss, and as a result there is very little difference between the friction losses for a standard motor and for an energy-efficient motor.

Stray load losses are the difference between the total motor losses and the sum of the rotor and stator iron losses and the windage and friction losses. The stray load losses are mainly a result of the air-gap flux not being sinusoidal, which is caused by the stator and rotor slots, and the flux saturation occurring at the teeth of the laminations. The effect of this nonsinusoidal flux distribution is to produce high-frequency currents in the rotor bars and also high-frequency iron losses in the rotor and stator teeth in the air gap. The high-frequency rotor currents are harmonic currents, which create an ohmic loss in the rotor conductors, but contribute very little useful torque to the output of the machine. Another source of iron loss is the presence of steel baffles controlling the air flow around the stator end connections. Also by carefully machining the rotor to obtain a uniform air gap the stray losses can be reduced.

14-19 INDUCTION GENERATOR

From Fig. 14-8 it was seen that if a polyphase induction motor is driven at speeds in excess of the synchronous speed, in the same direction as the stator field, by either an overhauling load or an external prime mover, the direction of the induced torque reverses and the machine acts as an *induction generator.* When acting as a generator it will produce a full-load output when the rotor speed exceeds the synchronous speed by the same percentage as it would when operating at full load as a motor, that is, the negative slip at full load as a generator is equal to the positive slip at full load as a motor. If the induction generator is driven by a torque in excess of this value, a peak torque known as the *pushover torque* is reached and the machine will overspeed (Fig. 14-34).

Since the induction generator does not have an excitation system, it cannot produce reactive power. Therefore a source of reactive power external to the machine must be provided to maintain the stator field. The reactive power is provided by connecting capacitors across the stator terminals (see Fig 14-35). The output voltage and frequency are controlled by the ac system to which the induction generator is connected.

Currently induction generators driven by variable-pitch waterwheel turbines are used as unattended hydroelectric generating station plants up to 100 kW. This approach permits the use of small rivers suitably dammed to utilize previously untapped power sources.

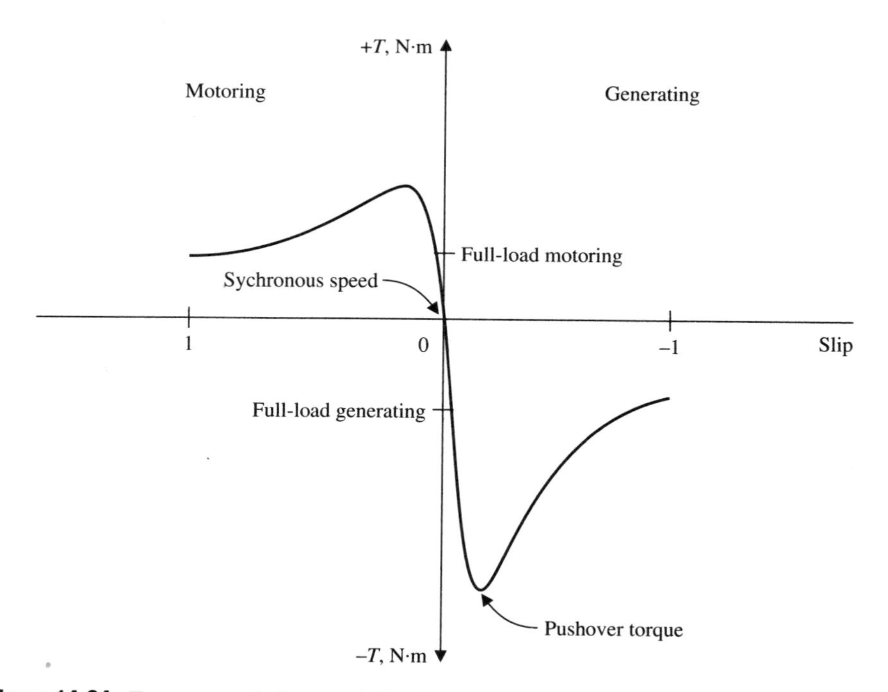

Figure 14-34 Torque-speed characteristic of polyphase induction motor acting as an induction generator.

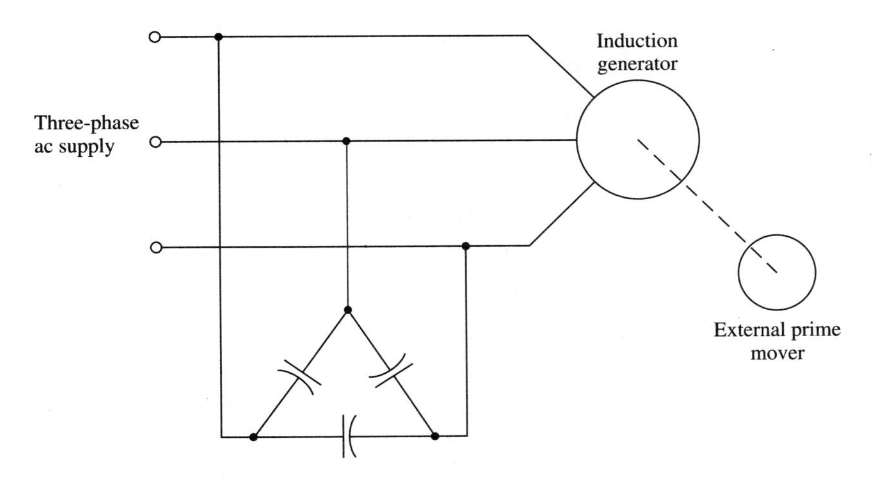

Figure 14-35 Induction generator driven by an external prime mover with a capacitor bank supplying reactive power for self-excitation.

QUESTIONS

14-1 Why is the polyphase squirrel-cage induction motor the most commonly used motor in industrial and commercial applications?

14-2 Discuss the two methods of constructing squirrel-cage rotors.

14-3 Discuss the construction of a wound rotor.

14-4 With the aid of sketches explain the principle of operation of a polyphase induction motor.

14-5 What is meant by the term *slip*?

14-6 Why must an induction motor operate with a slip less than 1?

14-7 When loaded lightly, why does a polyphase induction motor operate with a large lagging power factor?

14-8 Why does the power factor improve as a polyphase induction motor is loaded?

14-9 Explain with the aid of a sketch the shape of a typical polyphase induction motor torque-speed curve.

14-10 What is meant by breakdown torque?

14-11 What is meant by starting torque?

14-12 What are the conditions at which maximum power is developed by a polyphase induction motor?

14-13 What factor determines the slip at which breakdown torque occurs?

14-14 Describe how a no-load test is performed. What losses are measured?

14-15 Describe how you would perform a blocked-rotor test. What constants are obtained?

14-16 Why is the starting cycle the most severe period in the operation of an induction motor?

14-17 What are the advantages of using a wound-rotor induction motor?

14-18 How can the speed of using a wound-rotor induction motor be varied? What factors limit this method of speed control?

14-19 How are the inrush currents of a wound-rotor motor limited during the starting period?

14-20 Discuss how the starting and running characteristics of squirrel-cage induction motors can be modified.

14-21 Describe the construction of a deep-bar squirrel-cage rotor. Discuss the principle of operation.

14-22 Describe the construction of a double-cage rotor induction motor. Discuss the principle of operation and the torque-speed characteristics.

14-23 Discuss the characteristics and applications of NEMA design motors.

14-24 With the aid of a graph discuss the operating curves of a squirrel-cage induction motor.

14-25 Why is it necessary to use reduced-voltage starting with polyphase induction motors?

14-26 What are the objections to using full-voltage starting of polyphase induction motors?

14-27 What factors determine whether full-voltage starting should be employed?

14-28 With the aid of a sketch explain the operation of a full-voltage automatic starter.

14-29 Draw a schematic and explain the operation of a full-voltage reversing starter.

14-30 With the aid of a schematic explain the operation of a line-resistance reduced-voltage automatic starter. Why is the starter classified as a closed-transition starter? What is the advantage of using closed-transition starters?

14-31 When would a line-reactor starter be used?

14-32 With the aid of a sketch explain the operation of an autotransformer starter.

14-33 Why should closed-transition starters be used when starting higher power output induction motors?

14-34 With the aid of a sketch explain the operation of a star-delta starter.

14-35 With the aid of a sketch explain the operation of a three-step wound-rotor induction motor starter.

14-36 Discuss the principle of plug stopping of an induction motor.

14-37 Discuss the principle of dynamic braking as applied to a polyphase squirrel-cage induction motor.

14-38 With the aid of a sketch explain the principle of speed control using the consequent-pole method. What are the limitations of this method?

14-39 Why is the consequent-pole method of speed control restricted to squirrel-cage induction motors?

14-40 What are the advantages and disadvantages of using two-winding induction motors?

14-41 Discuss with the aid of waveforms the principle of pole-amplitude-modulation speed control of an induction motor.

14-42 Why is frequency control below base speed rarely used?

14-43 Why is the constant *V*/Hz control technique used in variable-voltage and variable-frequency speed control of an induction motor?

14-44 With the aid of a graph discuss the operation of induction motors in the constant-torque and constant-power output modes.

14-45 With the aid of a schematic explain the basic operation of a pulse-width modulated dc link converter driving a squirrel-cage induction motor. Why are these systems normally operated in open loop?

14-46 Discuss with the aid of a schematic the operation of a dual converter system for regenerative braking of a polyphase induction motor.

14-47 Discuss the advantages of using dc link converters.

14-48 Discuss with the aid of a schematic the operation of a slip power recovery speed control of a wound-rotor induction motor.

14-49 With the aid of a sketch discuss the use of ac voltage controllers to control the speed of a polyphase induction motor. What are the limitations of this method?

14-50 With the aid of a sketch explain the operation of an eddy current clutch variable-speed drive. What are the advantages and disadvantages of this type of drive?

14-51 In what ways do energy-efficient motors differ from the conventional polyphase squirrel-cage induction motor?

14-52 Discuss the operation of an induction generator.

PROBLEMS

14-1 How many stator poles per phase does a 60-Hz polyphase induction motor have if the nameplate speed is 695 r/min (72.78 rad/s)?

14-2 A three-phase 60-Hz eight-pole induction motor has a full-load slip of 0.05. What is the speed of the rotor with respect to the stator field?

14-3 A 208-V three-phase six-pole 60-Hz induction motor has a slip of 3.8%. Calculate: **(a)** the synchronous speed of the stator field; **(b)** the rotor speed; **(c)** the slip speed of the rotor; **(d)** the frequency of the rotor currents.

14-4 A 208-V three-phase 25-hp (18.65-kW) 60-Hz squirrel-cage induction motor runs at 1,150 r/min (120.43 rad/s) at full load and draws 75 A. If the power input is 20,500 W at full load, calculate: **(a)** the slip; **(b)** the speed regulation if the no-load speed is 1,185 r/min (124.09 rad/s); **(c)** the input power factor; **(d)** the output torque; **(e)** the efficiency.

14-5 A 460-V 60-Hz six-pole three-phase star-connected induction motor is rated at 150 hp (111.9 kW). Its equivalent circuit parameters are $R_S = 0.085\ \Omega$, $R_R' = 0.062\ \Omega$, $X_m = 7.5\ \Omega$, $X_S = 0.188\ \Omega$, $X_R' = 0.18\ \Omega$, $P_{fwc} = 2.45$ kW, and $P_{\text{stray}} = 1.25$ kW. For a slip of 0.04, calculate: **(a)** the line current; **(b)** the stator copper loss; **(c)** the rotor power input; **(d)** the gross developed output power; **(e)** the output torque; **(f)** the efficiency; **(g)** the rotor speed.

14-6 A 440-V 60-Hz three-phase star-connected four-pole induction motor develops 10 hp (7.46 kW) at 1,745 r/min (102.74 rad/s). The input power is 10 kW at 0.85 lagging power factor. The no-load input power is 0.6 kW. Calculate: **(a)** the full-load slip; **(b)** the line current; **(c)** the rotor copper loss; **(d)** the stator copper loss; **(e)** the output torque.

14-7 A 208-V 60-Hz six-pole three-phase star-connected squirrel-cage induction motor yielded the following test results. No load: 208 V, 8.2 A, and 550 W; blocked rotor: 138 V, 50 A, and 5,750 W. The dc resistance per phase is 0.45 Ω. Assume that the stator and referred rotor reactances are equal. Calculate the equivalent circuit parameters.

14-8 The total power input to a three-phase induction motor is 150 kW. The stator copper loss is 2.5 kW. If the slip is 3.2%, calculate: **(a)** the total power output; **(b)** the rotor copper loss.

14-9 A 125-hp (93.25-kW) 440-V 60-Hz six-pole, star-connected induction motor has the following equivalent circuit parameters: $R_S = 0.072\ \Omega$, $X_S = X_R' = 0.25\ \Omega$, $R_R' = 0.060\ \Omega$ and $X_m = 7.75\ \Omega$. The rotational losses are 2.75-kW. For a slip of 3.5%, calculate: **(a)** the line current; **(b)** the input power factor; **(c)** the mechanical power output; **(d)** the output torque; **(e)** the efficiency.

14-10 For the motor of Problem 14-9, calculate: **(a)** the slip at which maximum torque occurs; **(b)** the line current at maximum torque.

14-11 For the motor of Problem 14-9, calculate: **(a)** the starting torque; **(b)** the line current at start.

14-12 A 10-hp (7.46-kW) 440-V three-phase squirrel-cage induction motor draws a starting current of 100 A when rated voltage is applied. If the starting current is limited to 60 A, what voltage should be applied to the stator terminals?

14-13 A 5-hp (3.73-kW) 440-V three-phase induction motor with a full-load efficiency of 85% operates with a 0.8 lagging power factor and has a short-circuit current of 4.25 times rated full-load current. Calculate the inrush current at the instant of starting when connected to a 440-V supply if the motor is started using a star-delta starter. Neglect the magnetizing current.

CHAPTER

15

Single-Phase Induction Motors

15-1 INTRODUCTION

The single-phase induction motor is the motor in most common use in residential, commercial, and industrial applications. In terms of sheer numbers, the use of the single-phase induction motor far exceeds that of the polyphase induction motor. Single-phase induction motors perform many duties in the home, such as compressor motors for domestic refrigerators, freezers, and air-conditioners, sump-pump motors, furnace burner and blower motors, dishwasher motors, clothes washer and dryer motors, and exhaust fans.

In these applications the motor output power is usually less than 1 hp (0.746 kW), and as a result the term *small* or *fractional-horsepower* motors has become the common method of describing them. Nearly all single-phase induction motors can be described as small or fractional-horsepower, although some integral single-phase motors are manufactured in the following sizes: 1.5 hp (1.12 kW), 2 hp (1.19 kW), 3 hp (2.24 kW), 5 hp (3.73 kW), 7.5 hp (5.6 kW), and 10 hp (7.46 kW) for both 115- and 230-V single-phase supplies.

A small or fractional-horsepower motor is defined by the National Electrical Manufacturers Association (NEMA) and the American National Standards Institute (ANSI) as "a motor built in a frame smaller than that having a continuous rating of 1 hp, open type, at 1,700 to 1,800 r/min." A small motor is usually thought of as a fractional-horsepower motor, but in fact the classification is based on the frame size. The following two examples illustrate the significance of this definition:

1. A $\frac{3}{4}$-hp 900-r/min motor, if its frame is used for an 1,800-r/min motor, would have a rating of 0.75 hp $\times$ 1,800/900 r/min = 1.5 hp, and by definition this is

greater than 1 hp. Therefore the motor is considered to be an integral-horsepower machine.

2. A 1.5-hp 3,600-r/min motor, if its frame is used for an 1,800-r/min motor, would have a rating of 1.5 hp × 1,800/3,600 r/min = 0.75 hp, and by definition this is less than 1 hp. Therefore the motor is considered to be a *fractional-horsepower* machine.

15-2 CONSTRUCTION

Single-phase induction motors have squirrel-cage rotors that are identical to their three-phase counterparts. The laminated stator has winding slots distributed uniformly around the inner surface of the stator, in which are placed two windings, the *main* or running winding and the *auxiliary* or start winding. These two windings are displaced 90° electrically from each other, and there are an equal number of poles in each of the two windings. Usually each pole of the main and auxiliary windings is made up of groups of concentric coils connected in series. The main winding is placed at the bottom of the stator slots, and then the auxiliary windings are placed in the stator slots but displaced 90° electrical from the main winding. The two windings are displaced from each other, because provided a single-phase ac supply were connected to a single winding, such as the main winding, a pulsating magnetic field would be produced in the air gap, that is, a rotating magnetic field would not be produced, which must be present if a starting torque is to be developed. This problem is overcome by splitting the stator winding into two parts, the main and the auxiliary windings, each displaced 90°, so that the currents in them are displaced in both time and space. This combination produces a rotating magnetic field and hence a starting torque. The fact that a single-phase motor is not inherently self-starting has led to the development of a number of different ways of starting single-phase induction motors, which in turn has led to the classification of single-phase induction motors by their method of starting.

15-3 PRODUCTION OF A ROTATING MAGNETIC FIELD

Two theories have been developed that explain the production of torque by the rotor of a single-phase induction motor once it has started turning: the double revolving-field theory and the cross-field theory.

The speed of rotation of the rotor of a single-phase induction motor is determined by the rotational speed of the rotating magnetic field and the number of stator poles in the main winding. The rotational speed of the magnetic field is

$$S_S = \frac{120f}{P} \text{ r/min} \qquad \textbf{(15–1)(E)}$$

where S_s is the rotational speed of the rotating magnetic field in r/min, f the frequency of the applied voltage in hertz, and P the number of stator poles. In SI units,

$$\omega_S = \frac{4\pi f}{P} \text{ rad/s} \qquad \textbf{(15–1)(SI)}$$

where ω_s is the rotational speed of the rotating magnetic field in rad/s.

The rotor speed is always less than the speed of the rotating magnetic field, and the difference between the speed of the rotating magnetic field S_S or ω_S and the rotor speed S_R or ω_R is called the *slip*. The percentage slip is

$$\% \text{ slip } s = \frac{S_S - S_R}{S_S} \times 100\% \qquad \textbf{(15–2)(E)}$$

and the fractional slip is

$$s = \frac{S_S - S_R}{S_S} \qquad \textbf{(15–3)(E)}$$

In SI units the percentage slip is

$$\% \text{ slip } s = \frac{\omega_S - \omega_R}{\omega_S} \times 100\% \qquad \textbf{(15–2)(SI)}$$

and the fractional slip is

$$s = \frac{\omega_S - \omega_R}{\omega_S} \qquad \textbf{(15–3)(SI)}$$

Double Revolving-Field Theory

Applying a single-phase sinusoidal ac voltage to the two-pole stator winding shown in Fig. 15-1(a) produces a sinusoidally varying flux Φ_S, assuming that the iron of the magnetic circuit is operating on the straight-line portion of the *B–H* curve, acting along the vertical axis. This flux Φ_S in turn links with the conductors of the squirrel-cage rotor and by transformer action induces a voltage in the rotor. By Lenz's law the direction of the rotor flux Φ_R produced by the currents flowing in the short-circuited conductors of the rotor will oppose the stator flux Φ_S. The rotor flux is a pulsating flux along the vertical axis.

Ferraris showed that any sinusoidal magnetic field that was varying or pulsating with time along a fixed axis could be resolved into two equal sinusoidal magnetic fields rotating in opposite directions at ω or $2\pi f$ rad/s, each having a maximum amplitude equal to one-half that of the initial magnetic field. Figure 15-1(b) shows these two fields rotating at an angular velocity of ω rad/s in the air gap.

At time $t = 0$ the two phasors Φ_1 and Φ_2 are acting along the vertical axis and the resultant flux is at its maximum positive value, $\Phi = \Phi_1 + \Phi_2$. At the position shown, Φ_1 has turned counterclockwise through the angle ωt, simultaneously Φ_2 has turned clockwise through the angle ωt. Since any phasor can be resolved into two components at right angles to each other, Φ_1 and Φ_2 will both have vertical components which are additive, and horizontal components which cancel each other at every instant. Therefore, irrespective of the value of ωt, the angle turned through, the resultant of Φ_1 and Φ_2 will always lie along the vertical axis, and will assume all values from $+\Phi$ to $-\Phi$.

Each field Φ_1 and Φ_2 will act on the rotor in exactly the same manner as the rotating magnetic field of a polyphase induction motor, that is, Φ_1 tending to cause the

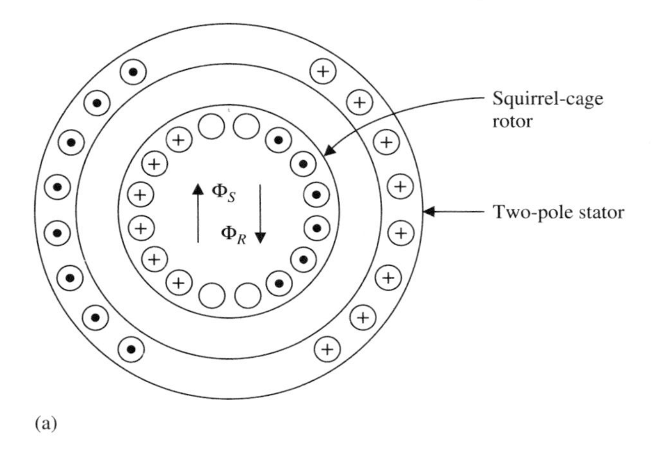

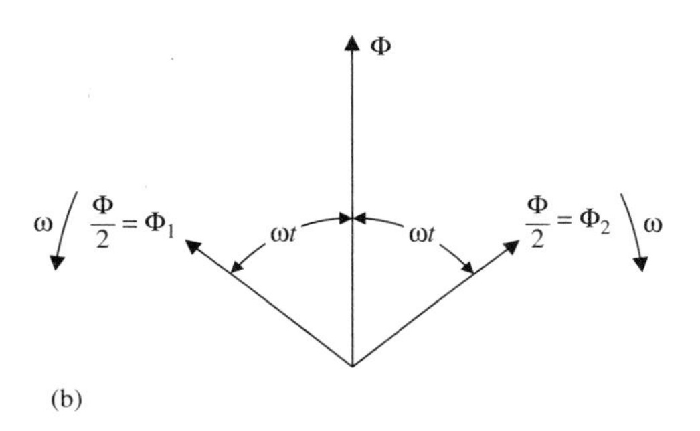

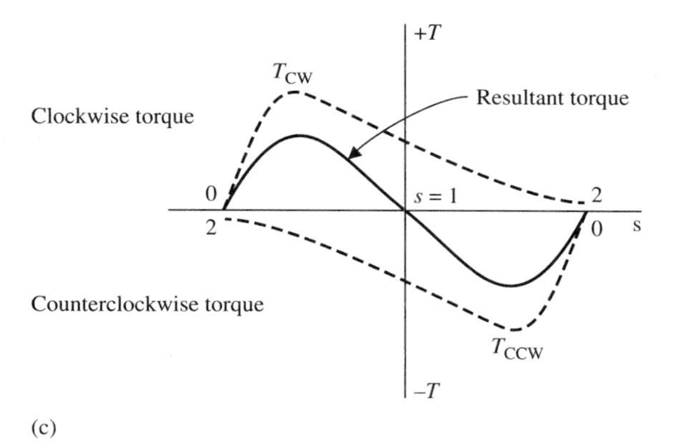

Figure 15-1 Double revolving-field theory. (a) Magnetic fields. (b) Double revolving fields. (c) Torque-slip characteristics.

rotor to turn in a counterclockwise direction and Φ_2 tending to cause the rotor to turn in a clockwise direction.

Figure 15-1(c) shows the torque-slip curves with T_{ccw} due to Φ_1 and T_{cw} due to Φ_2. As can be seen, the torques act in opposite directions. At standstill the slip $s = 1$, the two torques are equal and opposite, and the rotor will not turn. Provided the rotor is turned clockwise, T_{cw} will be greater than T_{ccw} and the rotor will accelerate in the clockwise direction and approach the synchronous speed of Φ_2. However, the rotor will not attain synchronous speed, but will run at a slip determined by the load connected to the motor shaft. There will always be a countertorque produced by T_{ccw}, but at normal operating speed it has a negligible effect. When the motor is operating at a small slip, its slip with respect to T_{ccw} is approaching $s = 2$. The rotating field component Φ_1 is moving past the rotor conductors at double the rate established by the ac supply. Therefore it induces double-frequency currents in the rotor. These currents will not produce a significant countertorque because the rotor will present a high impedance to them.

From the preceding explanation it can be seen that the rotor speed will build up to slip speed regardless of the direction in which it was originally started.

Cross-Field Theory

The cross-field theory provides an alternative explanation of torque production by a single-phase induction motor. Energizing the stator windings from the ac source (Fig. 15-2) creates a pulsating magnetic flux Φ_S acting horizontally. This flux induces electromotive forces in the rotor conductors by transformer action. Assuming that the rotor has been turned in a clockwise direction, then the rotor conductors cut the stator flux

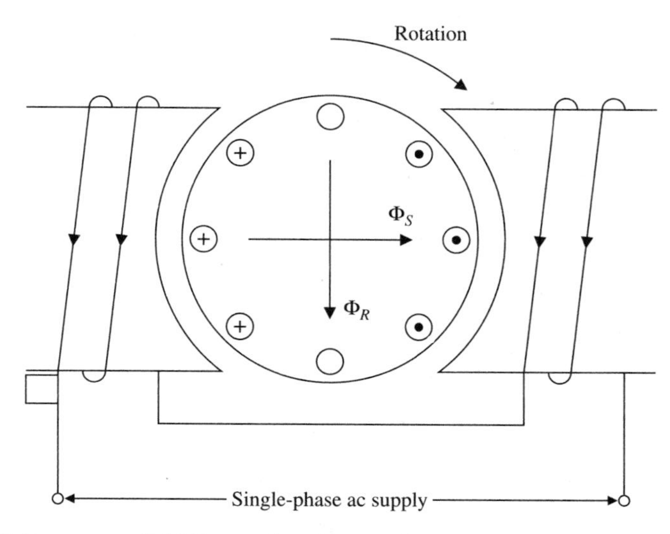

Figure 15-2 Rotor cross-field flux produced by rotation.

Φ_S, and a *speed voltage* is induced in the rotor conductors, which is the proportional to the rotor speed. Since the rotor is inductive, the rotor currents I_R produced in the conductors under the stator poles by the speed voltage will lag the speed voltage by nearly 90°. These rotor currents produce a rotor flux Φ_R, which lags the stator flux Φ_S by 90°.

The combination of Φ_S and Φ_R is to produce a synchronous clockwise rotating magnetic field. However, unlike the polyphase rotating magnetic field, it does not have a constant amplitude. This is because Φ_R varies with the rotor speed, being a minimum at low speeds and approximately equal to Φ_S as the rotor speed approaches synchronous speed. This is also consistent with the fact that the output torque increases with the rotor speed. The variation of the resultant magnetic field with variations of the rotor speed are shown in Fig. 15-3. It varies from being almost completely circular when the rotor is turning at close to synchronous speed to elliptical at half-speed and entirely pulsating when the rotor is stationary.

It can be seen that the single-phase induction motor is not self-starting. However, as soon as a quadrature flux is present, a synchronous rotating field will be developed to keep the rotor turning. To make a single-phase induction motor self-starting obviously requires that a quadrature field be present at the instant of applying power. This is the function of the auxiliary or start winding, which is placed at 90° electrical with respect to the main or run winding, the two windings being connected in parallel during the starting process.

15-4 SINGLE-PHASE INDUCTION MOTOR STARTING METHODS

There are three methods commonly used to make single-phase induction motors self-starting. The motors are classified by the way that starting torque is developed:

1. Split-phase motors
2. Capacitor-start motors
3. Shaded-pole motors

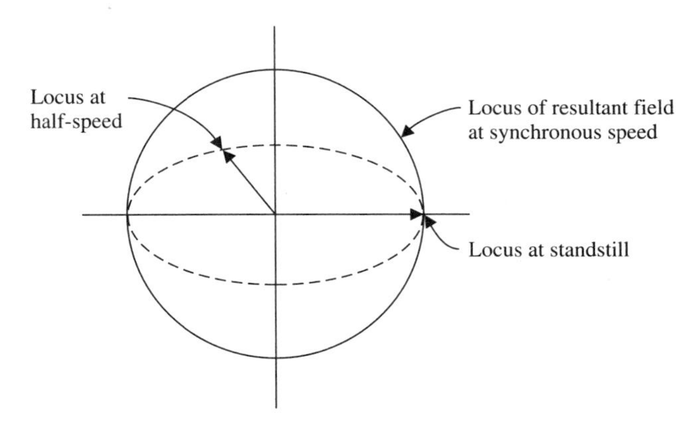

Figure 15-3 Variations of resultant flux amplitude with changes of rotor speed.

Split-Phase Motors

The split-phase or resistance-starting single-phase induction motor is shown diagrammatically in Fig. 15-4(a). It consists of a squirrel-cage rotor and two stator windings connected in parallel, but physically placed 90° electrical apart in the stator slots. The main or run winding has many turns of large-gauge insulated copper wire. This coil is placed in the bottom of the stator slot and has a low resistance and high inductive reactance. The auxiliary or start winding has fewer turns of smaller gauge insulated copper wire and is placed in the stator slots above, but displaced 90° electrical from the main winding. This coil has a high resistance but a low inductive reactance.

The main winding impedance is less than the auxiliary winding impedance. Therefore the main winding current I_m will be greater than the auxiliary winding current I_a. At the instant of starting with the centrifugal switch closed, the auxiliary winding current lags the supply voltage by approximately 15°, and the main winding current, because of the higher reactance, lags the supply voltage by approximately 40°, giving a displacement between the two currents of usually between 20 and 30°. The phasor diagram of the split-phase motor under starting conditions is shown in Fig. 15-4(b). The quadrature components of I_m and I_a, namely I_m' and I_a' are nearly equal. Since the quadrature components are displaced 90° in time, and the main and auxiliary windings are also displaced 90° in space, a two-phase rotating magnetic field is produced, which produces sufficient starting torque to be developed to accelerate the rotor against the connected load, in the direction of the rotating field.

In the process of accelerating, the rotor will also develop a field in accordance with the cross-field theory, and its direction of rotation is determined by the double revolving-field theory. However, at approximately 75–80% of the synchronous speed, the magnitude of the pulsating field produced by the main winding exceeds that produced by the combined main and auxiliary windings, that is, the rotor will develop less torque with both fields than it will with the main winding only. At this point the centrifugal switch is designed to open and remove power from the auxiliary winding. The motor continues accelerating until it reaches its operating speed under the influence of its own cross field.

The centrifugal switch has two functions: (1) to disconnect the auxiliary winding at the point where the torque produced by the main winding is greater than the torque produced by both the main and the auxiliary windings, and (2) to prevent the motor from drawing excessive power and destroying the auxiliary winding by overheating when it is running at rated speed under load. From the phasor diagram it can be seen that under operating conditions the motor current is I_m, which lags the supply voltage by ϕ_m, that is, the split-phase motor has a poor power factor under load conditions.

Centrifugal switches are the major weakness of single-phase motors, since under operating conditions the contacts arc and pit and eventually weld together, with the result that the auxiliary winding burns out because of the excessive I^2R losses. Modern practice, especially during repair, is to replace the centrifugal switch with a solid-state starting switch, which senses the induced voltage in the auxiliary winding and isolates the auxiliary winding at approximately 80% of the synchronous speed. These switches also have a restart capability, provided the motor speed drops below 50%, by cutting

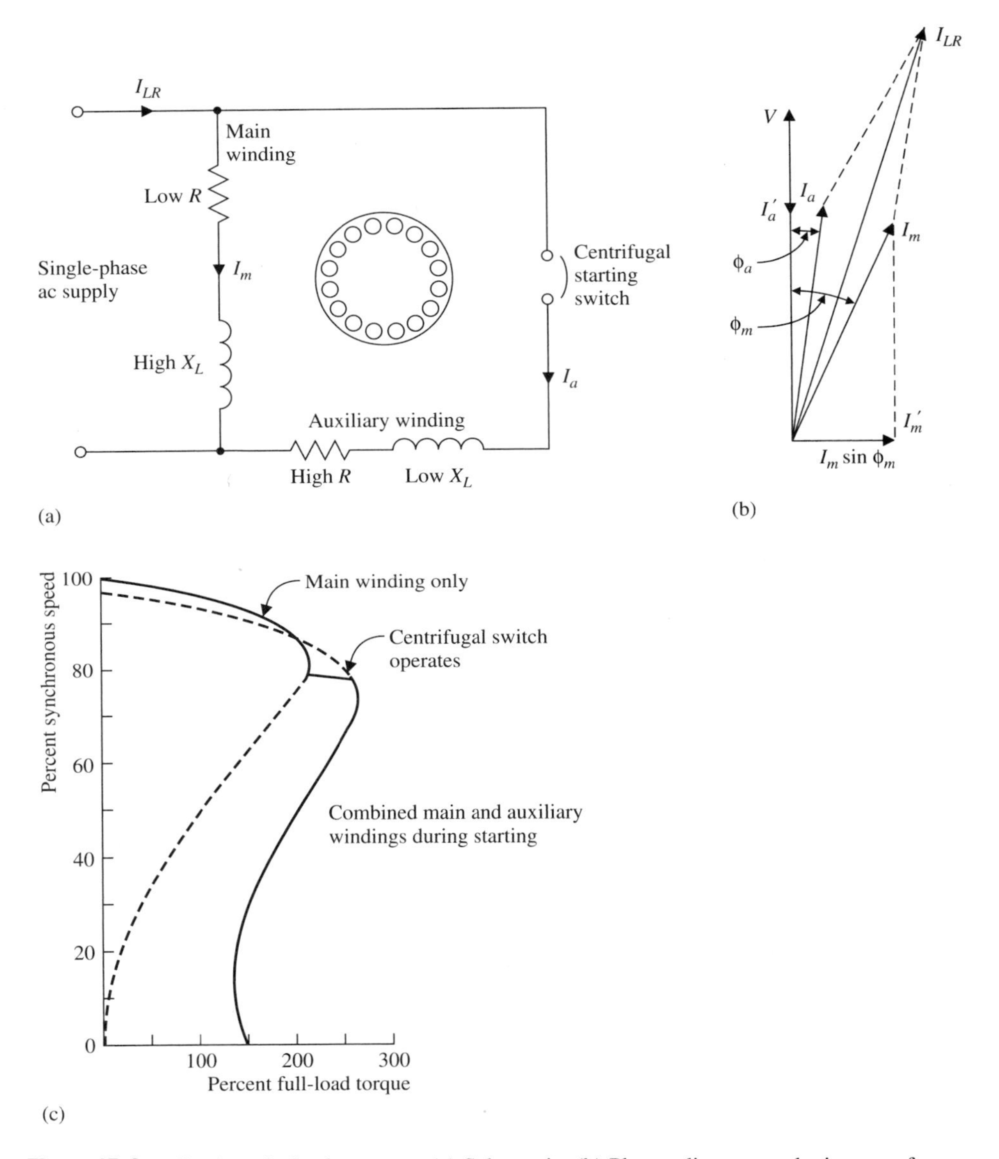

Figure 15-4 Split-phase induction motor. (a) Schematic. (b) Phasor diagram at the instant of starting. (c) Speed-torque characteristic.

the auxiliary winding back in and permitting the motor to reaccelerate back to rated speed. These units either fit inside the motor end bell or can be mounted externally.

Split-phase motors are readily available in sizes from $\frac{1}{20}$ hp (37 W) to $\frac{1}{3}$ hp (246 W), with starting torques varying from 150 to 200% of rated full-load torque.

The locked-rotor or starting current I_{LR} is usually between five and seven times rated full-load current, and the full-load slip is usually about 5%. Under full load the efficiency is approximately 65% at a power factor of about 0.7 lagging.

Split-phase motors can only be reversed at standstill by reversing either the main or the auxiliary winding connections. Two-speed and dual-voltage motors are available. Under heavy load conditions the elliptical torque (Fig. 15-3) produces a noisy vibration, which can be minimized by using rubber isolating mounts.

Speed control presents a problem because there are basically two methods: (1) changing the number of poles, or (2) changing the frequency. The number of poles can be changed by using consequent-pole arrangements of the main and auxiliary windings, which will give a 2:1 speed change, or alternately by using two entirely separate sets of stator windings. It is not practical to use variable-frequency speed control because of the limited range of control available without activating the starting switch.

Split-phase motors should be used in constant-torque applications, where low starting and accelerating torques are acceptable, and should not be used where frequent starting and high-inertia loads are encountered, because the auxiliary winding will be subjected to overheating. They are used in many applications such as oil-burner motors, blower fans, automatic clothes washers and dryers, and bench-mounted power tools.

Capacitor-Start Motors

To produce a rotating magnetic field it is necessary to generate two fluxes with an angular separation, that is, separated in both time and space. In the split-phase motor this was done by phase shifting the currents in the main and auxiliary windings, that is, making the main winding inductive and the auxiliary winding mainly resistive. This produces a phase shift between the two currents of approximately $\theta = 30°$. The starting or locked-rotor torque is

$$T = kI_m I_a \sin \theta \tag{15-4}$$

where T is the locked-rotor torque, I_m the main winding current, I_a the auxiliary winding current, both under locked-rotor conditions, and θ the phase angle between I_m and I_a, where $\theta = \phi_m - \phi_a$. k is a constant determined by the windings, the rotor resistance, and the synchronous speed of the motor.

From Eq. (15-4) it can be seen that the starting torque will be a maximum when $\theta = 90°$. Also the starting torque is proportional to $\sin \theta$. Therefore to increase the starting torque it is necessary to increase θ. This may be achieved by connecting a capacitor in series with the auxiliary winding. Usually θ is increased to approximately 80°.

Capacitor-start motors are used for general-purpose heavy-duty applications where high starting and running torques are required. There are three types of capacitor-start motors:

1. Capacitor-start motors, which have a capacitor connected in series with the auxiliary winding during the start cycle.
2. Two-value capacitor motors, which have a capacitor connected in series with the auxiliary winding during the starting cycle, and have a permanently connected capacitor in series with the auxiliary winding at all times.

3. Permanent-split capacitor motors, which have a capacitor permanently connected in series with the auxiliary winding and do not have a starting switch.

Capacitor-Start Motor. The schematic, phasor diagram, and speed-torque curve of a capacitor-start motor are shown in Fig. 15-5. The capacitors are selected to ensure that under starting conditions the angle between I_m and I_a is approximately 80°, that is, the capacitive reactance exceeds the inductive reactance of the auxiliary winding. Typically I_a leads I_m by 80°. From Eq. (15-4) the torque is proportional to sin θ. Therefore the starting torque of a capacitor-start motor as compared to a split-phase motor using identical main and auxiliary windings will be sin 80°/sin 25° = $0.98/0.42 \cong 2.33$, or 2.33 times greater than that of the split-phase motor. Capacitors also have the added benefit of improving the input power factor, thus reducing the size of the inrush starting current.

The use of capacitor-start induction motors has greatly increased with the introduction of small cylindrical, typically 1.5-in (3.8-cm)-diameter by 3.5-in (8.89-cm)-long dry-type ac electrolytic capacitors. They are assigned a voltage rating, usually 115 or 230 V, and a temperature rating. Though normally they are tested at 25°C (77°F), they are rated for operation at 65°C (150°F) and can be used up to 80°C (176°F). However, operation at the higher temperature reduces the life expectancy of the capacitor. Their duty cycle is based on 20 three-second operating periods per hour.

Typical values of starting capacitors for 115-V motors range from 135 μF for a $\frac{1}{4}$-hp (186-W) motor to 350 μF for a $\frac{3}{4}$-hp (560-W) motor.

The effect of the starting capacitor is to produce a starting torque of 3.5 to 4 times the rated full-load torque of the motor [Fig. 15-5(c)]. After the motor has run up to rated speed, it has the same operating characteristics as the split-phase motor. Unlike the split-phase motor, the capacitor-start motor is reversible. This is done by momentarily disconnecting the supply; then, when the starting switch recloses, the auxiliary winding connections are reversed and power is reapplied to the motor. The rotating magnetic field is now reversed, and the motor will decelerate to zero and then accelerate to full speed in the opposite direction. A typical application of a reversible capacitor-start motor is an electrically operated hoist.

The capacitor-start motor is more sensitive to centrifugal switch failures because of the duty cycle rating of the starting capacitor. For these reasons, especially with hermetically sealed units such as refrigerator compressors, it is common practice to use a current-sensitive relay mounted external to the unit; as an alternative, a solid-state starting switch, also externally mounted, can be used.

Capacitor-start motors are available in sizes ranging from $\frac{1}{6}$ hp (124 W) up to 10 hp (7.5 kW). The integral-size motors are usually dual voltage, that is, the windings are connected in parallel for 115-V and in series for 230-V operation. Because of their high starting torque, they are used for air-conditioners, refrigerators and deep freezers, electric hoists, large fans, pumps, and high-inertia loads.

Two-Value Capacitor Motors. The two-value or capacitor-start capacitor-run motor overcomes an undesirable feature of both split-phase and capacitor-start induction motors, which, under running conditions, both operate with a poor power factor. The poor power factor is caused by the large magnetizing component. As a result these motors draw higher input currents, which in turn means increased copper losses, which is the reason why these motors run relatively hot. Also because of the iron and copper losses, their overall efficiency is usually in the range of 65–75%.

This problem can be reduced by using an electrolytic capacitor, typically 300–500 μF, for starting, and an oil-filled capacitor, usually about 20 μF, to improve the power factor under running conditions. The connection arrangement is shown in Fig. 15-6(a). The capacitors are chosen so that the start capacitor gives the best starting torque

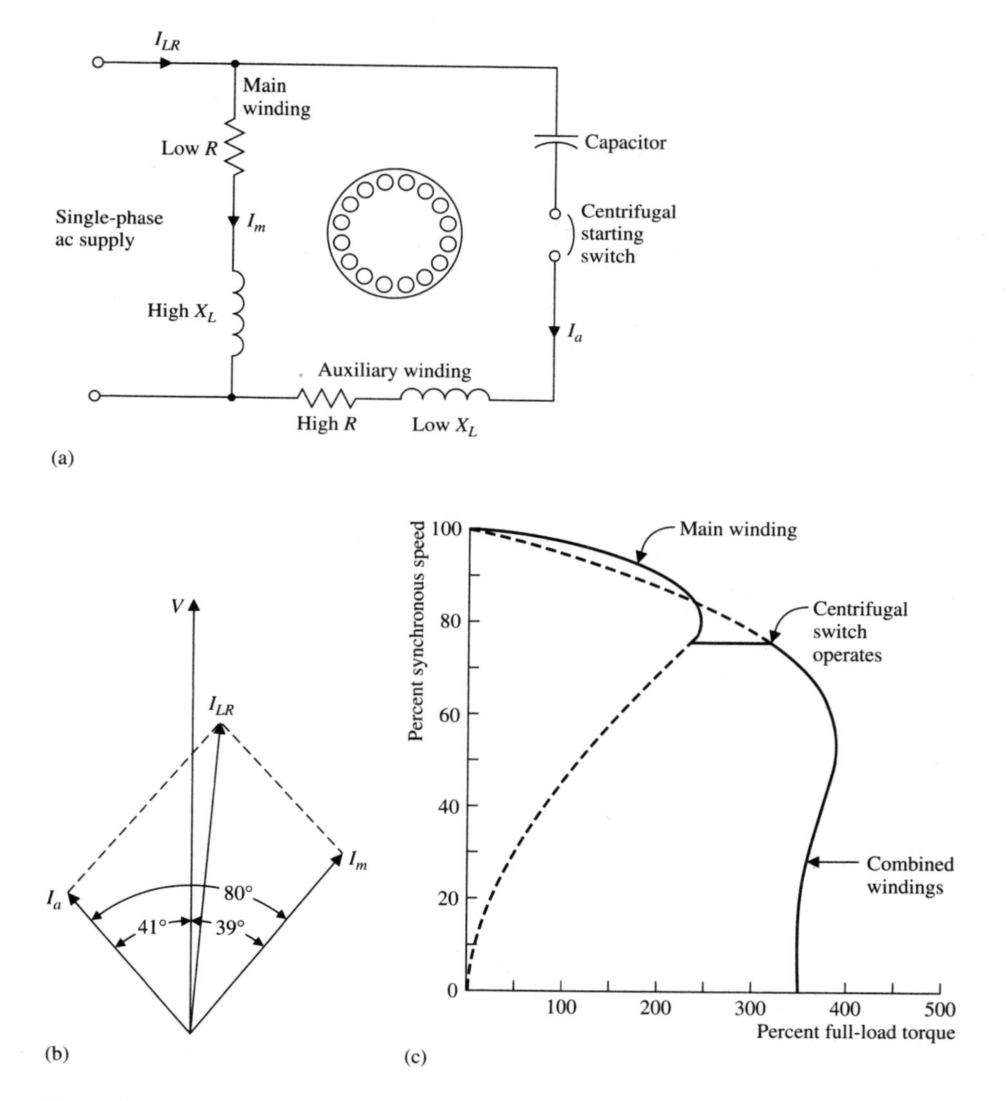

Figure 15-5 Capacitor-start induction motor. (a) Schematic. (b) Phasor diagram under starting conditions. (c) Speed-torque characteristic.

and the run capacitor improves the motor power factor to approximately unity, thus reducing the input line current and copper losses and therefore improving the overall efficiency.

An alternative method uses one oil-filled capacitor in conjunction with a tapped autotransformer [Fig. 15-6(b)]. This method depends on the principle that the reflected impedance of the secondary winding to the primary winding is proportional to the square of the ratio of the secondary turns to the primary turns, that is,

$$\alpha^2 = \left(\frac{N_S}{N_P}\right)^2 \tag{15–5}$$

Therefore if the autotransformer primary has 160 turns and the secondary has 30 turns, an 8-μF run capacitor appears as a $(160/30)^2 \times 8\mu F = 227$-μF start capacitor. The run capacitor must be able to withstand a step-up voltage of $(160/30) \times 115\text{ V} = 613\text{ V}$. It should be noted that a failure of the centrifugal starting switch can easily lead to the failure of the oil-filled capacitor which, since it is usually rated at 1,000 V, is relatively expensive to replace.

The two-value capacitor motor can be reversed by reversing the auxiliary winding connections. The autotransformer may first appear to be an expensive alternative to a capacitor, but an oil-filled capacitor of the required capacitance may be more costly.

The major benefits of using two-value capacitor motors are an improvement in the operating power factor to about 0.8 lagging, which in turn reduces the operating costs, although the overall efficiency remains basically unchanged. Also there is a slight increase in the breakdown torque. The two-value motor is considered to be an energy-efficient motor and is marketed by General Electric under the name Watt-Saver.

Permanent-Split Capacitor Motor. The permanent-split capacitor or single-value capacitor motor has a capacitor permanently connected in series with the auxiliary winding, and the main and auxiliary windings are energized during both starting and running conditions. The motor has a low locked-rotor torque and is designed for continuous-duty applications, such as fan and blower service.

Its major advantage is the total elimination of the centrifugal starting switch and its attendant problems. The motor is self-starting because a rotating magnetic field is produced by the main and auxiliary windings, which are displaced in both space and time. The resulting starting torque is low compared to the split-phase and capacitor-start motors [Fig. 15-7(a)]. Since the capacitor must be rated for continuous-duty operation, an oil-filled type will reduce leakage current and is selected on the basis of maximizing the motor running characteristics. However, it is smaller, usually by 5–10%, than that required for starting a capacitor-start motor. As a result the starting torque is usually about 60% of the rated full-load torque.

As can be seen from Fig. 15-7(a), the motor can be easily reversed by a simple switch. This motor is classified as a *reversing* motor since it can be reversed when operating at rated speed and load. Previously, reversing could only be done either by stopping the motor and switching the auxiliary winding connections and restarting the motor, or by switching the auxiliary winding connections after the centrifugal starting switch contacts closed. The permanent-split capacitor motor is classified as a *reversible* motor.

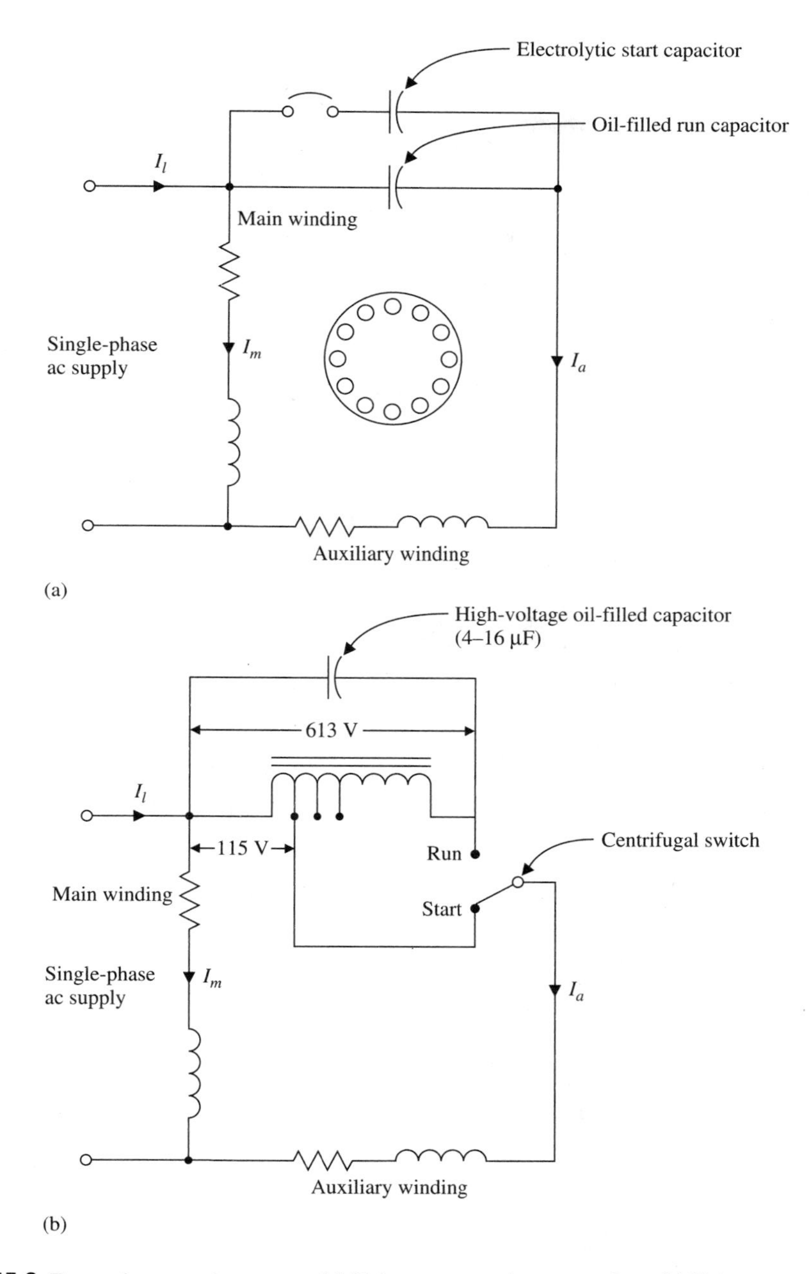

Figure 15-6 Two-value capacitor motor. (a) Using a start and run capacitor. (b) Using one capacitor and an autotransformer.

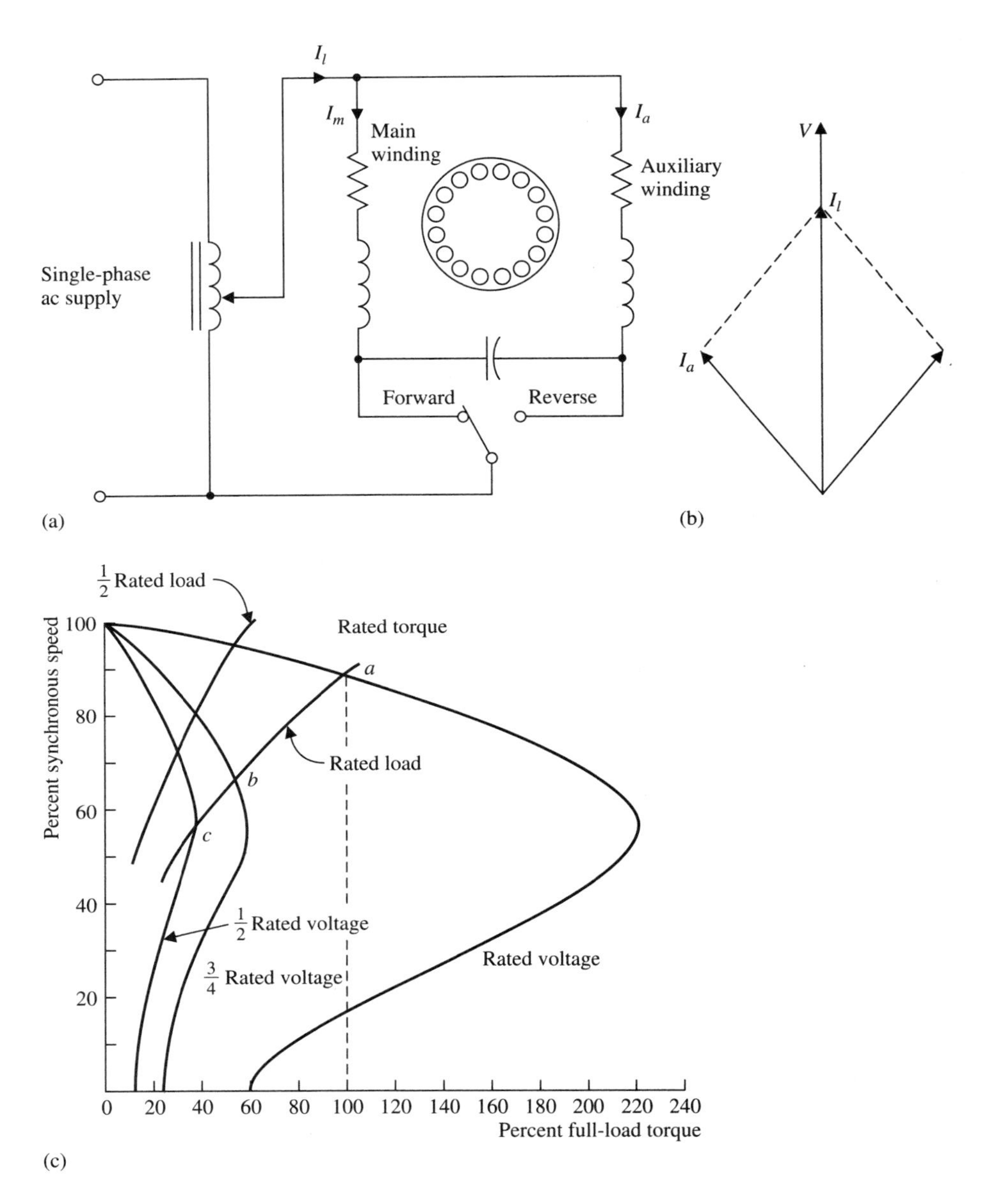

Figure 15-7 Permanent-split capacitor motor. (a) Schematic with adjustable voltage and reversing control. (b) Phasor diagram. (c) Speed-torque curves showing the effects of reduced voltage.

Since the torque T is proportional to the applied voltage squared, that is, $T \propto V^2$, and because of the low running torque, the speed of the permanent-split capacitor motor is easily controlled by varying the applied voltage. The voltage can be varied by using a tapped or a variable autotransformer, or even a phase-controlled TRIAC. Figure

15-7(c) clearly illustrates the speed changes that occur as the stator voltage is changed. For example, at rated load and voltage (point *a*) the rotor is developing rated torque at approximately 87% of synchronous speed. At point *b* with 75% of rated voltage, the rotor is developing 50% of rated torque at about 65% of synchronous speed. At point *c* at 50% of rated voltage, the rotor output is about 30% of rated full-load torque and the speed has been reduced to about 55% of the synchronous speed. When used for fan or blower loads at low-speed conditions, the speed regulation is erratic.

The advantages of using a permanent-split capacitor motor are: (1) it is a reversible motor, (2) speed can be controlled easily, and (3) it is quiet in operation. These motors are available in sizes ranging from $\frac{1}{20}$ hp (37 W) to $\frac{1}{4}$ hp (187 W), and are usually supplied as dual-voltage motors.

15-5 SINGLE-PHASE INDUCTION MOTOR TESTING

Single-phase induction motor testing is an essential element in all phases of the life cycle of a motor, starting with the initial design and development on through production-line testing, and finally as an essential tool in preventive maintenance and repair activities. Testing in the field aids maintenance personnel in determining the cause of a failure in a specific motor. It ensures that the repair action was carried out in accordance with repair specifications, and, most importantly, as a tool of preventive maintenance, it will reduce the chance of catastrophic breakdowns.

Tests carried out on single-phase induction motors are very similar to those for polyphase induction motors. The input power, current, and terminal voltage are measured with a wattmeter, an ammeter, and a voltmeter connected as shown in Fig. 15-8.

There are several ways of connecting the meters when testing a single-phase induction motor. In view of the low power levels that are encountered in fractional-horsepower motor testing, they all involve making allowances for losses in the voltmeter and wattmeter coils. This is required because the meters are connected so that the wattmeter not only measures the power input to the motor but also includes some meter losses.

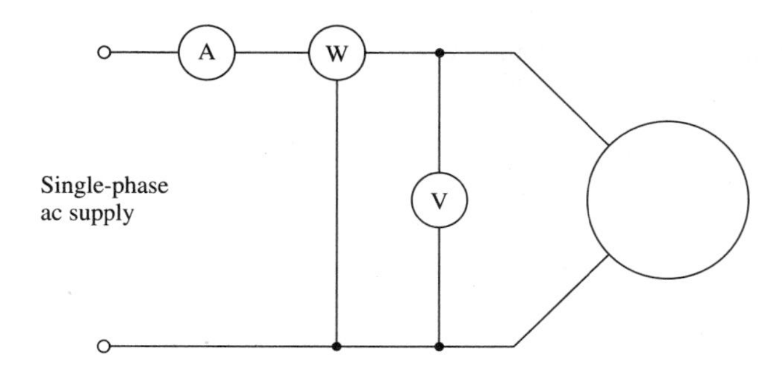

Figure 15-8 Meter connections for testing a single-phase induction motor.

Using the connections shown in Fig. 15-8, the correct power input to the motor is

$$W = W_W - \left(\frac{V_V^2}{R_V} + \frac{V_W^2}{R_W}\right) \tag{15–6}$$

where W_W is the wattmeter reading, V_V the voltage across the voltmeter terminals, V_W the voltage across the potential coil of the wattmeter ($V_V = V_W$), R_V the resistance of the voltmeter coil, and R_W the resistance of the potential coil of the wattmeter. These errors may be neglected when testing motors in excess of $\frac{1}{4}$ hp (187 W).

It is necessary to find the effective ac resistance of the main winding. Also if a two-value capacitor or permanent-split capacitor motor is to be tested, it is necessary to determine the resistance of the auxiliary winding so that its copper losses may be determined.

The effective ac resistance of the windings may be determined by the voltmeter-ammeter method, where $R_{dc} = V/I$, or it may also be determined by using a bridge megger. However, because of skin effect, eddy currents, and dielectric losses, the effective ac resistance R_{ac} will be greater than R_{dc}. It is usually assumed that $R_{ac} = 1.25\ R_{dc}$.

Also the motor temperature rise may be determined by the increase in the winding resistance, which is obtained by measuring the winding resistance before and immediately after load testing. The temperature rise is calculated from

$$\Delta T = \left(\frac{R_h}{R_c} - 1\right)(234.5 + T_a) \tag{15–7}$$

where ΔT is the temperature rise in °C above ambient temperature T_a, R_h the hot resistance, R_c the resistance at ambient temperature, and T_a the ambient temperature in °C.

Motors are usually given a commercial test, which is a simplified engineering test, before shipment by the manufacturer or after repair action. After the winding resistances have been measured and the ambient temperature recorded, the motor is usually run at no load for 10–30 min to allow the bearings to warm up. Then the no-load or rotational losses, with the motor running unloaded, are taken. They are

$$\begin{aligned} P_{rot} &= VI\cos\phi - I_l^2 R_r \\ &= W - I_l^2 R_r \end{aligned} \tag{15–8}$$

where P_{rot} denotes the no-load rotational losses and consists of the friction, windage, and iron losses, W is the wattmeter reading, I_l the line current, and R_r the effective ac resistance of the main winding for a split-phase or capacitor-start induction motor. For two-value capacitor and permanent-split capacitor motors, the copper losses of both the main and the auxiliary windings must be determined, and Eq. (15-8) becomes

$$P_{rot} = W - I_m^2 R_m - I_a^2 R_a \tag{15–9}$$

where I_m is the main winding current, I_a the auxiliary winding current, and R_m and R_a are the effective ac resistances of the main and auxiliary windings, respectively, that is, $R_m = 1.25 R_{m\,dc}$ and $R_a = 1.25 R_{a\,dc}$.

The full-load testing of the motor is usually done by using a Prony brake or a dynamometer. At rated full load the input power W, the line current I_l, the applied voltage V, and the shaft speed (r/min or rad/s) are measured and recorded. Locked-rotor readings on fractional-horsepower motors are usually taken at rated or near rated voltage. The ammeter, wattmeter, and voltmeter readings are taken simultaneously and as rapidly as possible to prevent the windings from heating up. From the locked-rotor test,

$$R_{es} = \frac{P_{LR}}{I_{LR}^2} \tag{15–10}$$

where R_{es} is the equivalent stator resistance, P_{LR} the input power under locked-rotor conditions, and I_{LR} the locked-rotor line current.

Normally with split-phase and capacitor-start motors the auxiliary winding is disconnected. When testing two-value capacitor motors, the capacitor voltage should also be measured and recorded. It should be noted that the locked-rotor torque varies with the relative position of the rotor, so several readings should be taken for different rotor positions, but only the maximum and minimum torque values need be recorded.

Another test which is often taken is the measurement of the breakdown torque. This is the maximum torque developed by the induction motor at rated voltage and frequency. It is measured by slowly increasing the load torque until there is either an abrupt speed change or the motor stalls, and then recording the torque at this point.

15-6 PERFORMANCE CALCULATIONS

➤ EXAMPLE 15-1

A $\frac{1}{3}$-hp (246-W) 115-V 60-Hz 1,730-r/min (181.17-rad/s) single-phase induction motor has an efficiency of 68% with a 0.6 power factor lagging when supplying rated load. Calculate: **(a)** the number of stator poles; **(b)** the slip; **(c)** the line current.

SOLUTION

(a)

$$S = \frac{120f}{P}$$

$$P = \frac{120f}{S} = \frac{120 \times 60\text{Hz}}{1{,}730 \text{ r/min}} = 4.16$$

Since the rotor speed is less than the speed of the rotating magnetic field, there must be an even number of poles; then $P = 4$ poles. In SI units,

$$\omega = \frac{4\pi f}{P}$$

Therefore

$$P = \frac{4\pi f}{\omega} = \frac{4\pi \times 60\text{Hz}}{181.17\text{rad/s}} = 4.16 = 4 \text{ poles}$$

(b)

$$S_S = \frac{120f}{P} = \frac{120 \times 60\text{Hz}}{4 \text{ poles}} = 1{,}800 \ r/\text{min}$$

$$\% \text{ slip} = \frac{S_S - S_R}{S_S} \times 100\% = \frac{1{,}800 - 1{,}730}{1{,}800} \times 100\% = 3.89\%$$

and the fractional slip is $s = 0.0389$. In SI units,

$$\omega_S = \frac{4\pi f}{P} = \frac{4\pi \times 60\,\text{Hz}}{4\text{ poles}} = 188.50\text{ rad/s}$$

$$\%\text{ slip} = \frac{\omega_S - \omega_R}{\omega_S} \times 100\% = 3.89\%$$

and the fractional slip is $s = 0.0389$.

(c)

$$P_{\text{out}} = \text{hp} \times 746\,\text{W/hp}$$
$$= 0.33\text{hp} \times 746\,\text{W/hp} = 246.18\,\text{W}$$
$$\eta = \frac{P_{\text{out}}}{P_{\text{in}}}$$
$$P_{\text{in}} = \frac{P_{\text{out}}}{\eta} = \frac{246.18\,\text{W}}{0.68} = 362.03\,\text{W}$$
$$= VI_l \cos\theta$$
$$I = \frac{P_{\text{in}}}{V\cos\theta} = \frac{362.03\,\text{W}}{115\,\text{V} \times 0.6} = 5.26\,\text{A}$$

➤ EXAMPLE 15-2

A $\frac{1}{4}$-hp (186-W) 1,725-r/min (180.64-rad/s) 115-V 60-Hz split-phase induction motor at the instant of starting draws a current of $11.2\angle{-15^\circ}$ A in its auxiliary winding and a current of $15.2\angle{40^\circ}$ in the main winding. At the instant of starting calculate: **(a)** the line current; **(b)** the power factor; **(c)** the in-phase components of the main and auxiliary winding currents with the line voltage.

SOLUTION

(a)

$$I_a = 11.2\angle{-15^\circ} = 10.82 - j2.90\,A$$
$$I_m = 11.64 - j9.77\,A$$
$$I_{LR} = I_a + I_m = 22.46 - j12.67 = 25.79\angle{-29.43^\circ}$$

(b)

$$\text{Power factor} = \cos(-29.43^\circ) = 0.87\text{ lagging}$$

(c) From **(a)** the in-phase components of I_a and I_m with respect to the line voltage are 10.82 and 11.64 A, respectively, that is, at the instant of starting these components are nearly equal.

➤ EXAMPLE 15-3

Repeat Example 15-2(**a**) and (**b**) if a capacitor is added in series with the auxiliary winding, which causes a current of $8.8\angle{40^\circ}$ to be drawn during starting, and calculate the value of the capacitor.

SOLUTION

(a)

$$I_a = 8.8\angle 40° = 6.74 + j5.66$$
$$I_m = 15.2\angle -40° = 11.64 - j9.77\ \text{A}$$
$$I_{LR} = I_a + I_m = 18.38 - j4.11 = 18.83\angle -12.61°\ \text{A}$$

The addition of the capacitor has reduced the inrush current, and at the same time the input power factor has been improved.

(b)

$$\text{Power factor} = \cos(-12.61°) = 0.98\ \text{lagging.}$$

(c) The auxiliary winding impedance without the capacitor is

$$Z_a = \frac{115\angle 0°\text{V}}{11.2\angle -15°\text{A}} = 10.27\angle 15°\,\Omega$$
$$= 9.92 + j2.66\,\Omega$$

The auxiliary winding impedance with the capacitor is

$$Z_a' = \frac{115\angle 0°\ \text{V}}{8.8\angle 40°\ \text{A}} = 12.95\angle -40°\,\Omega$$
$$= 9.92 - j8.32\,\Omega$$

The capacitive reactance is

$$X_c = -j8.32 - (+j2.66) = -j10.98\,\Omega$$
$$C = \frac{1}{2\pi f X_c} = \frac{1}{2\pi \times 60\,\text{Hz} \times 10.98\,\Omega} = 241.58\,\mu\text{F}$$

15-7 SOLID-STATE STARTING SWITCHES

As has been mentioned previously, the major cause of failure of split-phase capacitor-start and two-value capacitor induction motors has been the failure of the centrifugal starting switch. These switches are used extensively because of their low cost, and in the extremely competitive fractional-horsepower motor market it is essential to minimize manufacturing costs.

The major causes of centrifugal starting switch failures are:

1. Leakage of lubricating oil from sleeve bearings used in low-cost motors contaminates the switch contacts, leading to premature failure because of contact arcing.
2. High operating temperatures cause the spring material to lose its temper.
3. End play becomes excessive and breaks the switch assembly.
4. High-speed repetitive start-stop operating cycles.

The main problem is that the switch fails to open when the rotor has accelerated up to 75–80% of synchronous speed, which in turn causes the short-duty-cycle auxiliary winding to burn out. However, unless the motor is designed for a special application, it will probably be more economical to replace the motor.

It seems most logical to replace the failure-prone centrifugal switch by a solid-state switch. PT Components, Inc., Stearns Div., Milwaukee, have produced such a switch, called the Sinpac switch. This switch, which can be mounted either internally or externally to the motor, uses either a TRIAC or an SCR to satisfy the auxiliary-winding switching requirements.

Previous attempts to produce solid-state switches have assumed a constant acceleration time, sensed the main winding current, or with the use of a transducer have sensed the rotor speed. These approaches have proved unsatisfactory because timed acceleration devices are not responsive to load demand changes, and speed transducers have increased the cost of the switch significantly. The solution was found in a characteristic that is common to all single-phase motors, namely, that independent of the motor, there is a slight voltage dip in the voltage across the auxiliary winding, followed by an increase in the rotor speed at the point where switching should occur. This phenomenon is the secret of the Sinpac switch.

Two techniques of measuring and detecting the switching point have been developed. The first and simplest method is applicable to capacitor-start and capacitor start-run induction motors. With these motors the speed-sensitive voltage is monitored continuously. This method is called the continuously variable (CV) design. The speed-sensitive voltage is compared against the ac input voltage as a reference. The resulting error signal is then used to determine whether the auxiliary winding should be disconnected. When the motor has accelerated up to 75–80% of synchronous speed, the control logic initiates action to open the auxiliary winding circuit.

In the case of split-phase motors the speed-sensing technique has to be modified, since the voltage drop across both the main and the auxiliary windings is equal when the motor is first started. A pulsed-variable (PV) or sampling technique is used to measure the induced voltage in the auxiliary winding during the acceleration period. This voltage is sampled one or more times. After each sampling, if the motor is not up to switching speed, the TRIAC or SCR is triggered to reapply power to the auxiliary winding so that the power is applied in a series of steps until the motor can continue to accelerate up to normal operating speed.

Both the continuously variable and the pulsed-variable control systems enable the motor to restart if its speed drops to approximately 50–55% of synchronous speed, with the auxiliary winding being cut out at 75–80% of synchronous speed. These switches can be installed during manufacture or as a retrofit at a later date. They are currently available for 115-V and 115/230-V ratings, with maximum inrush current ratings ranging from 12 to 40 A for the pulsed-variable series and from 12 to 125 A for the continuously variable series of switches. Typical connection arrangements are revealed in Fig. 15-9.

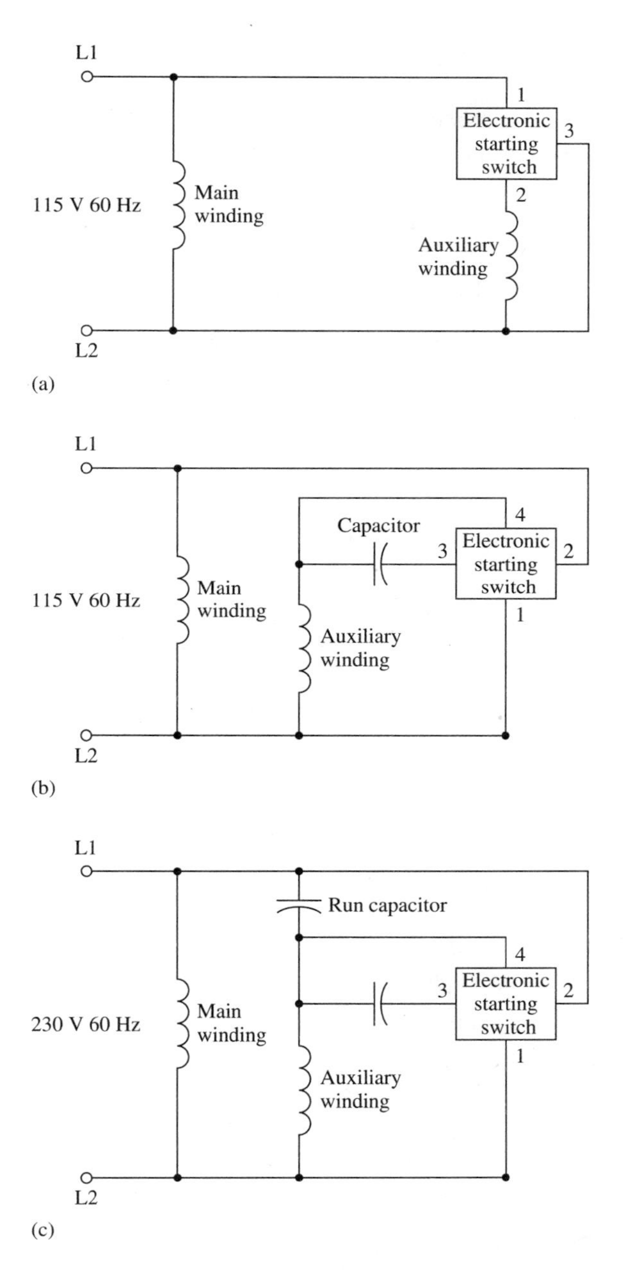

Figure 15-9 Typical connections for solid-state switch replacement for a mechanical centrifugal switch. (a) Split-phase motors. (b) Capacitor-start motors. (c) Capacitor-start capacitor-run motors. (Courtesy Power Transmission Design–A Penton Publication).

QUESTIONS

15-1 Discuss what is meant by the term *fractional-horsepower motor.*

15-2 Discuss the construction of single-phase induction motors.

15-3 With the aid of a sketch explain the double revolving-field theory.

15-4 With the aid of a sketch explain the cross-field theory.

15-5 Explain the operation of a split-phase induction motor.

15-6 Discuss the characteristics of split-phase induction motors.

15-7 How can a split-phase induction motor be reversed?

15-8 How can the speed of a split-phase induction motor be varied?

15-9 Discuss typical applications of split-phase induction motors.

15-10 What is the reason why capacitor motors are used?

15-11 With the aid of a schematic explain the operation of a capacitor-start induction motor. Why is the capacitor assigned a duty cycle?

15-12 How can a capacitor-start induction motor be reversed?

15-13 Discuss the characteristics and applications of capacitor-start induction motors.

15-14 What is the advantage of using two-value capacitor motors?

15-15 With the aid of schematics show and explain two methods of connecting capacitors in a two-value capacitor motor.

15-16 Is a two-value capacitor motor reversible?

15-17 Discuss with the aid of a sketch the construction and operation of a permanent-split capacitor motor.

15-18 Permanent-split capacitor motors are reversible; explain.

15-19 With the aid of a schematic explain how the speed of a permanent-split capacitor motor can be controlled.

15-20 Discuss the type of tests that should be conducted on a single-phase induction motor after repair action has taken place.

15-21 Discuss the reasons why centrifugal starting switches fail.

15-22 Discuss the two methods of controlling single-phase induction motors during the starting period by means of solid-state switches.

PROBLEMS

15-1 Determine which of the following motors are integral or fractional-horse-power motors: **(a)** $\frac{3}{4}$ hp, 3,600 r/min; **(b)** 1 hp, 3,600 r/min; **(c)** $\frac{1}{2}$ hp, 1,800 r/min; **(d)** $\frac{3}{4}$ hp, 1,200 r/min.

15-2 A $\frac{1}{4}$-hp 115-V 60-Hz split-phase induction motor has a 0.6 lagging power factor, and is 65% efficient at rated load. What is the line current at full-load?

15-3 A $\frac{1}{4}$-hp (187-W) 115-V 60-Hz 1,725-r/min (180.64-rad/s) split-phase induction motor draws an auxiliary winding current of $11.88\angle -16.3°$ and a main winding current of $14.72\angle -40.6°$ *A*. At the instant of starting, calculate: **(a)** the locked-rotor current; **(b)** the power factor; **(c)** the input power; **(d)** the running power factor and input with a full-load line current of $4.03\angle -40.6°$ *A*; **(e)** the full-load efficiency.

15-4 **(a)** Repeat Problem 15-3(**a**) and (**b**) when a capacitor is connected in series with the auxiliary winding, which causes the auxiliary winding to draw a leading current of $11.0\angle 40°$ *A*. **(b)** Compare the starting power factor and input power with that of the split-phase induction motor in Problem 15-3(**c**). **(c)** What is the value of the capacitor required to produce the leading current of $11.0\angle 40°$ *A*?

15-5 A $\frac{1}{3}$-hp (246-W) 115-V 60-Hz 1,725-r/min (180.6-rad/s) split-phase induction motor draws a line current of 6.0 A at a 0.7 lagging power factor at rated load. Calculate: **(a)** the full-load efficiency; **(b)** the full-load slip; **(c)** the output torque in lb·ft and N·m.

15-6 A 115-V two-value capacitor induction motor uses an autotransformer and a 6-μF oil-filled capacitor for both starting and running. Calculate the effective capacitor values when viewed from the primary side if the transformer ratios are 6:1 and 1.2:1 for the start and run connections. Also determine the minimum voltage rating of the capacitor.

15-7 A $\frac{1}{4}$-hp (187-W) 115-V 60-Hz split-phase induction motor operates at no load at 1,180 r/min (123.57 rad/s), and the full load speed is 1,120 r/min (117.29 rad/s). Calculate: **(a)** the slip at no load; **(b)** the slip at full load; **(c)** the speed regulation.

15-8 A $\frac{3}{4}$-hp (560-W) 115-V 60-Hz 1,750-r/min (183.26-rad/s) capacitor-start induction motor develops a starting torque of 350%. Calculate the starting torque in lb·ft and N·m.

CHAPTER

16

Single-Phase AC Motors

16-1 INTRODUCTION

Chapter 15 was devoted entirely to single-phase induction motors in which the starting torque was created by a rotating magnetic field produced by phase-splitting techniques. This chapter will concentrate on single-phase ac motors which use other techniques to produce torque. These motors are usually produced in fractional- and subfractional-horsepower sizes and are used widely in domestic and consumer products, business appliances, and power tools. These small-specialty motors can be grouped into (1) nonsynchronous: shaded pole and reluctance start; (2) synchronous: reluctance, hysteresis, and permanent magnet; and (3) variable speed: universal and permanent magnet.

16-2 NONSYNCHRONOUS MOTORS

Nonsynchronous motors are used in applications where their low efficiencies are more than compensated for by their low manufacturing cost. There are two major types: the shaded-pole motor and the reluctance-start motor.

Shaded-Pole Motor

The single-phase induction motors considered to date have all used slotted stators with concentrically wound main and auxiliary windings. Phase-splitting techniques were used to create an elliptical or near circular rotating magnetic field to produce starting torques.

The shaded-pole motor, on the other hand, used a stator with salient poles. The construction of a two-pole shaded-pole motor is shown in Fig. 16-1(a). The stator and salient-pole assembly is made from electrical steel laminations. The salient poles are divided into two unequal parts by a slot cut across the pole face. The smaller part, called the *shaded pole,* has a large-cross-sectional-area copper ring forming a short circuit placed around it. This copper ring is called the *shading coil.* When an ac supply is connected to the pole windings, a voltage is induced in the ring, which in turn produces a short-circuit current in the shading coils. This current, by Lenz's law, produces a magnetic field which opposes the changes in the magnetic flux that produced it. The shading coil therefore produces a delay in the magnetic flux produced in the shaded pole, that is, there is a phase delay in the production of the shaded-pole flux with respect to the unshaded-pole flux. This concept is illustrated in Fig. 16-1(b).

Referring to Fig. 16-1(b), as the sinusoidal flux is increasing (point 1), the current induced in the shaded coil produces a flux which opposes the flux buildup in the shaded pole. Hence the flux produced by the pole winding is concentrated in the unshaded pole. At point 2 the rate of change of current is zero. So there will be no voltage induced in the shading coil and hence no current flow, and therefore zero flux will be produced. Now the flux density in both the shaded and the unshaded

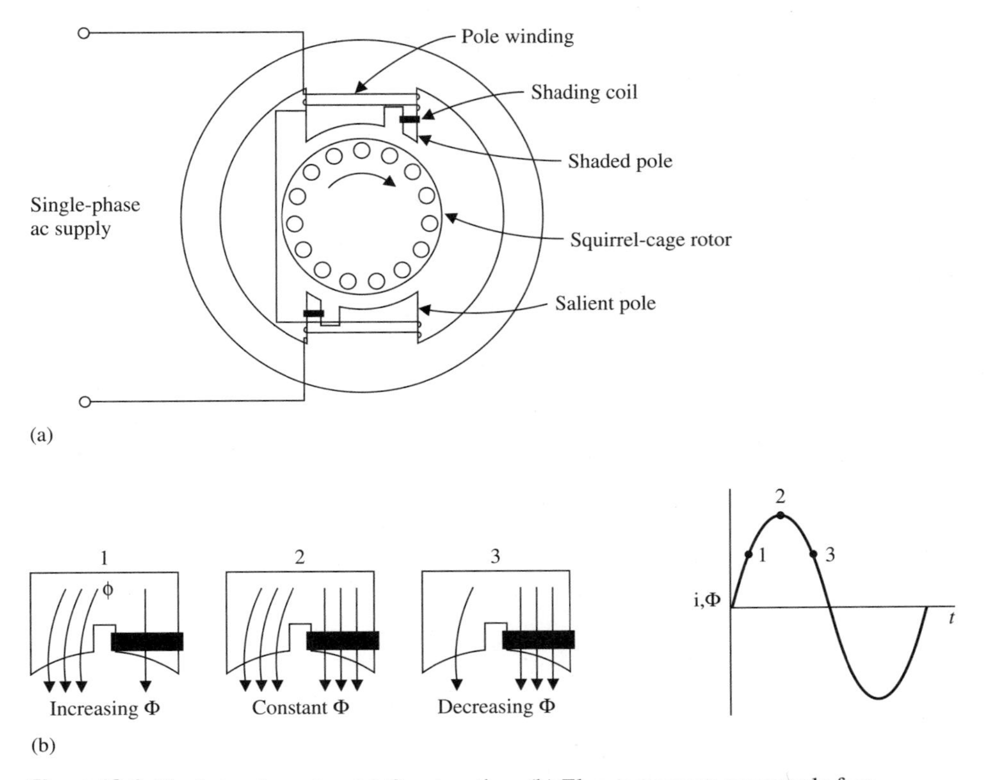

Figure 16-1 Shaded-pole motor. (a) Construction. (b) Flux movement across pole face.

poles will be about equal. At point 3 the sinusoidal flux produced by the pole winding is decreasing, and by Lenz's law, the shading coil will try to maintain the flux in the original direction. As a result the pole flux is now concentrated in the shaded pole. As can be seen from Fig. 16-1(b), there has been a net movement of the flux from the unshaded pole to the shaded pole across the salient pole face. The flux in the shaded pole is always lagging the flux produced by the pole winding by nearly 90°. The flux is not constant in amplitude and moves across all pole faces at the same time. This flux produces an unbalance in the rotor torques (double revolving-field theory) so that the clockwise torque is the stronger torque. As a result the squirrel-cage rotor will turn and accelerate in the clockwise direction. The flux movement across the pole faces continues as long as the pole windings are energized and produces an additional torque component.

The rotor of the shaded-pole motor shown in Fig. 16-1 will always revolve in a clockwise direction. However, some applications require that the motor be reversible. This can be accomplished by the arrangement shown in Fig. 16-2. The permanent copper shading rings are replaced by wound coils connected as shown. Depending on the desired direction of rotation, only two coils on the trailing side of the salient poles are short-circuited by the external switch, the other two coils remain open-circuited.

The speed of a shaded-pole motor may be varied by varying the input voltage to the pole windings, or by using taps brought out from the pole windings, in which case the induced voltage in the excited part of the pole winding is nearly equal to the

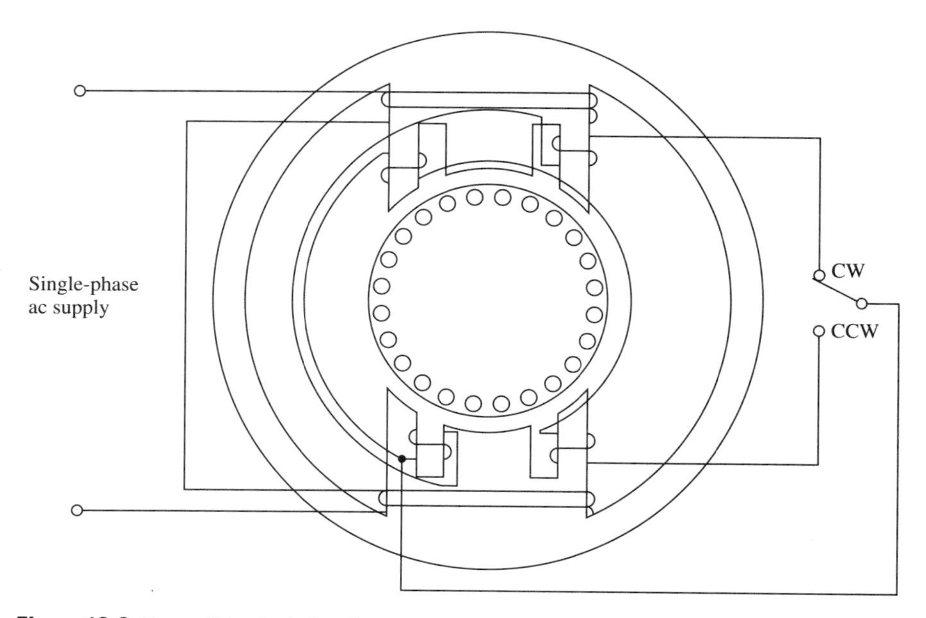

Figure 16-2 Reversible shaded-pole motor.

applied voltage. From this it can be seen that the fewer the number of pole winding turns involved, the greater will be the voltage per turn. Since the flux produced is proportional to the applied voltage, it then follows that the tap involving the least number of turns of the pole winding corresponds to the high-speed setting. Conversely when the supply voltage is applied to the maximum number of turns of the pole winding, the rotor will turn at its lowest speed.

The shaded-pole motor has a low starting torque, usually about 50% of the full-load rated torque, a low breakdown torque, low power factor, and efficiencies ranging from as low as 5% for the smallest motor rated at 1 mph (0.75 W) to between 30 and 40% for the largest motor rated at $\frac{1}{4}$ hp (187 W), and high full-load slips, usually between 7 and 10%. Standard speeds are 1,550 r/min (162.33 rad/s) for a four-pole motor, 1,050 r/min (109.96 rad/s) for a six-pole motor, and 800 r/min (83.78 rad/s) for an eight-pole motor.

Shaded-pole motors are produced in very large quantities and are cheap and simple. They are mainly used in applications where low efficiencies and power factors are unimportant, such as small fans and blowers, toys, rotisseries, and motion-picture projectors. They also need little maintenance and are trouble-free since there is no starting switch.

Reluctance-Start Motor

The reluctance-start motor uses a salient-pole stator with a nonuniform air gap. The rotor is a conventional squirrel-cage rotor [Fig. 16-3(a)].

As can be seen from Fig. 16-3(a), which shows the construction of a four-pole reluctance-start induction motor, the leading one-third of each salient pole has been cut back to increase the air gap between the pole and the rotor. From our studies of magnetic circuits it was established that the reluctance of a magnetic circuit was significantly increased by the presence of an air gap. The greater the air gap, the greater the reluctance of the magnetic circuit. Also, the self-inductance of a coil is dependent on the reluctance of the magnetic circuit; it is a minimum where the air gap is the greatest, and a maximum where the air gap is the smallest. This property causes the flux at the leading edge of the pole to lag the coil current by a small amount, whereas the flux produced at the trailing edge of the pole lags the current by nearly 90°. The overall effect is that the flux builds up rapidly in the region of the larger air gap, and is delayed in the region of the smaller air gap, that is, the flux is being delayed in time and displaced in space. The result is that the pole flux sweeps across the face of all salient poles at the same time [Fig. 16-3(b)], in a manner similar to the field produced by the shaded-pole motor.

The major difference as compared to the shaded-pole motor, is that the cross-field electromotive force, which depends on a uniform magnetic field to achieve its maximum effect, does not provide as great an effect as with the shaded-pole motor. As a result the running torque characteristics are inferior to those of the shaded-pole motor. In addition, the starting torque is usually less than 50% of the full-load rated torque, and the breakdown torque is very little greater than the full-load torque; the motor operates with a high slip.

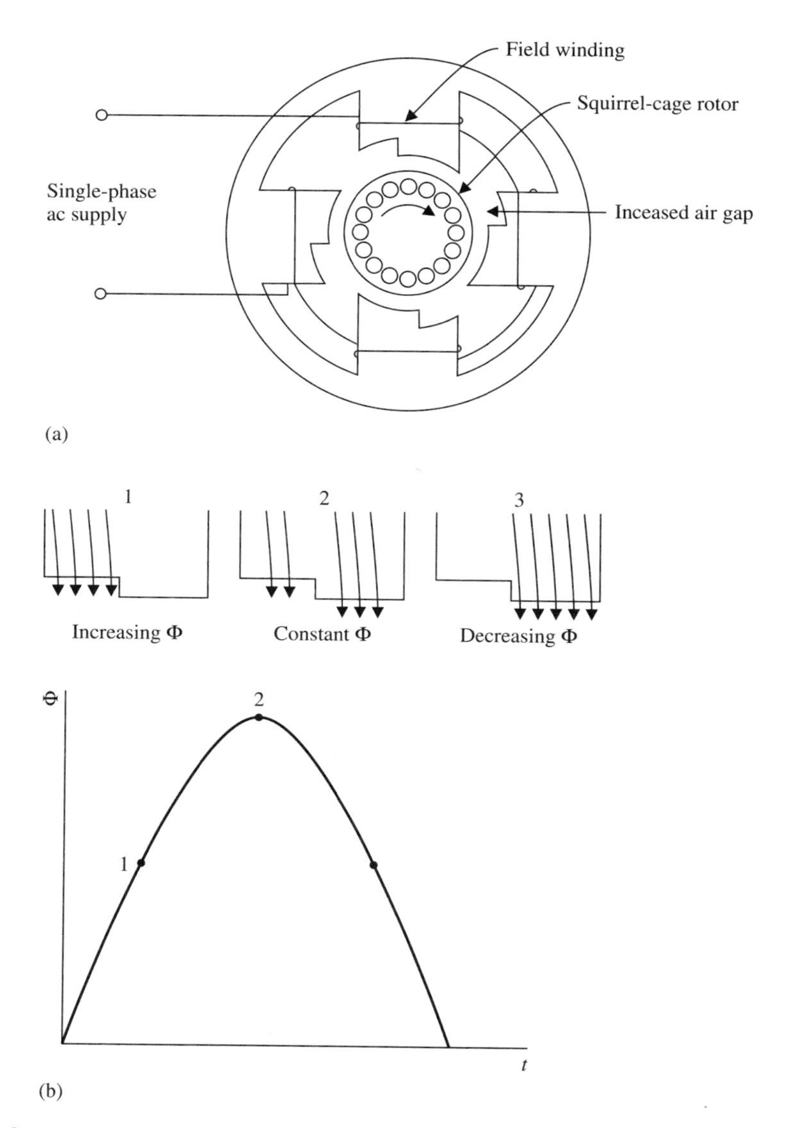

Figure 16-3 Reluctance-start motor. (a) Construction. (b) Development of moving magnetic field.

Speed control of the reluctance-start motor is achieved by controlling the voltage applied to the field windings in a manner similar to that used to control the shaded-pole motor. Unlike the shaded-pole motor, the reluctance-start motor cannot be reversed by electrical methods, since the direction of flux movement is always from the region of the large air gap to the region of the small air gap.

In summary, reluctance-start motors have poor starting and breakdown torques, with efficiencies even less than that of the shaded-pole motor, that is, less than 20%, and they are electrically nonreversible. Although they are cheap and simple to produce and call for little maintenance, there is little justification to their being used, and they most probably will be phased out of production.

16-3 SYNCHRONOUS MOTORS

Single-phase synchronous motors, just as is implied by their name, run at synchronous speed, which is solely determined by the number of stator poles and the supply frequency. They are used in applications where constant speed is the prime requirement. Just as with the single-phase induction motors, the stator windings produce a rotating magnetic field during starting. As a result the rotor accelerates and pulls into step with the stator field. However, unlike with the larger polyphase synchronous motors, the rotor is not dc excited. The main types of single-phase synchronous motors are the reluctance, hysteresis, permanent-magnet, and subsynchronous motors. The major applications of synchronous motors are electric clocks, timers, timing controls, appliances, fans and blowers, business machines, recorders, turntables, tape drives, and small drive motors.

Reluctance Motor

The reluctance motor is essentially a single-phase induction motor with a modified squirrel-cage rotor [Fig. 16-4(a)]. These single-phase reluctance motors use any one of the normal single-phase induction motor stator windings, that is, split-phase, capacitor-start, permanent-split capacitor, or shaded-pole arrangements. The rotor is modified by cutting away sections to form as many salient poles as there are stator poles. The cut-out sections increase the magnetic reluctance between the poles as compared to the low reluctance along the pole axis. The motor depends on this change of reluctance to produce a *reluctance torque,* hence the name reluctance motor. The reluctance torque is the torque that is produced by the low-reluctance rotor poles trying to line up with the stator field.

When first started up, the stator rotating magnetic field sweeps past the short-circuited rotor conductors on the pole tips and induces a rotor voltage, which in turn produces a rotor field. The interaction of the rotor and stator fields produces a starting torque of about 300–400% of rated full-load torque, depending on the relative position of the rotor poles with respect to the stator windings. At about 75% of synchronous speed, the centrifugal starting switch opens and disconnects the auxiliary winding. The rotor continues accelerating under the influence of the pulsating field produced by the main winding. As the rotor approaches synchronous speed, the reluctance torque pulls the rotor poles into synchronism with the stator field. This torque is known as the *pull-in* torque, and is usually about 120% of rated full-load torque. The rotor will continue to run at synchronous speed for all loads up to about 200% of full-load rated torque. At this point the rotor will pull out of synchronism. This is known as the *pull-out* torque. Provided the load torque exceeds the pull-out

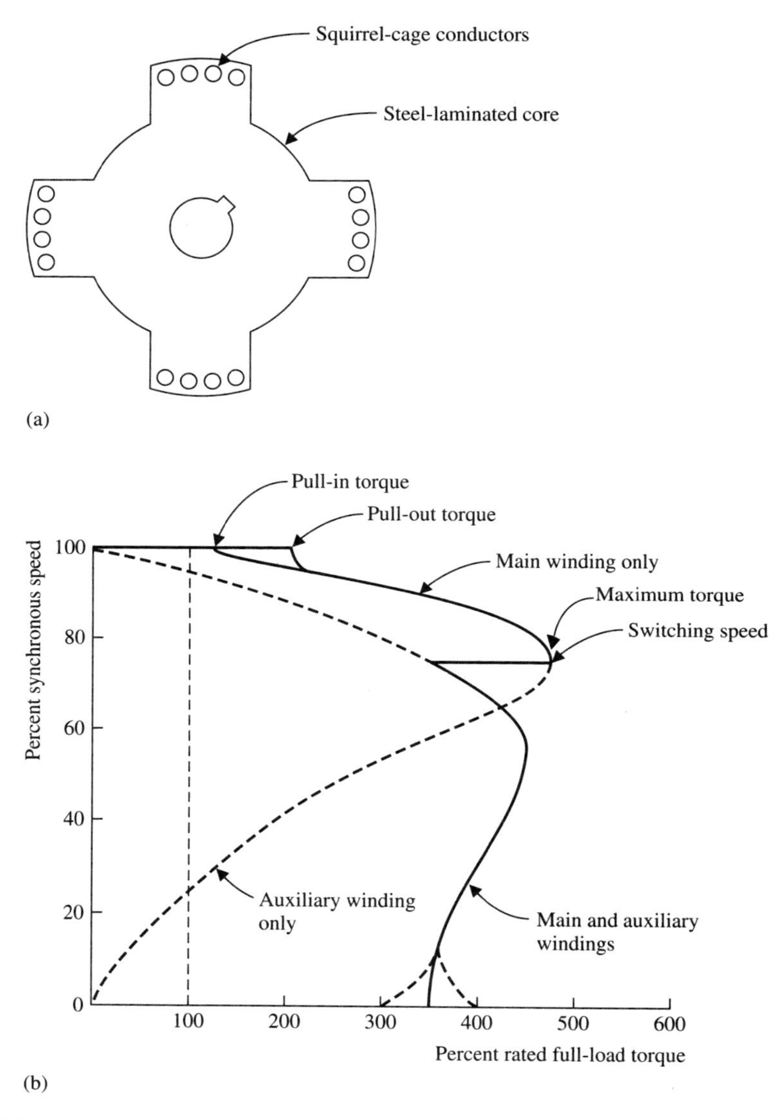

Figure 16-4 Reluctance motor. (a) Construction. (b) Speed-torque curves.

torque, the rotor will rotate at subsynchronous speed as a normal single-phase induction motor with a breakdown torque of about 500–550% of rated full-load torque.

The reluctance motor is normally used in continuous-duty applications, such as electric clocks, timing controls, fans and blowers, business machines, recorders, tape drives, and small drive systems. It calls for very little maintenance and can only be speed controlled by varying the frequency of the ac source. It is normally available in 60- and 400-Hz configurations, which gives it a speed range, depending on the source frequency, from 1,200 to 24,000 r/min (125.66 to 2,513.27 rad/s).

Hysteresis Motor

The stator winding of the hysteresis motor may be split phase, capacitor start, permanent-split capacitor, or shaded pole, although the permanent-split capacitor arrangement with a distributed winding is most commonly used. The capacitor is selected to produce a two-phase constant-amplitude rotating magnetic field.

The rotor shown in Fig. 16-5(a) is a smooth cylinder without teeth or windings, and is made up of laminations of a magnetically hard material, that is, a high-retentivity high-coercivity material such as cobalt-vanadium shrunk onto a nonmagnetic core. The hysteresis loops of an ideal material as well as of an acceptable material such as cobalt-vanadium are shown in Fig. 16-5(b).

Figure 16-5(c) illustrates the principle of torque production. *SS'* represents the axis of the rotating stator field, which is rotating counterclockwise at ω_s. Hysteresis causes the rotor magnetic field *RR'* to lag behind *SS'* by the *hysteretic angle* δ. The starting torque is proportional to the product of the stator and rotor magnetomotive forces and the sine of the hysteretic or torque angle δ. The greater the hysteresis loss of the rotor material, the greater will be the torque angle. At rotor speeds less than synchronous speed the stator field will induce eddy currents in the rotor, and in turn these currents will produce a magnetic field which will increase the rotor torque. The ideal speed-torque curve is shown in Fig. 16-5(d). The actual curve deviates from the ideal because the rotor material does not have an ideal hysteresis loop and also because the stator flux density varies slightly because of the elliptical shape of the rotating magnetic field. The torque component contributed by the eddy currents is proportional to the slip.

The hysteresis motor will achieve synchronous speed for any load that it can accelerate, no matter how great the inertia. It will also operate at synchronous speed irrespective of load torque changes by adjusting the torque angle. It is designed for continuous operation, needs little maintenance, and has the same speed ranges as the reluctance motor. The hysteresis motor produces an almost constant torque irrespective of supply voltage variations, and is the preferred motor for high-quality phonographs and recording equipment. The output power is fairly low for its frame size, and it has a low efficiency and power factor, especially at no load. It also has the added disadvantage that it is more costly than an equivalent output reluctance motor.

Permanent-Magnet Synchronous Motor

Permanent-magnet synchronous motors have rotors in which there are permanently magnetized poles. The interaction between the permanent-magnet rotor poles and the stator field produces an accelerating torque on the rotor. Under no-load conditions the rotor poles nearly line up with the rotating stator poles, the rotor only producing enough torque to maintain synchronous speed. As the load torque increases, the rotor poles fall back with respect to the stator poles to produce the needed

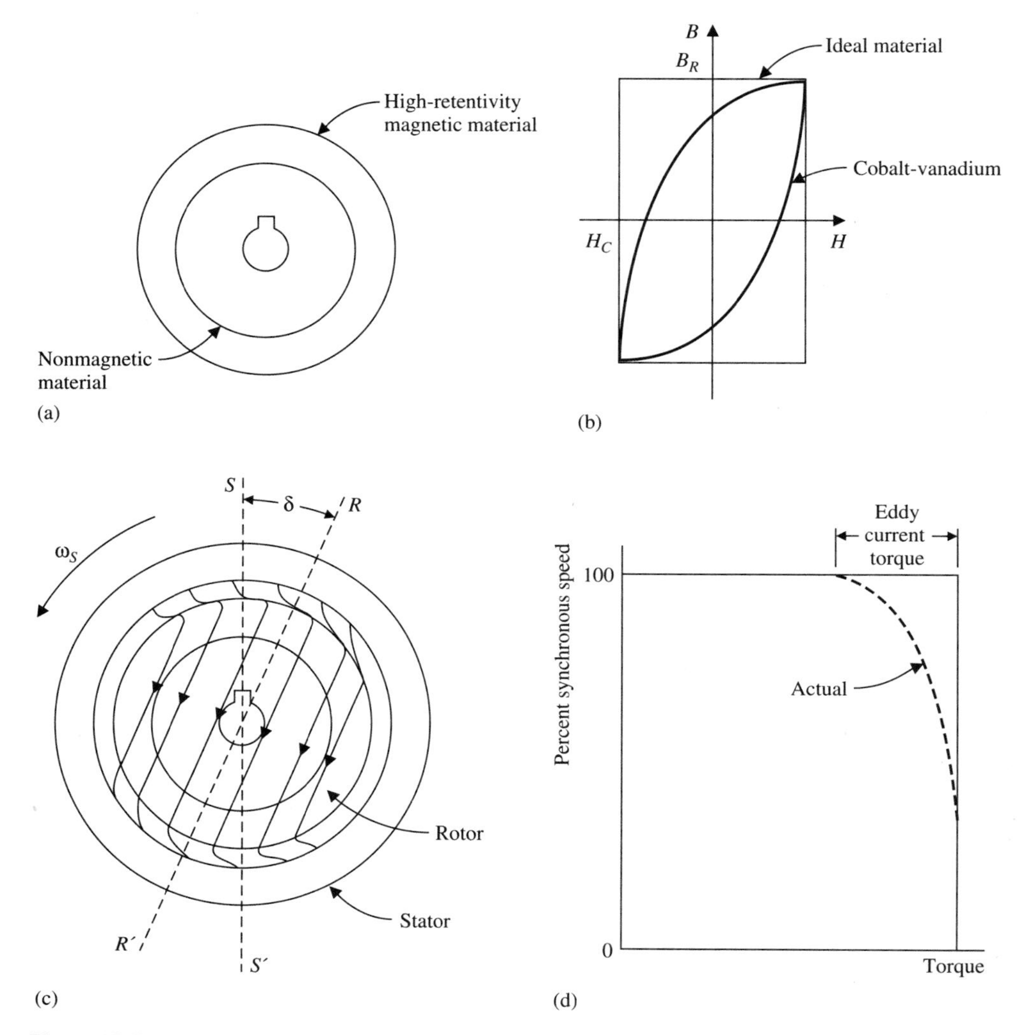

Figure 16-5 Hysteresis motor. (a) Rotor construction. (b) Hysteresis loops. (c) Torque production. (d) Speed-torque characteristics.

increase in torque. The rotor displacement must be less than half the angle between adjacent poles, that is, less than 90° electrical. Provided the load torque causes the rotor displacement angle to exceed 90° electrical, which corresponds to the pull-out torque, that is, the maximum torque developed at synchronous speed with the rated stator voltage and frequency, the rotor will pull out of synchronism and stall.

Permanent-magnet motors may be started against light-load torques. However, the starting torque cannot be greater than the pull-out torque. This limitation prevents the motor starting torque from exceeding about 25% of the pull-out torque.

Attempts have been made to overcome this problem by introducing a set of shaded poles midway between opposite-polarity stator poles. Another factor affecting the starting capability is the magnitude of the largest polar moment of inertia J that can be accelerated in one-quarter cycle of the ac source frequency, that is, too great a polar moment of inertia will prevent the rotor from achieving synchronous speed.

Subsynchronous Motor

The subsynchronous motor is a variation of the hysteresis motor. However, unlike the rotor of a hysteresis motor, the rotor of a subsynchronous motor is a toothed cylinder, and is formed by stacked laminations of magnetically hard material (Fig. 16-6).

Typical subsynchronous rotors have 16 salient poles or teeth, or 32 poles, which will give speeds of 450 r/min (47.12 rad/s) and 225 r/min (23.56 rad/s), respectively. The subsynchronous motor starts in exactly the same manner as the hysteresis motor. When running at synchronous speed under light-load conditions, the high-retentivity toothed rotor will have rotor poles induced on the rotor surface corresponding to the number of stator poles. When the load torque is increased, the rotor will slow down and rotate at the subsynchronous speed determined by the number of rotor poles. Because the output torque is inversely proportional to speed, the reduction in speed results in a greater torque output at subsynchronous speed.

The subsynchronous motor is self-starting since it uses either a shaded-pole or a permanent-split capacitor stator to produce a rotating magnetic field, and it will accelerate the rotor to synchronous speed under the influence of the hysteresis torque. This class of motor has a fairly high starting torque, but develops less torque at synchronous speed than is produced by a reluctance motor.

16-4 UNIVERSAL MOTORS

A dc series motor, provided that it has a laminated stator and field pole structure, will run when supplied from an ac source. Since the armature is in series with the series field winding, the cyclic reversals of current will produce cyclic reversals of flux, with the result that torque is always produced in the same direction. However, it will pulsate at twice the ac supply frequency. Since the motor has a high inductance, the current lags the voltage by an appreciable angle, that is, it has a poor power factor.

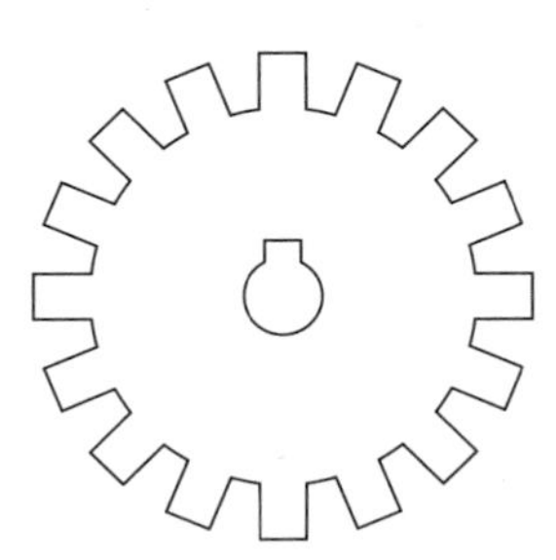

Figure 16-6 Subsynchronous motor rotor.

Several factors affect the motor's performance:

1. Excessive hysteresis and eddy current losses in the stator and field poles. These losses are minimized by laminating the stator and field poles. Two commonly used lamination shapes are shown in Fig. 16-7(a) and (b). The salient-pole lamination in Fig. 16-7(a) requires field windings on each pole, but because of its large mass and exterior surface area it has a high heat-dissipating capability. The C-type lamination in Fig. 16-7(b) requires only one field winding, but must have a directed air flow to help dissipate heat.
2. The series field winding has a high inductance, and therefore there is a high potential difference across the field winding, with the result that there is a much lower potential difference across the armature terminals. The distribution of the potential differences can be improved by reducing the number of field turns and increasing the number of armature turns. This solution is limited because the resulting increase in the armature magnetomotive force will badly distort the air-gap flux distribution. This distortion may be reduced by adding a compensating winding on the stator poles. However, this increases the manufacturing cost and is not commonly used.

At start-up the series field winding acts as a transformer primary, and the armature winding turns short-circuited by the brushes will carry heavy short-circuit currents. As the armature starts turning, the previously short-circuited turns are opened, and simultaneously other turns are short-circuited and then opened. This process causes heavy sparking and heating at the commutator segments in contact with the carbon brushes. The heating effect is reduced with increases in the armature speed. The sparking may be reduced by using high-resistance carbon brushes. A more practical solution is to run the motor at higher speeds, which in turn produces higher induced armature electromotive forces.

Because of the reactance voltage drops that occur with ac operation, the universal motor's speed will be slightly lower than it would be for dc operation with the same load. The drop in speed is lower than might be expected, because with ac operation the magnetic circuit is saturated at the current peaks, which reduces the air-gap flux density to less than that for dc operation. As a result it causes the armature speed to increase since the speed is inversely proportional to flux. The relative differences between ac and dc operation are illustrated in Fig. 16-7(c).

The universal motor operates under no load at speeds of 25,000–35,000 r/min (2,617.99–3,665.19 rad/s), and under normal load conditions at speeds of 8,000–10,000 r/min (837.76–1,047.20 rad/s). In some special applications, such as high-speed electric spindles used for precision milling, drilling, and grinding, continuously variable speeds up to 60,000 r/min (6,283.19 rad/s) are not uncommon. Typical performance characteristics are shown in Fig. 16-7(d).

There are many advantages for using universal motors:

1. Their ability to produce the highest horsepower output per pound (highest watt output per kilogram) and per-unit volume of any 60-Hz motor has led to their overwhelming adoption for portable tools, domestic appliances, and business machines.

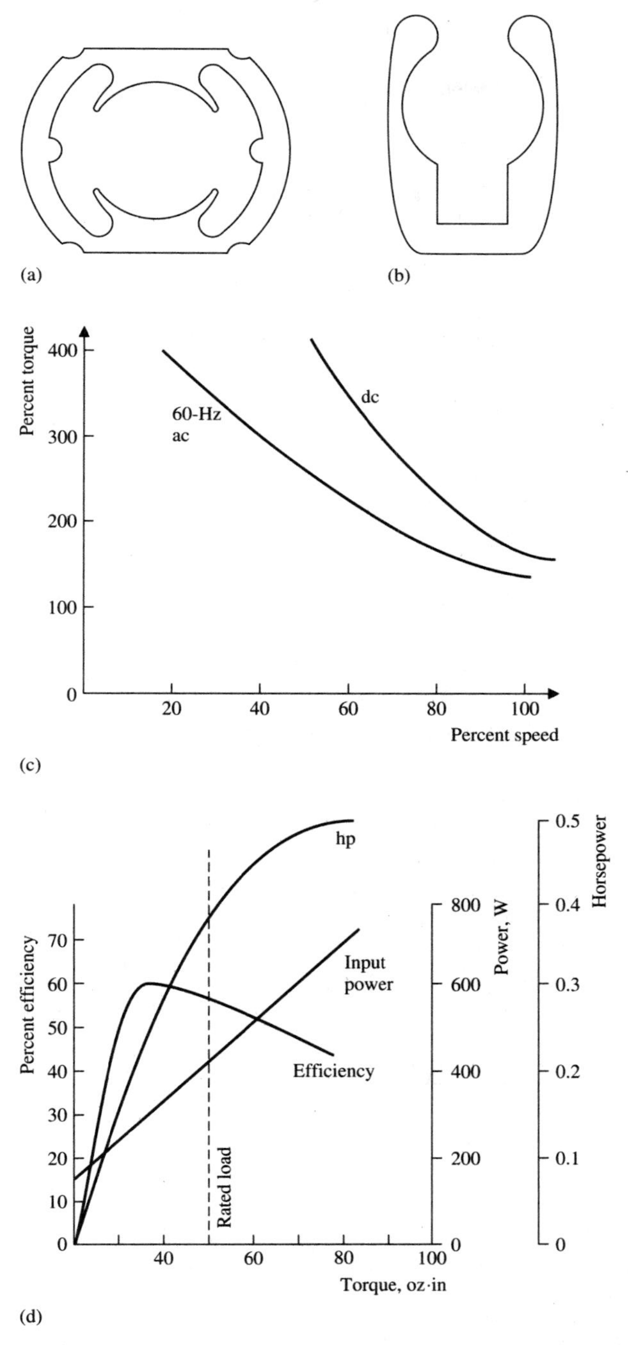

Figure 16-7 Universal motor. (a) Salient-pole stator lamination. (b) C-type stator lamination. (c) Torque-speed characteristics. (d) Performance characteristics.

2. They have high starting and stall torques.
3. They may be operated from either a dc or an ac source up to 60 Hz.
4. They are easily speed controlled and reversed by using two field windings, or by reversing the armature circuit.

The disadvantages to the use of universal motors are:

1. They have a high noise level caused by air movement, brush and commutator contact noise, and, in the case of portable drills, reduction gears, especially at high speeds.
2. There is no built-in protection against burnout if the motor stalls under load. This is caused by the increased current flow and the lack of cooling air.
3. They are not suitable for continuous-duty applications, mainly because of brush and commutator wear. Typical brush life ranges from 300 to 1,200 h.
4. They are a source of electromagnetic interference (EMI) over a wide frequency spectrum. This may be reduced by filtering.

The universal motor is widely used for portable tools such as sanders, saber saws, routers, and electric drills, and in household appliances such as vacuum cleaners, blenders, food processors, food mixers, and sewing machines.

Universal Motor Speed Control

From our study of the dc series motor it should be recalled that the speed of a series motor can be controlled by controlling the armature voltage or current. One method of speed control is to use a variable resistance in series with the armature and field, where an increase in resistance will decrease the speed. This method, while simple, is not very efficient because of the power dissipated in the resistance. In addition, there will be very poor speed regulation with changing loads and very little torque at low speed settings.

Another method is to use a tapped autotransformer, which in turn controls the voltage applied to the motor [Fig. 16-8(a)]. This method gives better speed regulation under changing loads with little power loss, and improved torque at low speed settings. However, the autotransformer is heavy and expensive. An alternative method of speed control is by using a tapped-field winding, which achieves control by varying the impedance of the motor [Fig. 16-8(b)].

Solid-state speed control methods are in common use because they are simple and relatively inexpensive. Since the universal motor produces positive torque during each half-cycle of the ac supply voltage, it can be operated under half- or full-wave control.

The simplest half-wave control scheme is shown in Fig. 16-9(a). It uses an *RC* phase shift control to vary the firing delay angle α, and a neon tube as the trigger device to supply the gate pulse signal to the SCR. There are several disadvantages to this circuit: (1) there is only one torque pulse per cycle, and (2) the speed regulation of the motor is very poor. The torque output of the universal motor can be greatly improved by using a TRIAC instead of the SCR to give two torque pulses per cycle

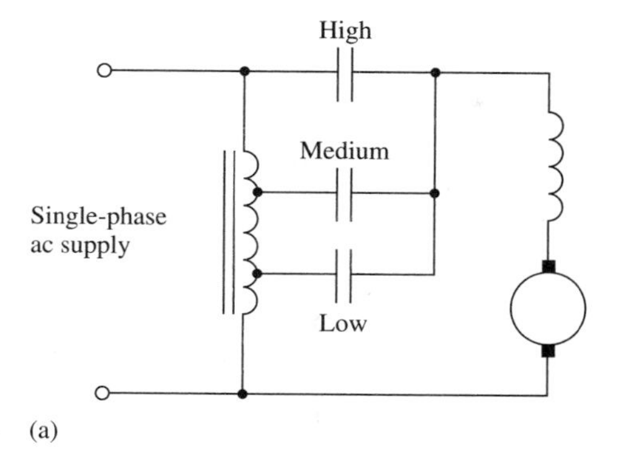

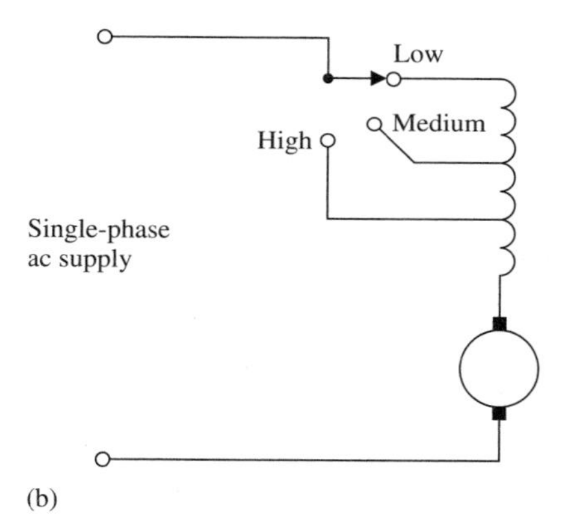

Figure 16-8 Universal motor speed control. (a) Tapped autotransformer. (b) Tapped field.

[Fig. 16-9(b)]. As before, the *RC* phase shift network will vary the firing delay angle α, with a DIAC being used as the trigger device to supply the gate pulse signal to the TRIAC. The major disadvantage of this circuit is that the speed regulation is still very poor. Both these circuits are classified as nonfeedback circuits. Speed regulation can be improved by using feedback techniques.

The circuit shown in Fig. 16-10 uses both half-cycles of the ac supply. The ac source voltage is rectified by the diode bridge consisting of diodes D1 to D4. The speed-adjusting circuit consisting of R1, R2, D5 and C1 controls the firing delay angle α. When C1 has charged up, it discharges via the trigger diode and D6 to supply a gate pulse to the SCR. This process is repeated twice per cycle. The combination of R3C2 in parallel with the universal motor prevents electrical noise (brush sparking) from being applied to the firing delay angle circuit, and it stabilizes the feedback signal. This circuit greatly improves the speed regulation, especially at low-speed operation.

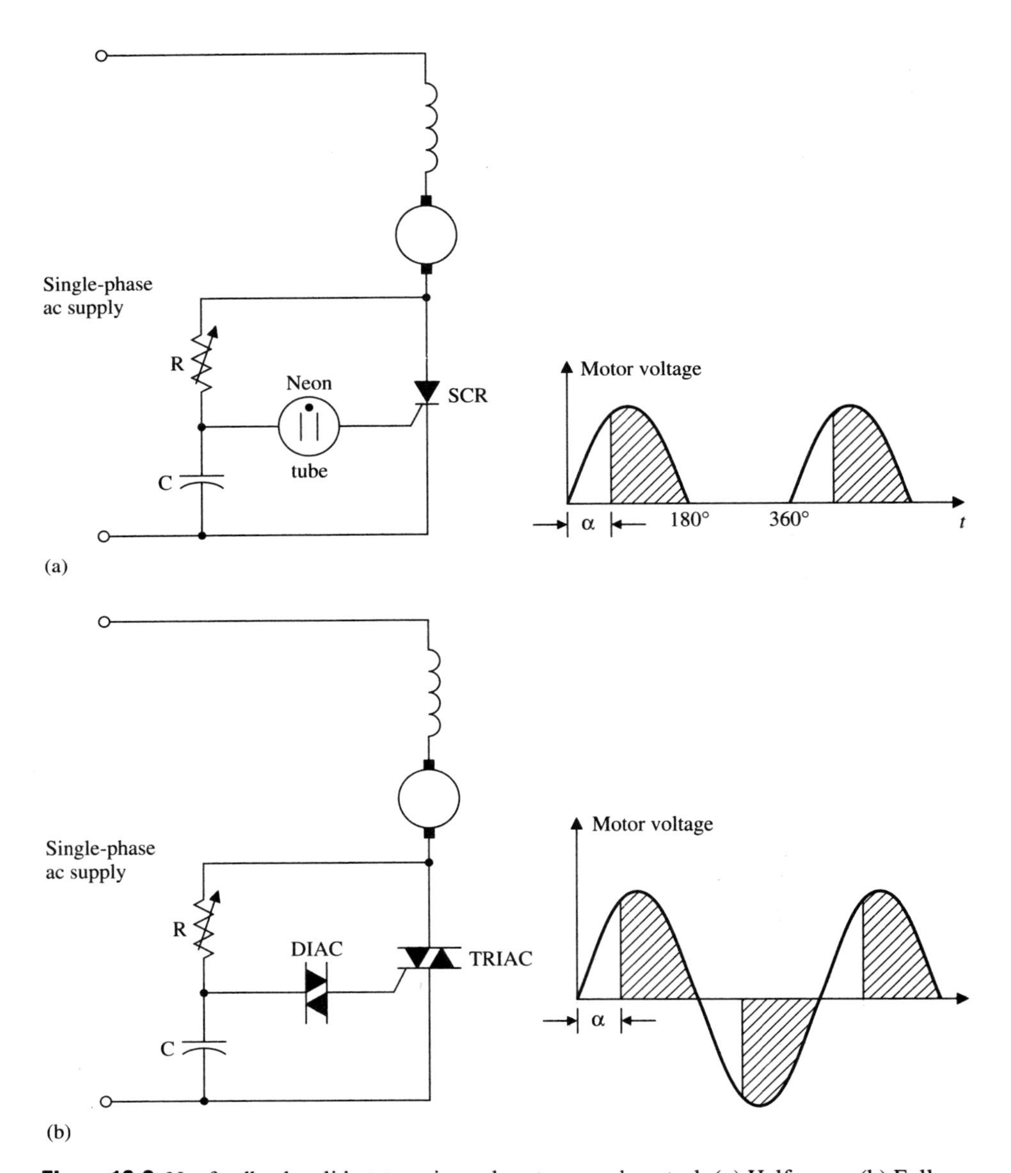

Figure 16-9 Nonfeedback solid-state universal motor speed control. (a) Half-wave. (b) Full wave.

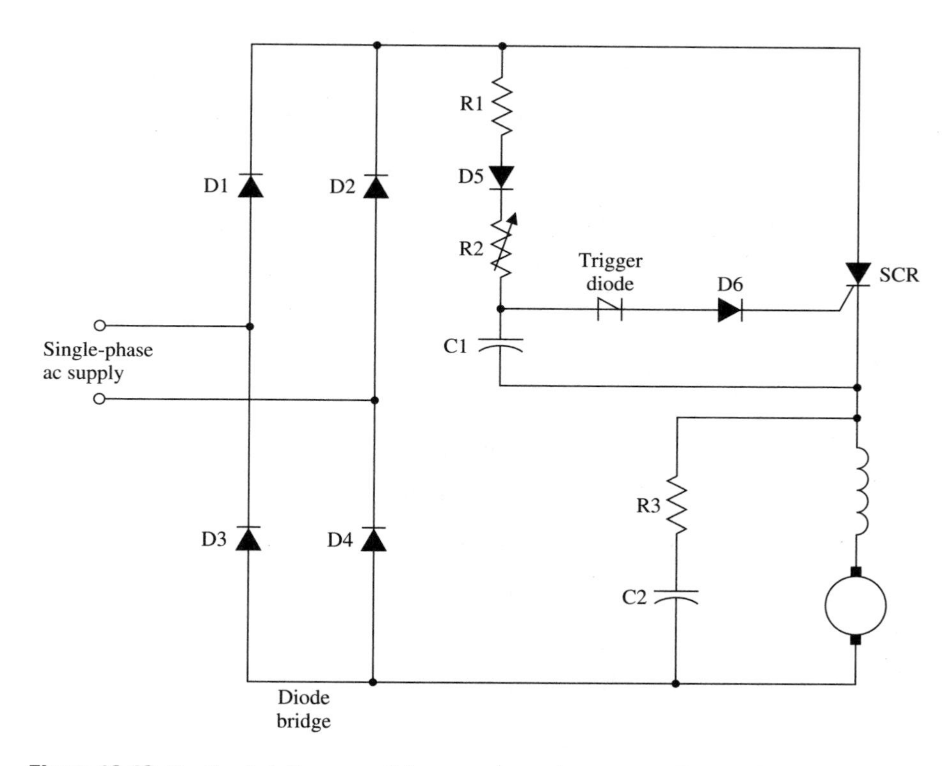

Figure 16-10 Feedback full-wave solid-state universal motor speed control.

QUESTIONS

16-1 Explain with the aid of a sketch how a shaded-pole motor produces a rotating magnetic field.

16-2 Explain with the aid of a sketch how a shaded-pole motor may be reversed.

16-3 Discuss the methods of controlling the speed of a shaded-pole motor.

16-4 What are the advantages of using shaded-pole motors?

16-5 Describe the construction of a shaded-pole motor.

16-6 Describe the construction of a reluctance-start motor.

16-7 Explain with the aid of sketches how a reluctance-start motor produces torque.

16-8 Why is a reluctance-start motor irreversible?

16-9 Discuss the methods of controlling the speed of a reluctance-start motor.

16-10 Why is the reluctance-start motor inferior to the shaded-pole motor?

16-11 Discuss the construction of the reluctance motor.

16-12 Explain how a reluctance motor develops a starting torque.

16-13 What is meant by pull-in and pull-out torque in a reluctance motor?

16-14 Why does a reluctance motor run at synchronous speed?

16-15 With the aid of speed-torque curves explain what happens when the pull-out torque of a reluctance motor is exceeded.

16-16 Describe the construction of a hysteresis motor.

16-17 How does a hysteresis motor develop torque?

16-18 For what reasons and applications would you select a hysteresis motor?

16-19 Describe the construction, characteristics, and limitations of a permanent-magnet synchronous motor.

16-20 Describe the construction and characteristics of a subsynchronous motor.

16-21 What is a universal motor?

16-22 What factors affect the performance of a universal motor, and how can they be minimized?

16-23 Why is the speed drop of a universal motor operating on ac greater than might be expected?

16-24 What are the advantages of a universal motor?

16-25 What are the disadvantages of a universal motor?

16-26 How may the speed of a universal motor be varied by conventional methods?

16-27 With the aid of sketches discuss half- and full-wave nonfeedback solid-state control of a universal motor. What are the limitations of these methods?

16-28 With the aid of a sketch discuss a method of controlling the speed of a universal motor using a solid-state feedback control.

CHAPTER

17

Special-Purpose Machines

17-1 INTRODUCTION

Our previous discussions have concentrated on polyphase synchronous and induction motors and single-phase motors. Considerable attention has been given in recent years to the development of small motors ($\frac{1}{4}$– $\frac{1}{2,000}$ hp; 186.5–0.37 W). These special-purpose machines have been developed mainly as a direct result of the need for precise speed and position control in automatic equipment, such as computerized numerically controlled (CNC) lathes, milling machines, and robotics. These devices are not solely confined to the very low end of the power spectrum, but are available in a wide range of power outputs.

17-2 STEPPER MOTORS

Stepper motors are electromagnetic devices which convert a pulsed electric input into a mechanical output in discrete equal angular increments called *steps*. There is a one-to-one correspondence between the input pulses and the angular steps. The amount of rotation is determined by the number of input pulses, the direction is controlled by applying the pulses to the clockwise (CW) or counterclockwise (CCW) drive circuits, and the rotational velocity is controlled by the frequency of the input pulses.

Stepper motors are used in industrial applications for the control of CNC machines, robots, plotters, printers, and the like. These applications are met by commercially available motors with rated output torques at 50 steps per second, ranging from 20 to 1,900 oz · in (0.14 to 13.4 N · m).

The major advantages for using stepper motors are:

1. Their performance is predictable and consistent, and they are easily adapted to computer control.
2. They are capable of rapid acceleration, deceleration, and reversal.

3. The rotor has a low moment of inertia.
4. They are easily controlled using digital electronics, minicomputers, microprocessors, and programmable logic controllers.
5. They may be used in open- or closed-loop control systems, although open-loop control is used more frequently.
6. They do not create electromagnetic interference.
7. They are relatively maintenance free since the only wear items are the bearings.
8. They may be supplied from ac or dc sources.

The main disadvantages are:

1. They have a low efficiency.
2. They require a variable pulse frequency or pulse rate source to control speed.
3. The load inertia must be limited to no more than four times the rotor inertia.
4. They require relatively complex control systems.
5. They must be carefully mated to the connected load in open-loop applications to prevent the gain or loss of steps because of variations in load inertia and friction.
6. They are more expensive than equivalently rated shaded-pole or vibrating-type ac motors.
7. They are not completely silent when running. An audible hum is present, which is a direct function of the control pulse rate.

Stepper Motor Terminology

The following definitions are commonly used to describe the specific characteristics of stepper motors and their controls:

Step angle (*SA*): The incremental amount that the motor shaft rotates each time the stator winding polarity is changed, that is, for each input pulse. The step angle is expressed in °/step or in degrees.

Steps per revolution (*SPR*): The total number of steps required for the motor shaft to turn through 360°, or one complete revolution,

$$\text{SPR} = \frac{360°}{\text{SA}} \tag{17–1}$$

Steps per second (*SPS*): The total number of angular steps accomplished by the motor shaft in 1 s. Steps per second is comparable to revolutions per second or radians per second of the conventional motor,

$$\text{SPS} = \frac{(\text{r/min}) \times \text{SPR}}{60} \text{ steps/s} \tag{17–2}$$

or

$$\omega = \frac{2\pi \times \text{SPS}}{\text{SPR}} \text{ rad/s} \tag{17-3}$$

Step accuracy: This is the positional accuracy tolerance. Step accuracy is usually expressed in percent, and represents the total error introduced by the stepper motor as a result of a single step movement. The error is noncumulative, that is, it will not accumulate or decrease regardless of the number of steps made.

Holding torque: With the motor shaft stationary, the holding or breakaway torque is the minimum amount of external torque that must be applied to cause the rotor to break away from its holding position with rated voltage and current applied to the stator.

Residual torque: With power removed from the motor, the residual torque is the amount of torque present as a result of the permanent-magnet flux acting on the stator poles. It is only present with permanent-magnet rotors.

Step response: The step response, or the time for a single step, is the time taken by the rotor to move a single step in response to a drive pulse. It is a function of the torque-to-inertia ratio of the motor and the characteristics of the drive cicuitry. Step response is defined under no-load conditions and is usually expressed in milliseconds.

Torque-to-inertia ratio (TIR): This is the ratio of the holding torque in oz·in (N·m) to the rotor inertia in $\text{oz}\cdot\text{in}\cdot\text{s}^2(10^{-3}\ \text{kg}\cdot\text{m}^2)$. The greater the ratio, the better the step response of the motor,

$$\text{TIR} = \frac{\text{holding torque (oz} \cdot \text{in)}}{\text{rotor inertia (oz} \cdot \text{in} \cdot \text{s}^2)} \tag{17–4)(E}$$

or

$$\text{TIR} = \frac{\text{holding torque (N} \cdot \text{m)}}{\text{rotor inertia } (10^{-3}\text{kg} \cdot \text{m}^2)} \tag{17–4)(SI}$$

Resonance: Stepper motors have a natural no-load frequency, usually between 90 and 160 steps per second, and if operated under load conditions at the natural or resonant frequency, there will be an increase in the vibration level and audible noise level. If it is necessary to operate the stepper motor in this range, the natural frequency can be lowered by increasing the inertia using a damping flywheel, but at the expense of a reduction in overall performance. A Lanchester damper is a viscous coupled inertia damper consisting of a lightweight aluminum cup driven by the motor shaft, which in turn surrounds a heavier flywheel, the whole arrangement being immersed in a liquid. As the lightweight aluminum cup rotates, the heavier flywheel starts to rotate because of the shear action of the fluid. This type of damper smooths out shaft oscillations, with very little reduction in overall performance.

Drives: The overall description given to the electronic circuitry that controls a stepper motor. It consists of a power supply, sequencing logic, and power output switching modules.

Translator: An electronic module that converts pulses into the correct switching sequence, producing one step for each received pulse.

Preset indexer: An electronic control module that includes the translator and the extra circuitry required to control the number of pulses, the pulse rate, and the direction of rotation of the motor.

Pulse rate: The rate at which the stator windings are switched. If one pulse produces one motor step, the pulse rate is also the motor stepping rate.

Ramping: The process of controlling the pulse frequency in order to control the acceleration rate of the motor from standstill to its maximum speed and to control the deceleration rate from the maximum speed to standstill. Ramping is necessary in order that high-inertia loads may be accelerated or decelerated without missing steps.

Slew rate: The maximum stepping speed where the stepping rate is in synchronism with the stepping pulses. However, the motor can only operate in this region after it has been accelerated by ramping up to maximum speed.

Stepper motors are classified in terms of the type of rotor, the step angle, the external diameter of the housing, and the mounting arrangements. In addition, the type and arrangement of the stator windings, torque output, number of stator phases, and operating voltage are also specified.

There are three main types of stepper motors: permanent magnet (PM), variable reluctance (VR), and hybrid (PM-VR).

The operation of a stepper motor is dependent on the magnetic characteristic—like magnetic poles repel each other, and unlike magnetic poles attract each other. Figure 17-1 illustrates the basic concept of stepper motor operation.

Considering Fig. 17-1(a), if the stator-pole windings are energized to produce the polarities shown, and the permanent-magnet rotor is positioned as shown between the stator poles, a torque will be produced by the interaction of the two magnetic fields, which causes the rotor to turn through 180°. The direction of rotation cannot be determined, that is, it may be CW or CCW. The direction of rotation may be controlled by adding an extra pair of stator poles *C* and *D* midway between the first pair, as shown in Fig. 17-1(b). With the stator windings energized to produce the magnetic polarities shown, the rotor will turn CCW through 135° until it comes to rest at the position shown in Fig. 17-1(c). The rotor turns in a series of steps which are determined by the magnetic polarities of the stator poles. As the rotor attempts to achieve an alignment between the rotor and stator pole fluxes, a torque is produced which causes the rotor to turn. The rotor can be either a permanent magnet or a nonretentive ferromagnetic material which is commonly used in variable-reluctance stepper motor rotors.

Permanent-Magnet Stepper Motors

The permanent-magnet stepper motor is so named because of the cylindrical ceramic permanent magnet which is used as the rotor. The rotor magnet is radially magnetized with the desired number of pairs of poles; the stator has an equal number

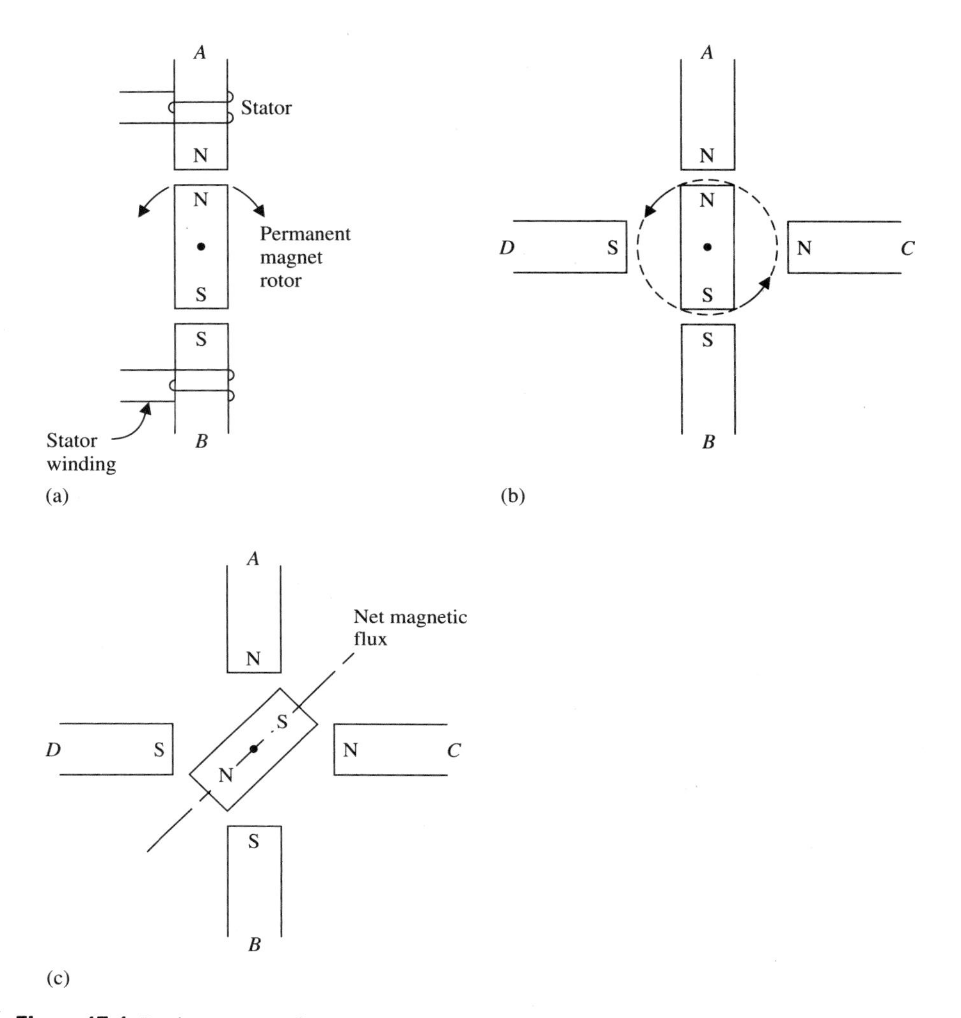

Figure 17-1 Basic concept of stepper motor operation.

of wound poles. The construction of a typical industrial tin-can or stamped-metal housing-type two-phase stepper motor is shown in Fig. 17-2(a). The motor consists of two sheet-metal cups with stator poles punched out around the inner surface surrounding the radially polarized magnetic rotor. Two center-tapped bifilar wound annular ring coils, coils *A* and *B*, and wound inside the cups. These two coils, depending on the way they are energized, determine the polarity of the stator poles by selectively controlling the direction of current flow in each coil. A rotating magnetic field is produced, which in conjunction with the permanent-magnet rotor field produces a torque that causes the rotor to step one-quarter pole pitch for each switching step, that is, 90°.

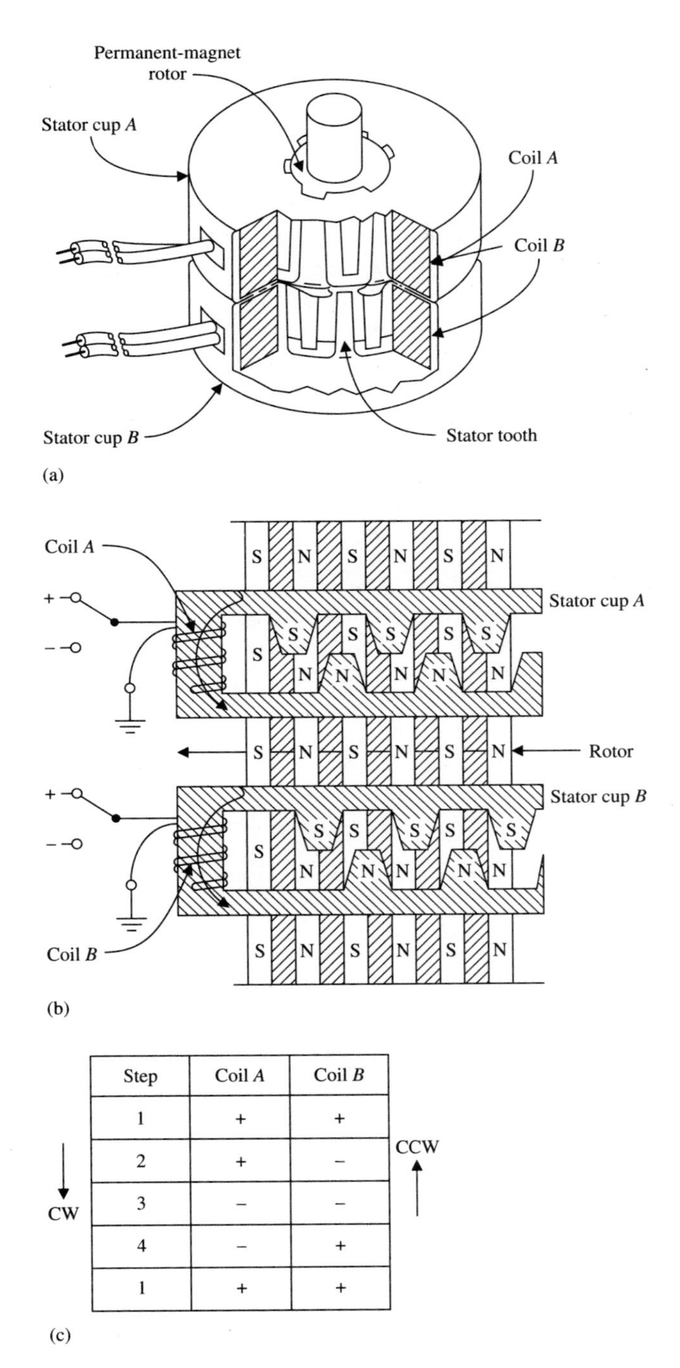

Step	Coil A	Coil B
1	+	+
2	+	–
3	–	–
4	–	+
1	+	+

Figure 17-2 Two-phase permanent-magnet stepper motor. (a) Sectionalized view. (b) Principle of operation. (c) Switching sequence.

This motor design is simple, and it is economical to produce. However, it is limited to a 90° step angle. The magnitude of the step can be reduced, but in this design it is not practical since there is a limit to the number of poles that can be accommodated around the periphery of the rotor. The circumferential length can be increased by increasing the diameter of the rotor so that more rotor poles can be accommodated. But this will increase the moment of inertia of the rotor, which in turn reduces the starting torque. An alternative arrangement is to use a stator with salient poles and a polyphase winding, with the pole faces of both the stator poles and the rotor periphery toothed. The teeth on the rotor surface and the stator-pole faces are offset, so that only a limited number of rotor poles will align with an energized stator pole. This design is, however, costly to build for small step angles. The permanent-magnet stepper motor resists movement when the stator windings are deenergized, because a static holding torque has developed, called the *detent torque*. This torque is useful because it holds the rotor in position when the power is off. The permanent-magnet stepper motor is bidirectional and is inherently self-damping, but it does have a tendency to overshoot when single stepped.

Variable-Reluctance Stepper Motors

A typical four-phase single-stack variable-reluctance stepper motor is illustrated in Fig. 17-3. This motor has eight stator poles and six rotor teeth or poles, a choice

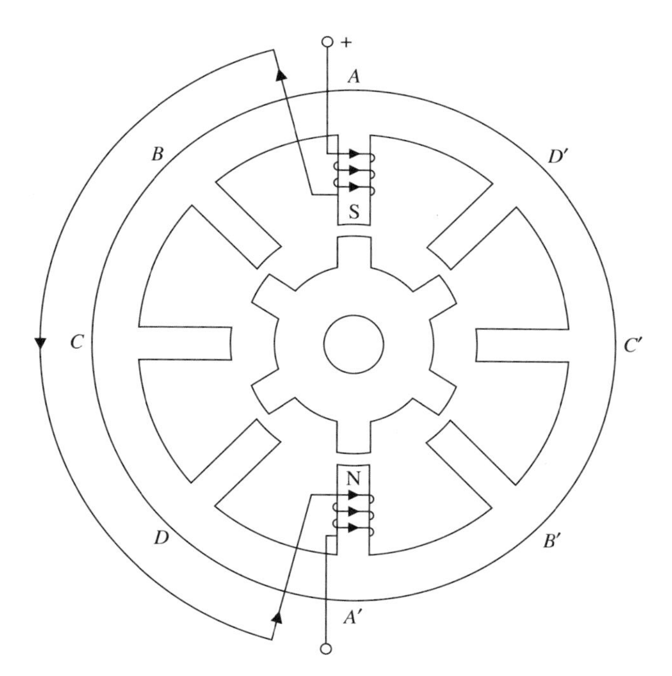

Figure 17-3 Four-phase variable-reluctance stepper motor. Only phase A windings are shown for clarity.

which ensures that when the adjoining phase is energized, the rotor will move to align itself with the energized stator poles. Variable-reluctance motors depend on the development of a reluctance torque, that is, the rotor being free to turn, it will move so as to minimize the length of the air gap.

Applying a dc voltage to phase ($A+$, $A'-$) will cause the six-pole soft-iron rotor to turn so that the flux through the pole teeth at the 12 o'clock and 6 o'clock positions is maximized, that is, a pair of diametrically opposite rotor poles have aligned themselves with the energized stator poles. If power is removed from the A phase (AA$'$), and power is applied to the B phase ($B+$, $B'-$), then the rotor will turn 15° CW, pulling the nearest set of rotor teeth into alignment with the B-phase stator poles. Similarly, as phases C and D are energized in succession, the rotor will rotate 15° CW as each phase is energized. If there were twice as many rotor teeth, that is, 12 teeth, then the rotor would turn 7.5° each time a stator phase is energized. If the sequence of phase excitation is reversed, that is, A', D, C, B', A, the direction of the rotor is reversed. The step angle SA is inversely related to the number of stator phases and rotor teeth by

$$\text{SA} = \frac{360^\circ}{\text{rotor teeth} \times \text{stator phases}} \tag{17–5}$$

For example, for the four-phase variable-reluctance stepper motor shown in Fig. 17-3,

$$\text{SA} = \frac{360^\circ}{6 \times 4} = 15^\circ$$

Similarly a 12-tooth rotor will have a step angle of 7.5°.

When the rotor and stator poles are aligned as shown in Fig. 17-3, the reluctance torque is zero. With phase A energized, if the rotor is mechanically displaced $\pm\theta°$, a reluctance torque is developed which attempts to return the rotor to its original position. This torque will be a maximum when the rotor has been displaced ±90°. This maximum torque is called the *holding* or *pull-in torque*. If the rotor is displaced past the ±90° position, the reluctance torque decreases, reaching zero when the rotor has turned through an angle of ±180°. The relationship of torque versus rotor displacement is shown in Fig. 17-4.

While the holding or pull-in torque occurs when the rotor is displaced ±90° electrical from the fully aligned position, the *maximum* or *dynamic* or *running* or *pull-out torque* occurs when the load on the rotor causes it to lag 45° electrical behind the energized stator poles. The running torque is at a maximum when the stepper motor is operating at slow stepping speeds; it falls off as the stepping speed is increased [Fig. 17-5(a)].

Figure 17-5(a) shows the relationship between the holding or pull-in torque and the running torque of a typical stepper motor. If the stepper motor is to be stopped or started accurately, it must operate in a *start-stop* running mode, that is, the rotor momentarily stops between drive pulses, or alternatively the motor is operating at a low number of steps per second. As the rotational velocity of the rotor increases, it

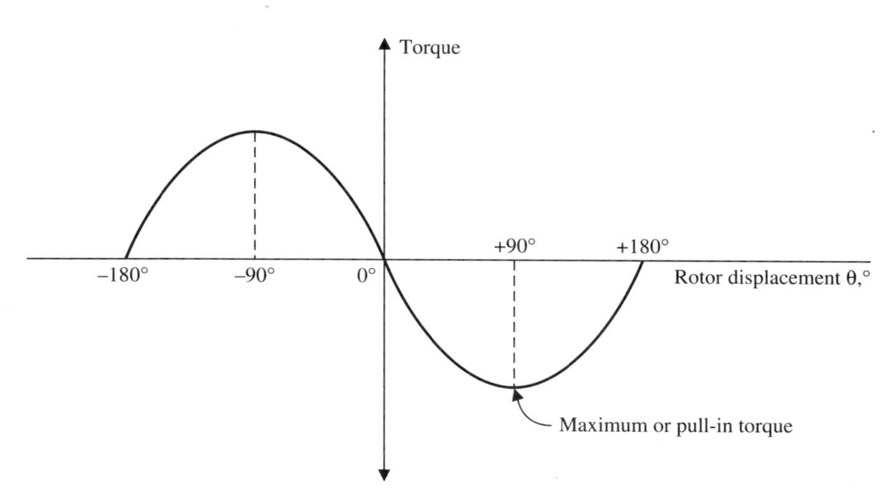

Figure 17-4 Stepper motor torque versus rotor displacement.

will attain *slew speed*, that is, the rotor is turning continuously without any momentary stops. At this point the rotor is rotating synchronously with the rotating field of the stators. To achieve slew speed without a loss of step, either from rest or when it is operating in the start-stop range, the rotor must be carefully ramped up to slew speed by slowly increasing the number of steps per second. In order to stop or reverse the motor without a loss of step, the rotor must be ramped down into the start-stop range before it is stopped or reversed. When the output torque is zero, that is, the point of maximum response, the rotor oscillates until the kinetic energy of the rotor and load have been damped. Permanent-magnet motors also experience this problem. However, it is damped out by the rotor's eddy currents and hysteresis losses. Variable-reluctance motors require an external damper, either electrical or mechanical, such as the Lanchester damper, to damp out the oscillations.

Multistack variable-reluctance stepper motors have the same number of stator sections as rotor sections, usually three or five stacks, and the stator and rotor sections have the same number of teeth. Each stator section is a phase, and the teeth of all stator phases are aligned, while the teeth of the corresponding rotor sections are offset.

As can be seen from Fig. 17-6, when phase *A* is energized, the teeth of rotor *A* align with the stator teeth, that is, the minimum-reluctance position. At the same time the teeth of rotor *B* are displaced by two-thirds of a tooth width in the CW direction, and the teeth of rotor *C* are displaced by two thirds of a tooth width in the CCW direction. If phase *B* is energized, the rotor will step CW. Energizing phase *C* will cause the rotor to step CCW, that is, energizing the phases in an *ABC* sequence will cause the rotor to step CW, and for CCW rotation the phases are energized in a *CBA* sequence.

Variable-reluctance motors do not use a permanent-magnet rotor, and as a result the rotor is smaller in diameter and has a low moment of inertia. Under light-load conditions these motors have a high slew rate. However, there are some restrictions,

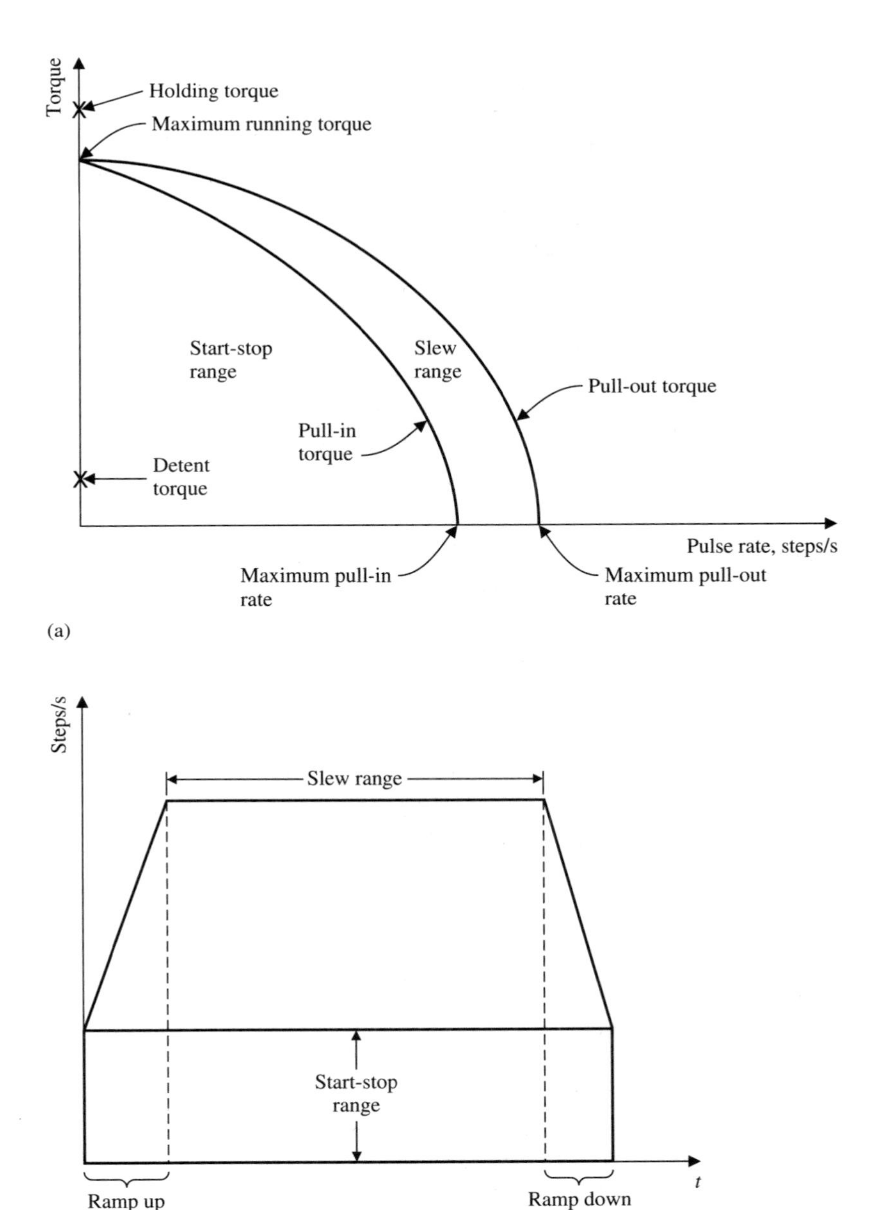

Figure 17-5 Stepper motor characteristics. (a) Torque versus pulse rate. (b) Velocity control.

such as only a limited range of step angles, usually 7.5 to 30°, are possible with single-stack variable-reluctance motors because the wound stator pole construction limits the number of poles that can be installed around the stator circumference.

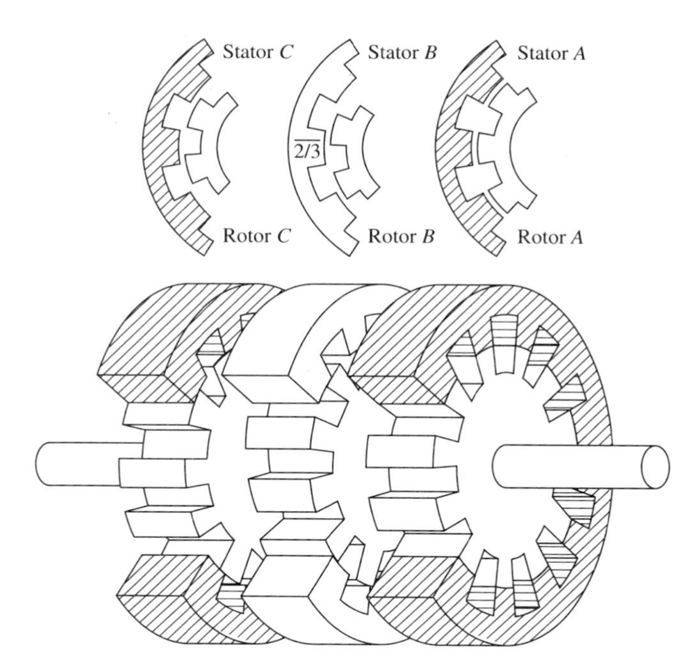

Figure 17-6 Three-phase multistack variable-reluctance stepper motor.

Multistack variable-reluctance motors have a low torque-to-stator-volume ratio when compared to hybrid stepper motors, and require sophisticated electronic drive controls because of the lack of inherent damping. Both single- and multistack variable-reluctance motors do not have a positive detent torque when power is removed, but this problem can be offset by reducing the power input and removing the drive pulses.

Hybrid Stepper Motors

The hybrid (permanent-magnet–variable-reluctance) stepper motor develops its torque by means of reluctance forces in exactly the same manner as the variable-reluctance motor, and has a stator similar to that of the single-stack variable-reluctance motor. However, the rotor is a combination of the variable-reluctance and permanent-magnet rotors [Fig. 17-7(b)].

The soft-iron rotor consists of two spools of soft iron shrunk onto the shaft. The outer surface of each spool is grooved, with the grooves on each spool being displaced with respect to each other by one-half of the tooth pitch. For example, for a 200-step or a 1.8° step-angle motor there are 50 teeth spaced 7.5° apart on one rotor spool. The teeth on the other spool are offset by 3.6° from those on the first spool. The two spools are magnetically polarized by an imbedded Alnico magnet, as shown in Fig. 17-7(b).

There are eight equidistant salient poles spaced around the inside circumference of the stator. The alternate poles A_1, A_2, A_1', and A_2', B_1, B_2, B_1', and B_2' are connected

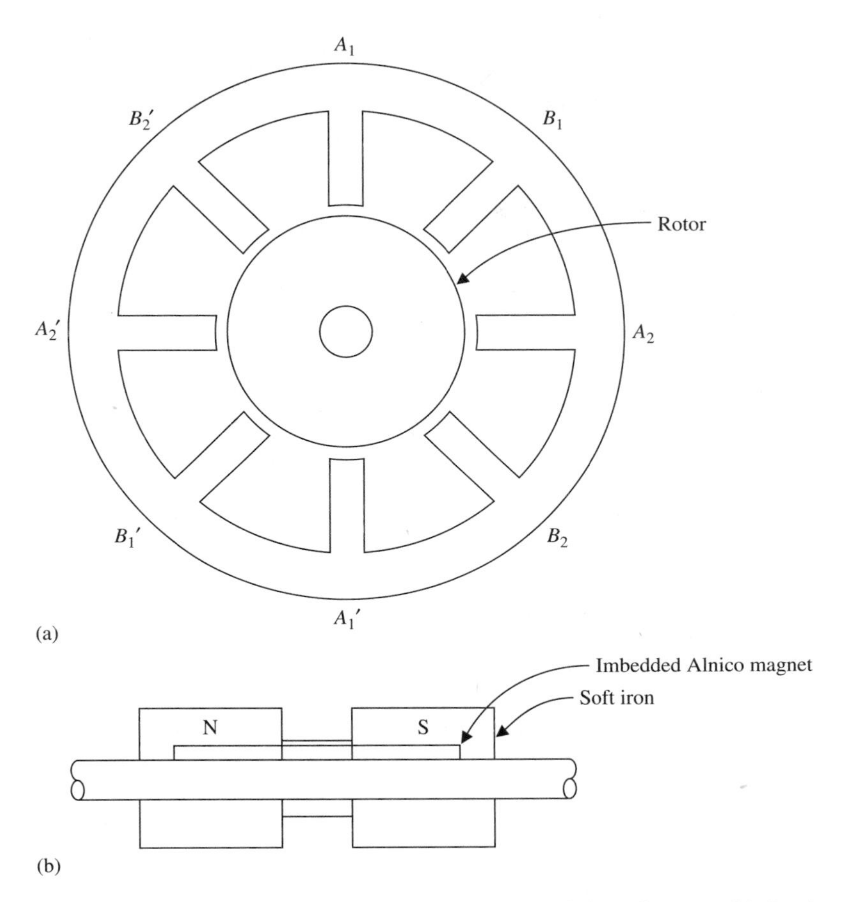

Figure 17-7 Conventional hybrid stepper motor. (a) Two-phase eight-pole stator. (b) Sectional view of rotor.

together electrically to form two phases. The pole faces are grooved with the same tooth pitch as the rotor. The teeth on each pole face are offset by one-quarter of a tooth pitch from those of the adjacent poles, with the offset being in the same direction around the whole stator, either a CW or a CCW offset.

When phase A is energized, poles A_1 and A_1' are N poles, and A_2 and A_2' are S poles, or vice versa, depending on the direction of current flow. The interaction of the stator flux and the rotor flux causes the rotor to turn so that the rotor teeth of one-half of the rotor line up with the A_1 pole, and the teeth of the other half of the rotor line up with the A_1' pole. At the same time the rotor teeth under the A_2 and A_2' are offset by one-half of a tooth pitch from those of the A_2 and A_2' stator poles. At the same time the teeth of the phase B poles, B_1, B_1', B_2, and B_2', are offset by one-quarter of a tooth pitch from those of the rotor.

When the stator windings are not excited, there will be a reluctance torque developed between the teeth of the magnetized rotor and the stator-pole teeth; when the teeth under a pole are aligned, the reluctance torque is zero. If the rotor is offset in either direction from this position, then a reluctance torque opposing the offset will be developed. The reluctance torque is a maximum when the teeth of the rotor and stator are offset by one-quarter of the tooth pitch. As the offset is increased, the reluctance torque decreases and becomes zero when the offset is equal to one-half of the tooth pitch. Increasing the offset beyond this point gives rise to an assisting torque, which becomes a maximum after the rotor has been offset three-quarters of a tooth pitch from its original position and decreases to zero as the offset is continued until the stator and rotor teeth are in alignment again. From this it can be seen that a small alternating detent torque is present.

When only phase A is excited [Fig. 17-7(a)], the rotor and stator teeth are aligned and there is zero torque output. When phase B only is excited, stator poles B_1 and B_1' will produce a CW torque, that is, the stator pole fluxes are aiding the rotor flux. The flux produced by stator poles B_1 and B_1' is proportional to the excitation current up to the point where the stator and rotor teeth saturate. As a result any further increases in the phase current will not significantly increase the torque. At the same time poles B_2 and B_2' are producing a CCW torque, which is quite appreciable when the excitation current is small, and the rotor permanent-magnet flux will be greater than the stator-pole flux. As the excitation current increases, the stator-pole flux increases and the CCW torque decreases, reaching a minimum when the magnetomotive forces of the rotor and stator teeth are approximately equal, which coincides with the saturation of the B_2 and B_2' poles. The actual motor torque is the sum of the CW and CCW torques. However, the effects of saturation and an increasing CCW torque limit the peak torque, and in fact even cause a reduction in output torque as the current increases.

The standard hybrid stepper motor is manufactured in torque ranges of 50 oz · in to in excess of 2,000 oz · in (0.35×10^{-1} to 14.12 N · m), with step angles from 0.5 to 15° and stepping rates in excess of 1,000 steps/s.

Recently the performance of the hybrid motor has been improved with the introduction of the *enhanced hybrid stepper motor*. The difference between the two motors is that small powerful rare-earth magnets are placed between the stator teeth. The permanent-magnet materials being used are samarium-cobalt or neodymium-iron-boron. These magnets minimize stator flux leakage and focus the air-gap flux into the stator teeth, as well as reversing the CCW field that opposed the CW field of the standard hybrid motor at or near load current.

The enhanced hybrid stepper motor has gains as high as 50% in running and holding torques, as well as having stepping rates in excess of 2,500 steps/s.

Stepper Motor Connections

The simplest stepper motor connection is the three-lead arrangement shown in Fig. 17-8. While there are two phase windings, obtaining reversed polarity in each of the phases requires a dual-polarity power supply and a complicated electronic switching sequence.

Most stepper motors use a bifilar connection for their stator windings. The bifilar winding connection shown in Fig. 17-9(a) has the advantage that it eliminates cou-

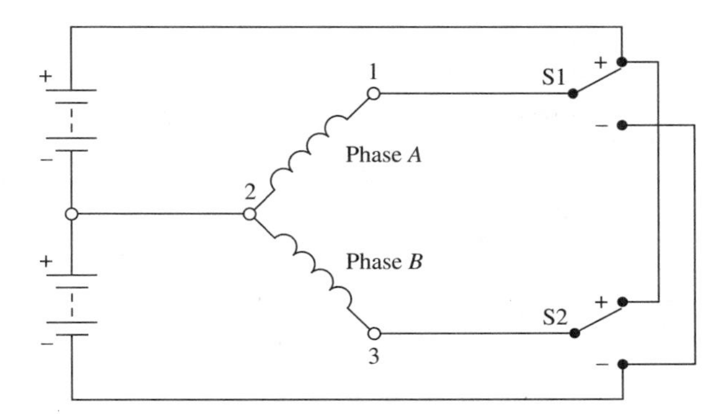

Figure 17-8 Standard three-lead stepper motor connections.

pling between adjacent windings. It has the center points connected to one side of a dc source, usually the negative. The electronic drive circuit then grounds the remaining leads either individually or in pairs, with the result that the control is greatly simplified since the polarities are not switched, which permits a higher pulsing rate to be achieved.

As compared to the standard three-lead connection, the bifilar wound stepper motor requires twice as many windings, and in the same frame size this can only be achieved by reducing the wire diameter, but at the expense of an increased winding resistance. A benefit of the increased resistance is that it reduces the winding time constant (L/R), which leads to faster response. As can be seen from Fig. 17-9(a), only single-ended power supply is required, which usually has an output voltage five times the required motor voltage. Series resistances R_{CL} are inserted in series with the common return lines to limit current, but they also further improve the motor's time constant. In comparison with the three-lead motor, the bifilar wound motor has an improved torque versus steps per second characteristic.

Normally bifilar motors use a *four-step* or *full-step* switching sequence [Fig. 17-9(b)]. Each time a switch (S1 through S4) is turned on, the rotor will step. After four steps the switching sequence is repeated. Each step causes the rotor to step one-fourth of a tooth pitch, that is, for every four steps, the rotor has moved one tooth pitch. Therefore a 50-tooth rotor will require 50 teeth × 4 steps/tooth = 200 steps to turn through 360°, that is, the step angle is 360°/200 steps = 1.8°/step. The angular velocity is determined by the switching rate.

An alternative switching sequence is the *eight-step* or *half-step* switching sequence [Fig. 17-9(c)]. This arrangement will cause the rotor to step 0.9°/step, that is, 400 steps per revolution. The main advantage to half-stepping is that a finer angular resolution is obtained as well as higher stepping rates; the penalty is a reduced output torque capability.

Stepper Motor Drive Schemes

The switching is accomplished by using power transistors or power MOSFETs in one of three types of stepper drives: series R (also sometimes called L/R); chopper, and bilevel. Each of these in turn is available in unipolar and bipolar versions. The major difference between power driver circuits is the technique used to limit current.

Series R (L/R). The series R (L/R) stepper motor driver (Fig. 17-10) is simple and inexpensive. Current-limiting resistances R_{CL} are connected in series with the stepper motor stator windings, so that

$$I_{\max} = \frac{V}{R_{CL} + R_{\text{WDG}}} \tag{17–6}$$

where R_{WDG} is the winding resistance.

An added benefit is the fact that the time constant is also reduced by the addition of the current-limit resistance, which improves the high-speed performance. The additional resistance requires an increase in the power supply voltage to achieve the

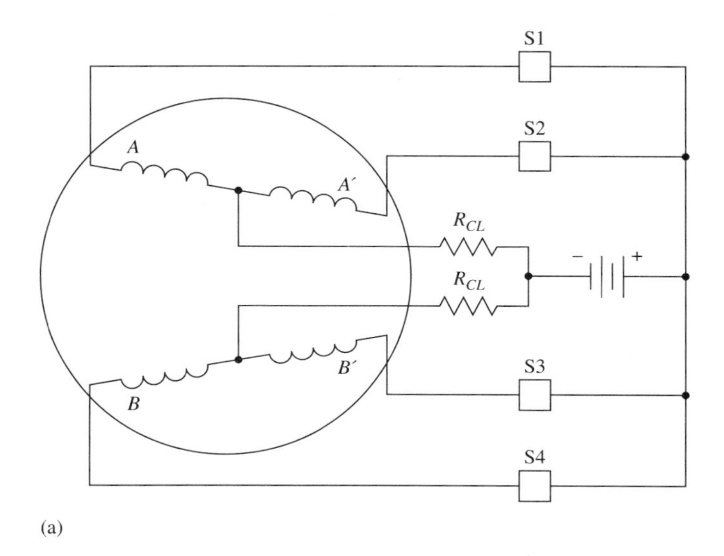

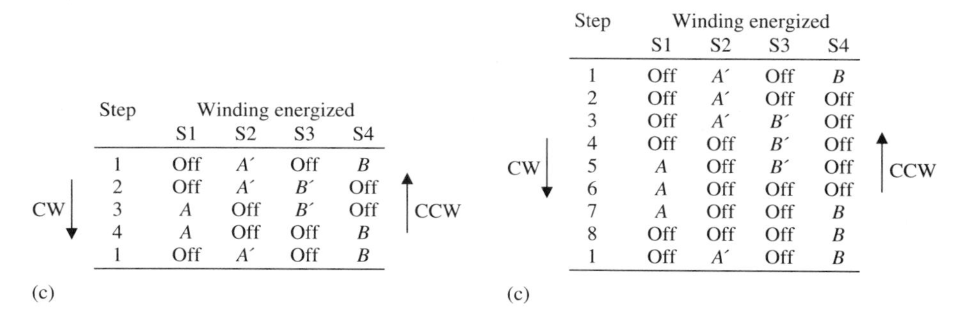

Step	Winding energized S1	S2	S3	S4
1	Off	A′	Off	B
2	Off	A′	B′	Off
3	A	Off	B′	Off
4	A	Off	Off	B
1	Off	A′	Off	B

CW ↓ ↑ CCW

(c)

Step	Winding energized S1	S2	S3	S4
1	Off	A′	Off	B
2	Off	A′	Off	Off
3	Off	A′	B′	Off
4	Off	Off	B′	Off
5	A	Off	B′	Off
6	A	Off	Off	Off
7	A	Off	Off	B
8	Off	Off	Off	B
1	Off	A′	Off	B

CW ↓ ↑ CCW

(c)

Figure 17-9 Bifilar stepper motor. (a) Schematic. (b) Full-step switching sequence. (c) Half-step switching sequence.

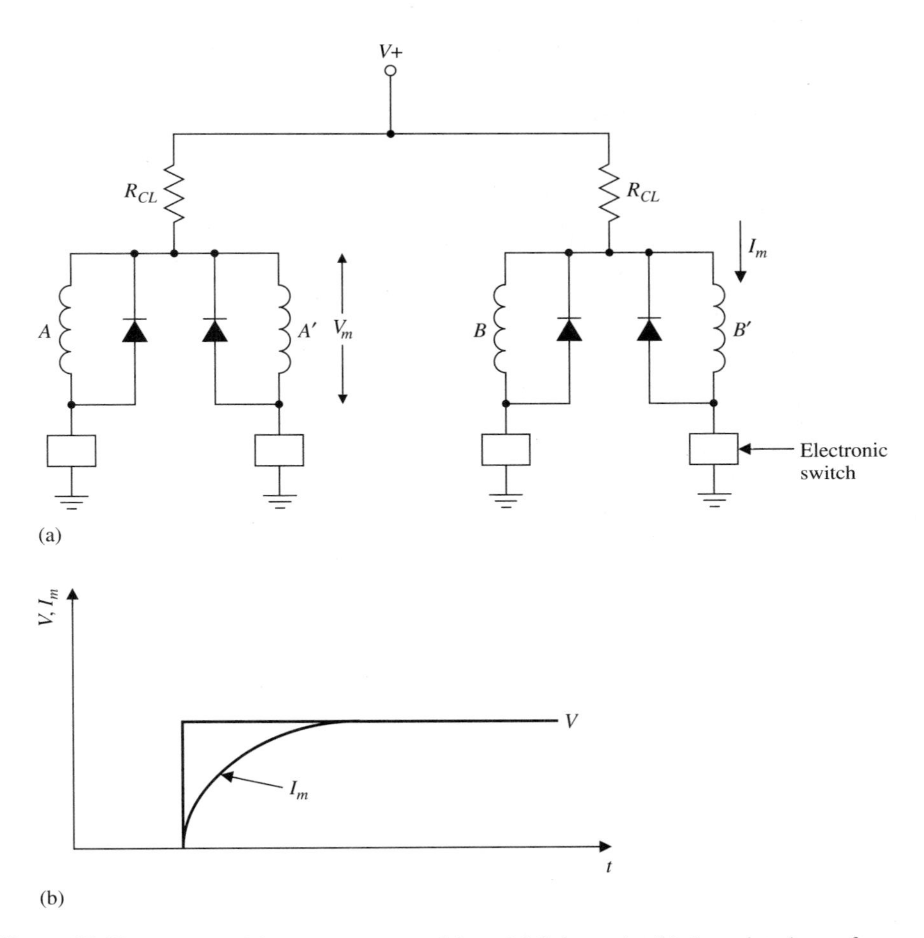

Figure 17-10 Series R (L/R) stepper motor driver. (a) Schematic. (b) Associated waveforms.

necessary stator current levels. Because of the high I^2R losses this type of driver is usually limited to low-cost small and simple systems.

Choppers. Chopper drivers use the principles discussed in Section 2-6 to control the current level to acceptable values. This scheme is illustrated in Fig. 17-11.

Since power loss in the series resistances is totally eliminated, as is the case in the series R driver, combined with accurate control of the stator current, it is possible to operate the stepper motor at higher current levels to provide a greater torque output. There is a tendency for the oscillating motor current to cause line resonance at some frequencies. This effect can be minimized by antiresonant circuitry.

Bilevel. An alternative method of controlling the current level is to use two separate input voltage levels. The motor is initially accelerated using the high-level voltage, and when the desired operating speed is achieved, the driver switches over to the lower voltage level source. This system eliminates the torque variation produced by the chopper system, but at the added expense of using a dual power supply, additional electronic switches, and increased complexity in system control.

Linear Stepper Motors

The linear stepper motor operates on exactly the same principle as the standard rotary hybrid stepper motor. The stationary part or member is called the *platen*, the moving part is the *forcer* (Fig. 17-12). The forcer consists of two electromagnets, which are driven by two phases of a pulse-width-modulated microstepping drive,

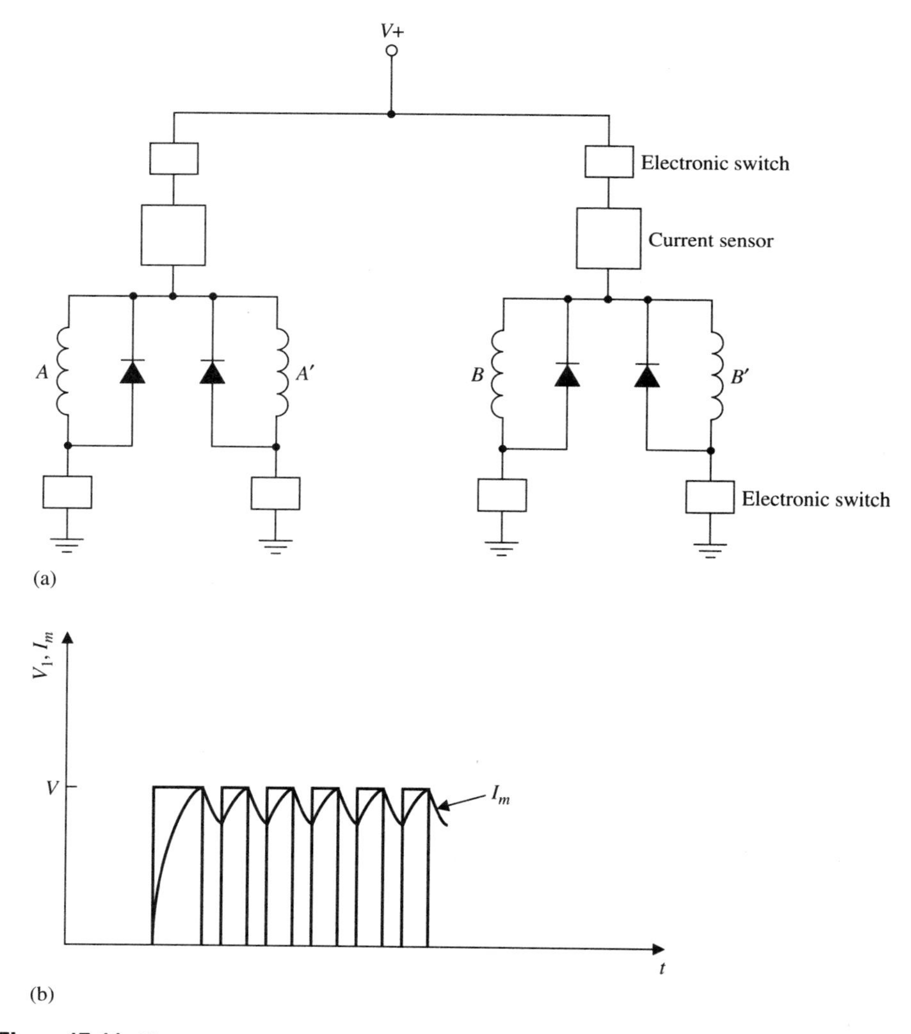

Figure 17-11 Chopper stepper motor driver. (a) Schematic. (b) Associated waveforms.

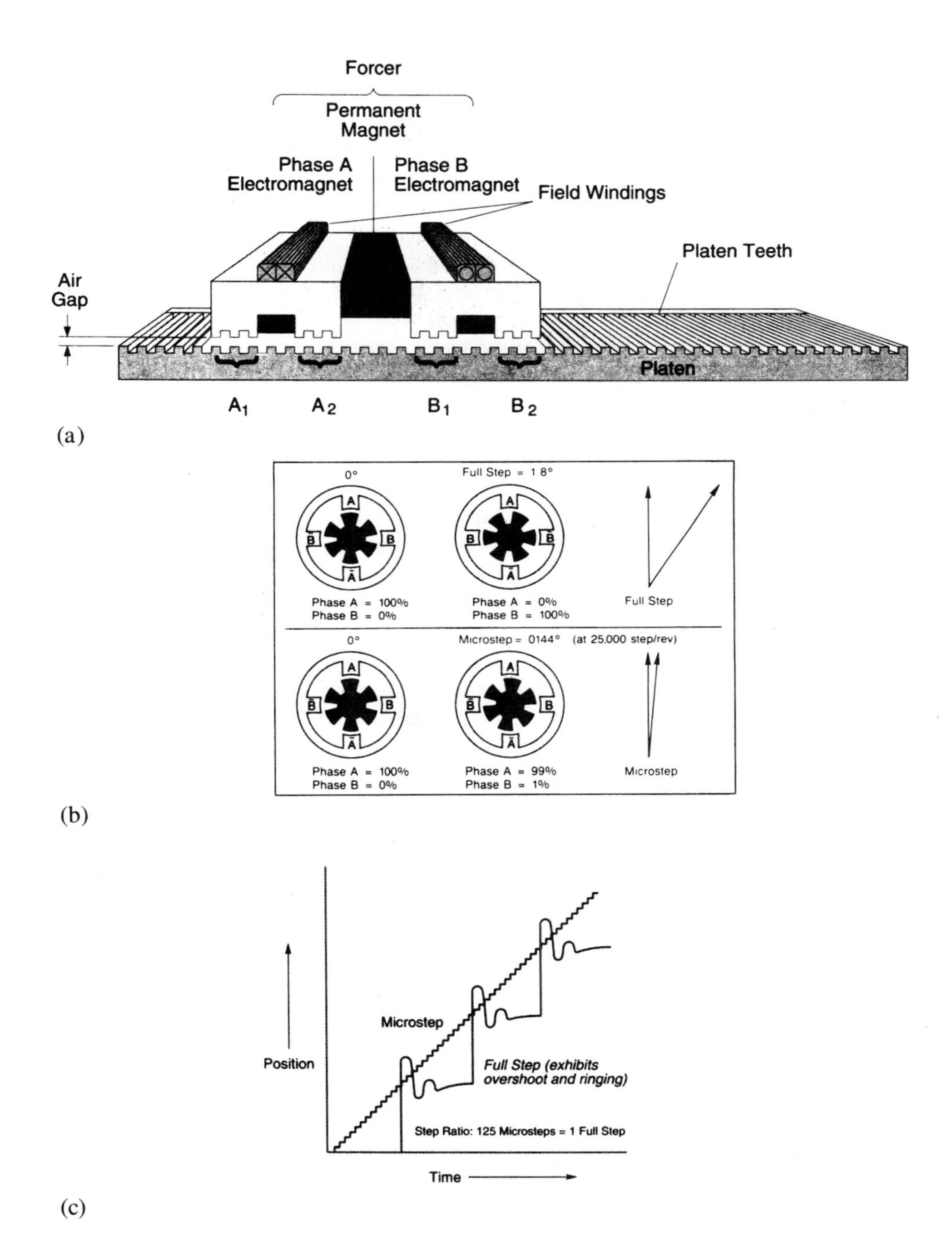

Figure 17-12 Linear stepper motor. (a) Principle of operation. (b) Microstep principle. (c) Comparison of microstepping and conventional stepping. *(Courtesy of Parker Hannifin Corporation)*

and a permanent magnet, aligned as shown in Fig. 17-12(a). The forcer unit is equipped with antifriction or air bearings, and when stepped it moves along the platen.

As can be seen from Fig. 17-12(a), when the electromagnets are deenergized, the permanent-magnet flux passes through the PHASE A electromagnet across the air gap at A_1 and A_2 along the platen and returns across the air gap at B_1 and B_2, through the PHASE B electromagnet to the opposite pole of the permanent magnet, with the flux being divided equally at the four poles. When the PHASE A electromagnet is energized, the flux path of both magnets is directed through the A_1 pole, that is, the path that is common to both the electromagnet and the permanent magnet, and there will be no flux passing through the A_2 pole because the two fields are canceled. The flux generated at the A_1 pole produces a force parallel to the platen, which causes the forcer to move along the platen until the teeth of the A_1 pole are aligned with the platen teeth. Alternately, switching the current between the two electromagnets moves the forcer a full step along the platen. Also changing the current direction each time an electromagnet is switched changes the polarity of the poles, so that the poles are aligned with the platen in the sequence A_2, B_1, A_1, and B_2 and the forcer continues in its original direction a distance of one-fourth of a platen tooth pitch for each step. The direction of movement is reversed by reversing the switching sequence.

Parker-Compumotor uses a combination of digital logic and bipolar pulse-width modulation to increase the resolution and smoothness of both their linear and their rotary stepper motors. A technique known as *microstepping* proportions the current levels between the two windings. Instead of switching the current on and off to produce the step sequence, the Parker-Compumotor drive decreases the current slightly in one phase and increases it slightly in the other. Each driver input pulse causes the motor to move one microstep; the motor usually requires between 5,000 and 12,500 steps/in. The actual number of microsteps per inch depends on the number of platen teeth per inch and the number of microsteps per full step provided by the driver. The linear speed is determined by the frequency of the input pulses to the drive controller. The microstep principle is illustrated in Fig. 17-12(b), and a comparison of microstepping and conventional full-step control is shown in Fig. 17-12(c). The Parker-Compumotor system provides smooth acceleration and deceleration through velocities in which half- and full-step systems would experience resonance problems with a consequent loss of force, providing at the same time a high-resolution digital open-loop control system.

The linear motor is best suited for high-speed low-mass moves, and has totally eliminated the need for mechanical rotary-to-linear conversion. The forcer is equipped with either antifriction or air bearings. Antifriction bearings minimize yaw, pitch, and roll, but still introduce friction into the system. The air bearing on the other hand permits a reduction in the air gap between the forcer and the platen, and at the same time increases the available force. Standard platen lengths range from 78 to 120 in (2 to 3.05 m) at linear speeds of 40–60 in/s (1.01–1.52 m/s). Parker-Compumotor also use the microstep principle for standard rotary stepper motors. Figure 17-13 shows representative linear microstepping motors and associated drivers.

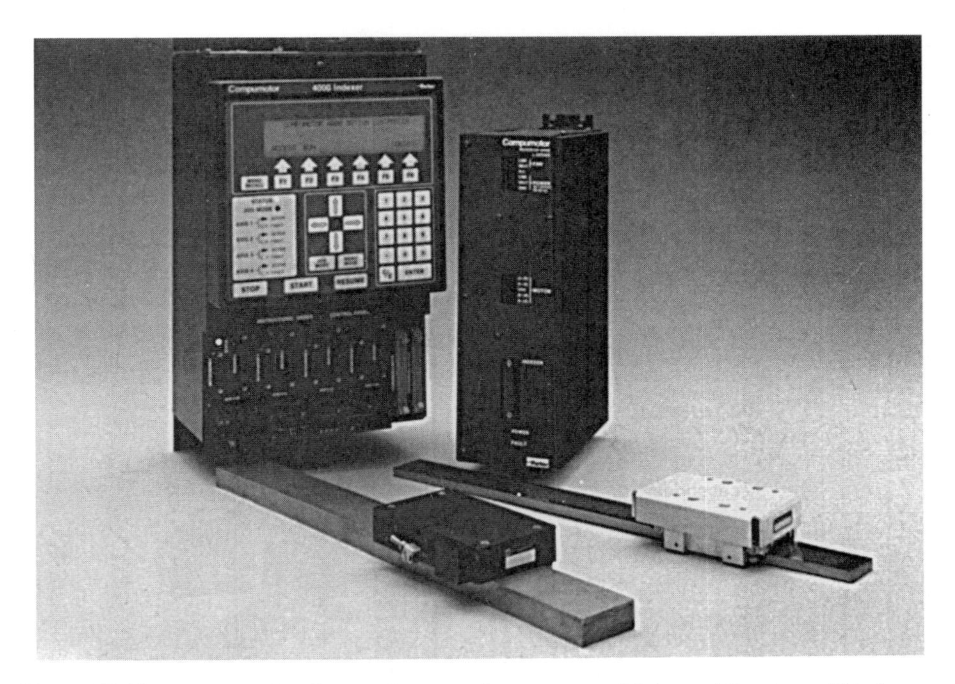

Figure 17-13 Representative linear microstepping motors and drivers. *(Courtesy of Parker Hannifin Corporation)*

The SynchroStep Motor

A revolutionary change in the design of stepper motors has been introduced by PMI Motion Technologies, Division of Kollmorgen Corporation. This motor, called the SynchroStep motor, consists of a thin disk of samarium cobalt, and has a pattern of alternative N and S poles permanently magnetized around the edge of the rotor disk [Fig. 17-14(a)].

The ironless rotor is free to rotate between the poles of two independent sets of C-shaped stator laminations [Fig. 17-14(b)]. Only one set of laminations is shown for clarity. Each set of laminations forms one phase of the motor. The two sets are offset from each other by one-half of a rotor pole. Energizing phase *A* causes the rotor poles to align with the phase *A* laminations, offset with respect to the phase *B* laminations [Fig. 17-15(a)]. Deenergizing phase *A* and energizing phase *B* causes the rotor to step forward one-half of a rotor pole width [Fig. 17-15(b)]. If phase *B* is now deenergized and phase *A* is energized with its polarity reversed, the rotor will once again step forward one-half of a rotor pole [Fig. 17-15(c)]. Once again phase *A* is deenergized and phase *B* is energized with reversed polarity, causing the rotor to step one-half of a rotor pole. [Fig. 17-15(d)]. Continuously repeating this sequence will cause the rotor to turn in a series of steps. It should be noted that the torque output is increased by energizing the two phases simultaneously. Reversing the sequence of phase energization will reverse the rotor; changing the stepping frequency controls the angular velocity of the rotor.

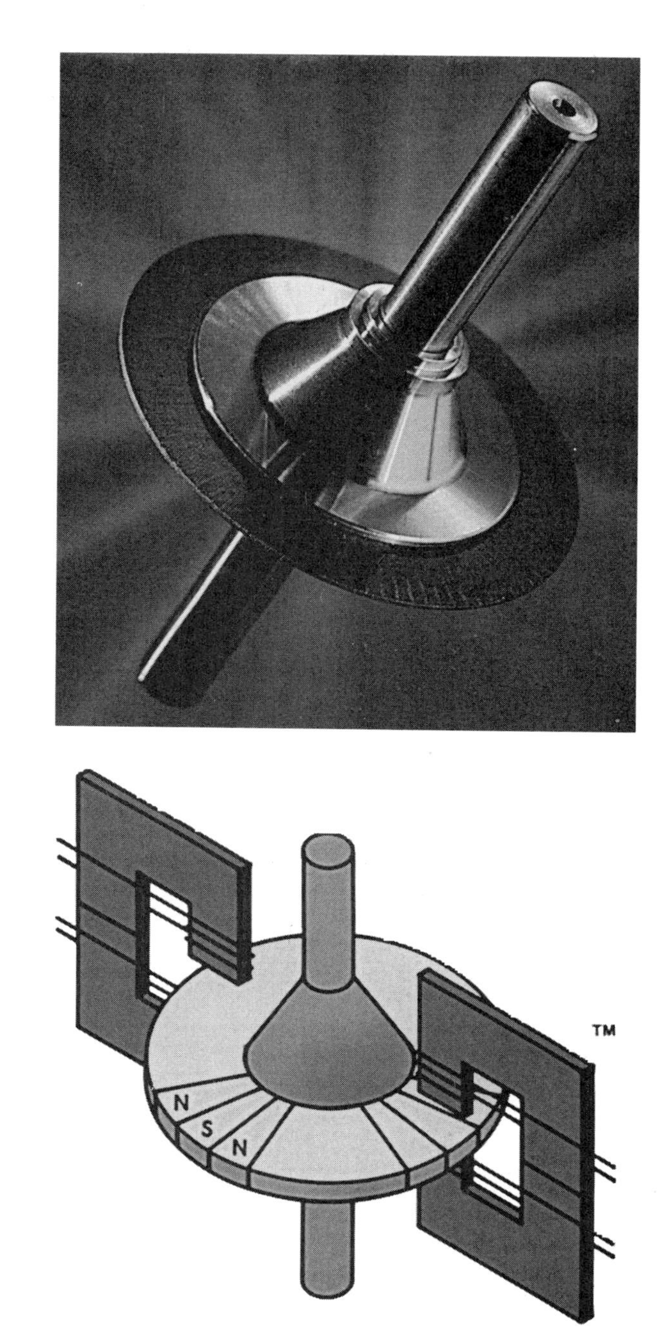

Figure 17-14 SynchroStep motor. (a) Ironless samarium cobalt disk rotor. (b) Axial air-gap arrangement of rotor and stator poles. *(Courtesy of PMI Motion Technologies, Division of Kollmorgen Corporation)*

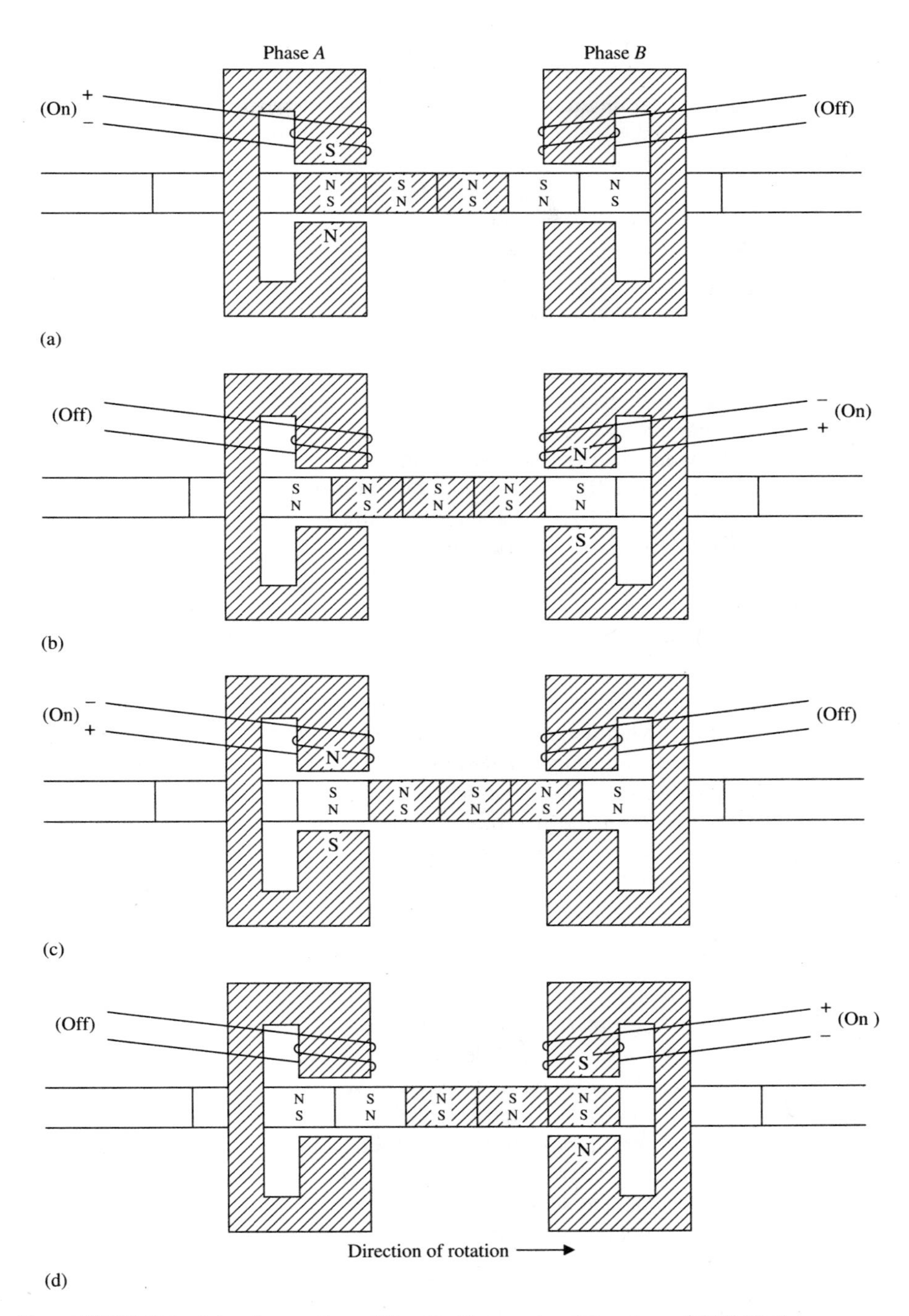

Figure 17-15 Principle of operation of SynchroStep motor. (*Courtesy of PMI Motion Technologies, Division of Kollmorgen Corporation*)

Iron losses, that is, hysteresis and eddy current losses, have been reduced significantly, as compared to the conventional stepper motor, by the use of grain-oriented silicon steel laminations. This is only possible in the SynchroStep motor because none of the phases share the stator iron, which precludes the use of oriented steel. This, combined with the absence of iron in the rotor, results in iron losses that are 20% of those of the hybrid stepper motor and 33% of those of a conventional dc servomotor. As a result the motor operates with a high efficiency. Also, for the same power output there is a corresponding reduction in the power input, which simplifies the design and cooling requirements of the associated electronic controls. Because of the low inertia of the rotor, very high acceleration rates are possible, up to 150,000 rad/s^2, which is seven times that of a hybrid stepper motor and three times that of most dc servomotors. Even though the torque output decreases as the angular velocity increases, there is a significant torque output up to 6,000 r/min (628.32 rad/s). These characteristics, combined with a high power-to-weight and power-to-size ratio, provide an excellent performance in a smaller and lighter package. These motors are especially suited to motion-control applications for robot wrists and grippers, and in small-size robots for multiaxis control, process control, computer peripherals, business machines, biomedical equipment, and instrumentation. Typical SynchroStep motors are shown in Fig. 17-16.

Stepper Motor Testing

Performance tests are carried out by the manufacturer. These include measurement of the holding torque, detent torque, and maximum response rate, which permit a check of the dynamic performance without having to carry out an operating test over the whole stepping range. The dc resistance of the stator winding is used as a check of the input power required at the operating voltage, as well as providing the application engineer with the maximum inrush current. Step-angle accuracy is also compared against factory standards. Finally, high potential and winding insulation tests are carried out.

17-3 SERVOMOTORS

With the continuing upsurge in electromechanical servomechanisms used in industrial robots, computerized numerical control, flexible manufacturing systems, automated warehouses, and the like, there is a continuing demand for servomotors. Not only is demand increasing, but there are very significant improvements in the capabilities and responses of the new generations of servomotors. Initially because ac squirrel-cage servomotors are rugged, require minimal maintenance, have high efficiency, and are relatively easy to control, they became the first choice for designers of electromechanical servomechanisms, even though the dc servomotor was capable of developing peak power outputs well in excess of their steady-state power output.

Increasing demands for higher efficiency and fast response, combined with the rapid growth of power electronics, have resulted in significant advances in high-response dc drives. In addition, the use of operational amplifiers has practically eliminated the drifting that was a characteristic of dc amplifiers. Improved cooling

Figure 17-16 Typical SynchroStep motors. (*Courtesy of PMI Motion Technologies, Division of Kollmorgen Corporation*)

techniques and an almost exclusive use of digital electronics have reduced both the size and the cost of these high-performance controllers.

The choice of whether to use an ac or a dc servomotor is not simple. The ac servomotor still finds a niche in low-power (≤100-W) instrument servomechanisms, especially when the control signals, the input power, the and excitation voltages are all derived from a single-frequency ac source, usually 60 or 400 Hz. If a variety of carrier frequencies are involved, it is simpler and less costly to rectify the signals and use a dc servomotor drive.

Both the ac and the dc servomotors depend on the interaction of two magnetic fields, the one produced by axial currents, and the radial field. If these fields are correctly oriented, the output torque is maximized in the dc machine. In the case of the ac servomotor the two fields must be in time phase; any space and time phase differences will produce a reduced torque output.

The polyphase induction motor produces a radial rotating field as a result of stator currents, and the rotor field is produced by currents flowing axially in the rotor conductors as a result of transformer action. An ac servomotor is effectively an inefficient transformer, and as a result is itself inefficient. Also because the peak flux density is $\sqrt{2}$ that of the mean flux density, it requires 57% more core iron. In the dc machine energy is transferred directly to the armature via the brushes and commutator, thus limiting the main source of energy loss to the brushes. In addition the iron losses are also significantly less than those of a comparably rated ac machine.

Both the two-phase servomotor and the dc servomotor can be assessed in terms of

$$T_{ST} = \frac{1{,}352 P_{LR}}{S_{nl}} \text{ oz} \cdot \text{in} \tag{17–7)(E}$$

or

$$T_{ST} = 91.17 \omega_{nl} P_{LR} \text{ N} \cdot \text{m} \tag{17–7)(SI}$$

where T_{ST} is the stall torque, S_{nl} or ω_{nl} the no-load angular velocity, and in the case of ac motors it is the synchronous speed, and P_{LR} is the locked-rotor input power at rated voltage.

From Eqs. (17-7) it can be seen that as the angular velocity decreases, the torque per watt increases. Low angular velocities are usually desirable in servo systems since less gear reduction will be required. In the case of ac servomotors there is a physical limit to the number of stator poles that can be wound in a given frame size. Furthermore the efficiency of transferring power to the rotor decreases sharply as the number of stator poles increases, whereas the dc motor speed is inversely proportional to the field pole flux, and with the use of rare-earth permanent magnets, the no-load angular velocity can be reduced well below that achievable with ac servomotors. Brush and radio frequency interference (RFI) problems traditionally associated with dc servomotors have been reduced considerably as a result of ongoing research and improved materials.

DC Servomotors

Dc servomotors are available in a number of configurations: (1) wound field, (2) moving-coil shell armature and printed circuit, and (3) torquer motors. Nearly all dc servomotors are used in position and velocity control applications. Position control applications usually require high torque at low speeds. Thus their torque-speed characteristics have a much steeper slope since their maximum torque occurs at standstill and their minimum torque at maximum speed. As a result these motors are frequently called torque motors. Velocity control applications demand a torque output that is relatively constant.

Separately Excited DC Servomotors. As you may recall, the rotational speed of a separately excited dc motor is proportional to the applied armature voltage V_a. Figure 17-17 shows the relationship between torque and rotational speed at a number of applied voltages. As can be seen, at any particular applied armature voltage there is a slight decrease in speed as the load torque increases. Since the motor is expected to operate over a wide range of speeds and loads, operating conditions must be specified so that the machine is not subjected to harmful operation. Figure 17-17 can be divided into three operating zones: (1) continuous operation without external cooling, (2) continuous operation with an external blower, and (3) intermittent operation. In the continuous operation zone the motor can be operated safely without temperature rises in excess of the designed values being reached. The specific information for these zones is readily available from manufacturers' data.

Similar torque-speed curves and operating zones for a permanent-magnet position control dc servomotor are shown in Fig. 17-18. The steepness of the torque-speed curves should be noted since the motor is designed to produce maximum torque at standstill or low speeds and minimum torque under high-speed operation.

In either case the range of safe operation is extended by using an externally mounted blower motor to provide sufficient cooling air to prevent the motor from overheating, especially at low speeds.

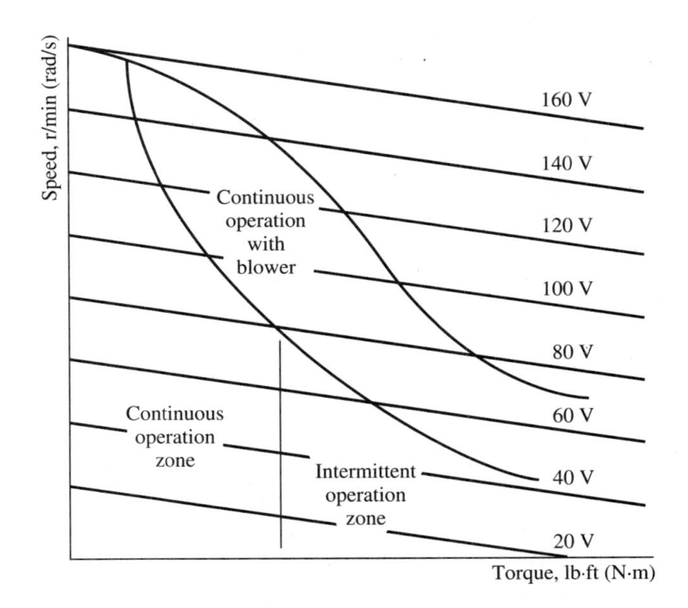

Figure 17-17 Torque-speed curves at varying applied armature voltages.

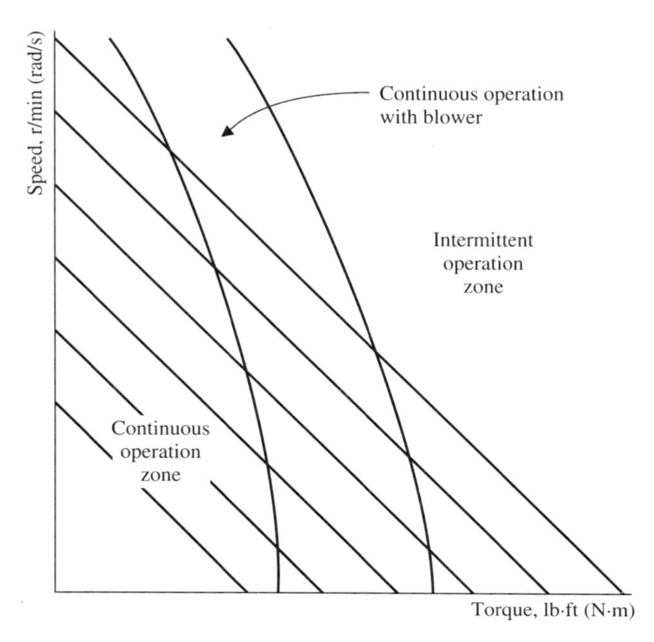

Figure 17-18 Torque-speed characteristics of a position control dc servomotor.

Moving-Coil Motors. Modern servomotors use permanent-magnet field structures. These magnet structures can be molded into any desired shape, and either Alnico or ceramic materials are used, although a number of machines using rare-earth magnets such as samarium-cobalt are now available. The major advantages to using permanent-magnet structures are: (1) optimized motor design, (2) increased motor efficiency since there are no field iron or copper losses, (3) reduced frame size for a given output, and (4) reduced cooling requirements. In turn the permanent-magnet motor has a number of advantages over the conventional wound field motor, such as (1) linear torque-speed characteristic, (2) high stall and accelerating torque, (3) reduced power input and higher efficiency. There are disadvantages to the use of permanent-magnet motors. (1) Speed control is limited to variable armature voltage control. (2) They may experience commutation problems because of the absence of interpoles. (3) Alnico permanent-magnet poles may be demagnetized under short-circuit conditions, although ceramic and rare-earth magnets are rarely affected.

In order to meet the requirements of high-performance systems where high acceleration and deceleration rates are required, it is necessary to reduce the inertia of the armature. Initial attempts included the slotless armature, which consisted of a small-diameter armature core with the conductors bonded to its surface in a very-high-density magnetic field. This design provided high mechanical strength and torsional rigidity, but suffered from a high armature inductance, which limited its response, and a low armature resistance and torque constant, which in turn required high armature current levels at low voltages. This concept has been replaced by the *moving-coil* armature, which is currently available in two forms: the shell armature and the disk or *printed-circuit* armature.

The shell armature has an armature that consists of a hollow-cup cylindrical rotor to which the armature coils are bonded, which are held in position by polymer resins and fiberglass, suitably attached to the rotor shaft. The resulting structure is small in diameter and the inertia is minimized; also, because there is no iron in the armature structure, the armature inductance is also minimized. As a result, the armature is extremely responsive to armature voltage control signals and this, combined with a very high torque-to-inertia ratio, enables acceleration rates of up to 1,000,000 rad/s^2 to be achieved. Yet at the same time it has the capability of completing several thousand start-stop cycles per second. This latter feature also permits their use in incremental-motion as well as continuous-motion applications.

A major disadvantage of moving-coil motors is their low thermal capacity and low thermal time constant, typically 20–30 s armature to frame, as against a frame to ambient thermal time constant of 30–60 min. The armature can overheat in less than 1 min. However, in addition to careful application the motor can be forced-air cooled, which is also necessary to prevent a reduction in the permanent-magnet flux if Alnico 5 or 5-7 magnets are used. Moving-coil motors are not usually affected by peak demagnetizing currents because of the large air gap and pole shoe design. Rather they are limited by current limitations of the armature and brushes. The construction of a typical shell armature moving-coil motor is shown in Fig. 17-19.

The printed-circuit or disk motor, which was originally patented in France in 1960, has been developed from an original concept put forward by Barlow in the 1820s. The first French design used printed-circuit techniques to produce a double-sided disk armature using photoetching. This technique was not adequate to meet the increasing demand, and it has now been replaced by a mechanical notching technique which permits greater uniformity of the armature conductors as well as permitting multilayered armatures to be produced.

The construction of a printed-circuit motor is illustrated in Fig. 17-20, where the disk armature rotates between the pole faces of eight or ten pairs of permanent magnets. These poles provide an axial magnetic field across a 0.1-in (2.54-mm) air gap. The permanent-magnet materials in common use are Alnico 5, which provides a high flux density. A lower flux density material such as barium ferrite is used in lower power output machines. The current flow in the armature conductors is radially across the disk. The winding end turns are at the outer radius of the disk and introduce inertia, since the moment of inertia is proportional to the fourth power of the disk diameter. The commutator consists of a cleaned track, and the power is usually supplied to the armature by two pairs of brushes bearing against the track. The brushes must be selected carefully to minimize wear, and pure carbon, graphite, or silver graphite brushes are used in cartridge-type brush holders. Conventional plain journal bearings are used, although in high-performance units high-quality antifriction bearings are preferred. High-performance printed-circuit motors also often incorporate an integral printed-circuit tachogenerator. The main advantages of the printed-circuit motor are the following:

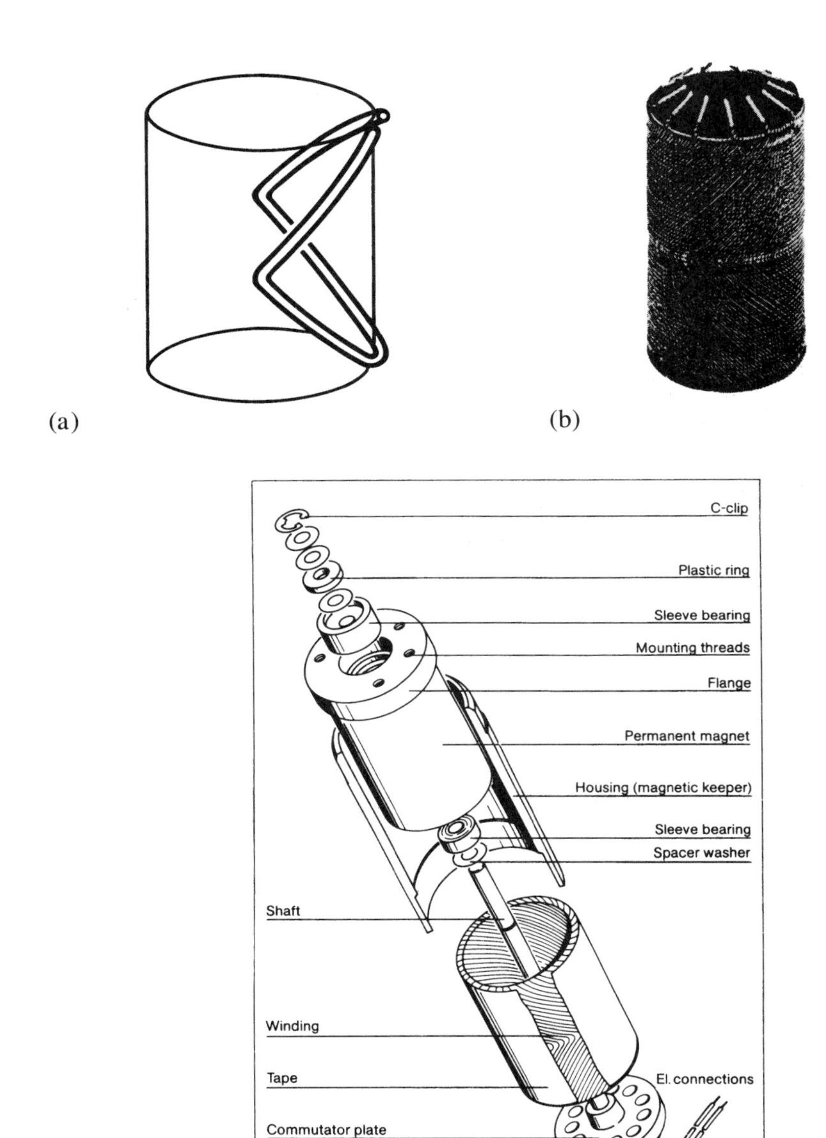

Figure 17-19 Shell type moving coil motor. (a) One turn of the armature winding. (b) Completed armature winding. (c) Exploded view of complete motor. *(Courtesy of Maxon Precision Motors, Inc.)*

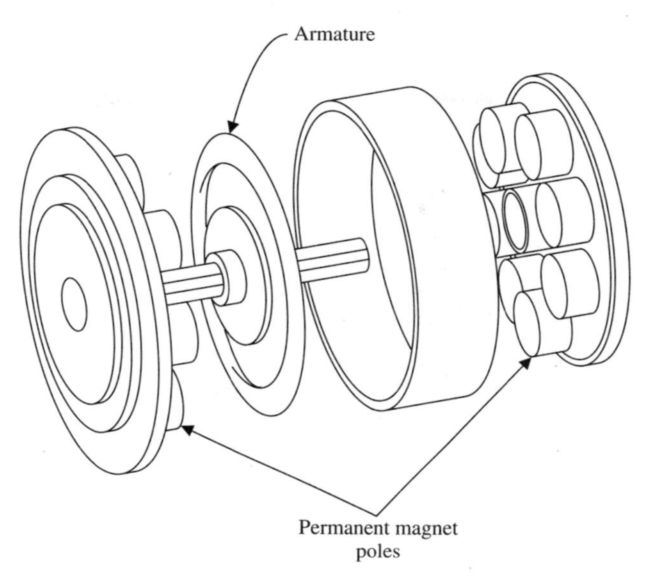

Figure 17-20 Construction of disk or printed-circuit motor.

1. Since there is no iron in the armature, it has low inertia, low mass, and a low armature inductance, typically 100 μH.
2. The output torque is not limited by saturation because of the absence of rotor iron.
3. Since the brushes bear against the coils, the number of commutator segments is effectively increased, which reduces torque ripple. Also the segment voltage is reduced and this, coupled with the low armature resistance, usually $\leq 1.2\ \Omega$, and a low reactance voltage, gives improved brush life and performance.
4. Because the armature conductors are directly exposed to the cooling medium, heat transfer is greatly improved, which in turn permits a smaller and lighter motor.
5. The large number of permanent-magnet field poles permits a higher current density in the armature conductors and improves the linearity between the armature current and the output torque.

The major applications of printed-circuit motors are where the frequent start-stop capabilities can be best utilized, such as magnetic tape drives, paper feeders for *X–Y* plotters, machine-tool positioning, and table feeds.

Direct-Drive Torque Motors. Direct-drive torque motors, or *torquers*, are frameless dc servomotors that are directly coupled to the load. They are thin in length compared to their diameter, and have axial mounting holes through the armature to permit them to be directly attached to shafts, hubs, and bosses on the load. These servomotors are especially useful where size, weight, power, and response time must be minimized. The direct connection of the motor armature to the load eliminates any need for gearing, and provides a high coupling stiffness as well as a high mechanical resonance. Operating the armature core in saturation, as well as the use of a large number of permanent-magnet field poles reduce the armature self-inductance, thus enabling the torquer to produce torque very rapidly at all rotational speeds. The direct coupling of the armature to the load enables the shaft to be positioned accurately, the only limit being the accuracy of the position transducer.

AC Servomotors

The most commonly used ac servomotor is the two-phase ac servomotor. It is simple and reliable and is frequently used in instrumentation servo systems. Its mechanical output ranges from 0.5 to 100 W. It differs from the more common polyphase squirrel-cage induction motor by having a high-resistance low-inertia rotor. Although the high-resistance rotor is essential for a linear torque-speed characteristic, it limits the power output capability as well as contributing to low overall efficiency.

The basic connections of a two-phase ac servomotor are shown in Fig. 17-21(a). This arrangement produces a rotating magnetic field. This rotating magnetic field induces electromotive forces and currents in the rotor conductors, and the interaction of the stator and rotor fields causes the rotor to rotate.

The two-phase ac servomotor differs from the conventional polyphase squirrel-cage induction motor because the magnitude of the rotating magnetic field, and thus the torque and speed, are controlled by the magnitude of the control-winding voltage. The direction of rotation is determined by the phase relationship between the reference and control-winding voltages. Also to reduce the load on the servo amplifier, the power factor of the control winding is adjusted to unity to minimize the current. Figure 17-21(b) shows the relationship between torque and speed.

17-4 SWITCHED-RELUCTANCE MOTORS

The switched-reluctance (SR) motor has been developed from the single-stack variable-reluctance stepper motor. Initially difficulties were experienced as a result of poor mechanical design, an incomplete understanding of the electromagnetic design of doubly excited salient-pole motors, and in addition the lack of suitable high-power electronic switching devices. Interest was rekindled in the 1960s with the development of the thyristor, and the motor was patented in the early 1970s. It has been applied commercially to battery-operated vehicles since the late 1970s.

The switched-reluctance motor is a dc brushless motor with a uniquely simple construction. The laminated rotor has no windings, and therefore no commutator is required. As a result the rotor is mechanically robust and has a low moment of inertia, which permits a fast response to control signals and is unaffected by temperature changes. The laminated salient-pole stator is also simple and robust, having stator windings that are very similar to those of a conventional dc machine, but with very short end turns. The motor is doubly salient and singly excited, that is, the stator and rotor both have salient poles, but only the stator poles are excited. The stator and rotor have a different number of poles, the usual combinations being 8:6 and 6:4. Although other combinations are possible, a 6:4 combination is shown in Fig. 17-22.

The windings on diagonally opposite pairs of stator poles are connected in series or parallel to produce radial magnetic fields of opposite magnetic polarities at the air gaps between the stator and rotor poles. The flux path shown in Fig. 17-22 is created by energizing the phase *A* windings with the rotor poles in position of minimum flux path reluctance. If phase *A* is deenergized and phase *B* energized, the rotor will turn CCW so that rotor poles 2 and 4 are aligned with the stator poles energized by phase *B*. Similarly, deenergizing phase *B* and energizing phase *C* will cause rotor poles 1 and 3 to be aligned with the stator poles energized by phase *C*. Energizing the stator phases in an *ABC* sequence will cause the rotor to turn in a CCW direction. Similarly if the stator phases are energized in an *ACB* sequence, the rotor will turn in a CW direction. It should be noted that if the motor is energized in a similar manner to a stepper motor, then

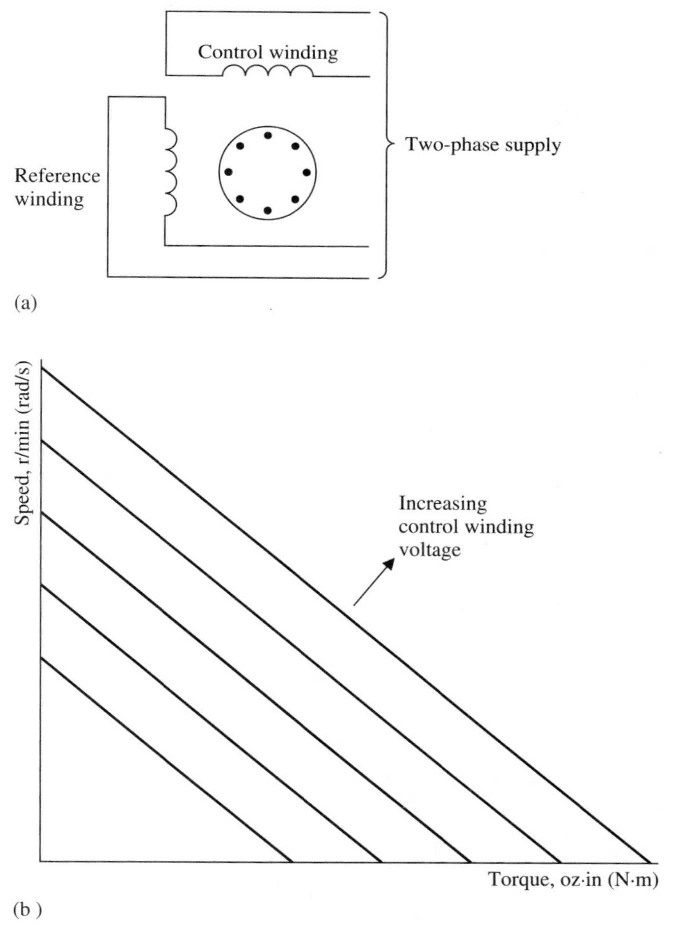

Figure 17-21 Two-phase ac servomotor. (a) Schematic. (b) Torque-speed characteristic.

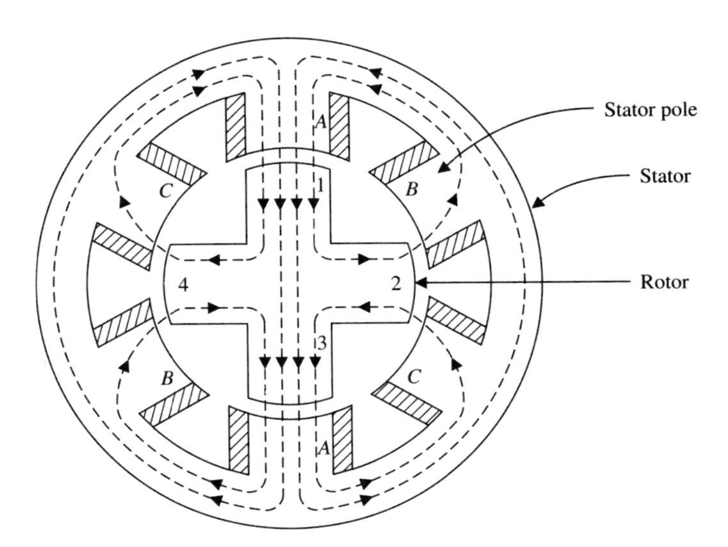

Figure 17-22 Cross section of switched-reluctance motor with six stator poles and four rotor poles showing magnetic flux distribution with phase A energized.

$$\text{Step length} = \frac{360°}{Np} \tag{17–8}$$

where N is the number of stator phases and p the number of rotor poles.

In practice, however, the phases are energized in synchronism with the rotor movement, giving a smooth rotation even at low speeds and low torque variation with rotor angle.

From Eq. (17-8) it can be seen that a switched-reluctance motor with a three-phase stator and a four-pole rotor has a step length of 30°. For the 6:4 switched-reluctance motor there are 12 steps per revolution (360°/30°), and the switching frequency is

$$\text{Switching frequency (Hz)} = \frac{\text{steps/r} \times \text{r/min}}{60N} \tag{17–9}$$

To run this motor at 3,000 r/min (314.16 rad/s) requires that the switching frequency be

$$\begin{aligned}\text{Switching frequency (Hz)} &= \frac{12 \times 3{,}000}{60 \times 3} \\ &= 200\text{Hz/phase}\end{aligned}$$

It should be noted that although the switched-reluctance motor is derived from the single-stack variable-reluctance stepper motor, the variable-reluctance stepper motor is a lower-power positioning device that is controlled by a square-wave pulse train applied to the stator windings. The switched-reluctance motor on the other hand is designed to be an efficient smooth running variable-speed drive in which the turn-on and the turn-off of the electronic switching devices are determined by the rotor position, the required torque, and the motor speed.

Figure 17-23(a) shows the variation of the inductance of a stator phase as the rotor poles move from a position of maximum reluctance and minimum stator inductance to a position of minimum reluctance and maximum stator inductance when the two poles are aligned. The motor only produces motoring torque if stator current is supplied as the stator inductance is increasing. In addition, because of the inductance of the stator windings, current flow for high output torque must be initiated before the stator inductance begins to increase so that the current will have peaked and largely decayed before the inductance becomes constant as the poles are fully over-

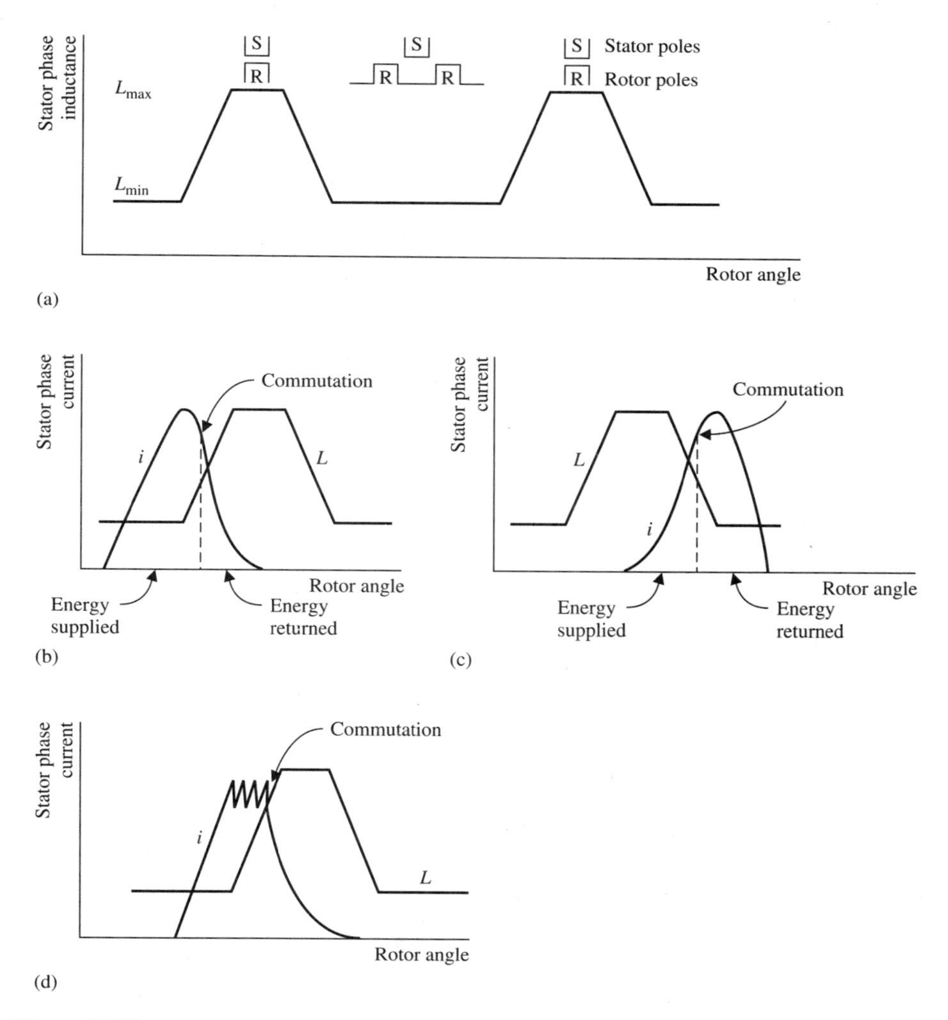

Figure 17-23 Switched-reluctance motor pole inductance profile. (a) Cyclic variation of stator phase inductance with respect to rotor poles. (b) Phase current variation, high-speed motoring. (c) Phase current variation, high-speed regeneration. (d) Phase current variation, low-speed motoring (chopper mode).

lapped. The stator phase current supplied up to the point of commutation supplies the energy converted to mechanical power output, the energy stored in the magnetic field, the stator copper losses, and the total iron losses of the machine. It should also be noted that the self-induced electromotive force in the stator winding can cause the stator current to start decreasing before commutation takes place [Fig. 17-23(b)]. After commutation the energy stored in the collapsing magnetic field is in part returned to the source and the remainder is converted to mechanical energy provided the stator phase inductance is still increasing.

If after commutation current is flowing in the stator phase winding on the negative slope of the inductance waveform, that is, the rotor pole is moving away from the stator pole, energy is returned to the source which may exceed that supplied by the source before commutation, in which case the motor is regenerating and being braked [Fig. 17-23(c)]. Figure 17-23(b) and (c) illustrates single-pulse operation. Figure 17-23(d) shows how the stator current is controlled by chopping to permit low-speed operation, which involves turning the switching device on and off rapidly during the normal conducting period.

Motoring torque production, unlike for the conventional dc motor, is entirely dependent on the variation of reluctance between the rotor and stator poles, that is, the rotor will always turn toward a position that minimizes the reluctance of the magnetic circuit. This means that the torque is totally independent of the polarity of the stator poles, and consequently the torque is independent of the current direction in the stator-pole windings. As a result only one switching device per phase is required, and this switching device must excite a pair of stator poles as they are being approached by a pair of rotor poles.

The rotor position is sensed by either digital Hall effect transducers, optical choppers, or high-frequency proximity detectors. This information is supplied to a logic circuit where it is encoded and used to control the stator phase switching devices (power transistors, MOSFETs, or GTOs) in the dc link converter.

A simple unipolar drive circuit controlling a three-phase switched-rotor motor is shown in Fig. 17-24. Each phase winding is controlled by a switching device. The low-power gating signals suitably conditioned are obtained from the drive control logic. The dc source voltage must be sufficient to establish the rated peak stator pole flux within half the switching period corresponding to maximum speed in the presence of the motor's self-induced electromotive force.

In addition, because of the high inductance of the stator phase windings when the stator and rotor pole are mostly overlapped, it is impossible to achieve an instantaneous reduction of the phase current when the switching device is turned off. An alternative path must be provided for the energy being returned from the stator phase winding by the collapsing magnetic field, and this consists of the diode D and the voltage sink V_{ds}.

The maximum voltage that will appear across the switching device is

$$V_{\max} = V_{do} + V_{ds} \tag{17–10}$$

To avoid serious losses and the consequent loss efficiency, the energy returned to the voltage sink V_{ds} via the diodes must be transferred to the voltage source V_{do}, either by an independent chopper circuit or by using one of the circuits shown in Fig. 17-25.

Figure 17-25 shows a number of different power circuits that have been used in conjunction with switched-reluctance motors. The first circuit [Fig. 17-25(a)] uses two switching devices per phase and is somewhat similar to that used in a dc link converter supplying a variable-frequency output to a polyphase induction motor. The major difference is that the stator windings are in series between the switching devices instead of being tapped off from the midpoint between the switching devices. The switching devices must be switched in pairs, with each device being voltage rated for V_{do} and being able to carry the peak phase current. If the two switching devices are turned off sequentially, a freewheeling path for the phase current is created as long as one switching device in on. This assists the control of the phase current when chopping.

Circuit 2 [Fig. 17-25(b)] is the preferred circuit, where the number of switching devices is to be kept to a minimum. The dc link filter capacitor is split to produce a dual voltage source. As can be seen, phase *A* current is drawn from the upper half of the voltage source when its switching device is turned on, and is returned to the bottom half of the voltage source when the switching device turns off, but always flows to the center tap of the split capacitor. The opposite applies to phase *B*. It is essential that the voltages across the capacitors remain equal. This is achieved by ensuring that the mean current drawn by phases *A* and *C* is equal to the mean current drawn by phases *B* and *D*. The penalties imposed by this configuration are: (1) there must be an even number of phases, and (2) it is impossible to retain independent control of the stator phase currents. But this only applies when operating at low speeds. The switching devices, now only one per phase, must be voltage rated for V_{do} and will carry twice the current of those in the first circuit for the same VA input to the motor.

The third circuit [Fig. 17-25(c)] uses a bifilar winding and one switching device per phase. This circuit, because of the imperfect coupling between the bifilar windings, has induced voltage spikes which require the voltage rating of the switching

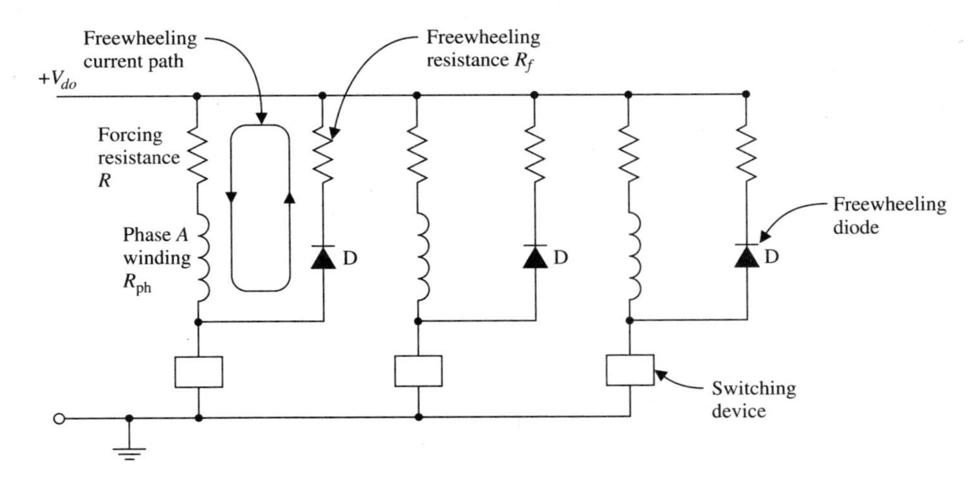

Figure 17-24 Three-phase unipolar drive circuit.

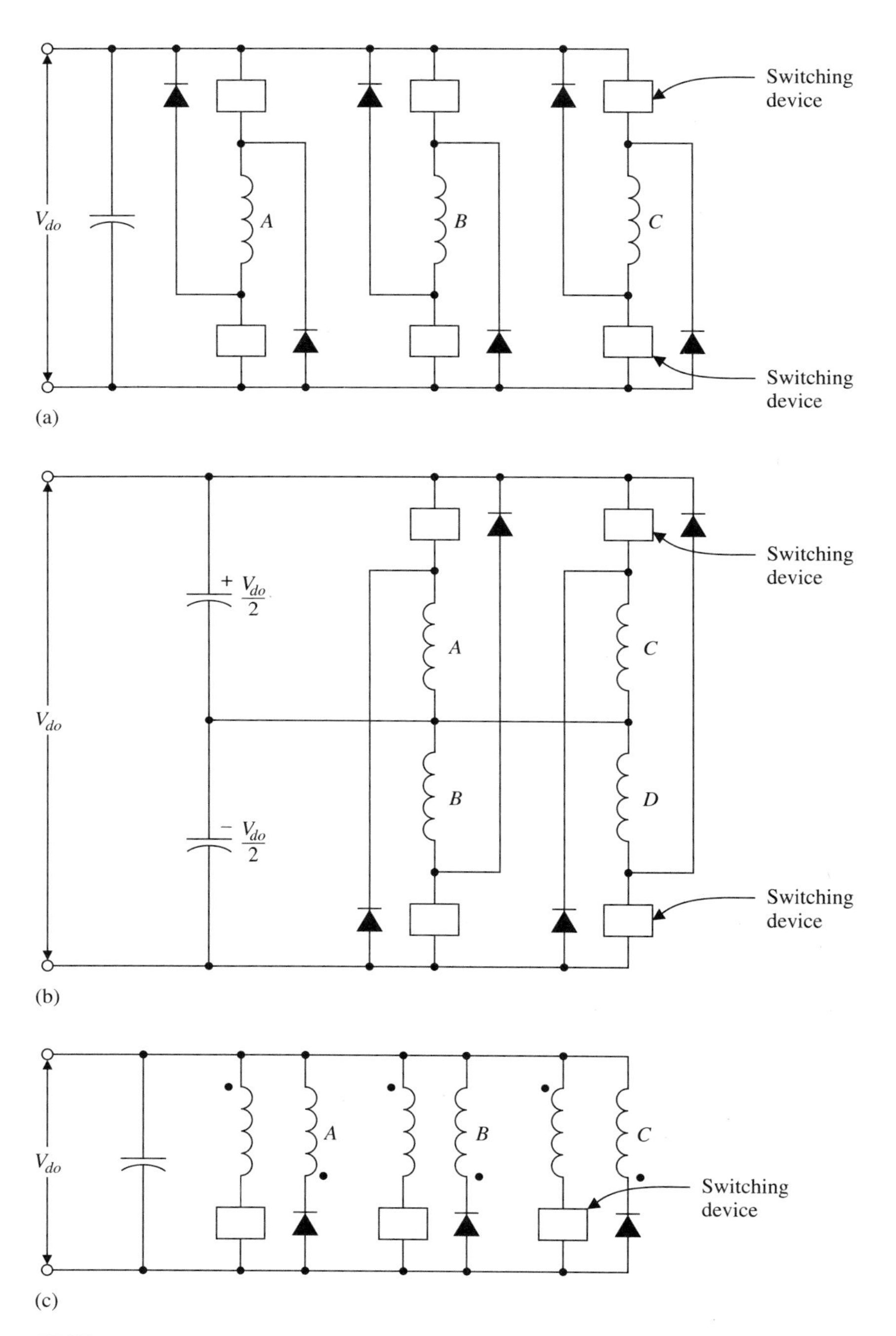

Figure 17-25 Switched-reluctance motor inverter circuits. (a) Requiring two switching devices per phase. (b) Requiring one switching device per phase with split supply. (c) Bifilar requiring one switching device per phase.

devices to be increased to $2V_{do} + \Delta V$, where ΔV is the amplitude of the voltage spike. Other negative factors to its use are the extra connections required between the converter and the motor as well as a reduction in the winding utilization in the motor. However, in spite of these problems it is appropriate for use in battery-powered vehicles. All of these configurations for switched-reluctance drives benefit from the absence of a path for a shoot-through fault current, as there is always a phase winding in series with a switching device.

The switched-reluctance motor has the ability to respond rapidly to step changes in load because of both its low rotor inertia and the individual control of the torque by each phase current pulse. In fact its performance is fully comparable to that of customized servomotors. The system has a very fast response, on the order of 40 ms, and is designed to provide a speed regulation of $\pm 0.01\%$ from no load to full load as well as maintaining a stability of approximately $\pm$ 1 r/min. In addition, because the switched-reluctance motor output torque can be rapidly altered by changing the switching point, typically within one-twelfth of a revolution, there is an extremely small time interval between the command being issued and the motor producing the desired output. The switched-reluctance motor, because of its high efficiency, controllability, and high torque-to-inertia ratio, obviously has excellent potential in servomotor applications associated with machine tools and robotics.

The choice of switching devices ranges from bipolar transistors and MOSFETs for low-power applications to GTOs and thyristors for medium- and high-power applications. The MOSFET permits chopping at above audible frequencies, but currently only for power levels of a few kilowatts per phase. Bipolar transistors and GTOs compete for the mid-power range, the GTO having the advantages of higher voltage ratings and better peak-to-mean-current ratios. The requirement for commutation circuitry makes the thyristor generally less attractive than GTOs.

Investigations at power outputs from a few watts to several megawatts have shown that the switched-reluctance motor has considerable potential in both low- and high-power applications. In fact, the limiting factor at high powers may be the practical realization of electronic switching devices to handle several kilovolts and kiloamperes with acceptable snubbers and adequate protection. This is seen as no more than a short-term limitation while high power switching technology is being developed.

17-5 TACHOMETER GENERATORS

A tachometer generator is an electromagnetic device that converts mechanical rotation to an electrical signal that is proportional to angular velocity, that is, it is linear over the normal operating range. These devices are used to provide a direct readout of shaft speed or, in closed-loop systems, to control the angular velocity of an ac or dc drive.

DC Permanent-Magnet Tachometer

The dc permanent-magnet tachometer, which has replaced the now obsolete wound field dc tachometer, consists of a permanent-magnet field structure and a conventional dc armature, although sometimes a moving-coil armature is also used. The dc tachometer differs from the conventional dc machine and has the following characteristics:

1. The output voltage is proportional to the shaft speed.
2. The output voltage is relatively free of voltage ripple.
3. The voltage gradient K_E is constant, regardless of ambient and device temperature changes. K_E is usually expressed as V/1,000 r/min or V/rad/s.

The requirement that the output voltage be relatively free of voltage ripple is met by increasing the number of armature coils and commutator segments. Transient voltage spikes caused by brush arcs can be minimized by using the correct brush materials and ensuring that the brushes are properly fitted in close-fitting brush holders during planned maintenance.

In closed-loop speed control applications it is essential that the coupling between the motor and the tachometer be torsionally stiff to prevent oscillations from being introduced into the drive system and producing instability. It is reasonable to expect a speed regulation of $\pm 1\%$ with a closed-loop dc drive.

AC Permanent-Magnet Tachometer

The ac permanent-magnet tachometer consists of a permanent magnet rotor revolving inside a polyphase wound stator. The output voltage and frequency are directly proportional to the angular velocity of the rotor. This type of tachometer eliminates the RFI problems inherent in the dc tachometer, as well as reducing the maintenance requirements. The main disadvantage is that its performance is poor at low angular velocities, and unlike the dc tachometer, the polarity of the output voltage does not reverse with reversed rotation. In addition, since most closed-loop speed controls use dc voltages, the output of the ac tachometer must be rectified and smoothed. Alternately since the output frequency is proportional to the angular velocity, it is possible to use frequency-to-voltage conversion techniques to produce a suitably conditioned speed signal.

AC Induction Tachometer

The ac induction tachometer shown in Fig. 17-26(a) has a laminated two-phase stator containing two electrically separate windings placed 90° electrical apart. These windings are the externally excited excitation or reference winding and the output winding. The rotor consists of a copper or aluminum cup, also known as a *drag cup*, attached to the rotor shaft. The drag cup rotates between stationary salient stator poles. The operation of the ac induction tachometer depends on transformer action inducing an eddy current flux in the rotor at right angles to the output winding. When the rotor is stationary, the alternating rotor flux will not induce a voltage in the output winding. As the rotor rotates, eddy currents are induced in the rotor cup at rights angles to the excitation winding flux, so that the resultant field is shifted in the direction of rotation. Assuming that the field produced by the excitation winding is constant, then the flux component at right angles to the excitation field will induce an alternative voltage of the same frequency as the excitation field supply in the output winding, the amplitude of the output voltage being proportional to the speed of rotation of the rotor. When the direction of rotation is reversed, the phase relationship of the output voltage with respect to the excitation or reference voltage is phase shifted by 180° [Fig. 17-26(b)]. As a result, by using a phase-sensitive rectifier, a dc voltage proportional to the angular velocity, and whose polarity is determined by the direction of rotation, can be obtained.

The major drawbacks of the ac induction tachometer are: (1) two air gaps are required in the magnetic circuit, and (2) a well-regulated voltage source operating at constant frequency is required to supply the excitation winding. The sensitivity of the ac induction tachometer ranges from 0.1 to 1.0 V/1,000 r/min (0.001 to 0.0095 V/rad/s).

17-6 DC BRUSHLESS MOTORS

Dc servomotors have far better characteristics than their ac counterparts and are used extensively. Their most important features are a relatively large power output compared to their physical size, a high starting torque, and their speed is easily con-

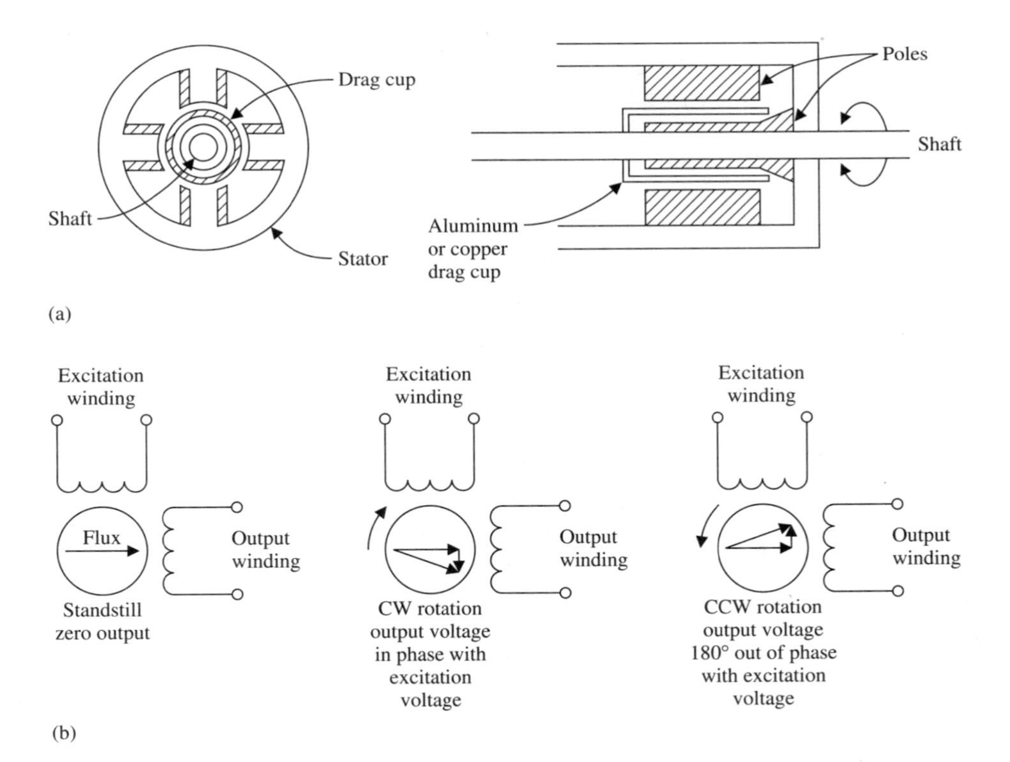

Figure 17-26 Ac induction tachometer. (a) Sectionalized view. (b) Principle of operation.

trolled. However, in spite of the excellent speed-torque characteristics of the conventional dc motor, the commutator and brushes have relatively short service lives and are a source of electrical noise. The increasing use of dc servomotors has resulted in considerable research being conducted to eliminate the commutator and brushes by using power semiconductor switches. The result is the brushless motor, which by definition has the speed-torque characteristics of a conventional dc permanent-magnet motor. At reduced speeds the dc brushless motor can be as much as 60% more efficient than an ac servomotor. When compared to variable-speed ac servomotors, the dc servomotor consistently maintains a 5–10% efficiency advantage. The modern brushless motor that utilizes microprocessor technology is being applied in all power output ranges from subfractional horsepower to 22 hp (16.4 kW), and is used where precise motor control is required, such as for Winchester drives, machine tools, and robotics. Even though this motor commands a price premium as compared to a variable-speed ac drive, its improved efficiency usually leads to a payback within one year.

Since the dc brushless motor must be designed to minimize the number of power semiconductor switching devices, the construction of the motor is the reverse of the conventional dc motor, that is, the armature windings are located in slots in the stator, which permits direct connection of the power supply to the armature windings via electronic switches and eliminates stationary to moving contacts. The field is provided by permanent magnets attached to the rotor shaft. The control electronics, which may be integral to the motor or remote, direct the current to the armature windings, creating a rotating magnetic field that the rotor permanent-magnet field will follow, causing the shaft to rotate. In order to distribute power to the armature windings at exactly the correct instant, logic-compatible control signals are developed from sensors which determine the position of the rotor. This is achieved by using Hall effect sensors, which sense the polarity and magnitude of the rotor field, and are usually mounted on the stator. However, since the Hall effect sensor is sensitive to temperatures in excess of 160°C and inductive transients, the sensor is often located away from the stator and is usually used in conjunction with an auxiliary magnet system mounted on the rotor shaft. An alternative arrangement is to use optoelectronic sensors, usually a light-emitting diode (LED) and a phototransistor or photoSCR. A segmented disk controls the transmission of light between the LED and the sensor. These signals are then processed to produce logic-compatible signals to trigger the power semiconductor switches supplying the stator windings.

The Hall effect was discovered by E. H. Hall in 1879. The concept of the Hall effect sensor is illustrated in Fig. 17-27. A constant current I_H is passed through a thin sheet of indium or antimide semiconducting material. Output connections are made to the element at right angles to the direction of current flow. When there is no magnetic field present, there will be no output Hall voltage V_H. When a perpendicular magnetic field is present, a Lorentz force is exerted on the distribution of the current across the element, and a Hall voltage is produced in a direction perpendicular to the current and magnetic field. The Hall voltage is

$$V_H = \frac{R_H}{t} \times I_H B \sin\theta \text{ V} \tag{17–11}$$

where V_H is the Hall voltage, I_H the input current, $B \sin \theta$ the component of the magnetic field perpendicular to the element, R_H the Hall coefficient, and t the thickness of the semiconductor element.

Hall Effect Controlled Motor

Dc brushless motor stators are connected in a number of different configurations: two-phase, three-phase star and delta, and four- and six-phase delta. The particular configuration chosen is determined by the peak torque and efficiency, and the number of semiconductor switches required. For example, a six-phase delta-connected stator used in conjunction with a four-pole rotor will require 12 commutations per revolution, that is, at 30° intervals, and 12 switching devices. The same peak torque can be obtained by using a four-phase stator with a four-pole rotor, which requires eight commutations per revolution at 45° intervals and eight switching devices.

Figure 17-28(a) shows the block diagram of a three-phase full-wave brushless motor control system. The stator winding is star connected with six transistors connected to the ends of the phase windings. The transistors conduct for 120° and are off for 240°. As can be seen from Fig. 17-28(b), when Q1 is turned on between 0° and

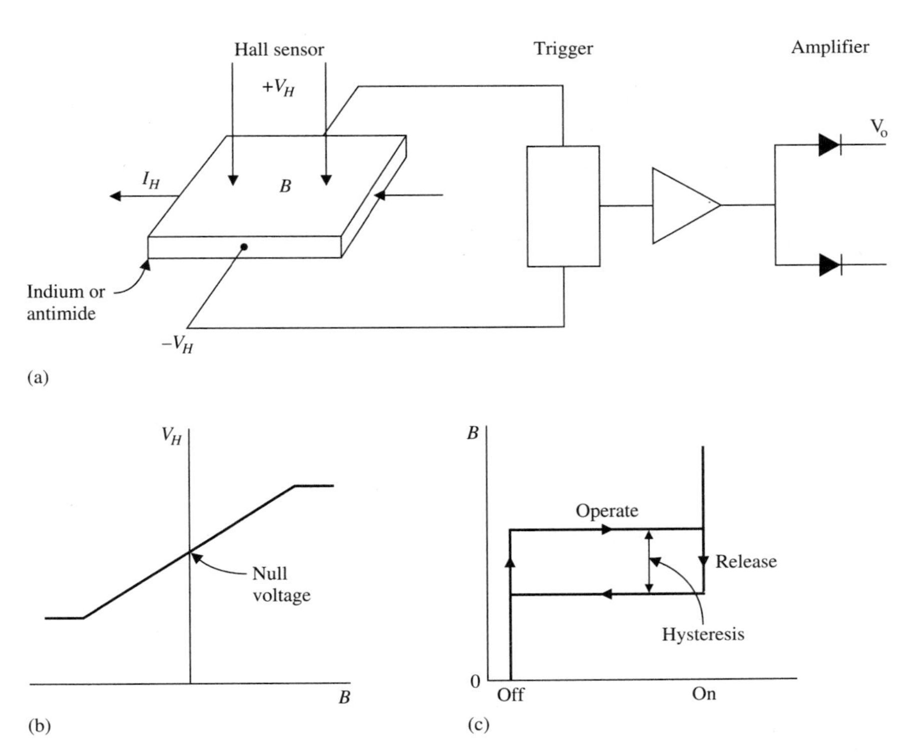

Figure 17-27 Hall effect transducer. (a) Block diagram. (b) Relationship between B and V_H. (c) Transfer function for digital-output Hall effect transducer.

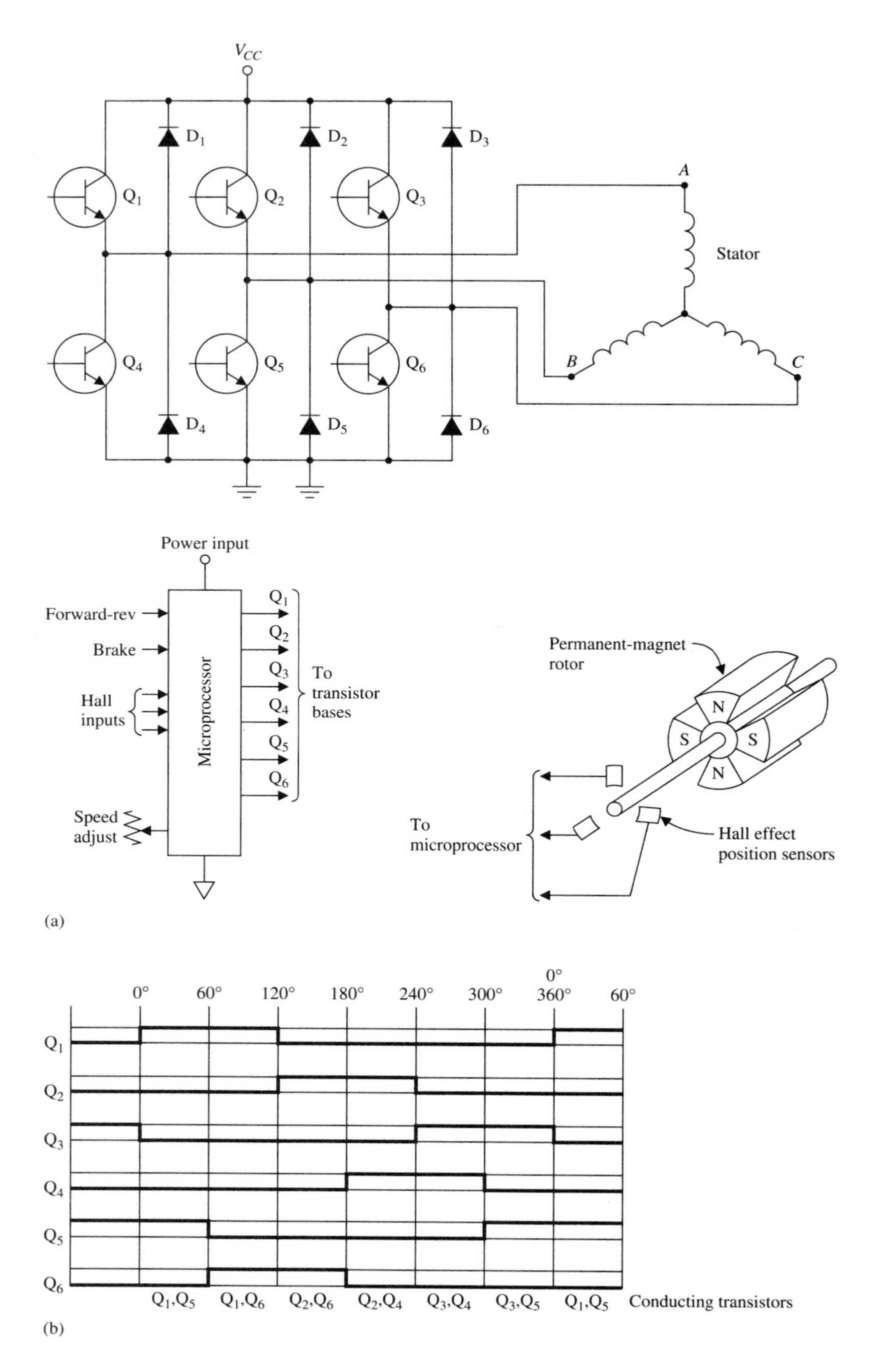

Figure 17-28a and b Three-phase full-wave motor. (a) Block diagram. (b) Transistor firing sequence.

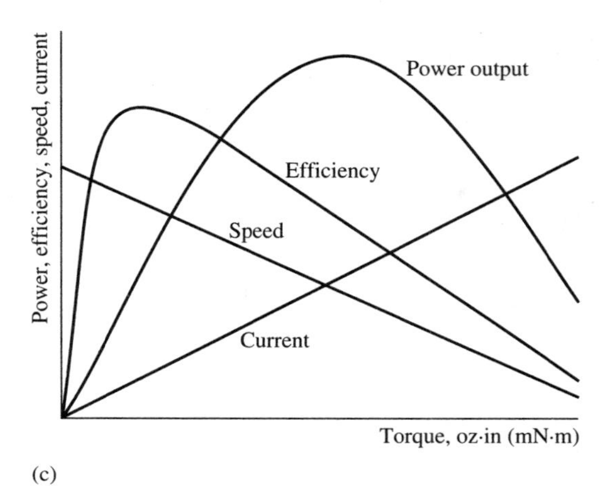

Figure 17-28c Three-phase full-wave motor. (c) Performance characteristics.

120°, Q5 is also turned on between 0° and 60° with the result that current is flowing in the stator winding from *A* to *B*. Between 60° and 120° Q6 is conducting, and the current is flowing in the stator winding between *A* and *C*. At the same time the stored energy in phase *B* is being returned to the power supply via the feedback diode D2. The process is continued with two transistors in conduction at the same time in the sequence Q1, Q5; Q1, Q6; Q2, Q6; Q2, Q4; Q3, Q4; Q3, Q5; and so on. The turn-on and turn-off points are being determined by the Hall effect position sensors, and the switching rate and thus the rotor speed by the speed adjust signal to the microprocessor. Additional controls such as forward, reverse, and braking are usually included. Reverse operation is achieved by reversing the firing sequence of the transistors. Typical performance curves are shown in Fig. 17-28(c).

It should be noted that the switching devices can be power transistors, SCRs, GTOs, and power MOSFETs, depending on the rating of the motor. Currently the major commercial applications of dc brushless motors are in the very small subfractional power output ratings, but they are being introduced in integral-horsepower (kW) sizes.

17-7 LINEAR INDUCTION MOTORS

The *linear induction motor* (LIM), which produces linear motion, was developed principally for the electric transportation industry. It is best understood by developing it from the polyphase squirrel-cage induction motor, as illustrated in Fig. 17-29. Figure 17-29(a) shows the squirrel-cage motor with a CCW rotating magnetic field created by the stator (primary) windings, with a CCW rotation of the rotor (secondary). Assuming that the stator is split and rolled out flat [Fig. 17-29(b)], the primary magnetic flux will now travel in a straight line from left to right. If the primary is fixed in position and the secondary free to move, the secondary will move

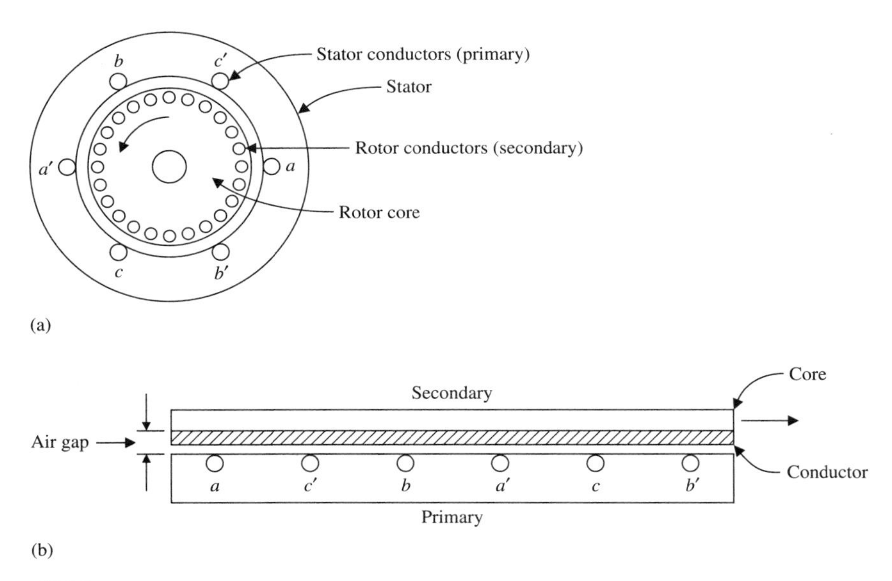

Figure 17-29 Development of linear induction motor from a polyphase squirrel-cage induction motor. (a) Squirrel-cage induction motor. (b) Linear induction motor.

from left to right at a velocity less than that of the primary magnetic flux. Conversely if the secondary is fixed in position, then the primary will move from right to left. Normally in transportation applications the secondary is fixed and the primary is attached to the vehicle, with the three-phase power being picked up by rolling contact shoes making contact with a stationary three-phase bus-bar system mounted parallel to the direction of the vehicle's movement.

The linear induction motor has been produced in two configurations: a double-sided structure (DLIM), where the primary is located on both sides of a conducting secondary, and a single-sided one (SLIM) with a single primary acting on one side of a conducting secondary. The DLIM is rarely encountered today, and most research is being conducted with the SLIM. The secondary is usually formed from a nonmagnetic material such as aluminum backed by a solid or laminated magnetic backing.

The operating characteristics of the linear induction motor are basically similar to those of a SCIM. However, its efficiency and power factor are usually poorer than those of a SCIM with a large air gap.

There has been a considerable interest in linear induction motors since the mid-1960s, and they are currently being used in urban rail systems. The system has the advantage that it has eliminated the motor-driven wheels, which permits a steerable truck design with the ability to negotiate tight turns as well as making propulsive and braking forces independent of adhesion between the rail and the vehicle. A system produced by the Urban Transportation Development Corporation Ltd. (UTDC) of Kingston, Ontario, Canada, incorporates a SLIM system and is currently being used in the United States and Canada. The UTDC system propels a 13,000-kg (14.30-ton) vehicle carrying 75 passengers at a cruising speed of 72 km/h (44.74 mi/h) on grades up to 6%. Speed control is achieved by using variable-frequency control. Linear induction motors are not limited to transportation applications. Some interesting examples are variable speed and force control of a steel mill drop hammer and hatch cover motors on merchant ships, to mention just a few applications.

17-8 SYNCHROS

The term *synchro* is a generic term covering a range of ac electromechanical devices which are used in angular data transmission. Synchros convert angular position information to an electrical output, or vice versa. Because of the ease with which this is accomplished, synchros are used extensively in industrial process control. Synchros have star- or delta-connected three-phase stator windings, and depending on the type of synchro, the rotors are either single-phase salient-pole rotors or three-phase wound rotors. There are two classes of synchros: *torque* and *control*. The torque synchros, which are less common, transmit angular data directly without the use of amplifiers. They are capable of driving light loads, but because the angular error is dependent on the connected load, and due to dynamic errors caused by internal friction and inertia, an overall error as great as 1.5% can be produced. The control synchro, which can only transmit data, achieves an angular accuracy of 10 minutes of arc. All synchros are subject to errors caused by a lack of uniformity in the stator windings, irregularities in the magnetic circuit, choice of slot combinations, uneven winding distribution, and rotor pole structures.

The usual operating voltages and frequencies of standard synchros are 115 V at 60 or 400 Hz, and 26 V at 400 Hz. Synchros are manufactured in a range of standard sizes and mounting arrangements, and vary in diameter from 0.5 to 3.7 in (12.7 to 94 mm).

Synchro Transmitter

The synchro control transmitter (CX) and torque transmitter (TX) have electromotive forces induced in the three-phase stators that are proportional to the angular position of the energized single-phase rotor [Fig. 17-30(a)]. From Fig. 17-30(b) it can be seen that there is a unique relationship between the amplitudes and the polarities of the induced voltages in the stator.

Synchro Receiver

The construction of the synchro control receiver (CR) and torque receiver (TR) is identical to that of the synchro transmitters, except that the rotor has an inertial damper mounter on the shaft to dampen out rotor oscillations when receiving angular signals, and also to prevent overspeeding during continuous operation. Synchro receivers may be used as transmitters, but transmitters should never be used as receivers.

The process of transmitting angular information is illustrated in Fig. 17-31. The amplitude of the voltages induced in the stator windings is determined by the rotor position. These voltages are transmitted over connecting lines between the transmitter and receiver stators, and cause currents to flow in the stator windings of the

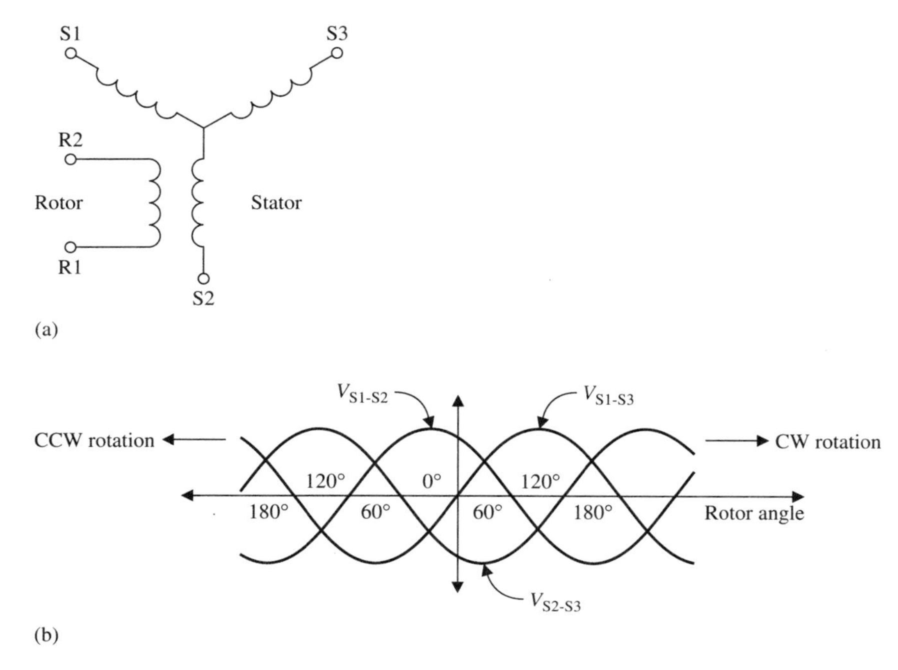

Figure 17-30 Synchro transmitter. (a) Schematic. (b) Stator terminal voltages versus rotor angle.

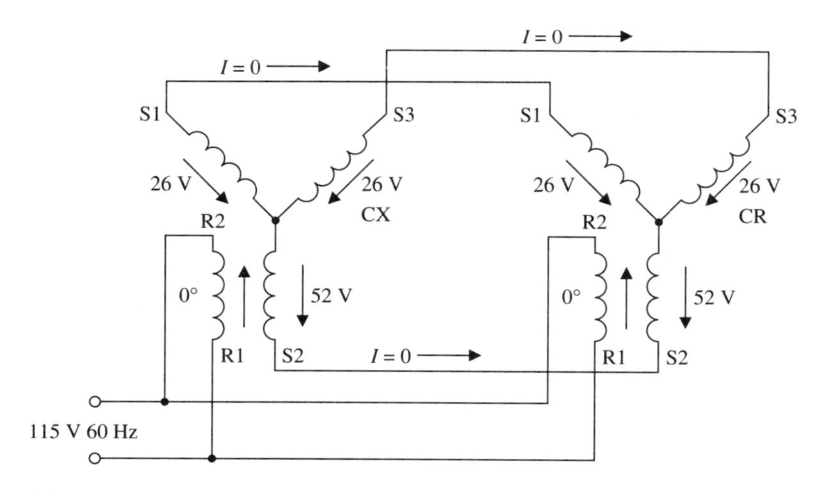

Figure 17-31 Synchro transmitter and receiver pair with both rotors at 0°.

receiver. In turn these currents create a magnetic field in the stator bore of the receiver, and since the receiver rotor is energized from the same single-phase source as the transmitter, there is an interaction between the stator and rotor magnetic fields, which creates a torque aligning the rotor field with the stator field. When the receiver field is aligned with the stator field, there is no current flow in the connecting lines between transmitter and receiver stators.

Synchro Differential Transmitter

Synchro differential transmitters (CDX) and (TDX) are connected between a synchro transmitter and receiver whenever it is necessary to add angular information to the system or subtract from it.

The stator of the differential transmitter is identical to those of the synchro transmitter and receiver. However, the rotor is cylindrical and has a three-phase star- or delta-connected winding which is connected via slip rings to the receiver stator. Normally the stator and rotor windings have a 1:1 transformation ratio. The electrical connections are shown in Fig. 17-32.

The stator of the synchro transmitter is directly connected to the stator of the differential transmitter, and the rotor connections of the differential transmitter are connected to the stator of the receiver. As can be seen, the output voltages of the synchro transmitter applied to the differential transmitter stator produce a magnetic field whose orientation is determined by the angular position of the transmitter rotor. In turn, these voltages induce, by transformer action, voltages in the differential transmitter stator with their amplitudes depending on the position of its rotor. These voltages are then applied to the stator of the receiver and are equal to the sum or difference of the angles represented by the mechanical position of the two inputting rotors. The electrical information used at the receiver to produce a change in the mechanical position of the rotor is determined by:

1. The physical movement of the transmitter and differential transmitter rotors
2. The connection of the wiring between the stators of the transmitter and the differential transmitter
3. The connection of the wiring between the rotor of the differential transmitter and the stator of the receiver

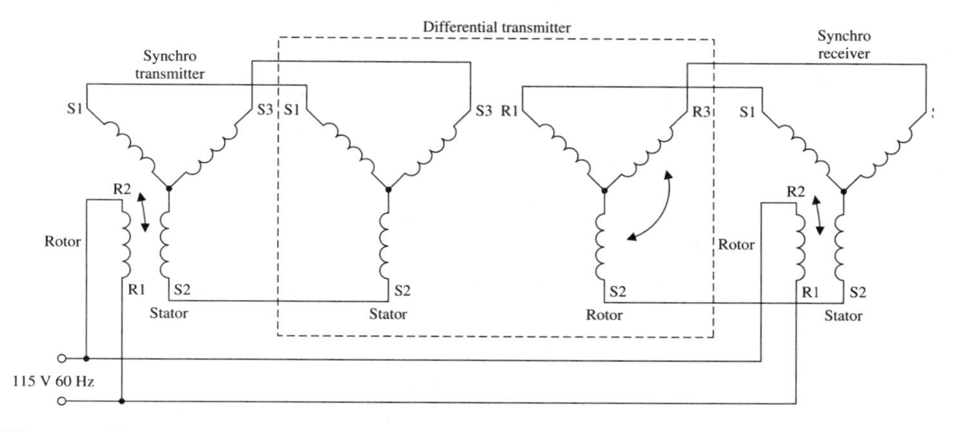

Figure 17-32 Synchro differential transmitter system.

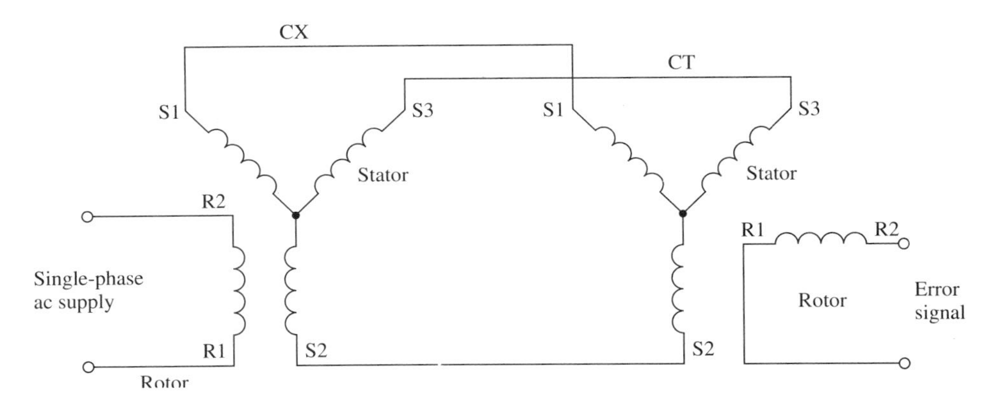

Figure 17-33 Synchro transmitter and control transformer system with rotors at 0°.

Synchro Differential Receiver

The synchro differential receiver (CDR) is similar to the differential transmitter, the only difference being that the rotor has a damper mounted on the rotor shaft to prevent oscillation. The role performed by the differential receiver is to produce a mechanical output at the rotor shaft, that is, the sum or difference of two synchro transmitters. The stator of one transmitter is connected to the stator of the differential receiver, and the stator of the other is connected to the rotor.

Synchro Control Transformer

The synchro control transformer (CT) develops an ac voltage at the rotor terminals that is proportional to the angular offset between the rotors of the synchro transmitter and the control transformer, when connected as shown in Fig. 17-33. The stator of the control transformer is similar to the stators of the other synchro units, but it has a much greater impedance so as to reduce the exciting current. The rotor is cylindrical, has a high impedance, and is aligned at right angles to the transmitter rotor. When connected as shown in Fig. 17-33, the stator of the control transformer is supplied from the stator of the synchro transmitter, and its rotor is mechanically coupled to the servomechanism whose position is being controlled. The stator field of the control transformer is established by the angular offset of the transmitters rotor. This magnetic field induces an error voltage in the rotor of the control transformer, which is proportional to the offset between the two rotors, and whose phase relationship with respect to the single-phase reference supplied to the transmitter rotor defines the direction of the error. This error signal in turn is used to initiate corrective action to reduce the positional error between the designated position, the position of the synchro transmitter rotor, and the actual position of the servomechanism being controlled, represented by the position of the control transformer rotor.

Figure 17-34 illustrates the use of the control transformer in a position control system. The desired position signal is produced by rotating the rotor of the synchro transmitter to the desired angular position of the servomechanism. This information

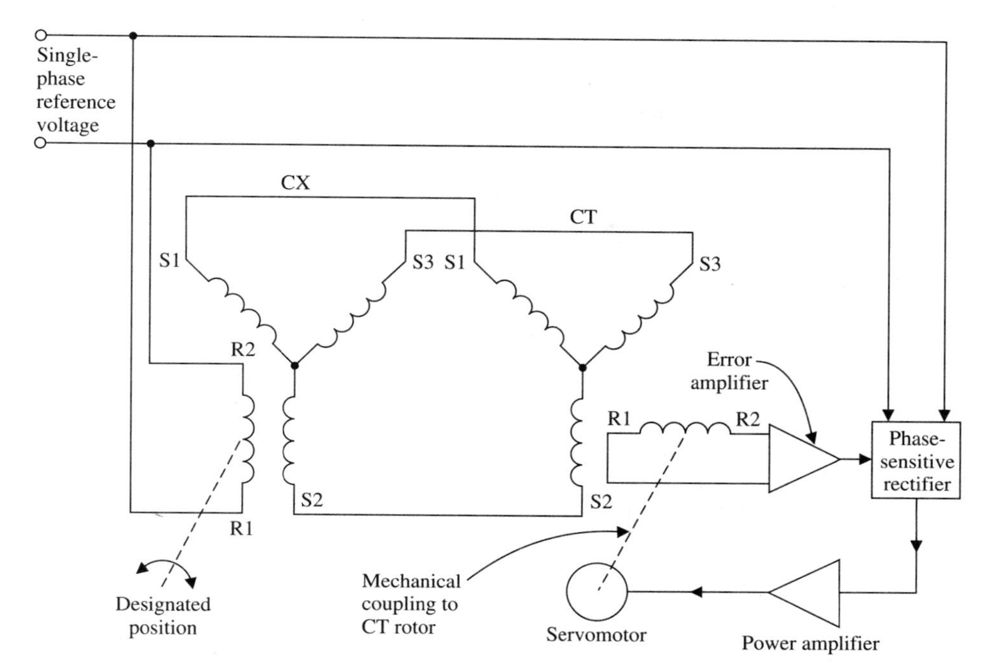

Figure 17-34 Schematic of simple position control system using a control transformer.

is transmitted to the control transformer over the interconnecting lines between the two stators. Assuming that the CT rotor is not at the designated position, then an error voltage will be induced in the rotor winding whose phase and amplitude are determined by the amount and direction of offset of the two rotors. This error voltage is amplified by the error amplifier. The amplified output is then supplied to the phase-sensitive rectifier, where it is compared against the reference voltage. The phase-sensitive rectifier will produce an output dc signal whose polarity and magnitude when applied to the armature of the dc servomotor via the power amplifier will cause it to turn in such a direction as to eliminate the error.

QUESTIONS

17-1 What is a stepper motor?

17-2 Discuss the principle of operation of a basic stepper motor.

17-3 Discuss the advantages and disadvantages of stepper motors.

17-4 Explain the following terms: **(a)** step angle; **(b)** steps per revolution; **(c)** steps per second; **(d)** step accuracy; **(e)** holding torque; **(f)** residual torque; **(g)** step response; **(h)** torque-to-inertia ratio; **(i)** resonance; **(j)** pulse rate; **(k)** ramping; **(l)** slew rate.

17-5 With the aid of a sketch explain the construction and operation of a permanent-magnet stepper motor.

17-6 What are the advantages and disadvantages of permanent-magnet stepper motors?

17-7 With the aid of a sketch explain the construction and operation of a single-stack variable-reluctance stepper motor.

17-8 What is the effect of increasing the number of rotor teeth of a variable-reluctance stepper motor?

17-9 With the aid of a graph explain the relationships between holding and running torque with respect to the stepping rate.

17-10 What is meant by start-stop running mode?

17-11 Why do variable-reluctance stepper motors require a damper?

17-12 With the aid of a sketch explain the operation of a three-phase multistack variable-reluctance stepper motor.

17-13 Explain with the aid of a sketch the operation of a conventional hybrid stepper motor.

17-14 What is an enhanced hybrid stepper motor? What are the advantages of using an enhanced hybrid stepper motor?

17-15 Why are stepper motor windings usually bifilar connected?

17-16 What is meant by full stepping and half stepping?

17-17 Discuss the relative merits of series R, chopper, and bilevel stepper motor drive systems.

17-18 Explain with the aid of sketches the construction and operation of a linear stepper motor.

17-19 What are the advantages of using microstepping?

17-20 What are the advantages of using linear stepper motors?

17-21 Explain with the aid of sketches the construction and operation of the SynchroStep motor.

17-22 What are the advantages of the SynchroStep motor over the conventional stepper motor?

17-23 Discuss the reasons that the dc servomotor is preferred to the ac servomotor.

17-24 Explain the difference between the characteristics of dc servomotors used in positional and velocity control applications.

17-25 Discuss the operating zones that are applied to the use of dc servomotors.

17-26 With the aid of a sketch explain the construction of a shell armature moving-coil dc servomotor.

17-27 With the aid of a sketch explain the construction of a printed-circuit armature dc servomotor.

17-28 Discuss the advantages and disadvantages of permanent-magnet moving-coil motors.

17-29 Discuss the advantages and disadvantages of permanent-magnet printed-circuit motors.

17-30 What is a torquer and where would it be used?

17-31 With the aid of sketches and graphs explain the operation of a two-phase ac servomotor.

17-32 With the aid of a sketch explain the operation of the switched-reluctance motor.

17-33 With the aid of pole inductance profile graphs explain the production of motoring and braking torque in a switched-reluctance motor.

17-34 Discuss the advantages and disadvantages of the switched-reluctance motor inverter circuits shown in Fig. 17-25.

17-35 Discuss the advantages and disadvantages of switched-reluctance motors. What factor is currently limiting the use of switched-reluctance motors in high-power-output applications?

17-36 What is a tachometer generator? Where is it used?

17-37 What are the characteristics of a dc permanent-magnet tachometer?

17-38 What are the advantages and disadvantages of ac permanent-magnet tachometers?

17-39 With the aid of sketches explain the operation of an ac induction tachometer.

17-40 What is a dc brushless motor? What are the advantages and disadvantages of dc brushless motors?

17-41 Explain with the aid of sketches the operation of a Hall effect position sensor.

17-42 Where and why are Hall effect sensors used?

17-43 Explain with the aid of a sketch the operation of a three-phase full-wave dc brushless motor using Hall effect position sensors.

17-44 With the aid of sketches explain the operation of a linear induction motor.

17-45 Discuss the applications of linear induction motors in electric traction.

17-46 What is a synchro and where would it be used?

17-47 Explain with the aid of a sketch the operation of a synchro transmitter and receiver transmitting angular information.

17-48 Discuss the function of a differential transmitter.

17-49 Discuss the function and operation of a synchro control transformer.

17-50 Discuss with the aid of a sketch the position control of a dc servomotor using a control transformer.

PROBLEMS

17-1 A bifilar stepper motor operating in the full-step mode has 100 rotor teeth. If it is controlled by a 1-kHz pulse train, calculate: **(a)** steps per revolution; **(b)** step angle; **(c)** rotor speed in rad/s.

17-2 A stepper motor is supplied from a 12-V source and draws 2.5 A. If it takes 30 ms for the current to build up to 2.5 A, calculate the resistance and inductance of the windings.

17-3 If a 200-step motor is supplied with a 1,000- step/s pulse train for 25 ms, calculate the angle that the rotor has turned through.

CHAPTER

18

AC Machine Selection

18-1 INTRODUCTION

Industrial and commercial installations use large numbers of single- and three-phase ac motors. However, in terms of electric power consumption, it is estimated that in excess of 90% of the energy consumed by electric motors is used to supply three-phase squirrel-cage induction motors (SCIMs). The main reasons for the bias toward the polyphase squirrel-cage induction motors are low cost, high reliability, high efficiency over a wide load range, and low to moderate maintenance costs. Energy-efficient squirrel-cage induction motors, although initially more expensive, usually recover the added cost within 18 months as a result of lower operating costs.

The remaining three-phase market is satisfied by the polyphase synchronous motor and the polyphase wound-rotor induction motor (WRIM). Generally the synchronous motor is used in low-speed high-power-output applications, and the wound-rotor induction motor is used in medium- to high-speed applications over a wide range of power outputs.

Both the synchronous motor and the squirrel-cage induction motor can also operate as generators, although the latter, acting as an induction generator, operates only as a generator in parallel with a large power system because it inherently has a poor voltage-regulating capability if operated by itself.

The major standard applying to motors and generators in the United States is NEMA MG-1, and its Canadian counterpart is the Canadian Standards Association (CSA) publication C 154. The two standards are practically identical. The National Electrical Code (NEC) is an ANSI standard issued by the National Fire Protection Association. This code is designed to protect people and buildings from electrical hazards resulting from the use of electricity for lighting, heating, power, and other applications. It covers wiring installation methods and materials, branch circuit protection, motors, and control, and it provides recommendations for suitable equipment for each classification.

Many factors are involved in selecting and applying polyphase squirrel-cage and wound-rotor induction motors and synchronous motors. The initial steps in motor selection include determination of the power supply, output rating (hp or kW), operating speed, duty cycle, service factor, mountings, enclosures, insulation class, frame size, and efficiency. In addition, the selection must include factors such as the operating environment, that is, temperature, humidity, altitude, atmospheric pollutants, and so on, the mounting requirements (vertical, inclined, or horizontal), the supply voltage and frequency variations, power factor, and unbalanced line voltages.

The principal factors to be discussed are:

1. Rated horsepower or kilowatt output
2. Load characteristics
3. Speed or angular velocity rating
4. Frame size
5. Ambient temperature
6. Temperature rise
7. Duty cycle
8. Voltage, current, and frequency ratings
9. Power factor
10. Enclosures
11. NEMA designations
12. Maintenance

18-2 RATED HORSEPOWER OR KILOWATT OUTPUT

The power output and speed are the two most important factors in determining the selection between a synchronous motor and a squirrel-cage induction motor. Low-speed synchronous motors generally are considered to be those that operate at less than 500–600 r/min (52.36–62.83 rad/s) and are manufactured in power outputs from 150 hp (111.9 kW) to in excess of 10,000 hp (7,460 kW). High-speed synchronous motors usually are used in applications ranging from 700 hp (522.2 kW) to in excess of 10,000 hp (7,460 kW) in the speed range of 500 or 600 r/min (52.36 or 62.83 rad/s) up to 1,200 r/min (125.56 rad/s). The initial cost of the synchronous motor, especially in the lower power output sizes, is greater than for the equivalent rated squirrel-cage induction motor. However, the synchronous motor has the higher efficiency, even at speeds as low as 72 r/min (7.54 rad/s), as well as having an excellent power factor at all loads and speeds.

The polyphase induction motor is especially suited for high-speed operation in all power output ratings from 1 hp (0.746 kW) up to 3,000 hp (2,238 kW), although the squirrel-cage induction motor is the preferred choice in the 100–200-hp (74.6–149.2-kW) power output range from 500–600 r/min (52.36–62.83 rad/s) to 3,600 r/min (376.99 rad/s). Above 3,000 hp (2,238 kW) at 3,600 r/min (376.99 rad/s) the choice may be either the synchronous or the squirrel-cage induction motor. At 1,800 r/min (188.50 rad/s) from 200 hp (149.2 kW) to in excess of 10,000 hp (7,460 kW), and in the range of 600–1,200 r/min (62.83–125.66 rad/s) from 200 to 700 hp

(149.2 to 522.2 kW) both the squirrel-cage induction motor and the synchronous motor find equal favor.

The squirrel-cage induction motor gives its best performance at high speeds with intermittent or fluctuating loads. But when lightly loaded it is the major source of poor system power factor as well as having an overall efficiency less than the synchronous motor. A rough rule of thumb suggests that the preferred choice should be a synchronous motor if the motor horsepower divided by the speed (r/min) is greater than 2, for example, 1,200 hp at 514.29 r/min (53.86 rad/s).

18-3 LOAD CHARACTERISTICS

The nature of the load and the speed regulation have a significant effect on the selection of an ac motor. A hard-starting load, such as an air or refrigeration compressor, demands a high-starting torque, which can be best met by a wound-rotor induction motor or a double-cage induction motor. Alternatively, if the load can be relieved during starting by venting to the atmosphere, as in the case of an air compressor, a synchronous motor could be used. Similarly, for a domestic refrigerator or freezer a single-phase capacitor-start motor is the ideal choice.

Operation at constant speed is best met by the synchronous motor, and the polyphase induction motor is best suited for applications where a speed regulation of 4–6% is the norm. It should be noted that the squirrel-cage induction motor can be operated at constant speed from a variable-frequency inverter by using closed-loop feedback techniques and varying the output frequency of the inverter to maintain constant motor speed.

18-4 SPEED RATING

The rotational speed of all polyphase ac motors is determined by

$$S = \frac{120f}{P} \text{ r/min} \qquad \textbf{(18–1)(E)}$$

or

$$\omega = \frac{4\pi f}{P} \text{ rad/s} \qquad \textbf{(18–1)(SI)}$$

where S and ω are the synchronous speed of the rotating magnetic field in r/min and rad/s, respectively, f is the power supply frequency in hertz, and P the number of stator poles per phase.

Synchronous, synchronous-reluctance, and synchronous induction motors rotate at synchronous speed; induction motors rotate at slightly less than synchronous speed. In general, synchronous motor speeds are 3,600, 1,800, 1,200, 900, . . . , r/min or 376.99, 188.50, 125.66, 94.25, . . . , rad/s. Polyphase induction motors, in order to develop their torque, operate at typically 3,500, 1,750, 1,160, 875, . . . , r/min or 366.52, 183.26, 121.47, 91.63, . . . , rad/s.

As can be seen, the rotational speeds of synchronous and polyphase induction motors operate from a constant-frequency source at speeds determined by the number of poles per phase, which does not permit smooth variation of the rotational speed. Various techniques have been developed which give a limited variation of speed, such as the consequent-pole connection, which gives a 2:1 change of speed; or a multiwinding stator winding can be used. However, this arrangement requires two separate stator windings, which means deeper stator slots, resulting in a larger and more costly machine, and a different power output at each speed. The only way to obtain a range of speed control comparable to that of the dc motor is by using a dc link converter to produce a variable-frequency output, usually variable over the range of 5–120 or 150 Hz. This output can in turn be used to control either a synchronous or a polyphase induction motor, although at very low speeds the cycloconverter is the preferred method of controlling a synchronous motor. It should be noted that if either a dc link converter or a cycloconverter is used, the requirement for an ac starter is eliminated.

In addition to the motor classifications discussed in Section 9-3, the following classifications specifically applicable to ac motors must be included:

1. *Multispeed motor:* Motors that can be reconnected to operate at more than one speed, so that the speed is controlled by changing the number of poles per phase. Multispeed motors can have one stator winding, so that reconnection provides two speeds in a ratio of 2:1, or two stator windings can be used to give speed ratios other than 2:1. These motors are designed to have an output capability that meets either of the following load characteristics:
   a. *Constant-torque,* that is, they develop the same torque at each speed, and as a result the output power is proportional to the rotational speed. For example, a two-speed motor rated at 20 hp (14.92 kW) at 3,600 r/min (376.99 rad/s) will produce 10 hp (7.46 kW) at 1,800 r/min (188.5 rad/s). Typical applications of constant-torque motors are conveyers, mixers, and positive-displacement compressors.
   b. *Variable-torque,* that is, they develop an output power that is proportional to the square of the rotational speed. For example, a two-speed motor rated at 5 hp (3.73 kW) at 900 r/min (94.25 rad/s) will develop 20 hp (14.92 kW) at 1,800 r/min (188.5 rad/s). Variable-torque motors are used to drive centrifugal pumps, fans, and blowers, where the load decreases as the square of the rotational speed.

   The use of a centrifugal starting switch in a single-phase induction motor precludes these motors from being used in multispeed applications.
2. *Nonreversible motor:* A motor that cannot be reversed electrically. Typical examples are reluctance-start and shaded-pole motors.
3. *Reversing motor:* A motor that can be reversed when running. All three-phase induction motors are reversing motors and are easily reversed by reversing any two phases. The prime requirement for a reversing motor is that the starting torque be the dominant torque. For this reason a

single-phase capacitor-start motor cannot be classified as a reversing motor, since when running at normal speed, the start or auxiliary winding is disconnected.

18-5 FRAME SIZE

Standard frame sizes have been designated by NEMA based on the developed torque and rotational speed. Torque is chosen as the parameter since it determines the winding and rotor configurations. As a result the same frame size may be common to several motors, because even though the torque is the same, the output power and rotational speeds are different.

NEMA frame size 680 (200 hp, 149.2 kW) and smaller polyphase induction motors by far form the largest group of polyphase induction motors used in industry. Also because they are standardized, they are available from warehouse stock. Large polyphase induction motors 680 frame size and larger, that is, 200 to 100,000 hp (149.2 to 74,600 kW), normally are horizontally or vertically mounted, and are used in applications such as air and gas compressors, coal pulverizers and crushers, high-volume fans, rolling mills, large-capacity pumps, ball mills, wood chippers, and so on.

18-6 AMBIENT TEMPERATURE

The ambient temperature conditions in the vicinity of the machine are important in determining the type of machine that should be selected. In addition, the ambient temperature also determines the insulation class and even the bearing lubricant. The standard motor is designed to operate in an atmosphere with an ambient temperature up to 40°C (104°F), and the bearings are filled with standard high-temperature grease. If the motor is to be used at altitudes greater than 3,330 ft (1,005 m), the motor may have to be derated, because the lower air density reduces the motor's ability to dissipate heat. Alternately any tendency for an excessive temperature rise can be corrected by reducing the service factor to 1.0 for motors rated with a service factor of 1.15 or greater. It should also be appreciated that at the higher altitudes the ambient temperature is usually lower.

18-7 TEMPERATURE RISE

The permissible temperature rise of a continuous-duty motor is determined by the insulation class and the type of enclosure. The insulation classes commonly used in polyphase motors are classes A, B, F, and H, which have hot-spot temperatures of 105, 130, 155, and 180°C. Class A insulation is rarely used today. Class B insulating material is the most commonly used, and the NEMA frame sizes are based on its use. Motors based on a 40°C ambient temperature then may safely operate with the following temperature rises measured by resistance change: class B 80°C, class F 105°C, and class H 125°C above ambient. The use of class F or H insulation material also increases the service factor, that is, it increases the motor's ability to operate in higher ambient temperature conditions.

18-8 DUTY CYCLE

The four duty cycles described in Section 9-8 also apply to ac machines. However, apart from the power output requirements of the motor, it is necessary to determine the frequency of starting, the inertia (Wk^2) or flywheel effect of the connected load, and the method of stopping the motor and connected load, that is, plugging, dc injection braking, or mechanical braking. The inertia of the connected load determines the acceleration time during the starting cycle and therefore the temperature rise. This characteristic is compounded if the motor is subjected to frequent starting or reversing.

The service factor specified on the motor nameplate is an indication of the amount of continuous overload to which the machine can be subjected without causing damage to the motor. Assuming that the motor is supplied at the voltage and frequency specified on the nameplate, then the service factor times the output power defines the safe output power level to which the machine may be loaded. It should be noted that the efficiency, power factor, and speed may deviate slightly from the nameplate values.

18-9 VOLTAGE, CURRENT, AND FREQUENCY RATINGS

Polyphase induction motors will perform satisfactorily at rated load with voltage variations of ±10% at rated frequency. An increase of 10% in voltage will increase the starting and breakdown torques by 21%, the efficiency by 0.5–1 point, and the starting current by 10–12%; the full-load current will decrease by approximately 7%, the power factor by 3 points, and the temperature rise by 3–4 °C. The rotor speed will increase by 1%. Similarly a 10% decrease in voltage will cause the starting and breakdown torques to decrease by 19%, the rotor speed by about 1.5%, and the efficiency by 2 points. The power factor improves by 1 point, the starting and full-load currents will increase by about 11%, and the temperature rise by 6–7°C.

Frequency variations, although less frequent, will affect the motor's performance. For example, a 5% decrease in frequency increases the starting and breakdown torques by 11%, the starting current by 5–6%, and there will also be a slight increase in the full-load current. Also the rotor speed will decrease by 5%, and there will be a slight decrease in the full-load efficiency and power factor.

Standard supply voltages for large polyphase induction and synchronous motors are 100–600 hp (74.6–447.6 kW) 460 or 575 V; 200–4,000 hp (149.2–2,984 kW) 2,300 V; 400–7,000 hp (298.4–5,222 kW) 4,000 or 4,600 V; 1,000–12,000 hp (746–8,952 kW) 6,600 V; and 3,500–25,000 hp (2,611–18,650 kW) 13,200 V. Smaller power output induction motors are also supplied from 200-, 230-, 460-, or 575-V three-phase sources.

The standard voltages for dc excitation voltages applied to the rotor of synchronous motors may be 62.5, 125, 250, 375, or 500 V, although 125 and 250 V are the most common voltages used in industrial applications.

The standard frequency for residential, commercial, and industrial users in North and South America is 60 Hz; 50 Hz is the standard in nearly all the rest of the world. In special applications some systems operate at 25 and $16\frac{2}{3}$ Hz, and in the aerospace industry 400 Hz is the standard frequency.

18-10 POWER FACTOR

The power factor of a polyphase induction motor is always lagging. However, as the connected load decreases, the motor power factor also decreases. In an industrial plant or a commercial facility the cumulative effect of a number of polyphase induction motors operating under lightly loaded conditions is to cause the whole plant electrical system to have a low power factor. In addition, a low-speed motor will have a poorer power factor than a high-speed motor. Also as was mentioned previously, an increase in the supply voltage above its rated value reduces the power factor of an induction motor. Apart from penalty costs imposed by the supply authority, poor power factors increase motor heating. The overall power factor can be improved by:

1. Using power factor correction capacitors connected across the incoming supply lines, or across the motor terminals of individual motors.
2. Using an overexcited synchronous motor to inject leading kVARs into the system to correct the overall power factor. This is the preferred method, since not only can the power factor be corrected automatically, but at the same time the synchronous motor can perform useful work.

18-11 ENCLOSURES

Enclosures were discussed in Chapter 3, but there are a few special applications that can be easily performed by polyphase induction motors, such as a submersible pump where the stator is suitably encapsulated and the rotor forms the pump impeller.

18-12 NEMA DESIGNATIONS

All ac motor nameplates with the exception of polyphase wound-rotor motors rated at $\frac{1}{20}$ hp (0.04 kW) or larger are marked with "Code" followed by a letter selected from Table 18-1, which shows the locked-rotor kVA per horsepower. Multispeed motor nameplates are marked with a code letter specifying the locked-rotor kVA per horsepower for the highest speed connection. Single-speed motors that use wye-delta starting are marked with a code letter corresponding to the locked-rotor kVA per horsepower on the wye or star connection. Dual-voltage motors usually have a different locked-rotor kVA for each voltage connection. They are marked with the code letter that gives the highest locked-rotor kVA per horsepower. Table 18-1 defines the NEMA code letter designations for locked-rotor kVA per horsepower.

Table 18-1 NEMA code letter designations for locked-rotor kVA/hp at rated voltage and frequency for large polyphase squirrel-cage induction motors

NEMA Code Letter	Locked-Rotor kVA/hp	NEMA Code Letter	Locked-Rotor kVA/hp
A	0–3.15	L	9.0–10.0
B	3.15–3.55	M	10.0–11.2
C	3.55–4.0	N	11.2–12.5
D	4.0–4.5	P	12.5–14.0
E	4.5–5.0	R	14.0–16.0
F	5.0–5.6	S	16.0–18.0
G	5.6–6.3	T	18.0–20.0
H	6.3–7.1	U	20.0–22.4
J	7.1–8.0	V	22.4 and up
K	8.0–9.0		

Table 18-2 lists NEMA frame sizes for open three-phase squirrel-cage induction motors, design classes A and B, either vertically or horizontally mounted, operating from a 60-Hz supply with a 1.15 service factor. It should be appreciated that this is only one table that has been selected from many to illustrate the NEMA classification system.

NEMA designations for starters and maximum ratings of motor branch circuit protective devices are listed in Tables 18-3 and 18-4, respectively.

In general NEMA standards are standards that have been voluntarily adopted by all manufacturers and represent the common practice of the electrical manufacturing industry. These standards define among other things dimensions, tolerances, operating characteristics, performance, quality, rating, and testing. In particular they cover frame sizes, torque classifications, and basis of rating.

18-13 MAINTENANCE

The maintenance requirements of polyphase induction and synchronous motors are much simpler than those required for dc machines. The stator windings are not subjected to centrifugal forces. However, because of the high surge currents that occur during the starting cycle, the resulting magnetic attractive and repulsive forces, which are proportional to the square of the inrush current, can lead to insulation breakdown if the slot wedges or the coil end turns can move. As a result all bolts, nuts, slot wedges, and the securing arrangements of the coil end turns must be carefully inspected on a regular basis.

The rotors of squirrel-cage induction motors, while very rugged in their construction, still may have a failure between a rotor bar or bars and the short-circuiting end rings. The result of an open circuit between a rotor bar and the end ring is that there is no torque produced by the bar(s). A single open circuit is hard to detect, but as the number of open bars increases, rotor noise and vibration will also increase.

Table 18-2 Three-phase squirrel-cage induction motor frame sizes, design A and B, either horizontal or vertical mounted, 60 Hz, class B insulation, general purpose–open type, service factor 1.15

	Rotational Speed (r/min)			
hp	**3,600**	**1,800**	**1,200**	**900**
$\frac{1}{2}$				143T
$\frac{3}{4}$			143T	145T
1		143T	145T	182T
$1\frac{1}{2}$	143T	145T	182T	184T
2	145T	145T	184T	213T
3	145T	182T	213T	215T
5	182T	184T	215T	254T
$7\frac{1}{2}$	184T	213T	254T	256T
10	213T	215T	256T	284T
15	215T	254T	284T	286T
20	254T	256T	286T	324T
25	256T	284T	324T	326T
30	284TS	286T	326T	364T
40	286TS	324T	364T	365T
50	324TS	326T	365T	404T
60	326TS	364TS*	404T	405T
75	364TS	365TS*	405T	444T
100	365TS	404TS*	444T	445T
125	404TS	405TS*	445T	
150	405TS	444TS*		
200	444TS	445TS*		

*In applications where V-belts or chain drives are used the correct frame size is as shown with the designator S omitted.

The rotors of wound-rotor induction motors are subjected to centrifugal forces. As a result they should be inspected for loose slot wedges, banding, and loose end turns by checking the varnish for any signs of cracking. In addition, the slip ring surfaces and slip ring brushgear should be inspected to ensure that the slip rings are smooth, and that the brushes are not worn excessively and are free in the brush holders with the correct brush pressure.

Salient-pole rotors of synchronous motors should be checked to ensure that the pole pieces are tightly wedged in their dovetail slots, and that the pole winding is not loose on the pole. If they are loose, additional insulation material should be added to ensure that there is no more movement. Usually a field coil failure can be spotted because some of the turns are flared. This flaring is a direct result of excessive heating caused by coil copper loss and aging, combined with the effects of centrifugal force. The cure is usually to have the field pole coils rewound. Damper or amortisseur windings installed in the pole faces of the salient poles should also be inspected for damage in exactly the same manner as for a squirrel-cage rotor. If slip rings are

Table 18-3 NEMA starter designations detailing maximum horsepower ratings at 230 and 460/575 V for polyphase motors

NEMA Size	Full-Voltage Starting		Autotransformer Starting		Wye-Delta Starting	
	230V	460/575V	230V	460/575V	230V	460/575V
00	1.5	2				
0	3	5				
1	7.5	10	7.5	10	10	15
2	15	25	15	25	25	40
3	30	50	30	50	50	75
4	50	100	50	100	75	150
5	100	200	100	200	150	300
6	200	400	200	400	350	700
7	300	600	300	600	500	1,000
8	450	900	450	900	800	1,500

Table 18-4 Maximum rating or setting of motor branch circuit protective devices for all single-phase and polyphase squirrel-cage and synchronous motors with full-voltage resistance or reactor starters

Code Letter	Percent of Full-Load Current			
	Non-time Delay Fuse	Dual-Element Time-Delay Fuse	Instant Trip Circuit Breaker	Time-Limit Circuit Breaker
No code	300	175	700	250
A	150	150	700	150
B to E	250	175	700	200
F to V	300	175	700	250

installed, the rings and brushgear should be inspected in exactly the same way as for the wound-rotor induction motor.

Normally annual maintenance routines for polyphase machines consists of the following procedures:

1. Blowing out accumulated dirt, carbon dust, and so on, using dry compressed air.
2. In the case of ball and roller bearings, the drain plug in the bearing housing is removed and new grease is added until the new grease appears in the drain hole. The motor should then be run for about 1 h to allow the grease to expand and the excess to drain before replacing the drain plug.

3. The insulation resistance of the stator and rotor windings as appropriate should be checked by means of a bridge megger. If the insulation reading is low (usually less than 1 MΩ), the machine should be stripped down to determine the cause.

QUESTIONS

18-1 Why is the polyphase squirrel-cage induction motor the preferred motor in industrial and commercial applications?

18-2 What factors determine the selection between a synchronous motor and an induction motor?

18-3 What types of motor should be considered where the load requires a high starting torque?

18-4 How can a squirrel-cage induction motor be operated at constant speed?

18-5 What motors operate at synchronous speed?

18-6 What are the disadvantages of using pole-changing techniques to obtain speed control of a polyphase motor?

18-7 What are the disadvantages of using a multispeed motor to obtain speed variation?

18-8 How can the speed of a polyphase squirrel-cage or wound-motor induction motor or a synchronous motor be made continuously variable above and below base speed?

18-9 What is meant by a constant-torque motor? Discuss typical applications.

18-10 What is meant by a variable-torque motor? Discuss typical applications.

18-11 What is the prime requirement of a reversing motor?

18-12 What factors are affected by the ambient temperature rating of a machine?

18-13 What effect does the altitude at which a motor operates have on the motor's performance?

18-14 What are the relative merits of using class F and H insulating materials?

18-15 What is meant by service factor?

18-16 Discuss the effects of a 10% increase of the terminal voltage applied to a squirrel-cage induction motor.

18-17 Discuss the effects of a 10% decrease of the terminal voltage applied to a squirrel-cage induction motor.

18-18 What operating conditions affect the temperature rise of an ac polyphase motor?

18-19 Discuss the effects of an increase and a decrease in frequency on the performance of a squirrel-cage induction motor.

18-20 What are the standard supply voltages for integral-horsepower squirrel-cage and wound-rotor induction motors and synchronous motors?

18-21 What are the standard excitation voltages for synchronous motors?

18-22 How may a low plant power factor be improved?

18-23 If you were responsible for the planned maintenance of a squirrel-cage induction motor, a wound-rotor induction motor, and a synchronous motor, what items should be checked?

Conversion Tables

Length

1 kilometer (km) = 1,000 m
1 meter (m) = 100 cm
1 centimeter (cm) = 10^{-2}m
1 millimeter (mm) = 10^{-3}m
1 micrometer (μm) = 10^{-6}m
1 inch (in) = 2.540 cm
= 25.4 mm
1 foot (ft) = 30.48 cm = 0.3048 m
1 yard (yd) = 0.9144 m
1 mile (statute mile, mi) = 1,609.344 m
1 cm = 0.3937 in
1 m = 39.37 in = 3.2808 ft = 1.0936 yd
1 kilometer (km) = 0.6214 m

Area

1 m^2 = 10^4 cm^2 = 10^6 mm
1 m^2 = 1,550.0 in^2 = 10.764 ft^2
1 in^2 = 645.16 mm^2 = 6.9444×10^{-3} ft^2
1 ft^2 = 144 in^2

Volume

1 m^3 = 6.1024×10^4 in^3 = 35.315 ft^3
1 liter (L) = 0.001 m^3

$1\ cm^3 = 6.1024 \times 10^{-2}\ in^3$
$1\ in^3 = 16.387\ cm^3$
$1\ ft^3 = 1{,}728\ in^3 = 2.8317 \times 10^{-2}\ m^3$
$1\ ft^3 = 7.481$ gal (US) = 6.229 gal (E)
1 US gallon (gal) = 231 in^3 = 0.833 gal (E) = 3.785 L
1 English gallon = 1.2009 gal (US) = 277.3 in^3 = 4.546 L

Linear Velocity

1 m/s = 10^3 mm/s = 3.6 km/h = 39.370 in/s = 3.2808 ft/s
1 in/s = 2.54×10^{-2} m/s = 25.4 mm/s
1 mi/h = 1.6093 km/h

Angular Velocity

1 rad/s = 0.15915 r/s = 9.5493 r/min
1 r/min = 10^{-3} kr/min = 1.6667×10^{-2} r/s = 6°/s = 0.10472 rad/s
1 r/s = 60 r/min/s = 360°/s = 6.2832 rad/s

Mass

1 kilogram (kg) = 0.0685 slug = 2.2046 lb (mass) = 35.274 oz (mass)
1 oz (mass) = 28.3495 g
1 lb (mass) = 16 oz (mass) = 0.45359 kg
1 slug = 14.59 kg

Force

1 newton (N) = 0.2248 lb (force) = 0.102 kg = 7.233 pdl
1 kilogram (kg) = 2.2046 lb (mass) = 9.807 N
1 pound (lb) = 4.448 N = 0.4536 kg = 32.17 pdl
1 poundal (pdl) = 0.138255 N
1 short ton = 2,000 lb
1 long ton = 2,240 lb
1 metric tonne = 2,205 lb

Torque

1 newton-meter (N · m) = 0.73756 lb · ft = 8.85075 lb · in = 141.612 oz · in
1 oz · in = 7.0615×10^{-3} N · m
1 lb · ft = 192 oz · in = 1.3558 N · m

Moment of Inertia

1 kg · cm^2 = 0.01416 oz · in · s^2
1 kg · m^2 = 10^7 g · cm^2 = 8.85075 lb · in · s^2 = 141.612 oz · in · s^2
1 oz · in · s^2 = 6.25×10^{-2} lb · in · s^2 = 7.06155×10^{-3} kg · m^2 = 70.6155 kg · cm^2

Energy (Work)

1 joule (J) = 1 N · m = 1 W · s = 2.7778 $\times$ 10^{-7} kWh = 2.38846 $\times$ 10^{-4} kcal = 9.4781 $\times$ 10^{-4} Btu
1 kcal = 10^3 cal = 4186.8 J = 1.1630 $\times$ 10^{-3} kWh = 3.9683 Btu
1 kWh = 3.6 $\times$ 10^6 J = 859.845 kcal = 3.4121 $\times$ 10^3 Btu
1 Btu = 1,055.06 J = 2.9307 $\times$ 10^{-4} kWh = 0.251997 kcal

Power

1 watt (W) = 10^{-3} Kw = 1 J/s = 0.73756 lb · ft/s = 1.3596 $\times$ 10^{-3} hp (metric) = 1.3410 $\times$ 10^{-3} hp (English)
1 hp (English) = 550 lb · ft/s = 745.7 W = 1.0139 hp (metric)
1 hp (metric) = 736.5 W = 0.98632 hp (English)

Temperature

°C = 5(°F − 32)/9
°F = [9(°C)/5] + 32
K = °C + 273.15

Transformer Efficiency Calculations

Computer Program

```
10 PRINT "TRANSFORMER EFFICIENCY CALCULATIONS"
20 PRINT
30 REM TYPE IN TRANSFORMER KVA
40 KVA=100
50 REM TYPE IN CORE LOSS
60 CL=800
70 REM TYPE IN FULL LOAD COPPER LOSS
80 CUL=1000
90 REM SPECIFY POWER FACTOR
100 PF=.7
110 PRINT
120 PRINT   " KW      COPPER      CORE         EFFICIENCY "
130 PRINT           "           LOSS        LOSS "
140 FOR I=1 TO 13
150 LET PCL=I/10
160 LET PCUL=PCL^2*CUL
170 LET ALOAD=KVA*PCL*PF
180 LET EFF=ALOAD/(ALOAD+(CL+PCUL)/1000)*100
190 PRINT USING  "###.#    ####   ####    ##.##" ; ALOAD,PCUL,CL,EFF
200 NEXT I
210 END
```

Power factor = 0.7

TRANSFORMER EFFICIENCY CALCULATIONS

kW	copper loss	core loss	efficiency
7.0	10	800	89.63
14.0	40	800	94.34
21.0	90	800	95.93
28.0	160	800	96.69
35.0	250	800	97.09
42.0	360	800	97.31
49.0	490	800	97.43
56.0	640	800	97.49
63.0	810	800	97.51
70.0	1000	800	97.49
77.0	1210	800	97.46
84.0	1440	800	97.40
91.0	1690	800	97.34

Power factor = 0.85

TRANSFORMER EFFICIENCY CALCULATIONS

kW	copper loss	core loss	efficiency
8.5	10	800	91.30
17.0	40	800	95.29
25.5	90	800	96.63
34.0	160	800	97.25
42.5	250	800	97.59
51.0	360	800	97.78
59.5	490	800	97.88
68.0	640	800	97.93
76.5	810	800	97.94
85.0	1000	800	97.93
93.5	1210	800	97.90
102.0	1440	800	97.85
110.5	1690	800	97.80

Power factor = 1.0

TRANSFORMER EFFICIENCY CALCULATIONS

kW	copper loss	core loss	efficiency
10.0	10	800	92.51
20.0	40	800	95.97
30.0	90	800	97.12
40.0	160	800	97.66
50.0	250	800	97.94
60.0	360	800	98.10
70.0	490	800	98.19
80.0	640	800	98.23
90.0	810	800	98.24
100.0	1000	800	98.23
110.0	1210	800	98.21
120.0	1440	800	98.17
130.0	1690	800	98.12

Bibliography

Bose, B. K.: *Power Electronics and AC Drives.* Englewood Cliffs, N.J.: Prentice-Hall, 1986.

Chapman, S. J.: *Electric Machinery Fundamentals.* New York: McGraw-Hill, 1985.

Electro-Craft Corp.: *DC Motors, Speed Controls, Servo Systems,* 3rd ed. Hopkins, Minn., 1975.

Fink, D. G., and H. W. Beaty: *Standard Handbook for Electrical Engineers,* 11th ed. New York: McGraw-Hill, 1978.

Gebert, K. L., and K. R. Edwards: *Transformers, Principles and Applications,* 2nd ed. Chicago, IL: American Technical Society, 1974.

Jordan, H. E.: *Energy Efficient Electric Motors and Their Applications.* New York: Van Nostrand Reinhold, 1983.

Kosow, I. L.: *Electric Machinery and Transformers.* Englewood Cliffs, N.J.: Prentice-Hall, 1972.

Parker, R. J., and R. J. Studders: *Permanent Magnets and Their Applications.* New York: Wiley, 1962.

Pearman, R. A.: *Power Electronics Solid-State Motor Control.* Reston, Va.: Reston Publishing, 1980.

Pearman, R.: *Solid-State Industrial Electronics.* Reston, Va.: Reston Publishing, 1984.

Pelley, B. R.: *Thyristor Phase-Controlled Converters and Cycloconverters.* New York: Wiley, 1971.

Smeaton, R. W.: *Motor Application and Maintenance,* 2nd ed. New York: McGraw-Hill, 1987.

Wildi, T.: *Electrical Power Technology.* New York: Wiley, 1981.

Answers to Selected Problems

Chapter 1

1-1. (a) 2 MΩ; (b) 1μF; (c) 2mH; (d) 2.5 kg; (e) 65 kPa; (f) 5 MW; (g) 500 kV; (h) 2 μA; (i) 2.5 kJ; (j) 1 pF.

1-3. (a) 52.3 N · m; (b) 25.4 mm; (c) 14.2 A; (d) 16 s; (e) 23 m; (f) 9.81 m/s^2.

1-5. (a) 2.28 kg; (b) 1.09 slug.

1-7. 999,756 ft-lb.

1-9. 36.86 m/s^2

1-11. 5.8 lb-ft, 8.01 N · m.

1-13. 15.33 kW.

1-15. 0.57 kg-m^2.

1-17. (a) 3,125 A/m; (b) 3.73 Wb/m; (c) 3.73×10^{-3} Wb.

1-19. (a) 500,000 A/Wb; (b) 500 A; (c) 1,326 A/m.

1-21. 43 mA, 3,979.

1-23. 1.86 A.

1-27. (a) 112,910 lines/m^2; (b) 2.07 A/m; (c) 7.50×10^{-4} Wb; (d) 0.85Wb/m^2 = 0.85 T.

1-29. 14.0 V/coil

Chapter 4

4-1. (a) 62.5 A; (b) 250 A.

4-3. 76.80 V

Chapter 5

5-1. (a) 252 V; (b) 264 V; (c) 224.40 V.

5-3. 0.2041Ω

5-5. (a) 73.53Ω; (b) 343 V.

5-7. (a) 554.35 V; (b) 200 A.

5-9. (a) 200 A; (b) 1.67 A; (c) 201.67 A; (d) 1.61%; (e) 418.34 W; (f) 813.42 W; (g) 0.95 or 95%.

Chapter 6

6-1. (a) 36,523.76 W or 36.52 kW; (b) 146.93 lb-ft; (c) 88.54%; (d) 1,843.18 r/min; (e) 5.36%.
6-3. 1,233.01 r/min.
6-5. 12.44 hp, 9.28 kW.
6-7. (a) 4,521.21 W; (b) 36.17 A; (c) 20.65 N.m.
6-9. $r_1 = 1.226\Omega$, $r_2 = 0.700\Omega$, $r_3 = 0.166\Omega$
6-11. 851.01 r/min.
6-13. (a) 210.63 V; (b) 22.12 kW.

Chapter 7

7-1. 89.33%
7-3. (a) 276.24 kW; (b) 4.04 kW.
7-7. (a) 81.77%; (b) 81.72%.
7-9. (a) 79.35%; (b) 22.64 kW; (c) 83.46%.
7-11. 82.06%.
7-13. Generator 87.06%, motor 86.85%.

Chapter 10

10-2. (a)3,983.72 V; (b) 144.93 A; (c) 1,000 kVA.
10-3. (a) 40 coils; (b) 5 coils per pole phase group.
10-5. 0.97.

Chapter 11

11-1. 90 poles.
11-3. 28.6%.
11-5. (a) $4{,}679.91 \angle 30.31°$ V/phase; (b) $5{,}588.27 \angle 51.63°$ V/phase.
11-7. 95.51%.
11-9. (a) 1.12Ω/phase; (b) 1.04Ω/phase; (c) 33.82%.

Chapter 12

12-1. (a) motor 6 poles, generator 40 poles; (b) 125.66 rad/s.
12-3. (a) $334.04 \angle 4.74°$; (b) 80.99 kW.
12-5. (a) 22.15°; (b) 213.45 A; (c) 0.99 lagging.

Chapter 13

13-1. (a) Primary 1,893 turns; (b) secondary 75 turns; (c) 0.0274 Wb.
13-3. (a) 0.0717; (b) 85.89°; (c) 0.9 A; (d) 12.47 A.
13-5. (a) 20.83 A; (b) 208.33 A.
13-7. (a) 300 turns; (b) 0.0688 m^2.
13-9. 10.42V
13-11. (a) 0.18Ω; (b) 1.70Ω; (c) 1.69Ω; (d) 31,464.73Ω; (e) 0.51 A; (f) 0.13 A; (g) 0.49 A; (h) 8,489.8Ω; (i) 1.7%.

13-13. (a) 118.2 W; (b) 12 W.

13-15. 91.46 kVA.

13-17. (a) 43.48 A; (b) 41.67 A; (c) 100 kVA.

13-19. (a) 26 kVA; (b) 21.06 kVA.

13-21. (a) V_p = 2,400V, V_s = 550 V, 46.27 kVA; (b) V_p = 4,160V, V_s = 317.54 V, 46.27 kVA; (c) V_p = 4,160V, V_s = 550V, 40.26 kVA.

Chapter 14

14-1. 10 poles.

14-3. (a) 1,200 r/min or 125.66 rad/s; (b) 1,154 r/min or 120.88 rad/s; (c) 46 r/min or 4.78 rad/s; (d) 2.28 Hz.

14-5. (a) 169.32 A; (b) 7.31 kW; (c) 4.64 kW; (d) 111.4 kW; (e) 851.03 N·m or 628.06 lb-ft; (f) 83.62%; (g) 1,152 r/min or 120.64 rad/s.

14-8. (a) 141.97 kW; (b) 5.53 kW.

14-9. (a) 148.92 A; (b) 0.88 lagging; (c) 120.59 hp or 89.96 kW; (d) 741.84 N·m or 547.18 lb-ft.

14-11. (a) 491.23 A; (b) 345.65 N·m or 254.95 lb-ft.

14-13. 10.2 A.

Chapter 15

15-3. (a) 26.01 A; (b) 0.87 lagging; (c) 2.60 kW.

15-5. (a) 51.48%; (b) 0.0417; (c) 1.0048 lb-ft or 0.74 N·m.

15-7. (a) 1.67%; (b) 6.67%; (c) 5.36%

Chapter 17

17-1. (a) 400 steps; (b) 0.9°/step; (c) 15.71 rad/s.

17-2. R = 4.8Ω and L = 0.03H.

17-3. 45°.

INDEX